SCHAUM'S OUTLINE OF

THEORY AND PROBLEMS

OF

THEORETICAL MECHANICS

with an introduction to

Lagrange's Equations and Hamiltonian Theory

•

BY

MURRAY R. SPIEGEL, Ph.D.

Professor of Mathematics

Rensselaer Polytechnic Institute

•

SCHAUM'S OUTLINE SERIES

McGRAW-HILL BOOK COMPANY

New York, St. Louis, San Francisco, Toronto, Sydney

60232

8 9 10 11 12 13 14 15 SH SH 7 5

Preface

In the 17th century, Sir Isaac Newton formulated his now famous laws of mechanics. These remarkably simple laws served to describe and predict the motions of observable objects in the universe, including those of the planets of our solar system.

Early in the 20th century it was discovered that various theoretical conclusions derived from Newton's laws were not in accord with certain conclusions deduced from theories of electromagnetism and atomic phenomena which were equally well founded experimentally. These discrepancies led to Einstein's *relativistic mechanics* which revolutionized the concepts of space and time, and to *quantum mechanics*. For objects which move with speeds much less than that of light and which have dimensions large compared with those of atoms and molecules Newtonian mechanics, also called *classical mechanics*, is nevertheless quite satisfactory. For this reason it has maintained its fundamental importance in science and engineering.

It is the purpose of this book to present an account of Newtonian mechanics and its applications. The book is designed for use either as a supplement to all current standard textbooks or as a textbook for a formal course in mechanics. It should also prove useful to students taking courses in physics, engineering, mathematics, astronomy, celestial mechanics, aerodynamics and in general any field which needs in its formulation the basic principles of mechanics.

Each chapter begins with a clear statement of pertinent definitions, principles and theorems together with illustrative and other descriptive material. This is followed by graded sets of solved and supplementary problems. The solved problems serve to illustrate and amplify the theory, bring into sharp focus those fine points without which the student continually feels himself on unsafe ground, and provide the repetition of basic principles so vital to effective learning. Numerous proofs of theorems and derivations of basic results are included in the solved problems. The large number of supplementary problems with answers serve as a complete review of the material of each chapter.

Topics covered include the dynamics and statics of a particle, systems of particles and rigid bodies. Vector methods, which lend themselves so readily to concise notation and to geometric and physical interpretations, are introduced early and used throughout the book. An account of vectors is provided in the first chapter and may either be studied at the beginning or referred to as the need arises. Added features are the chapters on Lagrange's equations and Hamiltonian theory which provide other equivalent formulations of Newtonian mechanics and which are of great practical and theoretical value.

Considerably more material has been included here than can be covered in most courses. This has been done to make the book more flexible, to provide a more useful book of reference and to stimulate further interest in the topics.

I wish to take this opportunity to thank the staff of the Schaum Publishing Company for their splendid cooperation.

M. R. Spiegel

Rensselaer Polytechnic Institute
February, 1967

CONTENTS

CONTENTS

CONTENTS

Chapter 1

VECTORS, VELOCITY and ACCELERATION

MECHANICS, KINEMATICS, DYNAMICS AND STATICS

Mechanics is a branch of physics concerned with motion or change in position of physical objects. It is sometimes further subdivided into:

1. *Kinematics,* which is concerned with the geometry of the motion,
2. *Dynamics,* which is concerned with the physical causes of the motion,
3. *Statics,* which is concerned with conditions under which no motion is apparent.

AXIOMATIC FOUNDATIONS OF MECHANICS

An axiomatic development of mechanics, as for any science, should contain the following basic ingredients:

1. *Undefined terms or concepts.* This is clearly necessary since ultimately any definition must be based on something which remains undefined.
2. *Unproved assertions.* These are fundamental statements, usually in mathematical form, which it is hoped will lead to valid descriptions of phenomena under study. In general these statements, called *axioms* or *postulates,* are based on experimental observations or abstracted from them. In such case they are often called *laws.*
3. *Defined terms or concepts.* These *definitions* are given by using the undefined terms or concepts.
4. *Proved assertions.* These are often called *theorems* and are proved from the definitions and axioms.

An example of the "axiomatic way of thinking" is provided by *Euclidean geometry* in which *point* and *line* are undefined concepts.

MATHEMATICAL MODELS

A mathematical description of physical phenomena is often simplified by replacing actual physical objects by suitable *mathematical models.* For example in describing the rotation of the earth about the sun we can for many practical purposes treat the earth and sun as points.

SPACE, TIME AND MATTER

From everyday experience, we all have some idea as to the meaning of each of the following terms or concepts. However, we would certainly find it difficult to formulate completely satisfactory definitions. We take them as undefined concepts.

1. *Space.* This is closely related to the concepts of *point, position, direction* and *displacement.* Measurement in space involves the concepts of *length* or *distance,* with which we assume familiarity. Units of length are feet, meters, miles, etc. In this book we assume that space is *Euclidean,* i.e. the space of *Euclid's geometry.*

2. *Time.* This concept is derived from our experience of having one *event* taking place after, before or simultaneous with another *event.* Measurement of time is achieved, for example, by use of *clocks.* Units of time are seconds, hours, years, etc.

3. *Matter.* Physical objects are composed of "small bits of matter" such as atoms and molecules. From this we arrive at the concept of a material object called a *particle* which can be considered as occupying a point in space and perhaps moving as time goes by. A measure of the "quantity of matter" associated with a particle is called its *mass.* Units of mass are grams, kilograms, etc. Unless otherwise stated we shall assume that the mass of a particle does not change with time.

Length, mass and time are often called *dimensions* from which other physical quantities are constructed. For a discussion of units and dimensions see Appendix A, Page 339.

SCALARS AND VECTORS

Various quantities of physics, such as length, mass and time, require for their specification a single real number (apart from units of measurement which are decided upon in advance). Such quantities are called *scalars* and the real number is called the *magnitude* of the quantity. A scalar is represented analytically by a letter such as t, m, etc.

Other quantities of physics, such as displacement, require for their specification a *direction* as well as magnitude. Such quantities are called *vectors.* A vector is represented analytically by a bold faced letter such as $\mathbf{A}$ in Fig. 1-1. Geometrically it is represented by an arrow PQ where P is called the *initial point* and Q is called the *terminal point.* The magnitude or length of the vector is then denoted by $|\mathbf{A}|$ or A.

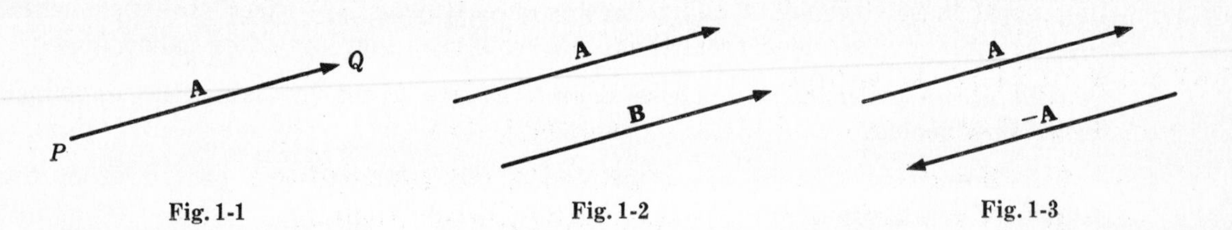

Fig. 1-1 Fig. 1-2 Fig. 1-3

VECTOR ALGEBRA

The operations of addition, subtraction and multiplication familiar in the algebra of real numbers are with suitable definition capable of extension to an algebra of vectors. The following definitions are fundamental.

1. Two vectors $\mathbf{A}$ and $\mathbf{B}$ are *equal* if they have the same magnitude and direction regardless of their initial points. Thus $\mathbf{A}=\mathbf{B}$ in Fig. 1-2 above.

2. A vector having direction opposite to that of vector $\mathbf{A}$ but with the same length is denoted by $-\mathbf{A}$ as in Fig. 1-3 above.

3. The *sum* or *resultant* of vectors $\mathbf{A}$ and $\mathbf{B}$ of Fig. 1-4(a) below is a vector $\mathbf{C}$ formed by placing the initial point of $\mathbf{B}$ on the terminal point of $\mathbf{A}$ and joining the initial point of $\mathbf{A}$ to the terminal point of $\mathbf{B}$ [see Fig. 1-4(b) below]. We write $\mathbf{C} = \mathbf{A}+\mathbf{B}$. This definition is equivalent to the *parallelogram law* for vector addition as indicated in Fig. 1-4(c) below.

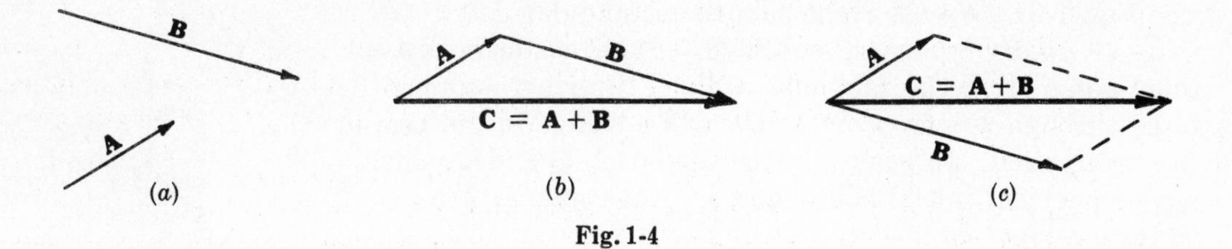

Fig. 1-4

Extensions to sums of more than two vectors are immediate. For example, Fig. 1-5 below shows how to obtain the sum or resultant **E** of the vectors **A, B, C** and **D**.

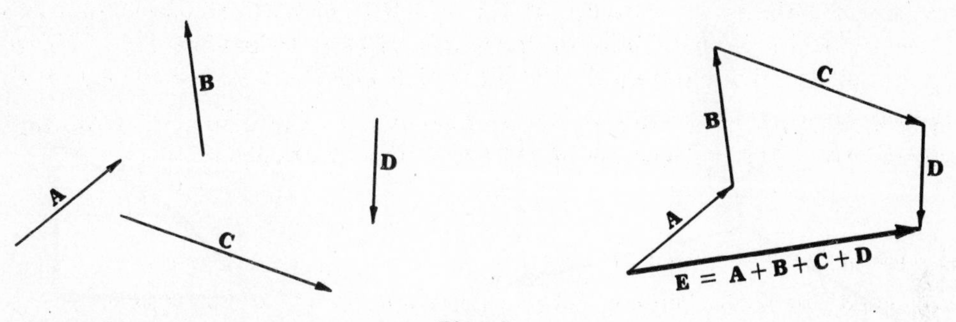

Fig. 1-5

4. The *difference* of vectors **A** and **B**, represented by $\mathbf{A}-\mathbf{B}$, is that vector **C** which when added to **B** gives **A**. Equivalently, $\mathbf{A}-\mathbf{B}$ may be defined as $\mathbf{A}+(-\mathbf{B})$. If $\mathbf{A}=\mathbf{B}$, then $\mathbf{A}-\mathbf{B}$ is defined as the *null* or *zero vector* represented by **0**. This has a magnitude of zero but its direction is not defined.
5. The *product* of a vector **A** by a scalar p is a vector $p\mathbf{A}$ or $\mathbf{A}p$ with magnitude $|p|$ times the magnitude of **A** and direction the same as or opposite to that of **A** according as p is positive or negative. If $p=0$, $p\mathbf{A}=\mathbf{0}$, the null vector.

LAWS OF VECTOR ALGEBRA

If **A, B** and **C** are vectors, and p and q are scalars, then

1. $\mathbf{A}+\mathbf{B} = \mathbf{B}+\mathbf{A}$ — Commutative Law for Addition
2. $\mathbf{A}+(\mathbf{B}+\mathbf{C}) = (\mathbf{A}+\mathbf{B})+\mathbf{C}$ — Associative Law for Addition
3. $p(q\mathbf{A}) = (pq)\mathbf{A} = q(p\mathbf{A})$ — Associative Law for Multiplication
4. $(p+q)\mathbf{A} = p\mathbf{A}+q\mathbf{A}$ — Distributive Law
5. $p(\mathbf{A}+\mathbf{B}) = p\mathbf{A}+p\mathbf{B}$ — Distributive Law

Note that in these laws only multiplication of a vector by one or more scalars is defined. On pages 4 and 5 we define products of vectors.

UNIT VECTORS

Vectors having unit length are called *unit vectors*. If **A** is a vector with length $A>0$, then $\mathbf{A}/A=\mathbf{a}$ is a unit vector having the same direction as **A** and $\mathbf{A}=A\mathbf{a}$.

RECTANGULAR UNIT VECTORS

The rectangular unit vectors **i**, **j** and **k** are mutually perpendicular unit vectors having directions of the positive x, y and z axes respectively of a rectangular coordinate system [see Fig. 1-6]. We use right-handed rectangular coordinate systems unless otherwise specified. Such systems derive their name from the fact that a right threaded screw rotated through 90° from Ox to Oy will advance in the positive z direction. In general three vectors **A**, **B** and **C** which have coincident initial points and are not coplanar are said to form a *right-handed system* or *dextral system* if a right threaded screw rotated through an angle less than 180° from **A** to **B** will advance in the direction **C** [see Fig. 1-7 below].

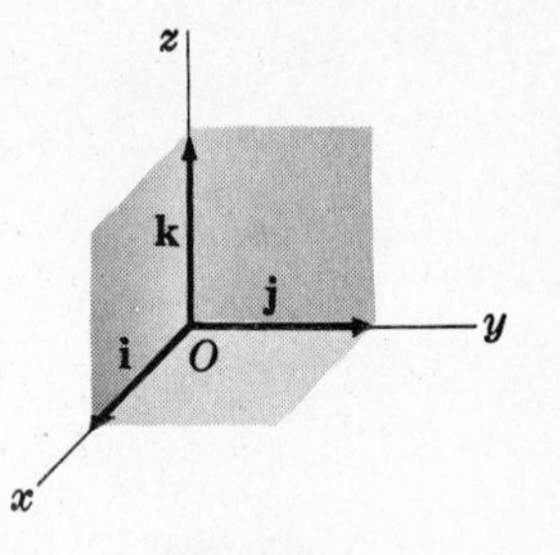

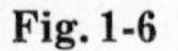
Fig. 1-6

Fig. 1-7

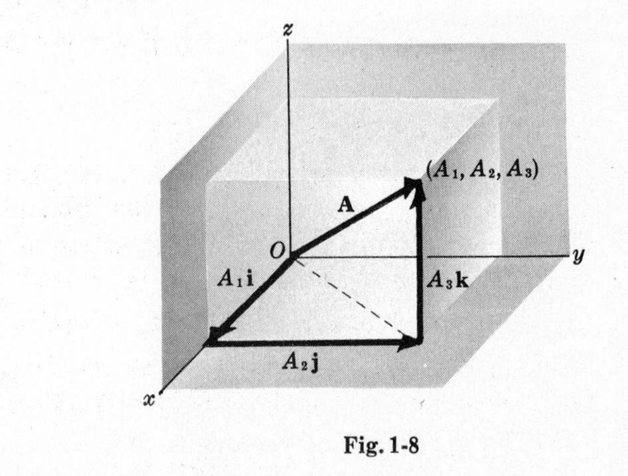

Fig. 1-8

COMPONENTS OF A VECTOR

Any vector **A** in 3 dimensions can be represented with initial point at the origin O of a rectangular coordinate system [see Fig. 1-8 above]. Let (A_1, A_2, A_3) be the rectangular coordinates of the terminal point of vector **A** with initial point at O. The vectors $A_1\mathbf{i}$, $A_2\mathbf{j}$ and $A_3\mathbf{k}$ are called the *rectangular component vectors*, or simply component vectors, of **A** in the x, y and z directions respectively. A_1, A_2 and A_3 are called the *rectangular components*, or simply *components*, of **A** in the x, y and z directions respectively.

The sum or resultant of $A_1\mathbf{i}$, $A_2\mathbf{j}$ and $A_3\mathbf{k}$ is the vector **A**, so that we can write

$$\mathbf{A} = A_1\mathbf{i} + A_2\mathbf{j} + A_3\mathbf{k} \tag{1}$$

The magnitude of **A** is

$$A = |\mathbf{A}| = \sqrt{A_1^2 + A_2^2 + A_3^2} \tag{2}$$

In particular, the *position vector* or *radius vector* **r** from O to the point (x, y, z) is written

$$\mathbf{r} = x\mathbf{i} + y\mathbf{j} + z\mathbf{k} \tag{3}$$

and has magnitude $r = |\mathbf{r}| = \sqrt{x^2 + y^2 + z^2}$.

DOT OR SCALAR PRODUCT

The dot or scalar product of two vectors **A** and **B**, denoted by $\mathbf{A}\cdot\mathbf{B}$ (read **A** dot **B**) is defined as the product of the magnitudes of **A** and **B** and the cosine of the angle between them. In symbols,

$$\mathbf{A}\cdot\mathbf{B} = AB\cos\theta, \qquad 0 \leqq \theta \leqq \pi \tag{4}$$

Note that $\mathbf{A}\cdot\mathbf{B}$ is a scalar and not a vector.

The following laws are valid:

1. $\mathbf{A}\cdot\mathbf{B} = \mathbf{B}\cdot\mathbf{A}$ Commutative Law for Dot Products
2. $\mathbf{A}\cdot(\mathbf{B}+\mathbf{C}) = \mathbf{A}\cdot\mathbf{B} + \mathbf{A}\cdot\mathbf{C}$ Distributive Law
3. $p(\mathbf{A}\cdot\mathbf{B}) = (p\mathbf{A})\cdot\mathbf{B} = \mathbf{A}\cdot(p\mathbf{B}) = (\mathbf{A}\cdot\mathbf{B})p$, where p is a scalar.
4. $\mathbf{i}\cdot\mathbf{i} = \mathbf{j}\cdot\mathbf{j} = \mathbf{k}\cdot\mathbf{k} = 1, \quad \mathbf{i}\cdot\mathbf{j} = \mathbf{j}\cdot\mathbf{k} = \mathbf{k}\cdot\mathbf{i} = 0$
5. If $\mathbf{A} = A_1\mathbf{i} + A_2\mathbf{j} + A_3\mathbf{k}$ and $\mathbf{B} = B_1\mathbf{i} + B_2\mathbf{j} + B_3\mathbf{k}$, then
$$\mathbf{A}\cdot\mathbf{B} = A_1B_1 + A_2B_2 + A_3B_3$$
$$\mathbf{A}\cdot\mathbf{A} = A^2 = A_1^2 + A_2^2 + A_3^2$$
$$\mathbf{B}\cdot\mathbf{B} = B^2 = B_1^2 + B_2^2 + B_3^2$$
6. If $\mathbf{A}\cdot\mathbf{B} = 0$ and $\mathbf{A}$ and $\mathbf{B}$ are not null vectors, then $\mathbf{A}$ and $\mathbf{B}$ are perpendicular.

CROSS OR VECTOR PRODUCT

The cross or vector product of **A** and **B** is a vector $\mathbf{C} = \mathbf{A}\times\mathbf{B}$ (read **A** cross **B**). The magnitude of $\mathbf{A}\times\mathbf{B}$ is defined as the product of the magnitudes of **A** and **B** and the sine of the angle between them. The direction of the vector $\mathbf{C} = \mathbf{A}\times\mathbf{B}$ is perpendicular to the plane of **A** and **B** and such that **A**, **B** and **C** form a right-handed system. In symbols,

$$\mathbf{A}\times\mathbf{B} = AB\sin\theta\,\mathbf{u}, \qquad 0 \leqq \theta \leqq \pi \tag{5}$$

where **u** is a unit vector indicating the direction of $\mathbf{A}\times\mathbf{B}$. If $\mathbf{A}=\mathbf{B}$ or if **A** is parallel to **B**, then $\sin\theta = 0$ and we define $\mathbf{A}\times\mathbf{B} = \mathbf{0}$.

The following laws are valid:

1. $\mathbf{A}\times\mathbf{B} = -\mathbf{B}\times\mathbf{A}$ (Commutative Law for Cross Products Fails)
2. $\mathbf{A}\times(\mathbf{B}+\mathbf{C}) = \mathbf{A}\times\mathbf{B} + \mathbf{A}\times\mathbf{C}$ Distributive Law
3. $p(\mathbf{A}\times\mathbf{B}) = (p\mathbf{A})\times\mathbf{B} = \mathbf{A}\times(p\mathbf{B}) = (\mathbf{A}\times\mathbf{B})p$, where p is a scalar.
4. $\mathbf{i}\times\mathbf{i} = \mathbf{j}\times\mathbf{j} = \mathbf{k}\times\mathbf{k} = \mathbf{0}, \quad \mathbf{i}\times\mathbf{j} = \mathbf{k},\ \mathbf{j}\times\mathbf{k} = \mathbf{i},\ \mathbf{k}\times\mathbf{i} = \mathbf{j}$
5. If $\mathbf{A} = A_1\mathbf{i} + A_2\mathbf{j} + A_3\mathbf{k}$ and $\mathbf{B} = B_1\mathbf{i} + B_2\mathbf{j} + B_3\mathbf{k}$, then
$$\mathbf{A}\times\mathbf{B} = \begin{vmatrix} \mathbf{i} & \mathbf{j} & \mathbf{k} \\ A_1 & A_2 & A_3 \\ B_1 & B_2 & B_3 \end{vmatrix}$$
6. $|\mathbf{A}\times\mathbf{B}|$ = the area of a parallelogram with sides **A** and **B**.
7. If $\mathbf{A}\times\mathbf{B} = \mathbf{0}$ and **A** and **B** are not null vectors, then **A** and **B** are parallel.

TRIPLE PRODUCTS

The *scalar triple product* is defined as

$$\mathbf{A}\cdot(\mathbf{B}\times\mathbf{C}) = \begin{vmatrix} A_1 & A_2 & A_3 \\ B_1 & B_2 & B_3 \\ C_1 & C_2 & C_3 \end{vmatrix} \tag{6}$$

where $\mathbf{A} = A_1\mathbf{i} + A_2\mathbf{j} + A_3\mathbf{k}$, $\mathbf{B} = B_1\mathbf{i} + B_2\mathbf{j} + B_3\mathbf{k}$, $\mathbf{C} = C_1\mathbf{i} + C_2\mathbf{j} + C_3\mathbf{k}$. It represents the volume of a parallelepiped having **A**, **B**, **C** as edges, or the negative of this volume according as **A**, **B**, **C** do or do not form a right handed system. We have $\mathbf{A}\cdot(\mathbf{B}\times\mathbf{C}) = \mathbf{B}\cdot(\mathbf{C}\times\mathbf{A}) = \mathbf{C}\cdot(\mathbf{A}\times\mathbf{B})$.

The *vector triple product* is defined as

$$\mathbf{A}\times(\mathbf{B}\times\mathbf{C}) = (\mathbf{A}\cdot\mathbf{C})\mathbf{B} - (\mathbf{A}\cdot\mathbf{B})\mathbf{C} \tag{7}$$

Since $(\mathbf{A}\times\mathbf{B})\times\mathbf{C} = (\mathbf{A}\cdot\mathbf{C})\mathbf{B} - (\mathbf{B}\cdot\mathbf{C})\mathbf{A}$, it is clear that $\mathbf{A}\times(\mathbf{B}\times\mathbf{C}) \neq (\mathbf{A}\times\mathbf{B})\times\mathbf{C}$.

DERIVATIVES OF VECTORS

If to each value assumed by a scalar variable u there corresponds a vector $\mathbf{A}(u)$, or briefly $\mathbf{A}$, then $\mathbf{A}(u)$ is called a (vector) function of u. The derivative of $\mathbf{A}(u)$ is defined as

$$\frac{d\mathbf{A}}{du} = \lim_{\Delta u \to 0} \frac{\mathbf{A}(u+\Delta u) - \mathbf{A}(u)}{\Delta u} \tag{8}$$

provided this limit exists. If $\mathbf{A}(u) = A_1(u)\mathbf{i} + A_2(u)\mathbf{j} + A_3(u)\mathbf{k}$, then

$$\frac{d\mathbf{A}}{du} = \frac{dA_1}{du}\mathbf{i} + \frac{dA_2}{du}\mathbf{j} + \frac{dA_3}{du}\mathbf{k} \tag{9}$$

Similarly we can define higher derivatives. For example the second derivative of $\mathbf{A}(u)$ if it exists is given by

$$\frac{d^2\mathbf{A}}{du^2} = \frac{d^2A_1}{du^2}\mathbf{i} + \frac{d^2A_2}{du^2}\mathbf{j} + \frac{d^2A_3}{du^2}\mathbf{k} \tag{10}$$

Example. If $\mathbf{A} = (2u^2 - 3u)\mathbf{i} + 5\cos u\,\mathbf{j} - 3\sin u\,\mathbf{k}$, then

$$\frac{d\mathbf{A}}{du} = (4u-3)\mathbf{i} - 5\sin u\,\mathbf{j} - 3\cos u\,\mathbf{k}, \qquad \frac{d^2\mathbf{A}}{du^2} = 4\mathbf{i} - 5\cos u\,\mathbf{j} + 3\sin u\,\mathbf{k}$$

The usual rules of differentiation familiar in the calculus can be extended to vectors, although order of factors in products may be important. For example if $\phi(u)$ is a scalar function while $\mathbf{A}(u)$ and $\mathbf{B}(u)$ are vector functions, then

$$\frac{d}{du}(\phi\mathbf{A}) = \phi\frac{d\mathbf{A}}{du} + \frac{d\phi}{du}\mathbf{A} \tag{11}$$

$$\frac{d}{du}(\mathbf{A}\cdot\mathbf{B}) = \mathbf{A}\cdot\frac{d\mathbf{B}}{du} + \frac{d\mathbf{A}}{du}\cdot\mathbf{B} \tag{12}$$

$$\frac{d}{du}(\mathbf{A}\times\mathbf{B}) = \mathbf{A}\times\frac{d\mathbf{B}}{du} + \frac{d\mathbf{A}}{du}\times\mathbf{B} \tag{13}$$

INTEGRALS OF VECTORS

Let $\mathbf{A}(u) = A_1(u)\mathbf{i} + A_2(u)\mathbf{j} + A_3(u)\mathbf{k}$ be a vector function of u. We define the *indefinite integral* of $\mathbf{A}(u)$ as

$$\int \mathbf{A}(u)\,du = \mathbf{i}\int A_1(u)\,du + \mathbf{j}\int A_2(u)\,du + \mathbf{k}\int A_3(u)\,du \tag{14}$$

If there exists a vector function $\mathbf{B}(u)$ such that $\mathbf{A}(u) = \frac{d}{du}\{\mathbf{B}(u)\}$, then

$$\int \mathbf{A}(u)\,du = \int \frac{d}{du}\{\mathbf{B}(u)\}\,du = \mathbf{B}(u) + \mathbf{c} \tag{15}$$

where $\mathbf{c}$ is an arbitrary constant vector independent of u. The *definite integral* between limits $u=\alpha$ and $u=\beta$ is in such case, as in elementary calculus, given by

$$\int_\alpha^\beta \mathbf{A}(u)\,du = \int_\alpha^\beta \frac{d}{du}\{\mathbf{B}(u)\}\,du = \mathbf{B}(u) + \mathbf{c}\Big|_\alpha^\beta = \mathbf{B}(\beta) - \mathbf{B}(\alpha) \tag{16}$$

The definite integral can also be defined as a limit of a sum analogous to that of elementary calculus.

VELOCITY

Suppose that a particle moves along a path or curve C [Fig. 1-9 below]. Let the position vector of point P at time t be $\mathbf{r} = \mathbf{r}(t)$ while the position vector of point Q at time $t + \Delta t$ is

$\mathbf{r}+\Delta\mathbf{r} = \mathbf{r}(t+\Delta t)$. Then the *velocity* (also called the *instantaneous velocity*) of the particle at P is given by

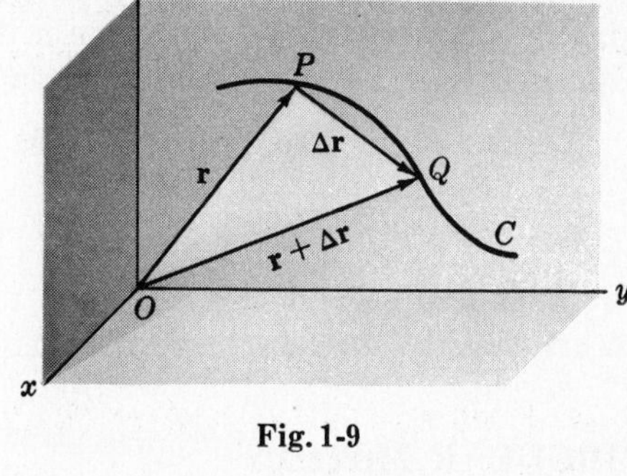

Fig. 1-9

$$\mathbf{v} = \frac{d\mathbf{r}}{dt} = \lim_{\Delta t \to 0} \frac{\Delta \mathbf{r}}{\Delta t} = \lim_{\Delta t \to 0} \frac{\mathbf{r}(t+\Delta t) - \mathbf{r}(t)}{\Delta t} \tag{17}$$

and is a vector tangent to C at P.

If $\mathbf{r} = \mathbf{r}(t) = x(t)\mathbf{i} + y(t)\mathbf{j} + z(t)\mathbf{k} = x\mathbf{i} + y\mathbf{j} + z\mathbf{k}$, we can write

$$\mathbf{v} = \frac{d\mathbf{r}}{dt} = \frac{dx}{dt}\mathbf{i} + \frac{dy}{dt}\mathbf{j} + \frac{dz}{dt}\mathbf{k} \tag{18}$$

The magnitude of the velocity is called the *speed* and is given by

$$v = |\mathbf{v}| = \left|\frac{d\mathbf{r}}{dt}\right| = \sqrt{\left(\frac{dx}{dt}\right)^2 + \left(\frac{dy}{dt}\right)^2 + \left(\frac{dz}{dt}\right)^2} = \frac{ds}{dt} \tag{19}$$

where s is the *arc length* along C measured from some initial point to P.

ACCELERATION

If $\mathbf{v} = d\mathbf{r}/dt$ is the velocity of the particle, we define the *acceleration* (also called the *instantaneous acceleration*) of the particle at P as

$$\mathbf{a} = \frac{d\mathbf{v}}{dt} = \lim_{\Delta t \to 0} \frac{\mathbf{v}(t+\Delta t) - \mathbf{v}(t)}{\Delta t} \tag{20}$$

In terms of $\mathbf{r} = x\mathbf{i} + y\mathbf{j} + z\mathbf{k}$ the acceleration is

$$\mathbf{a} = \frac{d^2\mathbf{r}}{dt^2} = \frac{d^2x}{dt^2}\mathbf{i} + \frac{d^2y}{dt^2}\mathbf{j} + \frac{d^2z}{dt^2}\mathbf{k} \tag{21}$$

and its magnitude is

$$a = |\mathbf{a}| = \sqrt{\left(\frac{d^2x}{dt^2}\right)^2 + \left(\frac{d^2y}{dt^2}\right)^2 + \left(\frac{d^2z}{dt^2}\right)^2} \tag{22}$$

RELATIVE VELOCITY AND ACCELERATION

If two particles P_1 and P_2 are moving with respective velocities $\mathbf{v}_1$ and $\mathbf{v}_2$ and accelerations $\mathbf{a}_1$ and $\mathbf{a}_2$, the vectors

$$\mathbf{v}_{P_2/P_1} = \mathbf{v}_2 - \mathbf{v}_1 \quad \text{and} \quad \mathbf{a}_{P_2/P_1} = \mathbf{a}_2 - \mathbf{a}_1 \tag{23}$$

are respectively called the *relative velocity* and *relative acceleration* of P_2 with respect to P_1.

TANGENTIAL AND NORMAL ACCELERATION

Suppose that particle P with position vector $\mathbf{r} = \mathbf{r}(t)$ moves along curve C [Fig. 1-10]. We can consider a rectangular coordinate system moving with the particle and defined by the *unit tangent vector* $\mathbf{T}$, the *unit principal normal* $\mathbf{N}$ and the *unit binormal* $\mathbf{B}$ to curve C where

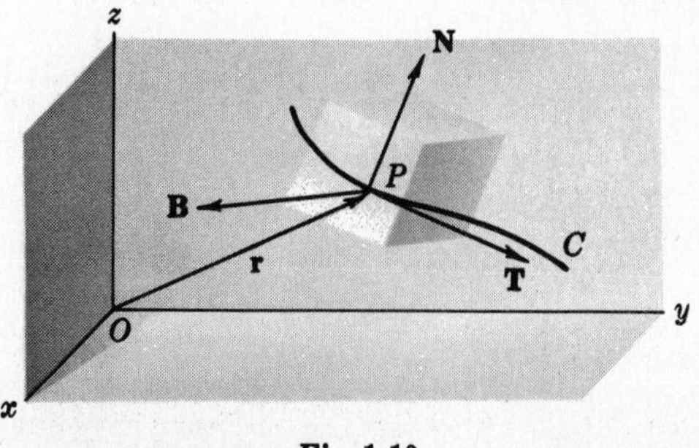

Fig. 1-10

$$\mathbf{T} = \frac{d\mathbf{r}}{ds}, \quad \mathbf{N} = R\frac{d\mathbf{T}}{ds}, \quad \mathbf{B} = \mathbf{T}\times\mathbf{N} \tag{24}$$

s being the arc length from some initial point to P and R the *radius of curvature* of C at P. The reciprocal of the radius of curvature is called the *curvature* and is given by $\kappa = 1/R$.

We can show [see Problem 1.35, page 20] that the acceleration along C is given by

$$\mathbf{a} = \frac{dv}{dt}\mathbf{T} + \frac{v^2}{R}\mathbf{N} \tag{25}$$

The first and second terms on the right are called the *tangential acceleration* and *normal* or *centripetal acceleration* respectively.

CIRCULAR MOTION

Suppose particle P moves on a circle C of radius R. If s is the arc length measured along C from A to P and θ is the corresponding angle subtended at the center O, then $s = R\theta$. Thus the magnitudes of the tangential velocity and acceleration are given respectively by

$$v = \frac{ds}{dt} = R\frac{d\theta}{dt} = R\omega \tag{26}$$

and

$$\frac{dv}{dt} = \frac{d^2s}{dt^2} = R\frac{d^2\theta}{dt^2} = R\alpha \tag{27}$$

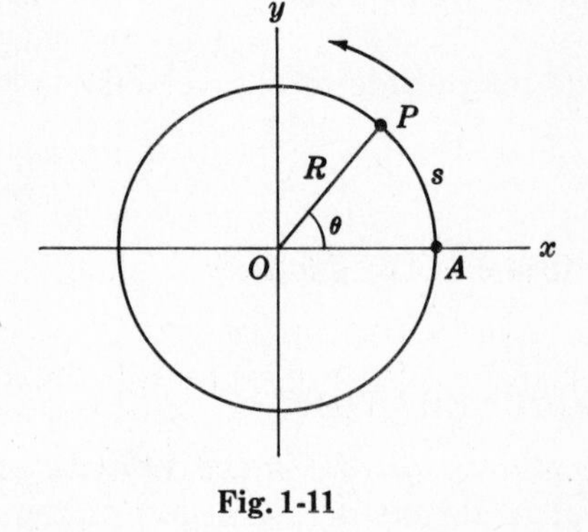

Fig. 1-11

We call $\omega = d\theta/dt$ and $\alpha = d^2\theta/dt^2$ the *angular speed* and *angular acceleration* respectively. The normal acceleration as seen from (25) is given by $v^2/R = \omega^2 R$.

NOTATION FOR TIME DERIVATIVES

We shall sometimes find it convenient to use dots placed over a symbol to denote derivatives with respect to time t, one dot for a first derivative, two dots for a second derivative, etc. Thus for example $\dot{\mathbf{r}} = d\mathbf{r}/dt$, $\ddot{\mathbf{r}} = d^2\mathbf{r}/dt^2$, $\dot{\mathbf{v}} = d\mathbf{v}/dt$, etc.

GRADIENT, DIVERGENCE AND CURL

If to each point (x, y, z) of a rectangular coordinate system there corresponds a vector $\mathbf{A}$, we say that $\mathbf{A} = \mathbf{A}(x, y, z)$ is a *vector function* of x, y, z. We also call $\mathbf{A}(x, y, z)$ a *vector field*. Similarly we call the (scalar) function $\phi(x, y, z)$ a *scalar field*.

It is convenient to consider a vector differential operator called *del* given by

$$\nabla = \mathbf{i}\frac{\partial}{\partial x} + \mathbf{j}\frac{\partial}{\partial y} + \mathbf{k}\frac{\partial}{\partial z} \tag{28}$$

Using this we define the following important quantities.

1. ***Gradient***
$$\nabla\phi = \left(\mathbf{i}\frac{\partial}{\partial x} + \mathbf{j}\frac{\partial}{\partial y} + \mathbf{k}\frac{\partial}{\partial z}\right)\phi = \mathbf{i}\frac{\partial\phi}{\partial x} + \mathbf{j}\frac{\partial\phi}{\partial y} + \mathbf{k}\frac{\partial\phi}{\partial z} \tag{29}$$
This is a vector called the *gradient* of ϕ and is also written grad ϕ.

2. ***Divergence***
$$\begin{aligned}\nabla\cdot\mathbf{A} &= \left(\mathbf{i}\frac{\partial}{\partial x} + \mathbf{j}\frac{\partial}{\partial y} + \mathbf{k}\frac{\partial}{\partial z}\right)\cdot(A_1\mathbf{i} + A_2\mathbf{j} + A_3\mathbf{k}) \\ &= \frac{\partial A_1}{\partial x} + \frac{\partial A_2}{\partial y} + \frac{\partial A_3}{\partial z}\end{aligned} \tag{30}$$
This is a scalar called the *divergence* of $\mathbf{A}$ and is also written div $\mathbf{A}$.

3. *Curl*

$$\nabla \times \mathbf{A} = \left(\mathbf{i}\frac{\partial}{\partial x} + \mathbf{j}\frac{\partial}{\partial y} + \mathbf{k}\frac{\partial}{\partial z}\right) \times (A_1\mathbf{i} + A_2\mathbf{j} + A_3\mathbf{k})$$

$$= \begin{vmatrix} \mathbf{i} & \mathbf{j} & \mathbf{k} \\ \frac{\partial}{\partial x} & \frac{\partial}{\partial y} & \frac{\partial}{\partial z} \\ A_1 & A_2 & A_3 \end{vmatrix} \tag{31}$$

$$= \left(\frac{\partial A_3}{\partial y} - \frac{\partial A_2}{\partial z}\right)\mathbf{i} + \left(\frac{\partial A_1}{\partial z} - \frac{\partial A_3}{\partial x}\right)\mathbf{j} + \left(\frac{\partial A_2}{\partial x} - \frac{\partial A_1}{\partial y}\right)\mathbf{k}$$

This is a vector called the *curl* of A and is also written curl $\mathbf{A}$.

Two important identities are

$$\text{div curl } \mathbf{A} = \nabla \cdot (\nabla \times \mathbf{A}) = 0 \tag{32}$$

$$\text{curl grad } \phi = \nabla \times (\nabla \phi) = 0 \tag{33}$$

LINE INTEGRALS

Let $\mathbf{r}(t) = x(t)\mathbf{i} + y(t)\mathbf{j} + z(t)\mathbf{k}$, where $\mathbf{r}(t)$ is the position vector of (x, y, z), define a curve C joining points P_1 and P_2 corresponding to $t = t_1$ and $t = t_2$ respectively. Let $\mathbf{A} = \mathbf{A}(x, y, z) = A_1\mathbf{i} + A_2\mathbf{j} + A_3\mathbf{k}$ be a vector function of position (vector field). The integral of the tangential component of $\mathbf{A}$ along C from P_1 to P_2, written as

$$\int_{P_1}^{P_2} \mathbf{A} \cdot d\mathbf{r} = \int_C \mathbf{A} \cdot d\mathbf{r} = \int_C A_1\,dx + A_2\,dy + A_3\,dz \tag{34}$$

is an example of a *line integral*.

If C is a closed curve (which we shall suppose is a *simple* closed curve, i.e. a curve which does not intersect itself anywhere) then the integral is often denoted by

$$\oint_C \mathbf{A} \cdot d\mathbf{r} = \oint_C A_1\,dx + A_2\,dy + A_3\,dz \tag{35}$$

In general, a line integral has a value which depends on the path. For methods of evaluation see Problems 1.39 and 1.40.

INDEPENDENCE OF THE PATH

The line integral (*34*) will be independent of the path joining P_1 and P_2 if and only if $\mathbf{A} = \nabla\phi$, or equivalently $\nabla \times \mathbf{A} = 0$. In such case its value is given by

$$\int_{P_1}^{P_2} \mathbf{A} \cdot d\mathbf{r} = \int_{P_1}^{P_2} d\phi = \phi(P_2) - \phi(P_1) = \phi(x_2, y_2, z_2) - \phi(x_1, y_1, z_1) \tag{36}$$

assuming that the coordinates of P_1 and P_2 are (x_1, y_1, z_1) and (x_2, y_2, z_2) respectively while $\phi(x, y, z)$ has continuous partial derivatives. The integral (*35*) in this case is zero.

FREE, SLIDING AND BOUND VECTORS

Up to now we have dealt with vectors which are specified by magnitude and direction only. Such vectors are called *free vectors*. Any two free vectors are equal as long as they have the same magnitude and direction [see Fig. 1-12(*a*) below].

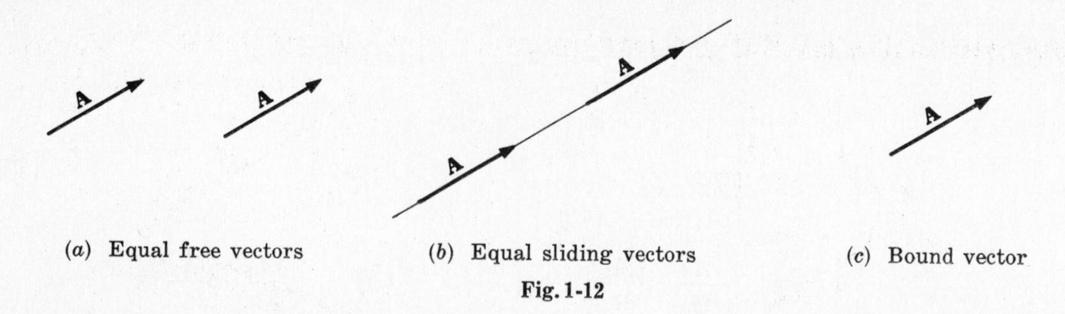

(a) Equal free vectors (b) Equal sliding vectors (c) Bound vector

Fig. 1-12

Sometimes in practice the particular *line of action* of a vector is important. In such case two vectors are equal if and only if they have the same magnitude, direction and line of action. Such vectors are often called *sliding vectors* [see Fig. 1-12(*b*)].

Sometimes it is important to specify the *point of action* of a vector. Such a vector [see Fig. 1-12(*c*)] is called a *bound vector*. In this case two vectors will be equal if and only if they are identical.

Most cases with which we shall deal involve free vectors. Cases where sliding vectors or bound vectors need to be employed will in general be clear from the context.

Solved Problems

VECTOR ALGEBRA

1.1. Show that addition of vectors is commutative, i.e. $\mathbf{A}+\mathbf{B} = \mathbf{B}+\mathbf{A}$. See Fig. 1-13 below.

$$\mathbf{OP}+\mathbf{PQ} = \mathbf{OQ} \quad \text{or} \quad \mathbf{A}+\mathbf{B} = \mathbf{C}$$

and

$$\mathbf{OR}+\mathbf{RQ} = \mathbf{OQ} \quad \text{or} \quad \mathbf{B}+\mathbf{A} = \mathbf{C}$$

Then $\mathbf{A}+\mathbf{B} = \mathbf{B}+\mathbf{A}$.

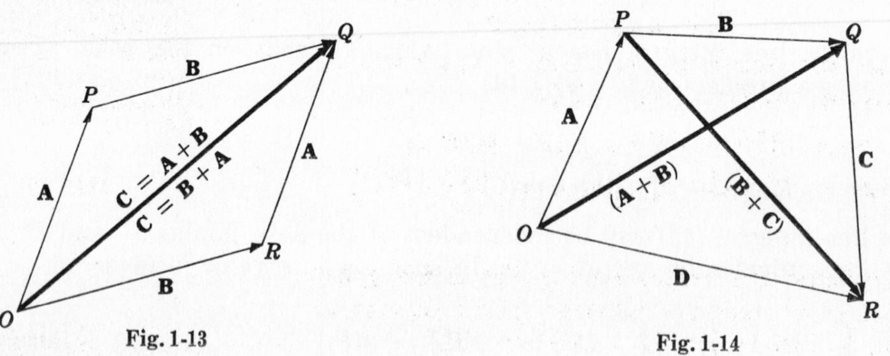

Fig. 1-13 Fig. 1-14

1.2. Show that the addition of vectors is associative, i.e. $\mathbf{A}+(\mathbf{B}+\mathbf{C}) = (\mathbf{A}+\mathbf{B})+\mathbf{C}$. See Fig. 1-14 above.

$$\mathbf{OP}+\mathbf{PQ} = \mathbf{OQ} = (\mathbf{A}+\mathbf{B}) \quad \text{and} \quad \mathbf{PQ}+\mathbf{QR} = \mathbf{PR} = (\mathbf{B}+\mathbf{C})$$

Since

$$\mathbf{OP}+\mathbf{PR} = \mathbf{OR} = \mathbf{D}, \quad \text{i.e.} \quad \mathbf{A}+(\mathbf{B}+\mathbf{C}) = \mathbf{D}$$

$$\mathbf{OQ}+\mathbf{QR} = \mathbf{OR} = \mathbf{D}, \quad \text{i.e.} \quad (\mathbf{A}+\mathbf{B})+\mathbf{C} = \mathbf{D}$$

we have $\mathbf{A}+(\mathbf{B}+\mathbf{C}) = (\mathbf{A}+\mathbf{B})+\mathbf{C}$.

Extensions of the results of Problems 1.1 and 1.2 show that the order of addition of any number of vectors is immaterial.

1.3. Given vectors **A**, **B** and **C** [Fig. 1-15(*a*)] construct (*a*) $\mathbf{A}-\mathbf{B}+2\mathbf{C}$, (*b*) $3\mathbf{C}-\frac{1}{2}(2\mathbf{A}-\mathbf{B})$.

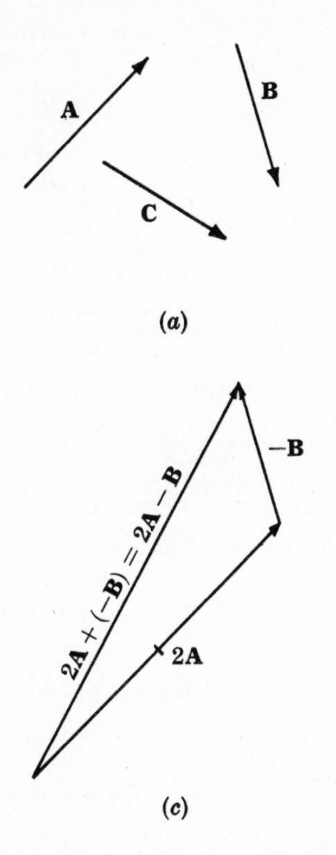

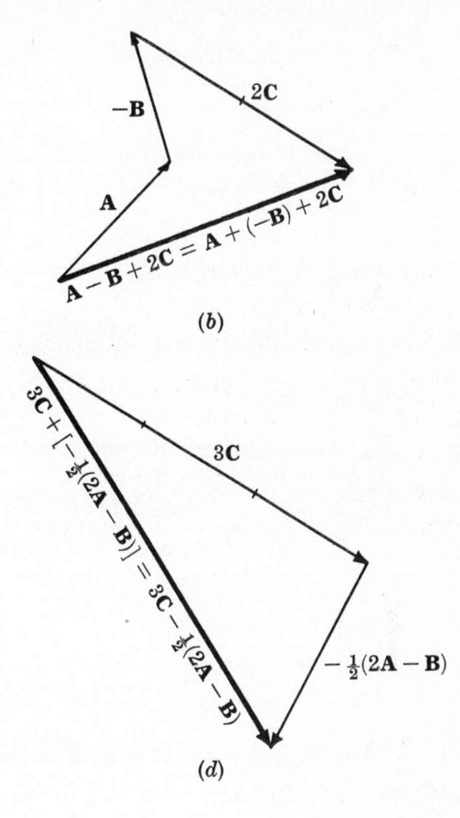

Fig. 1-15

1.4. Prove that the magnitude A of the vector $\mathbf{A} = A_1\mathbf{i} + A_2\mathbf{j} + A_3\mathbf{k}$ is $A = \sqrt{A_1^2 + A_2^2 + A_3^2}$. See Fig. 1-16.

By the Pythagorean theorem,

$$(\overline{OP})^2 = (\overline{OQ})^2 + (\overline{QP})^2$$

where $\overline{OP}$ denotes the magnitude of vector **OP**, etc. Similarly, $(\overline{OQ})^2 = (\overline{OR})^2 + (\overline{RQ})^2$.

Then $(\overline{OP})^2 = (\overline{OR})^2 + (\overline{RQ})^2 + (\overline{QP})^2$ or

$$A^2 = A_1^2 + A_2^2 + A_3^2, \quad \text{i.e.} \quad A = \sqrt{A_1^2 + A_2^2 + A_3^2}$$

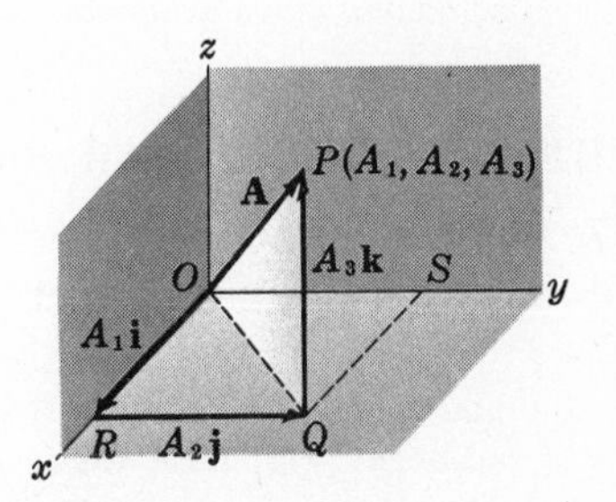

Fig. 1-16

1.5. Determine the vector having the initial point $P(x_1, y_1, z_1)$ and the terminal point $Q(x_2, y_2, z_2)$ and find its magnitude. See Fig. 1-17.

The position vector of P is $\mathbf{r}_1 = x_1\mathbf{i} + y_1\mathbf{j} + z_1\mathbf{k}$.

The position vector of Q is $\mathbf{r}_2 = x_2\mathbf{i} + y_2\mathbf{j} + z_2\mathbf{k}$.

$\mathbf{r}_1 + \mathbf{PQ} = \mathbf{r}_2$ or

$$\begin{aligned}\mathbf{PQ} = \mathbf{r}_2 - \mathbf{r}_1 &= (x_2\mathbf{i} + y_2\mathbf{j} + z_2\mathbf{k}) - (x_1\mathbf{i} + y_1\mathbf{j} + z_1\mathbf{k}) \\ &= (x_2 - x_1)\mathbf{i} + (y_2 - y_1)\mathbf{j} + (z_2 - z_1)\mathbf{k}\end{aligned}$$

$$\begin{aligned}\text{Magnitude of } \mathbf{PQ} &= \overline{PQ} \\ &= \sqrt{(x_2 - x_1)^2 + (y_2 - y_1)^2 + (z_2 - z_1)^2}\end{aligned}$$

Note that this is the distance between points P and Q.

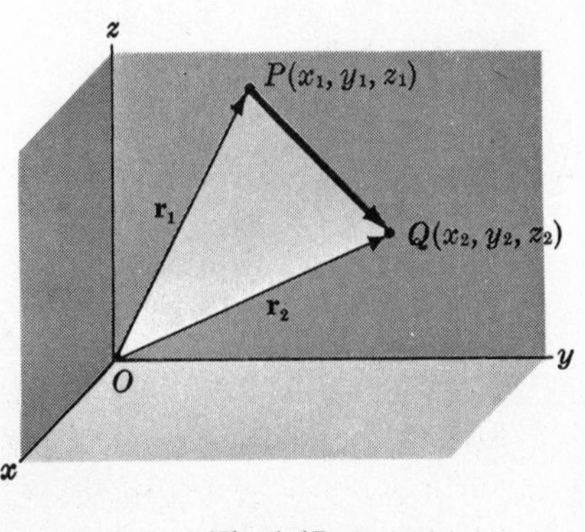

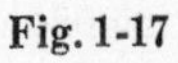
Fig. 1-17

1.6. Find (*a*) graphically and (*b*) analytically the sum or resultant of the following displacements:

A, 10 ft northwest; **B**, 20 ft 30° north of east; **C**, 35 ft due south. See Fig. 1-18.

Graphically.

At the terminal point of **A** place the initial point of **B**. At the terminal point of **B** place the initial point of **C**.

The resultant **D** is formed by joining the initial point of **A** to the terminal point of **C**, i.e. $\mathbf{D} = \mathbf{A} + \mathbf{B} + \mathbf{C}$.

The resultant is measured to have magnitude of 4.1 units = 20.5 ft and direction 60° south of east.

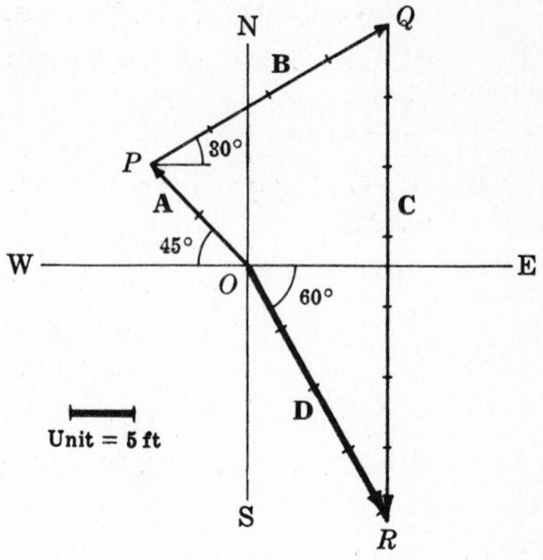

Fig. 1-18

Analytically.

From Fig. 1-18, if **i** and **j** are unit vectors in the E and N directions, we have

$$\mathbf{A} = -10 \cos 45^\circ \, \mathbf{i} + 10 \sin 45^\circ \, \mathbf{j}$$

$$\mathbf{B} = 20 \cos 30^\circ \, \mathbf{i} + 20 \sin 30^\circ \, \mathbf{j}$$

$$\mathbf{C} = -35\mathbf{j}$$

Then the resultant is

$$\begin{aligned} \mathbf{D} = \mathbf{A} + \mathbf{B} + \mathbf{C} &= (-10 \cos 45^\circ + 20 \cos 30^\circ)\mathbf{i} + (10 \sin 45^\circ + 20 \sin 30^\circ - 35)\mathbf{j} \\ &= (-5\sqrt{2} + 10\sqrt{3})\mathbf{i} + (5\sqrt{2} + 10 - 35)\mathbf{j} = 10.25\mathbf{i} - 17.93\mathbf{j} \end{aligned}$$

Thus the magnitude of **D** is $\sqrt{(10.25)^2 + (17.93)^2} = 20.65$ ft and the direction is

$$\tan^{-1} 17.93/10.25 = \tan^{-1} 1.749 = 60^\circ 45' \text{ south of east}$$

Note that although the graphical and analytical results agree fairly well, the analytical result is of course more accurate.

THE DOT OR SCALAR PRODUCT

1.7. Prove that the projection of **A** on **B** is equal to $\mathbf{A} \cdot \mathbf{b}$, where **b** is a unit vector in the direction of **B**.

Through the initial and terminal points of **A** pass planes perpendicular to **B** at G and H respectively as in the adjacent Fig. 1-19; then

$$\text{Projection of } \mathbf{A} \text{ on } \mathbf{B} = \overline{GH} = \overline{EF} = A \cos \theta = \mathbf{A} \cdot \mathbf{b}$$

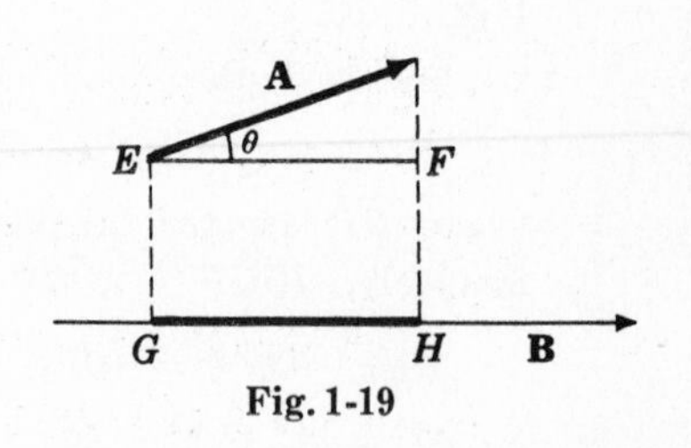

Fig. 1-19

1.8. Prove $\mathbf{A} \cdot (\mathbf{B} + \mathbf{C}) = \mathbf{A} \cdot \mathbf{B} + \mathbf{A} \cdot \mathbf{C}$.

Let **a** be a unit vector in the direction of **A**; then [see Fig. 1-20]

$$\text{Projection of } (\mathbf{B} + \mathbf{C}) \text{ on } \mathbf{A} = \text{projection of } \mathbf{B} \text{ on } \mathbf{A} + \text{projection of } \mathbf{C} \text{ on } \mathbf{A}$$

$$(\mathbf{B} + \mathbf{C}) \cdot \mathbf{a} = \mathbf{B} \cdot \mathbf{a} + \mathbf{C} \cdot \mathbf{a}$$

Multiplying by A,

$$(\mathbf{B} + \mathbf{C}) \cdot A\mathbf{a} = \mathbf{B} \cdot A\mathbf{a} + \mathbf{C} \cdot A\mathbf{a}$$

and

$$(\mathbf{B} + \mathbf{C}) \cdot \mathbf{A} = \mathbf{B} \cdot \mathbf{A} + \mathbf{C} \cdot \mathbf{A}$$

Then by the commutative law for dot products,

$$\mathbf{A} \cdot (\mathbf{B} + \mathbf{C}) = \mathbf{A} \cdot \mathbf{B} + \mathbf{A} \cdot \mathbf{C}$$

and the distributive law is valid.

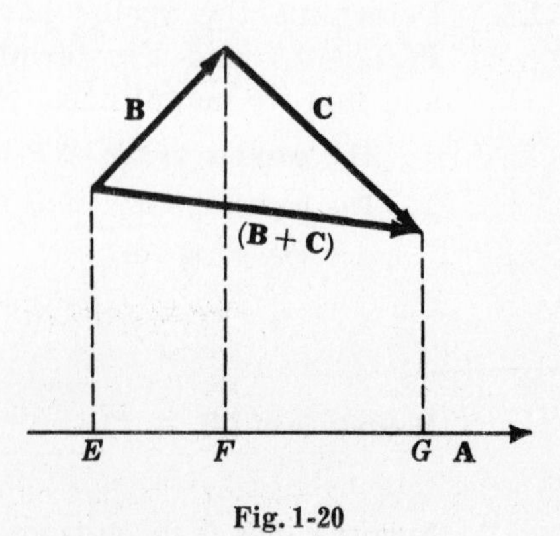

Fig. 1-20

1.9. Evaluate each of the following.

(a) $\mathbf{i}\cdot\mathbf{i} \;=\; |\mathbf{i}|\,|\mathbf{i}|\cos 0° \;=\; (1)(1)(1) \;=\; 1$

(b) $\mathbf{i}\cdot\mathbf{k} \;=\; |\mathbf{i}|\,|\mathbf{k}|\cos 90° \;=\; (1)(1)(0) \;=\; 0$

(c) $\mathbf{k}\cdot\mathbf{j} \;=\; |\mathbf{k}|\,|\mathbf{j}|\cos 90° \;=\; (1)(1)(0) \;=\; 0$

(d) $\mathbf{j}\cdot(2\mathbf{i}-3\mathbf{j}+\mathbf{k}) \;=\; 2\mathbf{j}\cdot\mathbf{i} - 3\mathbf{j}\cdot\mathbf{j} + \mathbf{j}\cdot\mathbf{k} \;=\; 0-3+0 \;=\; -3$

(e) $(2\mathbf{i}-\mathbf{j})\cdot(3\mathbf{i}+\mathbf{k}) \;=\; 2\mathbf{i}\cdot(3\mathbf{i}+\mathbf{k}) - \mathbf{j}\cdot(3\mathbf{i}+\mathbf{k}) \;=\; 6\mathbf{i}\cdot\mathbf{i} + 2\mathbf{i}\cdot\mathbf{k} - 3\mathbf{j}\cdot\mathbf{i} - \mathbf{j}\cdot\mathbf{k}$
$\;=\; 6+0-0-0 \;=\; 6$

1.10. If $\mathbf{A} = A_1\mathbf{i}+A_2\mathbf{j}+A_3\mathbf{k}$ and $\mathbf{B} = B_1\mathbf{i}+B_2\mathbf{j}+B_3\mathbf{k}$, prove that $\mathbf{A}\cdot\mathbf{B} = A_1B_1+A_2B_2+A_3B_3$.

$$\begin{aligned}
\mathbf{A}\cdot\mathbf{B} &= (A_1\mathbf{i}+A_2\mathbf{j}+A_3\mathbf{k})\cdot(B_1\mathbf{i}+B_2\mathbf{j}+B_3\mathbf{k})\\
&= A_1\mathbf{i}\cdot(B_1\mathbf{i}+B_2\mathbf{j}+B_3\mathbf{k}) + A_2\mathbf{j}\cdot(B_1\mathbf{i}+B_2\mathbf{j}+B_3\mathbf{k}) + A_3\mathbf{k}\cdot(B_1\mathbf{i}+B_2\mathbf{j}+B_3\mathbf{k})\\
&= A_1B_1\mathbf{i}\cdot\mathbf{i} + A_1B_2\mathbf{i}\cdot\mathbf{j} + A_1B_3\mathbf{i}\cdot\mathbf{k} + A_2B_1\mathbf{j}\cdot\mathbf{i} + A_2B_2\mathbf{j}\cdot\mathbf{j} + A_2B_3\mathbf{j}\cdot\mathbf{k}\\
&\quad + A_3B_1\mathbf{k}\cdot\mathbf{i} + A_3B_2\mathbf{k}\cdot\mathbf{j} + A_3B_3\mathbf{k}\cdot\mathbf{k}\\
&= A_1B_1 + A_2B_2 + A_3B_3
\end{aligned}$$

since $\mathbf{i}\cdot\mathbf{i} = \mathbf{j}\cdot\mathbf{j} = \mathbf{k}\cdot\mathbf{k} = 1$ and all other dot products are zero.

1.11. If $\mathbf{A} = A_1\mathbf{i}+A_2\mathbf{j}+A_3\mathbf{k}$, show that $A = \sqrt{\mathbf{A}\cdot\mathbf{A}} = \sqrt{A_1^2+A_2^2+A_3^2}$.

$\mathbf{A}\cdot\mathbf{A} \;=\; (A)(A)\cos 0° \;=\; A^2.$ Then $A = \sqrt{\mathbf{A}\cdot\mathbf{A}}$.

Also, $\mathbf{A}\cdot\mathbf{A} \;=\; (A_1\mathbf{i}+A_2\mathbf{j}+A_3\mathbf{k})\cdot(A_1\mathbf{i}+A_2\mathbf{j}+A_3\mathbf{k})$
$\;=\; (A_1)(A_1)+(A_2)(A_2)+(A_3)(A_3) \;=\; A_1^2+A_2^2+A_3^2$

by Problem 1.10, taking $\mathbf{B}=\mathbf{A}$.

Then $A \;=\; \sqrt{\mathbf{A}\cdot\mathbf{A}} \;=\; \sqrt{A_1^2+A_2^2+A_3^2}$ is the magnitude of $\mathbf{A}$. Sometimes $\mathbf{A}\cdot\mathbf{A}$ is written $\mathbf{A}^2$.

1.12. Find the acute angle between the diagonals of a quadrilateral having vertices at (0, 0, 0), (3, 2, 0), (4, 6, 0), (1, 3, 0) [Fig. 1-21].

We have $\mathbf{OA} = 3\mathbf{i}+2\mathbf{j}$, $\mathbf{OB} = 4\mathbf{i}+6\mathbf{j}$, $\mathbf{OC} = \mathbf{i}+3\mathbf{j}$ from which

$$\mathbf{CA} \;=\; \mathbf{OA}-\mathbf{OC} \;=\; 2\mathbf{i}-\mathbf{j}$$

Then

$$\mathbf{OB}\cdot\mathbf{CA} \;=\; |\mathbf{OB}|\,|\mathbf{CA}|\cos\theta$$

i.e.

$$(4\mathbf{i}+6\mathbf{j})\cdot(2\mathbf{i}-\mathbf{j}) \;=\; \sqrt{(4)^2+(6)^2}\,\sqrt{(2)^2+(-1)^2}\cos\theta$$

from which $\cos\theta = 2/(\sqrt{52}\,\sqrt{5}) = .1240$ and $\theta = 82°53'$.

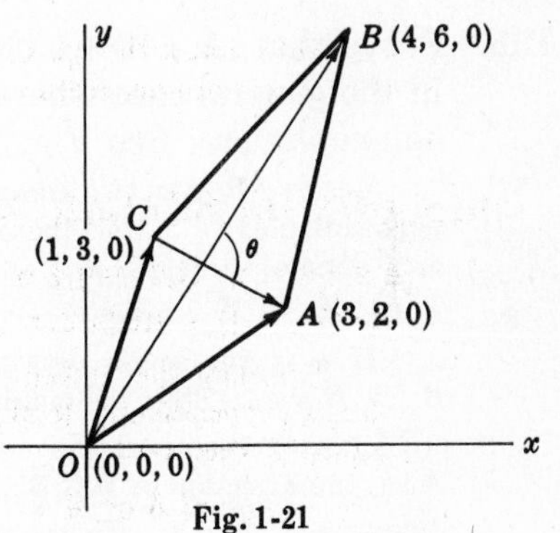

Fig. 1-21

THE CROSS OR VECTOR PRODUCT

1.13. Prove $\mathbf{A}\times\mathbf{B} = -\mathbf{B}\times\mathbf{A}$.

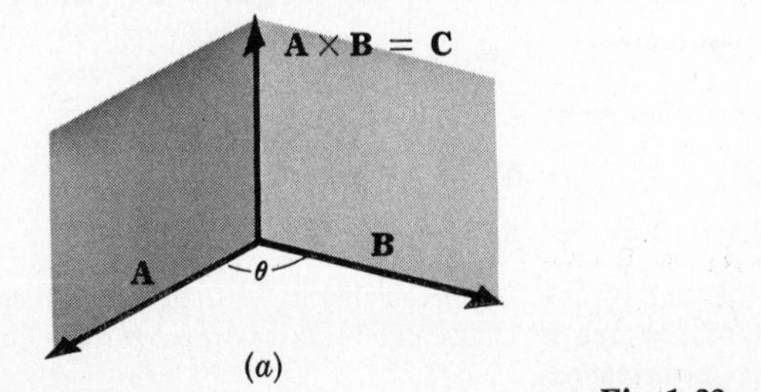

(a)

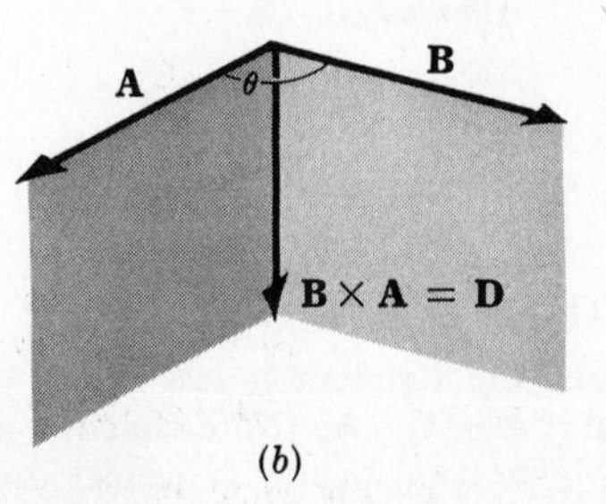

(b)

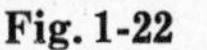

Fig. 1-22

$\mathbf{A}\times\mathbf{B} = \mathbf{C}$ has magnitude $AB\sin\theta$ and direction such that $\mathbf{A}$, $\mathbf{B}$ and $\mathbf{C}$ form a right-handed system [Fig. 1-22(a) above].

$\mathbf{B}\times\mathbf{A} = \mathbf{D}$ has magnitude $BA \sin\theta$ and direction such that **B**, **A** and **D** form a right-handed system [Fig. 1-22(*b*) above].

Then **D** has the same magnitude as **C** but is opposite in direction, i.e. $\mathbf{C} = -\mathbf{D}$ or $\mathbf{A}\times\mathbf{B} = -\mathbf{B}\times\mathbf{A}$.

The commutative law for cross products is not valid.

1.14. Prove that

$$\mathbf{A}\times(\mathbf{B}+\mathbf{C}) = \mathbf{A}\times\mathbf{B} + \mathbf{A}\times\mathbf{C}$$

for the case where **A** is perpendicular to **B** and also to **C**.

Since **A** is perpendicular to **B**, $\mathbf{A}\times\mathbf{B}$ is a vector perpendicular to the plane of **A** and **B** and having magnitude $AB \sin 90° = AB$ or magnitude of $A\mathbf{B}$. This is equivalent to multiplying vector **B** by A and rotating the resultant vector through 90° to the position shown in Fig. 1-23.

Similarly, $\mathbf{A}\times\mathbf{C}$ is the vector obtained by multiplying **C** by A and rotating the resultant vector through 90° to the position shown.

In like manner, $\mathbf{A}\times(\mathbf{B}+\mathbf{C})$ is the vector obtained by multiplying $\mathbf{B}+\mathbf{C}$ by A and rotating the resultant vector through 90° to the position shown.

Since $\mathbf{A}\times(\mathbf{B}+\mathbf{C})$ is the diagonal of the parallelogram with $\mathbf{A}\times\mathbf{B}$ and $\mathbf{A}\times\mathbf{C}$ as sides, we have $\mathbf{A}\times(\mathbf{B}+\mathbf{C}) = \mathbf{A}\times\mathbf{B} + \mathbf{A}\times\mathbf{C}$.

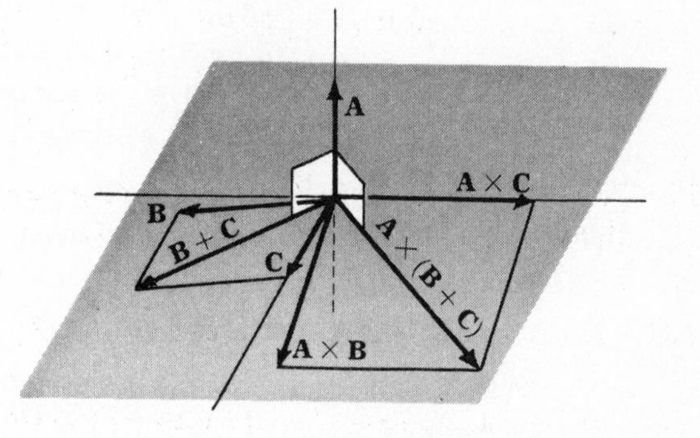

Fig. 1-23

1.15. Prove that $\mathbf{A}\times(\mathbf{B}+\mathbf{C}) = \mathbf{A}\times\mathbf{B} + \mathbf{A}\times\mathbf{C}$ in the general case where **A**, **B** and **C** are non-coplanar. See Fig. 1-24.

Resolve **B** into two component vectors, one perpendicular to **A** and the other parallel to **A**, and denote them by $\mathbf{B}_\perp$ and $\mathbf{B}_\parallel$ respectively. Then $\mathbf{B} = \mathbf{B}_\perp + \mathbf{B}_\parallel$.

If θ is the angle between **A** and **B**, then $B_\perp = B\sin\theta$. Thus the magnitude of $\mathbf{A}\times\mathbf{B}_\perp$ is $AB\sin\theta$, the same as the magnitude of $\mathbf{A}\times\mathbf{B}$. Also, the direction of $\mathbf{A}\times\mathbf{B}_\perp$ is the same as the direction of $\mathbf{A}\times\mathbf{B}$. Hence $\mathbf{A}\times\mathbf{B}_\perp = \mathbf{A}\times\mathbf{B}$.

Similarly if **C** is resolved into two component vectors $\mathbf{C}_\parallel$ and $\mathbf{C}_\perp$, parallel and perpendicular respectively to **A**, then $\mathbf{A}\times\mathbf{C}_\perp = \mathbf{A}\times\mathbf{C}$.

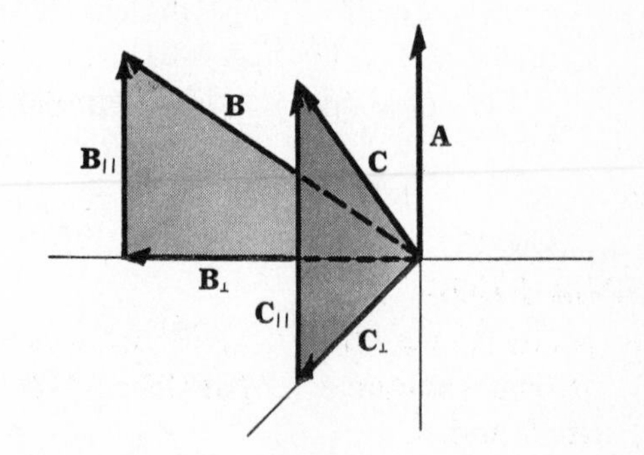

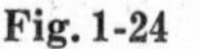

Fig. 1-24

Also, since $\mathbf{B}+\mathbf{C} = \mathbf{B}_\perp + \mathbf{B}_\parallel + \mathbf{C}_\perp + \mathbf{C}_\parallel = (\mathbf{B}_\perp + \mathbf{C}_\perp) + (\mathbf{B}_\parallel + \mathbf{C}_\parallel)$ it follows that

$$\mathbf{A}\times(\mathbf{B}_\perp + \mathbf{C}_\perp) = \mathbf{A}\times(\mathbf{B}+\mathbf{C})$$

Now $\mathbf{B}_\perp$ and $\mathbf{C}_\perp$ are vectors perpendicular to **A** and so by Problem 1.14,

$$\mathbf{A}\times(\mathbf{B}_\perp + \mathbf{C}_\perp) = \mathbf{A}\times\mathbf{B}_\perp + \mathbf{A}\times\mathbf{C}_\perp$$

Then

$$\mathbf{A}\times(\mathbf{B}+\mathbf{C}) = \mathbf{A}\times\mathbf{B} + \mathbf{A}\times\mathbf{C}$$

and the distributive law holds. Multiplying by -1, using Problem 1.13, this becomes $(\mathbf{B}+\mathbf{C})\times\mathbf{A} = \mathbf{B}\times\mathbf{A} + \mathbf{C}\times\mathbf{A}$. Note that the order of factors in cross products is important. The usual laws of algebra apply only if proper order is maintained.

1.16. If $\mathbf{A} = A_1\mathbf{i} + A_2\mathbf{j} + A_3\mathbf{k}$ and $\mathbf{B} = B_1\mathbf{i} + B_2\mathbf{j} + B_3\mathbf{k}$, prove that $\mathbf{A}\times\mathbf{B} = \begin{vmatrix} \mathbf{i} & \mathbf{j} & \mathbf{k} \\ A_1 & A_2 & A_3 \\ B_1 & B_2 & B_3 \end{vmatrix}$.

$$\begin{aligned}
\mathbf{A}\times\mathbf{B} &= (A_1\mathbf{i}+A_2\mathbf{j}+A_3\mathbf{k})\times(B_1\mathbf{i}+B_2\mathbf{j}+B_3\mathbf{k})\\
&= A_1\mathbf{i}\times(B_1\mathbf{i}+B_2\mathbf{j}+B_3\mathbf{k}) + A_2\mathbf{j}\times(B_1\mathbf{i}+B_2\mathbf{j}+B_3\mathbf{k}) + A_3\mathbf{k}\times(B_1\mathbf{i}+B_2\mathbf{j}+B_3\mathbf{k})\\
&= A_1B_1\mathbf{i}\times\mathbf{i} + A_1B_2\mathbf{i}\times\mathbf{j} + A_1B_3\mathbf{i}\times\mathbf{k} + A_2B_1\mathbf{j}\times\mathbf{i} + A_2B_2\mathbf{j}\times\mathbf{j} + A_2B_3\mathbf{j}\times\mathbf{k}\\
&\quad + A_3B_1\mathbf{k}\times\mathbf{i} + A_3B_2\mathbf{k}\times\mathbf{j} + A_3B_3\mathbf{k}\times\mathbf{k}\\
&= (A_2B_3-A_3B_2)\mathbf{i} + (A_3B_1-A_1B_3)\mathbf{j} + (A_1B_2-A_2B_1)\mathbf{k} = \begin{vmatrix}\mathbf{i} & \mathbf{j} & \mathbf{k}\\ A_1 & A_2 & A_3\\ B_1 & B_2 & B_3\end{vmatrix}
\end{aligned}$$

1.17. If $\mathbf{A} = 3\mathbf{i}-\mathbf{j}+2\mathbf{k}$ and $\mathbf{B} = 2\mathbf{i}+3\mathbf{j}-\mathbf{k}$, find $\mathbf{A}\times\mathbf{B}$.

$$\mathbf{A}\times\mathbf{B} = \begin{vmatrix}\mathbf{i} & \mathbf{j} & \mathbf{k}\\ 3 & -1 & 2\\ 2 & 3 & -1\end{vmatrix} = \mathbf{i}\begin{vmatrix}-1 & 2\\ 3 & -1\end{vmatrix} - \mathbf{j}\begin{vmatrix}3 & 2\\ 2 & -1\end{vmatrix} + \mathbf{k}\begin{vmatrix}3 & -1\\ 2 & 3\end{vmatrix}$$
$$= -5\mathbf{i} + 7\mathbf{j} + 11\mathbf{k}$$

1.18. Prove that the area of a parallelogram with sides $\mathbf{A}$ and $\mathbf{B}$ is $|\mathbf{A}\times\mathbf{B}|$.

$$\begin{aligned}
\text{Area of parallelogram} &= h\,|\mathbf{B}|\\
&= |\mathbf{A}|\sin\theta\,|\mathbf{B}|\\
&= |\mathbf{A}\times\mathbf{B}|
\end{aligned}$$

Note that the area of the triangle with sides $\mathbf{A}$ and $\mathbf{B}$ is $\frac{1}{2}|\mathbf{A}\times\mathbf{B}|$.

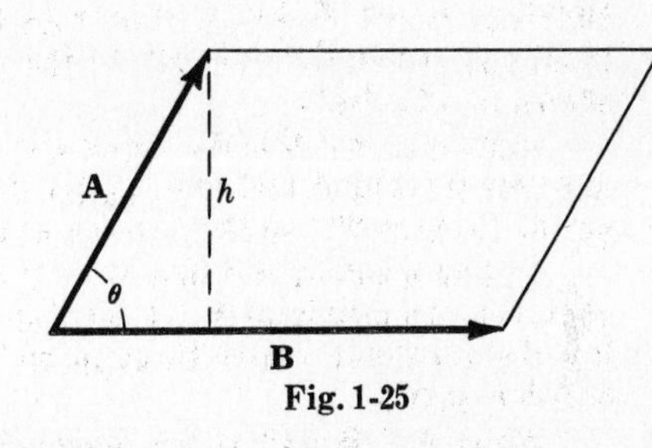

Fig. 1-25

1.19. Find the area of the triangle with vertices at $P(2,3,5)$, $Q(4,2,-1)$, $R(3,6,4)$.

$$\mathbf{PQ} = (4-2)\mathbf{i} + (2-3)\mathbf{j} + (-1-5)\mathbf{k} = 2\mathbf{i} - \mathbf{j} - 6\mathbf{k}$$
$$\mathbf{PR} = (3-2)\mathbf{i} + (6-3)\mathbf{j} + (4-5)\mathbf{k} = \mathbf{i} + 3\mathbf{j} - \mathbf{k}$$

$$\begin{aligned}
\text{Area of triangle} &= \tfrac{1}{2}|\mathbf{PQ}\times\mathbf{PR}| = \tfrac{1}{2}|(2\mathbf{i}-\mathbf{j}-6\mathbf{k})\times(\mathbf{i}+3\mathbf{j}-\mathbf{k})|\\
&= \tfrac{1}{2}\left|\begin{vmatrix}\mathbf{i} & \mathbf{j} & \mathbf{k}\\ 2 & -1 & -6\\ 1 & 3 & -1\end{vmatrix}\right| = \tfrac{1}{2}|19\mathbf{i}-4\mathbf{j}+7\mathbf{k}|\\
&= \tfrac{1}{2}\sqrt{(19)^2+(-4)^2+(7)^2} = \tfrac{1}{2}\sqrt{426}
\end{aligned}$$

TRIPLE PRODUCTS

1.20. Show that $\mathbf{A}\cdot(\mathbf{B}\times\mathbf{C})$ is in absolute value equal to the volume of a parallelepiped with sides $\mathbf{A}$, $\mathbf{B}$ and $\mathbf{C}$.

Let $\mathbf{n}$ be a unit normal to parallelogram I, having the direction of $\mathbf{B}\times\mathbf{C}$, and let h be the height of the terminal point of $\mathbf{A}$ above the parallelogram I.

Fig. 1-26

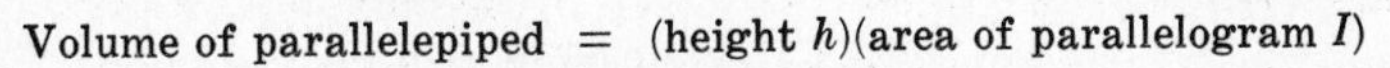

$$\begin{aligned}
\text{Volume of parallelepiped} &= (\text{height } h)(\text{area of parallelogram } I)\\
&= (\mathbf{A}\cdot\mathbf{n})(|\mathbf{B}\times\mathbf{C}|)\\
&= \mathbf{A}\cdot\{|\mathbf{B}\times\mathbf{C}|\,\mathbf{n}\} = \mathbf{A}\cdot(\mathbf{B}\times\mathbf{C})
\end{aligned}$$

If $\mathbf{A}$, $\mathbf{B}$ and $\mathbf{C}$ do not form a right-handed system, $\mathbf{A}\cdot\mathbf{n} < 0$ and the volume $= |\mathbf{A}\cdot(\mathbf{B}\times\mathbf{C})|$.

1.21. (a) If $\mathbf{A} = A_1\mathbf{i}+A_2\mathbf{j}+A_3\mathbf{k}$, $\mathbf{B} = B_1\mathbf{i}+B_2\mathbf{j}+B_3\mathbf{k}$, $\mathbf{C} = C_1\mathbf{i}+C_2\mathbf{j}+C_3\mathbf{k}$ show that

$$\mathbf{A}\cdot(\mathbf{B}\times\mathbf{C}) = \begin{vmatrix}A_1 & A_2 & A_3\\ B_1 & B_2 & B_3\\ C_1 & C_2 & C_3\end{vmatrix}$$

(*b*) Give a geometric significance of the case where $\mathbf{A}\cdot(\mathbf{B}\times\mathbf{C}) = 0$.

(*a*) $$\mathbf{A}\cdot(\mathbf{B}\times\mathbf{C}) = \mathbf{A}\cdot\begin{vmatrix} \mathbf{i} & \mathbf{j} & \mathbf{k} \\ B_1 & B_2 & B_3 \\ C_1 & C_2 & C_3 \end{vmatrix}$$

$$= (A_1\mathbf{i}+A_2\mathbf{j}+A_3\mathbf{k})\cdot[(B_2C_3-B_3C_2)\mathbf{i}+(B_3C_1-B_1C_3)\mathbf{j}+(B_1C_2-B_2C_1)\mathbf{k}]$$

$$= A_1(B_2C_3-B_3C_2)+A_2(B_3C_1-B_1C_3)+A_3(B_1C_2-B_2C_1) = \begin{vmatrix} A_1 & A_2 & A_3 \\ B_1 & B_2 & B_3 \\ C_1 & C_2 & C_3 \end{vmatrix}$$

(*b*) By Problem 1.20 if $\mathbf{A}\cdot(\mathbf{B}\times\mathbf{C}) = 0$ then $\mathbf{A}$, $\mathbf{B}$ and $\mathbf{C}$ are *coplanar*, i.e. are in the same plane, and conversely if $\mathbf{A}$, $\mathbf{B}$, $\mathbf{C}$ are coplanar then $\mathbf{A}\cdot(\mathbf{B}\times\mathbf{C}) = 0$.

1.22. Find the volume of a parallelepiped with sides $\mathbf{A} = 3\mathbf{i}-\mathbf{j}$, $\mathbf{B} = \mathbf{j}+2\mathbf{k}$, $\mathbf{C} = \mathbf{i}+5\mathbf{j}+4\mathbf{k}$.

By Problems 1.20 and 1.21, volume of parallelepiped $$= |\mathbf{A}\cdot(\mathbf{B}\times\mathbf{C})| = \left|\begin{vmatrix} 3 & -1 & 0 \\ 0 & 1 & 2 \\ 1 & 5 & 4 \end{vmatrix}\right| = |-20| = 20.$$

1.23. If $\mathbf{A} = \mathbf{i}+\mathbf{j}$, $\mathbf{B} = 2\mathbf{i}-3\mathbf{j}+\mathbf{k}$, $\mathbf{C} = 4\mathbf{j}-3\mathbf{k}$, find (*a*) $(\mathbf{A}\times\mathbf{B})\times\mathbf{C}$, (*b*) $\mathbf{A}\times(\mathbf{B}\times\mathbf{C})$.

(*a*) $$\mathbf{A}\times\mathbf{B} = \begin{vmatrix} \mathbf{i} & \mathbf{j} & \mathbf{k} \\ 1 & 1 & 0 \\ 2 & -3 & 1 \end{vmatrix} = \mathbf{i}-\mathbf{j}-5\mathbf{k}. \quad \text{Then } (\mathbf{A}\times\mathbf{B})\times\mathbf{C} = \begin{vmatrix} \mathbf{i} & \mathbf{j} & \mathbf{k} \\ 1 & -1 & -5 \\ 0 & 4 & -3 \end{vmatrix} = 23\mathbf{i}+3\mathbf{j}+4\mathbf{k}.$$

(*b*) $$\mathbf{B}\times\mathbf{C} = \begin{vmatrix} \mathbf{i} & \mathbf{j} & \mathbf{k} \\ 2 & -3 & 1 \\ 0 & 4 & -3 \end{vmatrix} = 5\mathbf{i}+6\mathbf{j}+8\mathbf{k}. \quad \text{Then } \mathbf{A}\times(\mathbf{B}\times\mathbf{C}) = \begin{vmatrix} \mathbf{i} & \mathbf{j} & \mathbf{k} \\ 1 & 1 & 0 \\ 5 & 6 & 8 \end{vmatrix} = 8\mathbf{i}-8\mathbf{j}+\mathbf{k}.$$

It follows that, in general, $(\mathbf{A}\times\mathbf{B})\times\mathbf{C} \neq \mathbf{A}\times(\mathbf{B}\times\mathbf{C})$.

DERIVATIVES AND INTEGRALS OF VECTORS

1.24. If $\mathbf{r} = (t^3+2t)\mathbf{i}-3e^{-2t}\mathbf{j}+2\sin 5t\,\mathbf{k}$, find (*a*) $\dfrac{d\mathbf{r}}{dt}$, (*b*) $\left|\dfrac{d\mathbf{r}}{dt}\right|$, (*c*) $\dfrac{d^2\mathbf{r}}{dt^2}$, (*d*) $\left|\dfrac{d^2\mathbf{r}}{dt^2}\right|$ at $t=0$.

(*a*) $$\frac{d\mathbf{r}}{dt} = \frac{d}{dt}(t^3+2t)\mathbf{i}+\frac{d}{dt}(-3e^{-2t})\mathbf{j}+\frac{d}{dt}(2\sin 5t)\mathbf{k} = (3t^2+2)\mathbf{i}+6e^{-2t}\mathbf{j}+10\cos 5t\,\mathbf{k}$$

At $t=0$, $d\mathbf{r}/dt = 2\mathbf{i}+6\mathbf{j}+10\mathbf{k}$.

(*b*) From (*a*), $|d\mathbf{r}/dt| = \sqrt{(2)^2+(6)^2+(10)^2} = \sqrt{140} = 2\sqrt{35}$ at $t=0$.

(*c*) $$\frac{d^2\mathbf{r}}{dt^2} = \frac{d}{dt}\left(\frac{d\mathbf{r}}{dt}\right) = \frac{d}{dt}\{(3t^2+2)\mathbf{i}+6e^{-2t}\mathbf{j}+10\cos 5t\,\mathbf{k}\} = 6t\mathbf{i}-12e^{-2t}\mathbf{j}-50\sin 5t\,\mathbf{k}$$

At $t=0$, $d^2\mathbf{r}/dt^2 = -12\mathbf{j}$.

(*d*) From (*c*), $|d^2\mathbf{r}/dt^2| = 12$ at $t=0$.

1.25. Prove that $\dfrac{d}{du}(\mathbf{A}\cdot\mathbf{B}) = \mathbf{A}\cdot\dfrac{d\mathbf{B}}{du}+\dfrac{d\mathbf{A}}{du}\cdot\mathbf{B}$, where $\mathbf{A}$ and $\mathbf{B}$ are differentiable functions of u.

Method 1.
$$\frac{d}{du}(\mathbf{A}\cdot\mathbf{B}) = \lim_{\Delta u\to 0}\frac{(\mathbf{A}+\Delta\mathbf{A})\cdot(\mathbf{B}+\Delta\mathbf{B}) - \mathbf{A}\cdot\mathbf{B}}{\Delta u}$$
$$= \lim_{\Delta u\to 0}\frac{\mathbf{A}\cdot\Delta\mathbf{B} + \Delta\mathbf{A}\cdot\mathbf{B} + \Delta\mathbf{A}\cdot\Delta\mathbf{B}}{\Delta u}$$
$$= \lim_{\Delta u\to 0}\left(\mathbf{A}\cdot\frac{\Delta\mathbf{B}}{\Delta u} + \frac{\Delta\mathbf{A}}{\Delta u}\cdot\mathbf{B} + \frac{\Delta\mathbf{A}}{\Delta u}\cdot\Delta\mathbf{B}\right) = \mathbf{A}\cdot\frac{d\mathbf{B}}{du} + \frac{d\mathbf{A}}{du}\cdot\mathbf{B}$$

Method 2. Let $\mathbf{A} = A_1\mathbf{i} + A_2\mathbf{j} + A_3\mathbf{k}$, $\mathbf{B} = B_1\mathbf{i} + B_2\mathbf{j} + B_3\mathbf{k}$. Then
$$\frac{d}{du}(\mathbf{A}\cdot\mathbf{B}) = \frac{d}{du}(A_1B_1 + A_2B_2 + A_3B_3)$$
$$= \left(A_1\frac{dB_1}{du} + A_2\frac{dB_2}{du} + A_3\frac{dB_3}{du}\right) + \left(\frac{dA_1}{du}B_1 + \frac{dA_2}{du}B_2 + \frac{dA_3}{du}B_3\right)$$
$$= \mathbf{A}\cdot\frac{d\mathbf{B}}{du} + \frac{d\mathbf{A}}{du}\cdot\mathbf{B}$$

1.26. If $\phi(x,y,z) = x^2yz$ and $\mathbf{A} = 3x^2y\mathbf{i} + yz^2\mathbf{j} - xz\mathbf{k}$, find $\dfrac{\partial^2}{\partial y\,\partial z}(\phi\mathbf{A})$ at the point $(1,-2,-1)$.

$$\phi\mathbf{A} = (x^2yz)(3x^2y\mathbf{i} + yz^2\mathbf{j} - xz\mathbf{k}) = 3x^4y^2z\mathbf{i} + x^2y^2z^3\mathbf{j} - x^3yz^2\mathbf{k}$$
$$\frac{\partial}{\partial z}(\phi\mathbf{A}) = \frac{\partial}{\partial z}(3x^4y^2z\mathbf{i} + x^2y^2z^3\mathbf{j} - x^3yz^2\mathbf{k}) = 3x^4y^2\mathbf{i} + 3x^2y^2z^2\mathbf{j} - 2x^3yz\mathbf{k}$$
$$\frac{\partial^2}{\partial y\,\partial z}(\phi\mathbf{A}) = \frac{\partial}{\partial y}(3x^4y^2\mathbf{i} + 3x^2y^2z^2\mathbf{j} - 2x^3yz\mathbf{k}) = 6x^4y\mathbf{i} + 6x^2yz^2\mathbf{j} - 2x^3z\mathbf{k}$$

If $x = 1$, $y = -2$, $z = -1$, this becomes $-12\mathbf{i} - 12\mathbf{j} + 2\mathbf{k}$.

1.27. Evaluate $\displaystyle\int_{u=1}^{2}\mathbf{A}(u)\,du$ if $\mathbf{A}(u) = (3u^2-1)\mathbf{i} + (2u-3)\mathbf{j} + (6u^2-4u)\mathbf{k}$.

The given integral equals
$$\int_{u=1}^{2}\{(3u^2-1)\mathbf{i} + (2u-3)\mathbf{j} + (6u^2-4u)\mathbf{k}\}\,du$$
$$= (u^3-u)\mathbf{i} + (u^2-3u)\mathbf{j} + (2u^3-2u^2)\mathbf{k}\Big|_{u=1}^{2}$$
$$= \{(8-2)\mathbf{i} + (4-6)\mathbf{j} + (16-8)\mathbf{k}\} - \{(1-1)\mathbf{i} + (1-3)\mathbf{j} + (2-2)\mathbf{k}\}$$
$$= 6\mathbf{i} + 8\mathbf{k}$$

VELOCITY AND ACCELERATION

1.28. A particle moves along a curve whose parametric equations are $x = 3e^{-2t}$, $y = 4\sin 3t$, $z = 5\cos 3t$ where t is the time.

(a) Find its velocity and acceleration at any time.

(b) Find the magnitudes of the velocity and acceleration at $t=0$.

(a) The position vector $\mathbf{r}$ of the particle is
$$\mathbf{r} = x\mathbf{i} + y\mathbf{j} + z\mathbf{k} = 3e^{-2t}\mathbf{i} + 4\sin 3t\,\mathbf{j} + 5\cos 3t\,\mathbf{k}$$
Then the velocity is
$$\mathbf{v} = d\mathbf{r}/dt = -6e^{-2t}\mathbf{i} + 12\cos 3t\,\mathbf{j} - 15\sin 3t\,\mathbf{k}$$
and the acceleration is
$$\mathbf{a} = d\mathbf{v}/dt = d^2\mathbf{r}/dt^2 = 12e^{-2t}\mathbf{i} - 36\sin 3t\,\mathbf{j} - 45\cos 3t\,\mathbf{k}$$

(b) At $t=0$, $\mathbf{v} = d\mathbf{r}/dt = -6\mathbf{i} + 12\mathbf{j}$ and $\mathbf{a} = d^2\mathbf{r}/dt^2 = 12\mathbf{i} - 45\mathbf{k}$. Then

magnitude of velocity at $t=0$ is $\sqrt{(-6)^2 + (12)^2} = 6\sqrt{5}$

magnitude of acceleration at $t=0$ is $\sqrt{(12)^2 + (-45)^2} = 3\sqrt{241}$

1.29. A particle travels so that its acceleration is given by

$$\mathbf{a} = 2e^{-t}\mathbf{i} + 5\cos t\,\mathbf{j} - 3\sin t\,\mathbf{k}$$

If the particle is located at $(1, -3, 2)$ at time $t = 0$ and is moving with a velocity given by $4\mathbf{i} - 3\mathbf{j} + 2\mathbf{k}$, find (*a*) the velocity and (*b*) the displacement of the particle at any time $t > 0$.

(*a*)
$$\mathbf{a} = \frac{d^2\mathbf{r}}{dt^2} = \frac{d\mathbf{v}}{dt} = 2e^{-t}\mathbf{i} + 5\cos t\,\mathbf{j} - 3\sin t\,\mathbf{k}$$

Integrating,
$$\mathbf{v} = \int (2e^{-t}\mathbf{i} + 5\cos t\,\mathbf{j} - 3\sin t\,\mathbf{k})\,dt$$
$$= -2e^{-t}\mathbf{i} + 5\sin t\,\mathbf{j} + 3\cos t\,\mathbf{k} + \mathbf{c}_1$$

Since $\mathbf{v} = 4\mathbf{i} - 3\mathbf{j} + 2\mathbf{k}$ at $t = 0$, we have

$$4\mathbf{i} - 3\mathbf{j} + 2\mathbf{k} = -2\mathbf{i} + 3\mathbf{k} + \mathbf{c}_1 \quad \text{or} \quad \mathbf{c}_1 = 6\mathbf{i} - 3\mathbf{j} - \mathbf{k}$$

Then
$$\mathbf{v} = -2e^{-t}\mathbf{i} + 5\sin t\,\mathbf{j} + 3\cos t\,\mathbf{k} + 6\mathbf{i} - 3\mathbf{j} - \mathbf{k}$$
$$= (6 - 2e^{-t})\mathbf{i} + (5\sin t - 3)\mathbf{j} + (3\cos t - 1)\mathbf{k} \qquad (1)$$

(*b*) Replacing $\mathbf{v}$ by $d\mathbf{r}/dt$ in (*1*) and integrating, we have

$$\mathbf{r} = \int [(6 - 2e^{-t})\mathbf{i} + (5\sin t - 3)\mathbf{j} + (3\cos t - 1)\mathbf{k}]\,dt$$
$$= (6t + 2e^{-t})\mathbf{i} - (5\cos t + 3t)\mathbf{j} + (3\sin t - t)\mathbf{k} + \mathbf{c}_2$$

Since the particle is located at $(1, -3, 2)$ at $t = 0$, we have $\mathbf{r} = \mathbf{i} - 3\mathbf{j} + 2\mathbf{k}$ at $t = 0$, so that

$$\mathbf{i} - 3\mathbf{j} + 2\mathbf{k} = 2\mathbf{i} - 5\mathbf{j} + \mathbf{c}_2 \quad \text{or} \quad \mathbf{c}_2 = -\mathbf{i} + 2\mathbf{j} + 2\mathbf{k}$$

Thus
$$\mathbf{r} = (6t + 2e^{-t} - 1)\mathbf{i} + (2 - 5\cos t - 3t)\mathbf{j} + (3\sin t - t + 2)\mathbf{k} \qquad (2)$$

RELATIVE VELOCITY AND ACCELERATION

1.30. An airplane moves in a northwesterly direction at 125 mi/hr relative to the ground, due to the fact that there is a westerly wind [i.e. from the west] of 50 mi/hr relative to the ground. Determine (*a*) graphically and (*b*) analytically how fast and in what direction the plane would have traveled if there were no wind.

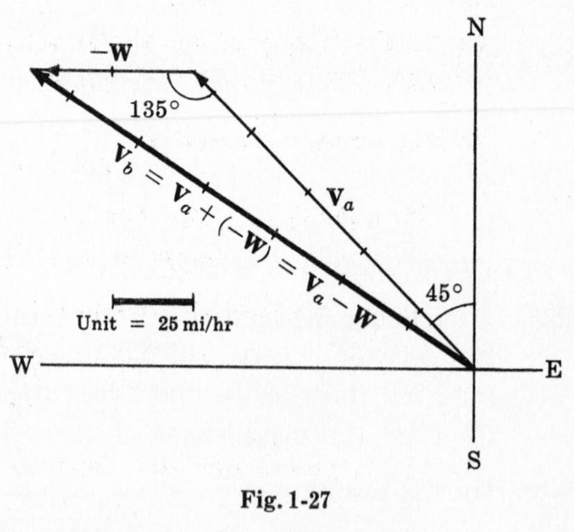

Fig. 1-27

(*a*) ***Graphically.***

Let $\mathbf{W}$ = wind velocity

$\mathbf{V}_a$ = velocity of plane with wind

$\mathbf{V}_b$ = velocity of plane without wind.

Then [see Fig. 1-27] $\quad \mathbf{V}_a = \mathbf{V}_b + \mathbf{W} \quad$ or $\quad \mathbf{V}_b = \mathbf{V}_a - \mathbf{W} = \mathbf{V}_a + (-\mathbf{W})$.

V_b has magnitude 6.5 units = 163 mi/hr and direction 33° north of west.

(*b*) ***Analytically.***

Letting $\mathbf{i}$ and $\mathbf{j}$ be unit vectors in directions E and N respectively, we see from Fig. 1-27 that

$$\mathbf{V}_a = -125\cos 45°\,\mathbf{i} + 125\sin 45°\,\mathbf{j} \quad \text{and} \quad \mathbf{W} = 50\mathbf{i}$$

Then $\mathbf{V}_b = \mathbf{V}_a - \mathbf{W} = (-125\cos 45° - 50)\mathbf{i} + 125\sin 45°\,\mathbf{j} = -138.39\mathbf{i} + 88.39\mathbf{j}$.

Thus the magnitude of $\mathbf{V}_b$ is $\sqrt{(-138.39)^2 + (88.39)^2} = 164.2$ mi/hr and the direction is $\tan^{-1} 88.39/138.39 = \tan^{-1} .6387 = 32°34'$ north of west.

1.31. Two particles have position vectors given by $\mathbf{r}_1 = 2t\mathbf{i} - t^2\mathbf{j} + (3t^2 - 4t)\mathbf{k}$ and $\mathbf{r}_2 = (5t^2 - 12t + 4)\mathbf{i} + t^3\mathbf{j} - 3t\mathbf{k}$. Find (*a*) the relative velocity and (*b*) the relative acceleration of the second particle with respect to the first at the instant where $t = 2$.

(*a*) The velocities of the particles at $t = 2$ are respectively

$$\mathbf{v}_1 = \dot{\mathbf{r}}_1 = 2\mathbf{i} - 2t\mathbf{j} + (6t - 4)\mathbf{k}\Big|_{t=2} = 2\mathbf{i} - 4\mathbf{j} + 8\mathbf{k}$$

$$\mathbf{v}_2 = \dot{\mathbf{r}}_2 = (10t - 12)\mathbf{i} + 3t^2\mathbf{j} - 3\mathbf{k}\Big|_{t=2} = 8\mathbf{i} + 12\mathbf{j} - 3\mathbf{k}$$

Relative velocity of particle 2 with respect to particle 1

$$= \mathbf{v}_2 - \mathbf{v}_1 = (8\mathbf{i} + 12\mathbf{j} - 3\mathbf{k}) - (2\mathbf{i} - 4\mathbf{j} + 8\mathbf{k}) = 6\mathbf{i} + 16\mathbf{j} - 11\mathbf{k}$$

(*b*) The accelerations of the particles at $t = 2$ are respectively

$$\mathbf{a}_1 = \dot{\mathbf{v}}_1 = \ddot{\mathbf{r}}_1 = -2\mathbf{j} + 6\mathbf{k}\Big|_{t=2} = -2\mathbf{j} + 6\mathbf{k}$$

$$\mathbf{a}_2 = \dot{\mathbf{v}}_2 = \ddot{\mathbf{r}}_2 = 10\mathbf{i} + 6t\mathbf{j}\Big|_{t=2} = 10\mathbf{i} + 12\mathbf{j}$$

Relative acceleration of particle 2 with respect to particle 1

$$= \mathbf{a}_2 - \mathbf{a}_1 = (10\mathbf{i} + 12\mathbf{j}) - (-2\mathbf{j} + 6\mathbf{k}) = 10\mathbf{i} + 14\mathbf{j} - 6\mathbf{k}$$

TANGENTIAL AND NORMAL ACCELERATION

1.32. Given a space curve C with position vector

$$\mathbf{r} = 3\cos 2t\,\mathbf{i} + 3\sin 2t\,\mathbf{j} + (8t - 4)\mathbf{k}$$

(*a*) Find a unit tangent vector $\mathbf{T}$ to the curve.

(*b*) If $\mathbf{r}$ is the position vector of a particle moving on C at time t, verify in this case that $\mathbf{v} = v\mathbf{T}$.

(*a*) A tangent vector to C is

$$d\mathbf{r}/dt = -6\sin 2t\,\mathbf{i} + 6\cos 2t\,\mathbf{j} + 8\mathbf{k}$$

The magnitude of this vector is

$$|d\mathbf{r}/dt| = ds/dt = \sqrt{(-6\sin 2t)^2 + (6\cos 2t)^2 + (8)^2} = 10$$

Then a unit tangent vector to C is

$$\mathbf{T} = \frac{d\mathbf{r}/dt}{|d\mathbf{r}/dt|} = \frac{d\mathbf{r}/dt}{ds/dt} = \frac{d\mathbf{r}}{ds} = \frac{-6\sin 2t\,\mathbf{i} + 6\cos 2t\,\mathbf{j} + 8\mathbf{k}}{10}$$
$$= -\tfrac{3}{5}\sin 2t\,\mathbf{i} + \tfrac{3}{5}\cos 2t\,\mathbf{j} + \tfrac{4}{5}\mathbf{k}$$

(*b*) This follows at once from (*a*) since

$$\mathbf{v} = d\mathbf{r}/dt = -6\sin 2t\,\mathbf{i} + 6\cos 2t\,\mathbf{j} + 8\mathbf{k}$$
$$= (10)(-\tfrac{3}{5}\sin 2t\,\mathbf{i} + \tfrac{3}{5}\cos 2t\,\mathbf{j} + \tfrac{4}{5}\mathbf{k}) = v\mathbf{T}$$

Note that in this case the speed of the particle along the curve is constant.

1.33. If $\mathbf{T}$ is a unit tangent vector to a space curve C, show that $d\mathbf{T}/ds$ is normal to $\mathbf{T}$.

Since $\mathbf{T}$ is a unit vector, we have $\mathbf{T}\cdot\mathbf{T} = 1$. Then differentiating with respect to s, we obtain

$$\mathbf{T}\cdot\frac{d\mathbf{T}}{ds} + \frac{d\mathbf{T}}{ds}\cdot\mathbf{T} = 2\mathbf{T}\cdot\frac{d\mathbf{T}}{ds} = 0 \quad \text{or} \quad \mathbf{T}\cdot\frac{d\mathbf{T}}{ds} = 0$$

which states that $d\mathbf{T}/ds$ is normal, i.e. perpendicular, to $\mathbf{T}$.

If $\mathbf{N}$ is a unit vector in the direction of $d\mathbf{T}/ds$, we have

$$d\mathbf{T}/ds = \kappa\mathbf{N}$$

and we call $\mathbf{N}$ the unit *principal normal* to C. The scalar $\kappa = |d\mathbf{T}/ds|$ is called the *curvature*, while $R = 1/\kappa$ is called the *radius of curvature*.

1.34. Find the (*a*) curvature, (*b*) radius of curvature and (*c*) unit principal normal $\mathbf{N}$ to any point of the space curve of Problem 1.32.

(*a*) From Problem 1.32, $\mathbf{T} = -\frac{3}{5}\sin 2t\,\mathbf{i} + \frac{3}{5}\cos 2t\,\mathbf{j} + \frac{4}{5}\mathbf{k}$. Then

$$\frac{d\mathbf{T}}{ds} = \frac{d\mathbf{T}/dt}{ds/dt} = \frac{(-6/5)\cos 2t\,\mathbf{i} - (6/5)\sin 2t\,\mathbf{j}}{10}$$
$$= -\tfrac{3}{25}\cos 2t\,\mathbf{i} - \tfrac{3}{25}\sin 2t\,\mathbf{j}$$

Thus the curvature is $\kappa = \left|\frac{d\mathbf{T}}{ds}\right| = \sqrt{(-\frac{3}{25}\cos 2t)^2 + (-\frac{3}{25}\sin 2t)^2} = \frac{3}{25}$

(*b*) Radius of curvature $= R = 1/\kappa = 25/3$

(*c*) From (*a*), (*b*) and Problem 1.33,

$$\mathbf{N} = \frac{1}{\kappa}\frac{d\mathbf{T}}{ds} = R\frac{d\mathbf{T}}{ds} = -\cos 2t\,\mathbf{i} - \sin 2t\,\mathbf{j}$$

1.35. Show that the acceleration $\mathbf{a}$ of a particle which travels along a space curve with velocity $\mathbf{v}$ is given by

$$\mathbf{a} = \frac{dv}{dt}\mathbf{T} + \frac{v^2}{R}\mathbf{N}$$

where $\mathbf{T}$ is the unit tangent vector to the space curve, $\mathbf{N}$ is its unit principal normal and R is the radius of curvature.

Velocity $\mathbf{v}$ = magnitude of $\mathbf{v}$ multiplied by unit tangent vector $\mathbf{T}$, or

$$\mathbf{v} = v\mathbf{T}$$

Differentiating, $$\mathbf{a} = \frac{d\mathbf{v}}{dt} = \frac{d}{dt}(v\mathbf{T}) = \frac{dv}{dt}\mathbf{T} + v\frac{d\mathbf{T}}{dt}$$

But $$\frac{d\mathbf{T}}{dt} = \frac{d\mathbf{T}}{ds}\frac{ds}{dt} = \kappa\mathbf{N}\frac{ds}{dt} = \kappa v\mathbf{N} = \frac{v\mathbf{N}}{R}$$

Then $$\mathbf{a} = \frac{dv}{dt}\mathbf{T} + v\left(\frac{v\mathbf{N}}{R}\right) = \frac{dv}{dt}\mathbf{T} + \frac{v^2}{R}\mathbf{N}$$

This shows that the component of the acceleration is dv/dt in a direction tangent to the path and v^2/R in the direction of the principal normal to the path. The latter acceleration is often called the *centripetal acceleration* or briefly *normal acceleration*.

CIRCULAR MOTION

1.36. A particle moves so that its position vector is given by $\mathbf{r} = \cos\omega t\,\mathbf{i} + \sin\omega t\,\mathbf{j}$ where ω is a constant. Show that (*a*) the velocity $\mathbf{v}$ of the particle is perpendicular to $\mathbf{r}$, (*b*) the acceleration $\mathbf{a}$ is directed toward the origin and has magnitude proportional to the distance from the origin, (*c*) $\mathbf{r}\times\mathbf{v}$ = a constant vector.

(*a*) $\mathbf{v} = \dfrac{d\mathbf{r}}{dt} = -\omega\sin\omega t\,\mathbf{i} + \omega\cos\omega t\,\mathbf{j}$. Then

$$\mathbf{r}\cdot\mathbf{v} = [\cos\omega t\,\mathbf{i} + \sin\omega t\,\mathbf{j}]\cdot[-\omega\sin\omega t\,\mathbf{i} + \omega\cos\omega t\,\mathbf{j}]$$
$$= (\cos\omega t)(-\omega\sin\omega t) + (\sin\omega t)(\omega\cos\omega t) = 0$$

and $\mathbf{r}$ and $\mathbf{v}$ are perpendicular.

(b) $\dfrac{d^2\mathbf{r}}{dt^2} = \dfrac{d\mathbf{v}}{dt} = -\omega^2 \cos \omega t\, \mathbf{i} - \omega^2 \sin \omega t\, \mathbf{j} = -\omega^2 [\cos \omega t\, \mathbf{i} + \sin \omega t\, \mathbf{j}] = -\omega^2 \mathbf{r}$

Then the acceleration is opposite to the direction of **r**, i.e. it is directed toward the origin. Its magnitude is proportional to $|\mathbf{r}|$ which is the distance from the origin.

(c) $\mathbf{r} \times \mathbf{v} = [\cos \omega t\, \mathbf{i} + \sin \omega t\, \mathbf{j}] \times [-\omega \sin \omega t\, \mathbf{i} + \omega \cos \omega t\, \mathbf{j}]$

$$= \begin{vmatrix} \mathbf{i} & \mathbf{j} & \mathbf{k} \\ \cos \omega t & \sin \omega t & 0 \\ -\omega \sin \omega t & \omega \cos \omega t & 0 \end{vmatrix} = \omega(\cos^2 \omega t + \sin^2 \omega t)\mathbf{k} = \omega \mathbf{k}, \quad \text{a constant vector.}$$

Physically, the motion is that of a particle moving on the circumference of a circle with constant angular speed ω. The acceleration, directed toward the center of the circle, is the *centripetal acceleration.*

GRADIENT, DIVERGENCE AND CURL

1.37. If $\phi = x^2yz^3$ and $\mathbf{A} = xz\mathbf{i} - y^2\mathbf{j} + 2x^2y\mathbf{k}$, find (a) $\nabla\phi$, (b) $\nabla \cdot \mathbf{A}$, (c) $\nabla \times \mathbf{A}$, (d) div $(\phi\mathbf{A})$, (e) curl $(\phi\mathbf{A})$.

(a) $$\nabla\phi = \left(\frac{\partial}{\partial x}\mathbf{i} + \frac{\partial}{\partial y}\mathbf{j} + \frac{\partial}{\partial z}\mathbf{k}\right)\phi = \frac{\partial\phi}{\partial x}\mathbf{i} + \frac{\partial\phi}{\partial y}\mathbf{j} + \frac{\partial\phi}{\partial z}\mathbf{k}$$

$$= \frac{\partial}{\partial x}(x^2yz^3)\mathbf{i} + \frac{\partial}{\partial y}(x^2yz^3)\mathbf{j} + \frac{\partial}{\partial z}(x^2yz^3)\mathbf{k} = 2xyz^3\mathbf{i} + x^2z^3\mathbf{j} + 3x^2yz^2\mathbf{k}$$

(b) $$\nabla \cdot \mathbf{A} = \left(\frac{\partial}{\partial x}\mathbf{i} + \frac{\partial}{\partial y}\mathbf{j} + \frac{\partial}{\partial z}\mathbf{k}\right) \cdot (xz\mathbf{i} - y^2\mathbf{j} + 2x^2y\mathbf{k})$$

$$= \frac{\partial}{\partial x}(xz) + \frac{\partial}{\partial y}(-y^2) + \frac{\partial}{\partial z}(2x^2y) = z - 2y$$

(c) $$\nabla \times \mathbf{A} = \left(\frac{\partial}{\partial x}\mathbf{i} + \frac{\partial}{\partial y}\mathbf{j} + \frac{\partial}{\partial z}\mathbf{k}\right) \times (xz\mathbf{i} - y^2\mathbf{j} + 2x^2y\mathbf{k})$$

$$= \begin{vmatrix} \mathbf{i} & \mathbf{j} & \mathbf{k} \\ \partial/\partial x & \partial/\partial y & \partial/\partial z \\ xz & -y^2 & 2x^2y \end{vmatrix}$$

$$= \left(\frac{\partial}{\partial y}(2x^2y) - \frac{\partial}{\partial z}(-y^2)\right)\mathbf{i} + \left(\frac{\partial}{\partial z}(xz) - \frac{\partial}{\partial x}(2x^2y)\right)\mathbf{j} + \left(\frac{\partial}{\partial x}(-y^2) - \frac{\partial}{\partial y}(xz)\right)\mathbf{k}$$

$$= 2x^2\mathbf{i} + (x - 4xy)\mathbf{j}$$

(d) $$\text{div}\,(\phi\mathbf{A}) = \nabla \cdot (\phi\mathbf{A}) = \nabla \cdot (x^3yz^4\mathbf{i} - x^2y^3z^3\mathbf{j} + 2x^4y^2z^3\mathbf{k})$$

$$= \frac{\partial}{\partial x}(x^3yz^4) + \frac{\partial}{\partial y}(-x^2y^3z^3) + \frac{\partial}{\partial z}(2x^4y^2z^3)$$

$$= 3x^2yz^4 - 3x^2y^2z^3 + 6x^4y^2z^2$$

(e) $$\text{curl}\,(\phi\mathbf{A}) = \nabla \times (\phi\mathbf{A}) = \nabla \times (x^3yz^4\mathbf{i} - x^2y^3z^3\mathbf{j} + 2x^4y^2z^3\mathbf{k})$$

$$= \begin{vmatrix} \mathbf{i} & \mathbf{j} & \mathbf{k} \\ \partial/\partial x & \partial/\partial y & \partial/\partial z \\ x^3yz^4 & -x^2y^3z^3 & 2x^4y^2z^3 \end{vmatrix}$$

$$= (4x^4yz^3 + 3x^2y^3z^2)\mathbf{i} + (4x^3yz^3 - 8x^3y^2z^3)\mathbf{j} - (2xy^3z^3 + x^3z^4)\mathbf{k}$$

1.38. (a) If $\mathbf{A} = (2xy + z^3)\mathbf{i} + (x^2 + 2y)\mathbf{j} + (3xz^2 - 2)\mathbf{k}$, show that $\nabla \times \mathbf{A} = 0$.
(b) Find a scalar function ϕ such that $\mathbf{A} = \nabla\phi$.

(a) $$\nabla \times \mathbf{A} = \begin{vmatrix} \mathbf{i} & \mathbf{j} & \mathbf{k} \\ \partial/\partial x & \partial/\partial y & \partial/\partial z \\ 2xy + z^3 & x^2 + 2y & 3xz^2 - 2 \end{vmatrix} = 0$$

(b) **Method 1.** If $\mathbf{A} = \nabla\phi = \frac{\partial\phi}{\partial x}\mathbf{i} + \frac{\partial\phi}{\partial y}\mathbf{j} + \frac{\partial\phi}{\partial z}\mathbf{k}$ then we must have

$$(1)\quad \frac{\partial\phi}{\partial x} = 2xy + z^3 \qquad (2)\quad \frac{\partial\phi}{\partial y} = x^2 + 2y \qquad (3)\quad \frac{\partial\phi}{\partial z} = 3xz^2 - 2$$

Integrating, we find

$$(4)\quad \phi = x^2y + xz^3 + F_1(y,z) \qquad (5)\quad \phi = x^2y + y^2 + F_2(x,z)$$

$$(6)\quad \phi = xz^3 - 2z + F_3(x,y)$$

Comparing these we must have $F_1(y,z) = y^2 - 2z$, $F_2(x,z) = xz^3 - 2z$, $F_3(x,y) = x^2y + y^2$ and so $\phi = x^2y + xz^3 + y^2 - 2z$.

Method 2. We have if $\mathbf{A} = \nabla\phi$,

$$\begin{aligned}\mathbf{A}\cdot d\mathbf{r} &= \left(\frac{\partial\phi}{\partial x}\mathbf{i} + \frac{\partial\phi}{\partial y}\mathbf{j} + \frac{\partial\phi}{\partial z}\mathbf{k}\right)\cdot(dx\,\mathbf{i} + dy\,\mathbf{j} + dz\,\mathbf{k}) \\ &= \frac{\partial\phi}{\partial x}dx + \frac{\partial\phi}{\partial y}dy + \frac{\partial\phi}{\partial z}dz = d\phi\end{aligned}$$

an exact differential. For this case,

$$\begin{aligned}d\phi = \mathbf{A}\cdot d\mathbf{r} &= (2xy + z^3)\,dx + (x^2 + 2y)\,dy + (3xz^2 - 2)\,dz \\ &= [(2xy + z^3)\,dx + x^2\,dy + 3xz^2\,dz] + 2y\,dy - 2\,dz \\ &= d(x^2y + xz^3) + d(y^2) + d(-2z) \\ &= d(x^2y + xz^3 + y^2 - 2z)\end{aligned}$$

Then $\phi = x^2y + xz^3 + y^2 - 2z$. Note that an arbitrary constant can also be added to ϕ.

LINE INTEGRALS AND INDEPENDENCE OF THE PATH

1.39. If $\mathbf{A} = (3x^2 - 6yz)\mathbf{i} + (2y + 3xz)\mathbf{j} + (1 - 4xyz^2)\mathbf{k}$, evaluate $\int_C \mathbf{A}\cdot d\mathbf{r}$ from $(0,0,0)$ to $(1,1,1)$ along the following paths C:

(a) $x = t,\ y = t^2,\ z = t^3$.

(b) the straight lines from $(0,0,0)$ to $(0,0,1)$, then to $(0,1,1)$, and then to $(1,1,1)$.

(c) the straight line joining $(0,0,0)$ and $(1,1,1)$.

$$\begin{aligned}\int_C \mathbf{A}\cdot d\mathbf{r} &= \int_C \{(3x^2 - 6yz)\mathbf{i} + (2y + 3xz)\mathbf{j} + (1 - 4xyz^2)\mathbf{k}\}\cdot(dx\,\mathbf{i} + dy\,\mathbf{j} + dz\,\mathbf{k}) \\ &= \int_C (3x^2 - 6yz)\,dx + (2y + 3xz)\,dy + (1 - 4xyz^2)\,dz\end{aligned}$$

(a) If $x = t$, $y = t^2$, $z = t^3$, points $(0,0,0)$ and $(1,1,1)$ correspond to $t = 0$ and $t = 1$ respectively. Then

$$\begin{aligned}\int_C \mathbf{A}\cdot d\mathbf{r} &= \int_{t=0}^{1} \{3t^2 - 6(t^2)(t^3)\}\,dt + \{2t^2 + 3(t)(t^3)\}\,d(t^2) + \{1 - 4(t)(t^2)(t^3)^2\}\,d(t^3) \\ &= \int_{t=0}^{1} (3t^2 - 6t^5)\,dt + (4t^3 + 6t^5)\,dt + (3t^2 - 12t^{11})\,dt = 2\end{aligned}$$

Another method.

Along C, $\mathbf{A} = (3t^2 - 6t^5)\mathbf{i} + (2t^2 + 3t^4)\mathbf{j} + (1 - 4t^9)\mathbf{k}$ and $\mathbf{r} = x\mathbf{i} + y\mathbf{j} + z\mathbf{k} = t\mathbf{i} + t^2\mathbf{j} + t^3\mathbf{k}$, $d\mathbf{r} = (\mathbf{i} + 2t\mathbf{j} + 3t^2\mathbf{k})\,dt$. Then

$$\int_C \mathbf{A}\cdot d\mathbf{r} = \int_0^1 (3t^2 - 6t^5)\,dt + (4t^3 + 6t^5)\,dt + (3t^2 - 12t^{11})\,dt = 2$$

(b) Along the straight line from $(0,0,0)$ to $(0,0,1)$, $x = 0$, $y = 0$, $dx = 0$, $dy = 0$ while z varies from 0 to 1. Then the integral over this part of the path is

$$\int_{z=0}^{1} \{3(0)^2 - 6(0)(z)\}0 + \{2(0) + 3(0)(z)\}0 + \{1 - 4(0)(0)(z^2)\}\,dz = \int_{z=0}^{1} dz = 1$$

Along the straight line from $(0,0,1)$ to $(0,1,1)$, $x=0$, $z=1$, $dx=0$, $dz=0$ while y varies from 0 to 1. Then the integral over this part of the path is

$$\int_{y=0}^{1} \{3(0)^2-6(y)(1)\}0 + \{2y+3(0)(1)\}\,dy + \{1-4(0)(y)(1)^2\}0 = \int_{y=0}^{1} 2y\,dy = 1$$

Along the straight line from $(0,1,1)$ to $(1,1,1)$, $y=1$, $z=1$, $dy=0$, $dz=0$ while x varies from 0 to 1. Then the integral over this part of the path is

$$\int_{x=0}^{1} \{3x^2-6(1)(1)\}\,dx + \{2(1)+3x(1)\}0 + \{1-4x(1)(1)^2\}0 = \int_{x=0}^{1} (3x^2-6)\,dx = -5$$

Adding, $\displaystyle\int_C \mathbf{A}\cdot d\mathbf{r} = 1+1-5 = -3.$

(c) Along the straight line joining $(0,0,0)$ and $(1,1,1)$ we have $x=t$, $y=t$, $z=t$. Then since $dx=dy=dz=dt$,

$$\int_C \mathbf{A}\cdot d\mathbf{r} = \int_C (3x^2-6yz)\,dx + (2y+3xz)\,dy + (1-4xyz^2)\,dz$$

$$= \int_{t=0}^{1} (3t^2-6t^2)\,dt + (2t+3t^2)\,dt + (1-4t^4)\,dt$$

$$= \int_{t=0}^{1} (2t+1-4t^4)\,dt = 6/5$$

Note that in this case the value of the integral depends on the particular path.

1.40. If $\mathbf{A} = (2xy+z^3)\mathbf{i} + (x^2+2y)\mathbf{j} + (3xz^2-2)\mathbf{k}$ show that (a) $\displaystyle\int_C \mathbf{A}\cdot d\mathbf{r}$ is independent of the path C joining the points $(1,-1,1)$ and $(2,1,2)$ and (b) find its value.

By Problem 1.38, $\nabla\times\mathbf{A} = \mathbf{0}$ or $\mathbf{A}\cdot d\mathbf{r} = d\phi = d(x^2y+xz^3+y^2-2z)$. Then the integral is independent of the path and its value is

$$\int_{(1,-1,1)}^{(2,1,2)} \mathbf{A}\cdot d\mathbf{r} = \int_{(1,-1,1)}^{(2,1,2)} d(x^2y+xz^3+y^2-2z)$$

$$= x^2y + xz^3 + y^2 - 2z \Big|_{(1,-1,1)}^{(2,1,2)} = 18$$

MISCELLANEOUS PROBLEMS

1.41. Prove that if $\mathbf{a}$ and $\mathbf{b}$ are non-collinear, then $x\mathbf{a}+y\mathbf{b} = \mathbf{0}$ implies $x=y=0$.

Suppose $x\neq 0$. Then $x\mathbf{a}+y\mathbf{b}=\mathbf{0}$ implies $x\mathbf{a}=-y\mathbf{b}$ or $\mathbf{a}=-(y/x)\mathbf{b}$, i.e. $\mathbf{a}$ and $\mathbf{b}$ must be parallel to the same line (collinear), contrary to hypothesis. Thus $x=0$; then $y\mathbf{b}=\mathbf{0}$, from which $y=0$.

1.42. Prove that the diagonals of a parallelogram bisect each other.

Let $ABCD$ be the given parallelogram with diagonals intersecting at P as shown in Fig. 1-28.

Since $\mathbf{BD}+\mathbf{a}=\mathbf{b}$, $\mathbf{BD}=\mathbf{b}-\mathbf{a}$. Then $\mathbf{BP}=x(\mathbf{b}-\mathbf{a})$.

Since $\mathbf{AC}=\mathbf{a}+\mathbf{b}$, $\mathbf{AP}=y(\mathbf{a}+\mathbf{b})$.

But $\mathbf{AB}=\mathbf{AP}+\mathbf{PB}=\mathbf{AP}-\mathbf{BP}$,
i.e. $\mathbf{a} = y(\mathbf{a}+\mathbf{b}) - x(\mathbf{b}-\mathbf{a}) = (x+y)\mathbf{a} + (y-x)\mathbf{b}$.

Since $\mathbf{a}$ and $\mathbf{b}$ are non-collinear we have by Problem 1.41, $x+y=1$ and $y-x=0$, i.e. $x=y=\frac{1}{2}$ and P is the midpoint of both diagonals.

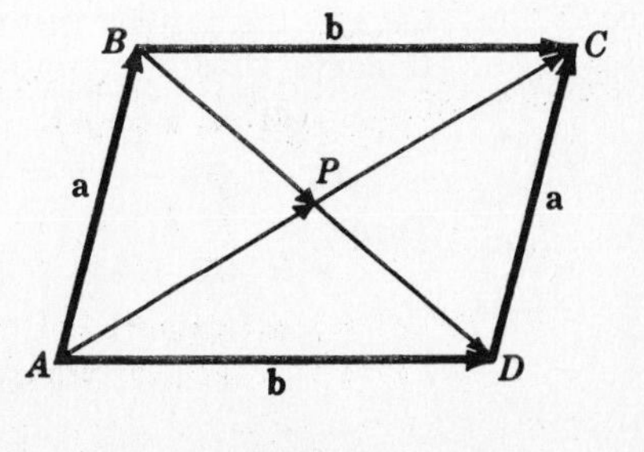

Fig. 1-28

1.43. Prove that for any vector $\mathbf{A}$,

$$(a) \quad \mathbf{A} = (\mathbf{A}\cdot\mathbf{i})\mathbf{i} + (\mathbf{A}\cdot\mathbf{j})\mathbf{j} + (\mathbf{A}\cdot\mathbf{k})\mathbf{k}$$

$$(b) \quad \mathbf{A} = A(\cos\alpha\,\mathbf{i} + \cos\beta\,\mathbf{j} + \cos\gamma\,\mathbf{k})$$

where α, β, γ are the angles which $\mathbf{A}$ makes with $\mathbf{i}, \mathbf{j}, \mathbf{k}$ respectively and $\cos\alpha$, $\cos\beta$, $\cos\gamma$ are called the *direction cosines* of $\mathbf{A}$.

(*a*) We have $\mathbf{A} = A_1\mathbf{i} + A_2\mathbf{j} + A_3\mathbf{k}$. Then

$$\mathbf{A}\cdot\mathbf{i} = (A_1\mathbf{i} + A_2\mathbf{j} + A_3\mathbf{k})\cdot\mathbf{i} = A_1$$
$$\mathbf{A}\cdot\mathbf{j} = (A_1\mathbf{i} + A_2\mathbf{j} + A_3\mathbf{k})\cdot\mathbf{j} = A_2$$
$$\mathbf{A}\cdot\mathbf{k} = (A_1\mathbf{i} + A_2\mathbf{j} + A_3\mathbf{k})\cdot\mathbf{k} = A_3$$

Thus
$$\mathbf{A} = (\mathbf{A}\cdot\mathbf{i})\mathbf{i} + (A\cdot\mathbf{j})\mathbf{j} + (\mathbf{A}\cdot\mathbf{k})\mathbf{k}$$

(*b*)
$$\mathbf{A}\cdot\mathbf{i} = |\mathbf{A}|\,|\mathbf{i}|\cos\alpha = A\cos\alpha$$
$$\mathbf{A}\cdot\mathbf{j} = |\mathbf{A}|\,|\mathbf{j}|\cos\beta = A\cos\beta$$
$$\mathbf{A}\cdot\mathbf{k} = |\mathbf{A}|\,|\mathbf{k}|\cos\gamma = A\cos\gamma$$

Then from part (*a*),

$$\mathbf{A} = (\mathbf{A}\cdot\mathbf{i})\mathbf{i} + (\mathbf{A}\cdot\mathbf{j})\mathbf{j} + (\mathbf{A}\cdot\mathbf{k})\mathbf{k} = A(\cos\alpha\,\mathbf{i} + \cos\beta\,\mathbf{j} + \cos\gamma\,\mathbf{k})$$

1.44. Prove that $\nabla\phi$ is a vector perpendicular to the surface $\phi(x,y,z) = c$, where c is a constant.

Let $\mathbf{r} = x\mathbf{i} + y\mathbf{j} + z\mathbf{k}$ be the position vector to any point $P(x,y,z)$ on the surface. Then $d\mathbf{r} = dx\,\mathbf{i} + dy\,\mathbf{j} + dz\,\mathbf{k}$ lies in the plane tangent to the surface at P. But

$$d\phi = \frac{\partial\phi}{\partial x}dx + \frac{\partial\phi}{\partial y}dy + \frac{\partial\phi}{\partial z}dz = 0 \quad\text{or}\quad \left(\frac{\partial\phi}{\partial x}\mathbf{i} + \frac{\partial\phi}{\partial y}\mathbf{j} + \frac{\partial\phi}{\partial z}\mathbf{k}\right)\cdot(dx\,\mathbf{i} + dy\,\mathbf{j} + dz\,\mathbf{k}) = 0$$

i.e. $\nabla\phi\cdot d\mathbf{r} = 0$ so that $\nabla\phi$ is perpendicular to $d\mathbf{r}$ and therefore to the surface.

1.45. Find a unit normal to the surface $2x^2 + 4yz - 5z^2 = -10$ at the point $P(3,-1,2)$.

By Problem 1.44, a vector normal to the surface is

$$\nabla(2x^2 + 4yz - 5z^2) = 4x\mathbf{i} + 4z\mathbf{j} + (4y - 10z)\mathbf{k} = 12\mathbf{i} + 8\mathbf{j} - 24\mathbf{k} \quad \text{at } (3,-1,2)$$

Then a unit normal to the surface at P is $\dfrac{12\mathbf{i} + 8\mathbf{j} - 24\mathbf{k}}{\sqrt{(12)^2 + (8)^2 + (-24)^2}} = \dfrac{3\mathbf{i} + 2\mathbf{j} - 6\mathbf{k}}{7}$.

Another unit normal to the surface at P is $-\dfrac{3\mathbf{i} + 2\mathbf{j} - 6\mathbf{k}}{7}$.

1.46. A ladder AB of length a rests against a vertical wall OA [Fig. 1-29]. The foot B of the ladder is pulled away with constant speed v_0. (*a*) Show that the midpoint of the ladder describes the arc of a circle of radius $a/2$ with center at O. (*b*) Find the velocity and speed of the midpoint of the ladder at the instant where B is distant $b < a$ from the wall.

(*a*) Let $\mathbf{r}$ be the position vector of midpoint M of AB. If angle $OBA = \theta$, we have

$$\mathbf{OB} = a\cos\theta\,\mathbf{i}, \quad \mathbf{OA} = a\sin\theta\,\mathbf{j}$$
$$\mathbf{AB} = \mathbf{OB} - \mathbf{OA} = a\cos\theta\,\mathbf{i} - a\sin\theta\,\mathbf{j}$$

Then
$$\mathbf{r} = \mathbf{OA} + \mathbf{AM} = \mathbf{OA} + \tfrac{1}{2}\mathbf{AB}$$
$$= a\sin\theta\,\mathbf{j} + \tfrac{1}{2}(a\cos\theta\,\mathbf{i} - a\sin\theta\,\mathbf{j})$$
$$= \tfrac{1}{2}a(\cos\theta\,\mathbf{i} + \sin\theta\,\mathbf{j})$$

Thus $|\mathbf{r}| = \frac{1}{2}a$, which is a circle of radius $a/2$ with center at O.

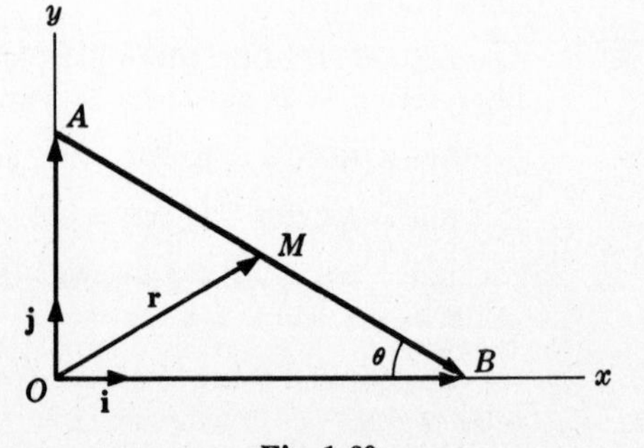

Fig. 1-29

(*b*) The velocity of the midpoint M is

$$\frac{d\mathbf{r}}{dt} = \frac{d}{dt}\{\tfrac{1}{2}a(\cos\theta\,\mathbf{i} + \sin\theta\,\mathbf{j})\} = \tfrac{1}{2}a(-\sin\theta\,\dot{\theta}\mathbf{i} + \cos\theta\,\dot{\theta}\mathbf{j}) \qquad (1)$$

where $\dot{\theta} = d\theta/dt$.

The velocity of the foot B of the ladder is

$$v_0\mathbf{i} = \frac{d}{dt}(\mathbf{OB}) = \frac{d}{dt}(a\cos\theta\,\mathbf{i}) = -a\sin\theta\,\dot{\theta}\mathbf{i} \quad \text{or} \quad a\sin\theta\,\dot{\theta} = -v_0 \qquad (2)$$

At the instant where B is distant b from the wall we have from (*2*),

$$\sin\theta = \frac{\sqrt{a^2-b^2}}{a}, \qquad \dot{\theta} = \frac{-v_0}{a\sin\theta} = \frac{-v_0}{\sqrt{a^2-b^2}}$$

Thus from (*1*) the required velocity of M at this instant is

$$\frac{d\mathbf{r}}{dt} = \tfrac{1}{2}v_0\left(\mathbf{i} - \frac{b}{\sqrt{a^2-b^2}}\mathbf{j}\right)$$

and its speed is $av_0/2\sqrt{a^2-b^2}$.

1.47. Let (r, θ) represent the polar coordinates describing the position of a particle. If $\mathbf{r}_1$ is a unit vector in the direction of the position vector $\mathbf{r}$ and $\boldsymbol{\theta}_1$ is a unit vector perpendicular to $\mathbf{r}$ and in the direction of increasing θ [see Fig. 1-30], show that

$$(a)\ \mathbf{r}_1 = \cos\theta\,\mathbf{i} + \sin\theta\,\mathbf{j}, \quad \boldsymbol{\theta}_1 = -\sin\theta\,\mathbf{i} + \cos\theta\,\mathbf{j}$$

$$(b)\ \mathbf{i} = \cos\theta\,\mathbf{r}_1 - \sin\theta\,\boldsymbol{\theta}_1, \quad \mathbf{j} = \sin\theta\,\mathbf{r}_1 + \cos\theta\,\boldsymbol{\theta}_1$$

(*a*) If $\mathbf{r}$ is the position vector of the particle at any time t, then $\partial\mathbf{r}/\partial r$ is a vector tangent to the curve $\theta = \text{constant}$, i.e. a vector in the direction of $\mathbf{r}$ (increasing r). A unit vector in this direction is thus given by

$$\mathbf{r}_1 = \frac{\partial\mathbf{r}}{\partial r}\bigg/\left|\frac{\partial\mathbf{r}}{\partial r}\right| \qquad (1)$$

Since

$$\mathbf{r} = x\mathbf{i} + y\mathbf{j} = r\cos\theta\,\mathbf{i} + r\sin\theta\,\mathbf{j} \qquad (2)$$

as seen from Fig. 1-30, we have

$$\frac{\partial\mathbf{r}}{\partial r} = \cos\theta\,\mathbf{i} + \sin\theta\,\mathbf{j}, \qquad \left|\frac{\partial\mathbf{r}}{\partial r}\right| = 1$$

so that

$$\mathbf{r}_1 = \cos\theta\,\mathbf{i} + \sin\theta\,\mathbf{j} \qquad (3)$$

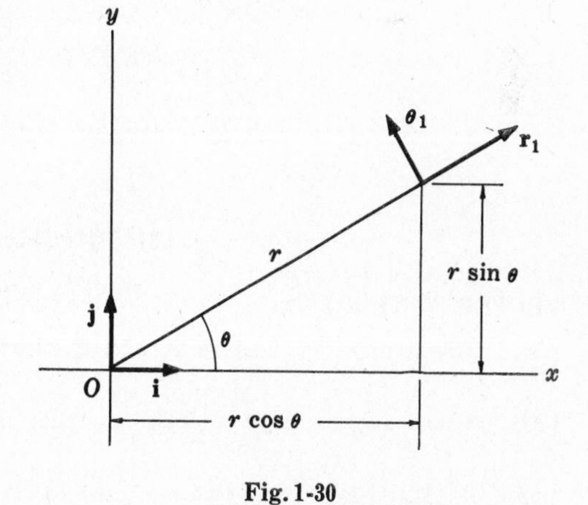

Fig. 1-30

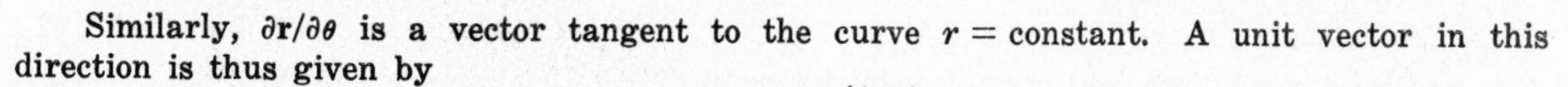

Similarly, $\partial\mathbf{r}/\partial\theta$ is a vector tangent to the curve $r = \text{constant}$. A unit vector in this direction is thus given by

$$\boldsymbol{\theta}_1 = \frac{\partial\mathbf{r}}{\partial\theta}\bigg/\left|\frac{\partial\mathbf{r}}{\partial\theta}\right| \qquad (4)$$

Now from (*2*),

$$\frac{\partial\mathbf{r}}{\partial\theta} = -r\sin\theta\,\mathbf{i} + r\cos\theta\,\mathbf{j}, \qquad \left|\frac{\partial\mathbf{r}}{\partial\theta}\right| = r$$

so that (*4*) yields

$$\boldsymbol{\theta}_1 = -\sin\theta\,\mathbf{i} + \cos\theta\,\mathbf{j} \qquad (5)$$

(*b*) These results follow by solving the simultaneous equations (*3*) and (*5*) for $\mathbf{i}$ and $\mathbf{j}$.

1.48. Prove that (*a*) $\dot{\mathbf{r}}_1 = \dot{\theta}\boldsymbol{\theta}_1$ (*b*) $\dot{\boldsymbol{\theta}}_1 = -\dot{\theta}\mathbf{r}_1$.

(*a*) From (*3*) of Problem 1.47 we have

$$\dot{\mathbf{r}}_1 = \frac{d\mathbf{r}_1}{dt} = \frac{\partial \mathbf{r}_1}{\partial r}\frac{dr}{dt} + \frac{\partial \mathbf{r}_1}{\partial \theta}\frac{d\theta}{dt}$$

$$= (\mathbf{0})(\dot{r}) + (-\sin\theta\,\mathbf{i} + \cos\theta\,\mathbf{j})(\dot{\theta}) = \dot{\theta}\boldsymbol{\theta}_1$$

(*b*) From (*5*) of Problem 1.47 we have

$$\dot{\boldsymbol{\theta}}_1 = \frac{d\boldsymbol{\theta}_1}{dt} = \frac{\partial \boldsymbol{\theta}_1}{\partial r}\frac{dr}{dt} + \frac{\partial \boldsymbol{\theta}_1}{\partial \theta}\frac{d\theta}{dt}$$

$$= (\mathbf{0})(\dot{r}) + (-\cos\theta\,\mathbf{i} - \sin\theta\,\mathbf{j})(\dot{\theta}) = -\dot{\theta}\mathbf{r}_1$$

1.49. Prove that in polar coordinates (*a*) the velocity is given by

$$\mathbf{v} = \dot{r}\mathbf{r}_1 + r\dot{\theta}\boldsymbol{\theta}_1$$

and (*b*) the acceleration is given by

$$\mathbf{a} = (\ddot{r} - r\dot{\theta}^2)\mathbf{r}_1 + (r\ddot{\theta} + 2\dot{r}\dot{\theta})\boldsymbol{\theta}_1$$

(*a*) We have $\mathbf{r} = r\mathbf{r}_1$ so that

$$\mathbf{v} = \frac{d\mathbf{r}}{dt} = \frac{dr}{dt}\mathbf{r}_1 + r\frac{d\mathbf{r}_1}{dt} = \dot{r}\mathbf{r}_1 + r\dot{\mathbf{r}}_1 = \dot{r}\mathbf{r}_1 + r\dot{\theta}\boldsymbol{\theta}_1$$

by Problem 1.48(*a*).

(*b*) From part (*a*) and Problem 1.48 we have

$$\mathbf{a} = \frac{d\mathbf{v}}{dt} = \frac{d}{dt}(\dot{r}\mathbf{r}_1 + r\dot{\theta}\boldsymbol{\theta}_1)$$

$$= \ddot{r}\mathbf{r}_1 + \dot{r}\dot{\mathbf{r}}_1 + \dot{r}\dot{\theta}\boldsymbol{\theta}_1 + r\ddot{\theta}\boldsymbol{\theta}_1 + r\dot{\theta}\dot{\boldsymbol{\theta}}_1$$

$$= \ddot{r}\mathbf{r}_1 + \dot{r}(\dot{\theta}\boldsymbol{\theta}_1) + \dot{r}\dot{\theta}\boldsymbol{\theta}_1 + r\ddot{\theta}\boldsymbol{\theta}_1 + (r\dot{\theta})(-\dot{\theta}\mathbf{r}_1)$$

$$= (\ddot{r} - r\dot{\theta}^2)\mathbf{r}_1 + (r\ddot{\theta} + 2\dot{r}\dot{\theta})\boldsymbol{\theta}_1$$

Supplementary Problems

VECTOR ALGEBRA

1.50. Given any two vectors **A** and **B**, illustrate geometrically the equality $4\mathbf{A} + 3(\mathbf{B} - \mathbf{A}) = \mathbf{A} + 3\mathbf{B}$.

1.51. Given vectors **A**, **B** and **C**, construct the vectors (*a*) $2\mathbf{A} - 3\mathbf{B} + \frac{1}{2}\mathbf{C}$, (*b*) $\mathbf{C} - \frac{1}{3}\mathbf{A} + \frac{1}{4}\mathbf{B}$.

1.52. If **A** and **B** are any two non-zero vectors which do not have the same direction, prove that $p\mathbf{A} + q\mathbf{B}$ is a vector lying in the plane determined by **A** and **B**.

1.53. (*a*) Determine the vector having initial point $(2, -1, 3)$ and terminal point $(3, 2, -4)$. (*b*) Find the distance between the two points in (*a*). *Ans.* (*a*) $\mathbf{i} + 3\mathbf{j} - 7\mathbf{k}$, (*b*) $\sqrt{59}$

1.54. A triangle has vertices at the points $A(2, 1, -1)$, $B(-1, 3, 2)$, $C(1, -2, 1)$. Find the length of the median to the side AB. *Ans.* $\frac{1}{2}\sqrt{66}$

1.55. A man travels 25 miles northeast, 15 miles due east and 10 miles due south. By using an appropriate scale determine (*a*) graphically and (*b*) analytically how far and in what direction he is from his starting position. *Ans.* 33.6 miles, 13.2° north of east.

1.56. Find a unit vector in the direction of the resultant of vectors $\mathbf{A} = 2\mathbf{i} - \mathbf{j} + \mathbf{k}$, $\mathbf{B} = \mathbf{i} + \mathbf{j} + 2\mathbf{k}$, $\mathbf{C} = 3\mathbf{i} - 2\mathbf{j} + 4\mathbf{k}$. *Ans.* $\pm(6\mathbf{i} - 2\mathbf{j} + 7\mathbf{k})/\sqrt{89}$

THE DOT OR SCALAR PRODUCT

1.57. Evaluate $|(\mathbf{A} + \mathbf{B}) \cdot (\mathbf{A} - \mathbf{B})|$ if $\mathbf{A} = 2\mathbf{i} - 3\mathbf{j} + 5\mathbf{k}$ and $\mathbf{B} = 3\mathbf{i} + \mathbf{j} - 2\mathbf{k}$. *Ans.* 24

1.58. Find a so that $2\mathbf{i} - 3\mathbf{j} + 5\mathbf{k}$ and $3\mathbf{i} + a\mathbf{j} - 2\mathbf{k}$ are perpendicular. *Ans.* $a = -4/3$

1.59. If $\mathbf{A} = 2\mathbf{i} + \mathbf{j} + \mathbf{k}$, $\mathbf{B} = \mathbf{i} - 2\mathbf{j} + 2\mathbf{k}$ and $\mathbf{C} = 3\mathbf{i} - 4\mathbf{j} + 2\mathbf{k}$, find the projection of $\mathbf{A} + \mathbf{C}$ in the direction of $\mathbf{B}$. *Ans.* 17/3

1.60. A triangle has vertices at $A(2,3,1)$, $B(-1,1,2)$, $C(1,-2,3)$. Find the acute angle which the median to side AC makes with side BC. *Ans.* $\cos^{-1}\sqrt{91}/14$

1.61. Prove the *law of cosines* for triangle ABC, i.e. $c^2 = a^2 + b^2 - 2ab\cos C$.
[*Hint.* Take the sides as $\mathbf{A}, \mathbf{B}, \mathbf{C}$ where $\mathbf{C} = \mathbf{A} - \mathbf{B}$. Then use $\mathbf{C} \cdot \mathbf{C} = (\mathbf{A} - \mathbf{B}) \cdot (\mathbf{A} - \mathbf{B})$.]

1.62. Prove that the diagonals of a rhombus are perpendicular to each other.

THE CROSS OR VECTOR PRODUCT

1.63. If $\mathbf{A} = 2\mathbf{i} - \mathbf{j} + \mathbf{k}$ and $\mathbf{B} = \mathbf{i} + 2\mathbf{j} - 3\mathbf{k}$, find $|(2\mathbf{A} + \mathbf{B}) \times (\mathbf{A} - 2\mathbf{B})|$. *Ans.* $25\sqrt{3}$

1.64. Find a unit vector perpendicular to the plane of the vectors $\mathbf{A} = 3\mathbf{i} - 2\mathbf{j} + 4\mathbf{k}$ and $\mathbf{B} = \mathbf{i} + \mathbf{j} - 2\mathbf{k}$.
Ans. $\pm(2\mathbf{j} + \mathbf{k})/\sqrt{5}$

1.65. Find the area of the triangle with vertices $(2,-3,1)$, $(1,-1,2)$, $(-1,2,3)$. *Ans.* $\frac{1}{2}\sqrt{3}$

1.66. Find the shortest distance from the point $(3,2,1)$ to the plane determined by $(1,1,0)$, $(3,-1,1)$, $(-1,0,2)$. *Ans.* 2

1.67. Prove the *law of sines* for triangle ABC, i.e. $\dfrac{\sin A}{a} = \dfrac{\sin B}{b} = \dfrac{\sin C}{c}$.
[*Hint.* Consider the sides to be $\mathbf{A}, \mathbf{B}, \mathbf{C}$ where $\mathbf{A} + \mathbf{B} + \mathbf{C} = \mathbf{0}$ and take the cross product of both sides with $\mathbf{A}$ and $\mathbf{B}$ respectively.]

TRIPLE PRODUCTS

1.68. If $\mathbf{A} = 2\mathbf{i} + \mathbf{j} - 3\mathbf{k}$, $\mathbf{B} = \mathbf{i} - 2\mathbf{j} + \mathbf{k}$ and $\mathbf{C} = -\mathbf{i} + \mathbf{j} - 4\mathbf{k}$, find (*a*) $\mathbf{A} \cdot (\mathbf{B} \times \mathbf{C})$, (*b*) $\mathbf{C} \cdot (\mathbf{A} \times \mathbf{B})$, (*c*) $\mathbf{A} \times (\mathbf{B} \times \mathbf{C})$, (*d*) $(\mathbf{A} \times \mathbf{B}) \times \mathbf{C}$. *Ans.* (*a*) 20, (*b*) 20, (*c*) $8\mathbf{i} - 19\mathbf{j} - \mathbf{k}$, (*d*) $25\mathbf{i} - 15\mathbf{j} - 10\mathbf{k}$

1.69. Prove that $\mathbf{A} \cdot (\mathbf{B} \times \mathbf{C}) = (\mathbf{A} \times \mathbf{B}) \cdot \mathbf{C}$, i.e. the dot and the cross can be interchanged.

1.70. Find the volume of a parallelepiped whose edges are given by $\mathbf{A} = 2\mathbf{i} + 3\mathbf{j} - \mathbf{k}$, $\mathbf{B} = \mathbf{i} - 2\mathbf{j} + 2\mathbf{k}$, $\mathbf{C} = 3\mathbf{i} - \mathbf{j} - 2\mathbf{k}$. *Ans.* 31

1.71. Find the volume of the tetrahedron with vertices at $(2,1,1)$, $(1,-1,2)$, $(0,1,-1)$, $(1,-2,1)$.
Ans. 4/3

1.72. Prove that (*a*) $\mathbf{A} \cdot (\mathbf{B} \times \mathbf{C}) = \mathbf{B} \cdot (\mathbf{C} \times \mathbf{A}) = \mathbf{C} \cdot (\mathbf{A} \times \mathbf{B})$,
(*b*) $\mathbf{A} \times (\mathbf{B} \times \mathbf{C}) = \mathbf{B}(\mathbf{A} \cdot \mathbf{C}) - \mathbf{C}(\mathbf{A} \cdot \mathbf{B})$.

1.73. (*a*) Let $\mathbf{r}_1, \mathbf{r}_2, \mathbf{r}_3$ be position vectors to three points P_1, P_2, P_3 respectively. Prove that the equation $(\mathbf{r} - \mathbf{r}_1) \cdot [(\mathbf{r} - \mathbf{r}_2) \times (\mathbf{r} - \mathbf{r}_3)] = 0$, where $\mathbf{r} = x\mathbf{i} + y\mathbf{j} + z\mathbf{k}$, represents an equation for the plane determined by P_1, P_2 and P_3. (*b*) Find an equation for the plane passing through $(2,-1,-2)$, $(-1,2,-3)$, $(4,1,0)$. *Ans.* (*b*) $2x + y - 3z = 9$

DERIVATIVES AND INTEGRALS OF VECTORS

1.74. Let $\mathbf{A} = 3t\mathbf{i} - (t^2 + t)\mathbf{j} + (t^3 - 2t^2)\mathbf{k}$. Find (*a*) $d\mathbf{A}/dt$ and (*b*) $d^2\mathbf{A}/dt^2$ at $t = 1$.
Ans. (*a*) $3\mathbf{i} - 3\mathbf{j} - \mathbf{k}$, (*b*) $-2\mathbf{j} + 2\mathbf{k}$

1.75. If $\mathbf{r} = \mathbf{a}\cos\omega t + \mathbf{b}\sin\omega t$, where $\mathbf{a}$ and $\mathbf{b}$ are any constant non-collinear vectors and ω is a constant scalar, prove that (*a*) $\mathbf{r} \times d\mathbf{r}/dt = \omega(\mathbf{a} \times \mathbf{b})$, (*b*) $d^2\mathbf{r}/dt^2 + \omega^2\mathbf{r} = \mathbf{0}$.

1.76. If $\mathbf{A} = t\mathbf{i} - \sin t\,\mathbf{k}$ and $\mathbf{B} = \cos t\,\mathbf{i} + \sin t\,\mathbf{j} + \mathbf{k}$, find $\dfrac{d}{dt}(\mathbf{A}\cdot\mathbf{B})$. *Ans.* $-t\sin t$

1.77. Prove that $\dfrac{d}{du}(\mathbf{A}\times\mathbf{B}) = \mathbf{A}\times\dfrac{d\mathbf{B}}{du} + \dfrac{d\mathbf{A}}{du}\times\mathbf{B}$ where $\mathbf{A}$ and $\mathbf{B}$ are differentiable functions of u.

1.78. If $\mathbf{A}(u) = 4(u-1)\mathbf{i} - (2u+3)\mathbf{j} + 6u^2\mathbf{k}$, evaluate (*a*) $\displaystyle\int_2^3 \mathbf{A}(u)\,du$, (*b*) $\displaystyle\int_1^2 (u\mathbf{i} - 2\mathbf{k})\cdot\mathbf{A}(u)\,du$.
Ans. (*a*) $6\mathbf{i} - 8\mathbf{j} + 38\mathbf{k}$, (*b*) -28

1.79. Find the vector $\mathbf{B}(u)$ such that $d^2\mathbf{B}/du^2 = 6u\mathbf{i} - 48u^2\mathbf{j} + 12\mathbf{k}$ where $\mathbf{B} = 2\mathbf{i} - 3\mathbf{k}$ and $d\mathbf{B}/du = \mathbf{i} + 5\mathbf{j}$ for $u = 0$. *Ans.* $(u^3 + u + 2)\mathbf{i} + (5u - 4u^4)\mathbf{j} + (6u^2 - 3)\mathbf{k}$

1.80. Prove that $\displaystyle\int \mathbf{A}\times\frac{d^2\mathbf{A}}{dt^2}\,dt = \mathbf{A}\times\frac{d\mathbf{A}}{dt} + \mathbf{c}$ where $\mathbf{c}$ is a constant vector.

1.81. If $\mathbf{R} = x^2y\mathbf{i} - 2y^2z\mathbf{j} + xy^2z^2\mathbf{k}$, find $\left|\dfrac{\partial^2\mathbf{R}}{\partial x^2}\times\dfrac{\partial^2\mathbf{R}}{\partial y^2}\right|$ at the point $(2,1,-2)$. *Ans.* $16\sqrt{5}$

1.82. If $\mathbf{A} = x^2\mathbf{i} - y\mathbf{j} + xz\mathbf{k}$ and $\mathbf{B} = y\mathbf{i} + x\mathbf{j} - xyz\mathbf{k}$, find $\dfrac{\partial^2}{\partial x\,\partial y}(\mathbf{A}\times\mathbf{B})$ at the point $(1,-1,2)$.
Ans. $-4\mathbf{i} + 8\mathbf{j}$

VELOCITY AND ACCELERATION

1.83. A particle moves along the space curve $\mathbf{r} = (t^2 + t)\mathbf{i} + (3t - 2)\mathbf{j} + (2t^3 - 4t^2)\mathbf{k}$. Find the (*a*) velocity, (*b*) acceleration, (*c*) speed or magnitude of velocity and (*d*) magnitude of acceleration at time $t = 2$. *Ans.* (*a*) $5\mathbf{i} + 3\mathbf{j} + 8\mathbf{k}$, (*b*) $2\mathbf{i} + 16\mathbf{k}$, (*c*) $7\sqrt{2}$, (*d*) $2\sqrt{65}$

1.84. A particle moves along the space curve defined by $x = e^{-t}\cos t$, $y = e^{-t}\sin t$, $z = e^{-t}$. Find the magnitude of the (*a*) velocity and (*b*) acceleration at any time t.
Ans. (*a*) $\sqrt{3}\,e^{-t}$, (*b*) $\sqrt{5}\,e^{-t}$

1.85. The position vector of a particle is given at any time t by $\mathbf{r} = a\cos\omega t\,\mathbf{i} + b\sin\omega t\,\mathbf{j} + ct^2\mathbf{k}$. (*a*) Show that although the speed of the particle increases with time the magnitude of the acceleration is always constant. (*b*) Describe the motion of the particle geometrically.

RELATIVE VELOCITY AND ACCELERATION

1.86. The position vectors of two particles are given respectively by $\mathbf{r}_1 = t\mathbf{i} - t^2\mathbf{j} + (2t + 3)\mathbf{k}$ and $\mathbf{r}_2 = (2t - 3t^2)\mathbf{i} + 4t\mathbf{j} - t^3\mathbf{k}$. Find (*a*) the relative velocity and (*b*) the relative acceleration of the second particle with respect to the first at $t = 1$. *Ans.* (*a*) $-5\mathbf{i} + 6\mathbf{j} - 5\mathbf{k}$, (*b*) $-6\mathbf{i} + 2\mathbf{j} - 6\mathbf{k}$

1.87. An automobile driver traveling northeast at 26 mi/hr notices that the wind appears to be coming from the northwest. When he drives southeast at 30 mi/hr the wind appears to be coming from 60° south of west. Find the velocity of the wind relative to the ground.
Ans. 52 mi/hr in a direction from 30° south of west

1.88. A man in a boat on one side of a river wishes to reach a point directly opposite him on the other side of the river. Assuming that the width of the river is D and that the speeds of the boat and current are V and $v < V$ respectively, show that (*a*) he should start his boat upstream at an angle of $\sin^{-1}(v/V)$ with the shore and (*b*) the time to cross the river is $D/\sqrt{V^2 - v^2}$.

TANGENTIAL AND NORMAL ACCELERATION

1.89. Show that the tangential and normal acceleration of a particle moving on a space curve are given by d^2s/dt^2 and $\kappa(ds/dt)^2$ where s is the arc length of the curve measured from some initial point and κ is the curvature.

1.90. Find the (*a*) unit tangent $\mathbf{T}$, (*b*) principal normal $\mathbf{N}$, (*c*) radius of curvature R and (*d*) curvature κ to the space curve $x = t$, $y = t^2/2$, $z = t$.
Ans. (*a*) $(\mathbf{i} + t\mathbf{j} + \mathbf{k})/\sqrt{t^2 + 2}$, (*b*) $(-t\mathbf{i} + 2\mathbf{j} - t\mathbf{k})/\sqrt{2t^2 + 4}$, (*c*) $(t^2 + 2)^{3/2}/\sqrt{2}$, (*d*) $\sqrt{2}/(t^2 + 2)^{3/2}$

1.91. A particle moves in such a way that its position vector at any time t is $\mathbf{r} = t\mathbf{i} + \frac{1}{2}t^2\mathbf{j} + t\mathbf{k}$. Find (*a*) the velocity, (*b*) the speed, (*c*) the acceleration, (*d*) the magnitude of the acceleration, (*e*) the magnitude of the tangential acceleration, (*f*) the magnitude of the normal acceleration.
Ans. (*a*) $\mathbf{i} + t\mathbf{j} + \mathbf{k}$, (*b*) $\sqrt{t^2+2}$, (*c*) $\mathbf{j}$, (*d*) 1, (*e*) $t/\sqrt{t^2+2}$, (*f*) $\sqrt{2}/\sqrt{t^2+2}$

1.92. Find the (*a*) tangential acceleration and (*b*) normal acceleration of a particle which moves on the ellipse $\mathbf{r} = a \cos \omega t\, \mathbf{i} + b \sin \omega t\, \mathbf{j}$.

Ans. (*a*) $\dfrac{\omega^2(a^2 - b^2) \sin \omega t \cos \omega t}{\sqrt{a^2 \sin^2 \omega t + b^2 \cos^2 \omega t}}$ (*b*) $\dfrac{\omega^2 ab}{\sqrt{a^2 \sin^2 \omega t + b^2 \cos^2 \omega t}}$

CIRCULAR MOTION

1.93. A particle moves in a circle of radius 20 cm. If its tangential speed is 40 cm/sec, find (*a*) its angular speed, (*b*) its angular acceleration, (*c*) its normal acceleration.
Ans. (*a*) 2 radians/sec, (*b*) 0 radians/sec^2, (*c*) 80 cm/sec^2

1.94. A particle moving on a circle of radius R has a constant angular acceleration α. If the particle starts from rest, show that after time t (*a*) its angular velocity is $\omega = \alpha t$, (*b*) the arc length covered is $s = \frac{1}{2}R\alpha t^2$.

1.95. A particle moves on a circle of radius R with constant angular speed ω_0. At time $t = 0$ it starts to slow down so that its angular acceleration is $-\alpha$ (or *deceleration* α). Show that (*a*) it comes to rest after a time ω_0/α and (*b*) has travelled a distance $R\omega_0^2/2\alpha$.

1.96. If the particle in Problem 1.95 is travelling at 3600 revolutions per minute in a circle of radius 100 cm and develops a constant deceleration of 5 radians/sec^2, (*a*) how long will it be before it comes to rest and (*b*) what distance will it have travelled? *Ans.* (*a*) 75.4 sec, (*b*) 1.42×10^6 cm

GRADIENT, DIVERGENCE AND CURL

1.97. If $\mathbf{A} = xz\mathbf{i} + (2x^2 - y)\mathbf{j} - yz^2\mathbf{k}$ and $\phi = 3x^2y + y^2z^3$, find (*a*) $\nabla\phi$, (*b*) $\nabla \cdot \mathbf{A}$ and (*c*) $\nabla \times \mathbf{A}$ at the point $(1, -1, 1)$. *Ans.* (*a*) $-6\mathbf{i} + \mathbf{j} + 3\mathbf{k}$, (*b*) 2, (*c*) $-\mathbf{i} + \mathbf{j} + 4\mathbf{k}$

1.98. If $\phi = xy + yz + zx$ and $\mathbf{A} = x^2y\mathbf{i} + y^2z\mathbf{j} + z^2x\mathbf{k}$, find (*a*) $\mathbf{A} \cdot \nabla\phi$, (*b*) $\phi\nabla \cdot \mathbf{A}$ and (*c*) $(\nabla\phi) \times \mathbf{A}$ at the point $(3, -1, 2)$. *Ans.* (*a*) 25, (*b*) 2, (*c*) $56\mathbf{i} - 30\mathbf{j} + 47\mathbf{k}$

1.99. Prove that if $U, V, \mathbf{A}, \mathbf{B}$ have continuous partial derivatives, then (*a*) $\nabla(U + V) = \nabla U + \nabla V$, (*b*) $\nabla \cdot (\mathbf{A} + \mathbf{B}) = \nabla \cdot \mathbf{A} + \nabla \cdot \mathbf{B}$, (*c*) $\nabla \times (\mathbf{A} + \mathbf{B}) = \nabla \times \mathbf{A} + \nabla \times \mathbf{B}$.

1.100. Show that $\nabla \times (r^2\mathbf{r}) = \mathbf{0}$ where $\mathbf{r} = x\mathbf{i} + y\mathbf{j} + z\mathbf{k}$ and $r = |\mathbf{r}|$.

1.101. Prove that (*a*) div curl $\mathbf{A} = 0$ and (*b*) curl grad $\phi = \mathbf{0}$ under suitable conditions on $\mathbf{A}$ and ϕ.

1.102. If $\mathbf{A} = (2x^2 - yz)\mathbf{i} + (y^2 - 2xz)\mathbf{j} + x^2z^3\mathbf{k}$ and $\phi = x^2y - 3xz^2 + 2xyz$, show directly that div curl $\mathbf{A} = 0$ and curl grad $\phi = \mathbf{0}$.

1.103. If $\mathbf{A} = 3xz^2\mathbf{i} - yz\mathbf{j} + (x + 2z)\mathbf{k}$, find curl curl $\mathbf{A}$. *Ans.* $-6x\mathbf{i} + (6z - 1)\mathbf{k}$

1.104. (*a*) Prove that $\nabla \times (\nabla \times \mathbf{A}) = -\nabla^2\mathbf{A} + \nabla(\nabla \cdot \mathbf{A})$. (*b*) Verify the result in (*a*) if $\mathbf{A}$ is as given in Problem 1.103.

1.105. Prove: (*a*) $\nabla \times (U\mathbf{A}) = (\nabla U) \times \mathbf{A} + U(\nabla \times \mathbf{A})$. (*b*) $\nabla \cdot (\mathbf{A} \times \mathbf{B}) = \mathbf{B} \cdot (\nabla \times \mathbf{A}) - \mathbf{A} \cdot (\nabla \times \mathbf{B})$.

LINE INTEGRALS AND INDEPENDENCE OF THE PATH

1.106. If $\mathbf{F} = (3x - 2y)\mathbf{i} + (y + 2z)\mathbf{j} - x^2\mathbf{k}$, evaluate $\int_C \mathbf{F} \cdot d\mathbf{r}$ from $(0,0,0)$ to $(1,1,1)$, where C is a path consisting of: (*a*) the curve $x = t$, $y = t^2$, $z = t^3$; (*b*) a straight line joining these points; (*c*) the straight lines from $(0,0,0)$ to $(0,1,0)$, then to $(0,1,1)$ and then to $(1,1,1)$; (*d*) the curve $x = z^2$, $z = y^2$. *Ans.* (*a*) 23/15, (*b*) 5/3, (*c*) 0, (*d*) 13/30

1.107. Evaluate $\int_C \mathbf{A} \cdot d\mathbf{r}$ where $\mathbf{A} = 3x^2\mathbf{i} + (2xz - y)\mathbf{j} + z\mathbf{k}$ along (*a*) the straight line from $(0,0,0)$ to $(2,1,3)$, (*b*) the space curve $x = 2t^2$, $y = t$, $z = 4t^2 - t$ from $t = 0$ to $t = 1$, (*c*) the curve defined by $x^2 = 4y$, $3x^3 = 8z$ from $x = 0$ to $x = 2$. *Ans.* (*a*) 16, (*b*) 14.2, (*c*) 16

1.108. Find $\oint_C \mathbf{F} \cdot d\mathbf{r}$ where $\mathbf{F} = (x - 3y)\mathbf{i} + (y - 2x)\mathbf{j}$ and C is the closed curve in the xy plane, $x = 2\cos t$, $y = 3\sin t$, $z = 0$ from $t = 0$ to $t = 2\pi$. *Ans.* 6π

1.109. (*a*) If $\mathbf{A} = (4xy - 3x^2z^2)\mathbf{i} + (4y + 2x^2)\mathbf{j} + (1 - 2x^3z)\mathbf{k}$, prove that $\int_C \mathbf{A} \cdot d\mathbf{r}$ is independent of the curve C joining two given points. (*b*) Evaluate the integral in (*a*) if C is the curve from the points $(1, -1, 1)$ to $(2, -2, -1)$. *Ans.* (*b*) -19

1.110. Determine whether $\int_C \mathbf{A} \cdot d\mathbf{r}$ is independent of the path C joining any two points if (*a*) $\mathbf{A} = 2xyz\mathbf{i} + x^2z\mathbf{j} + x^2y\mathbf{k}$, (*b*) $2xz\mathbf{i} + (x^2 - y)\mathbf{j} + (2z - x^2)\mathbf{k}$. In the case where it is independent of the path, determine ϕ such that $\mathbf{A} = \nabla\phi$.
Ans. (*a*) Independent of path, $\phi = x^2yz + c$; (*b*) dependent on path

1.111. Evaluate $\oint_C \mathbf{E} \cdot d\mathbf{r}$ where $\mathbf{E} = r\mathbf{r}$. *Ans.* 0

MISCELLANEOUS PROBLEMS

1.112. If $\mathbf{A} \times \mathbf{B} = 8\mathbf{i} - 14\mathbf{j} + \mathbf{k}$ and $\mathbf{A} + \mathbf{B} = 5\mathbf{i} + 3\mathbf{j} + 2\mathbf{k}$, find $\mathbf{A}$ and $\mathbf{B}$.
Ans. $\mathbf{A} = 2\mathbf{i} + \mathbf{j} - 2\mathbf{k}$, $\mathbf{B} = 3\mathbf{i} + 2\mathbf{j} + 4\mathbf{k}$

1.113. Let l_1, m_1, n_1 and l_2, m_2, n_2 be direction cosines of two vectors. Show that the angle θ between them is such that $\cos\theta = l_1l_2 + m_1m_2 + n_1n_2$.

1.114. Prove that the line joining the midpoints of two sides of a triangle is parallel to the third side and has half its length.

1.115. Prove that $(\mathbf{A} \times \mathbf{B})^2 + (\mathbf{A} \cdot \mathbf{B})^2 = \mathbf{A}^2\mathbf{B}^2$.

1.116. If $\mathbf{A}$, $\mathbf{B}$ and $\mathbf{C}$ are non-coplanar vectors [vectors which do not all lie in the same plane] and $x_1\mathbf{A} + y_1\mathbf{B} + z_1\mathbf{C} = x_2\mathbf{A} + y_2\mathbf{B} + z_2\mathbf{C}$, prove that necessarily $x_1 = x_2$, $y_1 = y_2$, $z_1 = z_2$.

1.117. Let $ABCD$ be any quadrilateral and points P, Q, R and S the midpoints of successive sides. Prove that (*a*) $PQRS$ is a parallelogram, (*b*) the perimeter of $PQRS$ is equal to the sum of the lengths of the diagonals of $ABCD$.

1.118. Prove that an angle inscribed in a semicircle is a right angle.

1.119. Find a unit normal to the surface $x^2y - 2xz + 2y^2z^4 = 10$ at the point $(2, 1, -1)$.
Ans. $\pm(3\mathbf{i} + 4\mathbf{j} - 6\mathbf{k})/\sqrt{61}$

1.120. Prove that $\mathbf{A} \cdot \frac{d\mathbf{A}}{dt} = A\frac{dA}{dt}$.

1.121. If $\mathbf{A}(u)$ is a differentiable function of u and $|\mathbf{A}(u)| = 1$, prove that $d\mathbf{A}/du$ is perpendicular to $\mathbf{A}$.

1.122. Prove $\nabla \cdot (\phi\mathbf{A}) = (\nabla\phi) \cdot \mathbf{A} + \phi(\nabla \cdot \mathbf{A})$.

1.123. If $\mathbf{A} \times \mathbf{B} = \mathbf{A} \times \mathbf{C}$, does $\mathbf{B} = \mathbf{C}$ necessarily? Explain.

1.124. A ship is traveling northeast at 15 miles per hour. A man on this ship observes that another ship located 5 miles west seems to be traveling south at 5 miles per hour. (*a*) What is the actual velocity of this ship? (*b*) At what distance will the two ships be closest together?

1.125. Prove that $(\mathbf{A} \times \mathbf{B}) \cdot (\mathbf{C} \times \mathbf{D}) + (\mathbf{B} \times \mathbf{C}) \cdot (\mathbf{A} \times \mathbf{D}) + (\mathbf{C} \times \mathbf{A}) \cdot (\mathbf{B} \times \mathbf{D}) = 0$.

1.126. Solve the equation $d^2\mathbf{r}/dt^2 = -g\mathbf{k}$ where g is a constant, given that $\mathbf{r} = 0$, $d\mathbf{r}/dt = v_0\mathbf{k}$ at $t = 0$.
Ans. $\mathbf{r} = (v_0t - \frac{1}{2}gt^2)\mathbf{k}$

1.127. If $\phi = (x^2 + y^2 + z^2)^{-1/2}$, show that $\nabla^2\phi = \nabla \cdot (\nabla\phi) = 0$ at all points except $(0,0,0)$.

1.128. The muzzle velocity of a gun is 60 mi/hr. How long does it take for a bullet to travel through the gun barrel which is 2.2 ft long, assuming that the bullet is uniformly accelerated? *Ans.* .05 sec

1.129. A 25 foot ladder AB rests against a vertical wall OA as in Fig. 1-29, page 24. If the foot of the ladder B is pulled away from the wall at 12 ft/sec, find (*a*) the velocity and (*b*) the acceleration of the top of the ladder A at the instant where B is 15 ft from the wall.
Ans. (*a*) 9 ft/sec downward, (*b*) 11.25 ft/sec² downward

1.130. Prove that (*a*) $|\mathbf{A}+\mathbf{B}| \leqq |\mathbf{A}| + |\mathbf{B}|$, (*b*) $|\mathbf{A}+\mathbf{B}+\mathbf{C}| \leqq |\mathbf{A}|+|\mathbf{B}|+|\mathbf{C}|$. Give a possible geometric interpretation.

1.131. A train starts from rest with uniform acceleration. After 10 seconds it has a speed of 20 mi/hr. (*a*) How far has it traveled from its starting point after 15 seconds and (*b*) what will be its speed in mi/hr? *Ans.* (*a*) 330 ft, (*b*) 30 mi/hr

1.132. Prove that the magnitude of the acceleration of a particle moving on a space curve is

$$\sqrt{(dv/dt)^2 + v^4/R^2}$$

where v is the tangential speed and R is the radius of curvature.

1.133. If $\mathbf{T}$ is the unit tangent vector to a curve C and $\mathbf{A}$ is a vector field, prove that

$$\int_C \mathbf{A} \cdot d\mathbf{r} = \int_C \mathbf{A} \cdot \mathbf{T}\, ds$$

where s is the arc length parameter.

1.134. If $\mathbf{A} = (2x - y + 4)\mathbf{i} + (5y + 3x - 6)\mathbf{j}$, evaluate $\oint \mathbf{A} \cdot d\mathbf{r}$ around a triangle with vertices at $(0,0,0)$, $(3,0,0)$, $(3,2,0)$. *Ans.* 12

1.135. An automobile driver starts at point A of a highway and stops at point B after traveling the distance D in time T. During the course of the trip he travels at a maximum speed V. Assuming that the acceleration is constant both at the beginning and end of the trip, show that the time during which he travels at the maximum speed is given by $2D/V - T$.

1.136. Prove that the medians of a triangle (*a*) can form a triangle, (*b*) meet in a point which divides the length of each median in the ratio two to one.

1.137. If a particle has velocity $\mathbf{v}$ and acceleration $\mathbf{a}$ along a space curve, prove that the radius of curvature of its path is given numerically by

$$R = \frac{v^3}{|\mathbf{v} \times \mathbf{a}|}$$

1.138. Prove that the area of a triangle formed by vectors $\mathbf{A}$, $\mathbf{B}$ and $\mathbf{C}$ is $\frac{1}{2}|\mathbf{A}\times\mathbf{B} + \mathbf{B}\times\mathbf{C} + \mathbf{C}\times\mathbf{A}|$.

1.139. (*a*) Prove that the equation $\mathbf{A}\times\mathbf{X} = \mathbf{B}$ can be solved for $\mathbf{X}$ if and only if $\mathbf{A}\cdot\mathbf{B} = 0$ and $\mathbf{A} \neq \mathbf{0}$. (*b*) Show that one solution is $\mathbf{X} = \mathbf{B}\times\mathbf{A}/A^2$. (*c*) Can you find the general solution?
Ans. (*c*) $\mathbf{X} = \mathbf{B}\times\mathbf{A}/A^2 + \lambda\mathbf{A}$ where λ is any scalar.

1.140. Find all vectors $\mathbf{X}$ such that $\mathbf{A}\cdot\mathbf{X} = p$.
Ans. $\mathbf{X} = p\mathbf{A}/A^2 + \mathbf{V}\times\mathbf{A}$ where $\mathbf{V}$ is an arbitrary vector.

1.141. Through any point inside a triangle three lines are constructed parallel respectively to each of the three sides of the triangle and terminating in the other two sides. Prove that the sum of the ratios of the lengths of these lines to the corresponding sides is 2.

1.142. If $\mathbf{T}$, $\mathbf{N}$ and $\mathbf{B} = \mathbf{T}\times\mathbf{N}$ are the unit tangent vector, unit principal normal and unit binormal to a space curve $\mathbf{r} = \mathbf{r}(u)$, assumed differentiable, prove that

$$\frac{d\mathbf{T}}{ds} = \kappa\mathbf{N}, \quad \frac{d\mathbf{B}}{ds} = -\tau\mathbf{N}, \quad \frac{d\mathbf{N}}{ds} = \tau\mathbf{B} - \kappa\mathbf{T}$$

These are called the *Frenet-Serret formulas*. In these formulas κ is called the *curvature*, τ is the *torsion* and their reciprocals $R = 1/\kappa$, $\sigma = 1/\tau$ are called the *radius of curvature* and *radius of torsion*.

1.143. In Fig. 1-31, AB is a piston rod of length l. If A moves along horizontal line CD while B moves with constant angular speed ω around the circle of radius a with center at O, find (a) the velocity and (b) the acceleration of A.

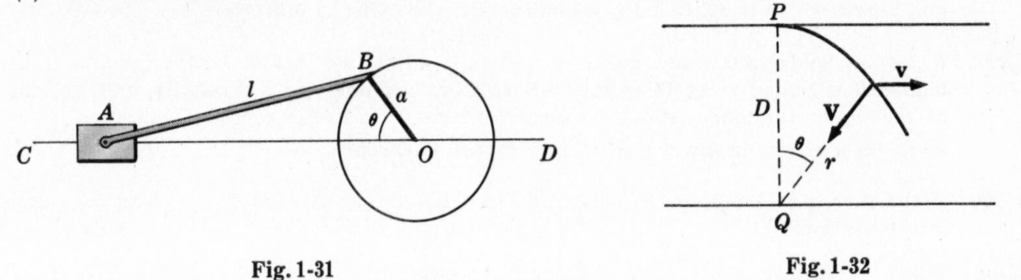

Fig. 1-31

Fig. 1-32

1.144. A boat leaves point P [see Fig. 1-32] on one side of a river bank and travels with constant velocity $\mathbf{V}$ in a direction toward point Q on the other side of the river directly opposite P and distance D from it. If r is the instantaneous distance from Q to the boat, θ is the angle between $\mathbf{r}$ and PQ, and the river travels with speed v, prove that the path of the boat is given by

$$r = \frac{D \sec\theta}{(\sec\theta + \tan\theta)^{v/V}}$$

1.145. If $v = V$ in Problem 1.144, prove that the path is an arc of a parabola.

1.146. (a) Prove that in cylindrical coordinates (ρ, ϕ, z) [see Fig. 1-33] the position vector is

$$\mathbf{r} = \rho\cos\phi\,\mathbf{i} + \rho\sin\phi\,\mathbf{j} + z\mathbf{k}$$

(b) Express the velocity in cylindrical coordinates.

(c) Express the acceleration in cylindrical coordinates.

Ans. (b) $\mathbf{v} = \dot{\rho}\boldsymbol{\rho}_1 + \rho\dot{\phi}\boldsymbol{\phi}_1 + \dot{z}\mathbf{k}$

(c) $\mathbf{a} = (\ddot{\rho} - \rho\dot{\phi}^2)\boldsymbol{\rho}_1 + (\rho\ddot{\phi} + 2\dot{\rho}\dot{\phi})\boldsymbol{\phi}_1 + \ddot{z}\mathbf{k}$

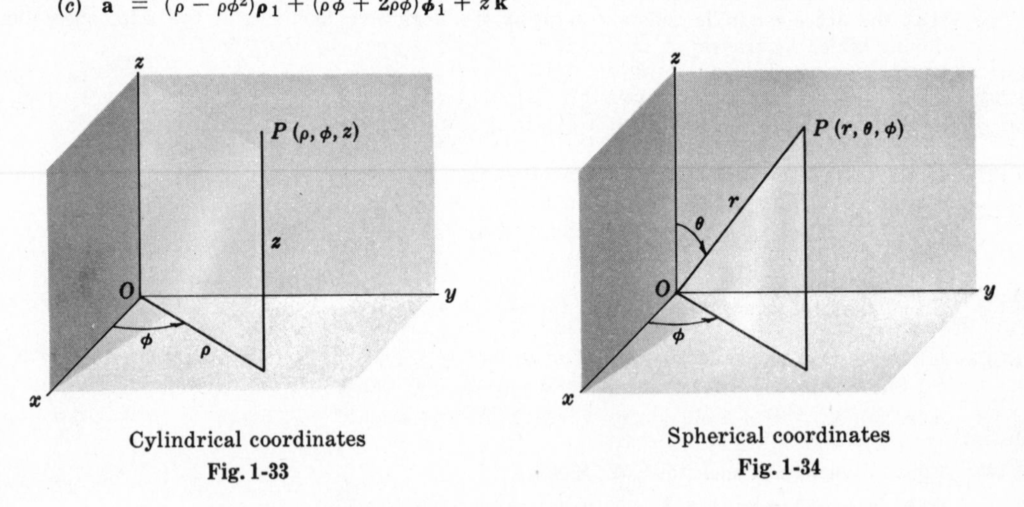

Cylindrical coordinates
Fig. 1-33

Spherical coordinates
Fig. 1-34

1.147. (a) Prove that in spherical coordinates (r, θ, ϕ) [see Fig. 1-34] the position vector is

$$\mathbf{r} = r\sin\theta\cos\phi\,\mathbf{i} + r\sin\theta\sin\phi\,\mathbf{j} + r\cos\theta\,\mathbf{k}$$

(b) Express the velocity in spherical coordinates.

(c) Express the acceleration in spherical coordinates.

Ans. (b) $\mathbf{v} = \dot{r}\mathbf{r}_1 + r\dot{\theta}\boldsymbol{\theta}_1 + r\dot{\phi}\sin\theta\,\boldsymbol{\phi}_1$

(c) $\mathbf{a} = (\ddot{r} - r\dot{\theta}^2 - r\dot{\phi}^2\sin^2\theta)\mathbf{r}_1 + (2\dot{r}\dot{\theta} + r\ddot{\theta} - r\dot{\phi}^2\sin\theta\cos\theta)\boldsymbol{\theta}_1$
$+ (2r\dot{\theta}\dot{\phi}\cos\theta + 2\dot{r}\dot{\phi}\sin\theta + r\ddot{\phi}\sin\theta)\boldsymbol{\phi}_1$

1.148. Show that if a particle moves in the xy plane the results of Problems 1.146 and 1.147 reduce to those of Problem 1.49.

Chapter 2 NEWTON'S LAWS of MOTION WORK, ENERGY and MOMENTUM

NEWTON'S LAWS

The following three laws of motion given by Sir Isaac Newton are considered the axioms of mechanics:

1. Every particle persists in a state of rest or of uniform motion in a straight line (i.e. with constant velocity) unless acted upon by a force.
2. If $\mathbf{F}$ is the (external) force acting on a particle of mass m which as a consequence is moving with velocity $\mathbf{v}$, then
$$\mathbf{F} = \frac{d}{dt}(m\mathbf{v}) = \frac{d\mathbf{p}}{dt} \tag{1}$$
where $\mathbf{p} = m\mathbf{v}$ is called the *momentum*. If m is independent of time t this becomes
$$\mathbf{F} = m\frac{d\mathbf{v}}{dt} = m\mathbf{a} \tag{2}$$
where $\mathbf{a}$ is the acceleration of the particle.
3. If particle 1 acts on particle 2 with a force $\mathbf{F}_{12}$ in a direction along the line joining the particles, while particle 2 acts on particle 1 with a force $\mathbf{F}_{21}$, then $\mathbf{F}_{21} = -\mathbf{F}_{12}$. In other words, to every *action* there is an equal and opposite *reaction*.

DEFINITIONS OF FORCE AND MASS

The concepts of *force* and *mass* used in the above axioms are as yet undefined, although intuitively we have some idea of mass as a measure of the "quantity of matter in an object" and force as a measure of the "push or pull on an object". We can however use the above axioms to develop definitions [see Problem 2.28, page 49].

UNITS OF FORCE AND MASS

Standard units of mass are the *gram* (gm) in the cgs (centimeter-gram-second) system, *kilogram* (kg) in the mks (meter-kilogram-second) system and *pound* (lb) in the fps (foot-pound-second) system. Standard units of force in these systems are the *dyne*, *newton* (nt) and *poundal* (pdl) respectively. A *dyne* is that force which will give a 1 gm mass an acceleration of 1 cm/sec^2. A *newton* is that force which will give a 1 kg mass an acceleration of 1 m/sec^2. A *poundal* is that force which will give a 1 lb mass an acceleration of 1 ft/sec^2. For relationships among these units see Appendix A, page 341.

INERTIAL FRAMES OF REFERENCE. ABSOLUTE MOTION

It must be emphasized that Newton's laws are postulated under the assumption that all measurements or observations are taken with respect to a coordinate system or *frame of reference* which is fixed in space, i.e. is absolutely at rest. This is the so-called assumption that space or motion is *absolute*. It is quite clear, however, that a particle can be at rest or in uniform motion in a straight line with respect to one frame of reference and be traveling in a curve and accelerating with respect to another frame of reference.

We can show that if Newton's laws hold in one frame of reference they also hold in any other frame of reference which is moving at constant velocity relative to it [see Problem 2.3]. All such frames of reference are called *inertial frames of reference* or *Newtonian frames of reference*. To all observers in such inertial systems the force acting on a particle will be the same, i.e. it will be *invariant*. This is sometimes called the *classical principle of relativity*.

The earth is not exactly an inertial system, but for many practical purposes can be considered as one so long as motion takes place with speeds which are not too large. For non-inertial systems we use the methods of Chapter 6. For speeds comparable with the speed of light (186,000 mi/sec), Newton's laws of mechanics must be replaced by *Einstein's laws of relativity* or *relativistic mechanics*.

WORK

If a force $\mathbf{F}$ acting on a particle gives it a displacement $d\mathbf{r}$, then the *work* done by the force on the particle is defined as

$$dW = \mathbf{F}\cdot d\mathbf{r} \qquad (3)$$

since only the component of $\mathbf{F}$ in the direction of $d\mathbf{r}$ is effective in producing the motion.

The total work done by a force field (vector field) $\mathbf{F}$ in moving the particle from point P_1 to point P_2 along the curve C of Fig. 2-1 is given by the line integral [see Chap. 1, page 9].

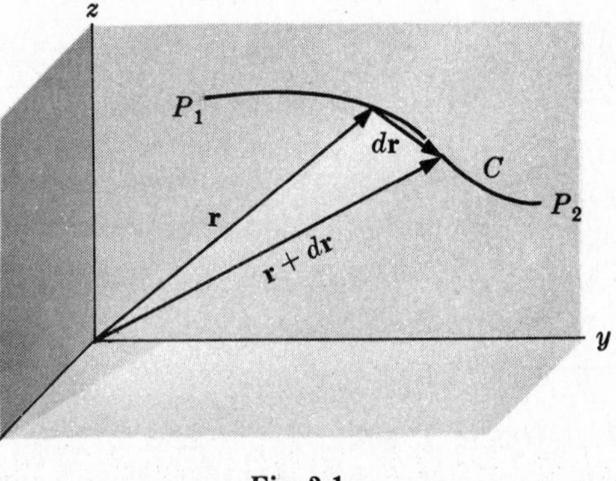

Fig. 2-1

$$W = \int_C \mathbf{F}\cdot d\mathbf{r} = \int_{P_1}^{P_2} \mathbf{F}\cdot d\mathbf{r} = \int_{\mathbf{r}_1}^{\mathbf{r}_2} \mathbf{F}\cdot d\mathbf{r} \qquad (4)$$

where $\mathbf{r}_1$ and $\mathbf{r}_2$ are the position vectors of P_1 and P_2 respectively.

POWER

The time rate of doing work on a particle is often called the *instantaneous power*, or briefly the *power*, applied to the particle. Using the symbols W and $\mathcal{P}$ for work and power respectively we have

$$\mathcal{P} = \frac{dW}{dt} \qquad (5)$$

If $\mathbf{F}$ is the force acting on a particle and $\mathbf{v}$ is the velocity of the particle, then we have

$$\mathcal{P} = \mathbf{F}\cdot\mathbf{v} \qquad (6)$$

KINETIC ENERGY

Suppose that the above particle has constant mass and that at times t_1 and t_2 it is located at P_1 and P_2 [Fig. 2-1] and moving with velocities $\mathbf{v}_1 = d\mathbf{r}_1/dt$ and $\mathbf{v}_2 = d\mathbf{r}_2/dt$ respectively. Then we can prove the following [see Problem 2.8].

Theorem 2.1. The total work done in moving the particle along C from P_1 to P_2 is given by

$$W = \int_C \mathbf{F} \cdot d\mathbf{r} = \tfrac{1}{2}m(v_2^2 - v_1^2) \tag{7}$$

If we call the quantity

$$T = \tfrac{1}{2}mv^2 \tag{8}$$

the *kinetic energy* of the particle, then Theorem 2.1 is equivalent to the statement

$$\begin{aligned}&\text{Total Work done from } P_1 \text{ to } P_2 \text{ along } C\\ &\quad = \text{Kinetic energy at } P_2 - \text{Kinetic energy at } P_1\end{aligned} \tag{9}$$

or, in symbols,

$$W = T_2 - T_1 \tag{10}$$

where $T_1 = \tfrac{1}{2}mv_1^2$, $T_2 = \tfrac{1}{2}mv_2^2$.

CONSERVATIVE FORCE FIELDS

Suppose there exists a scalar function V such that $\mathbf{F} = -\nabla V$. Then we can prove the following [see Problem 2.15].

Theorem 2.2. The total work done in moving the particle along C from P_1 to P_2 is

$$W = \int_{P_1}^{P_2} \mathbf{F} \cdot d\mathbf{r} = V(P_1) - V(P_2) \tag{11}$$

In such case the work done is *independent of the path C* joining points P_1 and P_2. If the work done by a force field in moving a particle from one point to another point is independent of the path joining the points, then the force field is said to be *conservative.*

The following theorems are valid.

Theorem 2.3. A force field $\mathbf{F}$ is conservative if and only if there exists a continuously differentiable scalar field V such that $\mathbf{F} = -\nabla V$ or, equivalently, if and only if

$$\nabla \times \mathbf{F} = \operatorname{curl} \mathbf{F} = \mathbf{0} \quad \text{identically} \tag{12}$$

Theorem 2.4. A continuously differentiable force field $\mathbf{F}$ is conservative if and only if for *any* closed non-intersecting curve C (simple closed curve)

$$\oint_C \mathbf{F} \cdot d\mathbf{r} = 0 \tag{13}$$

i.e. the total work done in moving a particle around any closed path is zero.

POTENTIAL ENERGY OR POTENTIAL

The scalar V such that $\mathbf{F} = -\nabla V$ is called the *potential energy,* also called the *scalar potential* or briefly the *potential,* of the particle in the conservative force field $\mathbf{F}$. In such case equation (*11*) of Theorem 2.2 can be written

$$\begin{aligned}&\text{Total Work done from } P_1 \text{ to } P_2 \text{ along } C\\ &\quad = \text{Potential energy at } P_1 - \text{Potential energy at } P_2\end{aligned} \tag{14}$$

or, in symbols,

$$W = V_1 - V_2 \tag{15}$$

where $V_1 = V(P_1)$, $V_2 = V(P_2)$.

It should be noted that the potential is defined within an arbitrary additive constant. We can express the potential as

$$V = -\int_{\mathbf{r}_0}^{\mathbf{r}} \mathbf{F} \cdot d\mathbf{r} \tag{16}$$

where we suppose that $V = 0$ when $\mathbf{r} = \mathbf{r}_0$.

CONSERVATION OF ENERGY

For a conservative force field we have from equations (*10*) and (*15*),

$$T_2 - T_1 = V_1 - V_2 \quad \text{or} \quad T_1 + V_1 = T_2 + V_2 \tag{17}$$

which can also be written

$$\tfrac{1}{2}mv_1^2 + V_1 = \tfrac{1}{2}mv_2^2 + V_2 \tag{18}$$

The quantity $E = T + V$, which is the sum of the kinetic energy and potential energy, is called the *total energy*. From (*18*) we see that the total energy at P_1 is the same as the total energy at P_2. We can state our results in the following

Theorem 2.5. In a conservative force field the total energy [i.e. the sum of kinetic energy and potential energy] is a constant. In symbols, $T + V = \text{constant} = E$.

This theorem is often called the principle of *conservation of energy*.

IMPULSE

Suppose that in Fig. 2-1 the particle is located at P_1 and P_2 at times t_1 and t_2 where it has velocities $\mathbf{v}_1$ and $\mathbf{v}_2$ respectively. The time integral of the force $\mathbf{F}$ given by

$$\int_{t_1}^{t_2} \mathbf{F}\, dt \tag{19}$$

is called the *impulse* of the force $\mathbf{F}$. The following theorem can be proved [see Problem 2.18].

Theorem 2.6. The impulse is equal to the change in momentum; or, in symbols,

$$\int_{t_1}^{t_2} \mathbf{F}\, dt = m\mathbf{v}_2 - m\mathbf{v}_1 = \mathbf{p}_2 - \mathbf{p}_1 \tag{20}$$

The theorem is true even when the mass is variable and the force is non-conservative.

TORQUE AND ANGULAR MOMENTUM

If a particle with position vector $\mathbf{r}$ moves in a force field $\mathbf{F}$ [Fig. 2-2], we define

$$\boldsymbol{\Lambda} = \mathbf{r} \times \mathbf{F} \tag{21}$$

as the *torque* or *moment* of the force $\mathbf{F}$ about O. The magnitude of $\boldsymbol{\Lambda}$ is a measure of the "turning effect" produced on the particle by the force. We can prove the following [see Problem 2.20]

Theorem 2.7.

$$\mathbf{r} \times \mathbf{F} = \frac{d}{dt}\{m(\mathbf{r} \times \mathbf{v})\} \tag{22}$$

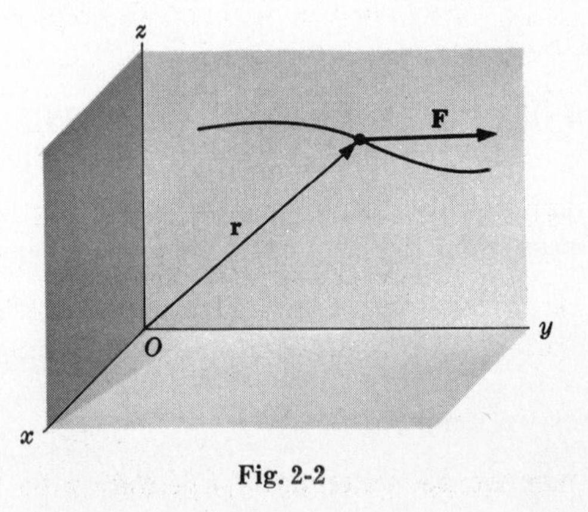

Fig. 2-2

The quantity

$$\boldsymbol{\Omega} = m(\mathbf{r} \times \mathbf{v}) = \mathbf{r} \times \mathbf{p} \tag{23}$$

is called the *angular momentum* or *moment of momentum* about O. In words the theorem states that the torque acting on a particle equals the time rate of change in its angular momentum, i.e.,

$$\boldsymbol{\Lambda} = \frac{d\boldsymbol{\Omega}}{dt} \tag{24}$$

This theorem is true even if the mass m is variable or the force non-conservative.

CONSERVATION OF MOMENTUM

If we let $\mathbf{F} = 0$ in Newton's second law, we find

$$\frac{d}{dt}(m\mathbf{v}) = 0 \quad \text{or} \quad m\mathbf{v} = \text{constant} \tag{25}$$

This leads to the following

Theorem 2.8. If the net external force acting on a particle is zero, its momentum will remain unchanged.

This theorem is often called the principle of *conservation of momentum.* For the case of constant mass it is equivalent to Newton's first law.

CONSERVATION OF ANGULAR MOMENTUM

If we let $\boldsymbol{\Lambda} = 0$ in (*24*), we find

$$\frac{d}{dt}\{m(\mathbf{r} \times \mathbf{v})\} = 0 \quad \text{or} \quad m(\mathbf{r} \times \mathbf{v}) = \text{constant} \tag{26}$$

This leads to the following

Theorem 2.9. If the net external torque acting on a particle is zero, the angular momentum will remain unchanged.

This theorem is often called the principle of *conservation of angular momentum.*

NON-CONSERVATIVE FORCES

If there is no scalar function V such that $\mathbf{F} = -\nabla V$ [or, equivalently, if $\nabla \times \mathbf{F} \neq \mathbf{0}$], then $\mathbf{F}$ is called a *non-conservative* force field. The results (*7*), (*20*) and (*24*) above hold for all types of force fields, conservative or not. However, (*11*) and (*17*) or (*18*) hold only for conservative force fields.

STATICS OR EQUILIBRIUM OF A PARTICLE

An important special case of motion of a particle occurs when the particle is, or appears to be, *at rest* or in *equilibrium* with respect to an inertial coordinate system or frame of reference. A necessary and sufficient condition for this is, from Newton's second law, that

$$\mathbf{F} = 0 \tag{27}$$

i.e. the net (external) force acting on the particle be zero.

If the force field is conservative with potential V, then a necessary and sufficient condition for a particle to be in equilibrium at a point is that

$$\nabla V = 0, \qquad \text{i.e.} \quad \frac{\partial V}{\partial x} = \frac{\partial V}{\partial y} = \frac{\partial V}{\partial z} = 0$$

at the point.

STABILITY OF EQUILIBRIUM

If a particle which is displaced slightly from an equilibrium point P tends to return to P, then we call P a *point of stability* or *stable point* and the equilibrium is said to be *stable.* Otherwise we say that the point is one of *instability* and the equilibrium is *unstable.* The following theorem is fundamental.

Theorem 2.10. A necessary and sufficient condition that an equilibrium point be one of stability is that the potential V at the point be a minimum.

Solved Problems

NEWTON'S LAWS

2.1. Due to a force field, a particle of mass 5 units moves along a space curve whose position vector is given as a function of time t by

$$\mathbf{r} = (2t^3 + t)\mathbf{i} + (3t^4 - t^2 + 8)\mathbf{j} - 12t^2\mathbf{k}$$

Find (*a*) the velocity, (*b*) the momentum, (*c*) the acceleration and (*d*) the force field at any time t.

(*a*) Velocity $= \mathbf{v} = \dfrac{d\mathbf{r}}{dt} = (6t^2 + 1)\mathbf{i} + (12t^3 - 2t)\mathbf{j} - 24t\mathbf{k}$

(*b*) Momentum $= \mathbf{p} = m\mathbf{v} = 5\mathbf{v} = (30t^2 + 5)\mathbf{i} + (60t^3 - 10t)\mathbf{j} - 120t\mathbf{k}$

(*c*) Acceleration $= \mathbf{a} = \dfrac{d\mathbf{v}}{dt} = \dfrac{d^2\mathbf{r}}{dt^2} = 12t\mathbf{i} + (36t^2 - 2)\mathbf{j} - 24\mathbf{k}$

(*d*) Force $= \mathbf{F} = \dfrac{d\mathbf{p}}{dt} = m\dfrac{d\mathbf{v}}{dt} = 60t\mathbf{i} + (180t^2 - 10)\mathbf{j} - 120\mathbf{k}$

2.2. A particle of mass m moves in the xy plane so that its position vector is

$$\mathbf{r} = a \cos \omega t\ \mathbf{i} + b \sin \omega t\ \mathbf{j}$$

where a, b and ω are positive constants and $a > b$. (*a*) Show that the particle moves in an ellipse. (*b*) Show that the force acting on the particle is always directed toward the origin.

(*a*) The position vector is

$$\mathbf{r} = x\mathbf{i} + y\mathbf{j} = a \cos \omega t\ \mathbf{i} + b \sin \omega t\ \mathbf{j}$$

and so $x = a \cos \omega t$, $y = b \sin \omega t$ which are the parametric equations of an ellipse having semi-major and semi-minor axes of lengths a and b respectively [see Fig. 2-3].

Since

$$(x/a)^2 + (y/b)^2 = \cos^2 \omega t + \sin^2 \omega t = 1$$

the ellipse is also given by $x^2/a^2 + y^2/b^2 = 1$.

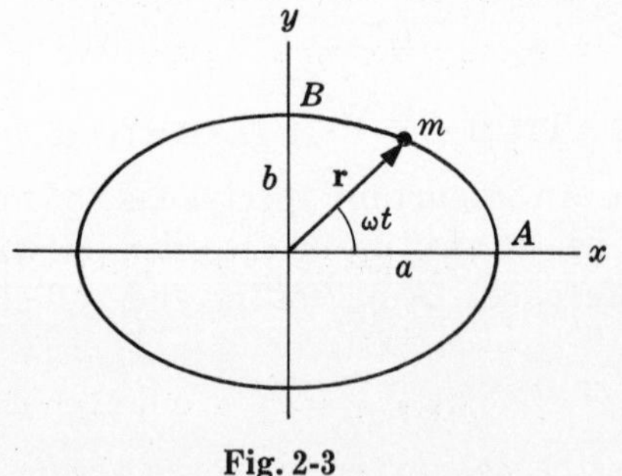

Fig. 2-3

(*b*) Assuming the particle has constant mass m, the force acting on it is

$$\begin{aligned} \mathbf{F} &= m\frac{d\mathbf{v}}{dt} = m\frac{d^2\mathbf{r}}{dt^2} = m\frac{d^2}{dt^2}[(a\cos\omega t)\mathbf{i} + (b\sin\omega t)\mathbf{j}] \\ &= m[-\omega^2 a\cos\omega t\,\mathbf{i} - \omega^2 b\sin\omega t\,\mathbf{j}] \\ &= -m\omega^2[a\cos\omega t\,\mathbf{i} + b\sin\omega t\,\mathbf{j}] = -m\omega^2\mathbf{r} \end{aligned}$$

which shows that the force is always directed toward the origin.

2.3. Two observers O and O', fixed relative to two coordinate systems $Oxyz$ and $O'x'y'z'$ respectively, observe the motion of a particle P in space [see Fig. 2-4]. Show that to both observers the particle appears to have the same force acting on it if and only if the coordinate systems are moving at constant velocity relative to each other.

Let the position vectors of the particle in the $Oxyz$ and $O'x'y'z'$ coordinate systems be $\mathbf{r}$ and $\mathbf{r}'$ respectively and let the position vector of O' with respect to O be $\mathbf{R} = \mathbf{r} - \mathbf{r}'$.

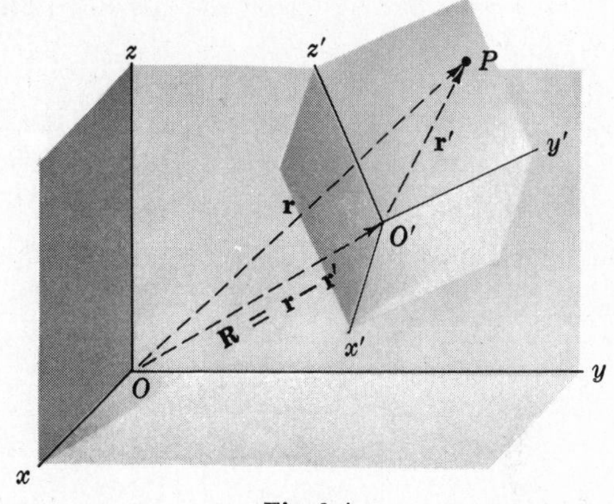

Fig. 2-4

Relative to observers O and O' the forces acting on P according to Newton's laws are given respectively by

$$\mathbf{F} = m\frac{d^2\mathbf{r}}{dt^2}, \qquad \mathbf{F}' = m\frac{d^2\mathbf{r}'}{dt^2}$$

The difference in observed forces is

$$\mathbf{F} - \mathbf{F}' = m\frac{d^2}{dt^2}(\mathbf{r} - \mathbf{r}') = m\frac{d^2\mathbf{R}}{dt^2}$$

and this will be zero if and only if

$$\frac{d^2\mathbf{R}}{dt^2} = 0 \quad \text{or} \quad \frac{d\mathbf{R}}{dt} = \text{constant}$$

i.e. the coordinate systems are moving at constant velocity relative to each other. Such coordinate systems are called *inertial coordinate systems.*

The result is sometimes called the *classical principle of relativity.*

2.4. A particle of mass 2 moves in a force field depending on time t given by

$$\mathbf{F} = 24t^2\mathbf{i} + (36t - 16)\mathbf{j} - 12t\mathbf{k}$$

Assuming that at $t = 0$ the particle is located at $\mathbf{r}_0 = 3\mathbf{i} - \mathbf{j} + 4\mathbf{k}$ and has velocity $\mathbf{v}_0 = 6\mathbf{i} + 15\mathbf{j} - 8\mathbf{k}$, find (*a*) the velocity and (*b*) the position at any time t.

(*a*) By Newton's second law,

$$2\,d\mathbf{v}/dt = 24t^2\mathbf{i} + (36t - 16)\mathbf{j} - 12t\mathbf{k}$$

or

$$d\mathbf{v}/dt = 12t^2\mathbf{i} + (18t - 8)\mathbf{j} - 6t\mathbf{k}$$

Integrating with respect to t and calling $\mathbf{c}_1$ the constant of integration, we have

$$\mathbf{v} = 4t^3\mathbf{i} + (9t^2 - 8t)\mathbf{j} - 3t^2\mathbf{k} + \mathbf{c}_1$$

Since $\mathbf{v} = \mathbf{v}_0 = 6\mathbf{i} + 15\mathbf{j} - 8\mathbf{k}$ at $t = 0$, we have $\mathbf{c}_1 = 6\mathbf{i} + 15\mathbf{j} - 8\mathbf{k}$ and so

$$\mathbf{v} = (4t^3 + 6)\mathbf{i} + (9t^2 - 8t + 15)\mathbf{j} - (3t^2 + 8)\mathbf{k}$$

(b) Since $\mathbf{v} = d\mathbf{r}/dt$, we have by part (a)

$$\frac{d\mathbf{r}}{dt} = (4t^3+6)\mathbf{i} + (9t^2-8t+15)\mathbf{j} - (3t^2+8)\mathbf{k}$$

Integrating with respect to t and calling $\mathbf{c}_2$ the constant of integration,

$$\mathbf{r} = (t^4+6t)\mathbf{i} + (3t^3-4t^2+15t)\mathbf{j} - (t^3+8t)\mathbf{k} + \mathbf{c}_2$$

Since $\mathbf{r} = \mathbf{r}_0 = 3\mathbf{i}-\mathbf{j}+4\mathbf{k}$ at $t=0$, we have $\mathbf{c}_2 = 3\mathbf{i}-\mathbf{j}+4\mathbf{k}$ and so

$$\mathbf{r} = (t^4+6t+3)\mathbf{i} + (3t^3-4t^2+15t-1)\mathbf{j} + (4-t^3-8t)\mathbf{k}$$

2.5. A constant force $\mathbf{F}$ acting on a particle of mass m changes the velocity from $\mathbf{v}_1$ to $\mathbf{v}_2$ in time τ.
(a) Prove that $\mathbf{F} = m(\mathbf{v}_2 - \mathbf{v}_1)/\tau$.
(b) Does the result in (a) hold if the force is variable? Explain.

(a) By Newton's second law,

$$m\frac{d\mathbf{v}}{dt} = \mathbf{F} \quad \text{or} \quad \frac{d\mathbf{v}}{dt} = \frac{\mathbf{F}}{m} \tag{1}$$

Then if $\mathbf{F}$ and m are constants we have on integrating,

$$\mathbf{v} = (\mathbf{F}/m)t + \mathbf{c}_1$$

At $t=0$, $\mathbf{v}=\mathbf{v}_1$ so that $\mathbf{c}_1=\mathbf{v}_1$ i.e.

$$\mathbf{v} = (\mathbf{F}/m)t + \mathbf{v}_1 \tag{2}$$

At $t=\tau$, $\mathbf{v}=\mathbf{v}_2$ so that $\quad \mathbf{v}_2 = (\mathbf{F}/m)\tau + \mathbf{v}_1$

i.e. $\quad \mathbf{F} = m(\mathbf{v}_2-\mathbf{v}_1)/\tau \qquad (3)$

Another method.

Write (1) as $m\,d\mathbf{v} = \mathbf{F}\,dt$. Then since $\mathbf{v}=\mathbf{v}_1$ at $t=0$ and $\mathbf{v}=\mathbf{v}_2$ at $t=\tau$, we have

$$\int_{\mathbf{v}_1}^{\mathbf{v}_2} m\,d\mathbf{v} = \int_0^\tau \mathbf{F}\,dt \quad \text{or} \quad m(\mathbf{v}_2-\mathbf{v}_1) = \mathbf{F}\tau$$

which yields the required result.

(b) No, the result does not hold in general if $\mathbf{F}$ is not a constant, since in such case we would not obtain the result of integration achieved in (a).

2.6. Find the constant force in the (a) cgs system and (b) mks system needed to accelerate a mass of 10,000 gm moving along a straight line from a speed of 54 km/hr to 108 km/hr in 5 minutes.

Assume the motion to be in the direction of the positive x axis. Then if $\mathbf{v}_1$ and $\mathbf{v}_2$ are the velocities, we have from the given data $\mathbf{v}_1 = 54\mathbf{i}$ km/hr, $\mathbf{v}_2 = 108\mathbf{i}$ km/hr, $m = 10{,}000$ gm, $t = 5$ min.

(a) In the cgs system

$$m = 10^4 \text{ gm}, \quad \mathbf{v}_1 = 54\mathbf{i}\text{ km/hr} = 1.5\times10^3\mathbf{i}\text{ cm/sec}, \quad \mathbf{v}_2 = 3.0\times10^3\mathbf{i}\text{ cm/sec}, \quad t = 300\text{ sec}$$

Then

$$\mathbf{F} = m\mathbf{a} = m\left(\frac{\mathbf{v}_2-\mathbf{v}_1}{t}\right) = (10^4\text{ gm})\left(\frac{1.5\times10^3\mathbf{i}\text{ cm/sec}}{3\times10^2\text{ sec}}\right)$$
$$= 0.5\times10^5\mathbf{i}\text{ gm cm/sec}^2 = 5\times10^4\mathbf{i}\text{ dynes}$$

Thus the magnitude of the force is 50,000 dynes in the direction of the positive x axis.

(b) In the mks system

$$m = 10\text{ kg}, \quad \mathbf{v}_1 = 54\mathbf{i}\text{ km/hr} = 15\mathbf{i}\text{ m/sec}, \quad \mathbf{v}_2 = 30\mathbf{i}\text{ m/sec}, \quad t = 300\text{ sec}$$

Then $$\mathbf{F} = m\mathbf{a} = m\left(\frac{\mathbf{v}_2 - \mathbf{v}_1}{t}\right) = (10 \text{ kg})\left(\frac{15\mathbf{i} \text{ m/sec}}{300 \text{ sec}}\right)$$

$$= 0.5\mathbf{i} \text{ kg m/sec}^2 = 0.5\mathbf{i} \text{ newtons}$$

Thus the magnitude is 0.5 newtons in the positive x direction. This result could also have been obtained from part (a) on noting that 1 newton $= 10^5$ dynes or 1 dyne $= 10^{-5}$ newtons.

In this simple problem the unit vector **i** is sometimes omitted, it being understood that the force **F** will have the direction of the positive x axis. However, it is good practice to work this and similar problems with the unit vector present so as to emphasize the vector character of force, velocity, etc. This is especially important in cases where velocities may change their directions. See, for example, Problem 2.46, page 56.

2.7. What constant force is needed to bring a 2000 lb mass moving at a speed of 60 mi/hr to rest in 4 seconds?

We shall assume that the motion takes place in a straight line which we choose as the positive direction of the x axis. Then using the English absolute system of units, we have

$$m = 2000 \text{ lb}, \quad \mathbf{v}_1 = 60\mathbf{i} \text{ mi/hr} = 88\mathbf{i} \text{ ft/sec}, \quad \mathbf{v}_2 = 0\mathbf{i} \text{ ft/sec}, \quad t = 4 \text{ sec}$$

Then $$\mathbf{F} = m\mathbf{a} = m\left(\frac{\mathbf{v}_2 - \mathbf{v}_1}{t}\right) = (2000 \text{ lb})\left(\frac{-88\mathbf{i} \text{ ft/sec}}{4 \text{ sec}}\right)$$

$$= -4.4 \times 10^4\mathbf{i} \text{ ft lb/sec}^2 = -4.4 \times 10^4\mathbf{i} \text{ poundals}$$

Thus the force has magnitude 4.4×10^4 poundals in the negative x direction, i.e. in a direction opposite to the motion. This is of course to be expected.

WORK, POWER, AND KINETIC ENERGY

2.8. A particle of constant mass m moves in space under the influence of a force field **F**. Assuming that at times t_1 and t_2 the velocity is $\mathbf{v}_1$ and $\mathbf{v}_2$ respectively, prove that the work done is the change in kinetic energy, i.e.,

$$\int_{t_1}^{t_2} \mathbf{F} \cdot d\mathbf{r} = \tfrac{1}{2}mv_2^2 - \tfrac{1}{2}mv_1^2$$

$$\text{Work done} = \int_{t_1}^{t_2} \mathbf{F} \cdot \frac{d\mathbf{r}}{dt}\,dt = \int_{t_1}^{t_2} \mathbf{F} \cdot \mathbf{v}\,dt$$

$$= \int_{t_1}^{t_2} m\frac{d\mathbf{v}}{dt} \cdot \mathbf{v}\,dt = m\int_{t_1}^{t_2} \mathbf{v} \cdot d\mathbf{v}$$

$$= \tfrac{1}{2}m\int_{t_1}^{t_2} d(\mathbf{v} \cdot \mathbf{v}) = \tfrac{1}{2}mv^2\Big|_{t_1}^{t_2} = \tfrac{1}{2}mv_2^2 - \tfrac{1}{2}mv_1^2$$

2.9. Find the work done in moving an object along a vector $\mathbf{r} = 3\mathbf{i} + 2\mathbf{j} - 5\mathbf{k}$ if the applied force is $\mathbf{F} = 2\mathbf{i} - \mathbf{j} - \mathbf{k}$. Refer to Fig. 2-5.

Work done = (magnitude of force in direction of motion)(distance moved)

$$= (F\cos\theta)(r) = \mathbf{F} \cdot \mathbf{r}$$

$$= (2\mathbf{i} - \mathbf{j} - \mathbf{k}) \cdot (3\mathbf{i} + 2\mathbf{j} - 5\mathbf{k})$$

$$= 6 - 2 + 5 = 9$$

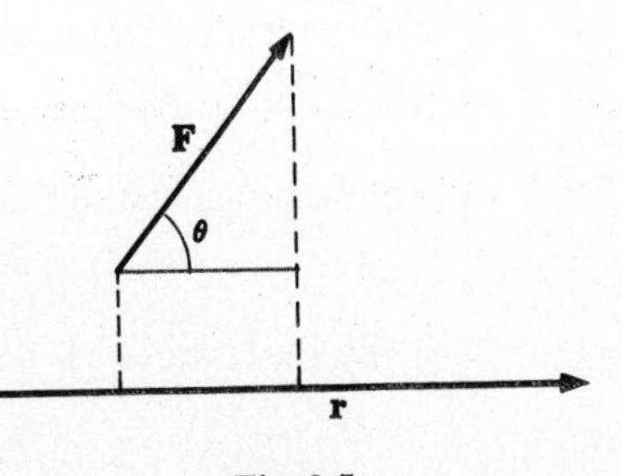

Fig. 2-5

2.10. Referring to Problem 2.2, (*a*) find the kinetic energy of the particle at points A and B, (*b*) find the work done by the force field in moving the particle from A to B, (*c*) illustrate the result of Problem 2.8 in this case and (*d*) show that the total work done by the field in moving the particle once around the ellipse is zero.

(*a*) Velocity $= \mathbf{v} = d\mathbf{r}/dt = -\omega a \sin \omega t\, \mathbf{i} + \omega b \cos \omega t\, \mathbf{j}$.

Kinetic energy $= \frac{1}{2}mv^2 = \frac{1}{2}m(\omega^2 a^2 \sin^2 \omega t + \omega^2 b^2 \cos^2 \omega t)$.

$$\text{Kinetic energy at } A \ \ [\text{where } \cos \omega t = 1,\ \sin \omega t = 0] \ = \ \tfrac{1}{2}m\omega^2 b^2$$
$$\text{Kinetic energy at } B \ \ [\text{where } \cos \omega t = 0,\ \sin \omega t = 1] \ = \ \tfrac{1}{2}m\omega^2 a^2$$

(*b*) **Method 1.** From part (*b*) of Problem 2.2,

$$\begin{aligned}\text{Work done} &= \int_A^B \mathbf{F} \cdot d\mathbf{r} = \int_A^B (-m\omega^2 \mathbf{r}) \cdot d\mathbf{r} = -m\omega^2 \int_A^B \mathbf{r} \cdot d\mathbf{r} \\ &= -\tfrac{1}{2}m\omega^2 \int_A^B d(\mathbf{r} \cdot \mathbf{r}) = -\tfrac{1}{2}m\omega^2 r^2 \Big|_A^B \\ &= \tfrac{1}{2}m\omega^2 a^2 - \tfrac{1}{2}m\omega^2 b^2 = \tfrac{1}{2}m\omega^2(a^2 - b^2)\end{aligned}$$

Method 2. We can assume that at A and B, $t = 0$ and $t = \pi/2\omega$ respectively. Then:

$$\begin{aligned}\text{Work done} &= \int_A^B \mathbf{F} \cdot d\mathbf{r} \\ &= \int_0^{\pi/2\omega} [-m\omega^2(a \cos \omega t\, \mathbf{i} + b \sin \omega t\, \mathbf{j})] \cdot [-\omega a \sin \omega t\, \mathbf{i} + \omega b \cos \omega t\, \mathbf{j}]\, dt \\ &= \int_0^{\pi/2\omega} m\omega^3(a^2 - b^2) \sin \omega t \cos \omega t\, dt \\ &= \tfrac{1}{2}m\omega^2(a^2 - b^2) \sin^2 \omega t \Big|_0^{\pi/2\omega} = \tfrac{1}{2}m\omega^2(a^2 - b^2)\end{aligned}$$

(*c*) From parts (*a*) and (*b*),

$$\begin{aligned}\text{Work done} &= \tfrac{1}{2}m\omega^2(a^2 - b^2) = \tfrac{1}{2}m\omega^2 a^2 - \tfrac{1}{2}m\omega^2 b^2 \\ &= \text{kinetic energy at } A - \text{kinetic energy at } B\end{aligned}$$

(*d*) Using Method 2 of part (*b*) we have, since t goes from 0 to $t = 2\pi/\omega$ for a complete circuit around the ellipse,

$$\begin{aligned}\text{Work done} &= \int_0^{2\pi/\omega} m\omega^3(a^2 - b^2) \sin \omega t \cos \omega t\, dt \\ &= \tfrac{1}{2}m\omega^2(a^2 - b^2) \sin^2 \omega t \Big|_0^{2\pi/\omega} = 0\end{aligned}$$

Method 1 can also be used to show the same result.

2.11. Prove that if $\mathbf{F}$ is the force acting on a particle and $\mathbf{v}$ is the (instantaneous) velocity of the particle, then the (instantaneous) power applied to the particle is given by

$$\mathcal{P} = \mathbf{F} \cdot \mathbf{v}$$

By definition the work done by a force $\mathbf{F}$ in giving a particle a displacement $d\mathbf{r}$ is

$$dW = \mathbf{F} \cdot d\mathbf{r}$$

Then the (instantaneous) power is given by

$$\mathcal{P} = \frac{dW}{dt} = \mathbf{F} \cdot \frac{d\mathbf{r}}{dt} = \mathbf{F} \cdot \mathbf{v}$$

as required.

2.12. Find the (instantaneous) power applied to the particle in Problem 2.1 by the force field.

By Problem 2.1, the velocity and force are given respectively by

$$\mathbf{v} = (6t^2+1)\mathbf{i} + (12t^3 - 2t)\mathbf{j} - 24t\mathbf{k}$$
$$\mathbf{F} = 60t\mathbf{i} + (180t^2 - 10)\mathbf{j} - 120\mathbf{k}$$

Then the power [by Problem 2.11] is given by

$$\mathcal{P} = \mathbf{F}\cdot\mathbf{v} = (60t)(6t^2+1) + (180t^2-10)(12t^3-2t) + (120)(24t)$$
$$= 2160t^5 - 120t^3 + 2960t$$

2.13. Find the work done by the force in (*a*) Problem 2.6, (*b*) Problem 2.7.

(*a*) In the cgs system: $v_1 = |\mathbf{v}_1| = 1.5\times 10^3$ cm/sec, $v_2 = |\mathbf{v}_2| = 3.0\times 10^3$ cm/sec, $m = 10^4$ gm. Then by Problem 2.8,

$$\begin{aligned}\text{Work done} &= \text{change in kinetic energy}\\ &= \tfrac{1}{2}m(v_2^2 - v_1^2)\\ &= \tfrac{1}{2}(10^4\text{ gm})(9.0\times 10^6 - 2.25\times 10^6)\,\frac{\text{cm}^2}{\text{sec}^2}\\ &= 3.38\times 10^{10}\,\frac{\text{gm cm}^2}{\text{sec}^2} = 3.38\times 10^{10}\left(\frac{\text{gm cm}}{\text{sec}^2}\right)(\text{cm})\\ &= 3.38\times 10^{10}\text{ dyne cm} = 3.38\times 10^{10}\text{ ergs}\end{aligned}$$

In the mks system we have similarly:

$$\begin{aligned}\text{Work done} &= \tfrac{1}{2}(10\text{ kg})(900-225)\,\frac{\text{m}^2}{\text{sec}^2}\\ &= 3.38\times 10^3\left(\frac{\text{kg m}}{\text{sec}^2}\right)(\text{m}) = 3.38\times 10^3\text{ newton meters}\end{aligned}$$

(*b*) As in part (*a*),

$$\begin{aligned}\text{Work done} &= \tfrac{1}{2}(2000\text{ lb})(88^2 - 0^2)\,\frac{\text{ft}^2}{\text{sec}^2}\\ &= 7.74\times 10^6(\text{ft})\left(\frac{\text{lb ft}}{\text{sec}^2}\right) = 7.74\times 10^6\text{ ft pdl}\end{aligned}$$

CONSERVATIVE FORCE FIELDS, POTENTIAL ENERGY, AND CONSERVATION OF ENERGY

2.14. Show that the force field $\mathbf{F}$ defined by

$$\mathbf{F} = (y^2z^3 - 6xz^2)\mathbf{i} + 2xyz^3\mathbf{j} + (3xy^2z^2 - 6x^2z)\mathbf{k}$$

is a conservative force field.

Method 1. The force field $\mathbf{F}$ is conservative if and only if curl $\mathbf{F} = \nabla\times\mathbf{F} = \mathbf{0}$. Now

$$\begin{aligned}\nabla\times\mathbf{F} &= \begin{vmatrix}\mathbf{i} & \mathbf{j} & \mathbf{k}\\ \partial/\partial x & \partial/\partial y & \partial/\partial z\\ y^2z^3 - 6xz^2 & 2xyz^3 & 3xy^2z^2 - 6x^2z\end{vmatrix}\\ &= \mathbf{i}\left[\frac{\partial}{\partial y}(3xy^2z^2 - 6x^2z) - \frac{\partial}{\partial z}(2xyz^3)\right]\\ &\quad + \mathbf{j}\left[\frac{\partial}{\partial z}(y^2z^3 - 6xz^2) - \frac{\partial}{\partial x}(3xy^2z^2 - 6x^2z)\right]\\ &\quad + \mathbf{k}\left[\frac{\partial}{\partial x}(2xyz^3) - \frac{\partial}{\partial y}(y^2z^3 - 6xz^2)\right]\\ &= \mathbf{0}\end{aligned}$$

Then the force field is conservative.

Method 2.

The force field $\mathbf{F}$ is conservative if and only if there exists a scalar function or potential $V(x,y,z)$ such that $\mathbf{F} = -\text{grad } V = -\nabla V$ Then

$$\begin{aligned}\mathbf{F} &= -\nabla V = -\frac{\partial V}{\partial x}\mathbf{i} - \frac{\partial V}{\partial y}\mathbf{j} - \frac{\partial V}{\partial z}\mathbf{k} \\ &= (y^2z^3 - 6xz^2)\mathbf{i} + 2xyz^3\mathbf{j} + (3xy^2z^2 - 6x^2z)\mathbf{k}\end{aligned}$$

Hence if $\mathbf{F}$ is conservative we must be able to find V such that

$$\partial V/\partial x = 6xz^2 - y^2z^3, \quad \partial V/\partial y = -2xyz^3, \quad \partial V/\partial z = 6x^2z - 3xy^2z^2 \tag{1}$$

Integrate the first equation with respect to x keeping y and z constant. Then

$$V = 3x^2z^2 - xy^2z^3 + g_1(y,z) \tag{2}$$

where $g_1(y,z)$ is a function of y and z.

Similarly integrating the second equation with respect to y (keeping x and z constant) and the third equation with respect to z (keeping x and y constant), we have

$$V = -xy^2z^3 + g_2(x,z) \tag{3}$$

$$V = 3x^2z^2 - xy^2z^3 + g_3(x,y) \tag{4}$$

Equations (2), (3) and (4) yield a common V if we choose

$$g_1(y,z) = c, \quad g_2(x,z) = 3x^2z^2 + c, \quad g_3(x,y) = c \tag{5}$$

where c is any arbitrary constant, and it follows that

$$V = 3x^2z^2 - xy^2z^3 + c$$

is the required potential.

Method 3.

$$\begin{aligned}V &= -\int_{\mathbf{r}_0}^{\mathbf{r}} \mathbf{F}\cdot d\mathbf{r} = -\int_{(x_0,y_0,z_0)}^{(x,y,z)} (y^2z^3 - 6xz^2)dx + 2xyz^3\,dy + (3xy^2z^2 - 6x^2z)dz \\ &= -\int_{(x_0,y_0,z_0)}^{(x,y,z)} d(xy^2z^3 - 3x^2z^2) = 3x^2z^2 - xy^2z^3 + c\end{aligned}$$

where $c = x_0y_0^2z_0^3 - 3x_0^2z_0^2$.

2.15. Prove Theorem 2.2, page 35: If the force acting on a particle is given by $\mathbf{F} = -\nabla V$, then the total work done in moving the particle along a curve C from P_1 to P_2 is

$$W = \int_{P_1}^{P_2} \mathbf{F}\cdot d\mathbf{r} = V(P_1) - V(P_2)$$

We have

$$W = \int_{P_1}^{P_2} \mathbf{F}\cdot d\mathbf{r} = \int_{P_1}^{P_2} -\nabla V\cdot d\mathbf{r} = \int_{P_1}^{P_2} -dV = -V\Big|_{P_1}^{P_2} = V(P_1) - V(P_2)$$

2.16. Find the work done by the force field $\mathbf{F}$ of Problem 2.14 in moving a particle from the point $A(-2,1,3)$ to $B(1,-2,-1)$.

$$\begin{aligned}\text{Work done} &= \int_A^B \mathbf{F}\cdot d\mathbf{r} = \int_A^B -\nabla V\cdot d\mathbf{r} \\ &= \int_{(-2,1,3)}^{(1,-2,-1)} -dV = -V(x,y,z)\Big|_{(-2,1,3)}^{(1,-2,-1)} \\ &= -3x^2z^2 + xy^2z^3 - c\Big|_{(-2,1,3)}^{(1,-2,-1)} = 155\end{aligned}$$

2.17. (*a*) Show that the force field of Problem 2.2 is conservative.
(*b*) Find the potential energy at points A and B of Fig. 2-3.
(*c*) Find the work done by the force in moving the particle from A to B and compare with Problem 2.10(*b*).
(*d*) Find the total energy of the particle and show that it is constant, i.e. demonstrate the principle of conservation of energy.

(*a*) From Problem 2(*b*), $\mathbf{F} = -m\omega^2\mathbf{r} = -m\omega^2(x\mathbf{i}+y\mathbf{j})$. Then

$$\nabla \times \mathbf{F} = \begin{vmatrix} \mathbf{i} & \mathbf{j} & \mathbf{k} \\ \partial/\partial x & \partial/\partial y & \partial/\partial z \\ -m\omega^2 x & -m\omega^2 y & 0 \end{vmatrix}$$

$$= \mathbf{i}\left[\frac{\partial}{\partial y}(0) - \frac{\partial}{\partial z}(-m\omega^2 y)\right] + \mathbf{j}\left[\frac{\partial}{\partial z}(-m\omega^2 x) - \frac{\partial}{\partial x}(0)\right] + \mathbf{k}\left[\frac{\partial}{\partial x}(-m\omega^2 y) - \frac{\partial}{\partial y}(-m\omega^2 x)\right]$$

$$= \mathbf{0}$$

Hence the field is conservative.

(*b*) Since the field is conservative there exists a potential V such that

$$\mathbf{F} = -m\omega^2 x\mathbf{i} - m\omega^2 y\mathbf{j} = -\nabla V = -\frac{\partial V}{\partial x}\mathbf{i} - \frac{\partial V}{\partial y}\mathbf{j} - \frac{\partial V}{\partial z}\mathbf{k}$$

Then $\qquad \partial V/\partial x = m\omega^2 x, \quad \partial V/\partial y = m\omega^2 y, \quad \partial V/\partial z = 0$

from which, omitting the constant, we have

$$V = \tfrac{1}{2}m\omega^2 x^2 + \tfrac{1}{2}m\omega^2 y^2 = \tfrac{1}{2}m\omega^2(x^2+y^2) = \tfrac{1}{2}m\omega^2 r^2$$

which is the required potential.

(*c*) Potential at point A of Fig. 2-3 [where $r=a$] $= \tfrac{1}{2}m\omega^2 a^2$.
Potential at point B of Fig. 2-3 [where $r=b$] $= \tfrac{1}{2}m\omega^2 b^2$. Then

$$\text{Work done from } A \text{ to } B = \text{Potential at } A - \text{Potential at } B = \tfrac{1}{2}m\omega^2 a^2 - \tfrac{1}{2}m\omega^2 b^2 = \tfrac{1}{2}m\omega^2(a^2-b^2)$$

agreeing with Problem 2.10(*b*).

(*d*) By Problems 2.10(*a*) and part (*b*),

$$\text{Kinetic energy at any point} = T = \tfrac{1}{2}mv^2 = \tfrac{1}{2}m\dot{\mathbf{r}}^2 = \tfrac{1}{2}m(\omega^2 a^2 \sin^2 \omega t + \omega^2 b^2 \cos^2 \omega t)$$

$$\text{Potential energy at any point} = V = \tfrac{1}{2}m\omega^2\mathbf{r}^2 = \tfrac{1}{2}m\omega^2(a^2\cos^2\omega t + b^2 \sin^2 \omega t)$$

Thus at any point we have on adding and using $\sin^2\omega t + \cos^2 \omega t = 1$,

$$T + V = \tfrac{1}{2}m\omega^2(a^2+b^2)$$

which is a constant.

IMPULSE, TORQUE, ANGULAR MOMENTUM, AND CONSERVATION OF MOMENTUM

2.18. Prove Theorem 2.6, page 36: The impulse of a force is equal to the change in momentum.

By definition of impulse [see (*19*), page 36] and Newton's second law, we have

$$\int_{t_1}^{t_2} \mathbf{F}\,dt = \int_{t_1}^{t_2} \frac{d}{dt}(m\mathbf{v})\,dt = \int_{t_1}^{t_2} d(m\mathbf{v}) = m\mathbf{v}\Big|_{t_1}^{t_2} = m\mathbf{v}_2 - m\mathbf{v}_1$$

2.19. A mass of 5000 kg moves on a straight line from a speed of 540 km/hr to 720 km/hr in 2 minutes. What is the impulse developed in this time?

Method 1.

Assume that the mass travels in the direction of the positive x axis. In the mks system,

$$\mathbf{v}_1 = 540\mathbf{i}\,\frac{\text{km}}{\text{hr}} = \frac{540\mathbf{i} \times 1000 \text{ m}}{3600 \text{ sec}} = 1.5 \times 10^2\mathbf{i}\,\frac{\text{m}}{\text{sec}}$$

$$\mathbf{v}_2 = 720\mathbf{i}\,\frac{\text{km}}{\text{hr}} = \frac{720\mathbf{i} \times 1000 \text{ m}}{3600 \text{ sec}} = 2.0 \times 10^2\mathbf{i}\,\frac{\text{m}}{\text{sec}}$$

Then from Problem 2.18,

$$\begin{aligned}\text{Impulse} &= m(\mathbf{v}_2 - \mathbf{v}_1) = (5000 \text{ kg})(0.5 \times 10^2\mathbf{i} \text{ m/sec})\\ &= 2.5 \times 10^5\mathbf{i} \text{ kg m/sec} = 2.5 \times 10^5\mathbf{i} \text{ newton sec}\end{aligned}$$

since 1 newton = 1 kg m/sec^2 or 1 newton sec = 1 kg m/sec.

Thus the impulse has magnitude 2.5×10^5 newton sec in the positive x direction.

Method 2.

Using the cgs system, $\mathbf{v}_1 = 540\mathbf{i}$ km/hr $= 1.5 \times 10^4\,\mathbf{i}$ cm/sec and $\mathbf{v}_2 = 720\mathbf{i}$ km/hr $= 2.0 \times 10^4\,\mathbf{i}$ cm/sec. Then

$$\begin{aligned}\text{Impulse} &= m(\mathbf{v}_2 - \mathbf{v}_1) = (5000 \times 10^3 \text{ gm})(0.5 \times 10^4\mathbf{i} \text{ cm/sec})\\ &= 2.50 \times 10^{10}\mathbf{i} \text{ gm cm/sec} = 2.50 \times 10^{10}\mathbf{i} \text{ dyne sec}\end{aligned}$$

since 1 dyne = 1 gm cm/sec^2 or 1 dyne sec = 1 gm cm/sec

Note that in finding the impulse we did not have to use the time 2 minutes as given in the statement of the problem.

2.20. Prove Theorem 2.7, page 36: The moment of force or torque about the origin O of a coordinate system is equal to the time rate of change of angular momentum.

The moment of force or torque about the origin O is

$$\boldsymbol{\Lambda} = \mathbf{r} \times \mathbf{F} = \mathbf{r} \times \frac{d}{dt}(m\mathbf{v})$$

The angular momentum or moment of momentum about O is

$$\boldsymbol{\Omega} = m(\mathbf{r} \times \mathbf{v}) = \mathbf{r} \times (m\mathbf{v})$$

Now we have

$$\begin{aligned}\frac{d\boldsymbol{\Omega}}{dt} &= \frac{d}{dt}(\mathbf{r} \times m\mathbf{v}) = \frac{d\mathbf{r}}{dt} \times (m\mathbf{v}) + \mathbf{r} \times \frac{d}{dt}(m\mathbf{v})\\ &= \mathbf{v} \times (m\mathbf{v}) + \mathbf{r} \times \frac{d}{dt}(m\mathbf{v}) = \mathbf{0} + \mathbf{r} \times \mathbf{F} = \boldsymbol{\Lambda}\end{aligned}$$

which gives the required result.

2.21. Determine (*a*) the torque and (*b*) the angular momentum about the origin for the particle of Problem 2.4 at any time t.

(*a*) Torque $\boldsymbol{\Lambda} = \mathbf{r} \times \mathbf{F}$

$$\begin{aligned}&= [(t^4 + 6t + 3)\mathbf{i} + (3t^3 - 4t^2 + 15t - 1)\mathbf{j} + (4 - t^3 - 8t)\mathbf{k}] \times [24t^2\mathbf{i} + (36t - 16)\mathbf{j} - 12t\mathbf{k}]\\ &= \begin{vmatrix} \mathbf{i} & \mathbf{j} & \mathbf{k} \\ t^4 + 6t + 3 & 3t^3 - 4t^2 + 15t - 1 & 4 - t^3 - 8t \\ 24t^2 & 36t - 16 & -12t \end{vmatrix}\\ &= (32t^3 + 108t^2 - 260t + 64)\mathbf{i} - (12t^5 + 192t^3 - 168t^2 - 36t)\mathbf{j}\\ &\quad - (36t^5 - 80t^4 + 360t^3 - 240t^2 - 12t + 48)\mathbf{k}\end{aligned}$$

(*b*) Angular momentum $\mathbf{\Omega} = \mathbf{r} \times (m\mathbf{v}) = m(\mathbf{r} \times \mathbf{v})$

$$= 2[(t^4 + 6t + 3)\mathbf{i} + (3t^3 - 4t^2 + 15t - 1)\mathbf{j} + (4 - t^3 - 8t)\mathbf{k}] \times [(4t^3 + 6)\mathbf{i} + (9t^2 - 8t + 15)\mathbf{j} - (3t^2 + 8)\mathbf{k}]$$

$$= 2\begin{vmatrix} \mathbf{i} & \mathbf{j} & \mathbf{k} \\ t^4 + 6t + 3 & 3t^3 - 4t^2 + 15t - 1 & 4 - t^3 - 8t \\ 4t^3 + 6 & 9t^2 - 8t + 15 & -3t^2 - 8 \end{vmatrix}$$

$$= (8t^4 + 36t^3 - 130t^2 + 64t - 104)\mathbf{i} - (2t^6 + 48t^4 - 56t^3 - 18t^2 - 96)\mathbf{j} - (6t^6 - 16t^5 + 90t^4 - 80t^3 - 6t^2 + 48t - 102)\mathbf{k}$$

Note that the torque is the derivative with respect to t of the angular momentum, illustrating the theorem of Problem 2.20.

2.22. A particle moves in a force field given by $\mathbf{F} = r^2\mathbf{r}$ where $\mathbf{r}$ is the position vector of the particle. Prove that the angular momentum of the particle is conserved.

The torque acting on the particle is

$$\mathbf{\Lambda} = \mathbf{r} \times \mathbf{F} = \mathbf{r} \times (r^2\mathbf{r}) = r^2(\mathbf{r} \times \mathbf{r}) = \mathbf{0}$$

Then by Theorem 2.9, page 37, the angular momentum is constant, i.e. the angular momentum is conserved.

NON-CONSERVATIVE FORCES

2.23. Show that the force field given by $\mathbf{F} = x^2yz\mathbf{i} - xyz^2\mathbf{k}$ is non-conservative.

We have

$$\nabla \times \mathbf{F} = \begin{vmatrix} \mathbf{i} & \mathbf{j} & \mathbf{k} \\ \partial/\partial x & \partial/\partial y & \partial/\partial z \\ x^2yz & 0 & -xyz^2 \end{vmatrix} = -xz^2\mathbf{i} + (x^2y + yz^2)\mathbf{j} - x^2z\mathbf{k}$$

Then since $\nabla \times \mathbf{F} \neq \mathbf{0}$, the field is non-conservative.

STATICS OF A PARTICLE

2.24. A particle P is acted upon by the forces $\mathbf{F}_1, \mathbf{F}_2, \mathbf{F}_3, \mathbf{F}_4, \mathbf{F}_5$ and $\mathbf{F}_6$ shown in Fig. 2-6. Represent geometrically the force needed to prevent P from moving.

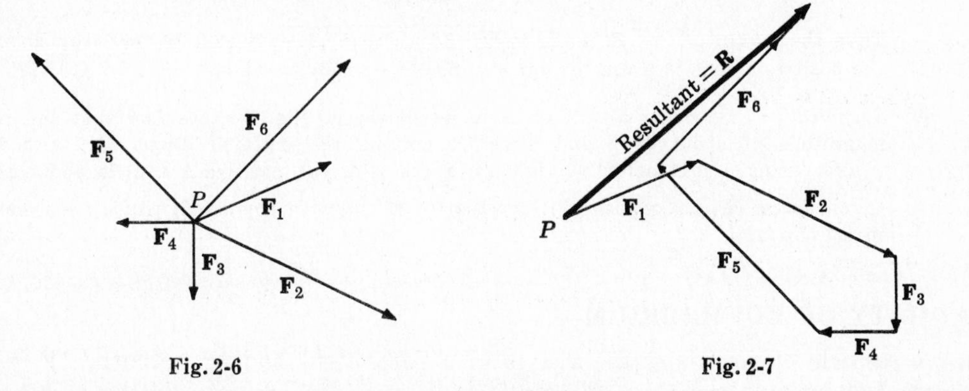

Fig. 2-6 Fig. 2-7

The resultant $\mathbf{R}$ of the forces $\mathbf{F}_1, \mathbf{F}_2, \mathbf{F}_3, \mathbf{F}_4, \mathbf{F}_5$ and $\mathbf{F}_6$ can be found by vector addition as indicated in Fig. 2-7. We have $\mathbf{R} = \mathbf{F}_1 + \mathbf{F}_2 + \mathbf{F}_3 + \mathbf{F}_4 + \mathbf{F}_5 + \mathbf{F}_6$. The force needed to prevent P from moving is $-\mathbf{R}$ which is a vector equal in magnitude to $\mathbf{R}$ but opposite in direction and sometimes called the *equilibrant*.

2.25. A particle is acted upon by the forces $\mathbf{F}_1 = 5\mathbf{i} - 10\mathbf{j} + 15\mathbf{k}$, $\mathbf{F}_2 = 10\mathbf{i} + 25\mathbf{j} - 20\mathbf{k}$ and $\mathbf{F}_3 = 15\mathbf{i} - 20\mathbf{j} + 10\mathbf{k}$. Find the force needed to keep the particle in equilibrium.

The resultant of the forces is

$$\mathbf{R} = \mathbf{F}_1 + \mathbf{F}_2 + \mathbf{F}_3 = (5\mathbf{i} - 10\mathbf{j} + 15\mathbf{k}) + (10\mathbf{i} + 25\mathbf{j} - 20\mathbf{k}) + (15\mathbf{i} - 20\mathbf{j} + 10\mathbf{k})$$
$$= 30\mathbf{i} - 5\mathbf{j} + 5\mathbf{k}$$

Then the force needed to keep the particle in equilibrium is $-\mathbf{R} = -30\mathbf{i} + 5\mathbf{j} - 5\mathbf{k}$.

2.26. The coplanar forces as indicated in Fig. 2-8 act on a particle P. Find the resultant of these forces (*a*) analytically and (*b*) graphically. What force is needed to keep the particle in equilibrium?

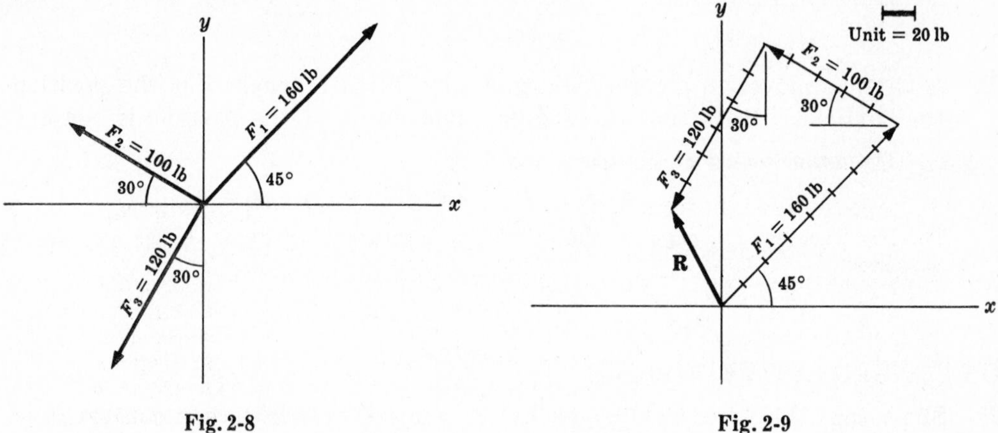

Fig. 2-8 Fig. 2-9

(*a*) ***Analytically.*** From Fig. 2-8 we have,

$$\mathbf{F}_1 = 160(\cos 45°\ \mathbf{i} + \sin 45°\ \mathbf{j}), \quad \mathbf{F}_2 = 100(-\cos 30°\ \mathbf{i} + \sin 30°\ \mathbf{j}),$$
$$\mathbf{F}_3 = 120(-\cos 60°\ \mathbf{i} - \sin 60°\ \mathbf{j})$$

Then the resultant $\mathbf{R}$ is

$$\mathbf{R} = \mathbf{F}_1 + \mathbf{F}_2 + \mathbf{F}_3$$
$$= (160 \cos 45° - 100 \cos 30° - 120 \cos 60°)\mathbf{i} + (160 \sin 45° + 100 \sin 30° - 120 \sin 60°)\mathbf{j}$$
$$= -33.46\mathbf{i} + 59.21\mathbf{j}$$

Writing $\mathbf{R} = R \cos\alpha\ \mathbf{i} + R \sin\alpha\ \mathbf{j}$ where α is the angle with the positive x axis measured counterclockwise, we see that

$$R \cos\alpha = -33.46, \qquad R \sin\alpha = 59.21$$

Thus the magnitude of $\mathbf{R}$ is $R = \sqrt{(-33.46)^2 + (59.21)^2} = 68.0$ lb, and the direction α with the positive x axis is given by $\tan\alpha = 59.21/(-33.46) = -1.770$ or $\alpha = 119° 28'$.

(*b*) ***Graphically.*** Choosing a unit of 20 lb as shown in Fig. 2-9, we find that the resultant has magnitude of about 68 lb and direction making an angle of about 61° with the negative x axis [using a protractor] so that the angle with the positive x axis is about 119°.

A force $-\mathbf{R}$, i.e. opposite in direction to $\mathbf{R}$ but with equal magnitude, is needed to keep P in equilibrium.

STABILITY OF EQUILIBRIUM

2.27. A particle moves along the x axis in a force field having potential $V = \frac{1}{2}\kappa x^2$, $\kappa > 0$. (*a*) Determine the points of equilibrium and (*b*) investigate the stability.

(*a*) Equilibrium points occur where $\nabla V = \mathbf{0}$ or in this case

$$dV/dx = \kappa x = 0 \quad \text{or} \quad x = 0$$

Thus there is only one equilibrium point, at $x = 0$.

(*b*) **Method 1.**

Since $d^2V/dx^2 = \kappa > 0$, it follows that at $x = 0$, V is a minimum. Thus by Theorem 2.10, page 38, $x = 0$ is a point of stability. This is also seen from Problem 2.36 where it is shown that the particle oscillates about $x = 0$.

Method 2.

We have $\mathbf{F} = -\nabla V = -\frac{dV}{dx}\mathbf{i} = -\kappa x\mathbf{i}$. Then when $x > 0$ the particle undergoes a force to the left, and when $x < 0$ the particle undergoes a force to the right. Thus $x = 0$ is a point of stability.

Method 3.

The fact that $x = 0$ is a minimum point can be seen from a graph of $V(x)$ vs x [Fig. 2-10].

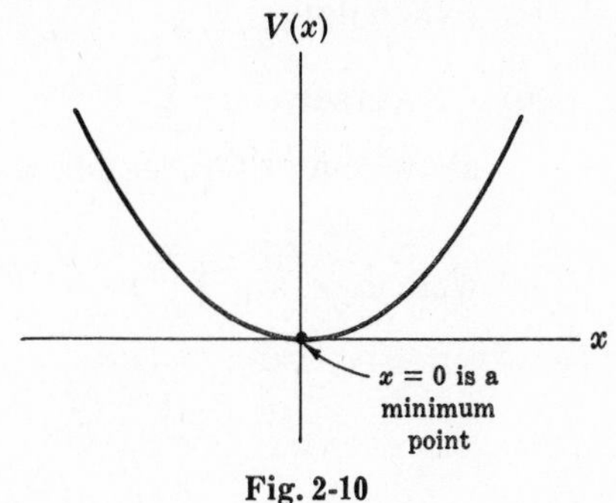

Fig. 2-10

MISCELLANEOUS PROBLEMS

2.28. Show how Newton's laws can be used to develop definitions of force and mass.

Let us first consider some given particle P, assuming for the present that its mass m_P is not defined but is simply some constant scalar quantity associated with P. Axiom 1 states that if P moves with constant velocity (which may be zero) then the force acting on it is zero. Axiom 2 states that if the velocity is not constant then there is a force acting on P given by $m_P\mathbf{a}_P$ where $\mathbf{a}_P$ is the acceleration of P. Thus force is defined by axioms 1 and 2 [although axiom 1 is unnecessary since it can in fact be deduced from axiom 2 by letting $\mathbf{F} = \mathbf{0}$]. It should be noted that force is a vector and thus has all the properties of vectors, in particular the parallelogram law for vector addition.

To define the mass m_P of particle P, let us now allow it to interact with some particular particle which we shall consider to be a *standard particle* and which we take to have *unit mass*. If $\mathbf{a}_P$ and $\mathbf{a}_S$ are the accelerations of particle P and the standard particle respectively, it follows from axioms 2 and 3 that $m_P\mathbf{a}_P = -\mathbf{a}_S$. Thus the mass m_P can be defined as $-\mathbf{a}_S/\mathbf{a}_P$.

2.29. Find the work done in moving a particle once around a circle C in the xy plane, if the circle has center at the origin and radius 3 and if the force field is given by

$$\mathbf{F} = (2x - y + z)\mathbf{i} + (x + y - z^2)\mathbf{j} + (3x - 2y + 4z)\mathbf{k}$$

In the plane $z = 0$, $\mathbf{F} = (2x - y)\mathbf{i} + (x + y)\mathbf{j} + (3x - 2y)\mathbf{k}$ and $d\mathbf{r} = dx\,\mathbf{i} + dy\,\mathbf{j}$ so that the work done is

$$\int_C \mathbf{F}\cdot d\mathbf{r} = \int_C [(2x - y)\mathbf{i} + (x + y)\mathbf{j} + (3x - 2y)\mathbf{k}]\cdot[dx\,\mathbf{i} + dy\,\mathbf{j}]$$

$$= \int_C (2x - y)\,dx + (x + y)\,dy$$

Choose the parametric equations of the circle as $x = 3\cos t$, $y = 3\sin t$ where t varies from 0 to 2π [see Fig. 2-11]. Then the line integral equals

$$\int_{t=0}^{2\pi} [2(3\cos t) - 3\sin t][-3\sin t]\,dt + [3\cos t + 3\sin t][3\cos t]\,dt$$

$$= \int_0^{2\pi} (9 - 9\sin t\cos t)\,dt = 9t - \frac{9}{2}\sin^2 t\Big|_0^{2\pi} = 18\pi$$

In traversing C we have chosen the counterclockwise direction indicated in Fig. 2-11. We call this the *positive* direction, or say that C has been traversed in the positive sense. If C were traversed in the clockwise (negative) direction the value of the integral would be -18π.

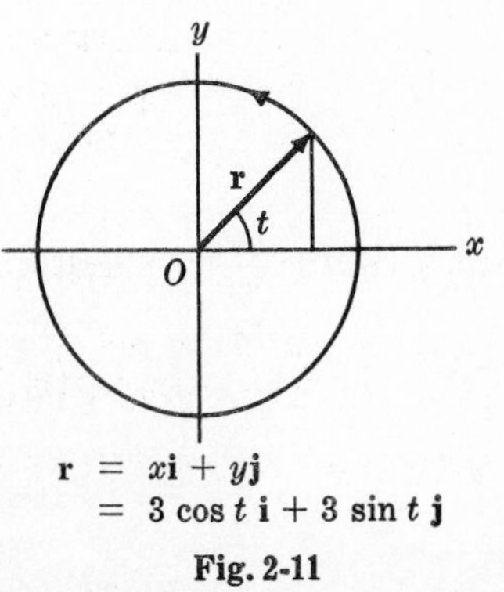

$$\mathbf{r} = x\mathbf{i} + y\mathbf{j} = 3\cos t\,\mathbf{i} + 3\sin t\,\mathbf{j}$$

Fig. 2-11

2.30. (*a*) If $\mathbf{F} = -\nabla V$, where V is single-valued and has continuous partial derivatives, show that the work done in moving a particle from one point $P_1 \equiv (x_1, y_1, z_1)$ in this field to another point $P_2 \equiv (x_2, y_2, z_2)$ is independent of the path joining the two points.

(*b*) Conversely, if $\int_C \mathbf{F} \cdot d\mathbf{r}$ is independent of the path C joining any two points, show that there exists a function V such that $\mathbf{F} = -\nabla V$.

(*a*) Work done

$$= \int_{P_1}^{P_2} \mathbf{F} \cdot d\mathbf{r} = -\int_{P_1}^{P_2} \nabla V \cdot d\mathbf{r}$$

$$= -\int_{P_1}^{P_2} \left(\frac{\partial V}{\partial x}\mathbf{i} + \frac{\partial V}{\partial y}\mathbf{j} + \frac{\partial V}{\partial z}\mathbf{k}\right) \cdot (dx\,\mathbf{i} + dy\,\mathbf{j} + dz\,\mathbf{k})$$

$$= -\int_{P_1}^{P_2} \frac{\partial V}{\partial x}dx + \frac{\partial V}{\partial y}dy + \frac{\partial V}{\partial z}dz$$

$$= -\int_{P_1}^{P_2} dV = V(P_1) - V(P_2) = V(x_1, y_1, z_1) - V(x_2, y_2, z_2)$$

Then the integral depends only on points P_1 and P_2 and not on the path joining them. This is true of course only if $V(x, y, z)$ is single-valued at all points P_1 and P_2.

(*b*) Let $\mathbf{F} = F_1\mathbf{i} + F_2\mathbf{j} + F_3\mathbf{k}$. By hypothesis, $\int_C \mathbf{F} \cdot d\mathbf{r}$ is independent of the path C joining any two points, which we take as (x_1, y_1, z_1) and (x, y, z) respectively. Then

$$V(x, y, z) = -\int_{(x_1, y_1, z_1)}^{(x, y, z)} \mathbf{F} \cdot d\mathbf{r} = -\int_{(x_1, y_1, z_1)}^{(x, y, z)} (F_1\,dx + F_2\,dy + F_3\,dz)$$

is independent of the path joining (x_1, y_1, z_1) and (x, y, z). Thus

$$V(x, y, z) = -\int_C [F_1(x, y, z)\,dx + F_2(x, y, z)\,dy + F_3(x, y, z)\,dz]$$

where C is a path joining (x_1, y_1, z_1) and (x, y, z). Let us choose as a particular path the straight line segments from (x_1, y_1, z_1) to (x, y_1, z_1) to (x, y, z_1) to (x, y, z) and call $V(x, y, z)$ the work done along this particular path. Then

$$V(x, y, z) = -\int_{x_1}^{x} F_1(x, y_1, z_1)\,dx - \int_{y_1}^{y} F_2(x, y, z_1)\,dy - \int_{z_1}^{z} F_3(x, y, z)\,dz$$

It follows that

$$\frac{\partial V}{\partial z} = -F_3(x, y, z)$$

$$\frac{\partial V}{\partial y} = -F_2(x, y, z_1) - \int_{z_1}^{z} \frac{\partial F_3}{\partial y}(x, y, z)\,dz = -F_2(x, y, z_1) - \int_{z_1}^{z} \frac{\partial F_2}{\partial z}(x, y, z)\,dz$$

$$= -F_2(x, y, z_1) - F_2(x, y, z)\Big|_{z_1}^{z}$$

$$= -F_2(x, y, z_1) - F_2(x, y, z) + F_2(x, y, z_1) = -F_2(x, y, z)$$

$$\frac{\partial V}{\partial x} = -F_1(x, y_1, z_1) - \int_{y_1}^{y} \frac{\partial F_2}{\partial x}(x, y, z_1)\,dy - \int_{z_1}^{z} \frac{\partial F_3}{\partial x}(x, y, z)\,dz$$

$$= -F_1(x, y_1, z_1) - \int_{y_1}^{y} \frac{\partial F_1}{\partial y}(x, y, z_1)\,dy - \int_{z_1}^{z} \frac{\partial F_1}{\partial z}(x, y, z)\,dz$$

$$= -F_1(x, y_1, z_1) - F_1(x, y, z_1)\Big|_{y_1}^{y} - F_1(x, y, z)\Big|_{z_1}^{z}$$

$$= -F_1(x, y_1, z_1) - F_1(x, y, z_1) + F(x, y_1, z_1) - F_1(x, y, z) + F(x, y, z_1) = -F_1(x, y, z)$$

Then $$\mathbf{F} = F_1\mathbf{i} + F_2\mathbf{j} + F_3\mathbf{k} = -\frac{\partial V}{\partial x}\mathbf{i} - \frac{\partial V}{\partial y}\mathbf{j} - \frac{\partial V}{\partial z}\mathbf{k} = -\nabla V$$

Thus a necessary and sufficient condition that a field $\mathbf{F}$ be conservative is that curl $\mathbf{F} = \nabla \times \mathbf{F} = \mathbf{0}$.

2.31. (a) Show that $\mathbf{F} = (2xy + z^3)\mathbf{i} + x^2\mathbf{j} + 3xz^2\mathbf{k}$ is a conservative force field. (b) Find the potential. (c) Find the work done in moving an object in this field from $(1, -2, 1)$ to $(3, 1, 4)$.

(a) A necessary and sufficient condition that a force will be conservative is that curl $\mathbf{F} = \nabla \times \mathbf{F} = \mathbf{0}$.

Now $$\nabla \times \mathbf{F} = \begin{vmatrix} \mathbf{i} & \mathbf{j} & \mathbf{k} \\ \partial/\partial x & \partial/\partial y & \partial/\partial z \\ 2xy + z^3 & x^2 & 3xz^2 \end{vmatrix} = \mathbf{0}.$$ Thus $\mathbf{F}$ is a conservative force field.

(b) As in Problem 2.14, Methods 2 or 3, we find $V = -(x^2y + xz^3)$.

(c) Work done $= -(x^2y + xz^3)\Big|_{(1,-2,1)}^{(3,1,4)} = -202$.

2.32. Prove that if $\int_{P_1}^{P_2} \mathbf{F}\cdot d\mathbf{r}$ is independent of the path joining any two points P_1 and P_2 in a given region, then $\oint \mathbf{F}\cdot d\mathbf{r} = 0$ for all closed paths in the region and conversely.

Let $P_1AP_2BP_1$ [see Fig. 2-12] be a closed curve. Then

$$\oint \mathbf{F}\cdot d\mathbf{r} = \int_{P_1AP_2BP_1} \mathbf{F}\cdot d\mathbf{r} = \int_{P_1AP_2} \mathbf{F}\cdot d\mathbf{r} + \int_{P_2BP_1} \mathbf{F}\cdot d\mathbf{r}$$
$$= \int_{P_1AP_2} \mathbf{F}\cdot d\mathbf{r} - \int_{P_1BP_2} \mathbf{F}\cdot d\mathbf{r} = 0$$

since the integral from P_1 to P_2 along a path through A is the same as that along a path through B, by hypothesis.

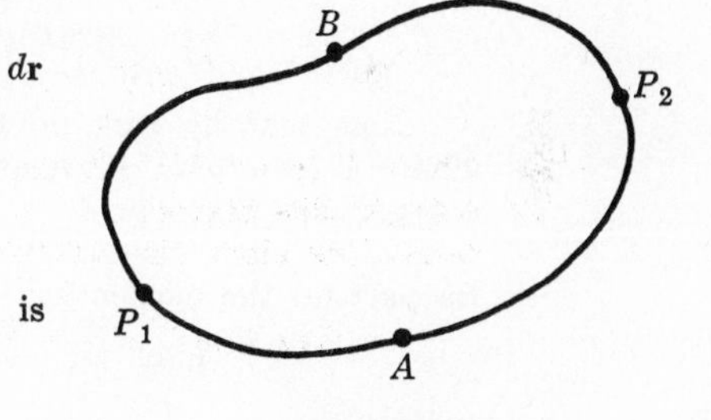

Fig. 2-12

Conversely if $\oint \mathbf{F}\cdot d\mathbf{r} = 0$, then

$$\int_{P_1AP_2BP_1} \mathbf{F}\cdot d\mathbf{r} = \int_{P_1AP_2} \mathbf{F}\cdot d\mathbf{r} + \int_{P_2BP_1} \mathbf{F}\cdot d\mathbf{r} = \int_{P_1AP_2} \mathbf{F}\cdot d\mathbf{r} - \int_{P_1BP_2} \mathbf{F}\cdot d\mathbf{r} = 0$$

so that, $$\int_{P_1AP_2} \mathbf{F}\cdot d\mathbf{r} = \int_{P_1BP_2} \mathbf{F}\cdot d\mathbf{r}.$$

2.33. (a) Show that a necessary and sufficient condition that $F_1\,dx + F_2\,dy + F_3\,dz$ be an exact differential is that $\nabla \times \mathbf{F} = \mathbf{0}$ where $\mathbf{F} = F_1\mathbf{i} + F_2\mathbf{j} + F_3\mathbf{k}$.

(b) Show that $(y^2z^3 \cos x - 4x^3z)\,dx + 2z^3y \sin x\,dy + (3y^2z^2 \sin x - x^4)\,dz$ is an exact differential of a function ϕ and find ϕ.

(a) Suppose $F_1\,dx + F_2\,dy + F_3\,dz = d\phi = \frac{\partial \phi}{\partial x}dx + \frac{\partial \phi}{\partial y}dy + \frac{\partial \phi}{\partial z}dz$, an exact differential. Then since x, y and z are independent variables,

$$F_1 = \frac{\partial\phi}{\partial x}, \quad F_2 = \frac{\partial\phi}{\partial y}, \quad F_3 = \frac{\partial\phi}{\partial z}$$

and so $\mathbf{F} = F_1\mathbf{i} + F_2\mathbf{j} + F_3\mathbf{k} = \frac{\partial\phi}{\partial x}\mathbf{i} + \frac{\partial\phi}{\partial y}\mathbf{j} + \frac{\partial\phi}{\partial z}\mathbf{k} = \nabla\phi$. Thus $\nabla \times \mathbf{F} = \nabla \times \nabla\phi = \mathbf{0}$.

Conversely if $\nabla \times \mathbf{F} = \mathbf{0}$, then $\mathbf{F} = \nabla\phi$ and so $\mathbf{F} \cdot d\mathbf{r} = \nabla\phi \cdot d\mathbf{r} = d\phi$, i.e. $F_1\,dx + F_2\,dy + F_3\,dz = d\phi$, an exact differential.

(*b*) $\mathbf{F} = (y^2z^3 \cos x - 4x^3z)\,\mathbf{i} + 2z^3y \sin x\,\mathbf{j} + (3y^2z^2 \sin x - x^4)\,\mathbf{k}$ and $\nabla \times \mathbf{F}$ is computed to be zero, so that by part (*a*) the required result follows.

2.34. Referring to Problem 2.4 find (*a*) the kinetic energy of the particle at $t=1$ and $t=2$, (*b*) the work done by the field in moving the particle from the point where $t=1$ to the point where $t=2$, (*c*) the momentum of the particle at $t=1$ and $t=2$ and (*d*) the impulse in moving the particle from $t=1$ to $t=2$.

(*a*) From part (*a*) of Problem 2.4,

$$\mathbf{v} = (4t^3+6)\mathbf{i} + (9t^2-8t+15)\mathbf{j} - (3t^2+8)\mathbf{k}$$

Then the velocities at $t=1$ and $t=2$ are

$$\mathbf{v}_1 = 10\mathbf{i} + 16\mathbf{j} - 11\mathbf{k}, \quad \mathbf{v}_2 = 38\mathbf{i} + 35\mathbf{j} - 20\mathbf{k}$$

and the kinetic energies at $t=1$ and $t=2$ are

$$T_1 = \tfrac{1}{2}mv_1^2 = \tfrac{1}{2}(2)[(10)^2+(16)^2+(-11)^2] = 477, \quad T_2 = \tfrac{1}{2}mv_2^2 = 3069$$

(*b*) Work done

$$= \int_{t=1}^{2} \mathbf{F}\cdot d\mathbf{r}$$

$$= \int_{t=1}^{2} [24t^2\mathbf{i} + (36t-16)\mathbf{j} - 12t\mathbf{k}] \cdot [(4t^3+6)\mathbf{i} + (9t^2-8t+15)\mathbf{j} - (3t^2+8)\mathbf{k}]\,dt$$

$$= \int_{t=1}^{2} [(24t^2)(4t^3+6) + (36t-16)(9t^2-8t+15) + (12t)(3t^2+8)]\,dt = 2592$$

Note that by part (*a*) this is the same as the difference or change in kinetic energies $3069 - 477 = 2592$, illustrating Theorem 2.1, page 35, that Work done = change in kinetic energy.

(*c*) By part (*a*) the momentum at any time t is

$$\mathbf{p} = m\mathbf{v} = 2\mathbf{v} = (8t^3+12)\mathbf{i} + (18t^2-16t+30)\mathbf{j} - (6t^2+16)\mathbf{k}$$

Then the momenta at $t=1$ and $t=2$ are

$$\mathbf{p}_1 = 20\mathbf{i} + 32\mathbf{j} - 22\mathbf{k}, \quad \mathbf{p}_2 = 76\mathbf{i} + 70\mathbf{j} - 40\mathbf{k}$$

(*d*) Impulse

$$= \int_{t=1}^{2} \mathbf{F}\,dt$$

$$= \int_{t=1}^{2} [24t^2\mathbf{i} + (36t-16)\mathbf{j} - 12t\mathbf{k}]\,dt = 56\mathbf{i} + 38\mathbf{j} - 18\mathbf{k}$$

Note that by part (*b*) this is the same as the difference or change in momentum, i.e. $\mathbf{p}_2 - \mathbf{p}_1 = (76\mathbf{i}+70\mathbf{j}-40\mathbf{k}) - (20\mathbf{i}+32\mathbf{j}-22\mathbf{k}) = 56\mathbf{i} + 38\mathbf{j} - 18\mathbf{k}$, illustrating Theorem 2.6, page 36, that Impulse = change in momentum.

2.35. A particle of mass m moves along the x axis under the influence of a conservative force field having potential $V(x)$. If the particle is located at positions x_1 and x_2 at respective times t_1 and t_2, prove that if E is the total energy,

$$t_2 - t_1 = \sqrt{\frac{m}{2}} \int_{x_1}^{x_2} \frac{dx}{\sqrt{E - V(x)}}$$

By the conservation of energy,

$$\text{Kinetic energy} + \text{Potential energy} = E$$

$$\tfrac{1}{2}m(dx/dt)^2 + V(x) = E$$

Then
$$(dx/dt)^2 = (2/m)\{E - V(x)\} \tag{1}$$

from which we obtain on considering the positive square root,

$$dt = \sqrt{m/2}\,(dx/\sqrt{E - V(x)})$$

Hence by integration,
$$\int_{t_1}^{t_2} dt = t_2 - t_1 = \sqrt{\frac{m}{2}} \int_{x_1}^{x_2} \frac{dx}{\sqrt{E - V(x)}} \tag{2}$$

2.36. (*a*) If the particle of Problem 2.35 has potential $V = \frac{1}{2}\kappa x^2$ and starts from rest at $x = a$, prove that $x = a \cos \sqrt{\kappa/m}\,t$ and (*b*) describe the motion.

(*a*) From (*1*) of Problem 2.35, $(dx/dt)^2 = (2/m)(E - \frac{1}{2}\kappa x^2)$. Since $dx/dt = 0$ where $x = a$, we find $E = \frac{1}{2}\kappa a^2$ so that

$$(dx/dt)^2 = (\kappa/m)(a^2 - x^2) \quad \text{or} \quad dx/\sqrt{a^2 - x^2} = \pm\sqrt{\kappa/m}\,dt$$

Integration yields $\sin^{-1}(x/a) = \pm\sqrt{\kappa/m}\,t + c_1$. Since $x = a$ at $t = 0$, $c_1 = \pi/2$. Then

$$\sin^{-1}(x/a) = \pm\sqrt{\kappa/m}\,t + \pi/2 \quad \text{or} \quad x = a \sin(\pi/2 \pm \sqrt{\kappa/m}\,t) = a \cos\sqrt{\kappa/m}\,t$$

(*b*) The particle oscillates back and forth along the x axis from $x = a$ to $x = -a$. The time for one complete vibration or oscillation from $x = a$ back to $x = a$ again is called the *period* of the oscillation and is given by $P = 2\pi\sqrt{m/\kappa}$.

2.37. A particle of mass 3 units moves in the xy plane under the influence of a force field having potential $V = 12x(3y - 4x)$. The particle starts at time $t = 0$ from rest at the point with position vector $10\mathbf{i} - 10\mathbf{j}$. (*a*) Set up the differential equations and conditions describing the motion. (*b*) Solve the equations in (*a*). (*c*) Find the position at any time. (*d*) Find the velocity at any time.

(*a*) Since $V = 12x(3y - 4x) = 36xy - 48x^2$, the force field is

$$\mathbf{F} = -\nabla V = -\frac{\partial V}{\partial x}\mathbf{i} - \frac{\partial V}{\partial y}\mathbf{j} - \frac{\partial V}{\partial z}\mathbf{k} = (-36y + 96x)\mathbf{i} - 36x\mathbf{j}$$

Then by Newton's second law,

$$3\frac{d^2\mathbf{r}}{dt^2} = (-36y + 96x)\mathbf{i} - 36x\mathbf{j}$$

or in component form, using $\mathbf{r} = x\mathbf{i} + y\mathbf{j}$,

$$d^2x/dt^2 = -12y + 32x, \quad d^2y/dt^2 = -12x \tag{1}$$

where
$$x = 10, \quad \dot{x} = 0, \quad y = -10, \quad \dot{y} = 0 \quad \text{at } t = 0 \tag{2}$$

using the fact that the particle starts at $\mathbf{r} = 10\mathbf{i} - 10\mathbf{j}$ with velocity $\mathbf{v} = \dot{\mathbf{r}} = 0$.

(*b*) From the second equation of (*1*), $x = -\frac{1}{12}\,d^2y/dt^2$. Substitution into the first equation of (*1*) yields

$$d^4y/dt^4 - 32\,d^2y/dt^2 - 144y = 0 \tag{3}$$

If α is constant then $y = e^{\alpha t}$ is a solution of (*3*) provided that

$$\alpha^4 - 32\alpha^2 - 144 = 0, \quad \text{i.e.} \quad (\alpha^2 + 4)(\alpha^2 - 36) = 0 \quad \text{or} \quad \alpha = \pm 2i,\ \alpha = \pm 6$$

Thus solutions are $e^{2it}, e^{-2it}, e^{6t}, e^{-6t}$ or $\cos 2t, \sin 2t, e^{6t}, e^{-6t}$ [in terms of real functions] and the general solution is

$$y = c_1 \cos 2t + c_2 \sin 2t + c_3 e^{6t} + c_4 e^{-6t} \tag{4}$$

Thus from $x = -\frac{1}{12} d^2y/dt^2$ we find, using (4),

$$x = \tfrac{1}{3}c_1 \cos 2t + \tfrac{1}{3}c_2 \sin 2t - 3c_3 e^{6t} - 3c_4 e^{-6t} \tag{5}$$

Using the conditions (2) in (4) and (5), we obtain

$$\tfrac{1}{3}c_1 - 3c_3 - 3c_4 = 10, \quad \tfrac{2}{3}c_2 - 18c_3 + 18c_4 = 0,$$

$$c_1 + c_3 + c_4 = -10, \quad 2c_2 + 6c_3 - 6c_4 = 0$$

Solving simultaneously, $c_1 = -6,\ c_2 = 0,\ c_3 = -2,\ c_4 = -2$ so that

$$x = -6 \cos 2t - 2e^{6t} - 2e^{-6t}, \quad y = -2 \cos 2t + 6e^{6t} + 6e^{-6t}$$

(c) The position at any time is

$$\mathbf{r} = x\mathbf{i} + y\mathbf{j} = (-6 \cos 2t - 2e^{6t} - 2e^{-6t})\mathbf{i} + (-2 \cos 2t + 6e^{6t} + 6e^{-6t})\mathbf{j}$$

(d) The velocity at any time is

$$\mathbf{v} = \dot{\mathbf{r}} = \dot{x}\mathbf{i} + \dot{y}\mathbf{j} = (12 \sin 2t - 12e^{6t} + 12e^{-6t})\mathbf{i} + (4 \sin 2t + 36e^{6t} - 36e^{-6t})\mathbf{j}$$

In terms of the hyperbolic functions

$$\sinh \alpha t = \tfrac{1}{2}(e^{\alpha t} - e^{-\alpha t}), \quad \cosh \alpha t = \tfrac{1}{2}(e^{\alpha t} + e^{-\alpha t})$$

we can also write

$$\mathbf{r} = (-6 \cos 2t - 4 \cosh 6t)\mathbf{i} + (-2 \cos 2t + 12 \cosh 6t)\mathbf{j}$$

$$\mathbf{v} = \dot{\mathbf{r}} = (12 \sin 2t - 24 \sinh 6t)\mathbf{i} + (4 \sin 2t + 72 \sinh 6t)\mathbf{j}$$

2.38. Prove that in polar coordinates (r, θ),

$$\nabla V = \frac{\partial V}{\partial r}\mathbf{r}_1 + \frac{1}{r}\frac{\partial V}{\partial \theta}\boldsymbol{\theta}_1$$

Let
$$\nabla V = G\mathbf{r}_1 + H\boldsymbol{\theta}_1 \tag{1}$$

where G and H are to be determined. Since $d\mathbf{r} = dx\,\mathbf{i} + dy\,\mathbf{j}$ we have on using $x = r\cos\theta$, $y = r\sin\theta$ and Problem 1.47(b), page 25,

$$d\mathbf{r} = (\cos\theta\, dr - r\sin\theta)(\cos\theta\,\mathbf{r}_1 - \sin\theta\,\boldsymbol{\theta}_1) + (\sin\theta\, dr + r\cos\theta\, d\theta)(\sin\theta\,\mathbf{r}_1 + \cos\theta\,\boldsymbol{\theta}_1)$$

or
$$d\mathbf{r} = dr\,\mathbf{r}_1 + r\,d\theta\,\boldsymbol{\theta}_1 \tag{2}$$

Now
$$\nabla V \cdot d\mathbf{r} = dV = \frac{\partial V}{\partial r}dr + \frac{\partial V}{\partial \theta}d\theta$$

Using (1) and (2) this becomes

$$(G\mathbf{r}_1 + H\boldsymbol{\theta}_1)\cdot(dr\,\mathbf{r}_1 + r\,d\theta\,\boldsymbol{\theta}_1) = G\,dr + Hr\,d\theta = \frac{\partial V}{\partial r}dr + \frac{\partial V}{\partial \theta}d\theta$$

so that
$$G = \frac{\partial V}{\partial r}, \quad H = \frac{1}{r}\frac{\partial V}{\partial \theta}$$

Then (1) becomes
$$\nabla V = \frac{\partial V}{\partial r}\mathbf{r}_1 + \frac{1}{r}\frac{\partial V}{\partial \theta}\boldsymbol{\theta}_1$$

2.39. According to the *theory of relativity*, the mass m of a particle is given by

$$m = \frac{m_0}{\sqrt{1 - v^2/c^2}} = \frac{m_0}{\sqrt{1 - \beta^2}}$$

where v is the speed, m_0 the rest mass, c the speed of light and $\beta = v/c$.

(a) Show that the time rate of doing work is given by

$$m_0c^2 \frac{d}{dt}(1-\beta^2)^{-1/2}$$

(b) Deduce from (a) that the kinetic energy is

$$T = (m-m_0)c^2 = m_0c^2\{(1-\beta^2)^{-1/2}-1\}$$

(c) If v is much less than c, show that $T = \frac{1}{2}mv^2$ approximately.

(a) By Newton's second law,
$$\mathbf{F} = \frac{d}{dt}(m\mathbf{v}) = \frac{d}{dt}\left(\frac{m_0\mathbf{v}}{\sqrt{1-\beta^2}}\right)$$

Then if W is the work done,
$$\frac{dW}{dt} = \mathbf{F}\cdot\mathbf{v} = v\frac{d}{dt}\left(\frac{m_0 v}{\sqrt{1-\beta^2}}\right) = m_0c^2\beta\frac{d}{dt}\left(\frac{\beta}{\sqrt{1-\beta^2}}\right) = m_0c^2\frac{d}{dt}\left(\frac{1}{\sqrt{1-\beta^2}}\right)$$
as proved by direct differentiation.

(b) Since Work done = change in kinetic energy, we have

Time rate of doing work = time rate of change in kinetic energy

or by part (a),
$$\frac{dW}{dt} = \frac{dT}{dt} = m_0c^2\frac{d}{dt}\left(\frac{1}{\sqrt{1-\beta^2}}\right)$$

Integrating,
$$T = \frac{m_0c^2}{\sqrt{1-\beta^2}} + c_1$$

To determine c_1 note that, by definition, $T=0$ when $v=0$ or $\beta=0$, so that $c_1 = -m_0c^2$. Hence we have, as required,
$$T = \frac{m_0c^2}{\sqrt{1-\beta^2}} - m_0c^2 = (m-m_0)c^2$$

(c) For $\beta < 1$ we have by the binomial theorem,
$$\frac{1}{\sqrt{1-\beta^2}} = (1-\beta^2)^{-1/2} = 1 + \frac{1}{2}\beta^2 + \frac{1\cdot 3}{2\cdot 4}\beta^4 + \frac{1\cdot 3\cdot 5}{2\cdot 4\cdot 6}\beta^6 + \cdots$$

Then
$$T = m_0c^2\left[1+\frac{1}{2}\frac{v^2}{c^2}+\cdots\right] - m_0c^2 = \frac{1}{2}mv^2 \quad \text{approximately}$$

Supplementary Problems

NEWTON'S LAWS

2.40. A particle of mass 2 units moves along the space curve defined by $\mathbf{r} = (4t^2-t^3)\mathbf{i} - 5t\mathbf{j} + (t^4-2)\mathbf{k}$. Find (a) the momentum and (b) the force acting on it at $t=1$.
Ans. (a) $10\mathbf{i}-10\mathbf{j}+8\mathbf{k}$, (b) $4\mathbf{i}+24\mathbf{k}$

2.41. A particle moving in a force field $\mathbf{F}$ has its momentum given at any time t by
$$\mathbf{p} = 3e^{-t}\mathbf{i} - 2\cos t\,\mathbf{j} - 3\sin t\,\mathbf{k}$$
Find $\mathbf{F}$. *Ans.* $-3e^{-t}\mathbf{i} + 2\sin t\,\mathbf{j} - 3\cos t\,\mathbf{k}$

2.42. Under the influence of a force field a particle of mass m moves along the ellipse
$$\mathbf{r} = a\cos\omega t\,\mathbf{i} + b\sin\omega t\,\mathbf{j}$$
If $\mathbf{p}$ is the momentum, prove that (a) $\mathbf{r}\times\mathbf{p} = mab\omega\mathbf{k}$, (b) $\mathbf{r}\cdot\mathbf{p} = \frac{1}{2}m(b^2-a^2)\sin 2\omega t$.

2.43. If $\mathbf{F}$ is the force acting on the particle of Problem 2.42, prove that $\mathbf{r} \times \mathbf{F} = \mathbf{0}$. Explain what this means physically.

2.44. A force of 100 dynes in the direction of the positive x axis acts on a particle of mass 2 gm for 10 minutes. What velocity does the particle acquire assuming that it starts from rest?
Ans. 3×10^4 cm/sec

2.45. Work Problem 2.44 if the force is 20 newtons and the mass is 10 kg. *Ans.* 1200 m/sec

2.46. (*a*) Find the constant force needed to accelerate a mass of 40 kg from the velocity $4\mathbf{i} - 5\mathbf{j} + 3\mathbf{k}$ m/sec to $8\mathbf{i} + 3\mathbf{j} - 5\mathbf{k}$ m/sec in 20 seconds. (*b*) What is the magnitude of the force in (*a*)?
Ans. (*a*) $8\mathbf{i} + 16\mathbf{j} - 16\mathbf{k}$ newtons or $(8\mathbf{i} + 16\mathbf{j} - 16\mathbf{k}) \times 10^5$ dynes
(*b*) 24 newtons or 24×10^5 dynes

2.47. An elevator moves from the top floor of a tall building to the ground floor without stopping. (*a*) Explain why a blindfolded person in the elevator may believe that the elevator is not moving at all. (*b*) Can the person tell when the motion begins or stops? Explain.

2.48. A particle of unit mass moves in a force field given in terms of time t by

$$\mathbf{F} = (6t - 8)\mathbf{i} - 60t^3\mathbf{j} + (20t^3 + 36t^2)\mathbf{k}$$

Its initial position and velocity are given respectively by $\mathbf{r}_0 = 2\mathbf{i} - 3\mathbf{k}$ and $\mathbf{v}_0 = 5\mathbf{i} + 4\mathbf{j}$. Find the (*a*) position and (*b*) velocity of the particle at $t = 2$.
Ans. (*a*) $4\mathbf{i} - 88\mathbf{j} + 77\mathbf{k}$, (*b*) $\mathbf{i} - 236\mathbf{j} + 176\mathbf{k}$

2.49. The force acting on a particle of mass m is given in terms of time t by

$$\mathbf{F} = a \cos \omega t \, \mathbf{i} + b \sin \omega t \, \mathbf{j}$$

If the particle is initially at rest at the origin, find its (*a*) position and (*b*) velocity at any later time.
Ans. (*a*) $\dfrac{a}{m\omega^2}(1 - \cos \omega t)\,\mathbf{i} + \dfrac{b}{m\omega^2}(\omega t - \sin \omega t)\,\mathbf{j}$, (*b*) $\dfrac{a}{m\omega} \sin \omega t\,\mathbf{i} + \dfrac{b}{m\omega}(1 - \cos \omega t)\,\mathbf{j}$

WORK, POWER AND KINETIC ENERGY

2.50. A particle is moved by a force $\mathbf{F} = 20\mathbf{i} - 30\mathbf{j} + 15\mathbf{k}$ along a straight line from point A to point B with position vectors $2\mathbf{i} + 7\mathbf{j} - 3\mathbf{k}$ and $5\mathbf{i} - 3\mathbf{j} - 6\mathbf{k}$ respectively. Find the work done.
Ans. 315

2.51. Find the kinetic energy of a particle of mass 20 moving with velocity $3\mathbf{i} - 5\mathbf{j} + 4\mathbf{k}$. *Ans.* 500

2.52. Due to a force field $\mathbf{F}$, a particle of mass 4 moves along the space curve $\mathbf{r} = (3t^2 - 2t)\mathbf{i} + t^3\mathbf{j} - t^4\mathbf{k}$. Find the work done by the field in moving the particle from the point where $t = 1$ to the point where $t = 2$. *Ans.* 2454

2.53. At one particular instant of time a particle of mass 10 is traveling along a space curve with velocity given by $4\mathbf{i} + 16\mathbf{k}$. At a later instant of time its velocity is $8\mathbf{i} - 20\mathbf{j}$. Find the work done on the particle between the two instants of time. *Ans.* 192

2.54. Verify Theorem 2.1, page 35 for the particle of Problem 2.52.

2.55. A particle of mass m moves under the influence of the force field given by $\mathbf{F} = a(\sin \omega t\,\mathbf{i} + \cos \omega t\,\mathbf{j})$. If the particle is initially at rest at the origin, prove that the work done on the particle up to time t is given by $(a^2/m\omega^2)(1 - \cos \omega t)$.

2.56. Prove that the instantaneous power applied to the particle in Problem 2.55 is $(a^2/m\omega) \sin \omega t$.

2.57. A particle moves with velocity $5\mathbf{i} - 3\mathbf{j} + 6\mathbf{k}$ under the influence of a constant force $\mathbf{F} = 20\mathbf{i} + 10\mathbf{j} + 15\mathbf{k}$. What is the instantaneous power applied to the particle? *Ans.* 160

CONSERVATIVE FORCE FIELDS, POTENTIAL ENERGY AND CONSERVATION OF ENERGY

2.58. (*a*) Prove that the force field $\mathbf{F} = (y^2 - 2xyz^3)\mathbf{i} + (3 + 2xy - x^2z^3)\mathbf{j} + (6z^3 - 3x^2yz^2)\mathbf{k}$ is conservative. (*b*) Find the potential V associated with the force field in (*a*).
Ans. (*b*) $xy^2 - x^2yz^3 + 3y + \frac{3}{2}z^4$

2.59. A particle moves in the force field of Problem 2.58 from the point $(2, -1, 2)$ to $(-1, 3, -2)$. Find the work done. *Ans.* 55

2.60. (*a*) Find constants a, b, c so that the force field defined by
$$\mathbf{F} = (x + 2y + az)\mathbf{i} + (bx - 3y - z)\mathbf{j} + (4x + cy + 2z)\mathbf{k}$$
is conservative.
(*b*) What is the potential associated with the force field in (*a*)?
Ans. (*a*) $a = 4$, $b = 2$, $c = -1$ (*b*) $V = -\frac{1}{2}x^2 + \frac{3}{2}y^2 - z^2 - 2xy - 4xz + yz$

2.61. Find the work done in moving a particle from the point $(1, -1, 2)$ to $(2, 3, -1)$ in a force field with potential $V = x^3 - y^3 + 2xy - y^2 + 4x$. *Ans.* 15

2.62. Determine whether the force field $\mathbf{F} = (x^2y - z^3)\mathbf{i} + (3xyz + xz^2)\mathbf{j} + (2x^2yz + yz^4)\mathbf{k}$ is conservative.
Ans. Not conservative

2.63. Find the work done in moving a particle in the force field $\mathbf{F} = 3x^2\mathbf{i} + (2xz - y)\mathbf{j} + z\mathbf{k}$ along (*a*) the straight line from $(0, 0, 0)$ to $(2, 1, 3)$, (*b*) the space curve $x = 2t^2$, $y = t$, $z = 4t^2 - t$ from $t = 0$ to $t = 1$. Is the work independent of the path? Explain. *Ans.* (*a*) 16, (*b*) 14.2

2.64. (*a*) Evaluate $\oint_C \mathbf{F} \cdot d\mathbf{r}$ where $\mathbf{F} = (x - 3y)\mathbf{i} + (y - 2x)\mathbf{j}$ and C is the closed curve in the xy plane $x = 2\cos t$, $y = 3\sin t$ from $t = 0$ to $t = 2\pi$. (*b*) Give a physical interpretation to the result in (*a*).
Ans. (*a*) 6π if C is traversed in the positive (counterclockwise) direction.

2.65. (*a*) Show that the force field $\mathbf{F} = -\kappa r^3 \mathbf{r}$ is conservative.
(*b*) Write the potential energy of a particle moving in the force field of (*a*).
(*c*) If a particle at mass m moves with velocity $\mathbf{v} = d\mathbf{r}/dt$ in this field, show that if E is the constant total energy then $\frac{1}{2}m(dr/dt)^2 + \frac{1}{5}\kappa r^5 = E$. What important physical principle does this illustrate?

2.66. A particle of mass 4 moves in the force field defined by $\mathbf{F} = -200\mathbf{r}/r^3$. (*a*) Show that the field is conservative and find the potential energy. (*b*) If a particle starts at $r = 1$ with speed 20, what will be its speed at $r = 2$? *Ans.* (*a*) $V = 200/r$, (*b*) $15\sqrt{2}$

IMPULSE, TORQUE AND ANGULAR MOMENTUM. CONSERVATION OF MOMENTUM

2.67. A particle of unit mass moves in a force field given by $\mathbf{F} = (3t^2 - 4t)\mathbf{i} + (12t - 6)\mathbf{j} + (6t - 12t^2)\mathbf{k}$ where t is the time. (*a*) Find the change in momentum of the particle from time $t = 1$ to $t = 2$. (*b*) If the velocity at $t = 1$ is $4\mathbf{i} - 5\mathbf{j} + 10\mathbf{k}$, what is the velocity at $t = 2$?
Ans. (*a*) $\mathbf{i} + 12\mathbf{j} - 19\mathbf{k}$, (*b*) $5\mathbf{i} + 7\mathbf{j} - 9\mathbf{k}$

2.68. A particle of mass m moves along a space curve defined by $\mathbf{r} = a\cos\omega t\,\mathbf{i} + b\sin\omega t\,\mathbf{j}$. Find (*a*) the torque and (*b*) the angular momentum about the origin. *Ans.* (*a*) 0, (*b*) $2mab\omega\mathbf{k}$

2.69. A particle moves in a force field given by $\mathbf{F} = \phi(r)\,\mathbf{r}$. Prove that the angular momentum of the particle about the origin is constant.

2.70 Find (*a*) the torque and (*b*) the angular momentum about the origin at the time $t = 2$ for the particle of Problem 2.67, assuming that at $t = 0$ it is located at the origin.
Ans. (*a*) $-(36\mathbf{i} + 128\mathbf{j} + 60\mathbf{k})$, (*b*) $-44\mathbf{i} + 52\mathbf{j} + 16\mathbf{k}$

2.71. Find the impulse developed by a force given by $\mathbf{F} = 4t\mathbf{i} + (6t^2 - 2)\mathbf{j} + 12\mathbf{k}$ from $t = 0$ to $t = 2$.
Ans. $8\mathbf{i} + 12\mathbf{j} + 24t\mathbf{k}$

2.72. What is the magnitude of the impulse developed by a mass of 200 gm which changes its velocity from $5\mathbf{i} - 3\mathbf{j} + 7\mathbf{k}$ m/sec to $2\mathbf{i} + 3\mathbf{j} + \mathbf{k}$ m/sec? *Ans.* 1.8×10^5 dyne sec or 1.8 newton sec

STATICS OF A PARTICLE

2.73. A particle is acted upon by the forces $\mathbf{F}_1 = 2\mathbf{i} + a\mathbf{j} - 3\mathbf{k}$, $\mathbf{F}_2 = 5\mathbf{i} + c\mathbf{j} + b\mathbf{k}$, $\mathbf{F}_3 = b\mathbf{i} - 5\mathbf{j} + 7\mathbf{k}$, $\mathbf{F}_4 = c\mathbf{i} - 6\mathbf{j} + a\mathbf{k}$. Find the values of the constants a, b, c in order that the particle will be in equilibrium. *Ans.* $a = 7, b = 11, c = 4$

2.74. Find (*a*) graphically and (*b*) analytically the resultant force acting on the mass m of Fig. 2-13 where all forces are in a plane.

Ans. (*b*) 19.5 dynes in a direction making an angle $85°22'$ with the negative x axis

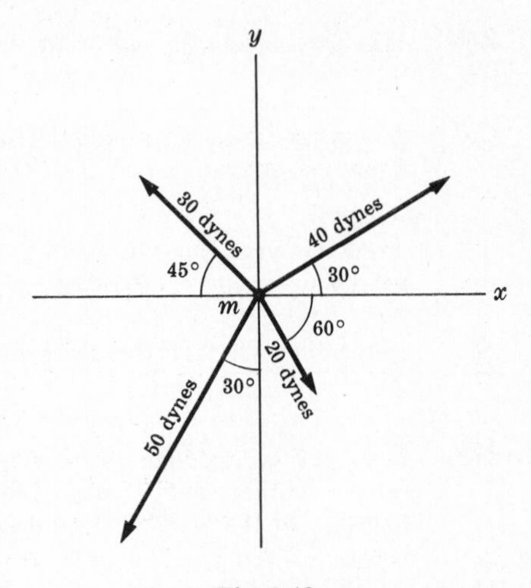

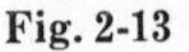
Fig. 2-13

2.75. The potential of a particle moving in the xy plane is given by $V = 2x^2 - 5xy + 3y^2 + 6x - 7y$. (*a*) Prove that there will be one and only one point at which a particle will remain in equilibrium and (*b*) find the coordinates of this point. *Ans.* (*b*) (1, 2)

2.76. Prove that a particle which moves in a force field of potential

$$V = x^2 + 4y^2 + z^2 - 4xy - 4yz + 2xz - 4x + 8y - 4z$$

can remain in equilibrium at infinitely many points and locate these points.

Ans. All points on the plane $x - 2y + z = 2$

STABILITY OF EQUILIBRIUM

2.77. A particle moves on the x axis in a force field having potential $V = x^2(6 - x)$.
(*a*) Find the points at equilibrium and (*b*) investigate their stability.

Ans. $x = 0$ is a point of stable equilibrium; $x = 4$ is a point of unstable equilibrium

2.78. Work Problem 2.77 if (*a*) $V = x^4 - 8x^3 - 6x^2 + 24x$, (*b*) $V = x^4$.

Ans. (*a*) $x = 1, 2$ are points of stable equilibrium; $x = -1$ is a point of unstable equilibrium.
(*b*) $x = 0$ is a point of stable equilibrium

2.79. Work Problem 2.77 if $V = \sin 2\pi x$.

Ans. If $n = 0, \pm 1, \pm 2, \pm 3, \ldots$ then $x = \frac{3}{4} + n$ are points of stable equilibrium, while $x = \frac{1}{4} + n$ are points of unstable equilibrium.

2.80. A particle moves in a force field with potential $V = x^2 + y^2 + z^2 - 8x + 16y - 4z$. Find the points of stable equilibrium. *Ans.* $(4, -8, 2)$

MISCELLANEOUS PROBLEMS

2.81. (*a*) Prove that $\mathbf{F} = (y^2 \cos x + z^3)\mathbf{i} + (2y \sin x - 4)\mathbf{j} + (3xz^2 + 2)\mathbf{k}$ is a conservative force field. (*b*) Find the potential corresponding to $\mathbf{F}$. (*c*) Find the work done in moving a particle in this field from $(0, 1, -1)$ to $(\pi/2, -1, 2)$.

Ans. (*a*) $V = y^2 \sin x + xz^3 - 4y + 2z + c$, (*b*) $15 + 4\pi$

2.82. A particle P is acted upon by 3 coplanar forces as indicated in Fig. 2-14. Find the force needed to prevent P from moving.

Ans. 323 lb in a direction opposite to 150 lb force

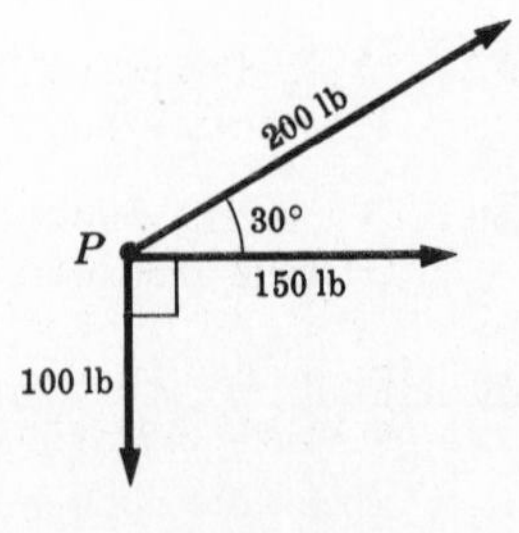

Fig. 2-14

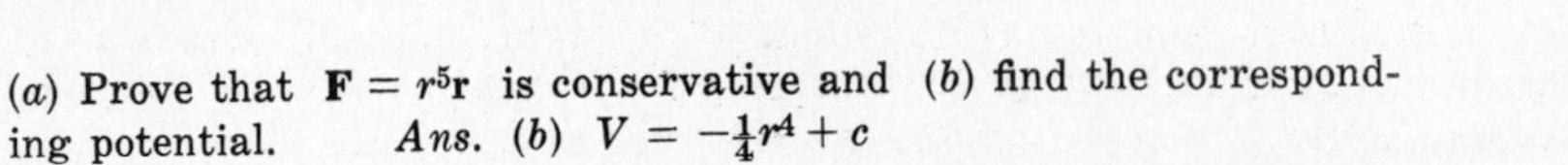

2.83. (*a*) Prove that $\mathbf{F} = r^5\mathbf{r}$ is conservative and (*b*) find the corresponding potential. *Ans.* (*b*) $V = -\frac{1}{4}r^4 + c$

2.84. Explain the following paradox: According to Newton's third law a trailer pulls back on an automobile to which it is attached with as much force as the auto pulls forward on the trailer. Therefore the auto cannot move.

2.85. Find the potential of a particle placed in a force field given by $\mathbf{F} = -\kappa r^{-n}\mathbf{r}$ where κ and n are constants. Treat all cases.

2.86. A waterfall 500 ft high has 440,000 ft^3 of water flowing over it per second. Assuming that the density of water is 62.5 lb/ft^3 and that 1 horsepower is 550 ft lb/sec, find the horsepower of the waterfall. *Ans.* 25×10^6 hp

2.87. The power applied to a particle by a force field is given as a function of time t by $\mathcal{P}(t) = 3t^2 - 4t + 2$. Find the work done in moving the particle from the point where $t = 2$ to the point where $t = 4$.
Ans. 36

2.88. Can the torque on a particle be zero without the force being zero? Explain.

2.89. Can the force on a particle be zero without the angular momentum being zero? Explain.

2.90. Under the influence of a force field $\mathbf{F}$ a particle of mass 2 moves along the space curve $\mathbf{r} = 6t^4\mathbf{i} - 3t^2\mathbf{j} + (4t^3 - 5)\mathbf{k}$. Find (*a*) the work done in moving the particle from the point where $t = 0$ to the point where $t = 1$, (*b*) the power applied to the particle at any time.
Ans. (*a*) 756 (*b*) $72t(48t^4 + 8t + 1)$

2.91. A force field moves a particle of mass m along the space curve $\mathbf{r} = a \cos \omega t\, \mathbf{i} + b \sin \omega t\, \mathbf{j}$. (*a*) What power is required? (*b*) Discuss physically the case $a = b$. *Ans.* (*a*) $m(a^2 - b^2)\omega^3 \sin \omega t \cos \omega t$

2.92. The angular momentum of a particle is given as a function of time t by

$$\mathbf{\Omega} = 6t^2\mathbf{i} - (2t+1)\mathbf{j} + (12t^3 - 8t^2)\mathbf{k}$$

Find the torque at the time $t = 1$. *Ans.* $12\mathbf{i} - 2\mathbf{j} + 20\mathbf{k}$

2.93. Find the constant force needed to give an object of mass 36,000 lb a speed of 10 mi/hr in 5 minutes starting from rest. *Ans.* 1760 poundals

2.94. A constant force of 100 newtons is applied for 2 minutes to a 20 kg mass which is initially at rest. (*a*) What is the speed achieved? (*b*) What is the distance traveled?
Ans. (*a*) 600 m/sec, (*b*) 36,000 m

2.95. A particle of mass m moves on the x axis under the influence of a force of attraction toward origin O given by $\mathbf{F} = -(\kappa/x^2)\mathbf{i}$. If the particle starts from rest at $x = a$, prove that it will arrive at O in a time given by $\frac{1}{2}\pi a\sqrt{ma/2\kappa}$.

2.96. Work Problem 2.95 if $\mathbf{F} = -(\kappa/x^3)\mathbf{i}$.

2.97. A particle of mass 2 units moves in the force field $\mathbf{F} = t^2\mathbf{i} - 3t\mathbf{j} + (t+2)\mathbf{k}$ where t is the time. (*a*) How far does the particle move from $t = 0$ to $t = 3$ if it is initially at rest at the origin? (*b*) Find the kinetic energy at times $t = 1$ and $t = 3$. (*c*) What is the work done on the particle by the field from $t = 1$ to $t = 3$? (*d*) What is the power applied to the particle at $t = 1$? (*e*) What is the impulse supplied to the particle at $t = 1$?

2.98. At $t = 0$ a particle of unit mass is at rest at the origin. If it is acted upon by a force $\mathbf{F} = 100te^{-2t}\mathbf{i}$, find (*a*) the change in momentum of the particle in going from time $t = 1$ to $t = 2$, (*b*) the velocity after a long time has elapsed. *Ans.* (*a*) $25e^{-2}(3 - 5e^{-2})\mathbf{i}$, (*b*) 25

2.99. A particle of mass 3 units moves in the xy plane under the influence of a force field having potential $V = 6x^3 + 12y^3 + 36xy - 48x^2$. Investigate the motion of the particle if it is displaced slightly from its equilibrium position.

[*Hint.* Near $x = 0, y = 0$ the potential is very nearly $36xy - 48x^2$ since $6x^3$ and $12y^3$ are negligible.]

2.100. A particle of unit mass moves on the x axis under the influence of a force field having potential $V = 6x(x-2)$. (*a*) Show that $x = 1$ is a position of stable equilibrium. (*b*) Prove that if the mass is displaced slightly from its position of equilibrium it will oscillate about it with period equal to $4\pi\sqrt{3}$.

[*Hint.* Let $x = 1 + u$ and neglect terms in u of degree higher than one.]

2.101. A particle of mass m moves in a force field $\mathbf{F} = -\kappa x\mathbf{i}$. (*a*) How much work is done in moving the particle from $x = x_1$ to $x = x_2$? (*b*) If a unit particle starts at $x = x_1$, with speed v_1, what is its speed on reaching $x = x_2$? *Ans.* (*a*) $\frac{1}{2}\kappa(x_1^2 - x_2^2)$, (*b*) $\sqrt{v_1^2 + (\kappa/m)(x_1^2 - x_2^2)}$

2.102. A particle of mass 2 moves in the xy plane under the influence of a force field having potential $V = x^2 + y^2$. The particle starts at time $t = 0$ from rest at the point $(2, 1)$. (*a*) Set up the differential equations and conditions describing the motion. (*b*) Find the position at any time t. (*c*) Find the velocity at any time t.

2.103. Work Problem 2.102 if $V = 8xy$.

2.104. Does Theorem 2.7, page 36, hold relative to a non-inertial frame of reference or coordinate system? Prove your answer.

2.105. (*a*) Prove that if a particle moves in the xy plane under the influence of a force field having potential $V = 12x(3y - 4x)$, then $x = 0$, $y = 0$ is a point of stable equilibrium. (*b*) Discuss the relationship of the result in (*a*) to Problem 2.37, page 53.

2.106. (*a*) Prove that a sufficient condition for the point (a, b) to be a minimum point of the function $V(x, y)$ is that at (a, b)

$$\text{(i)}\ \frac{\partial V}{\partial x} = \frac{\partial V}{\partial y} = 0, \qquad \text{(ii)}\ \Delta = \left(\frac{\partial^2 V}{\partial x^2}\right)\left(\frac{\partial^2 V}{\partial y^2}\right) - \left(\frac{\partial^2 V}{\partial x\, \partial y}\right)^2 > 0 \ \text{ and } \ \frac{\partial^2 V}{\partial x^2} > 0$$

(*b*) Use (*a*) to investigate the points of stability of a particle moving in a force field having potential $V = x^3 + y^3 - 3x - 12y$. *Ans.* (*b*) The point $(1, 2)$ is a point of stability

2.107. Suppose that a particle of unit mass moves in the force field of Problem 2.106. Find its speed at any time.

2.108. A particle moves once around the circle $\mathbf{r} = a(\cos\theta\, \mathbf{i} + \sin\theta\, \mathbf{j})$ in a force field

$$\mathbf{F} = (x\mathbf{i} - y\mathbf{j})/(x^2 + y^2)$$

(*a*) Find the work done. (*b*) Is the force field conservative? (*c*) Do your answers to (*a*) and (*b*) contradict Theorem 2.4, page 35? Explain.

2.109. It is sometimes stated that classical or Newtonian mechanics makes the assumption that space and time are both absolute. Discuss what is meant by this statement.

2.110. The quantity $\mathbf{F}_{av} = \dfrac{\int_{t_1}^{t_2} \mathbf{F}\, dt}{t_2 - t_1}$ is called the *average force* acting on a particle from time t_1 to t_2. Does the result (*3*) of Problem 2.5, page 40, hold if $\mathbf{F}$ is replaced by $\mathbf{F}_{av}$? Explain.

2.111. A particle of mass 2 gm moves in the force field $\mathbf{F} = 8xy\mathbf{i} + (4x^2 - 8z)\mathbf{j} - 8y\mathbf{k}$ dynes. If it has a speed of 4 cm/sec at the point $(-1, 2, -1)$, what is its speed at $(1, -1, 1)$? *Ans.* 6 cm/sec

2.112. (*a*) Find positions of stable equilibrium of a particle moving in a force field of potential $V = 18r^2e^{-2r}$.

(*b*) If the particle is released at $r = \frac{1}{4}$, find the speed when it reaches the equilibrium position.

(*c*) Find the period for small oscillations about the equilibrium position.

2.113. According to *Einstein's special theory of relativity* the mass m of a particle moving with speed v relative to an observer is given by $m = m_0/\sqrt{1 - v^2/c^2}$ where c is the *speed of light* [186,000 mi/sec] and m_0 is the *rest mass*. What is the percent increase in rest mass of (*a*) an airplane moving at 700 mi/hr, (*b*) a planet moving at 25,000 mi/hr, (*c*) an electron moving at half the speed of light? What conclusions do you draw from these results?

2.114. Prove that in cylindrical coordinates,

$$\nabla V = \frac{\partial V}{\partial \rho}\mathbf{e}_\rho + \frac{1}{\rho}\frac{\partial V}{\partial \phi}\mathbf{e}_\phi + \frac{\partial V}{\partial z}\mathbf{e}_z$$

where $\mathbf{e}_\rho, \mathbf{e}_\phi, \mathbf{e}_z$ are unit vectors in the direction of increasing ρ, ϕ and z respectively.

2.115. Prove that in spherical coordinates,

$$\nabla V = \frac{\partial V}{\partial r}\mathbf{e}_r + \frac{1}{r}\frac{\partial V}{\partial \theta}\mathbf{e}_\theta + \frac{1}{r \sin\theta}\frac{\partial V}{\partial \phi}\mathbf{e}_\phi$$

where $\mathbf{e}_r, \mathbf{e}_\theta, \mathbf{e}_\phi$ are unit vectors in the direction of increasing r, θ, ϕ respectively.

Chapter 3

MOTION in a UNIFORM FIELD FALLING BODIES and PROJECTILES

UNIFORM FORCE FIELDS

A force field which has constant magnitude and direction is called a *uniform* or *constant force field.* If the direction of this field is taken as the negative z direction as indicated in Fig. 3-1 and the magnitude is the constant $F_0 > 0$, then the force field is given by

$$\mathbf{F} = -F_0\,\mathbf{k} \tag{1}$$

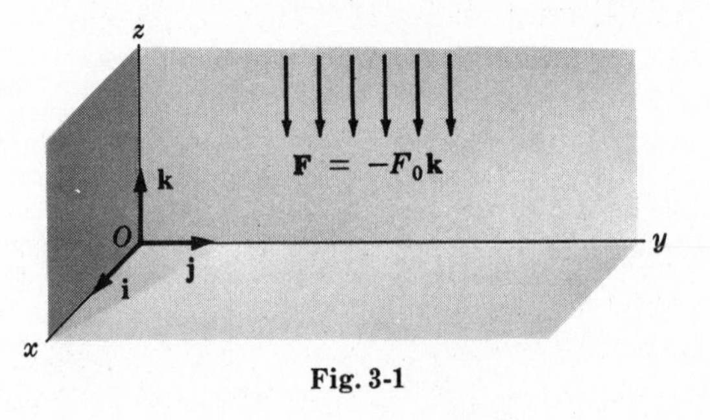

Fig. 3-1

UNIFORMLY ACCELERATED MOTION

If a particle of constant mass m moves in a uniform force field, then its acceleration is uniform or constant. The motion is then described as *uniformly accelerated motion.* Using $\mathbf{F} = m\mathbf{a}$ in (*1*), the acceleration of a particle of mass m moving in the uniform force field (*1*) is given by

$$\mathbf{a} = -\frac{F_0}{m}\mathbf{k} \tag{2}$$

WEIGHT AND ACCELERATION DUE TO GRAVITY

It is found experimentally that near the earth's surface objects fall with a vertical acceleration which is constant provided that air resistance is negligible. This acceleration is denoted by $\mathbf{g}$ and is called the *acceleration due to gravity* or the *gravitational acceleration.* The approximate magnitude of $\mathbf{g}$ is 980 cm/sec^2, 9.80 m/sec^2 or 32 ft/sec^2 according as the cgs, mks or fps system of units is used. This value varies at different parts of the earth's surface, increasing slightly as one goes from the equator to the poles.

Assuming the surface of the earth is represented by the xy plane of Fig. 3-2, the force acting on a particle of mass m is given by

$$\mathbf{W} = -mg\,\mathbf{k} \tag{3}$$

This force, which is called the *weight* of the particle, has magnitude $W = mg$.

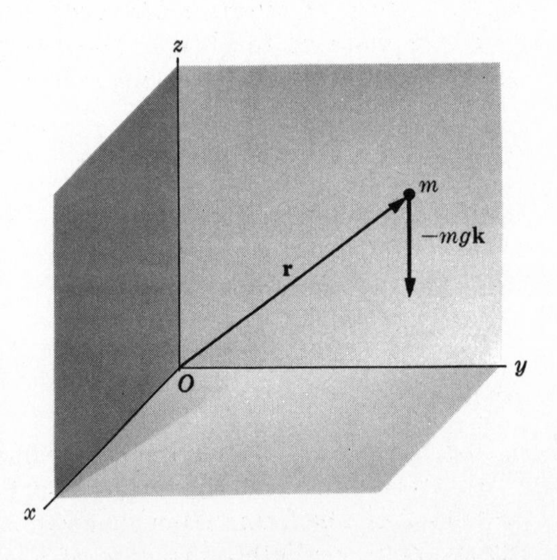

Fig. 3-2

GRAVITATIONAL SYSTEM OF UNITS

Because $W = mg$, it follows that $m = W/g$. This fact has led many scientists and engineers, who deal to a large extent with mechanics on the earth's surface, to rewrite the equations of motion with the fundamental mass quantity m replaced by the weight quantity W. Thus, for example, Newton's second law is rewritten as

$$\mathbf{F} = \frac{W}{g}\mathbf{a} \tag{4}$$

In this equation W and g can both vary while $m = W/g$ is constant. One system of units used in (4) is the *gravitational* or *English engineering system* where the unit of $\mathbf{F}$ or W is the *pound weight* (lb wt) while length is in feet and time is in seconds. In this case the unit of m is the *slug* and the system is often called the *foot-slug-second* (fss) *system*. Other systems are also possible. For example, we can take $\mathbf{F}$ or W in *kilograms weight* (kg wt) with length in meters and time in seconds.

ASSUMPTION OF A FLAT EARTH

Equation (3) indicates that the force acting on mass m has constant magnitude mg and is at each point directed perpendicular to the earth's surface represented by the xy plane. In reality this assumption, called the *assumption of the flat earth*, is not correct first because the earth is not flat and second because the force acting on mass m actually varies with the distance from the center of the earth, as shown in Chapter 5.

In practice the assumption of a flat earth is quite accurate for describing motions of objects at or near the earth's surface and will be used throughout this chapter. However, for describing the motion of objects far from the earth's surface the methods of Chapter 5 must be employed.

FREELY FALLING BODIES

If an object moves so that the only force acting upon it is its weight, or force due to gravity, then the object is often called a *freely falling body*. If $\mathbf{r}$ is the position vector and m is the mass of the body, then using Newton's second law the differential equation of motion is seen from equation (3) to be

$$m\frac{d^2\mathbf{r}}{dt^2} = -mg\mathbf{k} \quad \text{or} \quad \frac{d^2\mathbf{r}}{dt^2} = -g\mathbf{k} \tag{5}$$

Since this equation does not involve the mass m, the motion of a freely falling body is independent of its mass.

PROJECTILES

An object fired from a gun or dropped from a moving airplane is often called a *projectile*. If air resistance is negligible, a projectile can be considered as a freely falling body so that its motion can be found from equation (5) together with appropriate initial conditions. If air resistance is negligible the path of a projectile is an arc of a parabola (or a straight line which can be considered a degenerate parabola). See Problem 3.6.

POTENTIAL AND POTENTIAL ENERGY IN A UNIFORM FORCE FIELD

The potential of the uniform force field, or potential energy of a particle in this force field, is given by

$$V = F_0(z - z_0) \tag{6}$$

where z_0 is an arbitrary constant such that when $z = z_0$, $V = 0$. We call $z = z_0$ the *reference level.*

In particular for a constant gravitational field, $F_0 = mg$ and the potential energy of the particle is

$$V = mg(z - z_0) \tag{7}$$

This leads to

Theorem 3.1. The potential energy of a particle in a constant gravitational field is found by multiplying the magnitude of its weight by the height above some prescribed reference level. Note that the potential energy is the work done by the weight in moving through the distance $z - z_0$.

MOTION IN A RESISTING MEDIUM

In practice an object is acted upon not only by its weight but by other forces as well. An important class of forces are those which tend to oppose the motion of an object. Such forces, which generally arise because of motion in some medium such as air or water, are often called *resisting, damping* or *dissipative* forces and the corresponding medium is said to be a *resisting, damping* or *dissipative* medium.

It is found experimentally that for low speeds the resisting force is in magnitude proportional to the speed. In other cases it may be proportional to the square [or some other power] of the speed. If the resisting force is $\mathbf{R}$, then the motion of a particle of mass m in an otherwise uniform (gravitational) force field is given by

$$m\frac{d^2\mathbf{r}}{dt^2} = mg\mathbf{k} - \mathbf{R} \tag{8}$$

If $\mathbf{R} = \mathbf{0}$ this reduces to (5).

ISOLATING THE SYSTEM

In dealing with the dynamics or statics of a particle [or a system of particles, as we shall see later] it is extremely important to take into account all those forces which act *on* the particle [or *on* the system of particles]. This process is often called *isolating the system.*

CONSTRAINED MOTION

In some cases a particle P must move along some specified curve or surface as, for example, the inclined plane of Fig. 3-3 or the inner surface of a hemispherical bowl of Fig. 3-4 below. Such a curve or surface on which the particle must move is called a *constraint* and the resulting motion is called *constrained motion.*

Just as the particle exerts a force on the constraint, there will by Newton's third law be a *reaction force* of the constraint on the particle. This reaction force is often described by giving its components $\mathbf{N}$ and $\mathbf{f}$, normal to and parallel to the direction of motion respectively. In most cases which arise in practice, $\mathbf{f}$ is the force due to friction and is taken in a direction opposing the motion.

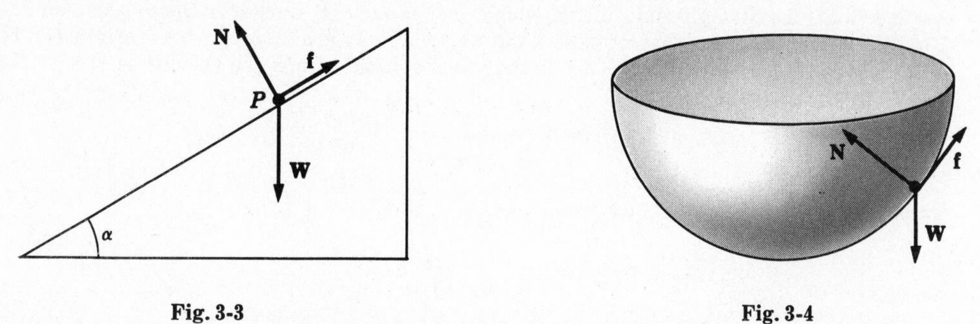

Fig. 3-3 Fig. 3-4

Problems involving constrained motion can be solved by using Newton's second law to arrive at differential equations for the motion and then solving these equations subject to initial conditions.

FRICTION

In the constrained motion of particles, one of the most important forces resisting motion is that due to *friction*. Referring to Fig. 3-5, let N be the magnitude of the normal component of the reaction of the constraint on the particle m. Then it is found experimentally that the magnitude of the force **f** due to friction is given by

$$f = \mu N \tag{9}$$

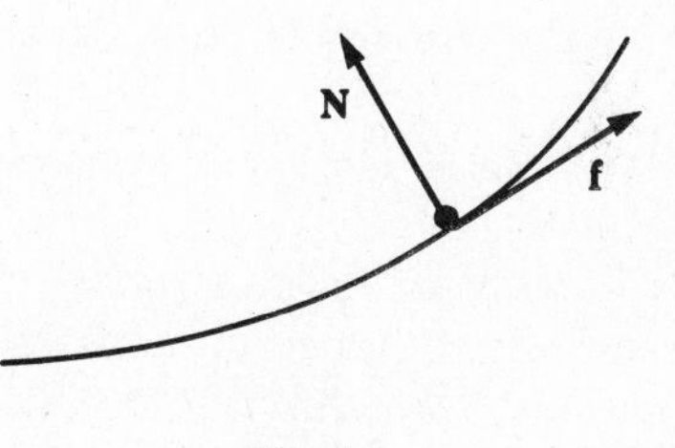

Fig. 3-5

where μ is called the *coefficient of friction*. The direction of **f** is always opposite to the direction of motion. The coefficient of friction, which depends on the material of both the particle and constraint, is taken as a constant in practice.

STATICS IN A UNIFORM GRAVITATIONAL FIELD

As indicated in Chapter 2, a particle is in equilibrium under the influence of a system of forces if and only if the net force acting on it is $\mathbf{F} = 0$.

Solved Problems

UNIFORM FORCE FIELDS AND UNIFORMLY ACCELERATED MOTION

3.1. A particle of mass m moves along a straight line under the influence of a constant force of magnitude F. If its initial speed is v_0, find (*a*) the speed, (*b*) the velocity and (*c*) the distance traveled after time t.

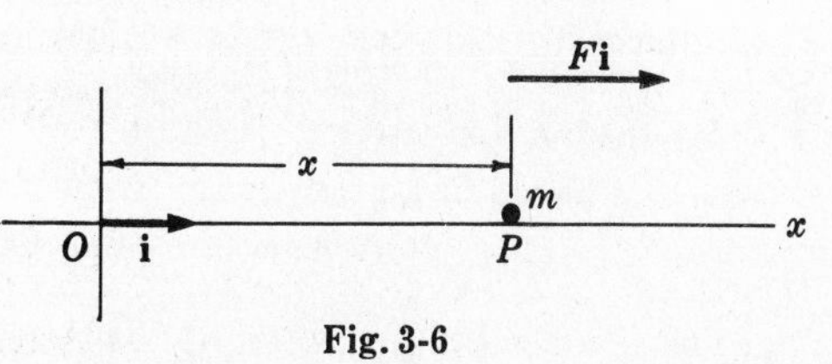

Fig. 3-6

(a) Assume that the straight line along which the particle P moves is the x axis, as shown in Fig. 3-6 above. Suppose that at time t the particle is at a distance x from origin O. If $\mathbf{i}$ is a unit vector in the direction OP and v is the speed at time t, then the velocity is $v\mathbf{i}$. By Newton's second law we have

$$\frac{d}{dt}(mv\mathbf{i}) = F\mathbf{i} \quad \text{or} \quad m\frac{dv}{dt} = F \tag{1}$$

Thus
$$dv = \frac{F}{m}dt \quad \text{or} \quad \int dv = \int \frac{F}{m}dt$$

i.e.
$$v = \frac{F}{m}t + c_1 \tag{2}$$

where c_1 is a constant of integration. To find c_1 we note the initial condition that $v = v_0$ at $t = 0$ so that from (2), $c_1 = v_0$ and

$$v = \frac{F}{m}t + v_0 \quad \text{or} \quad v = v_0 + \frac{F}{m}t \tag{3}$$

(b) From (3) the velocity at time t is

$$v\mathbf{i} = v_0\mathbf{i} + \frac{F}{m}t\mathbf{i} \quad \text{or} \quad \mathbf{v} = \mathbf{v}_0 + \frac{\mathbf{F}}{m}t$$

where $\mathbf{v} = v\mathbf{i}$, $\mathbf{v}_0 = v_0\mathbf{i}$ and $\mathbf{F} = F\mathbf{i}$.

(c) Since $v = dx/dt$ we have from (3),

$$\frac{dx}{dt} = v_0 + \frac{F}{m}t \quad \text{or} \quad dx = \left(v_0 + \frac{F}{m}t\right)dt$$

Then on integrating, assuming c_2 to be the constant of integration, we have

$$x = v_0 t + \left(\frac{F}{2m}\right)t^2 + c_2$$

Since $x = 0$ at $t = 0$, we find $c_2 = 0$. Thus

$$x = v_0 t + \left(\frac{F}{2m}\right)t^2 \tag{4}$$

3.2. Referring to Problem 3.1, show that the speed of the particle at any position x is given by $v = \sqrt{v_0^2 + (2F/m)x}$.

Method 1.

From (3) of Problem 3.1, we have $t = m(v - v_0)/F$. Substituting into (4) and simplifying, we find $x = (m/2F)(v^2 - v_0^2)$. Solving for v we obtain the required result.

Method 2.

From (1) of Problem 3.1, we have

$$\frac{dv}{dt} = \frac{F}{m}, \quad \text{i.e.} \quad \frac{dv}{dx}\frac{dx}{dt} = \frac{F}{m}$$

or since $v = dx/dt$,

$$v\frac{dv}{dx} = \frac{F}{m}, \quad \text{i.e.} \quad v\,dv = \frac{F}{m}dx$$

Integrating,
$$\frac{v^2}{2} = \frac{F}{m}x + c_3$$

Since $v = v_0$ when $x = 0$, we find $c_3 = v_0^2/2$ and hence $v = \sqrt{v_0^2 + (2F/m)x}$.

Method 3.

Change in kinetic energy from $t = 0$ to any time t

= Work done in moving particle from $x = 0$ to any position x

or $\frac{1}{2}mv^2 - \frac{1}{2}mv_0^2 = F(x - 0)$. Then $v = \sqrt{v_0^2 + (2F/m)x}$.

LINEAR MOTION OF FREELY FALLING BODIES

3.3. An object of mass m is thrown vertically upward from the earth's surface with speed v_0. Find (*a*) the position at any time, (*b*) the time taken to reach the highest point and (*c*) the maximum height reached.

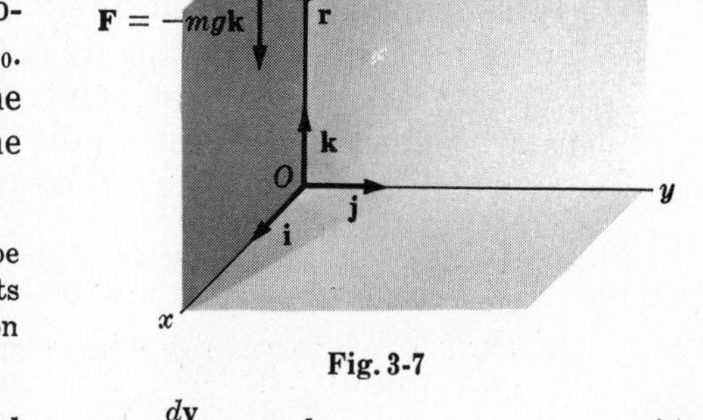

Fig. 3-7

(*a*) Let the position vector of m at any time t be $\mathbf{r} = x\mathbf{i} + y\mathbf{j} + z\mathbf{k}$. Assume that the object starts at $\mathbf{r} = \mathbf{0}$ when $t = 0$. Since the force acting on the object is $-mg\mathbf{k}$, we have by Newton's law,

$$m\frac{d^2\mathbf{r}}{dt^2} = m\frac{d\mathbf{v}}{dt} = -mg\mathbf{k} \quad \text{or} \quad \frac{d\mathbf{v}}{dt} = -g\mathbf{k} \tag{1}$$

where $\mathbf{v}$ is the velocity at time t. Integrating (*1*) once yields

$$\mathbf{v} = -gt\mathbf{k} + \mathbf{c}_1 \tag{2}$$

Since the velocity at $t = 0$ [i.e. the initial velocity] is $v_0\mathbf{k}$, we have from (*2*), $\mathbf{c}_1 = v_0\mathbf{k}$ so that

$$\mathbf{v} = -gt\mathbf{k} + v_0\mathbf{k} = (v_0 - gt)\mathbf{k} \tag{3}$$

or

$$\frac{d\mathbf{r}}{dt} = (v_0 - gt)\mathbf{k} \tag{4}$$

Integrating (*4*) yields

$$\mathbf{r} = (v_0t - \tfrac{1}{2}gt^2)\mathbf{k} + \mathbf{c}_2 \tag{5}$$

Then since $\mathbf{r} = \mathbf{0}$ when $t = 0$, $\mathbf{c}_2 = \mathbf{0}$. Thus the position vector is

$$\mathbf{r} = (v_0t - \tfrac{1}{2}gt^2)\mathbf{k} \tag{6}$$

or, equivalently,

$$x = 0, \quad y = 0, \quad z = v_0t - \tfrac{1}{2}gt^2 \tag{7}$$

(*b*) The highest point is reached when $\mathbf{v} = (v_0 - gt)\mathbf{k} = \mathbf{0}$, i.e. at time $t = v_0/g$.

(*c*) At time $t = v_0/g$ the maximum height reached is, from (*7*), $z = v_0^2/2g$.

Another method.

If we assume, as is physically evident, that the object must always be on the z axis, we may avoid vectors by writing Newton's law equivalently as [see equation (*1*) above and place $\mathbf{r} = z\mathbf{k}$]

$$d^2z/dt^2 = -g$$

from which, using $z = 0$, $dz/dt = v_0$ at $t = 0$, we find

$$z = v_0t - \tfrac{1}{2}gt^2$$

as above. The answers to (*b*) and (*c*) are then obtained as before.

3.4. Find the speed of the particle of Problem 3.3 in terms of its distance from origin O.

Method 1. From Problem 3.3, equations (*3*) and (*7*), we have

$$v = v_0 - gt, \quad z = v_0t - \tfrac{1}{2}gt^2$$

Solving for t in the first equation and substituting into the second equation, we find

$$z = v_0\left(\frac{v_0 - v}{g}\right) - \tfrac{1}{2}g\left(\frac{v_0 - v}{g}\right)^2 = \frac{v_0^2 - v^2}{2g} \quad \text{or} \quad v^2 = v_0^2 - 2gz$$

Method 2. From equation (*1*) of Problem 3.3 we have, since $\mathbf{v} = v\mathbf{k}$ and $v = dz/dt$,

$$\frac{dv}{dt} = -g, \quad \text{i.e.} \quad \frac{dv}{dz}\frac{dz}{dt} = -g \quad \text{or} \quad v\frac{dv}{dz} = -g$$

Then on integrating, $v^2/2 = -gz + c_3$. Since $v = v_0$ at $z = 0$, $c_3 = v_0^2/2$ and thus $v^2 = v_0^2 - 2gz$.

Method 3. See Problem 3.9 for a method using the principle of conservation of energy.

MOTION OF PROJECTILES

3.5. A projectile is launched with initial speed v_0 at an angle α with the horizontal. Find (*a*) the position vector at any time, (*b*) the time to reach the highest point, (*c*) the maximum height reached, (*d*) the time of flight back to earth and (*e*) the range.

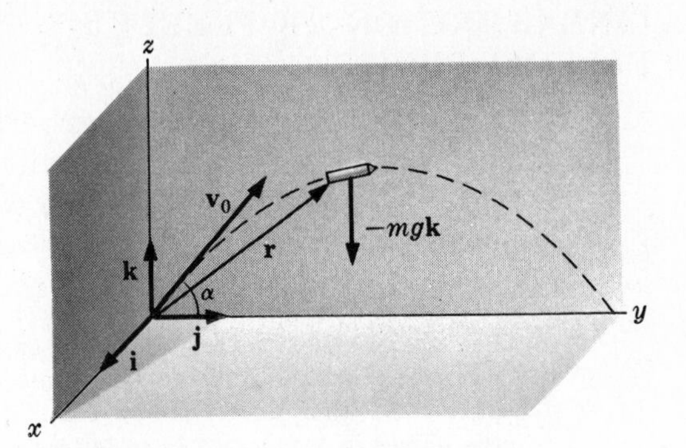

Fig. 3-8

(*a*) Let $\mathbf{r}$ be the position vector of the projectile and $\mathbf{v}$ the velocity at any time t. Then by Newton's law

$$m\frac{d^2\mathbf{r}}{dt^2} = -mg\mathbf{k} \tag{1}$$

i.e.,

$$\frac{d^2\mathbf{r}}{dt^2} = -g\mathbf{k} \quad \text{or} \quad \frac{d\mathbf{v}}{dt} = -g\mathbf{k} \tag{2}$$

Integration yields

$$\mathbf{v} = -gt\mathbf{k} + \mathbf{c}_1 \tag{3}$$

Assume the initial velocity of the projectile is in the yz plane so that the initial velocity is

$$\mathbf{v}_0 = v_0 \cos\alpha\,\mathbf{j} + v_0 \sin\alpha\,\mathbf{k} \tag{4}$$

Since $\mathbf{v} = \mathbf{v}_0$ at $t = 0$, we find from (*3*),

$$\mathbf{v} = v_0 \cos\alpha\,\mathbf{j} + (v_0 \sin\alpha - gt)\mathbf{k} \tag{5}$$

Replacing $\mathbf{v}$ by $d\mathbf{r}/dt$ in (*5*) and integrating, we obtain

$$\mathbf{r} = (v_0 \cos\alpha)t\,\mathbf{j} + \{(v_0 \sin\alpha)t - \tfrac{1}{2}gt^2\}\mathbf{k} \tag{6}$$

or, equivalently,

$$x = 0, \quad y = (v_0 \cos\alpha)t, \quad z = (v_0 \sin\alpha)t - \tfrac{1}{2}gt^2 \tag{7}$$

It follows that the projectile remains in the yz plane.

(*b*) At the highest point of the path the component of velocity $\mathbf{v}$ in the $\mathbf{k}$ direction is zero. Thus

$$v_0 \sin\alpha - gt = 0 \quad \text{and} \quad t = \frac{v_0 \sin\alpha}{g} \tag{8}$$

is the required time.

(*c*) Using the value of t obtained in (*b*), we find from (*7*) that

$$\text{Maximum height reached} = (v_0 \sin\alpha)\left(\frac{v_0 \sin\alpha}{g}\right) - \tfrac{1}{2}g\left(\frac{v_0 \sin\alpha}{g}\right)^2 = \frac{v_0^2 \sin^2\alpha}{2g} \tag{9}$$

(*d*) The time of flight back to earth is the time when $z = 0$, i.e. when

$$(v_0 \sin\alpha)t - \tfrac{1}{2}gt^2 = t[(v_0 \sin\alpha) - \tfrac{1}{2}gt] = 0$$

or since $t \neq 0$,

$$t = \frac{2v_0 \sin\alpha}{g} \tag{10}$$

Note that this is twice the time in (*b*).

(*e*) The range is the value of y at the time given by (*10*), i.e.,

$$\text{Range} = (v_0 \cos\alpha)\left(\frac{2v_0 \sin\alpha}{g}\right) = \frac{2v_0^2 \sin\alpha \cos\alpha}{g} = \frac{v_0^2 \sin 2\alpha}{g}$$

3.6. Show that the path of the projectile in Problem 3.5 is a parabola.

From the second equation of (*7*) in Problem 3.5, we have $t = y/(v_0 \cos\alpha)$. Substituting this into the third equation of (*7*) in Problem 3.5, we find

$$z = (v_0 \sin\alpha)(y/v_0 \cos\alpha) - \tfrac{1}{2}g(y/v_0 \cos\alpha)^2 \quad \text{or} \quad z = y \tan\alpha - (g/2v_0^2)y^2 \sec^2\alpha$$

which is a parabola in the yz plane.

3.7. Prove that the range of the projectile of Problem 3.5 is a maximum when the launching angle $\alpha = 45°$.

By Problem 3.5(*e*) the range is $(v_0^2 \sin 2\alpha)/g$. This is a maximum when $\sin 2\alpha = 1$, i.e. $2\alpha = 90°$ or $\alpha = 45°$.

POTENTIAL AND POTENTIAL ENERGY IN A UNIFORM FORCE FIELD

3.8. (*a*) Prove that a uniform force field is conservative, (*b*) find the potential corresponding to this field and (*c*) deduce the potential energy of a particle of mass m in a uniform gravitational force field.

(*a*) If the force field is as indicated in Fig. 3-1, then $\mathbf{F} = -F_0\mathbf{k}$. We have

$$\nabla \times \mathbf{F} = \begin{vmatrix} \mathbf{i} & \mathbf{j} & \mathbf{k} \\ \partial/\partial x & \partial/\partial y & \partial/\partial z \\ 0 & 0 & -F_0 \end{vmatrix} = \mathbf{0}$$

Thus the force field is conservative.

(*b*) $\mathbf{F} = -F_0\mathbf{k} = -\nabla V = -\frac{\partial V}{\partial x}\mathbf{i} - \frac{\partial V}{\partial y}\mathbf{j} - \frac{\partial V}{\partial z}\mathbf{k}$. Then $\frac{\partial V}{\partial x} = 0$, $\frac{\partial V}{\partial y} = 0$, $\frac{\partial V}{\partial z} = F_0$ from which $V = F_0 z + c$. If $V = 0$ at $z = z_0$, then $c = -F_0 z_0$ and so $V = F_0(z - z_0)$.

(*c*) For a uniform gravitational force field, $\mathbf{F} = -mg\mathbf{k}$ [see Fig. 3-2, page 62] and corresponds to $F_0 = mg$. Then by part (*b*) the potential or potential energy is $V = mg(z - z_0)$.

3.9. Work Problem 3.4 using the principle of conservation of energy.

According to the principle of conservation of energy, we have

$$\text{P.E. at } z = 0 + \text{K.E. at } z = 0 = \text{P.E. at } z + \text{K.E. at } z$$

$$0 + \tfrac{1}{2}mv_0^2 = mgz + \tfrac{1}{2}mv^2$$

Then $v^2 = v_0^2 - 2gz$.

MOTION IN A RESISTING MEDIUM

3.10. At time $t = 0$ a parachutist [Fig. 3-9] having weight of magnitude mg is located at $z = 0$ and is traveling vertically downward with speed v_0. If the force or air resistance acting on the parachute is proportional to the instantaneous speed, find the (*a*) speed, (*b*) distance traveled and (*c*) acceleration at any time $t > 0$.

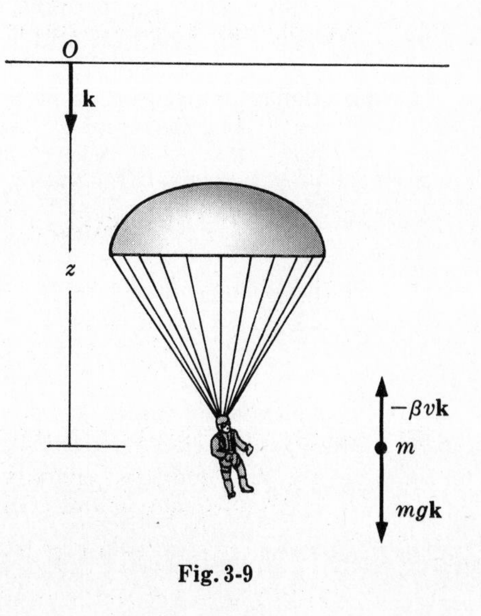

Fig. 3-9

(*a*) Assume the parachutist (considered as a particle of mass m) is located at distance z from origin O. If $\mathbf{k}$ is a unit vector in the vertically downward direction, then the weight is $mg\mathbf{k}$ while the force of air resistance is $-\beta v\mathbf{k}$ so that the net force is $(mg - \beta v)\mathbf{k}$.

Thus by Newton's law,

$$m\frac{dv}{dt}\mathbf{k} = (mg - \beta v)\mathbf{k} \qquad (1)$$

i.e.
$$m\frac{dv}{dt} = mg - \beta v \quad \text{or} \quad \frac{m\,dv}{mg - \beta v} = dt$$

Integrating,
$$-\frac{m}{\beta}\ln(mg - \beta v) = t + c_1 \tag{2}$$

Since $v = v_0$ at $t = 0$, $c_1 = -\frac{m}{\beta}\ln(mg - \beta v_0)$. Then from (2),

$$t = \frac{m}{\beta}\ln(mg - \beta v_0) - \frac{m}{\beta}\ln(mg - \beta v) = \frac{m}{\beta}\ln\left(\frac{mg - \beta v_0}{mg - \beta v}\right)$$

Thus
$$\frac{mg - \beta v_0}{mg - \beta v} = e^{\beta t/m} \quad \text{or} \quad v = \frac{mg}{\beta} + \left(v_0 - \frac{mg}{\beta}\right)e^{-\beta t/m} \tag{3}$$

(b) From (3), $dz/dt = mg/\beta + (v_0 - mg/\beta)e^{-\beta t/m}$. Then by integration,

$$z = \frac{mgt}{\beta} - \frac{m}{\beta}\left(v_0 - \frac{mg}{\beta}\right)e^{-\beta t/m} + c_2$$

Since $z = 0$ at $t = 0$, $c_2 = (m/\beta)(v_0 - mg/\beta)$ and thus

$$z = \frac{mgt}{\beta} + \frac{m}{\beta}\left(v_0 - \frac{mg}{\beta}\right)(1 - e^{-\beta t/m}) \tag{4}$$

(c) From (3), the acceleration is given by

$$a = \frac{dv}{dt} = -\frac{\beta}{m}\left(v_0 - \frac{mg}{\beta}\right)e^{-\beta t/m} = \left(g - \frac{\beta v_0}{m}\right)e^{-\beta t/m} \tag{5}$$

3.11. Show that the parachutist of Problem 3.10 approaches a limiting speed given by mg/β.

Method 1.

From equation (3) of Problem 3.10, $v = mg/\beta + (v_0 - mg/\beta)e^{-\beta t/m}$. Then as t increases, v approaches mg/β so that after a short time the parachutist is traveling with speed which is practically constant.

Method 2.

If the parachutist is to approach a limiting speed, the limiting acceleration must be zero. Thus from equation (1) of Problem 3.10 we have $mg - \beta v_{\text{lim}} = 0$ or $v_{\text{lim}} = mg/\beta$.

3.12. A particle of mass m is traveling along the x axis such that at $t = 0$ it is located at $x = 0$ and has speed v_0. The particle is acted upon by a force which opposes the motion and has magnitude proportional to the square of the instantaneous speed. Find the (a) speed, (b) position and (c) acceleration of the particle at any time $t > 0$.

(a) Suppose particle P is at a distance x from O at $t = 0$ and has speed v [see Fig. 3-10]. Then the force $\mathbf{F} = -\beta v^2\mathbf{i}$ where $\beta > 0$ is a constant of proportionality. By Newton's law,

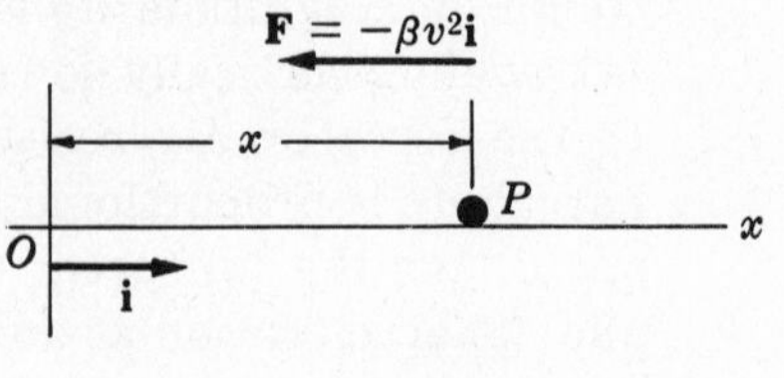

$$m\frac{dv}{dt}\mathbf{i} = -\beta v^2\mathbf{i} \quad \text{or} \quad \frac{dv}{v^2} = -\frac{\beta}{m}dt \tag{1}$$

$\mathbf{F} = -\beta v^2\mathbf{i}$; x; P; O; $\mathbf{i}$; x

Fig. 3-10

Integrating, $-1/v = -\beta t/m + c_1$. Since $v = v_0$ when $t = 0$, we have $c_1 = -1/v_0$. Thus

$$-\frac{1}{v} = -\frac{\beta t}{m} - \frac{1}{v_0} \quad \text{or} \quad v = \frac{mv_0}{\beta v_0 t + m} \tag{2}$$

which is the speed.

(b) From (2), $\frac{dx}{dt} = \frac{mv_0}{\beta v_0 t + m}$. Then $\int dx = \int \frac{mv_0}{\beta v_0 t + m}dt = \frac{mv_0}{\beta v_0}\int \frac{dt}{t + m/\beta v_0}$ or

$$x = \frac{m}{\beta}\ln\left(t + \frac{m}{\beta v_0}\right) + c_2$$

Since $x = 0$ at $t = 0$, $c_2 = -\frac{m}{\beta}\ln\left(\frac{m}{\beta v_0}\right)$. Thus

$$x = \frac{m}{\beta}\ln\left(t + \frac{m}{\beta v_0}\right) - \frac{m}{\beta}\ln\left(\frac{m}{\beta v_0}\right) = \frac{m}{\beta}\ln\left(1 + \frac{\beta v_0 t}{m}\right) \quad (3)$$

(c) From (a),

$$a = \frac{dv}{dt} = \frac{d}{dt}\left(\frac{mv_0}{\beta v_0 t + m}\right) = -\frac{\beta m v_0^2}{(\beta v_0 t + m)^2} \quad (4)$$

Note that although the speed of the particle continually decreases, it never comes to rest.

3.13. Determine the (a) speed and (b) acceleration of the particle of Problem 3.12 as a function of the distance x from O.

Method 1. From parts (a) and (b) of Problem 3.12,

$$x = \frac{m}{\beta}\ln\left(\frac{\beta v_0 t + m}{m}\right), \quad \text{and} \quad v = \frac{mv_0}{\beta v_0 t + m} \quad \text{or} \quad \frac{\beta v_0 t + m}{m} = \frac{v_0}{v}$$

Then

$$x = \frac{m}{\beta}\ln\left(\frac{v_0}{v}\right) \quad \text{or} \quad v = v_0 e^{-\beta x/m}$$

and the acceleration is given in magnitude by

$$a = \frac{dv}{dt} = -\frac{\beta v_0}{m}e^{-\beta x/m}\frac{dx}{dt} = -\frac{\beta v_0^2}{m}e^{-2\beta x/m}$$

which can also be obtained from equation (4) of Problem 3.12.

Method 2. From equation (1) of Problem 3.12 we have

$$m\frac{dv}{dt} = m\frac{dv}{dx}\frac{dx}{dt} = mv\frac{dv}{dx} = -\beta v^2$$

or since $v \neq 0$, $m\frac{dv}{dx} = -\beta v$ and $\frac{dv}{v} = -\frac{\beta}{m}x$. Integrating, $\ln v = -\beta x/m + c_3$. Since $v = v_0$ when $x = 0$, $c_3 = \ln v_0$. Thus $\ln(v/v_0) = -\beta x/m$ or $v = v_0 e^{-\beta x/m}$.

3.14. Suppose that in Problem 3.5 we assume that the projectile has acting upon it a force due to air resistance equal to $-\beta\mathbf{v}$ where β is a positive constant and $\mathbf{v}$ is the instantaneous velocity. Find (a) the velocity and (b) the position vector at any time.

(a) The equation of motion in this case is

$$m\frac{d^2\mathbf{r}}{dt^2} = -mg\mathbf{k} - \beta\mathbf{v} \quad \text{or} \quad m\frac{d\mathbf{v}}{dt} + \beta\mathbf{v} = -mg\mathbf{k} \quad (1)$$

Dividing by m and multiplying by the integrating factor $e^{\int \beta/m\,dt} = e^{\beta t/m}$, the equation can be written as

$$\frac{d}{dt}\{e^{\beta t/m}\mathbf{v}\} = -ge^{\beta t/m}\mathbf{k}$$

Integration yields

$$e^{\beta t/m}\mathbf{v} = -\frac{mg}{\beta}e^{\beta t/m}\mathbf{k} + \mathbf{c}_1 \quad (2)$$

The initial velocity or velocity at $t = 0$ is

$$\mathbf{v}_0 = v_0\cos\alpha\,\mathbf{j} + v_0\sin\alpha\,\mathbf{k} \quad (3)$$

Using this in (2) we find

$$\mathbf{c}_1 = v_0\cos\alpha\,\mathbf{j} + v_0\sin\alpha\,\mathbf{k} + \frac{mg}{\beta}\mathbf{k}$$

Thus (2) becomes on dividing by $e^{\beta t/m}$,

$$\mathbf{v} \;=\; (v_0 \cos\alpha\,\mathbf{j} + v_0 \sin\alpha\,\mathbf{k})e^{-\beta t/m} - \frac{mg}{\beta}(1 - e^{-\beta t/m})\mathbf{k} \tag{4}$$

(b) Replacing $\mathbf{v}$ by $d\mathbf{r}/dt$ in (4) and integrating, we find

$$\mathbf{r} \;=\; -\frac{m}{\beta}(v_0 \cos\alpha\,\mathbf{j} + v_0 \sin\alpha\,\mathbf{k})e^{-\beta t/m} - \frac{mg}{\beta}\left(t + \frac{m}{\beta}e^{-\beta t/m}\right)\mathbf{k} + \mathbf{c}_2 \tag{5}$$

Since $\mathbf{r} = \mathbf{0}$ at $t = 0$,

$$\mathbf{c}_2 \;=\; \frac{m}{\beta}(v_0 \cos\alpha\,\mathbf{j} + v_0 \sin\alpha\,\mathbf{k}) + \frac{m^2 g}{\beta^2}\mathbf{k} \tag{6}$$

Using (6) in (5), we find

$$\mathbf{r} \;=\; \frac{mv_0}{\beta}(\cos\alpha\,\mathbf{j} + \sin\alpha\,\mathbf{k})(1 - e^{-\beta t/m}) - \frac{mg}{\beta}\left(t + \frac{m}{\beta}e^{-\beta t/m} - \frac{m}{\beta}\right)\mathbf{k} \tag{7}$$

3.15. Prove that the projectile of Problem 3.14 attains a limiting velocity and find its value.

Method 1.

Refer to equation (4) of Problem 3.14. As t increases, $e^{-\beta t/m}$ approaches zero. Thus the velocity approaches a limiting value equal to $\mathbf{v}_{\text{lim}} = -(mg/\beta)\mathbf{k}$.

Method 2.

If the projectile is to approach a limiting velocity its limiting acceleration must be zero. Thus from equation (1) of Problem 3.14, $-mg\mathbf{k} - \beta\mathbf{v}_{\text{lim}} = 0$ or $\mathbf{v}_{\text{lim}} = -(mg/\beta)\mathbf{k}$.

CONSTRAINED MOTION

3.16. A particle P of mass m slides without rolling down a frictionless inclined plane AB of angle α [Fig. 3-11]. If it starts from rest at the top A of the incline, find (a) the acceleration, (b) the velocity and (c) the distance traveled after time t.

(a) Since there is no friction the only forces acting on P are the weight $\mathbf{W} = -mg\mathbf{k}$ and the reaction force of the incline which is given by the normal force $\mathbf{N}$.

Let $\mathbf{e}_1$ and $\mathbf{e}_2$ be unit vectors parallel and perpendicular to the incline respectively. If we denote by s the magnitude of the displacement from the top A of the inclined plane, we have by Newton's second law

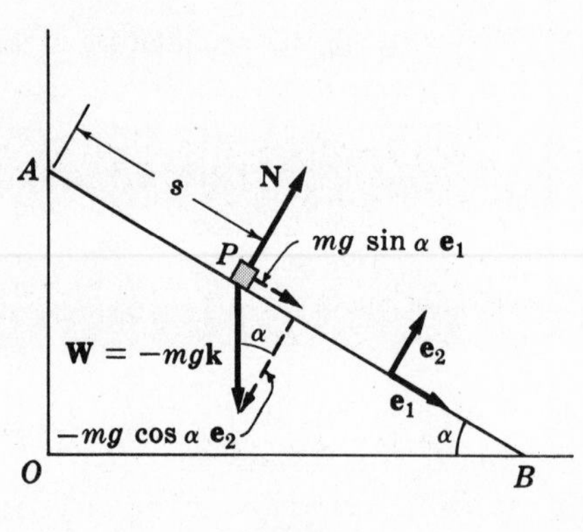

Fig. 3-11

$$m\frac{d^2}{dt^2}(s\mathbf{e}_1) \;=\; \mathbf{W} + \mathbf{N} \;=\; mg \sin\alpha\,\mathbf{e}_1 \tag{1}$$

since the resultant equal to $\mathbf{W} + \mathbf{N}$ is $mg \sin\alpha\,\mathbf{e}_1$, as indicated in Fig. 3-11. From (1) we have

$$d^2s/dt^2 \;=\; g \sin\alpha \tag{2}$$

Thus the acceleration down the incline at any time t is a constant equal to $g \sin\alpha$.

(b) Since $v = ds/dt$ is the speed, (2) can be written

$$dv/dt \;=\; g \sin\alpha \qquad \text{or} \qquad v \;=\; (g \sin\alpha)t + c_1$$

on integrating. Using the initial condition $v = 0$ at $t = 0$, we have $c_1 = 0$ so that the speed at any time t is

$$v \;=\; (g \sin\alpha)t \tag{3}$$

The velocity is $v\mathbf{e}_1 = (g \sin\alpha)t\mathbf{e}_1$ which has magnitude $(g \sin\alpha)t$ in the direction $\mathbf{e}_1$ down the incline.

(c) Since $v = ds/dt$, (3) can be written

$$ds/dt = (g \sin \alpha)t \quad \text{or} \quad s = \tfrac{1}{2}(g \sin \alpha)t^2 + c_2$$

on integrating. Using the initial condition $s = 0$ at $t = 0$, we find $c_2 = 0$ so that the required distance traveled is

$$s = \tfrac{1}{2}(g \sin \alpha)t^2 \tag{4}$$

3.17. If the length AB of the incline in Problem 3.16 is l, find (a) the time τ taken for the particle to reach the bottom B of the incline and (b) the speed at B.

(a) Since $s = l$ at B, the time τ to reach the bottom is from equation (4) of Problem 3.16 given by $l = \frac{1}{2}(g \sin \alpha)\tau^2$ or $\tau = \sqrt{2l/(g \sin \alpha)}$.

(b) The speed at B is given from (3) of Problem 3.16 by $v = (g \sin \alpha)\tau = \sqrt{2gl \sin \alpha}$.

MOTION INVOLVING FRICTION

3.18. Work Problem 3.16 if the inclined plane has a constant coefficient of friction μ.

(a) In this case there is, in addition to the forces $\mathbf{W}$ and $\mathbf{N}$ acting on P, a frictional force $\mathbf{f}$ [see Fig. 3-12] directed up the incline [in a direction opposite to the motion] and with magnitude

$$\mu N = \mu mg \cos \alpha \tag{1}$$

i.e.

$$\mathbf{f} = -\mu mg \cos \alpha \, \mathbf{e}_1 \tag{2}$$

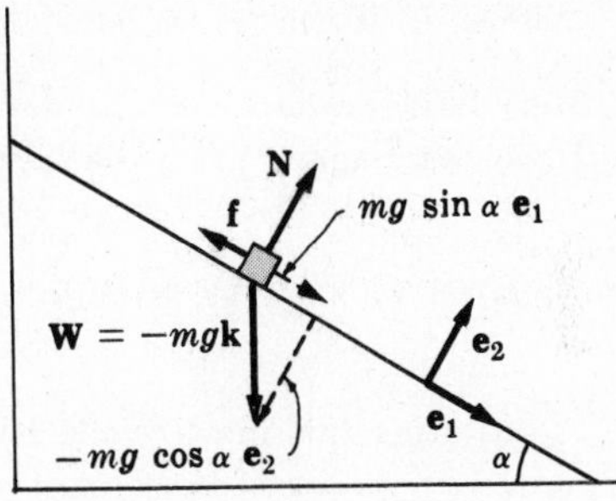

Fig. 3-12

Then equation (1) of Problem 3.16 is replaced by

$$m\frac{d^2(s\mathbf{e}_1)}{dt^2} = \mathbf{W} + \mathbf{N} + \mathbf{f} = mg \sin \alpha_1 \mathbf{e}_1 - \mu mg \cos \alpha \, \mathbf{e}_1 \tag{3}$$

or

$$d^2s/dt^2 = g(\sin \alpha - \mu \cos \alpha) \tag{4}$$

Thus the acceleration down the incline has the constant magnitude $g(\sin \alpha - \mu \cos \alpha)$ provided $\sin \alpha > \mu \cos \alpha$ or $\tan \alpha > \mu$ [otherwise the frictional force is so great that the particle will not move at all].

(b) Replacing d^2s/dt^2 by dv/dt in (4) and integrating as in part (b) of Problem 3.16, we find the speed at any time t to be

$$v = g(\sin \alpha - \mu \cos \alpha)t \tag{5}$$

(c) Replacing v by ds/dt in (5) and integrating as in part (c) of Problem 3.16, we find

$$s = \tfrac{1}{2}g(\sin \alpha - \mu \cos \alpha)t^2 \tag{6}$$

3.19. An object slides on a surface of ice along the horizontal straight line OA [Fig. 3-13]. At a certain point in its path the speed is v_0 and the object then comes to rest after traveling a distance x_0. Prove that the coefficient of friction is $v^2/2gx_0$.

Let x be the instantaneous distance of the object of mass m from O and suppose that at time $t = 0$, $x = 0$ and $dx/dt = v_0$.

Three forces act on the object, namely (1) the weight $\mathbf{W} = m\mathbf{g}$, (2) the normal force $\mathbf{N}$ of the ice surface on the object, and (3) the frictional force $\mathbf{f}$.

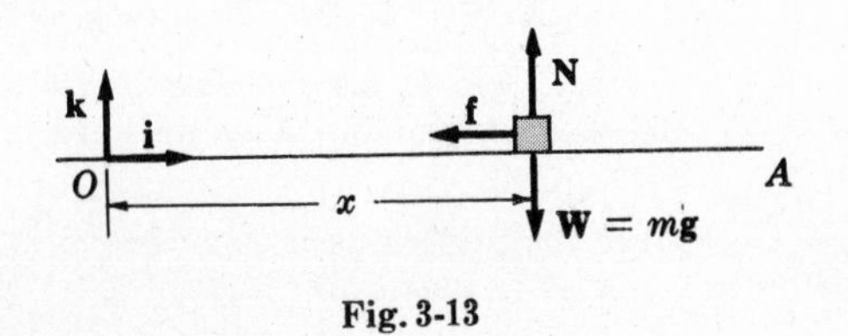

Fig. 3-13

By Newton's second law we have, if v is the instantaneous speed,

$$m\frac{dv}{dt}\mathbf{i} = \mathbf{W} + \mathbf{N} + \mathbf{f} \tag{1}$$

But since $\mathbf{N} = -\mathbf{W}$ and the magnitude of $\mathbf{f}$ is $f = \mu N = \mu mg$ so that $\mathbf{f} = -\mu mg\mathbf{i}$, (1) becomes

$$m\frac{dv}{dt}\mathbf{i} = -\mu mg\mathbf{i} \quad \text{or} \quad \frac{dv}{dt} = -\mu g \tag{2}$$

Method 1. Write (2) as

$$\frac{dv}{dx}\frac{dx}{dt} = -\mu g \quad \text{or} \quad v\frac{dv}{dx} = -\mu g \tag{3}$$

Then

$$v\,dv = -\mu g\,dx$$

Integrating, using the fact that $v = v_0$ at $x = 0$, we find

$$v^2/2 = -\mu gx + v_0^2/2 \tag{4}$$

Then since $v = 0$ when $x = x_0$, (4) becomes

$$-\mu gx_0 + v_0^2/2 = 0 \quad \text{or} \quad \mu = v_0^2/2gx_0 \tag{5}$$

Method 2. From (2) we have, on integrating and using the fact that $v = v_0$ at $t = 0$,

$$v = v_0 - \mu gt \quad \text{or} \quad dx/dt = v_0 - \mu gt \tag{6}$$

Integrating again, using the fact that $x = 0$ at $t = 0$, we find

$$x = v_0 t - \tfrac{1}{2}\mu g t^2 \tag{7}$$

From (7) we see that the object comes to rest (i.e., $v = 0$) when

$$v_0 - \mu gt = 0 \quad \text{or} \quad t = v_0/\mu g$$

Substituting this into (7) and noting that $x = x_0$, we obtain the required result.

STATICS IN A UNIFORM GRAVITATIONAL FIELD

3.20. A particle of mass m is suspended in equilibrium by two inelastic strings of lengths a and b from pegs A and B which are distant c apart. Find the tension in each string.

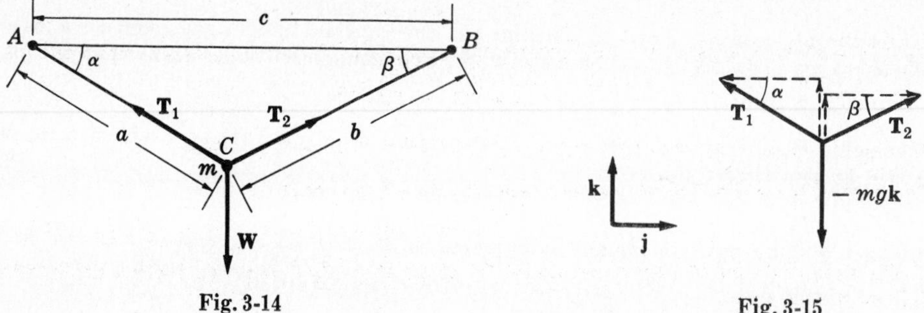

Fig. 3-14 Fig. 3-15

Let $\mathbf{W}$ denote the weight of the particle and $\mathbf{T}_1$ and $\mathbf{T}_2$ the respective tensions in the strings of lengths a and b as indicated in Fig. 3-14. These forces are also indicated in Fig. 3-15 and are assumed to lie in the plane of unit vectors $\mathbf{j}$ and $\mathbf{k}$. By resolving $\mathbf{T}_1$ and $\mathbf{T}_2$ into horizontal and vertical components it is clear that

$$\mathbf{T}_1 = T_1 \sin\alpha\,\mathbf{k} - T_1\cos\alpha\,\mathbf{j}, \qquad \mathbf{T}_2 = T_2\sin\beta\,\mathbf{k} + T_2\cos\beta\,\mathbf{j}$$

where T_1 and T_2 are the magnitudes of $\mathbf{T}_1$ and $\mathbf{T}_2$ respectively and where α and β are the respective angles at A and B. Also we have

$$\mathbf{W} = -mg\mathbf{k}$$

Since the particle is in equilibrium if and only if the net force acting on it is zero, we have

$$\begin{aligned}\mathbf{F} &= \mathbf{T}_1 + \mathbf{T}_2 + \mathbf{W}\\ &= T_1\sin\alpha\,\mathbf{k} - T_1\cos\alpha\,\mathbf{j} + T_2\sin\beta\,\mathbf{k} + T_2\cos\beta\,\mathbf{j} - mg\mathbf{k}\\ &= (T_2\cos\beta - T_1\cos\alpha)\mathbf{j} + (T_1\sin\alpha + T_2\sin\beta - mg)\mathbf{k}\\ &= \mathbf{0}\end{aligned}$$

From this we must have

$$T_2 \cos\beta - T_1 \cos\alpha = 0, \qquad T_1 \sin\alpha + T_2 \sin\beta - mg = 0$$

Solving simultaneously, we find

$$T_1 = \frac{mg\cos\beta}{\sin(\alpha+\beta)}, \qquad T_2 = \frac{mg\cos\alpha}{\sin(\alpha+\beta)}$$

The angles α and β can be determined from the law of cosines as

$$\alpha = \cos^{-1}\left(\frac{a^2+c^2-b^2}{2ac}\right), \qquad \beta = \cos^{-1}\left(\frac{b^2+c^2-a^2}{2bc}\right)$$

From these the tensions can be expressed in terms of a, b, c.

MISCELLANEOUS PROBLEMS

3.21. An inclined plane [Fig. 3-16] makes an angle α with the horizontal. A projectile is launched from the bottom A of the incline with speed v_0 in a direction making an angle β with the horizontal.

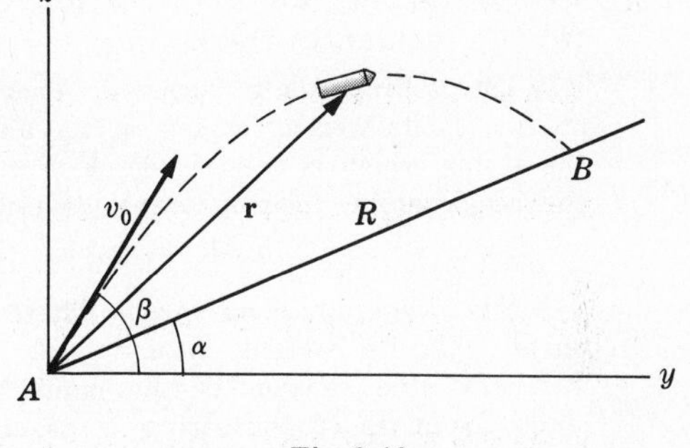

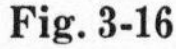

Fig. 3-16

(a) Prove that the range R up the incline is given by

$$R = \frac{2v_0^2 \sin(\beta-\alpha)\cos\beta}{g\cos^2\alpha}$$

(b) Prove that the maximum range up the incline is given by

$$R_{\max} = \frac{v_0^2}{g(1+\sin\alpha)}$$

and is achieved when $\beta = \pi/4 + \alpha/2$.

(a) As in Problem 3.5, equation (6), the position vector of the projectile at any time t is

$$\mathbf{r} = (v_0\cos\beta)t\mathbf{j} + \{(v_0\sin\beta)t - \tfrac{1}{2}gt^2\}\mathbf{k} \qquad (1)$$

or

$$y = (v_0\cos\beta)t, \qquad z = (v_0\sin\beta)t - \tfrac{1}{2}gt^2 \qquad (2)$$

The equation of the incline [which is a line in the yz plane] is

$$z = y\tan\alpha \qquad (3)$$

Using equations (2) in (3) we see that the projectile's path and the incline intersect for those values of t where

$$(v_0\sin\beta)t - \tfrac{1}{2}gt^2 = [(v_0\cos\beta)t]\tan\alpha$$

i.e.

$$t = 0 \quad \text{and} \quad t = \frac{2v_0(\sin\beta\cos\alpha - \cos\beta\sin\alpha)}{g\cos\alpha} = \frac{2v_0\sin(\beta-\alpha)}{g\cos\alpha}$$

The value $t=0$ gives the intersection point A. The second value of t yields point B which is the required point. Using this second value of t in the first equation of (2), we find that the required range R up the incline is

$$R = y\sec\alpha = (v_0\cos\beta)\left\{\frac{2v_0\sin(\beta-\alpha)}{g\cos\alpha}\right\}\sec\alpha = \frac{2v_0^2\sin(\beta-\alpha)\cos\beta}{g\cos^2\alpha}$$

(b) **Method 1.** The range R can be written by using the trigonometric identity

$$\sin A\cos B = \tfrac{1}{2}\{\sin(A+B) + \sin(A-B)\}$$

as

$$R = \frac{v_0^2}{g\cos^2\alpha}\{\sin(2\beta-\alpha) - \sin\alpha\}$$

This is a maximum when $\sin(2\beta - \alpha) = 1$, i.e. $2\beta - \alpha = \pi/2$ or $\beta = \alpha/2 + \pi/4$, and the value of this maximum is

$$R = \frac{v_0^2}{g\cos^2\alpha}(1 - \sin\alpha) = \frac{v_0^2}{g(1 - \sin^2\alpha)}(1 - \sin\alpha) = \frac{v_0^2}{g(1 + \sin\alpha)}$$

Method 2.

The required result can also be obtained by the methods of differential calculus for finding maxima and minima.

3.22. Two particles of masses m_1 and m_2 respectively are connected by an inextensible string of negligible mass which passes over a fixed frictionless pulley of negligible mass as shown in Fig. 3-17. Describe the motion by finding (*a*) the acceleration of the particles and (*b*) the tension in the string.

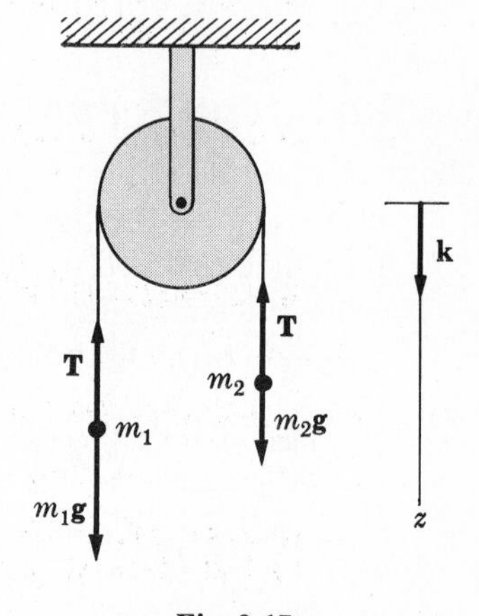

Fig. 3-17

Let us first isolate mass m_1. There are two forces acting on it: (1) its weight $m_1\mathbf{g} = m_1 g\mathbf{k}$, and (2) the force due to the string which is the tension $\mathbf{T} = -T\mathbf{k}$. If we call $\mathbf{a} = a\mathbf{k}$ the acceleration, then by Newton's law

$$m_1 a\mathbf{k} = m_1 g\mathbf{k} - T\mathbf{k} \qquad (1)$$

Next we isolate mass m_2. There are two forces acting on it: (1) its weight $m_2\mathbf{g} = m_2 g\mathbf{k}$, and (2) the tension $\mathbf{T} = -T\mathbf{k}$ [the tension is the same throughout the string since the mass of the string is assumed negligible and inextensible]. Since the string is inextensible, the acceleration of m_2 is $-\mathbf{a} = -a\mathbf{k}$. Then by Newton's law

$$-m_2 a\mathbf{k} = m_2 g\mathbf{k} - T\mathbf{k} \qquad (2)$$

From (*1*) and (*2*) we have

$$m_1 a = m_1 g - T, \qquad -m_2 a = m_2 g - T$$

Solving simultaneously, we find

$$a = \frac{m_1 - m_2}{m_1 + m_2}g, \qquad T = \frac{2m_1 m_2}{m_1 + m_2}g$$

Thus the particles move with constant acceleration, one particle rising and the other falling.

In this pulley system, sometimes called *Atwood's machine*, the pulley can rotate. However, since it is frictionless and has no mass [or negligible mass] the effect is the same as if the string passed over a smooth or frictionless peg instead of a pulley. In case the mass of the pulley is not negligible, rotational effects must be taken into account and are considered in Chapter 9.

3.23. A particle P of mass m rests at the top A of a frictionless fixed sphere of radius b. The particle is displaced slightly so that it slides (without rolling) down the sphere. (*a*) At what position will it leave the sphere and (*b*) what will its speed be at this position?

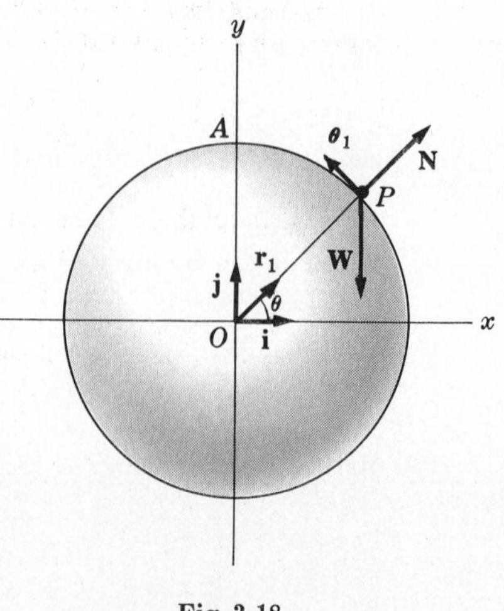

Fig. 3-18

The particle will slide down a circle of radius a which we choose to be in the xy plane as indicated in Fig. 3-18. The forces acting on the particle are: (1) its weight $\mathbf{W} = -mg\mathbf{j}$, and (2) the reaction force $\mathbf{N}$ of the sphere on the particle normal to the sphere.

Method 1.

(*a*) Let the position of the particle on the circle be measured by angle θ and let $\mathbf{r}_1$ and $\boldsymbol{\theta}_1$ be unit vectors. Resolving $\mathbf{W}$ into components in directions $\mathbf{r}_1$ and $\boldsymbol{\theta}_1$, we have as in Problem 1.43, page 24,

$$\begin{aligned}\mathbf{W} &= (\mathbf{W}\cdot\mathbf{r}_1)\mathbf{r}_1 + (\mathbf{W}\cdot\boldsymbol{\theta}_1)\boldsymbol{\theta}_1 \\ &= (-mg\mathbf{j}\cdot\mathbf{r}_1)\mathbf{r}_1 + (-mg\mathbf{j}\cdot\boldsymbol{\theta}_1)\boldsymbol{\theta}_1 = -mg\sin\theta\,\mathbf{r}_1 - mg\cos\theta\,\boldsymbol{\theta}_1\end{aligned}$$

Also,
$$\mathbf{N} = N\mathbf{r}_1$$

Using Newton's second law and the result of Problem 1.49, page 26, we have

$$\begin{aligned}\mathbf{F} = m\mathbf{a} &= m[(\ddot{r} - r\dot{\theta}^2)\mathbf{r}_1 + (r\ddot{\theta} + 2\dot{r}\dot{\theta})\boldsymbol{\theta}_1] \\ &= \mathbf{W} + \mathbf{N} = (N - mg\sin\theta)\mathbf{r}_1 - mg\cos\theta\,\boldsymbol{\theta}_1 \qquad (1)\end{aligned}$$

Thus
$$m(\ddot{r} - r\dot{\theta}^2) = N - mg\sin\theta, \qquad m(r\ddot{\theta} + 2\dot{r}\dot{\theta}) = -mg\cos\theta \qquad (2)$$

While the particle is on the circle (or sphere), we have $r = b$. Substituting this into (2),

$$-mb\dot{\theta}^2 = N - mg\sin\theta, \qquad b\ddot{\theta} = -g\cos\theta \qquad (3)$$

Multiplying the second equation by $\dot{\theta}$, we see that it can be written

$$b\frac{d}{dt}\left(\frac{\dot{\theta}^2}{2}\right) = -g\frac{d}{dt}(\sin\theta)$$

Integrating, $b\dot{\theta}^2/2 = -g\sin\theta + c_1$. Now when $\theta = \pi/2$, $\dot{\theta} = 0$ so that $c_1 = g$ and

$$b\dot{\theta}^2 = 2g(1 - \sin\theta) \qquad (4)$$

Substituting (4) into the first equation of (3), we find

$$N = mg(3\sin\theta - 2) \qquad (5)$$

Now as long as $N > 0$ the particle stays on the sphere; but when $N = 0$ the particle will be just about to leave the sphere. Thus the required angle is given by $3\sin\theta - 2 = 0$, i.e.,

$$\sin\theta = 2/3 \quad \text{or} \quad \theta = \sin^{-1}2/3 \qquad (6)$$

(b) Putting $\sin\theta = \frac{2}{3}$ into (4), we find

$$\dot{\theta}^2 = 2g/3b \qquad (7)$$

Then if v is the speed, we have $v = b\dot{\theta}$ so that (7) yields $v^2 = \frac{2}{3}bg$ or $v = \sqrt{\frac{2}{3}bg}$.

Method 2. By the conservation of energy, using the x axis as reference level, we have

$$\begin{aligned}\text{P.E. at } A + \text{K.E. at } A &= \text{P.E. at } P + \text{K.E. at } P \\ mgb + 0 &= mgb\sin\theta + \tfrac{1}{2}mv^2\end{aligned}$$

or
$$v^2 = 2gb(1 - \sin\theta) \qquad (8)$$

Using the result of Problem 1.35, page 20, together with Newton's second law, we have, since the radius of curvature is b,

$$\begin{aligned}\mathbf{F} = m\mathbf{a} &= \left(\frac{v^2}{b}\mathbf{r}_1 - \frac{dv}{dt}\boldsymbol{\theta}_1\right) = \mathbf{W} + \mathbf{N} \\ &= (N - mg\sin\theta)\mathbf{r}_1 - mg\cos\theta\,\boldsymbol{\theta}_1\end{aligned}$$

Using only the $\mathbf{r}_1$ component, we have

$$v^2/b = N - mg\sin\theta \qquad (9)$$

From (8) and (9) we find $N = mg(3\sin\theta - 2)$ which yields the required angle $\sin^{-1}(\frac{2}{3})$ as in Method 1. The speed is then found from (8).

Supplementary Problems

UNIFORM FORCE FIELDS AND LINEAR MOTION OF FREELY FALLING BODIES

3.24. An object of mass m is dropped from a height H above the ground. Prove that if air resistance is negligible, then it will reach the ground (a) in a time $\sqrt{2H/g}$ and (b) with speed $\sqrt{2gH}$.

3.25. Work Problem 3.24 if the object is thrown vertically downward with an initial velocity of magnitude v_0. *Ans.* (a) $(\sqrt{v_0^2 + 2gH} - v_0)/g$, (b) $\sqrt{v_0^2 + 2gH}$

3.26. Prove that the object of Problem 3.3, page 67, returns to the earth's surface (a) with the same speed as the initial speed and (b) in a time which is twice that taken to reach the maximum height.

3.27. A ball which is thrown upward reaches its maximum height of 100 ft and then returns to the starting point. (a) With what speed was it thrown? (b) How long does it take to return?
Ans. (a) 80 ft/sec, (b) 5 sec

3.28. A ball which is thrown vertically upward reaches a particular height H after a time τ_1 on the way up and a time τ_2 on the way down. Prove that (a) the initial velocity with which the ball was thrown has magnitude $\frac{1}{2}g(\tau_1 + \tau_2)$ and (b) the height $H = \frac{1}{2}g\tau_1\tau_2$.

3.29. In Problem 3.28, what is the maximum height reached? *Ans.* $\frac{1}{8}g(\tau_1 + \tau_2)^2$

3.30. Two objects are dropped from the top of a cliff of height H. The second is dropped when the first has traveled a distance D. Prove that at the instant when the first object has reached the bottom, the second object is at a distance above it given by $2\sqrt{DH} - D$.

3.31. An elevator starts from rest and attains a speed of 16 ft/sec in 2 sec. Find the weight of a 160 lb man in the elevator if the elevator is (a) moving up (b) moving down.
Ans. (a) 200 lb, (b) 120 lb

3.32. A particle of mass 3 kg moving in a straight line decelerates uniformly from a speed of 40 m/sec to 20 m/sec in a distance of 300 m. (a) Find the magnitude of the deceleration. (b) How much further does it travel before it comes to rest and how much longer will this take?
Ans. (a) 2 m/sec^2, (b) 100 m; 10 sec

3.33. In Problem 3.32, what is the total work done in bringing the particle to rest from the speed of 40 m/sec? *Ans.* 2400 newton meters (or joules)

MOTION OF PROJECTILES

3.34. A projectile is launched with a muzzle velocity of 1800 mi/hr at an angle of 60° with a horizontal and lands on the same plane. Find (a) the maximum height reached, (b) the time to reach the maximum height, (c) the total time of flight, (d) the range, (e) the speed after 1 minute of flight, (f) the speed at a height of 32,000 ft.
Ans. (a) 15.5 mi, (b) 71.4 sec, (c) 142.8 sec, (d) 35.7 mi, (e) 934 mi/hr, (f) 1558 mi/hr

3.35. (a) What is the maximum range possible for a projectile fired from a cannon having muzzle velocity 1 mi/sec and (b) what is the height reached in this case?
Ans. (a) 165 mi, (b) 41.25 mi

3.36 A cannon has its maximum range given by R_{max}. Prove that (a) the height reached in such case is $\frac{1}{4}R_{max}$ and (b) the time of flight is $\sqrt{R_{max}/2g}$.

3.37. It is desired to launch a projectile from the ground so as to hit a given point on the ground which is at a distance less than the maximum range. Prove that there are two possible angles for the launching, one which is less than 45° by a certain amount and the other greater than 45° by the same amount.

3.38. A projectile having horizontal range R reaches a maximum height H. Prove that it must have been launched with (a) an initial speed equal to $\sqrt{g(R^2 + 16H^2)/8H}$ and (b) at an angle with the horizontal given by $\sin^{-1}(4H/\sqrt{R^2 + 16H^2})$.

3.39. A projectile is launched at an angle α from a cliff of height H above sea level. If it falls into the sea at a distance D from the base of the cliff, prove that its maximum height above sea level is

$$H + \frac{D^2 \tan^2 \alpha}{4(H + D \tan \alpha)}$$

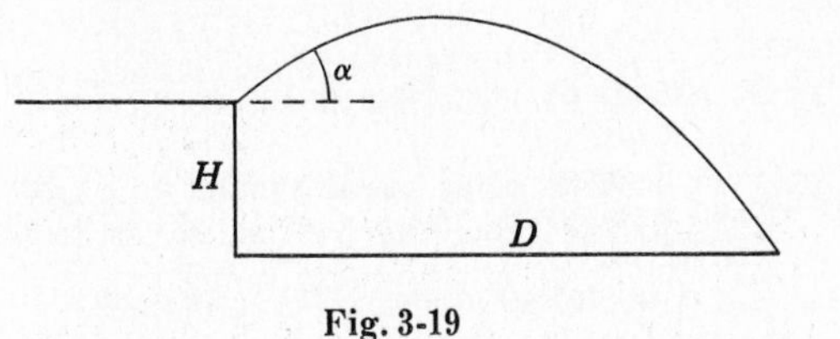

Fig. 3-19

MOTION IN A RESISTING MEDIUM

3.40. An object of weight W is thrown vertically upward with speed v_0. Assuming that air resistance is proportional to the instantaneous velocity and that the constant of proportionality is κ, prove that (a) the object will reach a maximum height of

$$\frac{W\kappa v_0}{\kappa^2 g} - \frac{W^2}{\kappa^2 g} \ln\left(1 + \frac{\kappa v_0}{W}\right)$$

and that (b) the time taken to reach this maximum height is

$$\frac{W}{\kappa g} \ln\left(1 + \frac{\kappa v_0}{W}\right)$$

3.41. A man on a parachute falls from rest and acquires a limiting speed of 15 mi/hr. Assuming that air resistance is proportional to the instantaneous speed, determine how long it takes to reach the speed of 14 mi/hr. *Ans.* 1.86 sec

3.42. A mass m moves along a straight line under the influence of a constant force F. Assuming that there is a resisting force numerically equal to κv^2 where v is the instantaneous speed and κ is a constant, prove that the distance traveled in going from speed v_1 to v_2 is $\dfrac{m}{2\kappa} \ln\left(\dfrac{F - \kappa v_1^2}{F - \kappa v_2^2}\right)$.

3.43. A particle of mass m moves in a straight line acted upon by a constant resisting force of magnitude F. If it starts with a speed of v_0, (a) how long will it take before coming to rest and (b) what distance will it travel in this time? *Ans.* (a) mv_0/F, (b) $mv_0^2/2F$

3.44. Can Problem 3.43 be worked by energy considerations? Explain.

3.45. A locomotive of mass m travels with constant speed v_0 along a horizontal track. (a) How long will it take for the locomotive to come to rest after the ignition is turned off, if the resistance to the motion is given by $\alpha + \beta v^2$ where v is the instantaneous speed and α and β are constants? (b) What is the distance traveled? *Ans.* (a) $\sqrt{m/\beta}\,\tan^{-1}(v_0\sqrt{\beta/\alpha})$, (b) $(m/2\beta)\ln(1 + \beta v_0^2/\alpha)$

3.46. A particle moves along the x axis acted upon only by a resisting force which is proportional to the cube of the instantaneous speed. If the initial speed is v_0 and after a time τ the speed is $\frac{1}{2}v_0$, prove that the speed will be $\frac{1}{4}v_0$ in time 5τ.

3.47. Find the total distance traveled by the particle of Problem 3.46 in reaching the speeds (a) $\frac{1}{2}v_0$, (b) $\frac{1}{4}v_0$. *Ans.* (a) $\frac{2}{3}v_0\tau$, (b) $v_0\tau$

3.48. Prove that for the projectile of Problem 3.14, page 71,

(a) the time to reach the highest point is $\dfrac{m}{\beta} \ln\left(1 + \dfrac{\beta v_0 \sin\alpha}{mg}\right)$,

(b) the maximum height is $\dfrac{mv_0 \sin\alpha}{\beta} - \dfrac{m^2 g}{\beta^2} \ln\left(1 + \dfrac{\beta v_0 \sin\alpha}{mg}\right)$.

CONSTRAINED MOTION AND FRICTION

3.49. A weight of 100 lb slides from rest down a 60° incline of length 200 ft starting from the top. Neglecting friction, (a) how long will it take to reach the bottom of the incline and (b) what is the speed with which it reaches the bottom? *Ans.* (a) 3.80 sec, (b) 105.3 ft/sec

3.50. Work Problem 3.49 if the coefficient of friction is 0.3. *Ans.* (a) 4.18 sec, (b) 95.7 ft/sec

3.51. (a) With what speed should an object be thrown up a smooth incline of angle α and length l, starting from the bottom, so as to just reach the top and (b) what is the time taken?
Ans. (a) $\sqrt{2gl \sin \alpha}$, (b) $\sqrt{2l/(g \sin \alpha)}$

3.52. If it takes a time τ for an object starting from speed v_0 on an icy surface to come to rest, prove that the coefficient of friction is $v_0/g\tau$.

3.53. What force is needed to move a 10 ton truck with uniform speed up an incline of 30° if the coefficient of friction is 0.1? *Ans.* 5.87 tons

3.54. A mass m rests on a horizontal piece of wood. The wood is tilted upward until the mass m just begins to slide. If the angle which the wood makes with the horizontal at that instant is α, prove that the coefficient of friction is $\mu = \tan \alpha$.

3.55. A 400 kg mass on a 30° inclined plane is acted upon by a force of 4800 newtons at angle 30° with the incline, as shown in Fig. 3-20. Find the acceleration of the mass if the incline (a) is frictionless, (b) has coefficient of friction 0.2. *Ans.* (a) 5.5 m/sec^2, (b) 5.0 m/sec^2

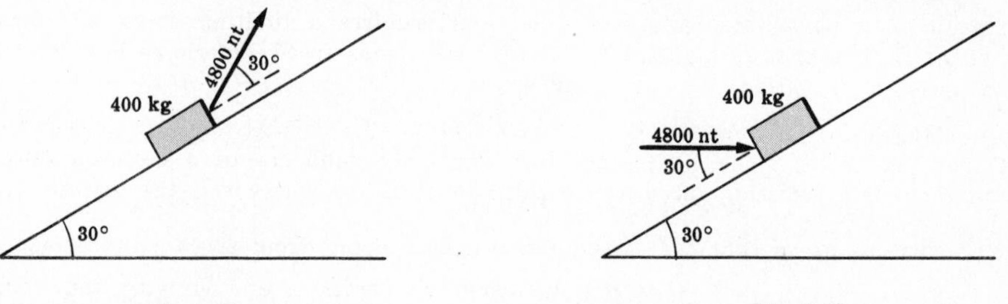

Fig. 3-20 Fig. 3-21

3.56. Work Problem 3.55 if the force of 4800 newtons acts as shown in Fig. 3-21.
Ans. (a) 5.5 m/sec^2, (b) 2.6 m/sec^2

STATICS IN A UNIFORM GRAVITATIONAL FIELD

3.57. A 100 kg weight is suspended vertically from the center of a rope as shown in Fig. 3-22. Determine the tension T in the rope. *Ans.* $T = 100$ kg wt $= 980$ nt

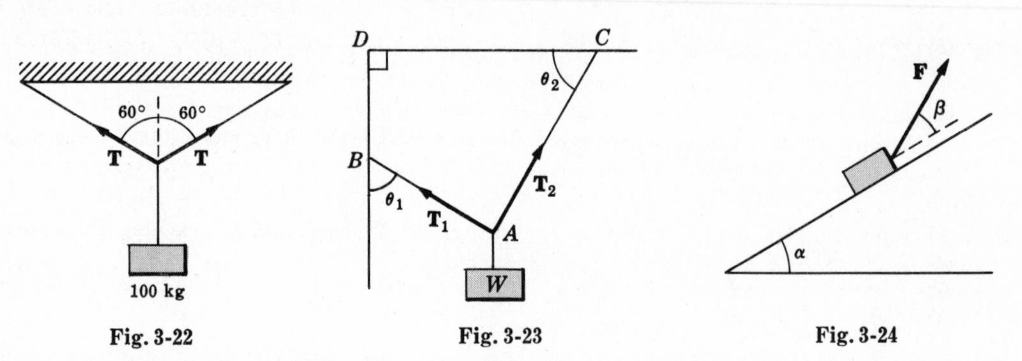

Fig. 3-22 Fig. 3-23 Fig. 3-24

3.58. In Fig. 3-23, AB and AC are ropes attached to the ceiling CD and wall BD at C and B respectively. A weight W is suspended from A. If the ropes AB and AC make angles θ_1 and θ_2 with the wall and ceiling respectively, find the tensions T_1 and T_2 in the ropes.

Ans. $T_1 = \dfrac{W \cos \theta_2}{\cos(\theta_1 - \theta_2)}$, $T_2 = \dfrac{W \sin \theta_1}{\cos(\theta_1 - \theta_2)}$

3.59. Find the magnitude of the force **F** needed to keep mass m in equilibrium on the inclined plane of Fig. 3-24 if (a) the plane is smooth, (b) the plane has coefficient of friction μ.

Ans. (a) $F = \dfrac{mg \sin \alpha}{\cos \beta}$, (b) $F = \dfrac{mg(\sin \alpha - \mu \cos \alpha)}{\cos \beta}$

3.60. How much force is needed to pull a train weighing 320 tons from rest to a speed of 15 mi/hr in 20 seconds if the coefficient of friction is 0.02 and (*a*) the track is horizontal, (*b*) the track is inclined at an angle of 10° with the horizontal and the train is going upward? [Use $\sin 10° = .1737$, $\cos 10° = .9848$.] *Ans.* (*a*) 17.4 tons, (*b*) 129.6 tons

3.61. Work Problem 3.60(*b*) if the train is going down the incline. *Ans.* 3.6 tons

3.62. A train of mass m is coasting down an inclined plane of angle α and coefficient of friction μ with constant speed v_0. Prove that the force needed to stop the train in a time τ is given by $mg(\sin\alpha - \mu\cos\alpha) + mv_0/\tau$.

MISCELLANEOUS PROBLEMS

3.63. A stone is dropped down a well and the sound of the splash is heard after time τ. Assuming the speed of sound is c, prove that the depth of the water level in the well is $(\sqrt{c^2 + 2gc\tau} - c)^2/2g$.

3.64. A projectile is launched downward from the top of an inclined plane of angle α in a direction making an angle γ with the incline. Assuming that the projectile hits the incline, prove that (*a*) the range is given by $R = \dfrac{2v_0^2 \sin\gamma \cos(\gamma - \alpha)}{g\cos^2\alpha}$ and that (*b*) the maximum range down the incline is $R_{max} = \dfrac{v_0^2}{g(1 - \sin\alpha)}$.

3.65. A cannon is located on a hill which has the shape of an inclined plane of angle α with the horizontal. A projectile is fired from this cannon in a direction up the hill and making an angle β with it. Prove that in order for the projectile to hit the hill horizontally we must have $\beta = \tan^{-1}\left(\dfrac{2\sin 2\alpha}{3 - \cos 2\alpha}\right)$.

3.66. Suppose that two projectiles are launched at angles α and β with the horizontal from the same place at the same time in the same vertical plane and with the same initial speed. Prove that during the course of the motion, the line joining the projectiles makes a constant angle with the vertical given by $\frac{1}{2}(\alpha + \beta)$.

3.67. Is it possible to solve equation (*1*), page 33, by the method of separation of variables? Explain.

3.68. When launched at angle θ_1 with the horizontal a projectile falls a distance D_1 short of its target, while at angle θ_2 it falls a distance D_2 beyond the target. Find the angle at which the projectile should be launched so as to hit the target.

3.69. An object was thrown vertically downward. During the tenth second of travel it fell twice as far as during the fifth second. With what speed was it thrown? *Ans.* 16 ft/sec

3.70. A gun of muzzle speed v_0 is situated at height h above a horizontal plane. Prove that the angle at which it must be fired so as to achieve the greatest range on the plane is given by $\theta = \frac{1}{2}\cos^{-1} gh/(v_0^2 + gh)$.

3.71. In Fig. 3-25, AB is a smooth table and masses m_1 and m_2 are connected by a string over the smooth peg at B. Find (*a*) the acceleration of mass m_2 and (*b*) the tension in the string.

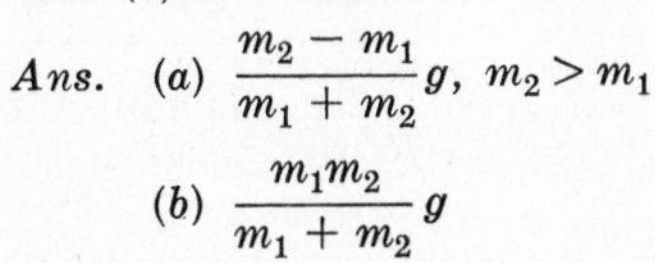

Ans. (*a*) $\dfrac{m_2 - m_1}{m_1 + m_2}g$, $m_2 > m_1$

(*b*) $\dfrac{m_1 m_2}{m_1 + m_2}g$

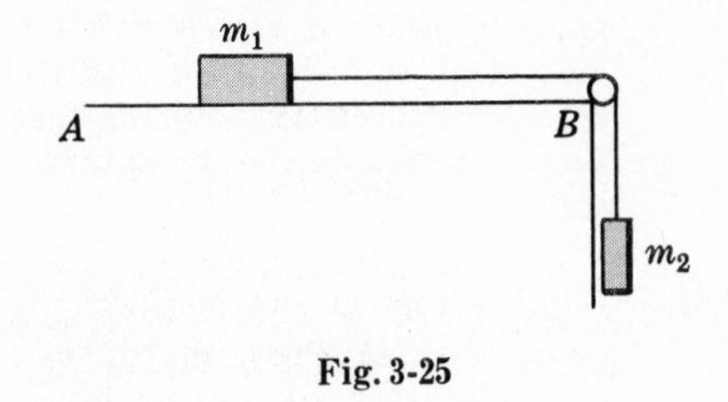

Fig. 3-25

3.72. Work Problem 3.71 if the table AB has coefficient of friction μ.

3.73. The maximum range of a projectile when fired down an inclined plane is twice the maximum range when fired up the inclined plane. Find the angle which the incline makes with the horizontal. *Ans.* $\sin^{-1} 1/3$

3.74. Masses m_1 and m_2 are located on smooth inclined planes of angles α_1 and α_2 respectively and are connected by an inextensible string of negligible mass which passes over a smooth peg at A [Fig. 3-26]. Find the accelerations of the masses.

Ans. The accelerations are in magnitude equal to

$$\frac{m_1 \sin \alpha_1 - m_2 \sin \alpha_2}{m_1 + m_2}\, g$$

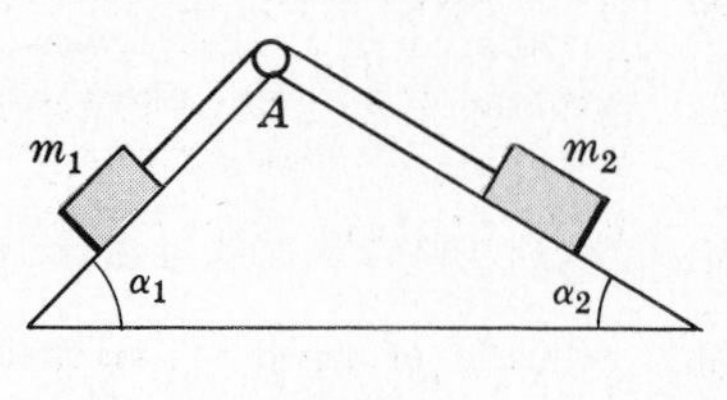

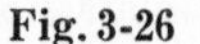
Fig. 3-26

3.75. Work Problem 3.74 if the coefficient of friction between the masses and the incline is μ.

Ans. $$\frac{m_1 \sin \alpha_1 - m_2 \sin \alpha_2 - \mu m_1 \cos \alpha_1 - \mu m_2 \cos \alpha_2}{m_1 + m_2}\, g$$

3.76. Prove that the least horizontal force $\mathbf{F}$ needed to pull a cylinder of radius a and weight W over an obstacle of height b [see Fig. 3-27] is given in magnitude by $W\sqrt{b(2a-b)}/(a-b)$.

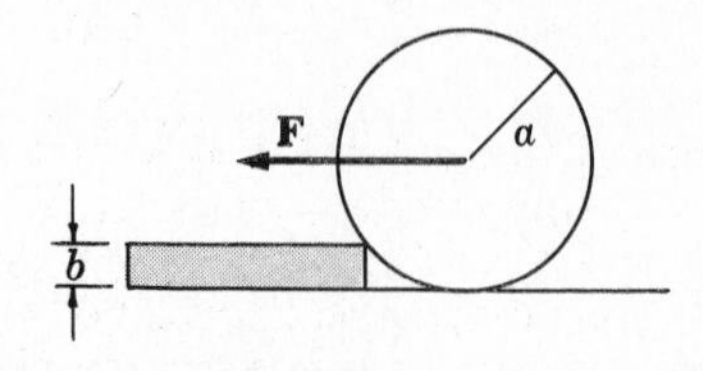

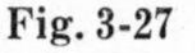
Fig. 3-27

3.77. Explain mathematically why a projectile fired from cannon A at the top of a cliff at height H above the ground can reach a cannon B located on the ground, while a projectile fired from cannon B with the same muzzle velocity will not be able to reach cannon A.

3.78. In Fig. 3-28 the mass m hangs from an inextensible string OA. It is pulled aside by a horizontal string AB so that OA makes an angle α with the vertical. Find the tension in each string.

Ans. Tension in $AB = mg \tan \alpha$; in $OA = mg \sec \alpha$

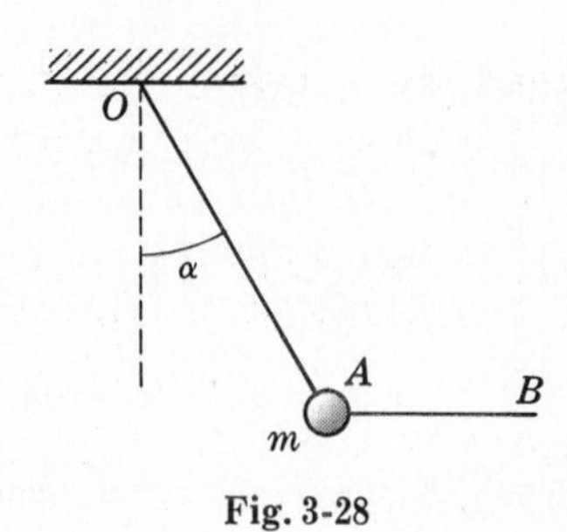

Fig. 3-28

3.79. A particle moving along the x axis is acted upon by a resisting force which is such that the time t for it to travel a distance x is given by $t = Ax^2 + Bx + C$ where A, B and C are constants. Prove that the magnitude of the resisting force is proportional to the cube of the instantaneous speed.

3.80. A projectile is to be launched so as to go from A to B [which are respectively at the bases of a double inclined plane having angles α and β as shown in Fig. 3-29] and just barely miss a pole of height H. If the distance between A and B is D, find the angle with the horizontal at which the projectile should be launched.

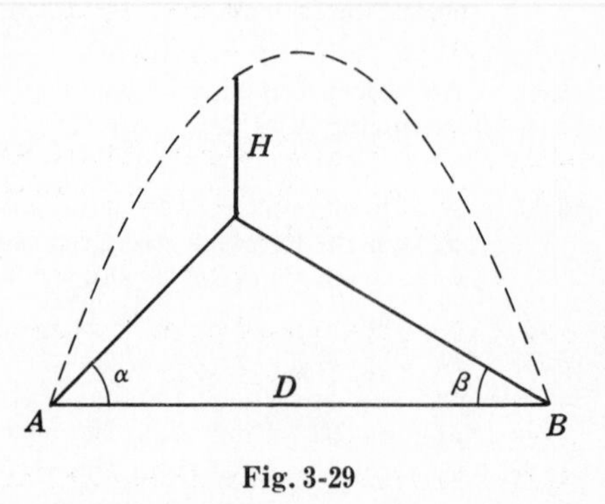

Fig. 3-29

3.81. A particle of mass m moves on a frictionless inclined plane of angle α and length l. If the particle starts from rest at the top of the incline, what will be its speed at the bottom assuming that air resistance is equal to κv where v is the instantaneous speed and κ is constant?

3.82. Suppose that in Problem 3.23 the particle P is given an initial speed v_0 at the top of the circle (or sphere). (*a*) Prove that if $v_0 \leqq \sqrt{gb}$, the angle θ at which the particle leaves the circle is given by $\sin^{-1}(\frac{2}{3} + v_0^2/3gb)$. (*b*) Discuss what happens if $v_0 > \sqrt{gb}$.

3.83. A cannon is situated at the top of a vertical cliff overlooking the sea at height H above sea level. What should be the least muzzle velocity of the cannon in order that a projectile fired from it will reach a ship at distance D from the foot of the cliff?

3.84. In Problem 3.83, (*a*) how long would it take the projectile to reach the ship and (*b*) what is the velocity on reaching the ship?

3.85. A uniform chain of total length a has a portion $0 < b < a$ hanging over the edge of a smooth table AB [see Fig. 3-30]. Prove that the time taken for the chain to slide off the table if it starts from rest is $\sqrt{a/g}\ \ln\,(a + \sqrt{a^2 - b^2})/b$.

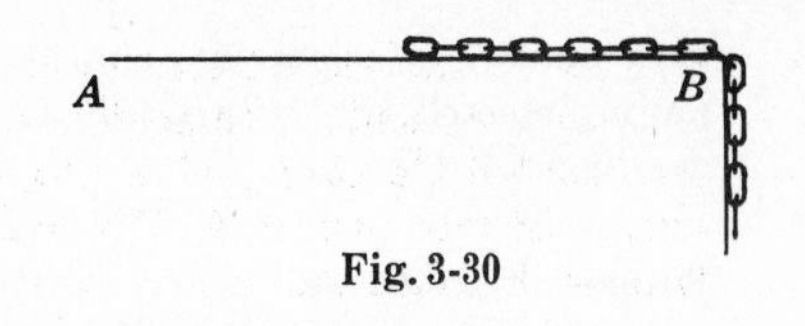

Fig. 3-30

3.86. If the table in Problem 3.85 has coefficient of friction μ, prove that the time taken is

$$\sqrt{\frac{a}{(1+\mu)g}}\ \ln\left\{\frac{a + \sqrt{a^2 - [b(1+\mu) - a\mu]^2}}{b(1+\mu) - a\mu}\right\}$$

3.87. A weight W_1 hangs on one side of a smooth fixed pulley of negligible mass [see Fig. 3-31]. A man of weight W_2 pulls himself up so that his acceleration relative to the fixed pulley is a. Prove that the weight W_1 moves upward with acceleration given by $[g(W_2 - W_1) - W_2 a]/W_1$.

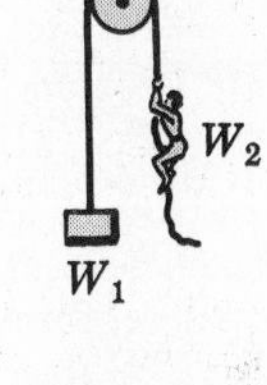

Fig. 3-31

3.88. Two monkeys of equal weight are hanging from opposite ends of a rope which passes over a smooth fixed pulley of negligible mass. The first monkey starts to climb the rope at a speed of 1 ft/sec while the other remains at rest relative to the rope. Describe the motion of the second monkey.

Ans. The second monkey moves up at the rate of 1 ft/sec.

3.89. Prove that the particle of Problem 3.23 will land at a distance from the base of the sphere given by $(4\sqrt{290} + 19\sqrt{5})b/81$.

3.90. Prove that if friction is negligible the time taken for a particle to slide down any chord of a vertical circle starting from rest at the top of the circle is the same regardless of the chord.

3.91. Given line AB of Fig. 3-32 and point P where AB and P are in the same vertical plane. Find a point Q on AB such that a particle starting from point P will reach Q in the shortest possible time.

[*Hint.* Use Problem 3.90.]

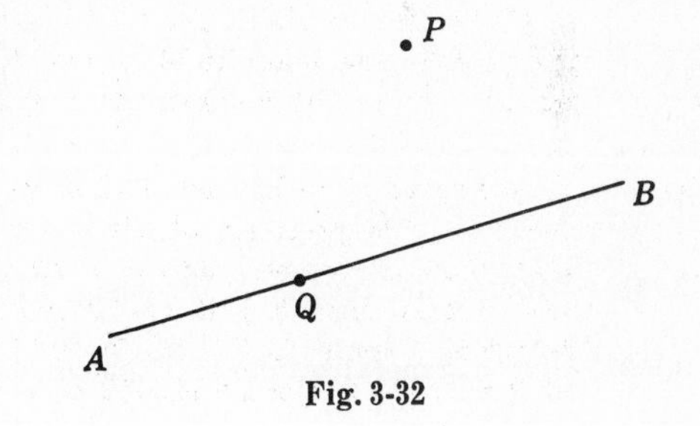

Fig. 3-32

3.92. Show how to work Problem 3.91 if line AB is replaced by a plane curve. Can it be done for a space curve? Explain.

3.93. Find the work done in moving the mass from the top of the incline of Problem 3.18 to the bottom.

Ans. $mgl(\sin\alpha - \mu\cos\alpha)$

3.94. The force on a particle having electrical charge q and which is moving in a magnetic field of intensity or strength $\mathbf{B}$ is given by $\mathbf{F} = q(\mathbf{v}\times\mathbf{B})$ where $\mathbf{v}$ is the instantaneous velocity. Prove that if the particle is given an initial speed v_0 in a plane perpendicular to a magnetic field $\mathbf{B}$ of constant strength, then it (*a*) will travel with constant speed v_0 and (*b*) will travel in a circular path of radius mv_0/qB. Assume that gravitational forces are negligible.

3.95. Prove that the *period*, i.e. the time for one complete vibration, of the particle of Problem 3.94 is independent of the speed of the particle and find its value. *Ans.* $2\pi m/qB$

3.96. Work Problem 3.94 if $\mathbf{B}$ is constant and the particle is given an initial speed v_0 in a plane which is not necessarily perpendicular to the magnetic field. Can we define a period in this case? Explain.

3.97. If a particle of electrical charge q and mass m moves with velocity $\mathbf{v}$ in an electromagnetic field having electric intensity $\mathbf{E}$ and magnetic intensity $\mathbf{B}$ the force acting on it, called the *Lorentz force*, is given by

$$\mathbf{F} = q(\mathbf{E} + \mathbf{v} \times \mathbf{B})$$

Suppose that $\mathbf{B}$ and $\mathbf{E}$ are constant and in the directions of the negative y and positive z axes respectively. Prove that if the particle starts from rest at the origin, then it will describe a *cycloid* in the yz plane whose equation is

$$y = b(\theta - \sin\theta), \qquad z = b(1 - \cos\theta)$$

where $\theta = qBt/m$, $b = mE/qB^2$ and t is the time.

3.98. (*a*) An astronaut of 80 kg wt on the earth takes off vertically in a space ship which achieves a speed of 2000 km/hr in 2 minutes. Assuming the acceleration to be constant, what is his apparent weight during this time? (*b*) Work part (*a*) if the astronaut has 180 lb wt on the earth and the space ship achieves a speed of 1280 mi/hr in 2 minutes. *Ans.* (*a*) 117 kg wt, (*b*) 268 lb wt

3.99. In Problem 3.82, how far from the base of the sphere will the particle land?

3.100. In Fig. 3-33 weight W_1 is on top of weight W_2 which is in turn on a horizontal plane. The coefficient of friction between W_1 and W_2 is μ_1 while that between W_2 and the plane is μ_2. Suppose that a force F inclined at angle α to the horizontal is applied to weight W_1. Prove that if $\cot\alpha \geqq \mu_1 > \mu_2$, then a necessary and sufficient condition that W_2 move relative to the plane while W_1 not move relative to W_2 is that

$$\frac{\mu_2(W_1 + W_2)}{\cos\alpha - \mu_2 \sin\alpha} < F \leqq \frac{\mu_1 W_1}{\cos\alpha - \mu_1 \sin\alpha}$$

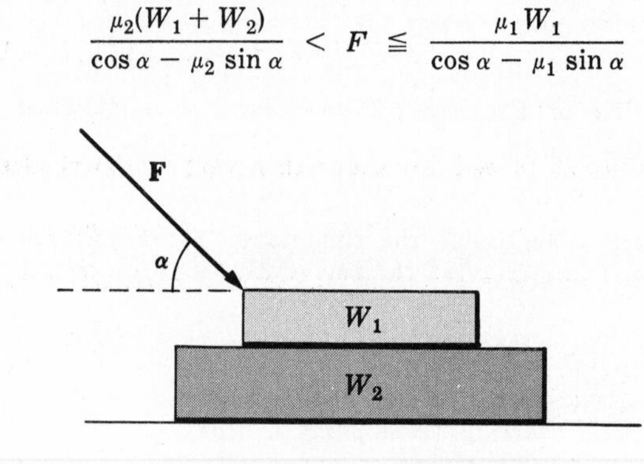

Fig. 3-33

3.101. Discuss the results in Problem 3.100 if any of the conditions are not satisfied.

3.102. Give a generalization of Problem 3.100.

3.103. Describe the motion of the particle of Problem 3.97 if $\mathbf{E}$ and $\mathbf{B}$ are constants, and have the same direction.

3.104. A bead of mass m is located on a parabolic wire with its axis vertical and vertex directed downward as in Fig. 3-34 and whose equation is $cz = x^2$. If the coefficient of friction is μ, find the highest distance above the x axis at which the particle will be in equilibrium. *Ans.* $\frac{1}{4}\mu^2 c$

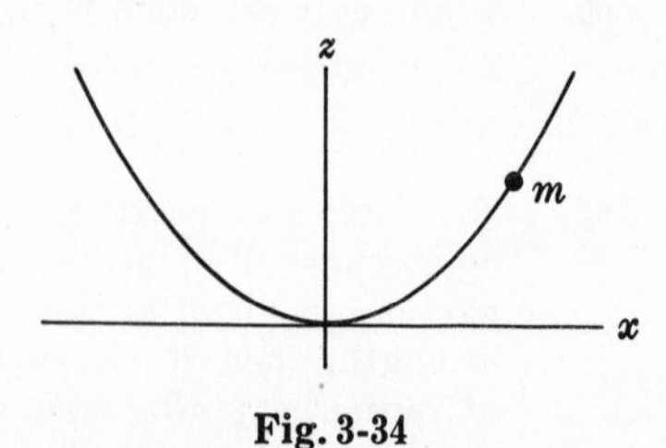

Fig. 3-34

3.105. Work Problem 3.104 if the parabola is replaced by a vertical circle of radius b which is tangent to the x axis.

3.106. A weight W is suspended from 3 equal strings of length l which are attached to the 3 vertices of a horizontal equilateral triangle of side s. Find the tensions in the strings.
Ans. $Wl/\sqrt{9l^2 - 3s^2}$

3.107. Work Problem 3.106 if there are n equal strings attached to the n vertices of a regular polygon having n sides.

3.108. A rope passes over a fixed pulley A of Fig. 3-35. At one end of this rope a mass M_1 is attached. At the other end of the rope there is a pulley of mass M_2 over which passes another rope with masses m_1 and m_2 attached. Prove that the acceleration of the mass m_1 is given by

$$\frac{3m_2M_2 - m_1M_1 - m_1M_2 - m_2M_1 - 4m_1m_2}{(m_1+m_2)(M_1+M_2) + 4m_1m_2}\,g$$

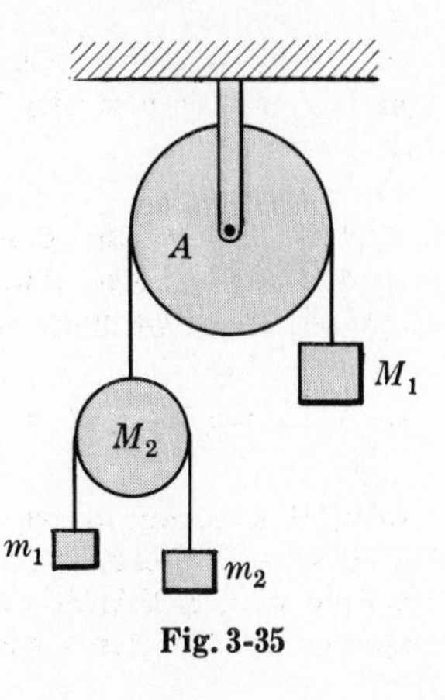

Fig. 3-35

3.109. An automobile of weight W with an engine having constant instantaneous power $\mathcal{P}$, travels up an incline of angle α. Assuming that resistance forces are r per unit weight, prove that the maximum speed which can be maintained up the incline is $\dfrac{\mathcal{P}}{W(r + \sin\alpha)}$.

3.110. An automobile of weight W moves up an incline of angle α, powered by an engine having constant instantaneous power $\mathcal{P}$. Assuming that the resistance to motion is equal to κv per unit weight where v is the instantaneous speed and κ is a constant, prove that the maximum speed which is possible on the incline is $(\sqrt{W^2\sin^2\alpha + 4\kappa W\mathcal{P}} - W\sin\alpha)/2\kappa W$.

3.111. A chain hangs over a smooth peg with length a on one side and length b, where $0 < b < a$, on the other side. Prove that the time taken for the chain to slide off is given by $\sqrt{\dfrac{a+b}{2g}}\ln\left(\dfrac{\sqrt{a}+\sqrt{b}}{\sqrt{a}-\sqrt{b}}\right)$.

3.112. Prove that a bead P which is placed anywhere on a vertical frictionless wire [see Fig. 3-36] in the form of a cycloid

$$x = b(\theta + \sin\theta), \qquad y = b(1 - \cos\theta)$$

will reach the bottom in the same time regardless of the starting point and find this time.
Ans. $\pi\sqrt{b/g}$

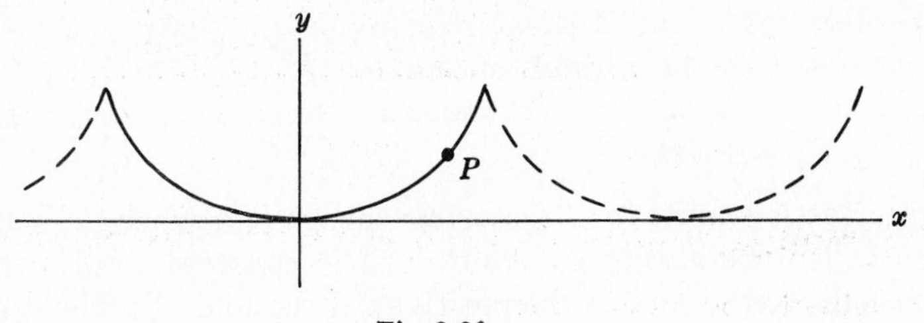

Fig. 3-36

Chapter 4

The SIMPLE HARMONIC OSCILLATOR and the SIMPLE PENDULUM

THE SIMPLE HARMONIC OSCILLATOR

In Fig. 4-1(*a*) the mass m lies on a frictionless horizontal table indicated by the x axis. It is attached to one end of a spring of negligible mass and unstretched length l whose other end is fixed at E.

If m is given a displacement along the x axis [see Fig. 4-1(*b*)] and released, it will vibrate or oscillate back and forth about the *equilibrium position* O.

To determine the equation of motion, note that at any instant when the spring has length $l + x$ [Fig. 4-1(*b*)] there is a force tending to restore m to its equilibrium position. According to *Hooke's law* this force, called the *restoring force*, is proportional to the stretch x and is given by

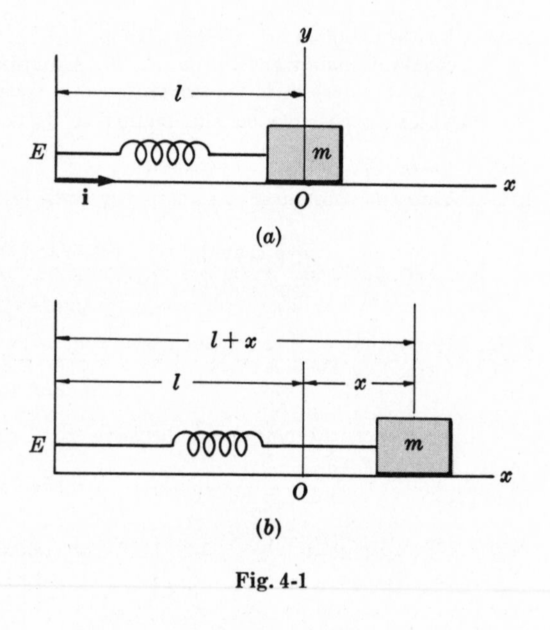

Fig. 4-1

$$\mathbf{F}_R = -\kappa x \mathbf{i} \qquad (1)$$

where the subscript R stands for "restoring force" and where κ is the constant of proportionality often called the *spring constant, elastic constant, stiffness factor* or *modulus of elasticity* and $\mathbf{i}$ is the unit vector in the positive x direction. By Newton's second law we have

$$m\frac{d^2(x\mathbf{i})}{dt^2} = -\kappa x\mathbf{i} \quad \text{or} \quad m\ddot{x} + \kappa x = 0 \qquad (2)$$

This vibrating system is called a *simple harmonic oscillator* or *linear harmonic oscillator*. This type of motion is often called *simple harmonic motion*.

AMPLITUDE, PERIOD AND FREQUENCY OF SIMPLE HARMONIC MOTION

If we solve the differential equation (*2*) subject to the initial conditions $x = A$ and $dx/dt = 0$ at $t = 0$, we find that

$$x = A\cos\omega t \quad \text{where} \quad \omega = \sqrt{\kappa/m} \qquad (3)$$

For the case where $A = 20$, $m = 2$ and $\kappa = 8$, see Problem 4.1.

Since $\cos\omega t$ varies between -1 and $+1$, the mass oscillates between $x = -A$ and $x = A$. A graph of x vs. t appears in Fig. 4-2.

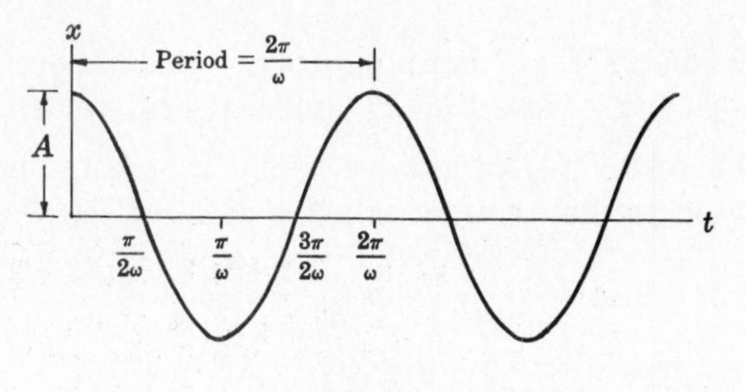

Fig. 4-2

The *amplitude* of the motion is the distance A and is the greatest distance from the equilibrium position.

The *period* of the motion is the time for one complete oscillation or vibration [sometimes called a *cycle*] such as, for example, from $x = A$ to $x = -A$ and then back to $x = A$ again. If P denotes the period, then

$$P = 2\pi/\omega = 2\pi\sqrt{m/\kappa} \tag{4}$$

The *frequency* of the motion, denoted by f, is the number of complete oscillations or cycles per unit time. We have

$$f = \frac{1}{P} = \frac{\omega}{2\pi} = \frac{1}{2\pi}\sqrt{\frac{\kappa}{m}} \tag{5}$$

In the general case, the solution of (*2*) is

$$x = A\cos\omega t + B\sin\omega t \qquad \text{where} \quad \omega = \sqrt{\kappa/m} \tag{6}$$

where A and B are determined from initial conditions. As seen in Problem 4.2, we can write (*6*) in the form

$$x = C\cos(\omega t - \phi) \qquad \text{where} \quad \omega = \sqrt{\kappa/m} \tag{7}$$

and where

$$C = \sqrt{A^2 + B^2} \quad \text{and} \quad \phi = \tan^{-1}(B/A) \tag{8}$$

The amplitude in this case is C while the period and frequency remain the same as in (*4*) and (*5*), i.e. they are unaffected by change of initial conditions. The angle ϕ is called the *phase angle* or *epoch* chosen so that $0 \leqq \phi \leqq \pi$. If $\phi = 0$, (*7*) reduces to (*3*).

ENERGY OF A SIMPLE HARMONIC OSCILLATOR

If T is the kinetic energy, V the potential energy and $E = T + V$ the total energy of a simple harmonic oscillator, then we have

$$T = \tfrac{1}{2}mv^2, \quad V = \tfrac{1}{2}\kappa x^2 \tag{9}$$

and

$$E = \tfrac{1}{2}mv^2 + \tfrac{1}{2}\kappa x^2 \tag{10}$$

See Problem 4.17.

THE DAMPED HARMONIC OSCILLATOR

In practice various forces may act on a harmonic oscillator, tending to reduce the magnitude of successive oscillations about the equilibrium position. Such forces are sometimes called *damping forces*. A useful approximate damping force is one which is proportional to the velocity and is given by

$$\mathbf{F}_D = -\beta\mathbf{v} = -\beta v\mathbf{i} = -\beta\frac{dx}{dt}\mathbf{i} \tag{11}$$

where the subscript D stands for "damping force" and where β is a positive constant called the *damping coefficient*. Note that $\mathbf{F}_D$ and $\mathbf{v}$ are in opposite directions.

If in addition to the restoring force we assume the damping force (*11*), the equation of motion of the harmonic oscillator, now called a *damped harmonic oscillator*, is given by

$$m\frac{d^2x}{dt^2} = -\kappa x - \beta\frac{dx}{dt} \quad \text{or} \quad m\frac{d^2x}{dt^2} + \beta\frac{dx}{dt} + \kappa x = 0 \tag{12}$$

on applying Newton's second law. Dividing by m and calling

$$\beta/m = 2\gamma, \quad \kappa/m = \omega^2 \tag{13}$$

this equation can be written

$$\ddot{x} + 2\gamma\dot{x} + \omega^2 x = 0 \tag{14}$$

where the dots denote, as usual, differentiation with respect to t.

OVER-DAMPED, CRITICALLY DAMPED AND UNDER-DAMPED MOTION

Three cases arise in obtaining solutions to the differential equation (*14*).

***Case 1,* Over-damped motion,** $\gamma^2 > \omega^2$, i.e. $\beta^2 > 4\kappa m$

In this case (*14*) has the general solution

$$x = e^{-\gamma t}(Ae^{\alpha t} + Be^{-\alpha t}) \quad \text{where} \quad \alpha = \sqrt{\gamma^2 - \omega^2} \tag{15}$$

and where the arbitrary constants A and B can be found from the initial conditions.

***Case 2,* Critically damped motion,** $\gamma^2 = \omega^2$, i.e. $\beta^2 = 4\kappa m$

In this case (*14*) has the general solution

$$x = e^{-\gamma t}(A + Bt) \tag{16}$$

where A and B are found from initial conditions.

***Case 3,* Under-damped or damped oscillatory motion,** $\gamma^2 < \omega^2$, i.e. $\beta^2 < 4\kappa m$

In this case (*14*) has the general solution

$$\begin{aligned} x &= e^{-\gamma t}(A \sin \lambda t + B \cos \lambda t) \\ &= Ce^{-\gamma t} \cos (\lambda t - \phi) \quad \text{where} \quad \lambda = \sqrt{\omega^2 - \gamma^2} \end{aligned} \tag{17}$$

and where $C = \sqrt{A^2 + B^2}$, called the *amplitude* and ϕ, called the *phase angle* or *epoch*, are determined from the initial conditions.

In Cases 1 and 2 damping is so large that no oscillation takes place and the mass m simply returns gradually to the equilibrium position $x = 0$. This is indicated in Fig. 4-3 where we have assumed the initial conditions $x = x_0$, $dx/dt = 0$. Note that in the critically damped case, mass m returns to the equilibrium position faster than in the over-damped case.

In Case 3, damping has been reduced to such an extent that oscillations about the equilibrium position do take place, although the magnitude of these oscillations tend to decrease with time as indicated in Fig. 4-3. The difference in times

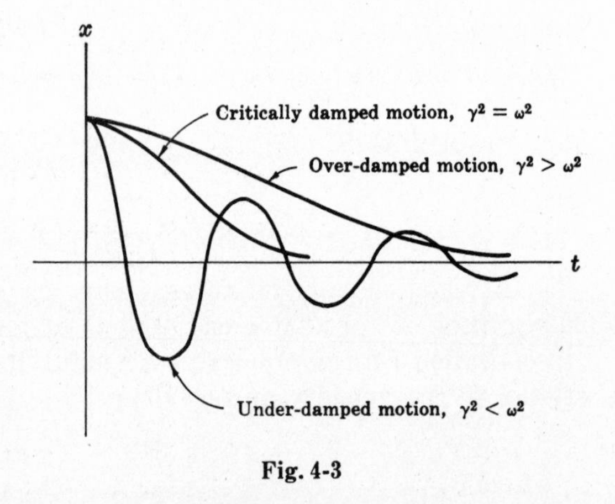

Fig. 4-3

between two successive maxima [or minima] in the under-damped [or damped oscillatory] motion of Fig. 4-3 is called the *period* of the motion and is given by

$$P = \frac{2\pi}{\lambda} = \frac{2\pi}{\sqrt{\omega^2 - \gamma^2}} = \frac{4\pi m}{\sqrt{4\kappa m - \beta^2}} \tag{18}$$

and the frequency, which is the reciprocal of the period, is given by

$$f = \frac{1}{P} = \frac{\lambda}{2\pi} = \frac{\sqrt{\omega^2 - \gamma^2}}{2\pi} = \frac{\sqrt{4\kappa m - \beta^2}}{4\pi m} \tag{19}$$

Note that if $\beta = 0$, *(18)* and *(19)* reduce to *(4)* and *(5)* respectively. The period and frequency corresponding to $\beta = 0$ are sometimes called the *natural period* and *natural frequency* respectively.

The period P given by *(18)* is also equal to two successive values of t for which $\cos(\lambda t - \phi) = 1$ [or $\cos(\lambda t - \phi) = -1$] as given in equation *(17)*. Suppose that the values of x corresponding to the two successive values t_n and $t_{n+1} = t_n + P$ are x_n and x_{n+1} respectively. Then

$$x_n/x_{n+1} = e^{-\gamma t_n}/e^{-\gamma(t_n+P)} = e^{\gamma P} \tag{20}$$

The quantity

$$\delta = \ln(x_n/x_{n+1}) = \gamma P \tag{21}$$

which is a constant, is called the *logarithmic decrement.*

FORCED VIBRATIONS

Suppose that in addition to the restoring force $-\kappa x\mathbf{i}$ and damping force $-\beta v\mathbf{i}$ we impress on the mass m a force $F(t)\mathbf{i}$ where

$$F(t) = F_0 \cos \alpha t \tag{22}$$

Then the differential equation of motion is

$$m\frac{d^2x}{dt^2} = -\kappa x - \beta\frac{dx}{dt} + F_0 \cos \alpha t \tag{23}$$

or

$$\ddot{x} + 2\gamma\dot{x} + \omega^2 x = f_0 \cos \alpha t \tag{24}$$

where

$$\gamma = \beta/2m, \quad \omega^2 = \kappa/m, \quad f_0 = F_0/m \tag{25}$$

The general solution of *(24)* is found by adding the general solution of

$$\ddot{x} + 2\gamma\dot{x} + \omega^2 x = 0 \tag{26}$$

[which has already been found and is given by *(15)*, *(16)* or *(17)*] to any particular solution of *(24)*. A particular solution of *(24)* is given by [see Problem 4.18]

$$x = \frac{f_0}{\sqrt{(\alpha^2 - \omega^2)^2 + 4\gamma^2\alpha^2}} \cos(\alpha t - \phi) \tag{27}$$

where

$$\tan \phi = \frac{2\gamma\alpha}{\alpha^2 - \omega^2} \qquad 0 \leqq \phi \leqq \pi \tag{28}$$

Now, as we have seen, the general solution of *(26)* approaches zero within a short time and we thus call this solution the *transient solution.* After this time has elapsed, the motion of the mass m is essentially given by *(27)* which is often called the *steady-state solution.* The vibrations or oscillations which take place, often called *forced vibrations* or *forced oscillations,* have a frequency which is equal to the frequency of the impressed force but lag behind by the phase angle ϕ.

RESONANCE

The amplitude of the steady-state oscillation (*27*) is given by

$$\mathcal{A} = \frac{f_0}{\sqrt{(\alpha^2 - \omega^2)^2 + 4\gamma^2\alpha^2}} \tag{29}$$

assuming $\gamma \neq 0$, i.e. $\beta \neq 0$, so that damping is assumed to be present. The maximum value of $\mathcal{A}$ in this case occurs where the frequency $\alpha/2\pi$ of the impressed force is such that

$$\alpha^2 = \alpha_R^2 = \omega^2 - 2\gamma^2 \tag{30}$$

assuming that $\gamma^2 < \frac{1}{2}\omega^2$ [see Problem 4.19]. Near this frequency very large oscillations may occur, sometimes causing damage to the system. The phenomenon is called *resonance* and the frequency $\alpha_R/2\pi$ is called the *frequency of resonance* or *resonant frequency.*

The value of the maximum amplitude at the resonant frequency is

$$\mathcal{A}_{\max} = \frac{f_0}{2\gamma\sqrt{\omega^2 - \gamma^2}} \tag{31}$$

The amplitude (*29*) can be written in terms of α_R as

$$\mathcal{A} = \frac{f_0}{\sqrt{(\alpha^2 - \alpha_R^2)^2 + 4\gamma^2(\omega^2 - \gamma^2)}} \tag{32}$$

A graph of $\mathcal{A}$ vs. α^2 is shown in Fig. 4-4. Note that the graph is symmetric around the resonant frequency and that the resonant frequency, frequency with damping and natural frequency (without damping) are all different. In case there is no damping, i.e. $\gamma = 0$ or $\beta = 0$, all of these frequencies are identical. In such case resonance occurs where the frequency of the impressed force equals the natural frequency of oscillation. The general solution for this case is

$$x = A\cos\omega t + B\sin\omega t + \frac{f_0 t}{2\omega}\sin\omega t \tag{33}$$

From the last term in (*33*) it is seen that the oscillations build up with time until finally the spring breaks. See Problem 4.20.

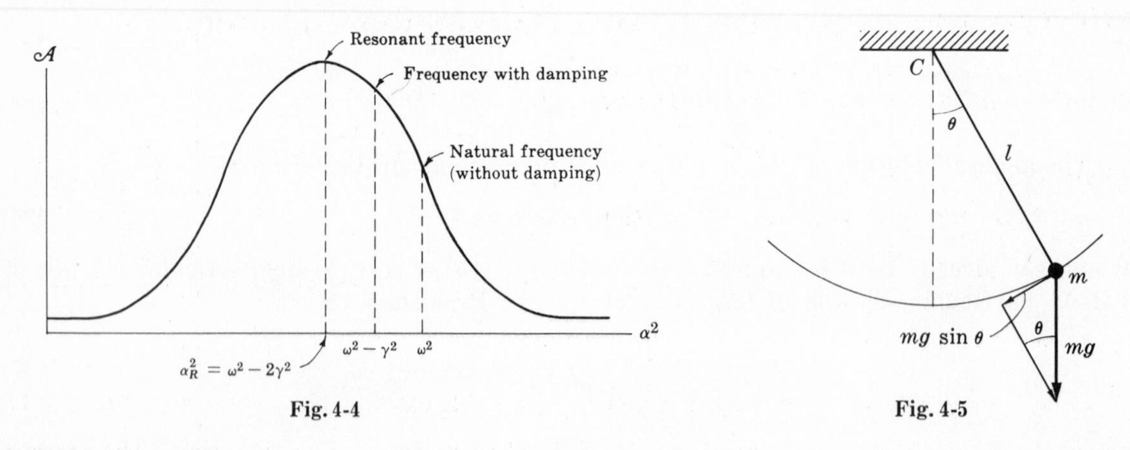

Fig. 4-4

Fig. 4-5

THE SIMPLE PENDULUM

A *simple pendulum* consists of a mass m [Fig. 4-5] at the end of a massless string or rod of length l [which always remains straight, i.e. rigid]. If the mass m, sometimes called the *pendulum bob,* is pulled aside and released, the resulting motion will be oscillatory.

Calling θ the instantaneous angle which the string makes with the vertical, the differential equation of motion is [see Problem 4.23]

$$\frac{d^2\theta}{dt^2} = -\frac{g}{l}\sin\theta \tag{34}$$

assuming no damping forces or other external forces are present.

For small angles [e.g. less than 5° with the vertical], $\sin\theta$ is very nearly equal to θ, where θ is in radians, and equation (34) becomes, to a high degree of approximation,

$$\frac{d^2\theta}{dt^2} = -\frac{g}{l}\theta \tag{35}$$

This equation has the general solution

$$\theta = A\cos\sqrt{g/l}\,t + B\sin\sqrt{g/l}\,t \tag{36}$$

where A and B are determined from initial conditions. For example, if $\theta = \theta_0$, $\dot{\theta} = 0$ at $t = 0$, then

$$\theta = \theta_0\cos\sqrt{g/l}\,t \tag{37}$$

In such case, the motion of the pendulum bob is that of simple harmonic motion. The period is given by

$$P = 2\pi\sqrt{l/g} \tag{38}$$

and the frequency is given by

$$f = \frac{1}{P} = \frac{1}{2\pi}\sqrt{g/l} \tag{39}$$

If the angles are not necessarily small, we can show [see Problems 4.29 and 4.30] that the period is equal to

$$\begin{aligned} P &= 4\sqrt{\frac{l}{g}}\int_0^{\pi/2}\frac{d\theta}{\sqrt{1-k^2\sin^2\theta}} \\ &= 2\pi\sqrt{\frac{l}{g}}\left\{1+\left(\frac{1}{2}\right)^2k^2+\left(\frac{1\cdot 3}{2\cdot 4}\right)^2k^4+\left(\frac{1\cdot 3\cdot 5}{2\cdot 4\cdot 6}\right)^2k^6+\cdots\right\} \end{aligned} \tag{40}$$

where $k = \sin(\theta_0/2)$. For small angles this reduces to (38).

For cases where damping and other external forces are considered, see Problems 4.25 and 4.114.

THE TWO AND THREE DIMENSIONAL HARMONIC OSCILLATOR

Suppose a particle of mass m moves in the xy plane under the influence of a force field $\mathbf{F}$ given by

$$\mathbf{F} = -\kappa_1 x\mathbf{i} - \kappa_2 y\mathbf{j} \tag{41}$$

where κ_1 and κ_2 are positive constants.

In this case the equations of motion of m are given by

$$m\frac{d^2x}{dt^2} = -\kappa_1 x, \qquad m\frac{d^2y}{dt^2} = -\kappa_2 y \tag{42}$$

and have solutions

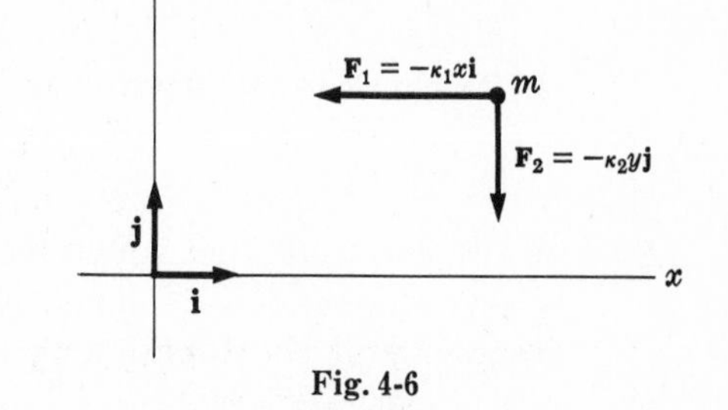

Fig. 4-6

$$x = A_1\cos\sqrt{\kappa_1/m}\,t + B_1\sin\sqrt{\kappa_1/m}\,t, \qquad y = A_2\cos\sqrt{\kappa_2/m}\,t + B_2\sin\sqrt{\kappa_2/m}\,t \tag{43}$$

where A_1, B_1, A_2, B_2 are constants to be determined from the initial conditions. The mass m subjected to the force field (41) is often called a *two-dimensional harmonic oscillator*. The various curves which m describes in its motion are often called *Lissajous curves* or *figures*.

These ideas are easily extended to a three dimensional harmonic oscillator of mass m which is subject to a force field given by

$$\mathbf{F} = -\kappa_1 x\mathbf{i} - \kappa_2 y\mathbf{j} - \kappa_3 z\mathbf{k} \tag{44}$$

where $\kappa_1, \kappa_2, \kappa_3$ are positive constants.

Solved Problems

SIMPLE HARMONIC MOTION AND THE SIMPLE HARMONIC OSCILLATOR

4.1. A particle P of mass 2 moves along the x axis attracted toward origin O by a force whose magnitude is numerically equal to $8x$ [see Fig. 4-7]. If it is initially at rest at $x = 20$, find (*a*) the differential equation and initial conditions describing the motion, (*b*) the position of the particle at any time, (*c*) the speed and velocity of the particle at any time, and (*d*) the amplitude, period and frequency of the vibration.

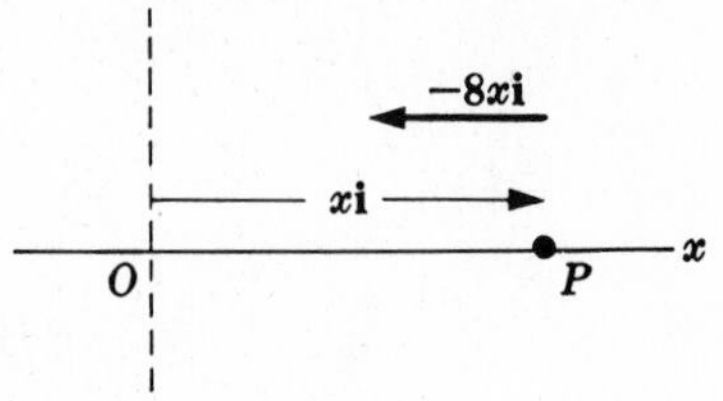

(*a*) Let $\mathbf{r} = x\mathbf{i}$ be the position vector of P. The acceleration of P is $\dfrac{d^2}{dt^2}(x\mathbf{i}) = \dfrac{d^2x}{dt^2}\mathbf{i}$. The net force acting on P is $-8x\mathbf{i}$. Then by Newton's second law,

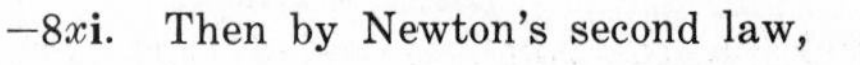

Fig. 4-7

$$2\frac{d^2x}{dt^2}\mathbf{i} = -8x\mathbf{i} \quad \text{or} \quad \frac{d^2x}{dt^2} + 4x = 0 \tag{1}$$

which is the required differential equation of motion. The initial conditions are

$$x = 20, \quad dx/dt = 0 \quad \text{at} \quad t = 0 \tag{2}$$

(*b*) The general solution of (*1*) is

$$x = A\cos 2t + B\sin 2t \tag{3}$$

When $t = 0$, $x = 20$ so that $A = 20$. Thus

$$x = 20\cos 2t + B\sin 2t \tag{4}$$

Then

$$dx/dt = -40\sin 2t + 2B\cos 2t \tag{5}$$

so that on putting $t = 0$, $dx/dt = 0$ we find $B = 0$. Thus (*3*) becomes

$$x = 20\cos 2t \tag{6}$$

which gives the position at any time.

(*c*) From (*6*) $dx/dt = -40\sin 2t$ which gives the speed at any time. The velocity is given by

$$\frac{dx}{dt}\mathbf{i} = -40\sin 2t\,\mathbf{i}$$

(*d*) Amplitude = 20. Period = $2\pi/2 = \pi$. Frequency = 1/period = $1/\pi$.

4.2. (*a*) Show that the function $A\cos\omega t + B\sin\omega t$ can be written as $C\cos(\omega t - \phi)$ where $C = \sqrt{A^2+B^2}$ and $\phi = \tan^{-1}(B/A)$. (*b*) Find the amplitude, period and frequency of the function in (*a*).

(*a*)
$$\begin{aligned} A\cos\omega t + B\sin\omega t &= \sqrt{A^2+B^2}\left(\frac{A}{\sqrt{A^2+B^2}}\cos\omega t + \frac{B}{\sqrt{A^2+B^2}}\sin\omega t\right) \\ &= \sqrt{A^2+B^2}\,(\cos\phi\cos\omega t + \sin\phi\sin\omega t) \\ &= \sqrt{A^2+B^2}\cos(\omega t - \phi) = C\cos(\omega t - \phi) \end{aligned}$$

where $\cos\phi = A/\sqrt{A^2+B^2}$ and $\sin\phi = B/\sqrt{A^2+B^2}$, i.e. $\tan\phi = B/A$ or $\phi = \tan^{-1}B/A$, and $C = \sqrt{A^2+B^2}$. We generally choose that value of ϕ which lies between 0° and 180°, i.e. $0 \leqq \phi \leqq \pi$.

(*b*) Amplitude = maximum value = $C = \sqrt{A^2+B^2}$. Period = $2\pi/\omega$. Frequency = $\omega/2\pi$.

4.3. Work Problem 4.1 if P is initially at $x = 20$ but is moving (*a*) to the right with speed 30, (*b*) to the left with speed 30. Find the amplitude, period and frequency in each case.

(*a*) The only difference here is that the condition $dx/dt = 0$ at $t = 0$ of Problem 4.1 is replaced by $dx/dt = 30$ at $t = 0$. Then from (*5*) of Problem 4.1 we find $B = 15$, and (*3*) of Problem 4.1 becomes

$$x = 20\cos 2t + 15\sin 2t \tag{1}$$

which gives the position of P at any time. This may be written [see Problem 4.2] as

$$\begin{aligned} x &= \sqrt{(20)^2 + (15)^2}\left\{\frac{20}{\sqrt{(20)^2+(15)^2}}\cos 2t + \frac{15}{\sqrt{(20)^2+(15)^2}}\sin 2t\right\} \\ &= 25\{\tfrac{4}{5}\cos 2t + \tfrac{3}{5}\sin 2t\} = 25\cos(2t - \phi) \end{aligned}$$

where
$$\cos\phi = \tfrac{4}{5}, \quad \sin\phi = \tfrac{3}{5} \tag{2}$$

The angle ϕ which can be found from (*2*) is often called the *phase angle* or *epoch*.

Since the cosine varies between -1 and $+1$, the amplitude $= 25$. The period and frequency are the same as before, i.e. period $= 2\pi/2 = \pi$ and frequency $= 2/2\pi = 1/\pi$.

(*b*) In this case the condition $dx/dt = 0$ at $t = 0$ of Problem 4.1 is replaced by $dx/dt = -30$ at $t = 0$. Then $B = -15$ and the position is given by

$$x = 20\cos 2t - 15\sin 2t$$

which as in part (*a*) can be written

$$\begin{aligned} x &= 25\{\tfrac{4}{5}\cos 2t - \tfrac{3}{5}\sin 2t\} \\ &= 25\{\cos\psi\cos 2t + \sin\psi\sin 2t\} = 25\cos(2t - \psi) \end{aligned}$$

where $\cos\psi = \frac{4}{5}$, $\sin\psi = -\frac{3}{5}$.

The amplitude, period and frequency are the same as in part (*a*). The only difference is in the phase angle. The relationship between ψ and ϕ is $\psi = \phi + \pi$. We often describe this by saying that the two motions are 180° *out of phase* with each other.

4.4. A spring of negligible mass, suspended vertically from one end, is stretched a distance of 20 cm when a 5 gm mass is attached to the other end. The spring and mass are placed on a horizontal frictionless table as in Fig. 4-1(*a*), page 86, with the suspension point fixed at E. The mass is pulled away a distance 20 cm beyond the equilibrium position O and released. Find (*a*) the differential equation and initial conditions describing the motion, (*b*) the position at any time t, and (*c*) the amplitude, period and frequency of the vibrations.

(*a*) The gravitational force on a 5 gm mass [i.e. the weight of a 5 gm mass] is $5g = 5(980)$ dynes $=$ 4900 dynes. Then since 4900 dynes stretches the spring 20 cm, the spring constant is $\kappa = 4900/20 = 245$ dynes/cm. Thus when the spring is stretched a distance x cm beyond the equilibrium position, the restoring force is $-245x\mathbf{i}$. Then by Newton's second law we have, if $\mathbf{r} = x\mathbf{i}$ is the position vector of the mass,

$$5\frac{d^2(x\mathbf{i})}{dt^2} = -245x\mathbf{i} \quad \text{or} \quad \frac{d^2x}{dt^2} + 49x = 0 \tag{1}$$

The initial conditions are
$$x = 20, \; dx/dt = 0 \quad \text{at } t = 0 \tag{2}$$

(*b*) The general solution of (*1*) is
$$x = A\cos 7t + B\sin 7t \tag{3}$$

Using the conditions (*2*) we find $A = 20$, $B = 0$ so that $x = 20\cos 7t$.

(*c*) From $x = 20\cos 7t$ we see that: amplitude $= 20$ cm; period $= 2\pi/7$ sec; frequency $= 7/2\pi$ vib/sec or $7/2\pi$ cycles/sec.

4.5. A particle of mass m moves along the x axis, attracted toward a fixed point O on it by a force proportional to the distance from O. Initially the particle is at distance x_0 from O and is given a velocity v_0 away from O. Determine (*a*) the position at any time, (*b*) the velocity at any time, and (*c*) the amplitude, period, frequency, and maximum speed.

(*a*) The force of attraction toward O is $-\kappa x\mathbf{i}$ where κ is a positive constant of proportionality. Then by Newton's second law,

$$m\frac{d^2x}{dt^2}\mathbf{i} = -\kappa x\mathbf{i} \quad \text{or} \quad \ddot{x} + \frac{\kappa x}{m} = 0 \tag{1}$$

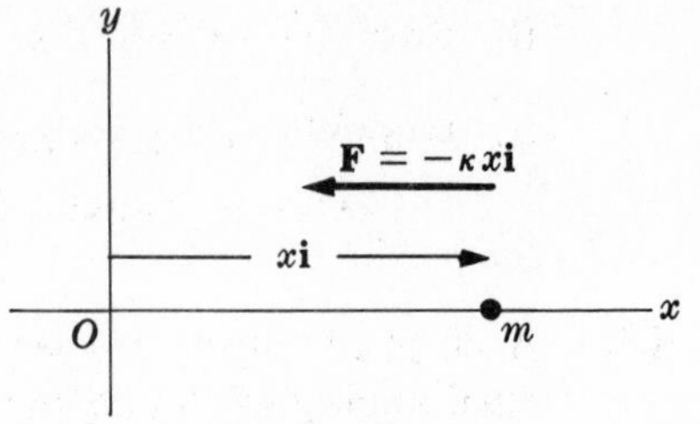

Solving (*1*), we find

$$x = A\cos\sqrt{\kappa/m}\,t + B\sin\sqrt{\kappa/m}\,t \tag{2}$$

y

F = −κxi

xi

O m x

Fig. 4-8

We also have the initial conditions

$$x = x_0,\ dx/dt = v_0 \quad \text{at } t = 0 \tag{3}$$

From $x = x_0$ at $t = 0$ we find, using (*2*), that $A = x_0$. Thus

$$x = x_0\cos\sqrt{\kappa/m}\,t + B\sin\sqrt{\kappa/m}\,t \tag{4}$$

so that

$$dx/dt = -x_0\sqrt{\kappa/m}\sin\sqrt{\kappa/m}\,t + B\sqrt{\kappa/m}\cos\sqrt{\kappa/m}\,t \tag{5}$$

From $dx/dt = v_0$ at $t = 0$ we find, using (*5*), that $B = v_0\sqrt{m/\kappa}$. Thus (*4*) becomes

$$x = x_0\cos\sqrt{\kappa/m}\,t + v_0\sqrt{m/\kappa}\sin\sqrt{\kappa/m}\,t \tag{6}$$

Using Problem 4.2, this can be written

$$x = \sqrt{x_0^2 + mv_0^2/\kappa}\,\cos(\sqrt{\kappa/m}\,t - \phi) \tag{7}$$

where

$$\phi = \tan^{-1}(v_0/x_0)\sqrt{m/\kappa} \tag{8}$$

(*b*) The velocity is, using (*6*) or (*7*),

$$\mathbf{v} = \frac{dx}{dt}\mathbf{i} = (-x_0\sqrt{\kappa/m}\sin\sqrt{\kappa/m}\,t + v_0\cos\sqrt{\kappa/m}\,t)\,\mathbf{i}$$

$$= -\sqrt{\kappa/m}\sqrt{x_0^2 + mv_0^2/\kappa}\,\sin(\sqrt{\kappa/m}\,t - \phi)\,\mathbf{i}$$

$$= -\sqrt{v_0^2 + \kappa x_0^2/m}\,\sin(\sqrt{\kappa/m}\,t - \phi)\,\mathbf{i} \tag{9}$$

(*c*) The amplitude is given from (*7*) by $\sqrt{x_0^2 + mv_0^2/\kappa}$.

From (*7*), the period is $P = 2\pi\sqrt{\kappa/m}$. The frequency is $f = 1/P = 2\pi\sqrt{m/\kappa}$.

From (*9*), the speed is a maximum when $\sin(\sqrt{\kappa/m}\,t - \phi) = \pm 1$; this speed is $\sqrt{v_0^2 + \kappa x_0^2/m}$.

4.6. An object of mass 20 kg moves with simple harmonic motion on the x axis. Initially ($t = 0$) it is located at the distance 4 meters away from the origin $x = 0$, and has velocity 15 m/sec and acceleration 100 m/sec² directed toward $x = 0$. Find (*a*) the position at any time, (*b*) the amplitude, period and frequency of the oscillations, and (*c*) the force on the object when $t = \pi/10$ sec.

(*a*) If x denotes the position of the object at time t, then the initial conditions are

$$x = 4,\ dx/dt = -15,\ d^2x/dt^2 = -100 \quad \text{at } t = 0 \tag{1}$$

Now for simple harmonic motion,

$$x = A\cos\omega t + B\sin\omega t \tag{2}$$

Differentiating, we find

$$dx/dt = -A\omega\sin\omega t + B\omega\cos\omega t \tag{3}$$

$$d^2x/dt^2 = -A\omega^2\cos\omega t - B\omega^2\sin\omega t \tag{4}$$

Using conditions (*1*) in (*2*), (*3*) and (*4*), we find $4 = A$, $-15 = B\omega$, $-100 = -A\omega^2$. Solving simultaneously, we find $A = 4$, $\omega = 5$, $B = -3$ so that

$$x = 4 \cos 5t - 3 \sin 5t \tag{5}$$

which can be written

$$x = 5 \cos (5t - \phi) \qquad \text{where} \quad \cos \phi = \tfrac{4}{5},\ \sin \phi = -\tfrac{3}{5} \tag{6}$$

(*b*) From (*6*) we see that: amplitude = 5 m, period = $2\pi/5$ sec, frequency = $5/2\pi$ vib/sec.

(*c*) Magnitude of acceleration = $d^2x/dt^2 = -100 \cos 5t + 75 \sin 5t = 75$ m/sec² at $t = \pi/10$.

Force on object = (mass)(acceleration) = (20 kg)(75 m/sec²) = 1500 newtons.

4.7. A 20 lb wt object suspended from the end of a vertical spring of negligible mass stretches it 6 inches. (*a*) Determine the position of the object at any time if initially it is pulled down 2 inches and then released. (*b*) Find the amplitude, period and frequency of the motion.

(*a*) Let D and E [Fig. 4-9] represent the position of the end of the spring before and after the object is put on the spring. Position E is the equilibrium position of the object.

Choose a coordinate system as shown in Fig. 4-9 so that the positive z axis is downward with origin at the equilibrium position.

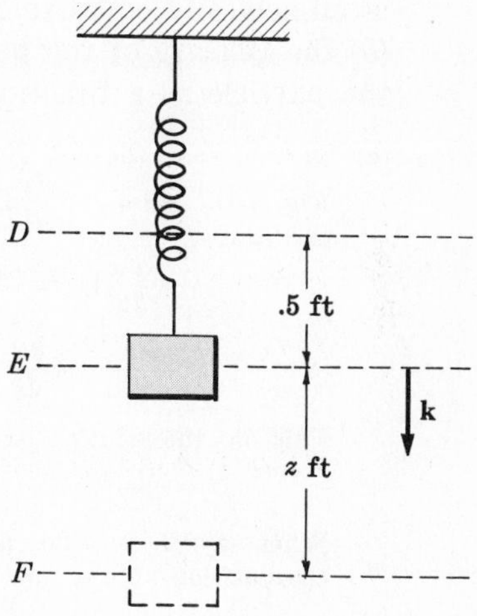

By Hooke's law, since 20 lb wt stretches the spring $\frac{1}{2}$ ft, 40 lb wt stretches it 1 ft; then $40(.5 + z)$ lb wt stretches it $(.5 + z)$ ft. Thus when the object is at position F there is an upward force acting on it of magnitude $40(.5 + z)$ and a downward force due to its weight of magnitude 20. By Newton's second law we thus have

D
E
F
.5 ft
z ft
k

Fig. 4-9

$$\frac{20}{32}\frac{d^2z}{dt^2}\mathbf{k} = 20\mathbf{k} - 40(.5 + z)\mathbf{k} \qquad \text{or} \qquad \frac{d^2z}{dt^2} + 64z = 0$$

Solving,
$$z = A \cos 8t + B \sin 8t \tag{1}$$

Now at $t = 0$, $z = \frac{1}{6}$ and $dz/dt = 0$; thus $A = \frac{1}{6}$, $B = 0$ and

$$z = \tfrac{1}{6} \cos 8t \tag{2}$$

(*b*) From (*2*): amplitude = $\frac{1}{6}$ ft, period = $2\pi/8 = \pi/4$ sec, frequency = $4/\pi$ vib/sec.

4.8. Work Problem 4.7 if initially the object is pulled down 3 inches (instead of 2 inches) and then given an initial velocity of 2 ft/sec downward.

In this case the solution (*1*) of Problem 4.7 still holds but the initial conditions are: at $t = 0$, $z = \frac{1}{4}$ and $dz/dt = 2$. From these we find $A = \frac{1}{4}$ and $B = \frac{1}{4}$, so that

$$z = \tfrac{1}{4} \cos 8t + \tfrac{1}{4} \sin 8t = \sqrt{2}/4 \cos (8t - \pi/4)$$

Thus amplitude = $\sqrt{2}/4$ ft, period = $2\pi/8 = \pi/4$ sec, frequency = $4/\pi$ vib/sec. Note that the period and frequency are unaffected by changing the initial conditions.

4.9. A particle travels with uniform angular speed ω around a circle of radius b. Prove that its projection on a diameter oscillates with simple harmonic motion of period $2\pi/\omega$ about the center.

Choose the circle in the xy plane with center at the origin O as in Fig. 4-10 below. Let Q be the projection of particle P on diameter AB chosen along the x axis.

If the particle is initially at B, then in time t we will have $\angle BOP = \theta = \omega t$. Then the position of P at time t is

$$\mathbf{r} = b \cos \omega t \,\mathbf{i} + b \sin \omega t \,\mathbf{j} \qquad (1)$$

The projection Q of P on the x axis is at distance

$$\mathbf{r} \cdot \mathbf{i} = x = b \cos \omega t \qquad (2)$$

from O at any time t. From (2) we see that the projection Q oscillates with simple harmonic motion of period $2\pi/\omega$ about the center O.

Fig. 4-10

THE DAMPED HARMONIC OSCILLATOR

4.10. Suppose that in Problem 4.1 the particle P has also a damping force whose magnitude is numerically equal to 8 times the instantaneous speed. Find (*a*) the position and (*b*) the velocity of the particle at any time. (*c*) Illustrate graphically the position of the particle as a function of time t.

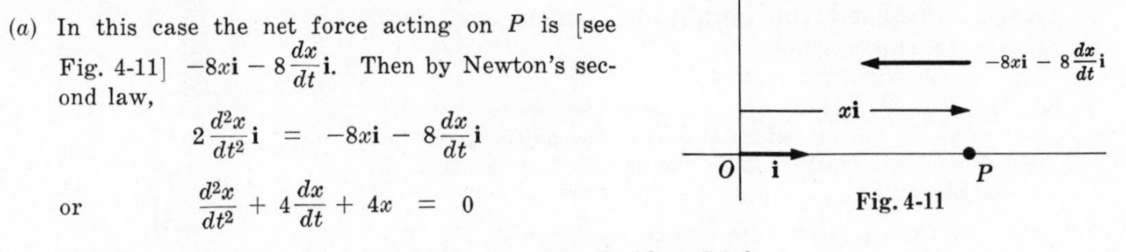

(*a*) In this case the net force acting on P is [see Fig. 4-11] $-8x\mathbf{i} - 8\dfrac{dx}{dt}\mathbf{i}$. Then by Newton's second law,

$$2\frac{d^2x}{dt^2}\mathbf{i} = -8x\mathbf{i} - 8\frac{dx}{dt}\mathbf{i}$$

or

$$\frac{d^2x}{dt^2} + 4\frac{dx}{dt} + 4x = 0$$

Fig. 4-11

This has the solution [see Appendix, page 352, Problem C.14]

$$x = e^{-2t}(A + Bt)$$

When $t = 0$, $x = 20$ and $dx/dt = 0$; thus $A = 20$, $B = 40$, and $x = 20e^{-2t}(1 + 2t)$ gives the position at any time t.

(*b*) The velocity is given by

$$\mathbf{v} = \frac{dx}{dt}\mathbf{i} = -80te^{-2t}\mathbf{i}$$

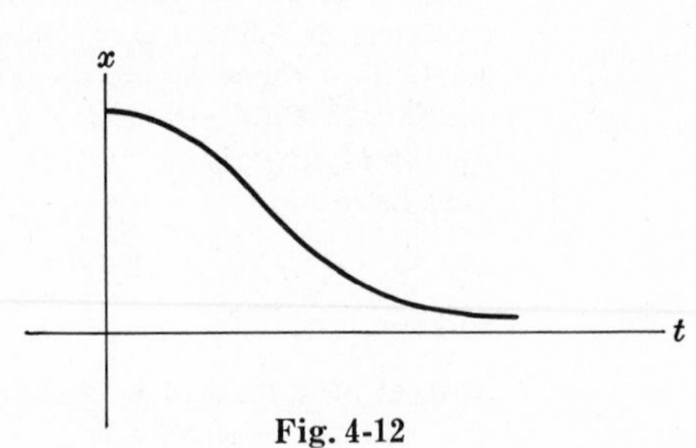

(*c*) The graph of x vs. t is shown in Fig. 4-12. It is seen that the motion is non-oscillatory. The particle approaches O slowly but never reaches it. This is an example where the motion is *critically damped.*

Fig. 4-12

4.11. A particle of mass 5 gm moves along the x axis under the influence of two forces: (i) a force of attraction to origin O which in dynes is numerically equal to 40 times the instantaneous distance from O, and (ii) a damping force proportional to the instantaneous speed such that when the speed is 10 cm/sec the damping force is 200 dynes. Assuming that the particle starts from rest at a distance 20 cm from O, (*a*) set up the differential equation and conditions describing the motion, (*b*) find the position of the particle at any time, (*c*) determine the amplitude, period and frequency of the damped oscillations, and (*d*) graph the motion.

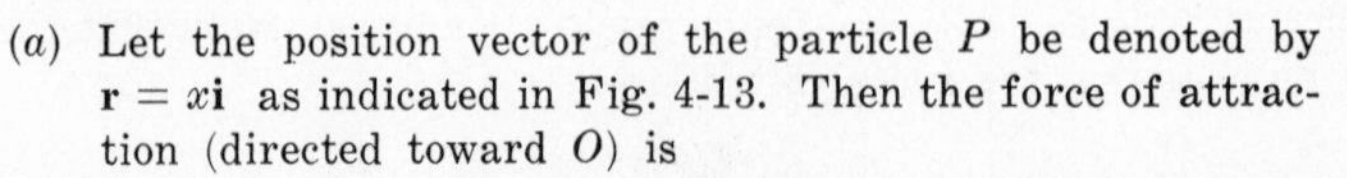

(*a*) Let the position vector of the particle P be denoted by $\mathbf{r} = x\mathbf{i}$ as indicated in Fig. 4-13. Then the force of attraction (directed toward O) is

$$-40x\mathbf{i} \qquad (1)$$

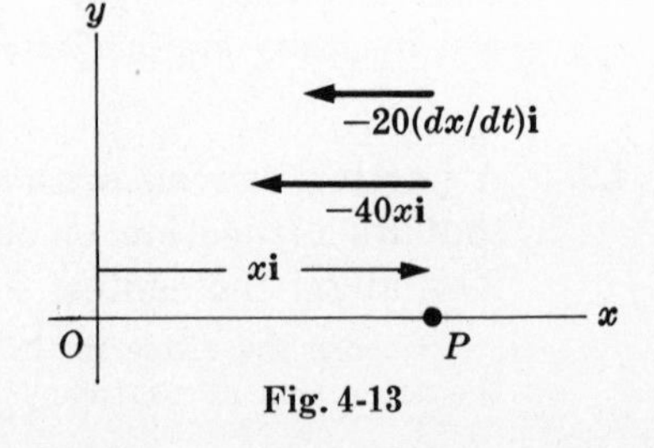

The magnitude of the damping force f is proportional to the speed, so that $f = \beta\, dx/dt$ where β is constant. Then since $f = 200$ when $dx/dt = 10$, we have $\beta = 20$ and $f = 20\, dx/dt$. To get $\mathbf{f}$, note that when $dx/dt > 0$ and $x > 0$ the particle is on the positive x axis and moving to

Fig. 4-13

the right. Thus the resistance force must be directed toward the left. This can only be accomplished if

$$\mathbf{f} = -20\frac{dx}{dt}\mathbf{i} \qquad (2)$$

This same form for $\mathbf{f}$ is easily shown to be correct if $x > 0,\ dx/dt < 0,\ x < 0,\ dx/dt > 0,\ x < 0,\ dx/dt < 0$ [see Problem 4.45].

Hence by Newton's second law we have

$$5\frac{d^2x}{dt^2}\mathbf{i} = -20\frac{dx}{dt}\mathbf{i} - 40x\mathbf{i} \qquad (3)$$

or

$$\frac{d^2x}{dt^2} + 4\frac{dx}{dt} + 8x = 0 \qquad (4)$$

Since the particle starts from rest at 20 cm from O, we have

$$x = 20,\ dx/dt = 0 \quad \text{at } t = 0 \qquad (5)$$

where we have assumed that the particle starts on the positive side of the x axis [we could just as well assume that the particle starts on the negative side, in which case $x = -20$].

(b) $x = e^{\alpha t}$ is a solution of (4) if

$$\alpha^2 + 4\alpha + 8 = 0 \quad \text{or} \quad \alpha = \tfrac{1}{2}(-4 \pm \sqrt{16-32}) = -2 \pm 2i$$

Then the general solution is

$$x = e^{-2t}(A\cos 2t + B\sin 2t) \qquad (6)$$

Since $x = 20$ at $t = 0$, we find from (6) that $A = 20$, i.e.,

$$x = e^{-2t}(20\cos 2t + B\sin 2t) \qquad (7)$$

Thus by differentiation,

$$dx/dt = (e^{-2t})(-40\sin 2t + 2B\cos 2t) + (-2e^{-2t})(20\cos 2t + B\sin 2t) \qquad (8)$$

Since $dx/dt = 0$ at $t = 0$, we have from (8), $B = 20$. Thus from (7) we obtain

$$x = 20e^{-2t}(\cos 2t + \sin 2t) = 20\sqrt{2}\,e^{-2t}\cos(2t - \pi/4) \qquad (9)$$

using Problem 4.2.

(c) From (9): amplitude $= 20\sqrt{2}\,e^{-2t}$ cm, period $= 2\pi/2 = \pi$ sec, frequency $= 1/\pi$ vib/sec.

(d) The graph is shown in Fig. 4-14. Note that the amplitudes of the oscillation decrease toward zero as t increases.

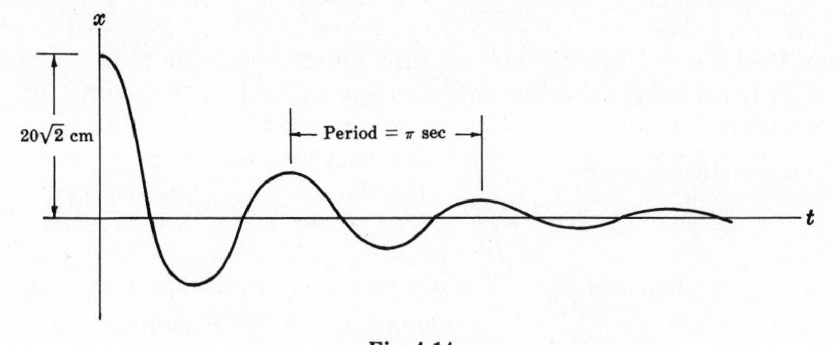

Fig. 4-14

4.12. Find the logarithmic decrement in Problem 4-11.

Method 1. The maxima (or minima) of x occur where $dx/dt = 0$. From (9) of Problem 4.11,

$$dx/dt = -80e^{-2t}\sin 2t = 0$$

when $t = 0, \pi/2, \pi, 3\pi/2, 2\pi, 5\pi/2, \ldots$. The maxima occur when $t = 0, \pi, 2\pi, \ldots$; the minima occur when $t = \pi/2, 3\pi/2, 5\pi/2, \ldots$. The ratio of two successive maxima is $e^{-2(0)}/e^{-2(\pi)}$ or $e^{-2(\pi)}/e^{-2(2\pi)}$, etc., i.e. $e^{2\pi}$. Then the logarithmic decrement is $\delta = \ln(e^{2\pi}) = 2\pi$.

Method 2.

From (9) of Problem 4.11, the difference between two successive values of t, denoted by t_n and t_{n+1}, for which $\cos(2t - \pi/4) = 1$ (or -1) is π, which is the period. Then

$$\frac{x_n}{x_{n+1}} = \frac{20\sqrt{2}\,e^{-2t_n}}{20\sqrt{2}\,e^{-2t_{n+1}}} = e^{2\pi} \quad \text{and} \quad \delta = \ln(x_n/x_{n+1}) = 2\pi$$

Method 3. From (13), (18) and (21), pages 88 and 89, we have

$$\delta = \gamma P = \left(\frac{\beta}{2m}\right)\left(\frac{4\pi m}{\sqrt{4\kappa m - \beta^2}}\right) = \frac{2\pi\beta}{\sqrt{4\kappa m - \beta^2}}$$

Then since $m = 5$, $\beta = 20$, $\kappa = 40$ [Problem 4.11, equation (3)], $\delta = 2\pi$.

4.13. Determine the natural period and frequency of the particle of Problem 4.11.

The natural period is the period when there is no damping. In such case the motion is given by removing the term involving dx/dt in equation (3) or (4) of Problem 4.11. Thus

$$d^2x/dt^2 + 8x = 0 \quad \text{or} \quad x = A\cos 2\sqrt{2}\,t + B\sin 2\sqrt{2}\,t$$

Then: natural period $= 2\pi/2\sqrt{2}$ sec $= \pi/\sqrt{2}$ sec; natural frequency $= \sqrt{2}/\pi$ vib/sec.

4.14. For what range of values of the damping constant in Problem 4.11 will the motion be (*a*) overdamped, (*b*) underdamped or damped oscillatory, (*c*) critically damped?

Denoting the damping constant by β, equation (3) of Problem 4.11 is replaced by

$$5\frac{d^2x}{dt^2}\mathbf{i} = -\beta\frac{dx}{dt}\mathbf{i} - 40x\mathbf{i} \quad \text{or} \quad \frac{d^2x}{dt^2} + \frac{\beta}{5}\frac{dx}{dt} + 8x = 0$$

Then the motion is:

(*a*) Overdamped if $(\beta/5)^2 > 32$, i.e. $\beta > 20\sqrt{2}$.

(*b*) Underdamped if $(\beta/5)^2 < 32$, i.e. $\beta < 20\sqrt{2}$.
[Note that this is the case for Problem 4.11 where $\beta = 20$.]

(*c*) Critically damped if $(\beta/5)^2 = 32$, i.e. $\beta = 20\sqrt{2}$.

4.15. Solve Problem 4.7 taking into account an external damping force given numerically in lb wt by βv where v is the instantaneous speed in ft/sec and (*a*) $\beta = 8$, (*b*) $\beta = 10$, (*c*) $\beta = 12.5$.

The equation of motion is

$$\frac{20}{32}\frac{d^2z}{dt^2}\mathbf{k} = 20\mathbf{k} - 40(.5 + z)\mathbf{k} - \beta\frac{dz}{dt}\mathbf{k} \quad \text{or} \quad \frac{d^2z}{dt^2} + \frac{8\beta}{5}\frac{dz}{dt} + 64z = 0$$

(*a*) If $\beta = 8$, then $d^2z/dt^2 + 12.8\,dz/dt + 64z = 0$. The solution is

$$z = e^{-6.4t}(A\cos 4.8t + B\sin 4.8t)$$

Using the conditions $z = 1/6$, $dz/dt = 0$ at $t = 0$, we find $A = 1/6$, $B = 2/9$ so that

$$z = \frac{1}{18}e^{-6.4t}(3\cos 4.8t + 4\sin 4.8t) = \frac{5}{18}e^{-6.4t}\cos(4.8t - 53°8')$$

The motion is *damped oscillatory* with period $2\pi/4.8 = 5\pi/12$ sec.

(*b*) If $\beta = 10$, then $d^2z/dt^2 + 16dz/dt + 64z = 0$. The solution is

$$z = e^{-4t}(A + Bt)$$

Solving subject to the initial conditions gives $A = \frac{1}{6}$, $B = \frac{2}{3}$; then $z = \frac{1}{6}e^{-4t}(1 + 4t)$.

The motion is *critically damped* since any decrease in β would produce oscillatory motion.

(c) If $\beta = 12.5$ then $d^2z/dt^2 + 20dz/dt + 64z = 0$. The solution is

$$z = Ae^{-4t} + Be^{-16t}$$

Solving subject to initial conditions gives $A = 1/6$, $B = -1/24$; then $z = \frac{1}{6}e^{-4t} - \frac{1}{24}e^{-16t}$.

The motion is *overdamped.*

ENERGY OF A SIMPLE HARMONIC OSCILLATOR

4.16. (a) Prove that the force $\mathbf{F} = -\kappa x\mathbf{i}$ acting on a simple harmonic oscillator is conservative. (b) Find the potential energy of a simple harmonic oscillator.

(a) We have

$$\nabla \times \mathbf{F} = \begin{vmatrix} \mathbf{i} & \mathbf{j} & \mathbf{k} \\ \partial/\partial x & \partial/\partial y & \partial/\partial z \\ -\kappa x & 0 & 0 \end{vmatrix} = \mathbf{0}$$

so that $\mathbf{F}$ is conservative.

(b) The potential or potential energy is given by V where $\mathbf{F} = -\nabla V$ or

$$-\kappa x\mathbf{i} = -\left(\frac{\partial V}{\partial x}\mathbf{i} + \frac{\partial V}{\partial y}\mathbf{j} + \frac{\partial V}{\partial z}\mathbf{k}\right)$$

Then $\partial V/\partial x = \kappa x$, $\partial V/\partial y = 0$, $\partial V/\partial z = 0$ from which $V = \frac{1}{2}\kappa x^2 + c$. Assuming $V = 0$ corresponding to $x = 0$, we find $c = 0$ so that $V = \frac{1}{2}\kappa x^2$.

4.17. Express in symbols the principle of conservation of energy for a simple harmonic oscillator.

By Problem 4.16(b), we have

$$\text{Kinetic energy} + \text{Potential energy} = \text{Total energy}$$

or

$$\tfrac{1}{2}mv^2 + \tfrac{1}{2}\kappa x^2 = E$$

which can also be written, since $v = dx/dt$, as $\frac{1}{2}m(dx/dt)^2 + \frac{1}{2}\kappa x^2 = E$.

Another method. The differential equation for the motion of a simple harmonic oscillator is

$$m\,d^2x/dt^2 = -\kappa x$$

Since $dx/dt = v$, this can also be written as

$$m\frac{dv}{dt} = -\kappa x \quad \text{or} \quad m\frac{dv}{dx}\frac{dx}{dt} = -\kappa x, \quad \text{i.e.} \quad mv\frac{dv}{dx} = -\kappa x$$

Integration yields $\frac{1}{2}mv^2 + \frac{1}{2}\kappa x^2 = E$.

FORCED VIBRATIONS AND RESONANCE

4.18. Derive the steady-state solution *(27)* corresponding to the differential equation *(24)* on page 89.

The differential equation is

$$\ddot{x} + 2\gamma\dot{x} + \omega^2 x = f_0 \cos \alpha t \tag{1}$$

Consider a particular solution having the form

$$x = c_1 \cos \alpha t + c_2 \sin \alpha t \tag{2}$$

where c_1 and c_2 are to be determined. Substituting (2) into (1), we find

$$(-\alpha^2 c_1 + 2\gamma\alpha c_2 + \omega^2 c_1)\cos\alpha t + (-\alpha^2 c_2 - 2\gamma\alpha c_1 + \omega^2 c_2)\sin\alpha t = f_0\cos\alpha t$$

from which

$$-\alpha^2 c_1 + 2\gamma\alpha c_2 + \omega^2 c_1 = f_0, \quad -\alpha^2 c_2 - 2\gamma\alpha c_1 + \omega^2 c_2 = 0 \tag{3}$$

or

$$(\alpha^2 - \omega^2)c_1 - 2\gamma\alpha c_2 = -f_0, \quad 2\gamma\alpha c_1 + (\alpha^2 - \omega^2)c_2 = 0 \tag{4}$$

Solving these simultaneously, we find

$$c_1 = \frac{f_0(\omega^2 - \alpha^2)}{(\alpha^2 - \omega^2)^2 + 4\gamma^2\omega^2}, \quad c_2 = \frac{2f_0\gamma\omega}{(\alpha^2 - \omega^2)^2 + 4\gamma^2\alpha^2} \tag{5}$$

Thus (2) becomes

$$x = \frac{f_0[(\omega^2 - \alpha^2)\cos\alpha t + 2\gamma\alpha\sin\alpha t]}{(\alpha^2 - \omega^2)^2 + 4\gamma^2\alpha^2} \tag{6}$$

Now by Problem 4.2, page 92,

$$(\omega^2 - \alpha^2)\cos\alpha t + 2\gamma\alpha\sin\alpha t = \sqrt{(\omega^2 - \alpha^2)^2 + 4\gamma^2\alpha^2}\cos(\alpha t - \phi) \tag{7}$$

where $\tan\phi = 2\gamma\alpha/(\alpha^2 - \omega^2)$, $0 \leqq \phi \leqq \pi$. Using (7) in (6), we find as required

$$x = \frac{f_0}{\sqrt{(\alpha^2 - \omega^2)^2 + 4\gamma^2\alpha^2}}\cos(\alpha t - \phi)$$

4.19. Prove (a) that the amplitude in Problem 4.18 is a maximum where the resonant frequency is determined from $\alpha = \sqrt{\omega^2 - 2\gamma^2}$ and (b) that the value of this maximum amplitude is $f_0/(2\gamma\sqrt{\omega^2 - \gamma^2})$.

Method 1. The amplitude in Problem 4.18 is

$$f_0/\sqrt{(\alpha^2 - \omega^2)^2 + 4\gamma^2\alpha^2} \tag{1}$$

It is a maximum when the denominator [or the square of the denominator] is a minimum. To find this minimum, write

$$\begin{aligned}(\alpha^2 - \omega^2)^2 + 4\gamma^2\alpha^2 &= \alpha^4 - 2(\omega^2 - 2\gamma^2)\alpha^2 + \omega^4 \\ &= \alpha^4 - 2(\omega^2 - 2\gamma^2)\alpha^2 + (\omega^2 - 2\gamma^2)^2 + \omega^4 - (\omega^2 - 2\gamma^2)^2 \\ &= [\alpha^2 - (\omega^2 - 2\gamma^2)]^2 + 4\gamma^2(\omega^2 - \gamma^2)\end{aligned}$$

This is a minimum where the first term on the last line is zero, i.e. when $\alpha^2 = \omega^2 - 2\gamma^2$, and the value is then $4\gamma^2(\omega^2 - \gamma^2)$. Thus the value of the maximum amplitude is given from (1) by $f_0/(2\gamma\sqrt{\omega^2 - \gamma^2})$.

Method 2. The function $U = (\alpha^2 - \omega^2)^2 + 4\gamma^2\alpha^2$ has a minimum or maximum when

$$\frac{dU}{d\alpha} = 2(\alpha^2 - \omega^2)2\alpha + 8\gamma^2\alpha = 0 \quad \text{or} \quad \alpha(\alpha^2 - \omega^2 + 2\gamma^2) = 0$$

i.e. $\alpha = 0$, $\alpha = \sqrt{\omega^2 - 2\gamma^2}$ where $\gamma^2 < \frac{1}{2}\omega^2$. Now

$$d^2U/d\alpha^2 = 12\alpha^2 - 4\omega^2 + 8\gamma^2$$

For $\alpha = 0$, $d^2U/d\alpha^2 = -4(\omega^2 - 2\gamma^2) < 0$. For $\alpha = \sqrt{\omega^2 - 2\gamma^2}$, $d^2U/d\alpha^2 = 8(\omega^2 - 2\gamma^2) > 0$. Thus $\alpha = \sqrt{\omega^2 - 2\gamma^2}$ gives the minimum value.

4.20. (a) Obtain the solution (33), page 90, for the case where there is no damping and the impressed frequency is equal to the natural frequency of the oscillation. (b) Give a physical interpretation.

(a) The case to be considered is obtained by putting $\gamma = 0$ or $\beta = 0$ and $\alpha = \omega$ in equations (23) or (24), page 89. We thus must solve the equation

$$\ddot{x} + \omega^2 x = f_0\cos\omega t \tag{1}$$

To find the general solution of this equation we add the general solution of

$$\ddot{x} + \omega^2 x = 0 \tag{2}$$

to a particular solution of (*1*).

Now the general solution of (*2*) is

$$x = A \cos \omega t + B \sin \omega t \tag{3}$$

To find a particular solution of (*1*) it would do no good to assume a particular solution of the form

$$x = c_1 \cos \omega t + c_2 \sin \omega t \tag{4}$$

since when we substitute (*4*) [which is identical in form to (*3*)] into the left side of (*1*), we would get zero. We must therefore modify the form of the assumed particular solution (*4*). As seen in Appendix C, the assumed particular solution has the form

$$x = t(c_1 \cos \omega t + c_2 \sin \omega t) \tag{5}$$

To see that this yields the required particular solution, let us differentiate (*5*) to obtain

$$\dot{x} = t(-\omega c_1 \sin \omega t + \omega c_2 \cos \omega t) + (c_1 \cos \omega t + c_2 \sin \omega t) \tag{6}$$

$$\ddot{x} = t(-\omega^2 c_1 \cos \omega t - \omega^2 c_2 \sin \omega t) + 2(-\omega c_1 \sin \omega t + \omega c_2 \cos \omega t) \tag{7}$$

Substituting (*5*), (*6*) and (*7*) into (*1*), we find after simplifying

$$-2\omega c_1 \sin \omega t + 2\omega c_2 \cos \omega t = f_0 \cos \omega t$$

from which $c_1 = 0$ and $c_2 = f_0/2\omega$. Thus the required particular solution (*5*) is $x = (f_0/2\omega)t \sin \omega t$. The general solution of (*1*) is therefore

$$x = A \cos \omega t + B \sin \omega t + (f_0/2\omega)t \sin \omega t \tag{8}$$

(*b*) The constants A and B in (*8*) are determined from the initial conditions. Unlike the case with damping, the terms involving A and B do not become small with time. However, the last term involving t increases with time to such an extent that the spring will finally break. A graph of the last term shown in Fig. 4-15 indicates how the oscillations build up in magnitude.

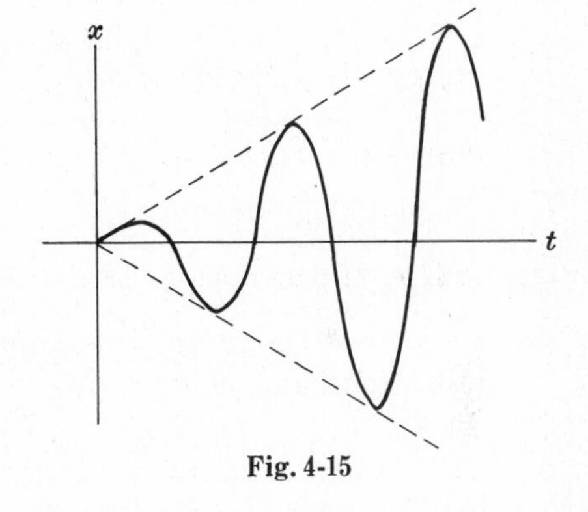

Fig. 4-15

4.21. **A vertical spring has a stiffness factor equal to 3 lb wt per ft. At $t = 0$ a force given in lb wt by $F(t) = 12 \sin 4t$, $t \geqq 0$ is applied to a 6 lb weight which hangs in equilibrium at the end of the spring. Neglecting damping, find the position of the weight at any later time t.**

Using the method of Problem 4.7, we have by Newton's second law,

$$\frac{6}{32}\frac{d^2z}{dt^2} = -3z + 12 \sin 4t$$

or

$$\frac{d^2z}{dt^2} + 16z = 64 \sin 4t \tag{1}$$

Solving,

$$z = A \cos 4t + B \sin 4t - 8t \cos 4t$$

When $t = 0$, $z = 0$ and $dz/dt = 0$; then $A = 0$, $B = 2$ and

$$z = 2 \sin 4t - 8t \cos 4t \tag{2}$$

As t gets larger the term $-8t \cos 4t$ increases numerically without bound, and physically the spring will ultimately break. The example illustrates the phenomenon of *resonance*. Note that the natural frequency of the spring $(4/2\pi = 2/\pi)$ equals the frequency of the impressed force.

4.22. Work Problem 4.21 if $F(t) = 30 \cos 6t,\ t \geqq 0$.

In this case the equation (*1*) of Problem 4.21 becomes

$$d^2z/dt^2 + 16z = 160 \cos 5t \tag{1}$$

and the initial conditions are

$$z = 0,\ dz/dt = 0 \quad \text{at} \quad t = 0 \tag{2}$$

The general solution of (*1*) is

$$z = A \cos 4t + B \sin 4t - 8 \cos 6t \tag{3}$$

Using conditions (*2*) in (*3*), we find $A = 8,\ B = 0$ and

$$z = 8(\cos 4t - \cos 6t) = 8\{\cos (5t - t) - \cos (5t + t)\} = 16 \sin t \sin 5t$$

The graph of z vs. t is shown by the heavy curve of Fig. 4-16. The dashed curves are the curves $z = \pm 16 \sin t$ obtained by placing $\sin 5t = \pm 1$. If we consider that $16 \sin t$ is the amplitude of $\sin 5t$, we see that the amplitude varies sinusoidally. The phenomenon is known as *amplitude modulation* and is of practical importance in communications and electronics.

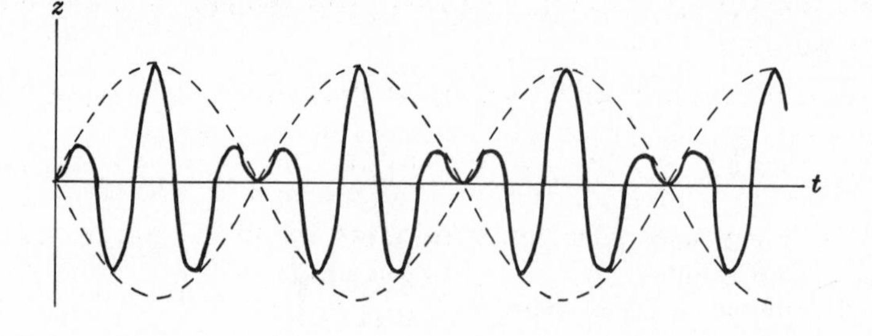

Fig. 4-16

THE SIMPLE PENDULUM

4.23. Determine the motion of a simple pendulum of length l and mass m assuming small vibrations and no resisting forces.

Let the position of m at any time be determined by s, the arclength measured from the equilibrium position O [see Fig. 4-17]. Let θ be the angle made by the pendulum string with the vertical.

If $\mathbf{T}$ is a unit tangent vector to the circular path of the pendulum bob m, then by Newton's second law

$$m \frac{d^2s}{dt^2} \mathbf{T} = -mg \sin \theta\, \mathbf{T} \tag{1}$$

or, since $s = l\theta$,

$$\frac{d^2\theta}{dt^2} = -\frac{g}{l} \sin \theta \tag{2}$$

For small vibrations we can replace $\sin \theta$ by θ so that to a high degree of accuracy equation (*2*) can be replaced by

$$\frac{d^2\theta}{dt^2} + \frac{g}{l}\theta = 0 \tag{3}$$

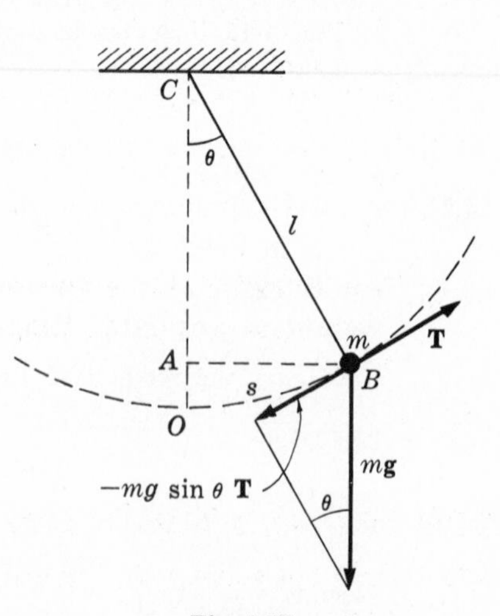

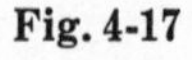

Fig. 4-17

which has solution

$$\theta = A \cos \sqrt{g/l}\, t + B \sin \sqrt{g/l}\, t$$

Taking as initial conditions $\theta = \theta_0,\ d\theta/dt = 0$ at $t = 0$, we find $A = \theta_0,\ B = 0$ and so

$$\theta = \theta_0 \cos \sqrt{g/l}\, t$$

From this we see that the period of the pendulum is $2\pi\sqrt{l/g}$.

4.24. Show how to obtain the equation (2) for the pendulum of Problem 4.23 by using the principle of conservation of energy.

We see from Fig. 4-17 that $OA = OC - AC = l - l\cos\theta = l(1-\cos\theta)$. Then by the conservation of energy [taking the reference level for the potential energy as a horizontal plane through the lowest point O] we have

$$\text{Potential energy at } B + \text{Kinetic energy at } B = \text{Total energy} = E = \text{constant}$$

$$mgl(1-\cos\theta) + \tfrac{1}{2}m(ds/dt)^2 = E \tag{1}$$

Since $s = l\theta$, this becomes

$$mgl(1-\cos\theta) + \tfrac{1}{2}ml^2(d\theta/dt)^2 = E \tag{2}$$

Differentiating both sides of (2) with respect to t, we find

$$mgl\sin\theta\,\dot{\theta} + ml^2\dot{\theta}\,\ddot{\theta} = 0 \quad \text{or} \quad \ddot{\theta} + (g/l)\sin\theta = 0$$

in agreement with equation (2) of Problem 4.23.

4.25. Work Problem 4.23 if a damping force proportional to the instantaneous velocity is taken into account.

In this case the equation of motion (1) of Problem 4.23 is replaced by

$$m\frac{d^2s}{dt^2}\mathbf{T} = -mg\sin\theta\,\mathbf{T} - \beta\frac{ds}{dt}\mathbf{T} \quad \text{or} \quad \frac{d^2s}{dt^2} = -g\sin\theta - \frac{\beta}{m}\frac{ds}{dt}$$

Using $s = l\theta$ and replacing $\sin\theta$ by θ for small vibrations, this becomes

$$\frac{d^2\theta}{dt^2} + \frac{\beta}{m}\frac{d\theta}{dt} + \frac{g}{l}\theta = 0$$

Three cases arise:

Case 1. $\beta^2/4m^2 < g/l$

$$\theta = e^{-\beta t/2m}(A\cos\omega t + B\sin\omega t) \qquad \text{where } \omega = \sqrt{g/l - \beta^2/4m^2}$$

This is the case of *damped oscillations* or *underdamped motion.*

Case 2. $\beta^2/4m^2 = g/l$

$$\theta = e^{-\beta t/2m}(A + Bt)$$

This is the case of *critically damped motion.*

Case 3. $\beta^2/4m^2 > g/l$

$$\theta = e^{-\beta t/2m}(Ae^{\lambda t} + Be^{-\lambda t}) \qquad \text{where } \lambda = \sqrt{\beta^2/4m^2 - g/l}$$

This is the case of *overdamped motion.*

In each case the constants A and B can be determined from the initial conditions. In Case 1 there are continually decreasing oscillations. In Cases 2 and 3 the pendulum bob gradually returns to the equilibrium position without oscillation.

THE TWO AND THREE DIMENSIONAL HARMONIC OSCILLATOR

4.26. Find the potential energy for (a) the two dimensional and (b) the three dimensional harmonic oscillator.

(a) In this case the force is given by

$$\mathbf{F} = -\kappa_1 x\mathbf{i} - \kappa_2 y\mathbf{j}$$

Since $\nabla \times \mathbf{F} = \mathbf{0}$, the force field is conservative. Thus a potential does exist, i.e. there exists a function V such that $\mathbf{F} = -\nabla V$. We thus have

$$\mathbf{F} = -\kappa_1 x\mathbf{i} - \kappa_2 y\mathbf{j} = -\nabla V = -\frac{\partial V}{\partial x}\mathbf{i} - \frac{\partial V}{\partial y}\mathbf{j} - \frac{\partial V}{\partial z}\mathbf{k}$$

from which $\partial V/\partial x = \kappa_1 x$, $\partial V/\partial y = \kappa_2 y$, $\partial V/\partial z = 0$ or

$$V = \tfrac{1}{2}\kappa_1 x^2 + \tfrac{1}{2}\kappa_2 y^2$$

choosing the arbitrary additive constant to be zero. This is the required potential energy.

(*b*) In this case we have $\mathbf{F} = -\kappa_1 x\mathbf{i} - \kappa_2 y\mathbf{j} - \kappa_3 z\mathbf{k}$ which is also conservative since $\nabla \times \mathbf{F} = \mathbf{0}$. We then find as in part (*a*), $\partial V/\partial x = \kappa_1 x$, $\partial V/\partial y = \kappa_2 y$, $\partial V/\partial z = \kappa_3 z$ from which the required potential energy is

$$V = \tfrac{1}{2}\kappa_1 x^2 + \tfrac{1}{2}\kappa_2 y^2 + \tfrac{1}{2}\kappa_3 z^2$$

4.27. A particle moves in the xy plane in a force field given by $\mathbf{F} = -\kappa x\mathbf{i} - \kappa y\mathbf{j}$. Prove that in general it will move in an elliptical path.

If the particle has mass m, its equation of motion is

$$m\frac{d^2\mathbf{r}}{dt^2} = \mathbf{F} = -\kappa x\mathbf{i} - \kappa y\mathbf{j} \tag{1}$$

or, since $\mathbf{r} = x\mathbf{i} + y\mathbf{j}$,

$$m\frac{d^2x}{dt^2}\mathbf{i} + m\frac{d^2y}{dt^2}\mathbf{j} = -\kappa x\mathbf{i} - \kappa y\mathbf{j}$$

Then

$$m\frac{d^2x}{dt^2} = -\kappa x, \qquad m\frac{d^2y}{dt^2} = -\kappa y \tag{2}$$

These equations have solutions given respectively by

$$x = A_1\cos\sqrt{\kappa/m}\,t + A_2\sin\sqrt{\kappa/m}\,t, \qquad y = B_1\cos\sqrt{\kappa/m}\,t + B_2\sin\sqrt{\kappa/m}\,t \tag{3}$$

Let us suppose that at $t = 0$ the particle is located at the point whose position vector is $\mathbf{r} = a\mathbf{i} + b\mathbf{j}$ and moving with velocity $d\mathbf{r}/dt = v_1\mathbf{i} + v_2\mathbf{j}$. Using these conditions, we find $A_1 = a$, $B_1 = b$, $A_2 = v_1\sqrt{m/\kappa}$, $B_2 = v_2\sqrt{m/\kappa}$ and so

$$x = a\cos\omega t + c\sin\omega t, \qquad y = b\cos\omega t + d\sin\omega t \tag{4}$$

where $c = v_1\sqrt{m/\kappa}$, $d = v_2\sqrt{m/\kappa}$. Solving for $\sin\omega t$ and $\cos\omega t$ in (*4*) we find, if $ad \neq bc$,

$$\cos\omega t = \frac{dx - cy}{ad - bc}, \qquad \sin\omega t = \frac{ay - bx}{ad - bc}$$

Squaring and adding, using the fact that $\cos^2\omega t + \sin^2\omega t = 1$, we find

$$(dx - cy)^2 + (ay - bx)^2 = (ad - bc)^2$$

or

$$(b^2 + d^2)x^2 - 2(cd + ab)xy + (a^2 + c^2)y^2 = (ad - bc)^2 \tag{5}$$

Now the equation

$$Ax^2 + Bxy + Cy^2 = D \qquad \text{where } A > 0,\ C > 0,\ D > 0$$

is an ellipse if $B^2 - 4AC < 0$, a parabola if $B^2 - 4AC = 0$, and a hyperbola if $B^2 - 4AC > 0$. To determine what (*5*) is, we see that $A = b^2 + d^2$, $B = -2(cd + ab)$, $C = a^2 + c^2$ so that

$$B^2 - 4AC = 4(cd + ab)^2 - 4(b^2 + d^2)(a^2 + c^2) = -4(ad - bc)^2 < 0$$

provided $ad \neq bc$. Thus in general the path is an ellipse, and if $A = C$ it is a circle. If $ad = bc$ the ellipse reduces to the straight line $ay = bx$.

MISCELLANEOUS PROBLEMS

4.28. A cylinder having axis vertical floats in a liquid of density σ. It is pushed down slightly and released. Find the period of the oscillation if the cylinder has weight W and cross sectional area A.

Let RS, the equilibrium position of the cylinder, be distant z from the liquid surface PQ at any time t. By *Archimedes' principle*, the buoyant force on the cylinder is $(Az)\sigma$. Then by Newton's second law,

$$\frac{W}{g}\frac{d^2z}{dt^2} = -Az\sigma$$

or

$$\frac{d^2z}{dt^2} + \frac{gA\sigma}{W}z = 0$$

Solving,

$$z = c_1 \cos\sqrt{gA\sigma/W}\,t + c_2 \sin\sqrt{gA\sigma/W}\,t$$

and the period of the oscillation is $2\pi\sqrt{W/gA\sigma}$.

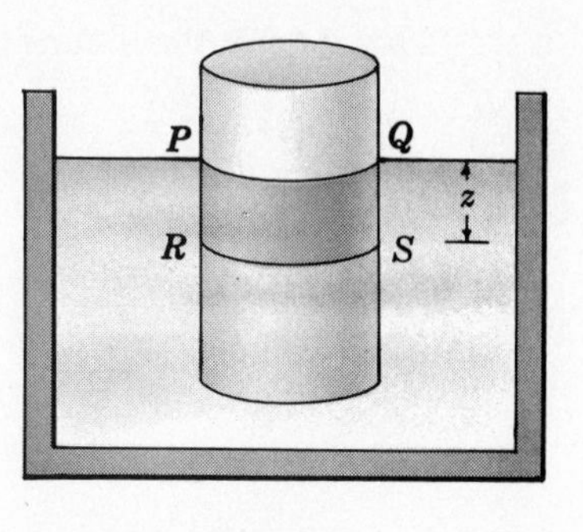

Fig. 4-18

4.29. Show that if the assumption of small vibrations is not made, then the period of a simple pendulum is

$$4\sqrt{\frac{l}{g}}\int_0^{\pi/2}\frac{d\phi}{\sqrt{1-k^2\sin^2\phi}} \qquad \text{where } k = \sin(\theta_0/2)$$

The equation of motion for a simple pendulum if small vibrations are not assumed is [equation (34), page 91]

$$\frac{d^2\theta}{dt^2} = -\frac{g}{l}\sin\theta \tag{1}$$

Let $d\theta/dt = u$. Then

$$\frac{d^2\theta}{dt^2} = \frac{du}{dt} = \frac{du}{d\theta}\frac{d\theta}{dt} = u\frac{du}{d\theta}$$

and (1) becomes

$$u\frac{du}{d\theta} = -\frac{g}{l}\sin\theta \tag{2}$$

Integrating (2) we obtain

$$\frac{u^2}{2} = \frac{g}{l}\cos\theta + c \tag{3}$$

Now when $\theta = \theta_0$, $u = 0$ so that $c = -(g/l)\cos\theta_0$. Thus (3) can be written

$$u^2 = (2g/l)(\cos\theta - \cos\theta_0) \quad \text{or} \quad d\theta/dt = \pm\sqrt{(2g/l)(\cos\theta - \cos\theta_0)} \tag{4}$$

If we restrict ourselves to that part of the motion where the bob goes from $\theta = \theta_0$ to $\theta = 0$, which represents a time equal to one fourth of the period, then we must use the minus sign in (4) so that it becomes

$$d\theta/dt = -\sqrt{(2g/l)(\cos\theta - \cos\theta_0)}$$

Separating the variables and integrating, we have

$$t = -\sqrt{\frac{l}{2g}}\int\frac{d\theta}{\sqrt{\cos\theta - \cos\theta_0}}$$

Since $t = 0$ at $\theta = \theta_0$ and $t = P/4$ at $\theta = 0$, where P is the period,

$$P = 4\sqrt{\frac{l}{2g}}\int_0^{\theta_0}\frac{d\theta}{\sqrt{\cos\theta - \cos\theta_0}} \tag{5}$$

Making use of the trigonometric identity $\cos\theta = 2\sin^2(\theta/2) - 1$, with a similar one replacing θ by θ_0, (5) can be written

$$P = 2\sqrt{\frac{l}{g}}\int_0^{\theta_0}\frac{d\theta}{\sqrt{\sin^2(\theta_0/2) - \sin^2(\theta/2)}} \tag{6}$$

Now let

$$\sin(\theta/2) = \sin(\theta_0/2)\sin\phi \tag{7}$$

Then taking the differential of both sides,

$$\tfrac{1}{2}\cos(\theta/2)\,d\theta \quad = \quad \sin(\theta_0/2)\cos\phi\,d\phi$$

or calling $k = \sin(\theta_0/2)$,

$$d\theta \quad = \quad \frac{2\sin(\theta_0/2)\cos\phi\,d\phi}{\sqrt{1-k^2\sin^2\phi}}$$

Also from (7) we see that when $\theta = 0$, $\phi = 0$; and when $\theta = \theta_0$, $\phi = \pi/2$. Hence (6) becomes, as required,

$$P \quad = \quad 4\sqrt{\frac{l}{g}}\int_0^{\pi/2}\frac{d\phi}{\sqrt{1-k^2\sin^2\phi}} \tag{8}$$

Note that if we have small vibrations, i.e. if k is equal to zero very nearly, then the period (8) becomes

$$P \quad = \quad 4\sqrt{\frac{l}{g}}\int_0^{\pi/2}d\phi \quad = \quad 2\pi\sqrt{\frac{l}{g}} \tag{9}$$

as we have already seen.

The integral in (8) is called an *elliptic integral* and cannot be evaluated exactly in terms of elementary functions. The equation of motion of the pendulum can be solved for θ in terms of *elliptic functions* which are generalizations of the trigonometric functions.

4.30. Show that period given in Problem 4.29 can be written as

$$P \quad = \quad 2\pi\sqrt{l/g}\left\{1 + \left(\frac{1}{2}\right)^2 k^2 + \left(\frac{1\cdot 3}{2\cdot 4}\right)^2 k^4 + \left(\frac{1\cdot 3\cdot 5}{2\cdot 4\cdot 6}\right)^2 k^6 + \cdots\right\}$$

The *binomial theorem* states that if $|x| < 1$, then

$$(1+x)^p \quad = \quad 1 + px + \frac{p(p-1)}{2\cdot 1}x^2 + \frac{p(p-1)(p-2)}{3\cdot 2\cdot 1}x^3 + \cdots$$

If $p = -\frac{1}{2}$, this can be written

$$(1+x)^{-1/2} \quad = \quad 1 - \frac{1}{2}x + \frac{1\cdot 3}{2\cdot 4}x^2 - \frac{1\cdot 3\cdot 5}{2\cdot 4\cdot 6}x^3 + \cdots$$

Letting $x = -k^2\sin^2\phi$ and integrating from 0 to $\pi/2$, we find

$$\begin{aligned} P \quad &= \quad 4\sqrt{l/g}\int_0^{\pi/2}\frac{d\phi}{\sqrt{1-k^2\sin^2\phi}} \\ &= \quad 4\sqrt{l/g}\int_0^{\pi/2}\left\{1 + \frac{1}{2}k^2\sin^2\phi + \frac{1\cdot 3}{2\cdot 4}k^4\sin^4\phi + \cdots\right\}d\phi \\ &= \quad 2\pi\sqrt{l/g}\left\{1 + \left(\frac{1}{2}\right)^2 k^2 + \left(\frac{1\cdot 3}{2\cdot 4}\right)^2 k^4 + \left(\frac{1\cdot 3\cdot 5}{2\cdot 4\cdot 6}\right)^2 k^6 + \cdots\right\} \end{aligned}$$

where we have made use of the integration formula

$$\int_0^{\pi/2}\sin^{2n}\phi\,d\phi \quad = \quad \frac{1\cdot 3\cdot 5\cdots(2n-1)}{2\cdot 4\cdot 6\cdots(2n)}\,\frac{\pi}{2}$$

The term by term integration is possible since $|k| < 1$.

4.31. A bead of mass m is constrained to move on a frictionless wire in the shape of a cycloid [Fig. 4-19 below] whose parametric equations are

$$x \;=\; a(\phi - \sin\phi), \qquad y \;=\; a(1-\cos\phi) \tag{1}$$

which lies in a vertical plane. If the bead starts from rest at point O, (*a*) find the speed at the bottom of the path and (*b*) show that the bead performs oscillations with period equivalent to that of a simple pendulum of length $4a$.

(a) Let P be the position of the bead at any time t and let s be the arclength along the cycloid measured from point O.

By the conservation of energy, measuring potential energy relative to line AB through the minimum point of the cycloid, we have

P.E. at P + K.E. at P = P.E. at O + K.E. at O

$$mg(2a - y) + \tfrac{1}{2}m(ds/dt)^2 = mg(2a) + 0 \qquad (2)$$

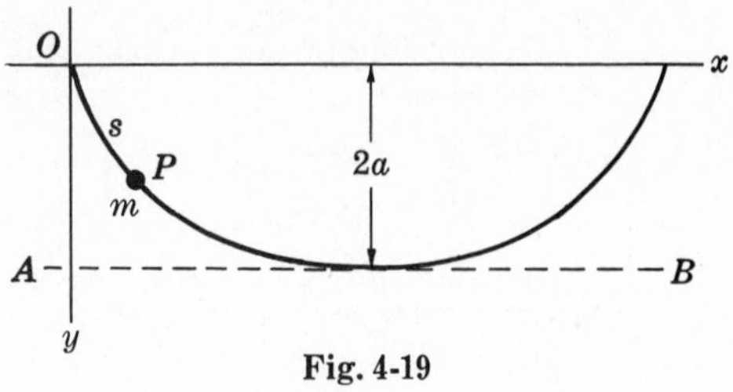

Fig. 4-19

Thus
$$v^2 = (ds/dt)^2 = 2gy \quad \text{or} \quad v = ds/dt = \sqrt{2gy} \qquad (3)$$

At the lowest point $y = 2a$ the speed is $v = \sqrt{2g(2a)} = 2\sqrt{ga}$.

(b) From part (a), $(ds/dt)^2 = 2gy$. But

$$(ds/dt)^2 = (dx/dt)^2 + (dy/dt)^2 = a^2(1-\cos\phi)^2\dot{\phi}^2 + a^2\sin^2\phi\,\dot{\phi}^2 = 2a^2(1-\cos\phi)\dot{\phi}^2$$

Then $2a^2(1-\cos\phi)\dot{\phi}^2 = 2ga(1-\cos\phi)$ or $\dot{\phi}^2 = g/a$. Thus

$$d\phi/dt = \sqrt{g/a} \quad \text{and} \quad \phi = \sqrt{g/a}\,t + c_1 \qquad (4)$$

When $\phi = 0$, $t = 0$; when $\phi = 2\pi$, $t = P/2$ where P is the period. Hence from the second equation of (4),

$$P = 4\pi\sqrt{a/g} = 2\pi\sqrt{4a/g}$$

and the period is the same as that of a simple pendulum of length $l = 4a$.

For some interesting applications see Problems 4.86-4.88.

4.32. A particle of mass m is placed on the inside of a smooth paraboloid of revolution having equation $cz = x^2 + y^2$ at a point P which is at height H above the horizontal [assumed as the xy plane]. Assuming that the particle starts from rest, (a) find the speed with which it reaches the vertex O, (b) find the time τ taken, and (c) find the period for small vibrations.

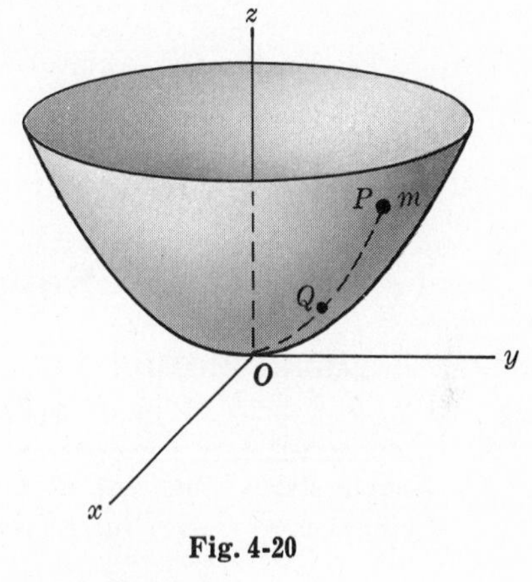

Fig. 4-20

It is convenient to choose the point P in the yz plane so that $x = 0$ and $cz = y^2$. By the principle of conservation of energy we have if Q is any point on the path PQO,

$$\text{P.E. at } P + \text{K.E. at } P = \text{P.E. at } Q + \text{K.E. at } Q$$

$$mgH + \tfrac{1}{2}m(0)^2 = mgz + \tfrac{1}{2}m(ds/dt)^2$$

where s is the arclength along OPQ measured from O. Thus

$$(ds/dt)^2 = 2g(H - z) \qquad (1)$$

or
$$ds/dt = -\sqrt{2g(H-z)} \qquad (2)$$

using the negative sign since s is decreasing with t.

(a) Putting $z = 0$, we see that the speed is $\sqrt{2gH}$ at the vertex.

(b) We have, since $x = 0$ and $cz = y^2$,

$$\left(\frac{ds}{dt}\right)^2 = \left(\frac{dx}{dt}\right)^2 + \left(\frac{dy}{dt}\right)^2 + \left(\frac{dz}{dt}\right)^2 = \left(\frac{dy}{dt}\right)^2 + \frac{4y^2}{c^2}\left(\frac{dy}{dt}\right)^2 = \left(1 + \frac{4y^2}{c^2}\right)\left(\frac{dy}{dt}\right)^2$$

Thus (1) can be written $(1 + 4y^2/c^2)(dy/dt)^2 = 2g(H - y^2/c)$. Then

$$\frac{dy}{dt} = -\sqrt{2gc}\,\frac{\sqrt{cH - y^2}}{\sqrt{c^2 + 4y^2}} \quad \text{or} \quad -\sqrt{2gc}\,dt = \frac{\sqrt{c^2+4y^2}}{\sqrt{cH - y^2}}\,dy$$

Integrating, using the fact that $z = H$ and thus $y = \sqrt{cH}$ at $t = 0$ while at $t = \tau$, $y = 0$, we have

$$\int_0^{\tau} -\sqrt{2gc}\, dt \quad = \quad \int_{\sqrt{cH}}^{0} \frac{\sqrt{c^2 + 4y^2}}{\sqrt{cH - y^2}}\, dy \qquad \text{or} \qquad \tau \quad = \quad \frac{1}{\sqrt{2gc}} \int_0^{\sqrt{cH}} \frac{\sqrt{c^2 + 4y^2}}{\sqrt{cH - y^2}}\, dy$$

Letting $y = \sqrt{cH}\cos\theta$, the integral can be written

$$\tau \quad = \quad \frac{1}{\sqrt{2gc}} \int_0^{\pi/2} \sqrt{c^2 + 4cH\cos^2\theta}\, d\theta \quad = \quad \frac{1}{\sqrt{2gc}} \int_0^{\pi/2} \sqrt{c^2 + 4cH - 4cH\sin^2\theta}\, d\theta$$

and this can be written

$$\tau \quad = \quad \sqrt{\frac{c + 4H}{2g}} \int_0^{\pi/2} \sqrt{1 - k^2\sin^2\theta}\, d\theta \tag{3}$$

where

$$k \quad = \quad \sqrt{4H/(c + 4H)} \quad < \quad 1 \tag{4}$$

The integral in (3) is an *elliptic integral* and cannot be evaluated in terms of elementary functions. It can, however, be evaluated in terms of series [see Problem 4.119].

(c) The particle oscillates back and forth on the inside of the paraboloid with period given by

$$P \quad = \quad 4\tau \quad = \quad 4\sqrt{\frac{c + 4H}{2g}} \int_0^{\pi/2} \sqrt{1 - k^2\sin^2\theta}\, d\theta \tag{5}$$

For small vibrations the value of k given by (4) can be assumed so small so as to be zero for practical purposes. Hence (5) becomes

$$P \quad = \quad 2\pi\sqrt{(c + 4H)/2g}$$

The length of the equivalent simple pendulum is $l = \frac{1}{2}(c + 4H)$.

Supplementary Problems

SIMPLE HARMONIC MOTION AND THE SIMPLE HARMONIC OSCILLATOR

4.33. A particle of mass 12 gm moves along the x axis attracted toward the point O on it by a force in dynes which is numerically equal to 60 times its instantaneous distance x cm from O. If the particle starts from rest at $x = 10$, find the (a) amplitude, (b) period and (c) frequency of the motion. *Ans.* (a) 10 cm, (b) $2\pi/\sqrt{5}$ sec, (c) $\sqrt{5}/2\pi$ vib/sec

4.34. (a) If the particle of Problem 4.33 starts at $x = 10$ with a speed toward O of 20 cm/sec, determine its amplitude, period and frequency. (b) Determine when the particle reaches O for the first time.
Ans. (a) Amplitude $= 6\sqrt{5}$ cm, period $= 2\pi/\sqrt{5}$ sec, frequency $= \sqrt{5}/2\pi$ vib/sec; (b) 0.33 sec

4.35. A particle moves on the x axis attracted toward the origin O on it with a force proportional to its instantaneous distance from O. If it starts from rest at $x = 5$ cm and reaches $x = 2.5$ cm for the first time after 2 sec, find (a) the position at any time t after it starts, (b) the speed at $x = 0$, (c) the amplitude, period and frequency of the vibration, (d) the maximum acceleration, (e) the maximum speed.
Ans. (a) $x = 5\cos(\pi t/6)$; (b) $5\pi/6$ cm/sec; (c) 5 cm, 12 sec, 1/12 vib/sec; (d) $5\pi^2/36$ cm/sec^2; (e) $5\pi/6$ cm/sec

4.36. If a particle moves with simple harmonic motion along the x axis, prove that (a) the acceleration is numerically greatest at the ends of the path, (b) the velocity is numerically greatest in the middle of the path, (c) the acceleration is zero in the middle of the path, (d) the velocity is zero at the ends of the path.

4.37. A particle moves with simple harmonic motion in a straight line. Its maximum speed is 20 ft/sec and its maximum acceleration is 80 ft/sec^2. Find the period and frequency of the motion.
Ans. $\pi/2$ sec, $2/\pi$ vib/sec

4.38. A particle moves with simple harmonic motion. If its acceleration at distance D from the equilibrium position is A, prove that the period of the motion is $2\pi\sqrt{D/A}$.

4.39. A particle moving with simple harmonic motion has speeds of 3 cm/sec and 4 cm/sec at distances 8 cm and 6 cm respectively from the equilibrium position. Find the period of the motion.
Ans. 4π sec

4.40. An 8 kg weight placed on a vertical spring stretches it 20 cm. The weight is then pulled down a distance of 40 cm and released. (*a*) Find the amplitude, period and frequency of the oscillations. (*b*) What is the position and speed at any time?
Ans. (*a*) 40 cm, $2\pi/7$ sec, $7/2\pi$ vib/sec
(*b*) $x = 40 \cos 7t$ cm, $v = -280 \sin 7t$ cm/sec

4.41. A mass of 200 gm placed at the lower end of a vertical spring stretches it 20 cm. When it is in equilibrium, the mass is hit and due to this goes up a distance of 8 cm before coming down again. Find (*a*) the magnitude of the velocity imparted to the mass when it is hit and (*b*) the period of the motion. *Ans.* (*a*) 56 cm/sec, (*b*) $2\pi/7$ sec

4.42. A 5 kg mass at the end of a spring moves with simple harmonic motion along a horizontal straight line with period 3 sec and amplitude 2 meters. (*a*) Determine the spring constant. (*b*) What is the maximum force exerted on the spring?
Ans. (*a*) 1140 dynes/cm or 1.14 newtons/meter
(*b*) 2.28×10^5 dynes or 2.28 newtons

4.43. When a mass M hanging from the lower end of a vertical spring is set into motion, it oscillates with period P. Prove that the period when mass m is added is $P\sqrt{1+m/M}$.

THE DAMPED HARMONIC OSCILLATOR

4.44. (*a*) Solve the equation $d^2x/dt^2 + 2\,dx/dt + 5x = 0$ subject to the conditions $x = 5$, $dx/dt = -3$ at $t = 0$ and (*b*) give a physical interpretation of the results.
Ans. (*a*) $x = \frac{1}{2}e^{-t}(10 \cos 2t - 5 \sin 2t)$

4.45. Verify that the damping force given by equation (*2*) of Problem 4.11 is correct regardless of the position and velocity of the particle.

4.46. A 60 lb weight hung on a vertical spring stretches it 2 ft. The weight is then pulled down 3 ft and released. (*a*) Find the position of the weight at any time if a damping force numerically equal to 15 times the instantaneous speed is acting. (*b*) Is the motion oscillatory damped, overdamped or critically damped? *Ans.* (*a*) $x = 3e^{-4t}(4t+1)$, (*b*) critically damped

4.47. Work Problem 4.46 if the damping force is numerically 18.75 times the instantaneous speed.
Ans. (*a*) $x = 4e^{-2t} - e^{-8t}$, (*b*) overdamped

4.48. In Problem 4.46, suppose that the damping force is numerically 7.5 times the instantaneous speed. (*a*) Prove that the motion is damped oscillatory. (*b*) Find the amplitude, period and frequency of the oscillations. (*c*) Find the logarithmic decrement.
Ans. (*b*) Amplitude $= 2\sqrt{3}\,e^{-2t}$ ft, period $= \pi/\sqrt{3}$ sec, frequency $= \sqrt{3}/\pi$ vib/sec; (*c*) $2\pi/\sqrt{3}$

4.49. Prove that the logarithmic decrement is the time required for the maximum amplitude during an oscillation to reduce to $1/e$ of this value.

4.50. The natural frequency of a mass vibrating on a spring is 20 vib/sec, while its frequency with damping is 16 vib/sec. Find the logarithmic decrement. *Ans.* 3/4

4.51. Prove that the difference in times corresponding to the successive maximum displacements of a damped harmonic oscillator with equation given by (*12*) of page 88 is constant and equal to $4\pi m/\sqrt{4\kappa m - \beta^2}$.

4.52. Is the difference in times between successive minimum displacements of a damped harmonic oscillator the same as in Problem 4.51? Justify your answer.

FORCED VIBRATIONS AND RESONANCE

4.53. The position of a particle moving along the x axis is determined by the equation $d^2x/dt^2 + 4dx/dt + 8x = 20 \cos 2t$. If the particle starts from rest at $x = 0$, find (*a*) x as a function of t, (*b*) the amplitude, period and frequency of the oscillation after a long time has elapsed.

Ans. (*a*) $x = \cos 2t + 2 \sin 2t - e^{-2t}(\cos 2t + 3 \sin 2t)$

(*b*) Amplitude $= \sqrt{5}$, period $= \pi$, frequency $= 1/\pi$

4.54. (*a*) Give a physical interpretation to Problem 4.53 involving a mass at the end of a vertical spring. (*b*) What is the natural frequency of such a vibrating spring? (*c*) What is the frequency of the impressed force? *Ans.* (*b*) $\sqrt{2}/\pi$, (*c*) $1/\pi$

4.55. The weight on a vertical spring undergoes forced vibrations according to the equation $d^2x/dt^2 + 4x = 8 \sin \omega t$ where x is the displacement from the equilibrium position and $\omega > 0$ is a constant. If at $t = 0$, $x = 0$ and $dx/dt = 0$, find (*a*) x as a function of t, (*b*) the period of the external force for which resonance occurs.

Ans. (*a*) $x = (8 \sin \omega t - 4\omega \sin 2t)/(4 - \omega^2)$ if $\omega \neq 2$; $x = \sin 2t - 2t \cos 2t$ if $\omega = 2$

(*b*) $\omega = 2$ or period $= \pi$

4.56. A vertical spring having constant 17 lb wt per ft has a 32 lb weight suspended from it. An external force given as a function of time t by $F(t) = 65 \sin 4t$, $t \geqq 0$ is applied. A damping force given numerically in lb wt by $2v$, where v is the instantaneous speed of the weight in ft/sec, is assumed to act. Initially the weight is at rest at the equilibrium position. (*a*) Determine the position of the weight at any time. (*b*) Indicate the transient and steady-state solutions, giving physical interpretations of each. (*c*) Find the amplitude, period and frequency of the steady-state solution. [Use $g = 32$ ft/sec^2.]

Ans. (*a*) $x = 4e^{-t} \cos 4t + \sin 4t - 4 \cos 4t$

(*b*) Transient, $4e^{-t} \cos 4t$; steady-state, $\sin 4t - 4 \cos 4t$

(*c*) Amplitude $= \sqrt{17}$ ft, period $= \pi/2$ sec, frequency $= 2/\pi$ vib/sec

4.57. A spring is stretched 5 cm by a force of 50 dynes. A mass of 10 gm is placed on the lower end of the spring. After equilibrium has been reached, the upper end of the spring is moved up and down so that the external force acting on the mass is given by $F(t) = 20 \cos \omega t$, $t \geqq 0$. (*a*) Find the position of the mass at any time, measured from its equilibrium position. (*b*) Find the value of ω for which resonance occurs.

Ans. (*a*) $x = (20 \cos \omega t)/(1 - \omega^2) - 20 \cos t$, (*b*) $\omega = 1$

4.58. A periodic external force acts on a 6 kg mass suspended from the lower end of a vertical spring having constant 150 newtons/meter. The damping force is proportional to the instantaneous speed of the mass and is 80 newtons when the speed is 2 meters/sec. Find the frequency at which resonance occurs. *Ans.* $5/6\pi$ vib/sec

THE SIMPLE PENDULUM

4.59. Find the length of a simple pendulum whose period is 1 second. Such a pendulum which registers seconds is called a *seconds pendulum*. *Ans.* 99.3 cm or 3.26 ft

4.60. Will a pendulum which registers seconds at one location lose or gain time when it is moved to another location where the acceleration due to gravity is greater? Explain.

Ans. Gain time

4.61. A simple pendulum whose length is 2 meters has its bob drawn to one side until the string makes an angle of 30° with the vertical. The bob is then released. (*a*) What is the speed of the bob as it passes through its lowest point? (*b*) What is the angular speed at the lowest point? (*c*) What is the maximum acceleration and where does it occur?

Ans. (*a*) 2.93 m/sec, (*b*) 1.46 rad/sec, (*c*) 2 m/sec^2

4.62. Prove that the tension in the string of a vertical simple pendulum of length l and mass m is given by $mg \cos\theta$ where θ is the instantaneous angle made by the string with the vertical.

4.63. A seconds pendulum which gives correct time at a certain location is taken to another location where it is found to lose T seconds per day. Determine the gravitational acceleration at the second location. *Ans.* $g(1 - T/86{,}400)^2$ where g is the gravitational acceleration at the first location

4.64. What is the length of a seconds pendulum on the surface of the moon where the acceleration due to gravity is approximately 1/6 that on the earth? *Ans.* 16.5 cm

4.65. A simple pendulum of length l and mass m hangs vertically from a fixed point O. The bob is given an initial horizontal velocity of magnitude v_0. Prove that the arc through which the bob swings in one period has a length given by $4l \cos^{-1}(1 - v_0^2/2gl)$

4.66. Find the minimum value of v_0 in Problem 4.65 in order that the bob will make a complete vertical circle with center at O. *Ans.* $2\sqrt{gl}$

THE TWO AND THREE DIMENSIONAL HARMONIC OSCILLATOR

4.67. A particle of mass 2 moves in the xy plane attracted to the origin with a force given by $\mathbf{F} = -18x\mathbf{i} - 50y\mathbf{j}$. At $t = 0$ the particle is placed at the point $(3, 4)$ and given a velocity of magnitude 10 in a direction perpendicular to the x axis. (*a*) Find the position and velocity of the particle at any time. (*b*) What curve does the particle describe?

Ans. (*a*) $\mathbf{r} = 3\cos 3t\,\mathbf{i} + [4\cos 5t + 2\sin 5t]\mathbf{j}$, $\mathbf{v} = -9\sin 3t\,\mathbf{i} + [10\cos 5t - 20\sin 5t]\mathbf{j}$

4.68. Find the total energy of the particle of Problem 4.67. *Ans.* 581

4.69. A two dimensional harmonic oscillator of mass 2 has potential energy given by $V = 8(x^2 + 4y^2)$. If the position vector and velocity of the oscillator at time $t = 0$ are given respectively by $\mathbf{r}_0 = 2\mathbf{i} - \mathbf{j}$ and $\mathbf{v}_0 = 4\mathbf{i} + 8\mathbf{j}$, (*a*) find its position and velocity at any time $t > 0$ and (*b*) determine the period of the motion.

Ans. (*a*) $\mathbf{r} = (2\cos 4t + \sin 4t)\mathbf{i} + (\sin 8t - \cos 8t)\mathbf{j}$, $\mathbf{v} = (4\cos 4t - 8\sin 4t)\mathbf{i} + (8\cos 8t + 8\sin 8t)\mathbf{j}$
(*b*) $\pi/8$

4.70. Work Problem 4.69 if $V = 8(x^2 + 2y^2)$. Is there a period defined for the motion in this case? Explain.

4.71. A particle of mass m moves in a 3 dimensional force field whose potential is given by $V = \frac{1}{2}\kappa(x^2 + 4y^2 + 16z^2)$. (*a*) Prove that if the particle is placed at an arbitrary point in space other than the origin, then it will return to the point after some period of time. Determine this time. (*b*) Is the velocity on returning to the starting point the same as the initial velocity? Explain.

4.72. Suppose that in Problem 4.71 the potential is $V = \frac{1}{2}\kappa(x^2 + 2y^2 + 5z^2)$. Will the particle return to the starting point? Explain.

MISCELLANEOUS PROBLEMS

4.73. A vertical spring of constant κ having natural length l is supported at a fixed point A. A mass m is placed at the lower end of the spring, lifted to a height h below A and dropped. Prove that the lowest point reached will be at a distance below A given by $l + mg/\kappa + \sqrt{m^2g^2/\kappa^2 + 2mgh/\kappa}$.

4.74. Work Problem 4.73 if damping proportional to the instantaneous velocity is taken into account.

4.75 Given the equation $m\ddot{x} + \beta\dot{x} + \kappa x = 0$ for damped oscillations of a harmonic oscillator. Prove that if $E = \frac{1}{2}m\dot{x}^2 + \frac{1}{2}\kappa x^2$, then $\dot{E} = -\beta\dot{x}^2$. Thus show that if there is damping the total energy E decreases with time. What happens to the energy lost? Explain.

4.76. (*a*) Prove that $$A_1 \cos(\omega t - \phi_1) + A_2 \cos(\omega t - \phi_2) \;=\; A \cos(\omega t - \phi)$$

where $$A \;=\; \sqrt{A_1^2 + A_2^2 + 2A_1A_2 \cos(\phi_1 - \phi_2)}, \quad \phi \;=\; \tan^{-1}\left(\frac{A_1 \sin\phi_1 + A_2 \sin\phi_2}{A_1 \cos\phi_1 + A_2 \cos\phi_2}\right).$$

(*b*) Use (*a*) to demonstrate that the sum of two simple harmonic motions of the same frequency and in the same straight line is simple harmonic of the same frequency.

4.77. Give a vector interpretation to the results of Problem 4.76.

4.78. Discuss Problem 4.76 in case the frequencies of the two simple harmonic motions are not equal. Is the resultant motion simple harmonic? Justify your answer.

4.79. A particle oscillates in a plane so that its distances x and y from two mutually perpendicular axes are given as functions of time t by

$$x \;=\; A \cos(\omega t + \phi_1), \qquad y \;=\; B \cos(\omega t + \phi_2)$$

(*a*) Prove that the particle moves in an ellipse inscribed in the rectangle defined by $x = \pm A$, $y = \pm B$. (*b*) Prove that the period of the particle in the elliptical path is $2\pi/\omega$.

4.80. Suppose that the particle of Problem 4.79 moves so that

$$x \;=\; A \cos(\omega t + \phi_1), \qquad y \;=\; B \cos(\omega t + \epsilon t + \phi_2)$$

where ϵ is assumed to be a positive constant which is assumed to be much smaller than ω. Prove that the particle oscillates in slowly rotating ellipses inscribed in the rectangle $x = \pm A$, $y = \pm B$.

4.81. Illustrate Problem 4.80 by graphing the motion of a particle which moves in the path

$$x \;=\; 3 \cos(2t + \pi/4), \qquad y \;=\; 4 \cos(2.4t)$$

4.82. In Fig. 4-21 a mass m which is on a frictionless table is connected to fixed points A and B by two springs of equal natural length, of negligible mass and spring constants κ_1 and κ_2 respectively. The mass m is displaced horizontally and then released. Prove that the period of oscillation is given by $P = 2\pi\sqrt{m/(\kappa_1 + \kappa_2)}$.

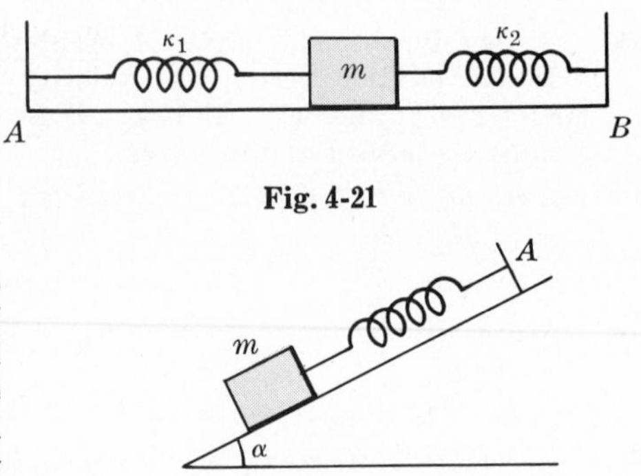

Fig. 4-21

4.83. A spring having constant κ and negligible mass has one end fixed at point A on an inclined plane of angle α and a mass m at the other end, as indicated in Fig. 4-22. If the mass m is pulled down a distance x_0 below the equilibrium position and released, find the displacement from the equilibrium position at any time if (*a*) the incline is frictionless, (*b*) the incline has coefficient of friction μ.

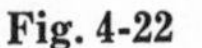
Fig. 4-22

4.84. A particle moves with simple harmonic motion along the x axis. At times t_0, $2t_0$ and $3t_0$ it is located at $x = a, b$ and c respectively. Prove that the period of oscillation is $\dfrac{4\pi t_0}{\cos^{-1}(a+c)/2b}$.

4.85. A seconds pendulum giving the correct time at one location is taken to another location where it loses 5 minutes per day. By how much must the pendulum rod be lengthened or shortened in order to give the correct time?

4.86. A vertical pendulum having a bob of mass m is suspended from the fixed point O. As it oscillates, the string winds up on the constraint curves ODA [or OC] as indicated in Fig. 4-23. Prove that if curve ABC is a cycloid, then the period of oscillation will be the same regardless of the amplitude of the oscillations. The pendulum in this case is called a *cycloidal pendulum*. The curves ODA and OC are constructed to be *evolutes* of the cycloid. [*Hint*. Use Problem 4.31.]

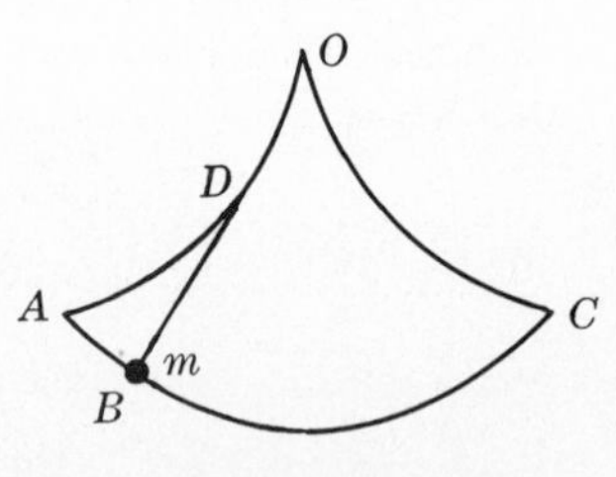

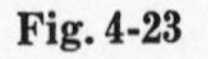
Fig. 4-23

4.87. A bead slides down a frictionless wire located in a vertical plane. It is desired to find the shape of the wire so that regardless of where the bead is placed on the wire it will slide under the influence of gravity to the bottom of the wire in the same time. This is often called the *tautochrone* problem. Prove that the wire must have the shape of a cycloid.
[*Hint.* Use Problem 4.31.]

4.88. Prove that the curves ODA and OC of Problem 4.86 are cycloids having the same shape as the cycloid ABC.

4.89. A simple pendulum of length l has its point of support moving back and forth on a horizontal line so that its distance from a fixed point on the line is $A \sin \omega t$, $t \geqq 0$. Find the position of the pendulum bob at any time t assuming that it is at rest at the equilibrium position at $t = 0$.

4.90. Work Problem 4.89 if the point of support moves vertically instead of horizontally and if at $t = 0$ the rod of the pendulum makes an angle θ_0 with the vertical.

4.91. A particle of mass m moves in a plane under the influence of forces of attraction toward fixed points which are directly proportional to its instantaneous distance from these points. Prove that in general the particle will describe an ellipse.

4.92. A vertical elastic spring of negligible weight and having its upper end fixed, carries a weight W at its lower end. The weight is lifted so that the tension in the spring is zero, and then it is released. Prove that the tension in the spring will not exceed $2W$.

4.93. A vertical spring having constant κ has a pan on top of it with a weight W on it [see Fig. 4-24]. Determine the largest frequency with which the spring can vibrate so that the weight will remain in the pan.

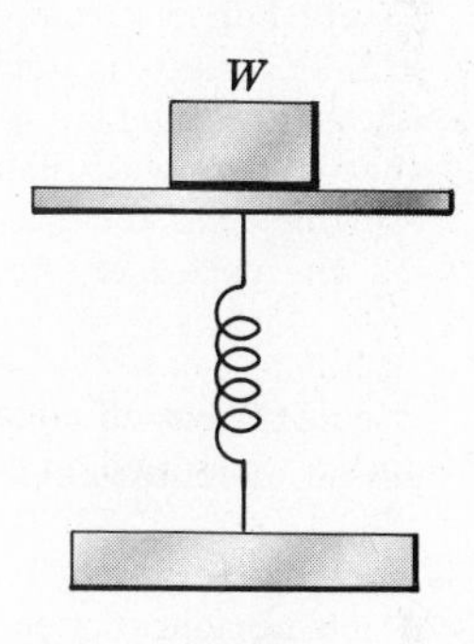

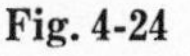
Fig. 4-24

4.94. A spring has a natural length of 50 cm and a force of 100 dynes is required to stretch it 25 cm. Find the work done in stretching the spring from 75 cm to 100 cm, assuming that the elastic limit is not exceeded so that the spring characteristics do not change.
Ans. 3750 ergs

4.95. A particle moves in the xy plane so that its position is given by $x = A \cos \omega t$, $y = B \cos 2\omega t$. Prove that it describes an arc of a parabola.

4.96. A particle moves in the xy plane so that its position is given by $x = A \cos(\omega_1 t + \phi_1)$, $y = B \cos(\omega_2 t + \phi_2)$. Prove that the particle describes a closed curve or not, according as ω_1/ω_2 is rational or not. In which cases is the motion periodic?

4.97. The position of a particle moving in the xy plane is described by the equations $d^2x/dt^2 = -4y$, $d^2y/dt^2 = -4x$. At time $t = 0$ the particle is at rest at the point $(6, 3)$. Find (*a*) its position and (*b*) its velocity at any later time t.

4.98. Find the period of a simple pendulum of length 1 meter if the maximum angle which the rod makes with the vertical is (*a*) 30°, (*b*) 60°, (*c*) 90°.

4.99. A simple pendulum of length 3 ft is suspended vertically from a fixed point. At $t = 0$ the bob is given a horizontal velocity of 8 ft/sec. Find (*a*) the maximum angle which the pendulum rod makes with the vertical, (*b*) the period of the oscillations.
Ans. (*a*) $\cos^{-1} 2/3 = 41° 48'$, (*b*) 1.92 sec

4.100. Prove that the time averages over a period of the potential energy and kinetic energy of a simple harmonic oscillator are equal to $2\pi^2 A^2/P^2$ where A is the amplitude and P is the period of the motion.

4.101. A cylinder of radius 10 ft with its axis vertical oscillates vertically in water of density 62.5 lb/ft³ with a period of 5 seconds. How much does it weigh? *Ans.* 3.98×10^5 lb wt

4.102. A particle moves in the xy plane in a force field whose potential is given by $V = x^2 + xy + y^2$. If the particle is initially at the point $(3, 4)$ and is given a velocity of magnitude 10 in a direction parallel to the positive x axis, (*a*) find the position at any time and (*b*) determine the period of the motion if one exists.

4.103. In Problem 4.96 suppose that ω_1/ω_2 is irrational and that at $t = 0$ the particle is at the particular point (x_0, y_0) inside the rectangle defined by $x = \pm A$, $y = \pm B$. Prove that the point (x_0, y_0) will never be reached again but that in the course of its motion the particle will come arbitrarily close to the point.

4.104. A particle oscillates on a vertical frictionless cycloid with its vertex downward. Prove that the projection of the particle on a vertical axis oscillates with simple harmonic motion.

4.105. A mass of 5 kg at the lower end of a vertical spring which has an elastic constant equal to 20 newtons/meter oscillates with a period of 10 seconds. Find (*a*) the damping constant, (*b*) the natural period and (*c*) the logarithmic decrement. *Ans.* (*a*) 19 nt sec/m, (*b*) 3.14 sec

4.106. A mass of 100 gm is supported in equilibrium by two identical springs of negligible mass having elastic constant equal to 50 dynes/cm. In the equilibrium position shown in Fig. 4-25 the springs make an angle of 30° with the horizontal and are 100 cm in length. If the mass is pulled down a distance of 2 cm and released, find the period of the resulting oscillation.

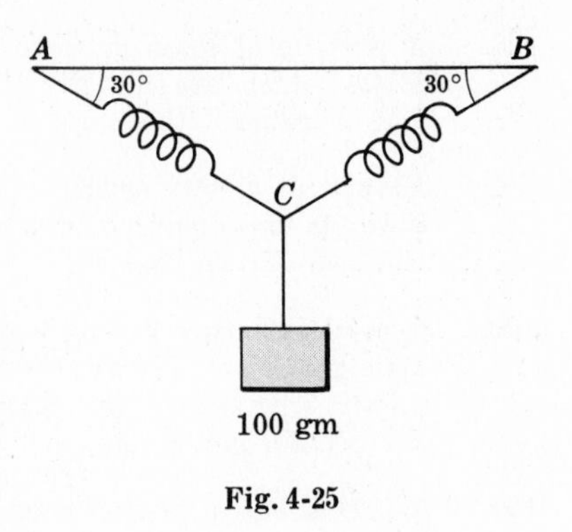

Fig. 4-25

4.107. A thin hollow circular cylinder of inner radius 10 cm is fixed so that its axis is horizontal. A particle is placed on the inner frictionless surface of the cylinder so that its vertical distance above the lowest point of the inner surface is 2 cm. Find (*a*) the time for the particle to reach the lowest point and (*b*) the period of the oscillations which take place.

4.108. A cubical box of side a and weight W vibrates vertically in water of density σ. Prove that the period of vibration is $(2\pi/a)\sqrt{\sigma g/W}$.

4.109. A spring vibrates so that its equation of motion is

$$m\,d^2x/dt^2 + \kappa x = F(t)$$

If $x = 0$, $dx/dt = 0$ at $t = 0$, find x as a function of time t.

Ans. $x = \dfrac{1}{\sqrt{m\kappa}} \displaystyle\int_0^t F(u) \sin \sqrt{\kappa/m}\,(t-u)\,du$

4.110. Work Problem 4.109 if damping proportional to dx/dt is taken into account.

4.111. A spring vibrates so that its equation of motion is

$$m\,d^2x/dt^2 + \kappa x = 5\cos\omega t + 2\cos 3\omega t$$

If $x = 0$, $\dot{x} = v_0$ at $t = 0$, (*a*) find x at any time t and (*b*) determine for what values of ω resonance will occur.

4.112. A vertical spring having elastic constant κ carries a mass m at its lower end. At $t = 0$ the spring is in equilibrium and its upper end is suddenly made to move vertically so that its distance from the original point of support is given by $A \sin \omega t$, $t \geqq 0$. Find (*a*) the position of the mass m at any time and (*b*) the values of ω for which resonance occurs.

4.113. (*a*) Solve $d^2x/dt^2 + x = t \sin t + \cos t$ where $x = 0$, $dx/dt = 0$ at $t = 0$, and (*b*) give a physical interpretation.

4.114. Discuss the motion of a simple pendulum for the case where damping and external forces are present.

4.115. Find the period of small vertical oscillations of a cylinder of radius a and height h floating with its axis horizontal in water of density σ.

4.116. A vertical spring having elastic constant 2 newtons per meter has a 50 gm weight suspended from it. A force in newtons which is given as a function of time t by $F(t) = 6\cos^4 t$, $t \geqq 0$ is applied. Assuming that the weight, initially at the equilibrium position, is given an upward velocity of 4 m/sec and that damping is negligible, determine the (a) position and (b) velocity of the weight at any time.

4.117. In Problem 4.55, can the answer for $\omega = 2$ be deduced from the answer for $\omega \neq 2$ by taking the limit as $\omega \to 2$? Justify your answer.

4.118. An oscillator has a restoring force acting on it whose magnitude is $-\kappa x - \epsilon x^2$ where ϵ is small compared with κ. Prove that the displacement of the oscillator [in this case often called an *anharmonic oscillator*] from the equilibrium position is given approximately by

$$x = A\cos(\omega t - \phi) + \frac{A\epsilon^2}{6\kappa}\{\cos 2(\omega t - \phi) - 3\}$$

where A and ϕ are determined from the initial conditions.

4.119. Prove that if the oscillations in Problem 4.32 are not necessarily small, then the period is given by

$$P = 2\pi\sqrt{\frac{c + 4H}{2g}}\left\{1 - \left(\frac{1}{2}\right)^2 k^2 - \left(\frac{1\cdot 3}{2\cdot 4}\right)^2 \frac{k^4}{3} - \left(\frac{1\cdot 3\cdot 5}{2\cdot 4\cdot 6}\right)^2 \frac{k^6}{5} - \cdots\right\}$$

Chapter 5

CENTRAL FORCES and PLANETARY MOTION

CENTRAL FORCES

Suppose that a force acting on a particle of mass m is such that [see Fig. 5-1]:

(a) it is always directed from m toward or away from a fixed point O,

(b) its magnitude depends only on the distance r from O.

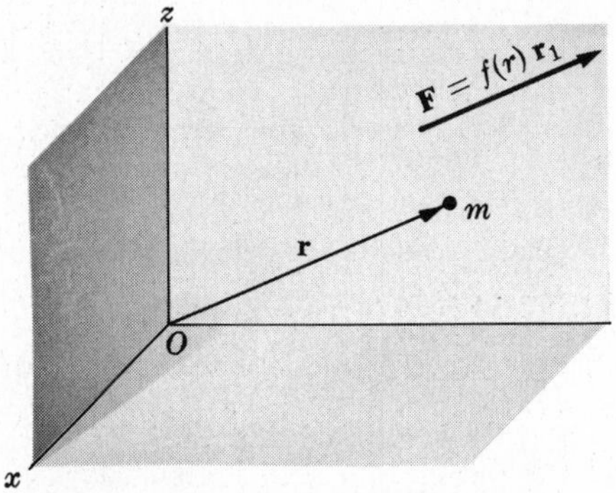

Then we call the force a *central force* or *central force field* with O as the *center of force*. In symbols $\mathbf{F}$ is a central force if and only if

$$\mathbf{F} = f(r)\,\mathbf{r}_1 = f(r)\,\mathbf{r}/r \qquad (1)$$

Fig. 5-1

where $\mathbf{r}_1 = \mathbf{r}/r$ is a unit vector in the direction of $\mathbf{r}$.

The central force is one of *attraction* toward O or *repulsion* from O according as $f(r) < 0$ or $f(r) > 0$ respectively.

SOME IMPORTANT PROPERTIES OF CENTRAL FORCE FIELDS

If a particle moves in a central force field, then the following properties are valid.

1. The path or *orbit* of the particle must be a plane curve, i.e. the particle moves in a plane. This plane is often taken to be the xy plane. See Problem 5.1.
2. The angular momentum of the particle is conserved, i.e. is constant. See Problem 5.2.
3. The particle moves in such a way that the position vector or radius vector drawn from O to the particle sweeps out equal areas in equal times. In other words, the time rate of change in area is constant. This is sometimes called the *law of areas*. See Problem 5.6.

EQUATIONS OF MOTION FOR A PARTICLE IN A CENTRAL FIELD

By Property 1, the motion of a particle in a central force field takes place in a plane. Choosing this plane as the xy plane and the coordinates of the particle as polar coordinates (r, θ), the equations of motion are found to be [see Problem 5.3]

$$m(\ddot{r} - r\dot{\theta}^2) = f(r) \qquad (2)$$

$$m(r\ddot{\theta} + 2\dot{r}\dot{\theta}) = 0 \qquad (3)$$

where dots denote differentiations with respect to time t.

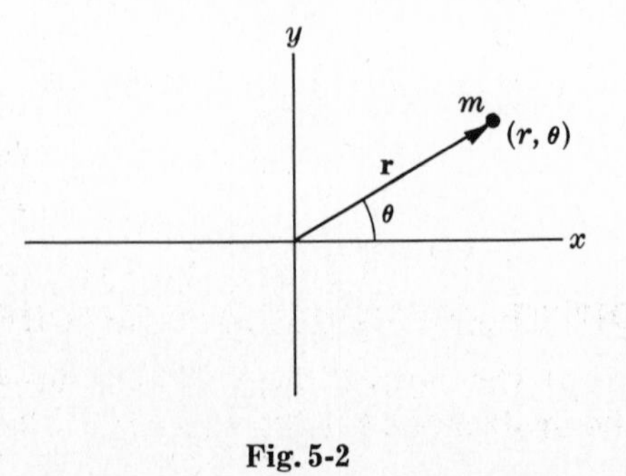

Fig. 5-2

From equation (3) we find

$$r^2\dot{\theta} = \text{constant} = h \tag{4}$$

This is related to Properties 2 and 3 above.

IMPORTANT EQUATIONS DEDUCED FROM THE EQUATIONS OF MOTION

The following equations deduced from the fundamental equations (2) and (3) often prove to be useful.

1.
$$\ddot{r} - \frac{h^2}{r^3} = \frac{f(r)}{m} \tag{5}$$

2.
$$\frac{d^2u}{d\theta^2} + u = -\frac{1}{mh^2u^2}f(1/u) \tag{6}$$

where $u = 1/r$.

3.
$$\frac{d^2r}{d\theta^2} - \frac{2}{r}\left(\frac{dr}{d\theta}\right)^2 - r = \frac{r^4 f(r)}{mh^2} \tag{7}$$

POTENTIAL ENERGY OF A PARTICLE IN A CENTRAL FIELD

A central force field is a conservative field, hence it can be derived from a potential. This potential which depends only on r is, apart from an arbitrary additive constant, given by

$$V(r) = -\int f(r)\,dr \tag{8}$$

This is also the potential energy of a particle in the central force field. The arbitrary additive constant can be obtained by assuming, for example, $V=0$ at $r=0$ or $V \to 0$ as $r \to \infty$.

CONSERVATION OF ENERGY

By using (8) and the fact that in polar coordinates the kinetic energy of a particle is $\frac{1}{2}m(\dot{r}^2 + r^2\dot{\theta}^2)$, the equation for conservation of energy can be written

$$\tfrac{1}{2}m(\dot{r}^2 + r^2\dot{\theta}^2) + V(r) = E \tag{9}$$

or

$$\tfrac{1}{2}m(\dot{r}^2 + r^2\dot{\theta}^2) - \int f(r)\,dr = E \tag{10}$$

where E is the total energy and is constant. Using (4), equation (10) can also be written as

$$\frac{mh^2}{2r^4}\left[\left(\frac{dr}{d\theta}\right)^2 + r^2\right] - \int f(r)\,dr = E \tag{11}$$

and also as

$$\frac{m}{2}\left(\dot{r}^2 + \frac{h^2}{r^2}\right) - \int f(r)\,dr = E \tag{12}$$

In terms of $u = 1/r$, we can also write equation (9) as

$$\left(\frac{du}{d\theta}\right)^2 + u^2 = \frac{2(E-V)}{mh^2} \tag{13}$$

DETERMINATION OF THE ORBIT FROM THE CENTRAL FORCE

If the central force field is prescribed, i.e. if $f(r)$ is given, it is possible to determine the orbit or path of the particle. This orbit can be obtained in the form

$$r = r(\theta) \tag{14}$$

i.e. r as a function of θ, or in the form

$$r = r(t), \quad \theta = \theta(t) \tag{15}$$

which are parametric equations in terms of the time parameter t.

To determine the orbit in the form (14) it is convenient to employ equations (6), (7) or (11). To obtain equations in the form (15), it is sometimes convenient to use (12) together with (4) or to use equations (4) and (5).

DETERMINATION OF THE CENTRAL FORCE FROM THE ORBIT

Conversely if we know the orbit or path of the particle, then we can find the corresponding central force. If the orbit is given by $r = r(\theta)$ or $u = u(\theta)$ where $u = 1/r$, the central force can be found from

$$f(r) = \frac{mh^2}{r^4}\left\{\frac{d^2r}{d\theta^2} - \frac{2}{r}\left(\frac{dr}{d\theta}\right)^2 - r\right\} \tag{16}$$

or

$$f(1/u) = -mh^2u^2\left\{\frac{d^2u}{d\theta^2} + u\right\} \tag{17}$$

which are obtained from equations (6) and (7) on page 117. The law of force can also be obtained from other equations, as for example equations (9)-(13).

It is important to note that given an orbit there may be infinitely many force fields for which the orbit is possible. However, if a central force field exists it is unique, i.e. it is the only one.

CONIC SECTIONS, ELLIPSE, PARABOLA AND HYPERBOLA

Consider a fixed point O and a fixed line AB distant D from O, as shown in Fig. 5-3. Suppose that a point P in the plane of O and AB moves so that the ratio of its distance from point O to its distance from line AB is always equal to the positive constant ϵ.

Then the curve described by P is given in polar coordinates (r, θ) by

$$r = \frac{p}{1 + \epsilon \cos\theta} \tag{18}$$

See Problem 5.16.

The point O is called a *focus*, the line AB is called a *directrix* and the ratio ϵ is called the *eccentricity*. The curve is often called a *conic section* since it can be obtained by intersecting a plane and a cone at different angles. Three possible types of curves exist, depending on the value of the eccentricity.

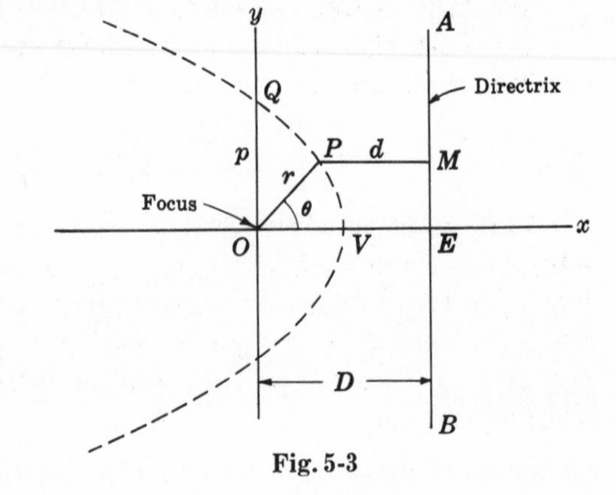

Fig. 5-3

1. **Ellipse:** $\epsilon < 1$ [See Fig. 5-4 below.]

   If C is the *center* of the ellipse and $CV = CU = a$ is the length of the *semi-major axis*, then the equation of the ellipse can be written as

$$r = \frac{a(1-\epsilon^2)}{1 + \epsilon\cos\theta} \tag{19}$$

Note that the *major axis* is the line joining the *vertices* V and U of the ellipse and has length $2a$.

If b is the length of the *semi-minor axis* [CW or CS in Fig. 5-4] and c is the distance CO from center to focus, then we have the important result

$$c = \sqrt{a^2 - b^2} = a\epsilon \qquad (20)$$

A circle can be considered as a special case of an ellipse with eccentricity equal to zero.

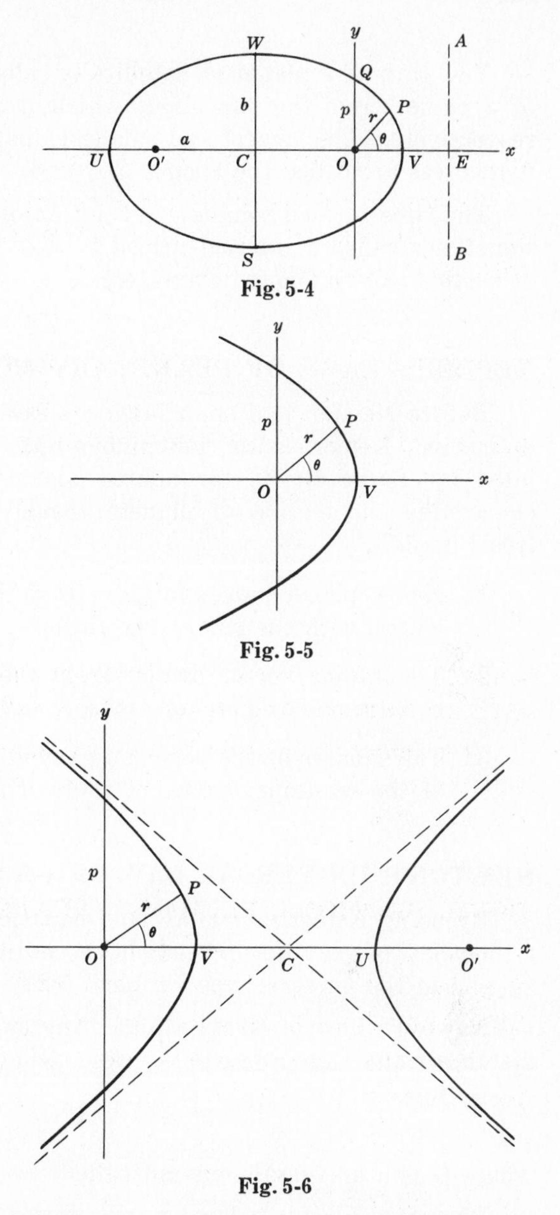

Fig. 5-4

Fig. 5-5

Fig. 5-6

2. **Parabola:** $\epsilon = 1$ [See Fig. 5-5.]

   The equation of the parabola is

$$r = \frac{p}{1 + \cos\theta} \qquad (21)$$

   We can consider a parabola to be a limiting case of the ellipse *(19)* where $\epsilon \to 1$, which means that $a \to \infty$ [i.e. the major axis becomes infinite] in such a way that $a(1 - \epsilon^2) = p$.

3. **Hyperbola:** $\epsilon > 1$ [See Fig. 5-6.]

   The hyperbola consists of two branches as indicated in Fig. 5-6. The branch on the left is the important one for our purposes. The hyperbola is asymptotic to the dashed lines of Fig. 5-6 which are called its *asymptotes.* The intersection C of the asymptotes is called the *center.* The distance $CV = a$ from the center C to vertex V is called the *semi-major axis* [the major axis being the distance between vertices V and U by analogy with the ellipse]. The equation of the hyperbola can be written as

$$r = \frac{a(\epsilon^2 - 1)}{1 + \epsilon\cos\theta} \qquad (22)$$

Various other alternative definitions for conic sections may be given. For example, an ellipse can be defined as the *locus* or *path* of all points the sum of whose distances from two fixed points is a constant. Similarly, a hyperbola can be defined as the locus of all points the difference of whose distances from two fixed points is a constant. In both these cases the two fixed points are the foci and the constant is equal in magnitude to the length of the major axis.

SOME DEFINITIONS IN ASTRONOMY

A *solar system* is composed of a *star* [such as our sun] and objects called *planets* which revolve around it. The star is an object which emits its own light, while the planets do not emit light but can reflect it. In addition there may be objects revolving about the planets. These are called *satellites.*

In our solar system, for example, the *moon* is a satellite of the earth which in turn is a planet revolving about our sun. In addition there are *artificial* or *man-made satellites* which can revolve about the planets or their moons.

The path of a planet or satellite is called its *orbit*. The largest and smallest distances of a planet from the sun about which it revolves are called the *aphelion* and *perihelion* respectively. The largest and smallest distances of a satellite around a planet about which it revolves are called the *apogee* and *perigee* respectively.

The time for one complete revolution of a body in an orbit is called its *period*. This is sometimes called a *sidereal period* to distinguish it from other periods such as the period of earth's motion about its axis, etc.

KEPLER'S LAWS OF PLANETARY MOTION

Before Newton had enunciated his famous laws of motion, Kepler, using voluminous data accumulated by Tycho Brahe formulated his three laws concerning the motion of planets around the sun [see Fig. 5-7].

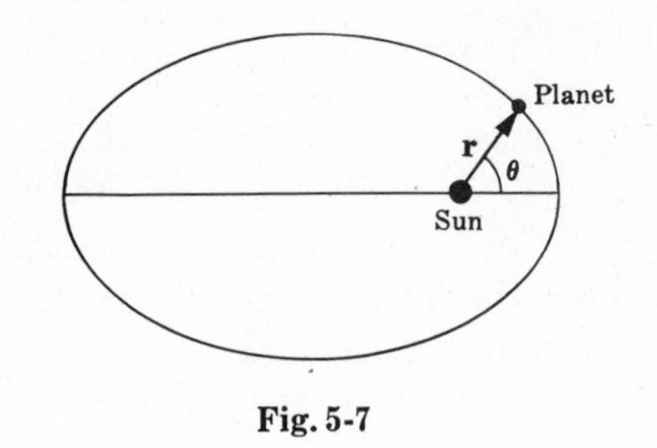

Fig. 5-7

1. Every planet moves in an orbit which is an ellipse with the sun at one focus.
2. The radius vector drawn from the sun to any planet sweeps out equal areas in equal times (the *law of areas*, as on page 116).
3. The squares of the periods of revolution of the planets are proportional to the cubes of the semi-major axes of their orbits.

NEWTON'S UNIVERSAL LAW OF GRAVITATION

By using Kepler's first law and equations (*16*) or (*17*), Newton was able to deduce his famous law of gravitation between the sun and planets, which he postulated as valid for any objects in the universe [see Problem 5.21].

Newton's Law of Gravitation. Any two particles of mass m_1 and m_2 respectively and distance r apart are attracted toward each other with a force

$$\mathbf{F} = -\frac{Gm_1m_2}{r^2}\mathbf{r}_1 \tag{23}$$

where G is a universal constant called the *gravitational constant*.

By using Newton's law of gravitation we can, conversely, deduce Kepler's laws [see Problems 5.13 and 5.23]. The value of G is shown in the table on page 342.

ATTRACTION OF SPHERES AND OTHER OBJECTS

By using Newton's law of gravitation, the forces of attraction between large objects such as spheres can be determined. To do this, we use the fact that each large object is composed of particles. We then apply the law of gravitation to find the forces between particles and sum over these forces, usually by methods of integration, to find the resultant force of attraction. An important application of this is given in the following

Theorem 5.1. Two solid or hollow uniform spheres of masses m_1 and m_2 respectively which do not intersect are attracted to each other as if they were particles of the same mass situated at their respective geometric centers.

Since the potential corresponding to

$$\mathbf{F} = -\frac{Gm_1m_2}{r^2}\mathbf{r}_1 \tag{24}$$

is
$$V = -\frac{Gm_1m_2}{r} \tag{25}$$
it is also possible to find the attraction between objects by first finding the potential and then using $\mathbf{F} = -\nabla V$. See Problems 5.26-5.33.

MOTION IN AN INVERSE SQUARE FORCE FIELD

As we have seen, the planets revolve in elliptical orbits about the sun which is at one focus of the ellipse. In a similar manner, satellites (natural or man-made) may revolve around planets in elliptical orbits. However, the motion of an object in an inverse square field of attraction need not always be elliptical but may be parabolic or hyperbolic. In such cases the object, such as a *comet* or *meteorite,* would enter the solar system and then leave but never return again.

The following simple condition in terms of the total energy E determines the path of an object.

(i) if $E < 0$ the path is an ellipse

(ii) if $E = 0$ the path is a parabola

(iii) if $E > 0$ the path is a hyperbola

Other conditions in terms of the speed of the object are also available. See Problem 5.37.

In this chapter we assume the sun to be fixed and the planets do not affect each other. Similarly in the motion of satellites around a planet such as the earth, for example, we assume the planet fixed and that the sun and all other planets have no effect.

Although such assumption is correct as a first approximation, the influence of other planets may have to be taken into account for more accurate purposes. The problems of dealing with the motions of two, three, etc., objects under their mutual attractions are often called the *two body problem, three body problem,* etc.

Solved Problems

CENTRAL FORCES AND IMPORTANT PROPERTIES

5.1. Prove that if a particle moves in a central force field, then its path must be a plane curve.

Let $\mathbf{F} = f(r)\,\mathbf{r}_1$ be the central force field. Then
$$\mathbf{r} \times \mathbf{F} = f(r)\,\mathbf{r} \times \mathbf{r}_1 = \mathbf{0} \tag{1}$$
since $\mathbf{r}_1$ is a unit vector in the direction of the position vector $\mathbf{r}$. Since $\mathbf{F} = m\,d\mathbf{v}/dt$, this can be written
$$\mathbf{r} \times d\mathbf{v}/dt = \mathbf{0} \tag{2}$$
or
$$\frac{d}{dt}(\mathbf{r} \times \mathbf{v}) = \mathbf{0} \tag{3}$$
Integrating, we find
$$\mathbf{r} \times \mathbf{v} = \mathbf{h} \tag{4}$$
where $\mathbf{h}$ is a constant vector. Multiplying both sides of (4) by $\mathbf{r}\cdot$,
$$\mathbf{r}\cdot\mathbf{h} = 0 \tag{5}$$
using the fact that $\mathbf{r}\cdot(\mathbf{r}\times\mathbf{v}) = (\mathbf{r}\times\mathbf{r})\cdot\mathbf{v} = 0$. Thus $\mathbf{r}$ is perpendicular to the constant vector $\mathbf{h}$, and so the motion takes place in a plane. We shall assume that this plane is taken to be the xy plane whose origin is at the center of force.

5.2. Prove that for a particle moving in a central force field the angular momentum is conserved.

From equation (4) of Problem 5.1, we have

$$\mathbf{r} \times \mathbf{v} = \mathbf{h}$$

where $\mathbf{h}$ is a constant vector. Then multiplying by mass m,

$$m(\mathbf{r} \times \mathbf{v}) = m\mathbf{h} \tag{1}$$

Since the left side of (1) is the angular momentum, it follows that the angular momentum is conserved, i.e. is always constant in magnitude and direction.

EQUATIONS OF MOTION FOR A PARTICLE IN A CENTRAL FIELD

5.3. Write the equations of motion for a particle in a central field.

By Problem 5.1 the motion of the particle takes place in a plane. Choose this plane to be the xy plane and the coordinates describing the position of the particle at any time t to be polar coordinates (r, θ). Using Problem 1.49, page 27, we have

$$\text{(mass)(acceleration)} = \text{net force}$$

$$m\{(\ddot{r} - r\dot{\theta}^2)\mathbf{r}_1 + (r\ddot{\theta} + 2\dot{r}\dot{\theta})\boldsymbol{\theta}_1\} = f(r)\,\mathbf{r}_1 \tag{1}$$

Thus the required equations of motion are given by

$$m(\ddot{r} - r\dot{\theta}^2) = f(r) \tag{2}$$

$$m(r\ddot{\theta} + 2\dot{r}\dot{\theta}) = 0 \tag{3}$$

5.4. Show that $r^2\dot{\theta} = h$, a constant.

Method 1. Equation (3) of Problem 5.3 can be written

$$m(r\ddot{\theta} + 2\dot{r}\dot{\theta}) = \frac{m}{r}(r^2\ddot{\theta} + 2r\dot{r}\dot{\theta}) = \frac{m}{r}\frac{d}{dt}(r^2\dot{\theta}) = 0$$

Thus $\frac{d}{dt}(r^2\dot{\theta}) = 0$ and so

$$r^2\dot{\theta} = h \tag{1}$$

where h is a constant.

Method 2. By Problem 1.49, page 27, the velocity in polar coordinates is

$$\mathbf{v} = \dot{r}\mathbf{r}_1 + r\dot{\theta}\boldsymbol{\theta}_1$$

Then from equation (4) of Problem 5.1

$$\mathbf{h} = \mathbf{r} \times \mathbf{v} = \dot{r}(\mathbf{r} \times \mathbf{r}_1) + r\dot{\theta}(\mathbf{r} \times \boldsymbol{\theta}_1) = r^2\dot{\theta}\mathbf{k} \tag{2}$$

since $\mathbf{r} \times \mathbf{r}_1 = \mathbf{0}$ and $\mathbf{r} \times \boldsymbol{\theta}_1 = r\mathbf{k}$ where $\mathbf{k}$ is the unit vector in a direction perpendicular to the plane of motion [the xy plane], i.e. in the direction $\mathbf{r} \times \mathbf{v}$. Using $\mathbf{h} = h\mathbf{k}$ in (2), we see that $r^2\dot{\theta} = h$.

5.5. Prove that $r^2\dot{\theta} = 2\dot{A}$ where $\dot{A}$ is the time rate at which area is swept out by the position vector $\mathbf{r}$.

Suppose that in time Δt the particle moves from M to N [see Fig. 5-8]. The area ΔA swept out by the position vector in this time is approximately half the area of a parallelogram with sides r and Δr or [see Problem 1.18, page 15]

$$\Delta A = \tfrac{1}{2}|\mathbf{r} \times \Delta\mathbf{r}|$$

Dividing by Δt and letting $\Delta t \to 0$,

$$\lim_{\Delta t \to 0} \frac{\Delta A}{\Delta t} = \lim_{\Delta t \to 0} \frac{1}{2}\left|\mathbf{r} \times \frac{\Delta\mathbf{r}}{\Delta t}\right| = \frac{1}{2}|\mathbf{r} \times \mathbf{v}|$$

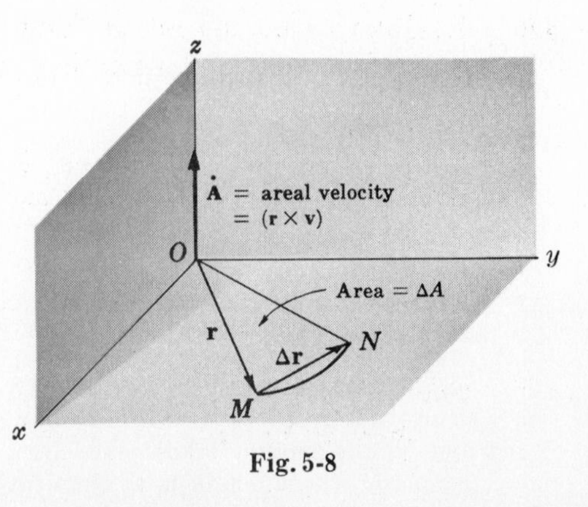

Fig. 5-8

i.e.,

$$\dot{A} = \tfrac{1}{2}|\mathbf{r}\times\mathbf{v}| = \tfrac{1}{2}r^2\dot{\theta}$$

using the result in Problem 5.4. Thus $r^2\dot{\theta} = 2\dot{A}$, as required. The vector quantity

$$\dot{\mathbf{A}} = \dot{A}\mathbf{k} = \tfrac{1}{2}(\mathbf{r}\times\mathbf{v}) = \tfrac{1}{2}(r^2\dot{\theta})\mathbf{k}$$

is often called the *areal velocity*.

5.6. Prove that for a particle moving in a central force field the areal velocity is constant.

By Problem 5.4, $r^2\dot{\theta} = h$ = a constant. Then the areal velocity is

$$\dot{\mathbf{A}} = \tfrac{1}{2}r^2\dot{\theta}\mathbf{k} = \tfrac{1}{2}h\mathbf{k} = \tfrac{1}{2}\mathbf{h}, \text{ a constant vector}$$

The result is often stated as follows: If a particle moves in a central force field with O as center, then the radius vector drawn from O to the particle sweeps out equal areas in equal times. This result is sometimes called the *law of areas.*

5.7. Show by means of the substitution $r = 1/u$ that the differential equation for the path of the particle in a central field is

$$\frac{d^2u}{d\theta^2} + u = -\frac{f(1/u)}{mh^2u^2}$$

From Problem 5.4 or equation (*3*) of Problem 5.3, we have

$$r^2\dot{\theta} = h \qquad \text{or} \qquad \dot{\theta} = h/r^2 = hu^2 \tag{1}$$

Substituting into equation (*2*) of Problem 5.3, we find

$$m(\ddot{r} - h^2/r^3) = f(r) \tag{2}$$

Now if $r = 1/u$, we have

$$\dot{r} = \frac{dr}{dt} = \frac{dr}{d\theta}\frac{d\theta}{dt} = \frac{h}{r^2}\frac{dr}{d\theta} = -h\frac{du}{d\theta} \tag{3}$$

$$\ddot{r} = \frac{d\dot{r}}{dt} = \frac{d}{dt}\left(-h\frac{du}{d\theta}\right) = \frac{d}{d\theta}\left(-h\frac{du}{d\theta}\right)\frac{d\theta}{dt} = -h^2u^2\frac{d^2u}{d\theta^2} \tag{4}$$

From this we see that (*2*) can be written

$$m(-h^2u^2\, d^2u/d\theta^2 - h^2u^3) = f(1/u) \tag{5}$$

or, as required,

$$\frac{d^2u}{d\theta^2} + u = -\frac{f(1/u)}{mh^2u^2} \tag{6}$$

POTENTIAL ENERGY AND CONSERVATION OF ENERGY FOR CENTRAL FORCE FIELDS

5.8. (*a*) Prove that a central force field is conservative and (*b*) find the corresponding potential energy of a particle in this field.

Method 1.

If we can find the potential or potential energy, then we will have also incidentally proved that the field is conservative. Now if the potential V exists, it must be such that

$$\mathbf{F}\cdot d\mathbf{r} = -dV \tag{1}$$

where $\mathbf{F} = f(r)\,\mathbf{r}_1$ is the central force. We have

$$\mathbf{F}\cdot d\mathbf{r} = f(r)\,\mathbf{r}_1\cdot d\mathbf{r} = f(r)\frac{\mathbf{r}}{r}\cdot d\mathbf{r} = f(r)\,dr$$

since $\mathbf{r}\cdot d\mathbf{r} = r\,dr$.

Since we can determine V such that

$$-dV = f(r)\,dr$$

for example,
$$V = -\int f(r)\,dr \tag{2}$$
it follows that the field is conservative and that (2) represents the potential or potential energy.

Method 2.

We can show that $\nabla \times \mathbf{F} = \mathbf{0}$ directly, but this method is tedious although straightforward.

5.9. Write the conservation of energy for a particle of mass m in a central force field.

Method 1. The velocity of a particle expressed in polar coordinates is [Problem 1.49, page 27]
$$\mathbf{v} = \dot{r}\mathbf{r}_1 + r\dot{\theta}\boldsymbol{\theta}_1 \qquad \text{so that} \qquad v^2 = \mathbf{v}\cdot\mathbf{v} = \dot{r}^2 + r^2\dot{\theta}^2$$
Then the principle of conservation of energy can be expressed as
$$\tfrac{1}{2}mv^2 + V = E \qquad \text{or} \qquad \tfrac{1}{2}m(\dot{r}^2 + r^2\dot{\theta}^2) - \int f(r)\,dr = E$$
where E is a constant.

Method 2. The equations of motion for a particle in a central field are, by Problem 5.3,
$$m(\ddot{r} - r\dot{\theta}^2) = f(r) \tag{1}$$
$$m(r\ddot{\theta} + 2\dot{r}\dot{\theta}) = 0 \tag{2}$$
Multiply equation (1) by $\dot{r}$, equation (2) by $r\dot{\theta}$ and add to obtain
$$m(\dot{r}\ddot{r} + r^2\dot{\theta}\ddot{\theta} + r\dot{r}\dot{\theta}^2) = f(r)\dot{r} \tag{3}$$
This can be written
$$\tfrac{1}{2}m\frac{d}{dt}(\dot{r}^2 + r^2\dot{\theta}^2) = \frac{d}{dt}\int f(r)\,dr \tag{4}$$
Then integrating both sides, we obtain
$$\tfrac{1}{2}m(\dot{r}^2 + r^2\dot{\theta}^2) - \int f(r)\,dr = E \tag{5}$$

5.10. Show that the differential equation describing the motion of a particle in a central field can be written as
$$\frac{mh^2}{2r^4}\left[\left(\frac{dr}{d\theta}\right)^2 + r^2\right] - \int f(r)\,dr = E$$

From Problem 5.9 we have by the conservation of energy,
$$\tfrac{1}{2}m(\dot{r}^2 + r^2\dot{\theta}^2) - \int f(r)\,dr = E \tag{1}$$
We also have
$$\dot{r} = \frac{dr}{dt} = \frac{dr}{d\theta}\frac{d\theta}{dt} = \frac{dr}{d\theta}\dot{\theta} \tag{2}$$
Substituting (2) into (1), we find
$$\tfrac{1}{2}m\left[\left(\frac{dr}{d\theta}\right)^2 + r^2\right]\dot{\theta}^2 - \int f(r)\,dr = E \qquad \text{or} \qquad \frac{mh^2}{2r^4}\left[\left(\frac{dr}{d\theta}\right)^2 + r^2\right] - \int f(r)\,dr = E$$
since $\dot{\theta} = h/r^2$.

5.11. (a) If $u = 1/r$, prove that $v^2 = \dot{r}^2 + r^2\dot{\theta}^2 = h^2\{(du/d\theta)^2 + u^2\}$.

(b) Use (a) to prove that the conservation of energy equation becomes
$$(du/d\theta)^2 + u^2 = 2(E - V)/mh^2$$

(a) From equations (1) and (3) of Problem 5.7 we have $\dot{\theta} = hu^2$, $\dot{r} = -h\,du/d\theta$. Thus

$$v^2 = \dot{r}^2 + r^2\dot{\theta}^2 = h^2(du/d\theta)^2 + (1/u^2)(hu^2)^2 = h^2\{(du/d\theta)^2 + u^2\}$$

(b) From the conservation of energy [Problem 5.9] and part (a),

$$\tfrac{1}{2}mv^2 = \tfrac{1}{2}m(\dot{r}^2 + r\dot{\theta}^2) = E - V \qquad \text{or} \qquad (du/d\theta)^2 + u^2 = 2(E - V)/mh^2$$

DETERMINATION OF ORBIT FROM CENTRAL FORCE, OR CENTRAL FORCE FROM ORBIT

5.12. Show that the position of the particle as a function of time t can be determined from the equations

$$t = \int [G(r)]^{-1/2}\,dr, \qquad t = \frac{1}{h}\int r^2\,d\theta$$

where
$$G(r) = \frac{2E}{m} + \frac{2}{m}\int f(r)\,dr - \frac{h^2}{2m^2r^2}$$

Placing $\dot{\theta} = h/r^2$ in the equation for conservation of energy of Problem 5.9,

$$\tfrac{1}{2}m(\dot{r}^2 + h^2/r^2) - \int f(r)\,dr = E$$

or
$$\dot{r}^2 = \frac{2E}{m} + \frac{2}{m}\int f(r)\,dr - \frac{h^2}{r^2} = G(r)$$

Then assuming the positive square root, we have

$$dr/dt = \sqrt{G(r)}$$

and so separating the variables and integrating, we find

$$t = \int [G(r)]^{-1/2}\,dr$$

The second equation follows by writing $\dot{\theta} = h/r^2$ as $dt = r^2\,d\theta/h$ and integrating.

5.13. Show that if the law of central force is defined by

$$f(r) = -K/r^2, \qquad K > 0$$

i.e. an inverse square law of attraction, then the path of the particle is a conic.

Method 1.

In this case $f(1/u) = -Ku^2$. Substituting into the differential equation of motion in Problem 5.7, we find

$$d^2u/d\theta^2 + u = K/mh^2 \tag{1}$$

This equation has the general solution

$$u = A\cos\theta + B\sin\theta + K/mh^2 \tag{2}$$

or using Problem 4.2, page 92,

$$u = K/mh^2 + C\cos(\theta - \phi) \tag{3}$$

i.e.,
$$r = \frac{1}{K/mh^2 + C\cos(\theta - \phi)} \tag{4}$$

It is always possible to choose the axes so that $\phi = 0$, in which case we have

$$r = \frac{1}{K/mh^2 + C\cos\theta} \tag{5}$$

This has the general form of the conic [see Problem 5.16]

$$r = \frac{p}{1 + \epsilon \cos \theta} = \frac{1}{1/p + (\epsilon/p) \cos \theta} \tag{6}$$

Then comparing (5) and (6) we see that

$$1/p = K/mh^2, \qquad \epsilon/p = C \tag{7}$$

or

$$p = mh^2/K, \qquad \epsilon = mh^2C/K \tag{8}$$

Method 2. Since $f(r) = -K/r^2$, we have

$$V = -\int f(r)\, dr = -K/r + c_1 \tag{9}$$

where c_1 is a constant. If we assume that $V \to 0$ as $r \to \infty$, then $c_1 = 0$ and so

$$V = -K/r \tag{10}$$

Using Problem 5.10, page 124, we find

$$\frac{mh^2}{2r^4}\left[\left(\frac{dr}{d\theta}\right)^2 + r^2\right] = E + \frac{K}{r} \tag{11}$$

from which

$$\frac{dr}{d\theta} = \pm r \sqrt{\frac{2Er^2}{mh^2} + \frac{2Kr}{mh^2} - 1} \tag{12}$$

By separating variables and integrating [see Problem 5.66] we find the solution (5) where C is expressed in terms of the energy E.

5.14. (a) Obtain the constant C of Problem 5.13 in terms of the total energy E and (b) thus show that the conic is an ellipse, parabola or hyperbola according as $E < 0$, $E = 0$, $E > 0$ respectively.

Method 1.

(a) The potential energy is

$$V = -\int f(r)\, dr = \int (K/r^2)\, dr = -K/r = -Ku \tag{1}$$

where we use $u = 1/r$ and choose the constant of integration so that $\lim_{r\to\infty} V = 0$. Now from equation (5) of Problem 5.13,

$$u = 1/r = K/mh^2 + C \cos \theta \tag{2}$$

Thus from Problem 5.11(b) together with (1), we have

$$(C \sin \theta)^2 + \left(\frac{K}{mh^2} + C \cos \theta\right)^2 = \frac{2E}{mh^2} + \frac{2K}{mh^2}\left(\frac{K}{mh^2} + C \cos \theta\right)$$

or

$$C^2 = \frac{K^2}{m^2h^4} + \frac{2E}{mh^2} \qquad \text{or} \qquad C = \sqrt{\frac{K^2}{m^2h^4} + \frac{2E}{mh^2}} \tag{3}$$

assuming $C > 0$.

(b) Using the value of C in part (a), the equation of the conic becomes

$$u = \frac{1}{r} = \frac{K}{mh^2}\left\{1 + \sqrt{1 + \frac{2Emh^2}{K^2}} \cos \theta\right\}$$

Comparing this with (4) of Problem 5.16, we see that the eccentricity is

$$\epsilon = \sqrt{1 + \frac{2Emh^2}{K^2}} \tag{4}$$

From this we see that the conic is an ellipse if $E < 0$ [but greater than $-K^2/2mh^2$], a parabola if $E = 0$ and a hyperbola if $E > 0$, since in such cases $\epsilon < 1$, $\epsilon = 1$ and $\epsilon > 1$ respectively.

Method 2. The value of C can also be obtained as in the second method of Problem 5.13.

5.15. Under the influence of a central force at point O, a particle moves in a circular orbit which passes through O. Find the law of force.

Method 1.

In polar coordinates the equation of a circle of radius a passing through O is [see Fig. 5-9]

$$r = 2a\cos\theta$$

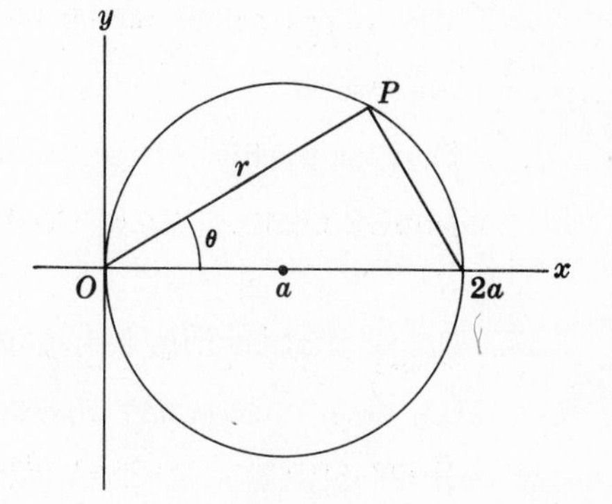

Fig. 5-9

Then since $u = 1/r = (\sec\theta)/2a$, we have

$$\frac{du}{d\theta} = \frac{\sec\theta\tan\theta}{2a}$$

$$\frac{d^2u}{d\theta^2} = \frac{(\sec\theta)(\sec^2\theta) + (\sec\theta\tan\theta)(\tan\theta)}{2a} = \frac{\sec^3\theta + \sec\theta\tan^2\theta}{2a}$$

Thus by Problem 5.7,

$$f(1/u) = -mh^2u^2\left(\frac{d^2u}{d\theta^2} + u\right) = -mh^2u^2\left(\frac{\sec^3\theta + \sec\theta\tan^2\theta + \sec\theta}{2a}\right)$$

$$= -\frac{mh^2u^2}{2a}\{\sec^3\theta + \sec\theta(\tan^2\theta + 1)\} = -\frac{mh^2u^2}{2a}\cdot 2\sec^3\theta$$

$$= -8mh^2a^2u^5$$

or

$$f(r) = -\frac{8mh^2a^2}{r^5}$$

Thus the force is one of attraction varying inversely as the fifth power of the distance from O.

Method 2. Using $r = 2a\cos\theta$ in equation (*16*), page 118, we have

$$f(r) = \frac{mh^2}{r^4}\left\{-2a\cos\theta - \frac{2}{2a\cos\theta}(-2a\sin\theta)^2 - 2a\cos\theta\right\}$$

$$= -\frac{4amh^2}{r^4\cos\theta} = -\frac{8a^2mh^2}{r^5}$$

CONIC SECTIONS. ELLIPSE, PARABOLA AND HYPERBOLA

5.16. Derive equation (*18*), page 118, for a conic section.

Referring to Fig. 5-3, page 118, by definition of a conic section we have for any point P on it,

$$r/d = \epsilon \quad \text{or} \quad d = r/\epsilon \tag{1}$$

Corresponding to the particular point Q, we have

$$p/D = \epsilon \quad \text{or} \quad p = \epsilon D \tag{2}$$

But

$$D = d + r\cos\theta = \frac{r}{\epsilon} + r\cos\theta = \frac{r}{\epsilon}(1 + \epsilon\cos\theta) \tag{3}$$

Then from (*2*) and (*3*), we have on eliminating D,

$$p = r(1 + \epsilon\cos\theta) \quad \text{or} \quad r = \frac{p}{1 + \epsilon\cos\theta} \tag{4}$$

The equation is a circle if $\epsilon = 0$, an ellipse if $0 < \epsilon < 1$, a parabola if $\epsilon = 1$ and a hyperbola if $\epsilon > 1$.

5.17. Derive equation (*19*), page 118, for an ellipse.

Referring to Fig. 5-4, page 119, we see that when $\theta = 0$, $r = OV$ and when $\theta = \pi$, $r = OU$. Thus using equation (*4*) of Problem 5.16,

$$OV = p/(1+\epsilon), \qquad OU = p/(1-\epsilon) \tag{1}$$

But since $2a$ is the length of the major axis,

$$OV + OU = 2a \qquad \text{or} \qquad p/(1+\epsilon) + p/(1-\epsilon) = 2a \tag{2}$$

from which

$$p = a(1-\epsilon^2) \tag{3}$$

Thus the equation of the ellipse is

$$r = \frac{a(1-\epsilon^2)}{1+\epsilon\cos\theta} \tag{4}$$

5.18. Prove that in Fig. 5-4, page 119, (*a*) $OV = a(1-\epsilon)$, (*b*) $OU = a(1+\epsilon)$.

(*a*) From Problem 5.17, equation (*3*) and the first equation of (*1*),

$$OV = \frac{p}{1+\epsilon} = \frac{a(1-\epsilon^2)}{1+\epsilon} = a(1-\epsilon) \tag{1}$$

(*b*) From Problem 5.17, equation (*3*) and the second equation of (*1*),

$$OU = \frac{p}{1-\epsilon} = \frac{a(1-\epsilon^2)}{1-\epsilon} = a(1+\epsilon) \tag{2}$$

5.19. Prove that $c = a\epsilon$ where c is the distance from the center to the focus of the ellipse. a is the length of the semi-major axis and ϵ is the eccentricity.

From Fig. 5-4, page 119, we have $c = CO = CV - OV = a - a(1-\epsilon) = a\epsilon$.

An analogous result holds for the hyperbola [see Problem 5.73(*c*), page 139].

5.20. If a and c are as in Problem 5.19 and b is the length of the semi-minor axis, prove that (*a*) $c = \sqrt{a^2 - b^2}$, (*b*) $b = a\sqrt{1-\epsilon^2}$.

(*a*) From Fig. 5-4, page 119, and the definition of an ellipse, we have

$$\epsilon = \frac{OV}{VE} = \frac{CV - CO}{VE} = \frac{a-c}{VE} \qquad \text{or} \qquad VE = \frac{a-c}{\epsilon} \tag{1}$$

Also since the eccentricity is the distance from O to W divided by the distance from W to the directrix AB [which is equal to CE], we have

$$OW/CE = \epsilon$$

or, using (*1*) and the result of Problem 5.19,

$$OW = \epsilon\, CE = \epsilon(CV + VE) = \epsilon[a + (a-c)/\epsilon] = \epsilon a + a - c = a$$

Then $(OW)^2 = (OC)^2 + (CW)^2$ or $a^2 = b^2 + c^2$, i.e. $c = \sqrt{a^2 - b^2}$.

(*b*) From Problem 5.19 and part (*a*), $a^2 = b^2 + a^2\epsilon^2$ or $b = a\sqrt{1-\epsilon^2}$.

KEPLER'S LAWS OF PLANETARY MOTION AND NEWTON'S UNIVERSAL LAW OF GRAVITATION

5.21. Prove that if a planet is to revolve around the sun in an elliptical path with the sun at a focus [Kepler's first law], then the central force necessary varies inversely as the square of the distance of the planet from the sun.

If the path is an ellipse with the sun at a focus, then calling r the distance from the sun, we have by Problem 5.16,

$$r = \frac{p}{1+\epsilon\cos\theta} \qquad \text{or} \qquad u = \frac{1}{r} = \frac{1}{p} + \frac{\epsilon}{p}\cos\theta \tag{1}$$

where $\epsilon < 1$. Then the central force is given as in Problem 5.7 by

$$f(1/u) = -mh^2u^2(d^2u/d\theta^2 + u) = -mh^2u^2/p \tag{2}$$

on substituting the value of u in (1). From (2) we have on replacing u by $1/r$,

$$f(r) = -mh^2/pr^2 = -K/r^2 \tag{3}$$

5.22. Discuss the connection of Newton's universal law of gravitation with Problem 5.21.

Historically, Newton arrived at the inverse square law of force for planets by using Kepler's first law and the method of Problem 5.21. He was then led to the idea that perhaps all objects of the universe were attracted to each other with a force which was inversely proportional to the square of the distance r between them and directly proportional to the product of their masses. This led to the fundamental postulate

$$\mathbf{F} = -\frac{GMm}{r^2}\mathbf{r}_1 \tag{1}$$

where G is the universal gravitational constant. Equivalently, the law of force (3) of Problem 5.21 is the same as (1) where

$$K = GMm \tag{2}$$

5.23. Prove Kepler's third law: The squares of the periods of the various planets are proportional to the cubes of their corresponding semi-major axes.

If a and b are the lengths of the semi-major and semi-minor axes, then the area of the ellipse is πab. Since the areal velocity has magnitude $h/2$ [Problem 5.6], the time taken to sweep over area πab, i.e. the period, is

$$P = \frac{\pi ab}{h/2} = \frac{2\pi ab}{h} \tag{1}$$

Now by Problem 5.17 equation (3), Problem 5.20(b), and Problem 5.13 equation (8), we have

$$b = a\sqrt{1-\epsilon^2}, \qquad p = a(1-\epsilon^2) = mh^2/K \tag{2}$$

Then from (1) and (2) we find

$$P = 2\pi m^{1/2}a^{3/2}/K^{1/2} \qquad \text{or} \qquad P^2 = 4\pi^2 ma^3/K$$

Thus the squares of the periods are proportional to the cubes of the semi-major axes.

5.24. Prove that $GM = gR^2$.

On the earth's surface, i.e. $r = R$ where R is the radius, the force of attraction of the earth on an object of mass m is equal to the weight mg of the object. Thus if M is the mass of the earth,

$$GMm/R^2 = mg \qquad \text{or} \qquad GM = gR^2$$

5.25. Calculate the mass of the earth.

From Problem 5.24, $GM = gR^2$ or $M = gR^2/G$. Taking the radius of the earth as 6.38×10^8 cm, $g = 980$ cm/sec^2 and $G = 6.67 \times 10^{-8}$ cgs units, we find $M = 5.98 \times 10^{27}$ gm $= 1.32 \times 10^{25}$ lb.

ATTRACTION OF OBJECTS

5.26. Find the force of attraction of a thin uniform rod of length $2a$ on a particle of mass m placed at a distance b from its midpoint.

Choose the x axis along the rod and the y axis perpendicular to the rod and passing through its center O, as shown in Fig. 5-10. Let σ be the mass per unit length of the rod. The force of attraction $d\mathbf{F}$ between an element of mass $\sigma\,dx$ of the rod and m is, by Newton's universal law of gravitation,

$$d\mathbf{F} = \frac{Gm\sigma\,dx}{x^2+b^2}(\sin\theta\,\mathbf{i} - \cos\theta\,\mathbf{j})$$

$$= \frac{Gm\sigma x\,dx}{(x^2+b^2)^{3/2}}\,\mathbf{i} - \frac{Gm\sigma b\,dx}{(x^2+b^2)^{3/2}}\,\mathbf{j}$$

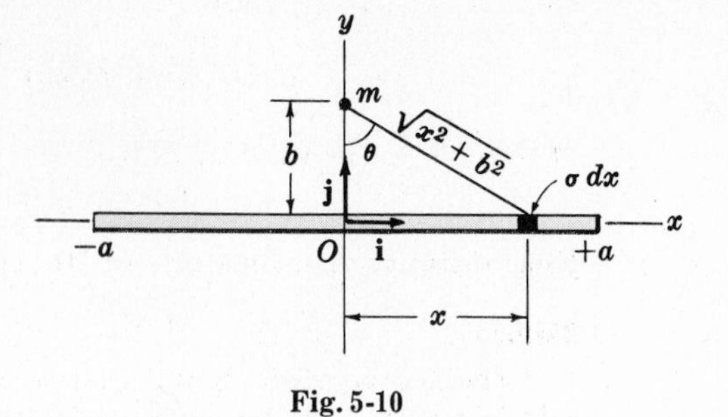

Fig. 5-10

since from Fig. 5-10, $\sin\theta = x/\sqrt{x^2+b^2}$, $\cos\theta = b/\sqrt{x^2+b^2}$. Then the total force of attraction is

$$\mathbf{F} = \mathbf{i}\int_{x=-a}^{a}\frac{Gm\sigma x\,dx}{(x^2+b^2)^{3/2}} - \mathbf{j}\int_{x=-a}^{a}\frac{Gm\sigma b\,dx}{(x^2+b^2)^{3/2}}$$

$$= 0 - 2\mathbf{j}\int_0^a \frac{Gm\sigma b\,dx}{(x^2+b^2)^{3/2}} = -2Gm\sigma b\,\mathbf{j}\int_0^a\frac{dx}{(x^2+b^2)^{3/2}}$$

Let $x = b\tan\theta$ in this integral. Then when $x=0$, $\theta=0$; and when $x=a$, $\theta = \tan^{-1}(a/b)$. Thus the integral becomes

$$\mathbf{F} = -2Gm\sigma b\,\mathbf{j}\int_0^{\tan^{-1}(a/b)}\frac{b\sec^2\theta\,d\theta}{(b^2\sec^2\theta)^{3/2}} = -\frac{2Gm\sigma a}{b\sqrt{a^2+b^2}}\,\mathbf{j}$$

Since the mass of the rod is $M = 2a\sigma$, this can also be written as

$$\mathbf{F} = -\frac{GMm}{b\sqrt{a^2+b^2}}\,\mathbf{j}$$

Thus we see that the force of attraction is directed from m to the center of the rod and of magnitude $2Gm\sigma a/b\sqrt{a^2+b^2}$ or $GMm/b\sqrt{a^2+b^2}$.

5.27. A mass m lies on the perpendicular through the center of a uniform thin circular plate of radius a and at distance b from the center. Find the force of attraction between the plate and the mass m.

Method 1.

Let $\mathbf{n}$ be a unit vector drawn from point P where m is located to the center O of the plate. Subdivide the circular plate into circular rings [such as ABC in Fig. 5-11] of radius r and thickness dr. If σ is the mass per unit area, then the mass of the ring is $\sigma(2\pi r\,dr)$. Since all points of the ring are at the same distance $\sqrt{r^2+b^2}$ from P, the force of attraction of the ring on m will be

$$d\mathbf{F} = \frac{G\sigma(2\pi r\,dr)m}{r^2+b^2}\cos\phi\,\mathbf{n}$$

$$= \frac{G\sigma\,2\pi r\,dr\,mb}{(r^2+b^2)^{3/2}}\,\mathbf{n} \qquad (1)$$

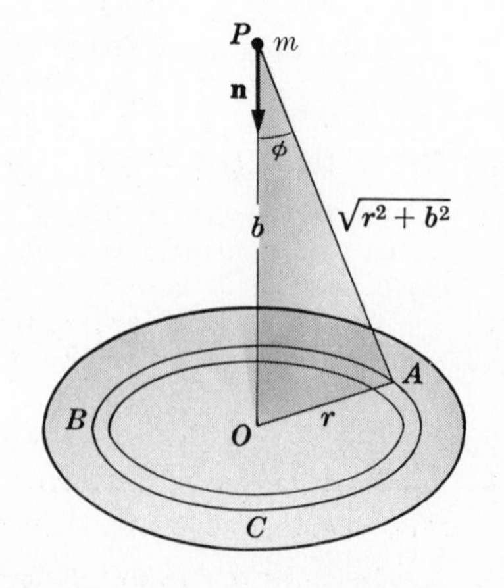

Fig. 5-11

where we have used the fact that due to symmetry the resultant force of attraction is in the direction $\mathbf{n}$. By integrating over all rings from $r=0$ to $r=a$, we find that the total attraction is

$$\mathbf{F} = 2\pi G\sigma mb\,\mathbf{n}\int_0^a\frac{r\,dr}{(r^2+b^2)^{3/2}} \qquad (2)$$

To evaluate the integral, let $r^2+b^2 = u^2$ so that $r\,dr = u\,du$. Then since $u=b$ when $r=0$ and $u=\sqrt{a^2+b^2}$ when $r=a$, the result is

$$\mathbf{F} \;=\; 2\pi G\sigma m b\,\mathbf{n} \int_b^{\sqrt{a^2+b^2}} \frac{u\,du}{u^3} \;=\; 2\pi G\sigma m\,\mathbf{n}\left(1 - \frac{b}{\sqrt{a^2+b^2}}\right)$$

If we let α be the value of ϕ when $r = a$, this can be written

$$\mathbf{F} \;=\; 2\pi G\sigma m\,\mathbf{n}\,(1 - \cos\alpha) \tag{3}$$

Thus the force is directed from m to the center O of the plate and has magnitude $2\pi G\sigma m(1 - \cos\alpha)$.

Method 2.

The method of double integration can also be used. In such case the element of area at A is $r\,dr\,d\theta$ where θ is the angle measured from a line [taken as the x axis] in the plane of the circular plate and passing through the center O. Then we have as in equation (*1*),

$$d\mathbf{F} \;=\; \frac{G\sigma(r\,dr\,d\theta)mb}{(r^2+b^2)^{3/2}}\,\mathbf{n}$$

and by integrating over the circular plate

$$\mathbf{F} \;=\; G\sigma m b\,\mathbf{n}\int_{r=0}^{a}\int_{\theta=0}^{2\pi}\frac{r\,dr\,d\theta}{(r^2+b^2)^{3/2}} \;=\; G\sigma m b\,\mathbf{n}\int_{r=0}^{a}\frac{2\pi r\,dr}{(r^2+b^2)^{3/2}} \;=\; 2\pi G\sigma m\,\mathbf{n}\,(1-\cos\alpha)$$

5.28. A uniform plate has its boundary consisting of two concentric half circles of inner and outer radii a and b respectively, as shown in Fig. 5-12. Find the force of attraction of the plate on a mass m located at the center O.

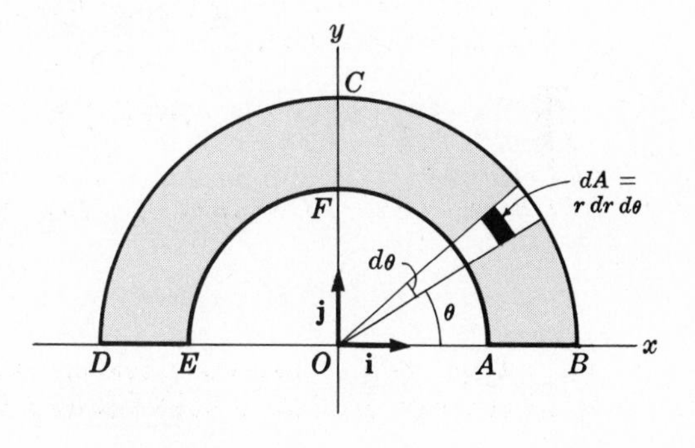

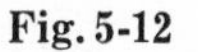
Fig. 5-12

It is convenient to use polar coordinates (r, θ). The element of area of the plate [shaded in Fig. 5-12] is $dA = r\,dr\,d\theta$, and the mass is $\sigma r\,dr\,d\theta$. Then the force of attraction between dA and O is

$$d\mathbf{F} \;=\; \frac{G(\sigma r\,dr\,d\theta)m}{r^2}\,(\cos\theta\,\mathbf{i} + \sin\theta\,\mathbf{j})$$

Thus the total force of attraction is

$$\begin{aligned}\mathbf{F} \;&=\; \int_{\theta=0}^{\pi}\int_{r=a}^{b}\frac{G(\sigma r\,dr\,d\theta)m}{r^2}\,(\cos\theta\,\mathbf{i}+\sin\theta\,\mathbf{j})\\ &=\; G\sigma m \ln\left(\frac{b}{a}\right)\int_{\theta=0}^{\pi}(\cos\theta\,\mathbf{i}+\sin\theta\,\mathbf{j})\,d\theta \;=\; 2G\sigma m\ln\left(\frac{b}{a}\right)\mathbf{j}\end{aligned}$$

Since $M = \sigma(\frac{1}{2}\pi b^2 - \frac{1}{2}\pi a^2)$, we have $\sigma = 2M/\pi(b^2 - a^2)$ and the force can be written

$$\mathbf{F} \;=\; \frac{4GMm}{\pi(b^2-a^2)}\ln\left(\frac{b}{a}\right)\mathbf{j}$$

The method of single integration can also be used by dividing the region between $r = a$ and $r = b$ into circular rings as in Problem 5.27.

5.29. Find the force of attraction of a thin spherical shell of radius a on a particle P of mass m at a distance $r > a$ from its center.

Let O be the center of the sphere. Subdivide the surface of the sphere into circular elements such as $ABCDA$ of Fig. 5-13 below by using parallel planes perpendicular to OP.

The area of the surface element $ABCDA$ as seen from Fig. 5-13 is

$$2\pi(a\sin\theta)(a\,d\theta) \;=\; 2\pi a^2\sin\theta\,d\theta$$

since the radius is $a\sin\theta$ [so that the perimeter is $2\pi(a\sin\theta)$] and the thickness is $a\,d\theta$. Then if σ is the mass per unit area, the mass of $ABCDA$ is $2\pi a^2\sigma\sin\theta\,d\theta$.

Since all points of $ABCDA$ are at the same distance $w = AP$ from P, the force of attraction of the element $ABCDA$ on m is

$$d\mathbf{F} = \frac{G(2\pi a^2\sigma \sin\theta\, d\theta)m}{w^2} \cos\phi\, \mathbf{n} \qquad (1)$$

where we have used the fact that from symmetry the net force will be in the direction of the unit vector $\mathbf{n}$ from P toward O. Now from Fig. 5-13,

$$\cos\phi = \frac{PE}{AP} = \frac{PO - EO}{AP} = \frac{r - a\cos\theta}{w} \qquad (2)$$

Using (2) in (1) together with the fact that by the cosine law

$$w^2 = a^2 + r^2 - 2ar\cos\theta \qquad (3)$$

we find

$$d\mathbf{F} = \frac{G(2\pi a^2\sigma \sin\theta\, d\theta)m(r - a\cos\theta)}{(a^2 + r^2 - 2ar\cos\theta)^{3/2}}\,\mathbf{n}$$

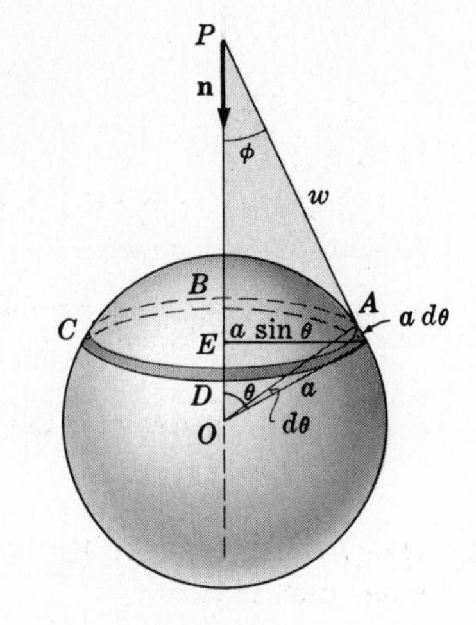

Fig. 5-13

Then the total force is

$$\mathbf{F} = 2\pi G a^2 \sigma m\, \mathbf{n} \int_{\theta=0}^{\pi} \frac{(r - a\cos\theta)\sin\theta}{(a^2 + r^2 - 2ar\cos\theta)^{3/2}}\, d\theta \qquad (4)$$

We can evaluate the integral by using the variable w given by (3) in place of θ. When $\theta = 0$, $w^2 = a^2 - 2ar + r^2 = (r-a)^2$ so that $w = r - a$ if $r > a$. Also when $\theta = \pi$, $w^2 = a^2 + 2ar + r^2 = (r+a)^2$ so that $w = r + a$. In addition, we have

$$2w\, dw = 2ar \sin\theta\, d\theta$$

$$r - a\cos\theta = r - a\left(\frac{a^2 + r^2 - w^2}{2ar}\right) = \frac{w^2 - a^2 + r^2}{2r}$$

Then (4) becomes

$$\mathbf{F} = \frac{\pi G a \sigma m\, \mathbf{n}}{r^2} \int_{r-a}^{r+a} \left(1 + \frac{r^2 - a^2}{w^2}\right) dw = \frac{4\pi G a^2 \sigma m\, \mathbf{n}}{r^2}$$

5.30. Work Problem 5.29 if $r < a$.

In this case the force is also given by (4) of Problem 5.29. However, in evaluating the integral we note that on making the substitution (3) of Problem 5.29 that $\theta = 0$ yields $w^2 = (a - r)^2$ or $w = a - r$ if $r < a$. Then the result (4) of Problem 5.29 becomes

$$\mathbf{F} = \frac{\pi G a \sigma m\, \mathbf{n}}{r^2} \int_{a-r}^{a+r} \left(1 - \frac{a^2 - r^2}{w^2}\right) dw = \mathbf{0}$$

Thus there will be no force of attraction of a spherical shell on any mass placed inside. This means that in such case a particle will be in equilibrium inside of the shell.

5.31. Prove that the force of attraction in Problem 5.29 is the same as if all the mass of the spherical shell were concentrated at its center.

The mass of the shell is $M = 4\pi a^2 \sigma$. Thus the force is $\mathbf{F} = (GMm/r^2)\mathbf{n}$, which proves the required result.

5.32. (a) Find the force of attraction of a solid uniform sphere on a mass m placed outside of it and (b) prove that the force is the same as if all the mass were concentrated at its center.

(a) We can subdivide the solid sphere into thin concentric spherical shells. If ρ is the distance of any of these shells from the center and $d\rho$ is the thickness, then by Problem 5.29 the force of attraction of this shell on the mass m is

$$d\mathbf{F} = \frac{G\sigma(4\pi\rho^2\,d\rho)m}{r^2}\,\mathbf{n} \tag{1}$$

where σ is the mass per unit volume. Then the total force obtained by integrating from $r = 0$ to $r = a$ is

$$\mathbf{F} = \frac{4\pi G\sigma m\,\mathbf{n}}{r^2}\int_0^a \rho^2\,d\rho = \frac{G(\frac{4}{3}\pi a^3)\sigma m\,\mathbf{n}}{r^2} \tag{2}$$

(b) Since the mass of the sphere is $M = \frac{4}{3}\pi a^3\sigma$, (2) can be written as $\mathbf{F} = (GMm/r^2)\mathbf{n}$, which shows that the force of attraction is the same as if all the mass were concentrated at the center.

We can also use triple integration to obtain this result [see Problem 5.130].

5.33. Derive the result of Problems 5.29 and 5.30 by first finding the potential due to the mass distribution.

The potential dV due to the element $ABCDA$ is

$$dV = -\frac{G(2\pi a^2\sigma\sin\theta\,d\theta)m}{w} = -\frac{G(2\pi a^2\sigma\sin\theta\,d\theta)m}{\sqrt{a^2+r^2-2ar\cos\theta}}$$

Then the total potential is

$$V = -2\pi G a^2\sigma m\int_0^\pi \frac{\sin\theta\,d\theta}{\sqrt{a^2+r^2-2ar\cos\theta}}$$
$$= -\frac{2\pi G a\sigma m}{r}\{\sqrt{(a+r)^2}-\sqrt{(a-r)^2}\}$$

If $r > a$ this yields $\quad V = -\frac{4\pi G a^2\sigma m}{r} = -\frac{GMm}{r}$

If $r < a$ it yields $\quad V = -4\pi G a\sigma m$

Then if $r > a$ the force is

$$\mathbf{F} = -\nabla V = -\nabla\left(-\frac{GMm}{r}\right) = -\frac{GMm}{r^2}\mathbf{r}_1$$

and if $r < a$ the force is

$$\mathbf{F} = -\nabla V = -\nabla(-4\pi G a\sigma m) = \mathbf{0}$$

in agreement with Problems 5.29 and 5.30.

MISCELLANEOUS PROBLEMS

5.34. An object is projected vertically upward from the earth's surface with initial speed v_0. Neglecting air resistance, (a) find the speed at a distance H above the earth's surface and (b) the smallest velocity of projection needed in order that the object never return.

(a) Let r denote the radial distance of the object at time t from the center of the earth, which we assume is fixed [see Fig. 5-14]. If M is the mass of the earth and R is its radius, then by Newton's universal law of gravitation and Problem 5.29, the force between m and M is

$$\mathbf{F} = -\frac{GMm}{r^2}\mathbf{r}_1 \tag{1}$$

where $\mathbf{r}_1$ is a unit vector directed radially outward from the earth's center in the direction of motion of the object.

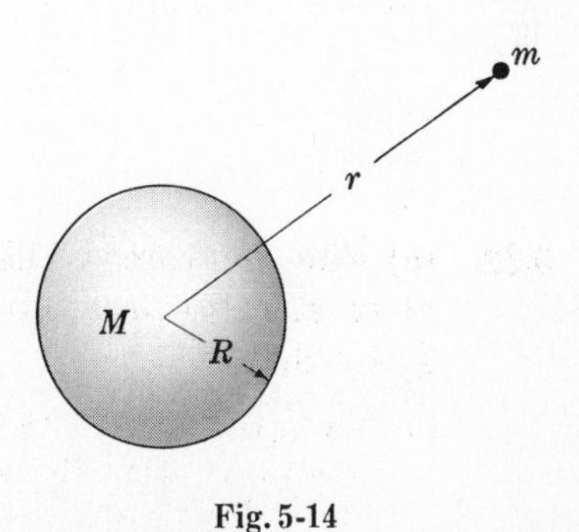

Fig. 5-14

If v is the speed at time t, we have by Newton's second law,

$$m\frac{dv}{dt}\mathbf{r}_1 = -\frac{GMm}{r^2}\mathbf{r}_1 \qquad \text{or} \qquad \frac{dv}{dt} = -\frac{GM}{r^2} \tag{2}$$

This can be written as

$$\frac{dv}{dr}\frac{dr}{dt} = -\frac{GM}{r^2} \qquad \text{or} \qquad v\frac{dv}{dr} = -\frac{GM}{r^2} \tag{3}$$

Then by integrating, we find $\qquad v^2/2 = GM/r + c_1 \qquad (4)$

Since the object starts from the earth's surface with speed v_0, we have $v = v_0$ when $r = R$ so that $c_1 = v_0^2/2 - GM/R$. Then (4) becomes

$$v^2 = 2GM\left(\frac{1}{r} - \frac{1}{R}\right) + v_0^2 \tag{5}$$

Thus when the object is at height H above the earth's surface, i.e. $r = R + H$,

$$v^2 = 2GM\left(\frac{1}{R+H} - \frac{1}{R}\right) + v_0^2 = v_0^2 - \frac{2GMH}{R(R+H)}$$

i.e.,

$$v = \sqrt{v_0^2 - \frac{2GMH}{R(R+H)}}$$

Using Problem 5.24, this can be written

$$v = \sqrt{v_0^2 - \frac{2gRH}{R+H}} \tag{6}$$

(b) As $H \to \infty$, the limiting speed (6) becomes

$$\sqrt{v_0^2 - 2GM/R} \qquad \text{or} \qquad \sqrt{v_0^2 - 2gR} \tag{7}$$

since $\lim\limits_{H\to\infty} \dfrac{H}{(R+H)} = 1$. The minimum initial speed occurs where (7) is zero or where

$$v_0 = \sqrt{2GM/R} = \sqrt{2gR} \tag{8}$$

This minimum speed is called the *escape speed* and the corresponding velocity is called the *escape velocity* from the earth's surface.

5.35. Show that the magnitude of the escape velocity of an object from the earth's surface is about 7 mi/sec.

From equation (8) of Problem 5.34, $v_0 = \sqrt{2gR}$. Taking $g = 32$ ft/sec^2 and $R = 4000$ mi, we find $v_0 = 6.96$ mi/sec.

5.36. Prove, by using vector methods primarily, that the path of a planet around the sun is an ellipse with the sun at one focus.

Since the force $\mathbf{F}$ between the planet and sun is

$$\mathbf{F} = m\frac{d\mathbf{v}}{dt} = -\frac{GMm}{r^2}\mathbf{r}_1 \tag{1}$$

we have

$$\frac{d\mathbf{v}}{dt} = -\frac{GM}{r^2}\mathbf{r}_1 \tag{2}$$

Also, by Problem 5.1, equation (4), we have

$$\mathbf{r} \times \mathbf{v} = \mathbf{h} \tag{3}$$

Now since $\mathbf{r} = r\mathbf{r}_1$, $\mathbf{v} = \dfrac{d\mathbf{r}}{dt} = r\dfrac{d\mathbf{r}_1}{dt} + \dfrac{dr}{dt}\mathbf{r}_1$. Thus from (3),

$$\mathbf{h} = \mathbf{r} \times \mathbf{v} = r\mathbf{r}_1 \times \left(r\frac{d\mathbf{r}_1}{dt} + \frac{dr}{dt}\mathbf{r}_1\right) = r^2\mathbf{r}_1 \times \frac{d\mathbf{r}_1}{dt} \tag{4}$$

From (2),

$$\frac{d\mathbf{v}}{dt} \times \mathbf{h} = -\frac{GM}{r^2}\mathbf{r}_1 \times \mathbf{h} = -GM\,\mathbf{r}_1 \times \left(\mathbf{r}_1 \times \frac{d\mathbf{r}_1}{dt}\right)$$

$$= -GM\left\{\left(\mathbf{r}_1 \cdot \frac{d\mathbf{r}_1}{dt}\right)\mathbf{r}_1 - (\mathbf{r}_1 \cdot \mathbf{r}_1)\frac{d\mathbf{r}_1}{dt}\right\} = GM\frac{d\mathbf{r}_1}{dt}$$

using equation (4) above and equation (7), page 5.

But since $\mathbf{h}$ is a constant vector, $\frac{d\mathbf{v}}{dt} \times \mathbf{h} = \frac{d}{dt}(\mathbf{v} \times \mathbf{h})$ so that

$$\frac{d}{dt}(\mathbf{v} \times \mathbf{h}) = GM\frac{d\mathbf{r}_1}{dt}$$

Integrating,

$$\mathbf{v} \times \mathbf{h} = GM\,\mathbf{r}_1 + \mathbf{c}$$

from which

$$\mathbf{r} \cdot (\mathbf{v} \times \mathbf{h}) = GM\,\mathbf{r} \cdot \mathbf{r}_1 + \mathbf{r} \cdot \mathbf{c} = GMr + r\mathbf{r}_1 \cdot \mathbf{c} = GMr + rc\cos\theta$$

where $\mathbf{c}$ is an arbitrary constant vector having magnitude c, and θ is the angle between $\mathbf{c}$ and $\mathbf{r}_1$.

Since $\mathbf{r} \cdot (\mathbf{v} \times \mathbf{h}) = (\mathbf{r} \times \mathbf{v}) \cdot \mathbf{h} = \mathbf{h} \cdot \mathbf{h} = h^2$ [see Problem 1.72(a), page 27],

$$h^2 = GMr + rc\cos\theta$$

and so

$$r = \frac{h^2}{GM + c\cos\theta} = \frac{h^2/GM}{1 + (c/GM)\cos\theta}$$

which is the equation of a conic. Since the only conic which is a closed curve is an ellipse, the required result is proved.

5.37. Prove that the speed v of a particle moving in an elliptical path in an inverse square field is given by

$$v^2 = \frac{K}{m}\left(\frac{2}{r} - \frac{1}{a}\right)$$

where a is the semi-major axis.

By (8) of Problem 5.13, (4) of Problem 5.14 and (3) of Problem 5.17, we have

$$p = \frac{mh^2}{K} = a(1 - \epsilon^2) = a\left(-\frac{2Emh^2}{K^2}\right) \tag{1}$$

from which

$$E = -K/2a \tag{2}$$

Thus by the conservation of energy we have, using $V = -K/r$,

$$\tfrac{1}{2}mv^2 = E - V = -\frac{K}{2a} + \frac{K}{r}$$

or

$$v^2 = \frac{K}{m}\left(\frac{2}{r} - \frac{1}{a}\right) \tag{3}$$

We can similarly show that for a hyperbola,

$$v^2 = \frac{K}{m}\left(\frac{2}{r} + \frac{1}{a}\right) \tag{4}$$

while for a parabola [which corresponds to letting $a \to \infty$ in either (3) or (4)],

$$v^2 = 2K/mr$$

5.38. An artificial (man-made) satellite revolves about the earth at height H above the surface. Determine the (a) orbital speed and (b) orbital period so that a man in the satellite will be in a state of weightlessness.

(a) Assume that the earth is spherical and has radius R. Weightlessness will result when the centrifugal force [equal and opposite to the centripetal force, i.e. the force due to the centripetal acceleration] acting on the man due to rotation of the satellite just balances his attraction to the earth. Then if v_0 is the orbital speed,

$$\frac{mv_0^2}{R+H} = \frac{GMm}{(R+H)^2} = \frac{gR^2m}{(R+H)^2} \qquad \text{or} \qquad v_0 = \frac{R}{R+H}\sqrt{(R+H)g}$$

If H is small compared with R, this is $\sqrt{Rg}$ approximately.

(*b*)

$$\text{Orbital speed} = \frac{\text{distance traveled in one revolution}}{\text{time for one revolution, or period}}$$

Thus

$$v_0 = \frac{2\pi(R+H)}{P}$$

Then from part (*a*)

$$P = \frac{2\pi(R+H)}{v_0} = 2\pi\left(\frac{R+H}{R}\right)\sqrt{\frac{R+H}{g}}$$

If H is small compared with R, this is $2\pi\sqrt{R/g}$ approximately.

5.39. Calculate the (*a*) orbital speed and (*b*) period in Problem 5.38 assuming that the height H above the earth's surface is small compared with the earth's radius.

Taking the earth's radius as 4000 miles and $g = 32$ ft/sec^2, we find (*a*) $v_0 = \sqrt{Rg} =$ 4.92 mi/sec and (*b*) $P = 2\pi\sqrt{R/g} = 1.42$ hr $= 85$ minutes, approximately.

5.40. Find the force of attraction of a solid sphere of radius a on a particle of mass m at a distance $b < a$ from its center.

By Problem 5.30 the force of attraction of any spherical shell containing m in its interior [such as the spherical shell shown dashed in Fig. 5-15] is zero.

Thus the force of attraction on m is the force due to a sphere of radius $b < a$ with center at O. If σ is the mass per unit volume, the force of attraction is

$$G(\tfrac{4}{3}\pi b^3)\sigma m/b^2 = (\tfrac{4}{3}\pi G\sigma m)b$$

Thus the force varies as the distance b from the mass to the center.

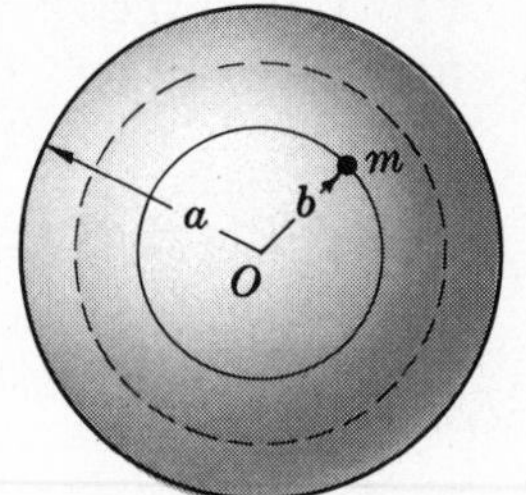

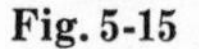
Fig. 5-15

Supplementary Problems

CENTRAL FORCES AND EQUATIONS OF MOTION

5.41. Indicate which of the following central force fields are attractive toward origin O and which are repulsive from O. (*a*) $\mathbf{F} = -4r^3\mathbf{r}_1$; (*b*) $\mathbf{F} = K\mathbf{r}_1/\sqrt{r}$, $K > 0$; (*c*) $\mathbf{F} = r(r-1)\mathbf{r}_1/(r^2+1)$; (*d*) $\mathbf{F} = \sin \pi r\, \mathbf{r}_1$.

Ans. (*a*) attractive; (*b*) repulsive; (*c*) attractive if $0 < r < 1$, repulsive if $r > 1$; (*d*) repulsive for $2n < r < 2n+1$, attractive for $2n+1 < r < 2n+2$ where $n = 0, 1, 2, 3 \ldots$.

5.42. Prove that in rectangular coordinates the magnitude of the areal velocity is $\frac{1}{2}(x\dot{y} - y\dot{x})$.

5.43. Give an example of a force field directed toward a fixed point which is not a central force field.

5.44. Derive equation (*7*), page 117.

5.45. If a particle moves in a circular orbit under the influence of a central force at its center, prove that its speed around the orbit must be constant.

5.46. A particle of mass m moves in a force field defined by $\mathbf{F} = -K\mathbf{r}_1/r^3$. If it starts on the positive x axis at distance a away from the origin and moves with speed v_0 in direction making angle α with the positive x axis, prove that the differential equation for the radial position r of the particle at any time t is

$$\frac{d^2r}{dt^2} = -\frac{(K - ma^2v_0^2 \sin^2 \alpha)}{mr^3}$$

5.47. (*a*) Show that the differential equation for the orbit in Problem 5.46 is given in terms of $u = 1/r$ by

$$\frac{d^2u}{d\theta^2} + (1-\gamma)u = 0 \qquad \text{where} \quad \gamma = \frac{K}{ma^2v_0^2 \sin^2 \alpha}$$

(*b*) Solve the differential equation in (*a*) and interpret physically.

5.48. A particle is to move under the influence of a central force field so that its orbital speed is always constant and equal to v_0. Determine all possible orbits.

POTENTIAL ENERGY AND CONSERVATION OF ENERGY

5.49. Find the potential energy or potential corresponding to the central force fields defined by (*a*) $\mathbf{F} = -K\mathbf{r}_1/r^3$, (*b*) $\mathbf{F} = (\alpha/r^2 + \beta/r^3)\mathbf{r}_1$, (*c*) $\mathbf{F} = Kr\mathbf{r}_1$, (*d*) $\mathbf{F} = \mathbf{r}_1/\sqrt{r}$, (*e*) $\mathbf{F} = \sin \pi r \, \mathbf{r}_1$.
Ans. (*a*) $-K/2r^2$, (*b*) $\alpha/r + \beta/2r^2$, (*c*) $\frac{1}{2}Kr^2$, (*d*) $2\sqrt{r}$, (*e*) $(\cos \pi r)/\pi$

5.50. (*a*) Find the potential energy for a particle which moves in the force field $\mathbf{F} = -K\mathbf{r}_1/r^2$. (*b*) How much work is done by the force field in (*a*) in moving the particle from a point on the circle $r = a > 0$ to another point on the circle $r = b > 0$? Does the work depend on the path? Explain.
Ans. (*a*) $-K/r$, (*b*) $K(a-b)/ab$

5.51. Work Problem 5.50 for the force field $\mathbf{F} = -K\mathbf{r}_1/r$. *Ans.* (*a*) $-K \ln r$, (*b*) $-K \ln (a/b)$

5.52. A particle of mass m moves in a central force field defined by $\mathbf{F} = -K\mathbf{r}_1/r^3$. (*a*) Write an equation for the conservation of energy. (*b*) Prove that if E is the total energy supplied to the particle, then its speed is given by $v = \sqrt{K/mr^2 + 2E/m}$.

5.53. A particle moves in a central force field defined by $\mathbf{F} = -Kr^2\mathbf{r}_1$. It starts from rest at a point on the circle $r = a$. (*a*) Prove that when it reaches the circle $r = b$ its speed will be $\sqrt{2K(a^3 - b^3)/3m}$ and that (*b*) the speed will be independent of the path.

5.54. A particle of mass m moves in a central force field $\mathbf{F} = K\mathbf{r}_1/r^n$ where K and n are constants. It starts from rest at $r = a$ and arrives at $r = 0$ with finite speed v_0. (*a*) Prove that we must have $n < 1$ and $K > 0$. (*b*) Prove that $v_0 = \sqrt{2Ka^{1-n}/m(n-1)}$. (*c*) Discuss the physical significance of the results in (*a*).

5.55. By differentiating both sides of equation (*13*), page 117, obtain equation (*6*).

DETERMINATION OF ORBIT FROM CENTRAL FORCE OR CENTRAL FORCE FROM ORBIT

5.56. A particle of mass m moves in a central force field given in magnitude by $f(r) = -Kr$ where K is a positive constant. If the particle starts at $r = a$, $\theta = 0$ with a speed v_0 in a direction perpendicular to the x axis, determine its orbit. What type of curve is described?

5.57. (*a*) Work Problem 5.56 if the speed is v_0 in a direction making angle α with the positive x axis. (*b*) Discuss the cases $\alpha = 0$, $\alpha = \pi$ and give the physical significance.

5.58. A particle moving in a central force field located at $r = 0$ describes the spiral $r = e^{-\theta}$. Prove that the magnitude of the force is inversely proportional to r^3.

5.59. Find the central force necessary to make a particle describe the lemniscate $r^2 = a^2 \cos 2\theta$ [see Fig. 5-16]. *Ans.* A force proportional to r^{-7}.

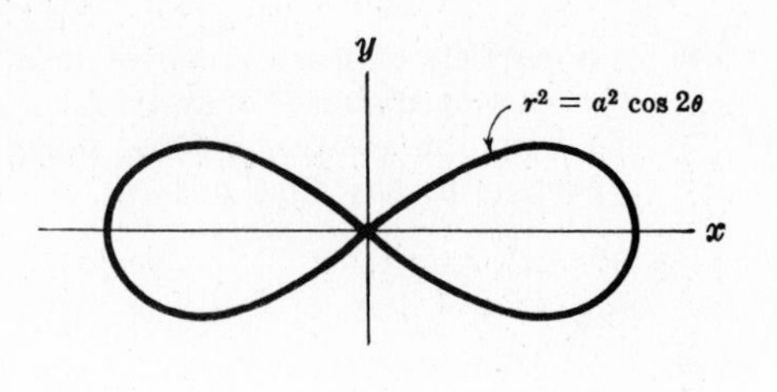

Fig. 5-16

5.60. Obtain the orbit for the particle of Problem 5.46 and describe physically.

5.61. Prove that the orbits $r = e^{-\theta}$ and $r = 1/\theta$ are both possible for the case of an inverse cube field of force. Explain physically how this is possible.

5.62. (*a*) Show that if the law of force is given by

$$\mathbf{F} = \frac{A\mathbf{r}_1}{r^4 \cos\theta} \qquad \text{or} \qquad \mathbf{F} = \frac{B\mathbf{r}_1}{r^2 \cos^3\theta}$$

then a particle can move in the circular orbit $r = 2a\cos\theta$. (*b*) What can you conclude about the uniqueness of forces when the orbit is specified? (*c*) Answer part (*b*) when the forces are central forces.

5.63. (*a*) What central force at the origin O is needed to make a particle move around O with a speed which is inversely proportional to the distance from O. (*b*) What types of orbits are possible in such case? *Ans.* (*a*) Inverse cube force.

5.64. Discuss the motion of a particle moving in a central force field given by $\mathbf{F} = (\alpha/r^2 + \beta/r^3)\mathbf{r}_1$.

5.65. Prove that there is no central force which will enable a particle to move in a straight line.

5.66. Complete the integration of equation (*12*) of Problem 5.13, page 125 and thus arrive at equation (*5*) of the same problem. [*Hint.* Let $r = 1/u$.]

5.67. Suppose that the orbit of a particle moving in a central force field is given by $\theta = \theta(r)$. Prove that the law of force is $-\dfrac{mh^2[2\theta' + r\theta'' + r^2(\theta')^3]}{r^5(\theta')^3}$ where primes denote differentiations with respect to r.

5.68. (*a*) Use Problem 5.67 to show that if $\theta = 1/r$, the central force is one of attraction and varies inversely as r^3. (*b*) Graph the orbit in (*a*) and explain physically.

CONIC SECTIONS. ELLIPSE, PARABOLA AND HYPERBOLA

5.69. The equation of a conic is $r = \dfrac{12}{3 + \cos\theta}$. Graph the conic, finding (*a*) the foci, (*b*) the vertices, (*c*) the length of the major axis, (*d*) the length of the minor axis, (*e*) the distance from the center to the directrix.

5.70. Work Problem 5.69 for the conic $r = \dfrac{24}{3 + 5\cos\theta}$.

5.71. Show that the equation of a parabola can be written as $r = p\sec^2(\theta/2)$.

5.72. Find an equation for an ellipse which has one focus at the origin, its center at the point $(-4, 0)$, and its major axis of length 10. *Ans.* $r = 9/(5 + 4\cos\theta)$

5.73. In Fig. 5-17, SR or TN is called the *minor axis* of the hyperbola and its length is generally denoted by $2b$. The length of the *major axis* VU is $2a$, while the distance between the foci O and O' is $2c$ [i.e. the distance from the center C to a focus O or O' is C].

(a) Prove that $c^2 = a^2 + b^2$.

(b) Prove that $b = a\sqrt{\epsilon^2 - 1}$ where ϵ is the eccentricity.

(c) Prove that $c = a\epsilon$. Compare with results for the ellipse.

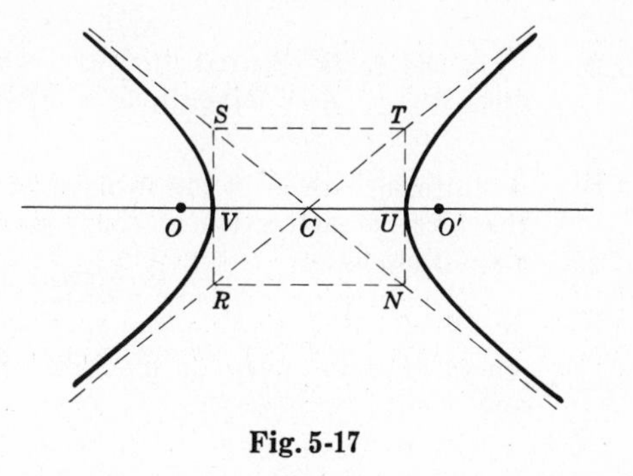

Fig. 5-17

5.74. Derive equation (22), page 119, for a hyperbola.

5.75. In rectangular coordinates the equations for an ellipse and hyperbola in standard form are given by

$$\frac{x^2}{a^2} + \frac{y^2}{b^2} = 1 \quad \text{and} \quad \frac{x^2}{a^2} - \frac{y^2}{b^2} = 1$$

respectively, where a and b are the lengths of the semi-major and semi-minor axes. Graph these equations, locating vertices, foci and directrices, and explain the relation of these equations to equations (19), page 118, and (22), page 119.

5.76. Using the alternative definitions for an ellipse and hyperbola given on pages 118-119, obtain the equations (19) and (22).

5.77. Prove that the angle between the asymptotes of a hyperbola is $2 \cos^{-1}(1/\epsilon)$.

KEPLER'S LAWS AND NEWTON'S LAW OF GRAVITATION

5.78. Assuming that the planet Mars has a period about the sun equal to 687 earth days approximately, find the mean distance of Mars from the sun. Take the distance of the earth from the sun as 93 million miles. *Ans.* 140 million miles

5.79. Work Problem 5.78 for (a) Jupiter and (b) Venus which have periods of 4333 earth days and 225 earth days respectively. *Ans.* (a) 484 million miles, (b) 67 million miles

5.80. Suppose that a small spherical planet has a radius of 10 km and a mean density of 5 gm/cm^3.
(a) What would be the acceleration due to gravity at its surface? (b) What would a man weigh on this planet if he weighed 80 kg wt on earth?

5.81. If the acceleration due to gravity on the surface of a spherically shaped planet P is g_P while its mean density and radius are given by σ_P and R_P respectively, prove that $g_P = \frac{4}{3}\pi G R_P \sigma_P$ where G is the universal gravitational constant.

5.82. If L, M, T represent the dimensions of length, mass and time, find the dimensions of the universal gravitational constant. *Ans.* $L^3M^{-1}T^{-2}$

5.83. Calculate the mass of the sun using the fact that the earth is approximately 150×10^6 kilometers from it and makes one complete revolution about it in approximately 365 days. *Ans.* 2×10^{30} kg

5.84. Calculate the force between the sun and the earth if the distance between the earth and the sun is taken as 150×10^6 kilometers and the masses of the earth and sun are 6×10^{24} kg and 2×10^{30} kg respectively. *Ans.* 1.16×10^{24} newtons

ATTRACTION OF OBJECTS

5.85. Find the force of attraction of a thin uniform rod of length a on a mass m outside the rod but on the same line as the rod and distance b from an end. *Ans.* $GMm/b(a+b)$

5.86. In Problem 5.85 determine where the mass of the rod should be concentrated so as to give the same force of attraction. *Ans.* At a point in the rod a distance $\sqrt{b(a+b)} - b$ from the end

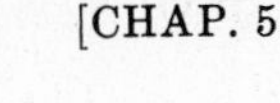

5.87. Find the force of attraction of an infinitely long thin uniform rod on a mass m at distance b from it. *Ans.* Magnitude is $2Gm\sigma/b$

5.88. A uniform wire is in the form of an arc of a circle of radius b and central angle ψ. Prove that the force of attraction of the wire on a mass m placed at the center of the circle is given in magnitude by

$$\frac{2GMm \sin(\psi/2)}{b^2\psi} \quad \text{or} \quad \frac{2G\sigma m \sin(\psi/2)}{b}$$

where M is the mass of the wire and σ is the mass per unit length. Discuss the cases $\psi = \pi/2$ and $\psi = \pi$.

5.89. In Fig. 5-18, AB is a thin rod of length $2a$ and m is a mass located at point C a distance b from the rod. Prove that the force of attraction of the rod on m has magnitude

$$\frac{GMm}{ab} \sin \tfrac{1}{2}(\alpha + \beta)$$

in a direction making an angle with the rod given by

$$\tan^{-1}\left(\frac{\cos\beta + \cos\alpha}{\sin\beta - \sin\alpha}\right)$$

Discuss the case $\alpha = \beta$ and compare with Problem 5.26.

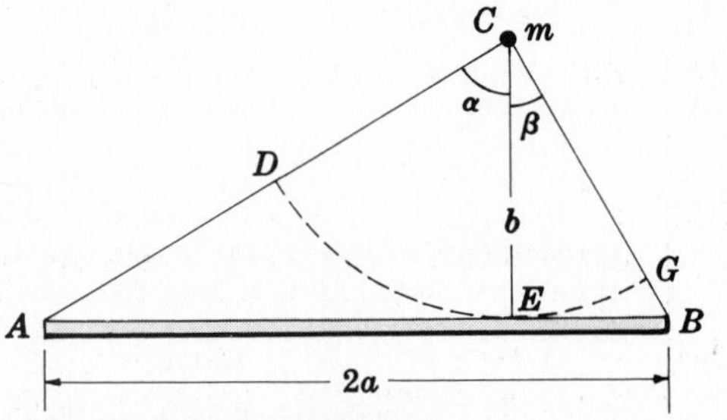

Fig. 5-18

5.90. By comparing Problem 5.89 with Problem 5.88, prove that the rod of Problem 5.89 can be replaced by a wire in the form of circular arc DEG [shown dashed in Fig. 5-18] which has its center at C and is tangent to the rod at E. Prove that the direction of the attraction is toward the midpoint of this arc.

5.91. A hemisphere of mass M and radius a has a particle of mass m located at its center. Find the force of attraction if (a) the hemisphere is a thin shell, (b) the hemisphere is solid.
Ans. (a) $GMm/2a^2$, (b) $3GMm/2a^2$

5.92. Work Problem 5.91 if the hemisphere is a shell having outer radius a and inner radius b.

5.93. Deduce from Kepler's laws that if the force of attraction between sun and planets is given in magnitude by $\gamma m/r^2$, then γ must be independent of the particular planet.

5.94. A cone has height H and radius a. Prove that the force of attraction on a particle of mass m placed at its vertex has magnitude $\dfrac{6GMm}{a^2}\left(1 - \dfrac{H}{\sqrt{a^2+H^2}}\right)$.

5.95. Find the force of attraction between two non-intersecting spheres.

5.96. A particle of mass m is placed outside of a uniform solid hemisphere of radius a at a distance a on a line perpendicular to the base through its center. Prove that the force of attraction is given in magnitude by $GMm(\sqrt{2}-1)/a^2$.

5.97. Work (a) Problem 5.26, (b) Problem 5.27, and (c) Problem 5.94 by first finding the potential.

MISCELLANEOUS PROBLEMS

5.98. A particle is projected vertically upward from the earth's surface with initial speed v_0.

(a) Prove that the maximum height H reached above the earth's surface is $H = v_0^2 R/(2gR - v_0^2)$.

(b) Discuss the significance of the case where $v_0^2 = 2gR$.

(c) Prove that if H is small, then it is equal to $v_0^2/2g$ very nearly.

5.99. (*a*) Prove that the time taken to reach the maximum height of Problem 5.98 is

$$\sqrt{\frac{R+H}{2g}}\left\{\sqrt{\frac{H}{R}}+\frac{R+H}{2R}\cos^{-1}\left(\frac{R-H}{R+H}\right)\right\}$$

(*b*) Prove that if H is very small compared with R, then the time in (*a*) is very nearly $\sqrt{2H/g}$.

5.100. (*a*) Prove that if an object is dropped to the earth's surface from a height H, then if air resistance is negligible it will hit the earth with a speed $v = \sqrt{2gRH/(R+H)}$ where R is the radius of the earth.

(*b*) Calculate the speed in part (*a*) for the cases where $H = 100$ miles and $H = 10{,}000$ miles respectively. Take the radius of the earth as 4000 miles.

5.101. Find the time taken for the object of Problem 5.100 to reach the earth's surface in each of the two cases.

5.102. What must be the law of force if the speed of a particle in a central force field is to be proportional to r^{-n} where n is a constant?

5.103. What velocity must a space ship have in order to keep it in an orbit around the earth at a distance of (*a*) 200 miles, (*b*) 2000 miles above the earth's surface?

5.104. An object is thrown upward from the earth's surface with velocity v_0. Assuming that it returns to earth and that air resistance is negligible, find its velocity on returning.

5.105. (*a*) What is the work done by a space ship of mass m in moving from a distance a above the earth's surface to a distance b?

(*b*) Does the work depend on the path? Explain. *Ans.* (*a*) $GmM(a-b)/ab$

5.106. (*a*) Prove that it is possible for a particle to move in a circle of radius a in any central force field whose law of force is $f(r)$.

(*b*) Suppose the particle of part (*a*) is displaced slightly from its circular orbit. Prove that it will return to the orbit, i.e. the motion is *stable*, if

$$a\,f'(a) + 3\,f(a) > 0$$

but is unstable otherwise.

(*c*) Illustrate the result in (*b*) by considering $f(r) = 1/r^n$ and deciding for which values of n stability can occur. *Ans.* (*c*) For $n < 3$ there is stability.

5.107. If the moon were suddenly stopped in its orbit, how long would it take to fall to the earth assuming that the earth remained at rest? *Ans.* About 4 days 18 hours

5.108. If the earth were suddenly stopped in its orbit, how long would it take for it to fall into the sun? *Ans.* About 65 days

5.109. Work Problem 5.34, page 133, by using energy methods.

5.110. Find the velocity of escape for an object on the surface of the moon. Use the fact that the acceleration due to gravity on the moon's surface is approximately 1/6 that on the earth and that the radius of the moon is approximately 1/4 of the earth's radius. *Ans.* 1.5 mi/sec

5.111. An object is dropped through a hole bored through the center of the earth. Assuming that the resistance to motion is negligible, show that the speed of the particle as it passes through the center of the earth is slightly less than 5 mi/sec.
[*Hint.* Use Problem 5.40, page 136.]

5.112. In Problem 5.111 show that the time taken for the object to return is about 85 minutes.

5.113. Work Problems 5.111 and 5.112 if the hole is straight but does not pass through the center of the earth.

5.114. Discuss the relationship between the results of Problems 5.111 and 5.112 and that of Problem 5.39.

5.115. How would you explain the fact that the earth has an atmosphere while the moon has none?

5.116. Prove Theorem 5.1, page 120.

5.117. Discuss Theorem 5.1 if the spheres intersect.

5.118. Explain how you could use the result of Problem 5.27 to find the force of attraction of a solid sphere on a particle.

5.119. Find the force of attraction between a uniform circular ring of outer radius a and inner radius b and a mass m located on its axis at a distance b from its center.

5.120. Two space ships move about the earth on the same elliptical path of eccentricity ϵ. If they are separated by a small distance D at perigee, prove that at apogee they will be separated by the distance $D(1-\epsilon)/(1+\epsilon)$.

5.121. (*a*) Explain how you could calculate the velocity of escape from a planet. (*b*) Use your method to calculate the velocity of escape from Mars. *Ans.* (*b*) 5 km/sec, or about 3 mi/sec

5.122. Work Problem 5.121 for (*a*) Jupiter, (*b*) Venus. *Ans.* (*a*) about 38 mi/sec, (*b*) about 6.3 mi/sec

5.123. Three infinitely long thin uniform rods having the same mass per unit length lie in the same plane and form a triangle. Prove that force of attraction on a particle will be zero if and only if the particle is located at the intersection of the medians of the triangle.

5.124. Find the force of attraction between a uniform rod of length a and a sphere of radius b if they do not intersect and the line of the rod passes through the center.

5.125. Work Problem 5.124 if the rod is situated so that a line drawn from the center perpendicular to the line of the rod bisects the rod.

5.126. A satellite of radius a revolves in a circular orbit about a planet of radius b with period P. If the shortest distance between their surfaces is c, prove that the mass of the planet is $4\pi^2(a+b+c)^3/GP^2$.

5.127. Given that the moon is approximately 240,000 miles from the earth and makes one complete revolution about the earth in $27\frac{1}{3}$ days approximately, find the mass of the earth.
Ans. 6×10^{24} kg

5.128. Discuss the relationship of Problem 5.126 with Kepler's third law.

5.129. Prove that the only central force field **F** whose divergence is zero is an inverse square force field.

5.130. Work Problem 5.32, page 132, by using triple integration.

5.131. A uniform solid right circular cylinder has radius a and height H. A particle of mass m is placed on the extended axis of the cylinder so that it is at a distance D from one end. Prove that the force of attraction is directed along the axis and given in magnitude by

$$\frac{2GMm}{a^2H}\{H+\sqrt{a^2+D^2}-\sqrt{a^2+(D+H)^2}\}$$

5.132. Suppose that the cylinder of Problem 5.131 has a given volume. Prove that the force of attraction when the particle is at the center of one end of the cylinder is a maximum when $a/H = \frac{1}{8}(9-\sqrt{17})$.

5.133. Work (*a*) Problem 5.26 and (*b*) Problem 5.27 assuming an inverse cube law of attraction.

5.134. Do the results of Problems 5.29 and 5.30 apply if there is an inverse cube law of attraction? Explain.

5.135. What would be the velocity of escape from the small planet of Problem 5.80?

5.136. A spherical shell of inner radius a and outer radius b has constant density σ. Prove that the gravitational potential $V(r)$ at distance r from the center is given by

$$V(r) = \begin{cases} 2\pi\sigma(b^2 - a^2) & r < a \\ 2\pi\sigma(b^2 - \frac{1}{3}r^2) - 4\pi\sigma a^3/3r & a < r < b \\ 4\pi\sigma(b^3 - a^3)/3r & r > b \end{cases}$$

5.137. If *Einstein's theory of relativity* is taken into account, the differential equation for the orbit of a planet becomes

$$\frac{d^2u}{d\theta^2} + u = \frac{K}{mh^2} + \gamma u^2$$

where $\gamma = 3K/mc^2$, c being the speed of light. (*a*) Prove that if axes are suitably chosen, then the position r of the planet can be determined approximately from

$$r = \frac{mh^2/K}{1 + \epsilon \cos \alpha\theta} \qquad \text{where} \quad \alpha = 1 - \gamma K/mh^2$$

(*b*) Use (*a*) to show that a planet actually moves in an elliptical path but that this ellipse slowly rotates in space, the rate of angular rotation being $2\pi\gamma K/mh^2$. (*c*) Show that in the case of Mercury this rotation amounts to 43 seconds of arc per century. This was actually observed, thus offering experimental proof of the validity of the theory of relativity.

5.138. Find the position of a planet in its orbit around the sun as a function of time t measured from where it is furthest from the sun.

5.139. At apogee of 200 miles from the earth's surface, two space ships in the same elliptical path are 500 feet apart. How far apart will they be at perigee 150 miles assuming that they drift without altering their path in any way?

5.140. A particle of mass m is located on a perpendicular line through the center of a rectangular plate of sides $2a$ and $2b$ at a distance D from this center. Prove that the force of attraction of the plate on the particle is given in magnitude by

$$\frac{GMm}{ab} \sin^{-1}\left(\frac{ab}{\sqrt{(a^2 + D^2)(b^2 + D^2)}}\right)$$

5.141. Find the force of attraction of a uniform infinite plate of negligible thickness and density σ on a particle at distance D from it. *Ans.* $2\pi\sigma Gm$

5.142. Points where $\dot{r} = 0$ are called *apsides* [singular, *apsis*]. (*a*) Prove that apsides for a central force field with potential $V(r)$ and total energy E are roots of the equation $V(r) + h^2/2r^2 = E$. (*b*) Find the apsides corresponding to an inverse square field of force, showing that there are two, one or none according as the orbit is an ellipse, hyperbola or parabola.

5.143. A particle moving in a central force field travels in a path which is the cycloid $r = a(1 - \cos\theta)$. Find the law of force. *Ans.* Inverse fourth power of r.

5.144. Set up equations for the motion of a particle in a central force field if it takes place in a medium where the resistance is proportional to the instantaneous speed of the particle.

5.145. A satellite has its largest and smallest orbital speeds given by v_{max} and v_{min} respectively. Prove that the eccentricity of the orbit in which the satellite moves is equal to $\dfrac{v_{max} - v_{min}}{v_{max} + v_{min}}$.

5.146. Prove that if the satellite of Problem 5.145 has a period equal to τ, then it moves in an elliptical path having major axis whose length is $\frac{\tau}{\pi}\sqrt{v_{max}\, v_{min}}$.

Chapter 6

MOVING COORDINATE SYSTEMS

NON-INERTIAL COORDINATE SYSTEMS

In preceding chapters the coordinate systems used to describe the motions of particles were assumed to be inertial [see page 33]. In many instances of practical importance, however, this assumption is not warranted. For example, a coordinate system fixed in the earth is not an inertial system since the earth itself is rotating in space. Consequently if we use this coordinate system to describe the motion of a particle relative to the earth we obtain results which may be in error. We are led therefore to consider the motion of particles relative to moving coordinate systems.

ROTATING COORDINATE SYSTEMS

In Fig. 6-1 let XYZ denote an inertial coordinate system with origin O which we shall consider fixed in space. Let the coordinate system xyz having the same origin O be rotating with respect to the XYZ system.

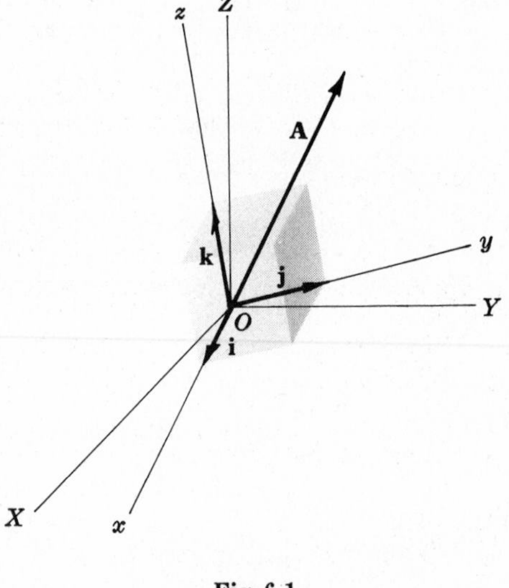

Fig. 6-1

Consider a vector $\mathbf{A}$ which is changing with time. To an observer fixed relative to the xyz system the time rate of change of $\mathbf{A} = A_1\mathbf{i} + A_2\mathbf{j} + A_3\mathbf{k}$ is found to be

$$\left.\frac{d\mathbf{A}}{dt}\right|_M = \frac{dA_1}{dt}\mathbf{i} + \frac{dA_2}{dt}\mathbf{j} + \frac{dA_3}{dt}\mathbf{k} \qquad (1)$$

where subscript M indicates the derivative in the moving (xyz) system.

However, the time rate of change of $\mathbf{A}$ relative to the fixed XYZ system symbolized by the subscript F is found to be [see Problem 6.1]

$$\left.\frac{d\mathbf{A}}{dt}\right|_F = \left.\frac{d\mathbf{A}}{dt}\right|_M + \boldsymbol{\omega} \times \mathbf{A} \qquad (2)$$

where $\boldsymbol{\omega}$ is called the *angular velocity* of the xyz system with respect to the XYZ system.

DERIVATIVE OPERATORS

Let D_F and D_M represent time derivative operators in the fixed and moving systems. Then we can write the operator equivalence

$$D_F \equiv D_M + \boldsymbol{\omega}\times \qquad (3)$$

This result is useful in relating higher order time derivatives in the fixed and moving systems. See Problem 6.6.

VELOCITY IN A MOVING SYSTEM

If, in particular, vector $\mathbf{A}$ is the position vector $\mathbf{r}$ of a particle, then (2) gives

$$\left.\frac{d\mathbf{r}}{dt}\right|_F = \left.\frac{d\mathbf{r}}{dt}\right|_M + \boldsymbol{\omega}\times\mathbf{r} \tag{4}$$

or

$$D_F\mathbf{r} = D_M\mathbf{r} + \boldsymbol{\omega}\times\mathbf{r} \tag{5}$$

Let us write

$$\mathbf{v}_{P|F} = d\mathbf{r}/dt\,|_F = D_F\mathbf{r} = \text{velocity of particle } P \text{ relative to fixed system}$$

$$\mathbf{v}_{P|M} = d\mathbf{r}/dt\,|_M = D_M\mathbf{r} = \text{velocity of particle } P \text{ relative to moving system}$$

$$\mathbf{v}_{M|F} = \boldsymbol{\omega}\times\mathbf{r} = \text{velocity of moving system relative to fixed system.}$$

Then (4) or (5) can be written

$$\mathbf{v}_{P|F} = \mathbf{v}_{P|M} + \mathbf{v}_{M|F} \tag{6}$$

ACCELERATION IN A MOVING SYSTEM

If $D_F^2 = d^2/dt^2\,|_F$ and $D_M^2 = d^2/dt^2\,|_M$ are second derivative operators with respect to t in the fixed and moving systems, then application of (3) yields [see Problem 6.6]

$$D_F^2\mathbf{r} = D_M^2\mathbf{r} + (D_M\boldsymbol{\omega})\times\mathbf{r} + 2\boldsymbol{\omega}\times D_M\mathbf{r} + \boldsymbol{\omega}\times(\boldsymbol{\omega}\times\mathbf{r}) \tag{7}$$

Let us write

$$\mathbf{a}_{P|F} = d^2\mathbf{r}/dt^2\,|_F = D_F^2\mathbf{r} = \text{acceleration of particle } P \text{ relative to fixed system}$$

$$\mathbf{a}_{P|M} = d^2\mathbf{r}/dt^2\,|_M = D_M^2\mathbf{r} = \text{acceleration of particle } P \text{ relative to moving system}$$

$$\begin{aligned}\mathbf{a}_{M|F} &= (D_M\boldsymbol{\omega})\times\mathbf{r} + 2\boldsymbol{\omega}\times D_M\mathbf{r} + \boldsymbol{\omega}\times(\boldsymbol{\omega}\times\mathbf{r})\\ &= \text{acceleration of moving system relative to fixed system}\end{aligned}$$

Then (7) can be written

$$\mathbf{a}_{P|F} = \mathbf{a}_{P|M} + \mathbf{a}_{M|F} \tag{8}$$

CORIOLIS AND CENTRIPETAL ACCELERATION

The last two terms on the right of (7) are called the *Coriolis acceleration* and *centripetal acceleration* respectively, i.e.,

$$\text{Coriolis acceleration} = 2\boldsymbol{\omega}\times D_M\mathbf{r} = 2\boldsymbol{\omega}\times\mathbf{v}_M \tag{9}$$

$$\text{Centripetal acceleration} = \boldsymbol{\omega}\times(\boldsymbol{\omega}\times\mathbf{r}) \tag{10}$$

The second term on the right of (7) is sometimes called the *linear acceleration,* i.e.,

$$\text{Linear acceleration} = (D_M\boldsymbol{\omega})\times\mathbf{r} = \left(\left.\frac{d\boldsymbol{\omega}}{dt}\right|_M\right)\times\mathbf{r} \tag{11}$$

and $D_M\boldsymbol{\omega}$ is called the *angular acceleration.* For many cases of practical importance [e.g. in the rotation of the earth] $\boldsymbol{\omega}$ is constant and $D_M\boldsymbol{\omega} = 0$.

The quantity $-\boldsymbol{\omega}\times(\boldsymbol{\omega}\times\mathbf{r})$ is often called the *centrifugal acceleration.*

MOTION OF A PARTICLE RELATIVE TO THE EARTH

Newton's second law is strictly applicable only to inertial systems. However, by using (7) we obtain a result valid for non-inertial systems. This has the form

$$mD_M^2\mathbf{r} = \mathbf{F} - m(D_M\boldsymbol{\omega})\times\mathbf{r} - 2m(\boldsymbol{\omega}\times D_M\mathbf{r}) - m\boldsymbol{\omega}\times(\boldsymbol{\omega}\times\mathbf{r}) \tag{12}$$

where $\mathbf{F}$ is the resultant of all forces acting on the particle as seen by the observer in the fixed or inertial system.

From (2) and (6) we find, as required,

$$\left.\frac{d\mathbf{A}}{dt}\right|_F = \left.\frac{d\mathbf{A}}{dt}\right|_M + \boldsymbol{\omega} \times \mathbf{A}$$

The vector quantity $\boldsymbol{\omega}$ is the *angular velocity* of the moving system relative to the fixed system.

6.2. Let D_F and D_M be symbolic time derivative operators in the fixed and moving systems respectively. Demonstrate the operator equivalence

$$D_F \equiv D_M + \boldsymbol{\omega}\times$$

By definition

$$D_F\mathbf{A} = \left.\frac{d\mathbf{A}}{dt}\right|_F = \text{derivative in fixed system}$$

$$D_M\mathbf{A} = \left.\frac{d\mathbf{A}}{dt}\right|_M = \text{derivative in moving system}$$

Then from Problem 6.1,

$$D_F\mathbf{A} = D_M\mathbf{A} + \boldsymbol{\omega}\times\mathbf{A} = (D_M + \boldsymbol{\omega}\times)\mathbf{A}$$

which shows the equivalence of the operators $D_F \equiv D_M + \boldsymbol{\omega}\times$.

6.3. Prove that the angular acceleration is the same in both XYZ and xyz coordinate systems.

Let $\mathbf{A} = \boldsymbol{\omega}$ in Problem 6.1. Then

$$\left.\frac{d\boldsymbol{\omega}}{dt}\right|_F = \left.\frac{d\boldsymbol{\omega}}{dt}\right|_M + \boldsymbol{\omega}\times\boldsymbol{\omega} = \left.\frac{d\boldsymbol{\omega}}{dt}\right|_M$$

Since $d\boldsymbol{\omega}/dt$ is the angular acceleration, the required statement is proved.

VELOCITY AND ACCELERATION IN MOVING SYSTEMS

6.4. Determine the velocity of a moving particle as seen by the two observers in Problem 6.1.

Replacing $\mathbf{A}$ by the position vector $\mathbf{r}$ of the particle, we have

$$\left.\frac{d\mathbf{r}}{dt}\right|_F = \left.\frac{d\mathbf{r}}{dt}\right|_M + \boldsymbol{\omega}\times\mathbf{r} \tag{1}$$

If $\mathbf{r}$ is expressed in terms of the unit vectors $\mathbf{i},\mathbf{j},\mathbf{k}$ of the moving coordinate system, then the velocity of the particle relative to this system is, on dropping the subscript M,

$$\frac{d\mathbf{r}}{dt} = \frac{dx}{dt}\mathbf{i} + \frac{dy}{dt}\mathbf{j} + \frac{dz}{dt}\mathbf{k} \tag{2}$$

and the velocity of the particle relative to the fixed system is from (1)

$$\left.\frac{d\mathbf{r}}{dt}\right|_F = \frac{d\mathbf{r}}{dt} + \boldsymbol{\omega}\times\mathbf{r} \tag{3}$$

The velocity (3) is sometimes called the *true velocity*, while (2) is the *apparent velocity*.

6.5. An xyz coordinate system is rotating with respect to an XYZ coordinate system having the same origin and assumed to be fixed in space [i.e. it is an inertial system]. The angular velocity of the xyz system relative to the XYZ system is given by $\boldsymbol{\omega} = 2t\mathbf{i} - t^2\mathbf{j} + (2t+4)\mathbf{k}$ where t is the time. The position vector of a particle at time t as observed in the xyz system is given by $\mathbf{r} = (t^2+1)\mathbf{i} - 6t\mathbf{j} + 4t^3\mathbf{k}$. Find (a) the apparent velocity and (b) the true velocity at time $t = 1$.

(a) The apparent velocity at any time t is

$$d\mathbf{r}/dt = 2t\mathbf{i} - 6\mathbf{j} + 12t^2\mathbf{k}$$

At time $t = 1$ this is $2\mathbf{i} - 6\mathbf{j} + 12\mathbf{k}$.

(b) The true velocity at any time t is

$$d\mathbf{r}/dt + \boldsymbol{\omega} \times \mathbf{r} = (2t\mathbf{i} - 6\mathbf{j} + 12t^2\mathbf{k}) + [2t\mathbf{i} - t^2\mathbf{j} + (2t+4)\mathbf{k}] \times [(t^2+1)\mathbf{i} - 6t\mathbf{j} + 4t^3\mathbf{k}]$$

At time $t = 1$ this is

$$2\mathbf{i} - 6\mathbf{j} + 12\mathbf{k} + \begin{vmatrix} \mathbf{i} & \mathbf{j} & \mathbf{k} \\ 2 & -1 & 6 \\ 2 & -6 & 4 \end{vmatrix} = 34\mathbf{i} - 2\mathbf{j} + 2\mathbf{k}$$

6.6. Determine the acceleration of a moving particle as seen by the two observers in Problem 6.1.

The acceleration of the particle as seen by the observer in the fixed XYZ system is $D_F^2\mathbf{r} = D_F(D_F\mathbf{r})$. Using the operator equivalence established in Problem 6.2, we have

$$\begin{aligned} D_F(D_F\mathbf{r}) &= D_F(D_M\mathbf{r} + \boldsymbol{\omega} \times \mathbf{r}) \\ &= (D_M + \boldsymbol{\omega} \times\ )(D_M\mathbf{r} + \boldsymbol{\omega} \times \mathbf{r}) \\ &= D_M(D_M\mathbf{r} + \boldsymbol{\omega} \times \mathbf{r}) + \boldsymbol{\omega} \times (D_M\mathbf{r} + \boldsymbol{\omega} \times \mathbf{r}) \\ &= D_M^2\mathbf{r} + D_M(\boldsymbol{\omega} \times \mathbf{r}) + \boldsymbol{\omega} \times D_M\mathbf{r} + \boldsymbol{\omega} \times (\boldsymbol{\omega} \times \mathbf{r}) \end{aligned}$$

or since $D_M(\boldsymbol{\omega} \times \mathbf{r}) = (D_M\boldsymbol{\omega}) \times \mathbf{r} + \boldsymbol{\omega} \times (D_M\mathbf{r})$,

$$D_F^2\mathbf{r} = D_M^2\mathbf{r} + (D_M\boldsymbol{\omega}) \times \mathbf{r} + 2\boldsymbol{\omega} \times (D_M\mathbf{r}) + \boldsymbol{\omega} \times (\boldsymbol{\omega} \times \mathbf{r}) \tag{1}$$

If $\mathbf{r}$ is the position vector expressed in terms of $\mathbf{i}, \mathbf{j}, \mathbf{k}$ of the moving coordinate system, then the acceleration of the particle relative to this system is, on dropping the subscript M,

$$\frac{d^2\mathbf{r}}{dt^2} = \frac{d^2x}{dt^2}\mathbf{i} + \frac{d^2y}{dt^2}\mathbf{j} + \frac{d^2z}{dt^2}\mathbf{k} \tag{2}$$

The acceleration of the particle relative to the fixed system is given from (1) as

$$\left.\frac{d^2\mathbf{r}}{dt^2}\right|_F = \frac{d^2\mathbf{r}}{dt^2} + \frac{d\boldsymbol{\omega}}{dt} \times \mathbf{r} + 2\boldsymbol{\omega} \times \left(\frac{d\mathbf{r}}{dt}\right) + \boldsymbol{\omega} \times (\boldsymbol{\omega} \times \mathbf{r}) \tag{3}$$

The acceleration (3) is sometimes called the *true acceleration*, while (2) is the *apparent acceleration*.

6.7. Find (a) the apparent acceleration and (b) the true acceleration of the particle in Problem 6.5.

(a) The apparent acceleration at any time t is

$$\frac{d^2\mathbf{r}}{dt^2} = \frac{d}{dt}\left(\frac{d\mathbf{r}}{dt}\right) = \frac{d}{dt}(2t\mathbf{i} - 6\mathbf{j} + 12t^2\mathbf{k}) = 2\mathbf{i} + 24t\mathbf{k}$$

At time $t = 1$ this is $2\mathbf{i} + 24\mathbf{k}$.

(b) The true acceleration at any time t is

$$\frac{d^2\mathbf{r}}{dt^2} + 2\boldsymbol{\omega} \times \frac{d\mathbf{r}}{dt} + \frac{d\boldsymbol{\omega}}{dt} \times \mathbf{r} + \boldsymbol{\omega} \times (\boldsymbol{\omega} \times \mathbf{r})$$

At time $t = 1$ this equals

$$\begin{aligned} &2\mathbf{i} + 24\mathbf{k} + (4\mathbf{i} - 2\mathbf{j} + 12\mathbf{k}) \times (2\mathbf{i} - 6\mathbf{j} + 12\mathbf{k}) \\ &\quad + (2\mathbf{i} - 2\mathbf{j} + 2\mathbf{k}) \times (2\mathbf{i} - 6\mathbf{j} + 4\mathbf{k}) \\ &\quad\quad + (2\mathbf{i} - \mathbf{j} + 6\mathbf{k}) \times \{(2\mathbf{i} - \mathbf{j} + 6\mathbf{k}) \times (2\mathbf{i} - 6\mathbf{j} + 4\mathbf{k})\} \\ &= 2\mathbf{i} + 24\mathbf{k} + (48\mathbf{i} - 24\mathbf{j} - 20\mathbf{k}) + (4\mathbf{i} - 4\mathbf{j} - 8\mathbf{k}) + (-14\mathbf{i} + 212\mathbf{j} + 40\mathbf{k}) \\ &= 40\mathbf{i} + 184\mathbf{j} + 36\mathbf{k} \end{aligned}$$

CORIOLIS AND CENTRIPETAL ACCELERATION

6.8. Referring to Problem 6.5, find (*a*) the Coriolis acceleration, (*b*) the centripetal acceleration and (*c*) their magnitudes at time $t = 1$.

(*a*) From Problem 6.5 we have,

$$\text{Coriolis acceleration} = 2\boldsymbol{\omega} \times d\mathbf{r}/dt = (4\mathbf{i} - 2\mathbf{j} + 12\mathbf{k}) \times (2\mathbf{i} - 6\mathbf{j} + 12\mathbf{k})$$
$$= 48\mathbf{i} - 24\mathbf{j} - 20\mathbf{k}$$

(*b*) From Problem 6.5 we have,

$$\text{Centripetal acceleration} = \boldsymbol{\omega} \times (\boldsymbol{\omega} \times \mathbf{r}) = (2\mathbf{i} - \mathbf{j} + 6\mathbf{k}) \times (32\mathbf{i} + 4\mathbf{j} - 10\mathbf{k})$$
$$= -14\mathbf{i} + 212\mathbf{j} + 40\mathbf{k}$$

(*c*) From parts (*a*) and (*b*) we have

$$\text{Magnitude of Coriolis acceleration} = \sqrt{(48)^2 + (-24)^2 + (-20)^2} = 4\sqrt{205}$$

$$\text{Magnitude of centripetal acceleration} = \sqrt{(-14)^2 + (212)^2 + (40)^2} = 2\sqrt{11{,}685}$$

MOTION OF A PARTICLE RELATIVE TO THE EARTH

6.9. (*a*) Express Newton's second law for the motion of a particle relative to an XYZ coordinate system fixed in space (inertial system). (*b*) Use (*a*) to find an equation of motion for the particle relative to an xyz system having the same origin as the XYZ system but rotating with respect to it.

(*a*) If m is the mass of the particle (assumed constant), $d^2\mathbf{r}/dt^2\,|_F$ its acceleration in the fixed system and $\mathbf{F}$ the resultant of all forces acting on the particle as viewed in the fixed system, then Newton's second law states that

$$m\frac{d^2\mathbf{r}}{dt^2}\bigg|_F = \mathbf{F} \tag{1}$$

(*b*) Using subscript M to denote quantities as viewed in the moving system, we have from Problem 6.6,

$$\frac{d^2\mathbf{r}}{dt^2}\bigg|_F = \frac{d^2\mathbf{r}}{dt^2}\bigg|_M + \dot{\boldsymbol{\omega}} \times \mathbf{r} + 2\boldsymbol{\omega} \times \frac{d\mathbf{r}}{dt}\bigg|_M + \boldsymbol{\omega} \times (\boldsymbol{\omega} \times \mathbf{r}) \tag{2}$$

Substituting this into (*1*), we find the required equation

$$m\frac{d^2\mathbf{r}}{dt^2}\bigg|_M = \mathbf{F} - m(\dot{\boldsymbol{\omega}} \times \mathbf{r}) - 2m\left(\boldsymbol{\omega} \times \frac{d\mathbf{r}}{dt}\bigg|_M\right) - m[\boldsymbol{\omega} \times (\boldsymbol{\omega} \times \mathbf{r})] \tag{3}$$

We can drop the subscript M provided it is clear that all quantities except $\mathbf{F}$ are as determined by an observer in the moving system. The quantity $\mathbf{F}$, it must be emphasized, is the resultant force as observed in the fixed or inertial system. If we do remove the subscript M and write $d\mathbf{r}/dt = \mathbf{v}$, then (*3*) can be written

$$m\frac{d^2\mathbf{r}}{dt^2} = \mathbf{F} - m(\dot{\boldsymbol{\omega}} \times \mathbf{r}) - 2m(\boldsymbol{\omega} \times \mathbf{v}) - m[\boldsymbol{\omega} \times (\boldsymbol{\omega} \times \mathbf{r})] \tag{4}$$

6.10. Calculate the angular speed of the earth about its axis.

Since the earth makes one revolution [2π radians] about its axis in approximately 24 hours = 86,400 sec, the angular speed is

$$\omega = \frac{2\pi}{86{,}400} = 7.27 \times 10^{-5} \text{ rad/sec}$$

The actual time for one revolution is closer to 86,164 sec and the angular speed 7.29×10^{-5} rad/sec.

MOVING COORDINATE SYSTEMS IN GENERAL

6.11. Work Problem 6.4 if the origins of the XYZ and xyz systems do not coincide.

Let $\mathbf{R}$ be the position vector of origin Q of the xyz system relative to origin O of the fixed (or inertial) XYZ system [see Fig. 6-3]. The velocity of the particle P relative to the moving system is, as before,

$$\left.\frac{d\mathbf{r}}{dt}\right|_M = \frac{d\mathbf{r}}{dt} = \frac{dx}{dt}\mathbf{i} + \frac{dy}{dt}\mathbf{j} + \frac{dz}{dt}\mathbf{k} \qquad (1)$$

Now the position vector of P relative to O is $\boldsymbol{\rho} = \mathbf{R} + \mathbf{r}$ and thus the velocity of P as viewed in the XYZ system is

$$\frac{d\boldsymbol{\rho}}{dt} = \left.\frac{d}{dt}(\mathbf{R}+\mathbf{r})\right|_F = \left.\frac{d\mathbf{R}}{dt}\right|_F + \left.\frac{d\mathbf{r}}{dt}\right|_F$$

$$= \dot{\mathbf{R}} + \frac{d\mathbf{r}}{dt} + \boldsymbol{\omega} \times \mathbf{r} \qquad (2)$$

using equation (3) of Problem 6.4. Note that $\dot{\mathbf{R}}$ is the velocity of Q with respect to O. If $\mathbf{R} = \mathbf{0}$ this reduces to the result of Problem 6.4.

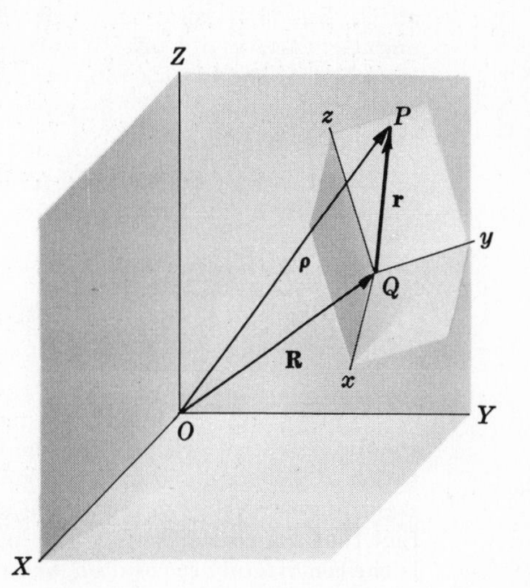

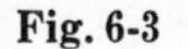

Fig. 6-3

6.12. Work Problem 6.6 if the origins of the XYZ and xyz systems do not coincide.

Referring to Fig. 6-3, the acceleration of the particle P relative to the moving system is, as before,

$$\left.\frac{d^2\mathbf{r}}{dt^2}\right|_M = \frac{d^2\mathbf{r}}{dt^2} = \frac{d^2x}{dt^2}\mathbf{i} + \frac{d^2y}{dt^2}\mathbf{j} + \frac{d^2z}{dt^2}\mathbf{k} \qquad (1)$$

Since the position vector of P relative to O is $\boldsymbol{\rho} = \mathbf{R} + \mathbf{r}$, the acceleration of P as viewed in the XYZ system is

$$\left.\frac{d^2\boldsymbol{\rho}}{dt^2}\right|_F = \left.\frac{d^2}{dt^2}(\mathbf{R}+\mathbf{r})\right|_F = \left.\frac{d^2\mathbf{R}}{dt^2}\right|_F + \left.\frac{d^2\mathbf{r}}{dt^2}\right|_F$$

$$= \ddot{\mathbf{R}} + \frac{d^2\mathbf{r}}{dt^2} + \frac{d\boldsymbol{\omega}}{dt} \times \mathbf{r} + 2\boldsymbol{\omega} \times \frac{d\mathbf{r}}{dt} + \boldsymbol{\omega} \times (\boldsymbol{\omega} \times \mathbf{r}) \qquad (2)$$

using equation (3) of Problem 6.6. Note that $\ddot{\mathbf{R}}$ is the acceleration of Q with respect to O. If $\mathbf{R} = \mathbf{0}$ this reduces to the result of Problem 6.6.

6.13. Work Problem 6.9 if the origins of the XYZ and xyz systems do not coincide.

(a) The position vector of the particle relative to the fixed (XYZ) system is $\boldsymbol{\rho}$. Then the required equation of motion is

$$m\left.\frac{d^2\boldsymbol{\rho}}{dt^2}\right|_F = \mathbf{F} \qquad (1)$$

(b) Using the result (2) of Problem 6.12 in (1), we obtain

$$m\frac{d^2\mathbf{r}}{dt^2} = \mathbf{F} - m\ddot{\mathbf{R}} - m(\dot{\boldsymbol{\omega}} \times \mathbf{r}) - 2m(\boldsymbol{\omega} \times \mathbf{v}) - m[\boldsymbol{\omega} \times (\boldsymbol{\omega} \times \mathbf{r})] \qquad (2)$$

where $\mathbf{F}$ is the force acting on m as viewed in the inertial system and where $\mathbf{v} = \dot{\mathbf{r}}$.

6.14. Find the equation of motion of a particle relative to an observer on the earth's surface.

We assume the earth to be a sphere with center at O [Fig. 6-4] rotating about the Z axis with angular velocity $\boldsymbol{\omega} = \omega\mathbf{K}$. We also use the fact that the effect of the earth's rotation around the sun is negligible, so that the XYZ system can be taken as an inertial system.

Then we can use equation (2) of Problem 6.12. For the case of the earth, we have

$$\dot{\boldsymbol{\omega}} = \mathbf{0} \tag{1}$$

$$\ddot{\mathbf{R}} = \boldsymbol{\omega} \times (\boldsymbol{\omega} \times \mathbf{R}) \tag{2}$$

$$\mathbf{F} = -\frac{GMm}{\rho^3}\boldsymbol{\rho} \tag{3}$$

the first equation arising from the fact that the rotation of the earth about its axis proceeds with constant angular velocity, the second arising from the fact that the acceleration of origin Q relative to O is the centripetal acceleration, and the third arising from Newton's law of gravitation. Using these in (2) of Problem 6.12 yields the required equation,

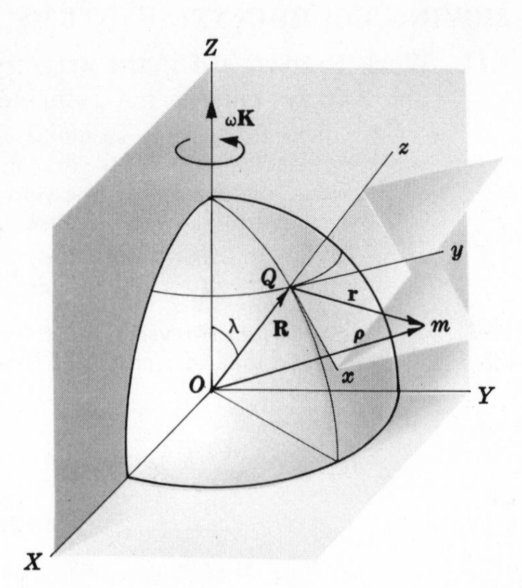

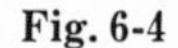
Fig. 6-4

$$\frac{d^2\mathbf{r}}{dt^2} = -\frac{GM}{\rho^3}\boldsymbol{\rho} - \boldsymbol{\omega} \times (\boldsymbol{\omega} \times \mathbf{R}) - 2(\boldsymbol{\omega} \times \mathbf{v}) - \boldsymbol{\omega} \times (\boldsymbol{\omega} \times \mathbf{r}) \tag{4}$$

assuming that other forces acting on m [such as air resistance, etc.] are neglected.

We can define

$$\mathbf{g} = -\frac{GM}{\rho^3}\boldsymbol{\rho} - \boldsymbol{\omega} \times (\boldsymbol{\omega} \times \mathbf{R}) \tag{5}$$

as the acceleration due to gravity, so that (4) becomes

$$\frac{d^2\mathbf{r}}{dt^2} = \mathbf{g} - 2(\boldsymbol{\omega} \times \mathbf{v}) - \boldsymbol{\omega} \times (\boldsymbol{\omega} \times \mathbf{r}) \tag{6}$$

Near the earth's surface the last term in (6) can be neglected, so that to a high degree of approximation,

$$\frac{d^2\mathbf{r}}{dt^2} = \mathbf{g} - 2(\boldsymbol{\omega} \times \mathbf{v}) \tag{7}$$

In practice we choose $\mathbf{g}$ as constant in magnitude although it varies slightly over the earth's surface. If other external forces act, we must add them to the right side of equations (6) or (7).

6.15. Show that if the particle of Problem 6.14 moves near the earth's surface, then the equations of motion are given by

$$\ddot{x} = 2\omega \cos\lambda\, \dot{y}$$

$$\ddot{y} = -2(\omega \cos\lambda\, \dot{x} + \omega \sin\lambda\, \dot{z})$$

$$\ddot{z} = -g + 2\omega \sin\lambda\, \dot{y}$$

where the angle λ is the *colatitude* [see Fig. 6-4] and $90° - \lambda$ is the *latitude*.

From Fig. 6-4 we have

$$\mathbf{K} = (\mathbf{K}\cdot\mathbf{i})\mathbf{i} + (\mathbf{K}\cdot\mathbf{j})\mathbf{j} + (\mathbf{K}\cdot\mathbf{k})\mathbf{k}$$

$$= (-\sin\lambda)\mathbf{i} + 0\mathbf{j} + (\cos\lambda)\mathbf{k} = -\sin\lambda\,\mathbf{i} + \cos\lambda\,\mathbf{k}$$

and so

$$\boldsymbol{\omega} = \omega\mathbf{K} = -\omega \sin\lambda\,\mathbf{i} + \omega \cos\lambda\,\mathbf{k}$$

Then
$$\boldsymbol{\omega}\times\mathbf{v} = \boldsymbol{\omega}\times(\dot{x}\mathbf{i}+\dot{y}\mathbf{j}+\dot{z}\mathbf{k})$$
$$= \begin{vmatrix} \mathbf{i} & \mathbf{j} & \mathbf{k} \\ -\omega\sin\lambda & 0 & \omega\cos\lambda \\ \dot{x} & \dot{y} & \dot{z} \end{vmatrix}$$
$$= (-\omega\cos\lambda\,\dot{y})\mathbf{i} + (\omega\cos\lambda\,\dot{x} + \omega\sin\lambda\,\dot{z})\mathbf{j} - (\omega\sin\lambda\,\dot{y})\mathbf{k}$$

Thus from equation (7) of Problem 6.14 we have
$$\frac{d^2\mathbf{r}}{dt^2} = \mathbf{g} - 2(\boldsymbol{\omega}\times\mathbf{v})$$
$$= -g\mathbf{k} + 2\omega\cos\lambda\,\dot{y}\,\mathbf{i} - 2(\omega\cos\lambda\,\dot{x} + \omega\sin\lambda\,\dot{z})\mathbf{j} + 2\omega\sin\lambda\,\dot{y}\,\mathbf{k}$$

Equating corresponding coefficients of $\mathbf{i},\mathbf{j},\mathbf{k}$ on both sides of this equation, we find, as required,
$$\ddot{x} = 2\omega\cos\lambda\,\dot{y} \quad (1)$$
$$\ddot{y} = -2(\omega\cos\lambda\,\dot{x} + \omega\sin\lambda\,\dot{z}) \quad (2)$$
$$\ddot{z} = -g + 2\omega\sin\lambda\,\dot{y} \quad (3)$$

6.16. An object of mass m initially at rest is dropped to the earth's surface from a height which is small compared with the earth's radius. Assuming that the angular speed of the earth about its axis is a constant ω, prove that after time t the object is deflected east of the vertical by the amount $\frac{1}{3}\omega g t^3 \sin\lambda$.

Method 1.

We assume that the object is located on the z axis at $x=0,\ y=0,\ z=h$ [see Fig. 6-4]. From equations (1) and (2) of Problem 6.15 we have on integrating,
$$\dot{x} = 2\omega\cos\lambda\,y + c_1, \qquad \dot{y} = -2(\omega\cos\lambda\,x + \omega\sin\lambda\,z) + c_2$$

Since at $t=0,\ \dot{x}=0,\ \dot{y}=0,\ x=0,\ y=0,\ z=h$ we have $c_1=0,\ c_2=2\omega\sin\lambda\,h$. Thus
$$\dot{x} = 2\omega\cos\lambda\,y, \qquad \dot{y} = -2(\omega\cos\lambda\,x + \omega\sin\lambda\,z) + 2\omega\sin\lambda\,h \quad (1)$$

Then (3) of Problem 6.15 becomes
$$\ddot{z} = -g + 2\omega\sin\lambda\,\dot{y} = -g - 4\omega^2\sin\lambda\,[\cos\lambda\,x + \sin\lambda\,(z-h)]$$

But since the terms on the right involving ω^2 are very small compared with $-g$ we can neglect them and write $\ddot{z}=-g$. Integration yields $\dot{z}=-gt+c_3$. Since $\dot{z}=0$ at $t=0$, we have $c_3=0$ or
$$\dot{z} = -gt \quad (2)$$

Using equation (2) and the first equation of (1) in equation (2) of Problem 6.15 we find
$$\ddot{y} = (-2\omega\cos\lambda)(2\omega\cos\lambda\,y) + (-2\omega\sin\lambda)(-gt)$$
$$= -4\omega^2\cos^2\lambda\,y + 2\omega\sin\lambda\,gt$$

Then neglecting the first term, we have $\ddot{y} = 2\omega\sin\lambda\,gt$. Integrating,
$$\dot{y} = \omega g\sin\lambda\,t^2 + c_4$$

Since $\dot{y}=0$ at $t=0$, we have $c_4=0$ and $\dot{y}=\omega g\sin\lambda\,t^2$. Integrating again,
$$y = \tfrac{1}{3}\omega g\sin\lambda\,t^3 + c_5$$

Then since $y=0$ at $t=0$, $c_5=0$ so that, as required,
$$y = \tfrac{1}{3}\omega g\sin\lambda\,t^3 \quad (3)$$

Method 2.

Integrating equations (1), (2) and (3) of Problem 6.15, we have
$$\dot{x} = 2\omega\cos\lambda\,y + c_1$$
$$\dot{y} = -2(\omega\cos\lambda\,x + \omega\sin\lambda\,z) + c_2$$
$$\dot{z} = -gt + 2\omega\sin\lambda\,y + c_3$$

Using the fact that at $t = 0$, $\dot{x} = \dot{y} = \dot{z} = 0$ and $x = 0$; $y = 0$, $z = h$, we have $c_1 = 0$, $c_2 = 2\omega h \sin\lambda$, $c_3 = 0$. Thus

$$\begin{aligned} \dot{x} &= 2\omega \cos\lambda\, y \\ \dot{y} &= -2(\omega\cos\lambda\, x + \omega\sin\lambda\, z) + 2\omega h \sin\lambda \\ \dot{z} &= -gt + 2\omega\sin\lambda\, y \end{aligned}$$

Integrating these we find, using the above conditions,

$$x = 2\omega\cos\lambda \int_0^t y\,du \tag{4}$$

$$y = 2\omega h t \sin\lambda - 2\omega\cos\lambda \int_0^t x\,du - 2\omega\sin\lambda \int_0^t z\,du \tag{5}$$

$$z = h - \tfrac{1}{2}gt^2 + 2\omega\sin\lambda \int_0^t y\,du \tag{6}$$

Since the unknowns are under the integral sign, these equations are called *integral equations.* We shall use a method called the *method of successive approximations* or *method of iteration* to obtain a solution to any desired accuracy. The method consists of using a first guess for x, y, z under the integral signs in (4), (5) and (6) to obtain a better guess. As a first guess we can try $x = 0$, $y = 0$, $z = 0$ under the integral signs. Then we find as a second guess

$$x = 0, \qquad y = 2\omega h t \sin\lambda, \qquad z = h - \tfrac{1}{2}gt^2$$

Substituting these in (4), (5) and (6) and neglecting terms involving ω^2, we find the third guess

$$x = 0, \qquad y = 2\omega h t\sin\lambda - 2\omega\sin\lambda(ht - \tfrac{1}{6}gt^3) = \tfrac{1}{3}\omega g t^3 \sin\lambda, \qquad z = h - \tfrac{1}{2}gt^2$$

Using these in (4), (5) and (6) and again neglecting terms involving ω^2, we find the fourth guess

$$x = 0, \qquad y = \tfrac{1}{3}\omega g t^3 \sin\lambda, \qquad z = h - \tfrac{1}{2}gt^2$$

Since this fourth guess is identical with the third guess, these results are accurate up to terms involving ω^2, and no further guesses need be taken. It is thus seen that the deflection is $y = \frac{1}{3}\omega g t^3 \sin\lambda$, as required.

6.17. Referring to Problem 6.16, show that an object dropped from height h above the earth's surface hits the earth at a point east of the vertical at a distance $\frac{2}{3}\omega h \sin\lambda \sqrt{2h/g}$.

From (2) of Problem 6.16 we have on integrating, $z = -\frac{1}{2}gt^2 + c$. Since $z = h$ at $t = 0$, $c = h$ and $z = h - \frac{1}{2}gt^2$. Then at $z = 0$, $h = \frac{1}{2}gt^2$ or $t = \sqrt{2h/g}$. Substituting this value of t into (3) of Problem 6.16, we find the required distance.

THE FOUCAULT PENDULUM

6.18. Derive an equation of motion for a simple pendulum, taking into account the earth's rotation about its axis.

Choose the xyz coordinate system of Fig. 6-5. Suppose that the origin O is the equilibrium position of the bob B, A is the point of suspension and the length of string AB is l. If the tension in the string is $\mathbf{T}$, then we have

$$\begin{aligned} \mathbf{T} &= (\mathbf{T}\cdot\mathbf{i})\mathbf{i} + (\mathbf{T}\cdot\mathbf{j})\mathbf{j} + (\mathbf{T}\cdot\mathbf{k})\mathbf{k} \\ &= T\cos\alpha\,\mathbf{i} + T\cos\beta\,\mathbf{j} + T\cos\gamma\,\mathbf{k} \\ &= -T\left(\frac{x}{l}\right)\mathbf{i} - T\left(\frac{y}{l}\right)\mathbf{j} + T\left(\frac{l-z}{l}\right)\mathbf{k} \end{aligned} \tag{1}$$

Since the net force acting on B is $\mathbf{T} + m\mathbf{g}$, the equation of motion of B is given by [see Problem 6.14]

$$m\frac{d^2\mathbf{r}}{dt^2} = \mathbf{T} + m\mathbf{g} - 2m(\boldsymbol{\omega}\times\mathbf{v}) - m\boldsymbol{\omega}\times(\boldsymbol{\omega}\times\mathbf{r}) \tag{2}$$

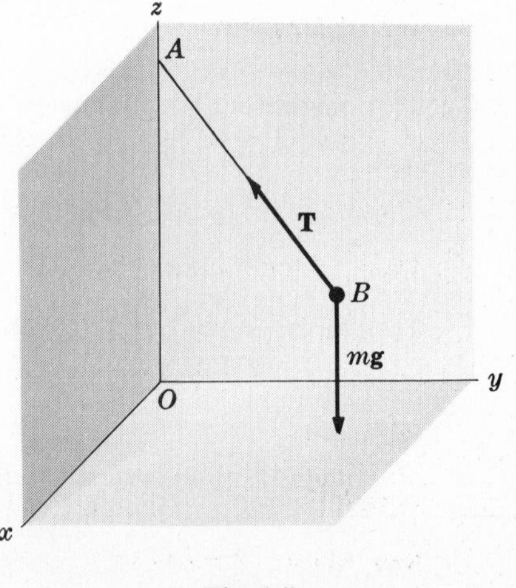

Fig. 6-5

If we neglect the last term in (2), put $\mathbf{g} = -g\mathbf{k}$ and use (1), then (2) can be written in component form as

$$m\ddot{x} = -T(x/l) + 2m\omega\dot{y}\cos\lambda \tag{3}$$

$$m\ddot{y} = -T(y/l) - 2m\omega(\dot{x}\cos\lambda + \dot{z}\sin\lambda) \tag{4}$$

$$m\ddot{z} = T(l-z)/l - mg + 2m\omega\dot{y}\sin\lambda \tag{5}$$

6.19. **By assuming that the bob of the simple pendulum in Problem 6.18 undergoes small oscillations about the equilibrium position so that its motion can be assumed to take place in a horizontal plane, simplify the equations of motion.**

Making the assumption that the motion of the bob takes place in a horizontal plane amounts to assuming that $\ddot{z}$ and $\dot{z}$ are zero. For small vibrations $(l-z)/l$ is very nearly equal to one. Then equation (5) of Problem 6.18 yields

$$0 = T - mg + 2m\omega\dot{y}\sin\lambda$$

or

$$T = mg - 2m\omega\dot{y}\sin\lambda \tag{1}$$

Substituting (1) into equations (3) and (4) of Problem 6.18 and simplifying, we obtain

$$\ddot{x} = -\frac{gx}{l} + \frac{2\omega x\dot{y}\sin\lambda}{l} + 2\omega\dot{y}\cos\lambda \tag{2}$$

$$\ddot{y} = -\frac{gy}{l} + \frac{2\omega y\dot{y}\sin\lambda}{l} - 2\omega\dot{x}\cos\lambda \tag{3}$$

These differential equations are non-linear because of the presence of the terms involving $x\dot{y}$ and $y\dot{y}$. However, these terms are negligible compared with the others since ω, x and y are small. Upon neglecting them we obtain the linear differential equations

$$\ddot{x} = -gx/l + 2\omega\dot{y}\cos\lambda \tag{4}$$

$$\ddot{y} = -gy/l - 2\omega\dot{x}\cos\lambda \tag{5}$$

6.20. **Solve the equations of motion of the pendulum obtained in Problem 6.19, assuming suitable initial conditions.**

Suppose that initially the bob is in the yz plane and is given a displacement from the z axis of magnitude $A > 0$, after which it is released. Then the initial conditions are

$$x = 0,\ \dot{x} = 0,\ y = A,\ \dot{y} = 0 \quad \text{at } t = 0 \tag{1}$$

To find the solution of equations (4) and (5) of Problem 6.19, it is convenient to place

$$K^2 = g/l, \quad \alpha = \omega\cos\lambda \tag{2}$$

so that they become

$$\ddot{x} = -K^2x + 2\alpha\dot{y} \tag{3}$$

$$\ddot{y} = -K^2y - 2\alpha\dot{x} \tag{4}$$

It is also convenient to use complex numbers. Multiplying equation (4) by i and adding to (3), we find

$$\ddot{x} + i\ddot{y} = -K^2(x+iy) + 2\alpha(\dot{y} - i\dot{x}) = -K^2(x+iy) - 2i\alpha(\dot{x}+i\dot{y})$$

Then calling $u = x + iy$, this can be written

$$\ddot{u} = -K^2u - 2i\alpha\dot{u} \quad \text{or} \quad \ddot{u} + 2i\alpha\dot{u} + K^2u = 0 \tag{5}$$

If $u = Ce^{\gamma t}$ where C and γ are constants, this becomes

$$\gamma^2 + 2i\alpha\gamma + K^2 = 0$$

so that

$$\gamma = (-2i\alpha \pm \sqrt{-4\alpha^2 - 4K^2})/2 = -i\alpha \pm i\sqrt{\alpha^2 + K^2} \tag{6}$$

Now since $\alpha^2 = \omega^2\cos^2\lambda$ is small compared to $K^2 = g/l$, we can write

$$\gamma = -i\alpha \pm iK \tag{7}$$

Then solutions of the equation are (allowing for complex coefficients)

$$(C_1 + iC_2)e^{-i(\alpha-K)t} \quad \text{and} \quad (C_3 + iC_4)e^{-i(\alpha+K)t}$$

and the general solution is

$$u = (C_1 + iC_2)e^{-i(\alpha-K)t} + (C_3 + iC_4)e^{-i(\alpha+K)t} \tag{8}$$

where C_1, C_2, C_3, C_4 are assumed real. Using Euler's formulas

$$e^{i\theta} = \cos\theta + i\sin\theta, \qquad e^{-i\theta} = \cos\theta - i\sin\theta \tag{9}$$

and the fact that $u = x + iy$, *(8)* can be written

$$x + iy = (C_1 + iC_2)\{\cos(\alpha - K)t - i\sin(\alpha - K)t\} + (C_3 + iC_4)\{\cos(\alpha + K)t - i\sin(\alpha + K)t\}$$

Equating real and imaginary parts, we find

$$x = C_1\cos(\alpha - K)t + C_2\sin(\alpha - K)t + C_3\cos(\alpha + K)t + C_4\sin(\alpha + K)t \tag{10}$$

$$y = -C_1\sin(\alpha - K)t + C_2\cos(\alpha - K)t - C_3\sin(\alpha + K)t + C_4\cos(\alpha + K)t \tag{11}$$

Using the initial condition $x = 0$ at $t = 0$, we find from *(10)* that $C_1 + C_3 = 0$ or $C_3 = -C_1$. Similarly, using $\dot{x} = 0$ at $t = 0$, we find from *(10)* that

$$C_4 = C_2\left(\frac{K - \alpha}{K + \alpha}\right) = C_2\left(\frac{\sqrt{g/l} - \omega\cos\lambda}{\sqrt{g/l} + \omega\cos\lambda}\right)$$

Now since $\omega\cos\lambda$ is small compared with $\sqrt{g/l}$, we have, to a high degree of approximation, $C_4 = C_2$.

Thus equations *(10)* and *(11)* become

$$x = C_1\cos(\alpha - K)t + C_2\sin(\alpha - K)t - C_1\cos(\alpha + K)t + C_2\sin(\alpha + K)t \tag{12}$$

$$y = -C_1\sin(\alpha - K)t + C_2\cos(\alpha - K)t + C_1\sin(\alpha + K)t + C_2\cos(\alpha + K)t \tag{13}$$

Using the initial condition $\dot{y} = 0$, *(13)* yields $C_1 = 0$. Similarly using $y = A$ at $t = 0$, we find $C_2 = \frac{1}{2}A$. Thus *(12)* and *(13)* become

$$x = \tfrac{1}{2}A\sin(\alpha - K)t + \tfrac{1}{2}A\sin(\alpha + K)t$$

$$y = \tfrac{1}{2}A\cos(\alpha - K)t + \tfrac{1}{2}A\cos(\alpha + K)t$$

or

$$\left.\begin{aligned} x &= A\cos Kt\sin\alpha t \\ y &= A\cos Kt\cos\alpha t \end{aligned}\right\} \tag{14}$$

i.e.,

$$\left.\begin{aligned} x &= A\cos\sqrt{g/l}\,t\,\sin(\omega\cos\lambda\ t) \\ y &= A\cos\sqrt{g/l}\,t\,\cos(\omega\cos\lambda\ t) \end{aligned}\right\} \tag{15}$$

6.21. Give a physical interpretation to the solution *(15)* of Problem 6.20.

In vector form, *(15)* can be written

$$\mathbf{r} = x\mathbf{i} + y\mathbf{j} = A\cos\sqrt{g/l}\,t\,\mathbf{n}$$

where

$$\mathbf{n} = \mathbf{i}\sin(\omega\cos\lambda)t + \mathbf{j}\cos(\omega\cos\lambda)t$$

is a unit vector.

The period of $\cos\sqrt{g/l}\,t$ [namely, $2\pi\sqrt{l/g}$] is very small compared with the period of $\mathbf{n}$ [namely, $2\pi/(\omega\cos\lambda)$]. It follows that $\mathbf{n}$ is a very slowly turning vector. Thus physically the pendulum oscillates in a plane through the z axis which is slowly rotating (or *precessing*) about the z axis.

Now at $t = 0$, $\mathbf{n} = \mathbf{j}$ and the bob is at $y = A$. After a time $t = 2\pi/(4\omega\cos\lambda)$, for example, $\mathbf{n} = \frac{1}{2}\sqrt{2}\,\mathbf{i} + \frac{1}{2}\sqrt{2}\,\mathbf{j}$ so that the rotation of the plane is proceeding in the clockwise direction as viewed from above the earth's surface in the northern hemisphere [where $\cos\lambda > 0$]. In the southern hemisphere the rotation of the plane is counterclockwise.

The rotation of the plane was observed by Foucault in 1851 and served to provide laboratory evidence of the rotation of the earth about its axis.

MISCELLANEOUS PROBLEMS

6.22. The vertical rod AB of Fig. 6-6 is rotating with constant angular velocity $\boldsymbol{\omega}$. A light inextensible string of length l has one end attached at point O of the rod while the other end P of the string has a mass m attached. Find (*a*) the tension in the string and (*b*) the angle which string OP makes with the vertical when equilibrium conditions prevail.

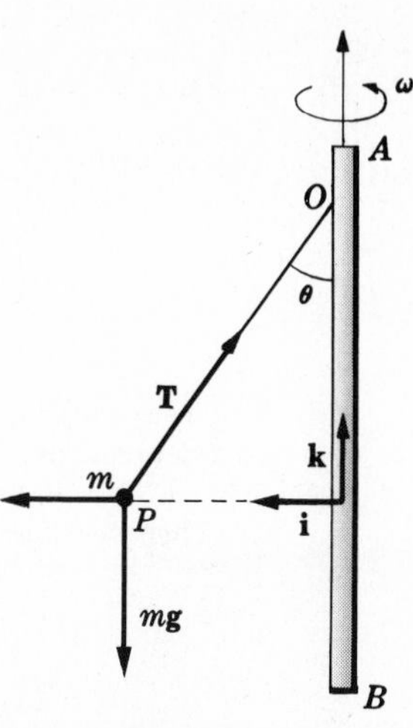

Fig. 6-6

Choose unit vectors $\mathbf{i}$ and $\mathbf{k}$ perpendicular and parallel respectively to the rod and rotating with it. The unit vector $\mathbf{j}$ can be chosen perpendicular to the plane of $\mathbf{i}$ and $\mathbf{k}$. Let

$$\mathbf{r} = l \sin\theta\,\mathbf{i} - l\cos\theta\,\mathbf{k}$$

be the position vector of m with respect to O.

Three forces act on particle m

(i) The weight, $m\mathbf{g} = -mg\mathbf{k}$

(ii) The centrifugal force,

$$-m\{\boldsymbol{\omega}\times(\boldsymbol{\omega}\times\mathbf{r})\} = -m\{[\omega\mathbf{k}]\times([\omega\mathbf{k}]\times[l\sin\theta\,\mathbf{i} - l\cos\theta\,\mathbf{k}])\}$$
$$= -m\{[\omega\mathbf{k}]\times(\omega l\sin\theta\,\mathbf{j})\} = m\omega^2 l\sin\theta\,\mathbf{i}$$

(iii) The tension, $\mathbf{T} = -T\sin\theta\,\mathbf{i} + T\cos\theta\,\mathbf{k}$

When the particle is in equilibrium, the resultant of all these forces is zero. Then

$$-mg\mathbf{k} + m\omega^2 l\sin\theta\,\mathbf{i} - T\sin\theta\,\mathbf{i} + T\cos\theta\,\mathbf{k} = \mathbf{0}$$

i.e.,

$$(m\omega^2 l\sin\theta - T\sin\theta)\mathbf{i} + (T\cos\theta - mg)\mathbf{k} = \mathbf{0}$$

or

$$m\omega^2 l\sin\theta - T\sin\theta = 0 \qquad (1)$$

$$T\cos\theta - mg = 0 \qquad (2)$$

Solving (*1*) and (*2*) simultaneously, we find (*a*) $T = m\omega^2 l$, (*b*) $\theta = \cos^{-1}(g/\omega^2 l)$.

Since the string OP with mass m at P describe the surface of a cone the system is sometimes called a *conical pendulum*.

6.23. A rod AOB [Fig. 6-7] rotates in a vertical plane [the yz plane] about a horizontal axis through O perpendicular to this plane [the x axis] with constant angular velocity $\boldsymbol{\omega}$. Assuming no frictional forces, determine the motion of a particle P of mass m which is constrained to move along the rod. An equivalent problem exists when the rod AOB is replaced by a thin hollow tube inside which the particle can move.

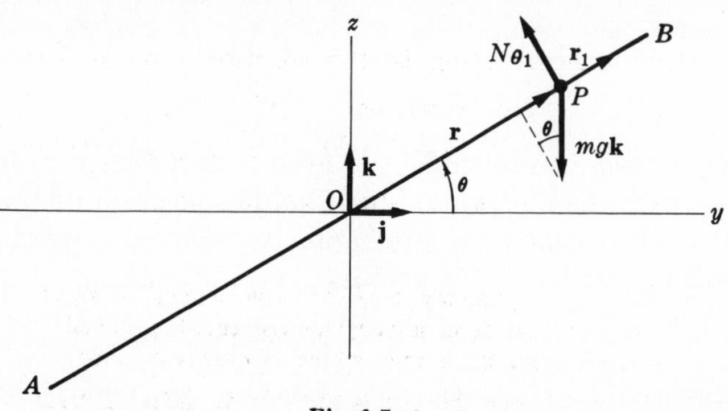

Fig. 6-7

At time t let $\mathbf{r}$ be the position vector of the particle and θ the angle made by the rod with the y axis. Choose unit vectors $\mathbf{j}$ and $\mathbf{k}$ in the y and z directions respectively and unit vector $\mathbf{i} = \mathbf{j}\times\mathbf{k}$. Let $\mathbf{r}_1$ be a unit vector in the direction $\mathbf{r}$ and $\boldsymbol{\theta}_1$ a unit vector in the direction of increasing θ.

There are three forces acting on P:

(i) The weight, $m\mathbf{g} = -mg\mathbf{k} = -mg \sin\theta\, \mathbf{r}_1 - mg\cos\theta\, \boldsymbol{\theta}_1$

(ii) The centrifugal force,

$$\begin{aligned} -m[\boldsymbol{\omega} \times (\boldsymbol{\omega} \times \mathbf{r})] &= -m[\omega\mathbf{i} \times (\omega\mathbf{i} \times r\mathbf{r}_1)] \\ &= -m[\omega\mathbf{i}(\omega\mathbf{i} \cdot r\mathbf{r}_1) - r\mathbf{r}_1(\omega\mathbf{i} \cdot \omega\mathbf{i})] \\ &= -m[\mathbf{0} - \omega^2 r\mathbf{r}_1] \;=\; m\omega^2 r\mathbf{r}_1 \end{aligned}$$

(iii) The reaction force $\mathbf{N} = N\boldsymbol{\theta}_1$ of the rod which is perpendicular to the rod since there are no frictional or resistance forces.

Then by Newton's second law,

$$m\frac{d^2\mathbf{r}}{dt^2} = -mg\mathbf{k} + m\omega^2 r\mathbf{r}_1 + N\boldsymbol{\theta}_1$$

or

$$\begin{aligned} m\frac{d^2r}{dt^2}\mathbf{r}_1 &= -mg\sin\theta\,\mathbf{r}_1 - mg\cos\theta\,\boldsymbol{\theta}_1 + m\omega^2 r\mathbf{r}_1 + N\boldsymbol{\theta}_1 \\ &= (m\omega^2 r - mg\sin\theta)\mathbf{r}_1 + (N - mg\cos\theta)\boldsymbol{\theta}_1 \end{aligned}$$

It follows that $N = mg\cos\theta$ and

$$d^2r/dt^2 = \omega^2 r - g\sin\theta \tag{1}$$

Since $\dot{\theta} = \omega$, a constant, we have $\theta = \omega t$ if we assume $\theta = 0$ at $t = 0$. Then (1) becomes

$$d^2r/dt^2 - \omega^2 r = -g\sin\omega t \tag{2}$$

If we assume that at $t = 0$, $r = r_0$, $dr/dt = v_0$, we find

$$r = \left(\frac{r_0}{2} + \frac{v_0}{2\omega} - \frac{g}{4\omega^2}\right)e^{\omega t} + \left(\frac{r_0}{2} - \frac{v_0}{2\omega} + \frac{g}{4\omega^2}\right)e^{-\omega t} + \frac{g}{2\omega^2}\sin\omega t \tag{3}$$

or in terms of hyperbolic functions,

$$r = r_0\cosh\omega t + \left(\frac{v_0}{\omega} - \frac{g}{2\omega^2}\right)\sinh\omega t + \frac{g}{2\omega^2}\sin\omega t \tag{4}$$

6.24. (a) Show that under suitable conditions the particle of Problem 6.23 can oscillate along the rod with simple harmonic motion and find these conditions. (b) What happens to the particle if the conditions of (a) are not satisfied?

(a) The particle will oscillate with simple harmonic motion along the rod if and only if $r_0 = 0$ and $v_0 = g/2\omega$. In this case, $r = (g/2\omega^2)\sin\omega t$. Thus the amplitude and period of the simple harmonic motion in such case are given by $g/2\omega^2$ and $2\pi/\omega$ respectively.

(b) If $v_0 = (g/2\omega) - \omega r_0$ then $r = r_0 e^{-\omega t} + (g/2\omega^2)\sin\omega t$ and the motion is approximately simple harmonic after some time. Otherwise the mass will ultimately fly off the rod if it is finite.

6.25. A projectile located at colatitude λ is fired with velocity v_0 in a southward direction at an angle α with the horizontal. (a) Find the position of the projectile after time t. (b) Prove that after time t the projectile is deflected toward the east of the original vertical plane of motion by the amount

$$\tfrac{1}{3}\omega g\sin\lambda\, t^3 - \omega v_0\cos(\alpha - \lambda)\, t^2$$

(a) We use the equations of Problem 6.15. Assuming the projectile starts at the origin, we have

$$x = 0,\; y = 0,\; z = 0 \qquad \text{at } t = 0 \tag{1}$$

Also, the initial velocity is $\mathbf{v}_0 = v_0\cos\alpha\,\mathbf{i} + v_0\sin\alpha\,\mathbf{k}$ so that

$$\dot{x} = v_0\cos\alpha,\; \dot{y} = 0,\; \dot{z} = v_0\sin\alpha \qquad \text{at } t = 0 \tag{2}$$

Integrating equations (*1*), (*2*) and (*3*) of Problem 6.15, we obtain on using conditions (*2*),

$$\dot{x} = 2\omega \cos\lambda\, y + v_0 \cos\alpha \tag{3}$$

$$\dot{y} = -2(\omega\cos\lambda\, x + \omega\sin\lambda\, z) \tag{4}$$

$$\dot{z} = -gt + 2\omega\sin\lambda\, y + v_0\sin\alpha \tag{5}$$

Instead of attempting to solve these equations directly we shall use the *method of iteration* or *successive approximations* as in Method 2 of Problem 6.16. Thus by integrating and using conditions (*1*), we find

$$x = 2\omega\cos\lambda \int_0^t y\,du + (v_0\cos\alpha)t \tag{6}$$

$$y = -2\omega\cos\lambda \int_0^t x\,du - 2\omega\sin\lambda\int_0^t z\,du \tag{7}$$

$$z = (v_0\sin\alpha)t - \tfrac{1}{2}gt^2 + 2\omega\sin\lambda\int_0^t y\,du \tag{8}$$

As a first guess we use $x = 0$, $y = 0$, $z = 0$ under the integral signs. Then (*6*), (*7*) and (*8*) become, neglecting terms involving ω^2,

$$x = (v_0\cos\alpha)t \tag{9}$$

$$y = 0 \tag{10}$$

$$z = (v_0\sin\alpha)t - \tfrac{1}{2}gt^2 \tag{11}$$

To obtain a better guess we now use (*9*), (*10*) and (*11*) under the integral signs in (*6*), (*7*) and (*8*), thus arriving at

$$x = (v_0\cos\alpha)t \tag{12}$$

$$y = -\omega v_0\cos(\alpha-\lambda)\,t^2 + \tfrac{1}{3}\omega g t^3\sin\lambda \tag{13}$$

$$z = (v_0\sin\alpha)t - \tfrac{1}{2}gt^2 \tag{14}$$

where we have again neglected terms involving ω^2. Further guesses again produce equations (*12*), (*13*) and (*14*), so that these equations are accurate up to terms involving ω^2.

(*b*) From equation (*13*) we see that the projectile is deflected toward the east of the xz plane by the amount $\frac{1}{3}\omega g t^3\sin\lambda - \omega v_0\cos(\alpha-\lambda)\,t^2$. If $v_0 = 0$ this agrees with Problem 6.16.

6.26. Prove that when the projectile of Problem 6.25 returns to the horizontal, it will be at the distance

$$\frac{\omega v_0^3\sin^2\alpha}{3g^2}(3\cos\alpha\cos\lambda + \sin\alpha\sin\lambda)$$

to the west of that point where it would have landed assuming no axial rotation of the earth.

The projectile will return to the horizontal when $z = 0$, i.e.,

$$(v_0\sin\alpha)t - \tfrac{1}{2}gt^2 = 0 \quad\text{or}\quad t = (2v_0\sin\alpha)/g$$

Using this value of t in equation (*13*) of Problem 6.25, we find the required result.

Supplementary Problems

ROTATING COORDINATE SYSTEMS. VELOCITY AND ACCELERATION

6.27. An xyz coordinate system moves with angular velocity $\boldsymbol{\omega} = 2\mathbf{i} - 3\mathbf{j} + 5\mathbf{k}$ relative to a fixed or inertial XYZ coordinate system having the same origin. If a vector relative to the xyz system is given as a function of time t by $\mathbf{A} = \sin t\,\mathbf{i} - \cos t\,\mathbf{j} + e^{-t}\,\mathbf{k}$, find (a) $d\mathbf{A}/dt$ relative to the fixed system, (b) $d\mathbf{A}/dt$ relative to the moving system.

Ans. (a) $(6\cos t - 3e^{-t})\mathbf{i} + (6\sin t - 2e^{-t})\mathbf{j} + (3\sin t - 2\cos t - e^{-t})\mathbf{k}$

(b) $\cos t\,\mathbf{i} + \sin t\,\mathbf{j} - e^{-t}\mathbf{k}$

6.28. Find $d^2\mathbf{A}/dt^2$ for the vector $\mathbf{A}$ of Problem 6.27 relative to (a) the fixed system and (b) the moving system.

Ans. (a) $(6\cos t - 45\sin t + 16e^{-t})\mathbf{i} + (40\cos t - 6\sin t - 11e^{-t})\mathbf{j}$
$+ (10\sin t - 23\cos t + 16e^{-t})\mathbf{k}$

(b) $-\sin t\,\mathbf{i} + \cos t\,\mathbf{j} + e^{-t}\mathbf{k}$

6.29. An xyz coordinate system is rotating with angular velocity $\boldsymbol{\omega} = 5\mathbf{i} - 4\mathbf{j} - 10\mathbf{k}$ relative to a fixed XYZ coordinate system having the same origin. Find the velocity of a particle fixed in the xyz system at the point $(3, 1, -2)$ as seen by an observer fixed in the XYZ system.

Ans. $18\mathbf{i} - 20\mathbf{j} + 17\mathbf{k}$

6.30. Discuss the physical interpretation of replacing $\boldsymbol{\omega}$ by $-\boldsymbol{\omega}$ in (a) Problem 6.4, page 148, and (b) Problem 6.6, page 149.

6.31. Explain from a physical point of view why you would expect the result of Problem 6.3, page 148, to be correct.

6.32. An xyz coordinate system rotates with angular velocity $\boldsymbol{\omega} = \cos t\,\mathbf{i} + \sin t\,\mathbf{j} + \mathbf{k}$ with respect to a fixed XYZ coordinate system having the same origin. If the position vector of a particle is given by $\mathbf{r} = \sin t\,\mathbf{i} - \cos t\,\mathbf{j} + t\mathbf{k}$, find (a) the apparent velocity and (b) the true velocity at any time t. *Ans.* (a) $\cos t\,\mathbf{i} + \sin t\,\mathbf{j} + \mathbf{k}$ (b) $(t\sin t + 2\cos t)\mathbf{i} + (2\sin t - t\cos t)\mathbf{j}$

6.33. Determine (a) the apparent acceleration and (b) the true acceleration of the particle of Problem 6.32.

Ans. (a) $-\sin t\,\mathbf{i} + \cos t\,\mathbf{j}$ (b) $(2t\cos t - 3\sin t)\mathbf{i} + (3\cos t + 2t\sin t)\mathbf{j} + (1 - t)\mathbf{k}$

CORIOLIS AND CENTRIPETAL ACCELERATIONS AND FORCES

6.34. A ball is thrown horizontally in the northern hemisphere. (a) Would the path of the ball, if the Coriolis force is taken into account, be to the right or to the left of the path when it is not taken into account as viewed by the person throwing the ball? (b) What would be your answer to (a) if the ball were thrown in the southern hemisphere? *Ans.* (a) to the right, (b) to the left

6.35. What would be your answer to Problem 6.34 if the ball were thrown at the north or south poles?

6.36. Explain why water running out of a vertical drain will swirl counterclockwise in the northern hemisphere and clockwise in the southern hemisphere. What happens at the equator?

6.37. Prove that the centrifugal force acting on a particle of mass m on the earth's surface is a vector (a) directed away from the earth and perpendicular to the angular velocity vector $\boldsymbol{\omega}$ and (b) of magnitude $m\omega^2 R \sin\lambda$ where λ is the colatitude.

6.38. In Problem 6.37, where would the centrifugal force be (a) a maximum, (b) a minimum?

Ans. (a) at the equator, (b) at the north and south poles.

6.39. Find the centrifugal force acting on a train of mass 100,000 kg at (a) the equator (b) colatitude 30°.

Ans. (a) 35.0 kg wt, (b) 17.5 kg wt

6.40. (a) A river of width D flows northward with a speed v_0 at colatitude λ. Prove that the left bank of the river will be higher than the right bank by an amount equal to

$$(2D\omega v_0 \cos\lambda)(g^2 + 4\omega^2 v^2 \cos^2\lambda)^{-1/2}$$

where ω is the angular speed of the earth about its axis.

(b) Prove that the result in part (a) is for all practical purposes equal to $(2D\omega v_0 \cos\lambda)/g$.

6.41. If the river of Problem 6.40 is 2 km wide and flows at a speed of 5 km/hr at colatitude 45°, how much higher will the left bank be than the right bank? *Ans.* 2.9 cm

6.42. An automobile rounds a curve whose radius of curvature is ρ. If the coefficient of friction is μ, prove that the greatest speed with which it can travel so as not to slip on the road is $\sqrt{\mu\rho g}$.

6.43. Determine whether the automobile of Problem 6.42 will slip if the speed is 60 mi/hr, $\mu = .05$ and (*a*) $\rho = 500$ ft, (*b*) $\rho = 50$ ft. Discuss the results physically.

MOTION OF A PARTICLE RELATIVE TO THE EARTH

6.44. An object is dropped at the equator from a height of 400 meters. If air resistance is neglected, how far will the point where it hits the earth's surface be from the point vertically below the initial position? *Ans.* 17.6 cm toward the east

6.45. Work Problem 6.44 if the object is dropped (*a*) at colatitude 60° and (*b*) at the north pole.
Ans. (*a*) 15.2 cm toward the east

6.46. An object is thrown vertically upward at colatitude λ with speed v_0. Prove that when it returns it will be at a distance westward from its starting point equal to $(4\omega v_0^3 \sin\lambda)/3g^2$.

6.47. An object at the equator is thrown vertically upward with a speed of 60 mi/hr. How far from its initial position will it land? *Ans.* .78 inches

6.48. With what speed must the object of Problem 6.47 be thrown in order that it return to a point on the earth which is 20 ft from its original position? *Ans.* 406 mi/hr

6.49. An object is thrown downward with initial speed v_0. Prove that after time t the object is deflected east of the vertical by the amount

$$\omega v_0 \sin\lambda\ t^2 + \tfrac{1}{3}\omega g \sin\lambda\ t^3$$

6.50. Prove that if the object of Problem 6.49 is thrown downward from height h above the earth's surface, then it will hit the earth at a point east of the vertical at a distance

$$\frac{\omega \sin\lambda}{3g^2}(\sqrt{v_0^2 + 2gh} - v_0)^2(\sqrt{v_0^2 + 2gh} + 2v_0)$$

6.51. Suppose that the mass m of a conical pendulum of length l moves in a horizontal circle of radius a. Prove that (*a*) the speed is $a\sqrt{g}/\sqrt[4]{l^2 - a^2}$ and (*b*) the tension in the string is $mgl\sqrt{l^2 - a^2}$.

6.52. If an object is dropped to the earth's surface prove that its path is a semicubical parabola.

THE FOUCAULT PENDULUM

6.53. Explain physically why the plane of oscillation of a Foucault pendulum should rotate clockwise when viewed from above the earth's surface in the northern hemisphere but counterclockwise in the southern hemisphere.

6.54. How long would it take the plane of oscillation of a Foucault pendulum to make one complete revolution if the pendulum is located at (*a*) the north pole, (*b*) colatitude 45°, (*c*) colatitude 85°?
Ans. (*a*) 23.94 hr, (*b*) 33.86 hr, (*c*) 92.50 hr

6.55. Explain physically why a Foucault pendulum situated at the equator would not detect the rotation of the earth about its axis. Is this physical result supported mathematically? Explain.

MOVING COORDINATE SYSTEMS IN GENERAL

6.56. An xyz coordinate system rotates about the z axis with angular velocity $\boldsymbol{\omega} = \cos t\,\mathbf{i} + \sin t\,\mathbf{j}$ relative to a fixed XYZ coordinate system where t is the time. The origin of the xyz system has position vector $\mathbf{R} = t\mathbf{i} - \mathbf{j} + t^2\mathbf{k}$ with respect to the XYZ system. If the position vector of a particle is given by $\mathbf{r} = (3t+1)\mathbf{i} - 2t\mathbf{j} + 5\mathbf{k}$ relative to the moving system, find the (*a*) apparent velocity and (*b*) true velocity at any time.

6.57. Determine (*a*) the apparent acceleration and (*b*) the true acceleration of the particle in Problem 6.56.

6.58. Work (*a*) Problem 6.5, page 148, and (*b*) Problem 6.7, page 149, if the position vector of the xyz system relative to the origin of the fixed XYZ system is $\mathbf{R} = t^2\mathbf{i} - 2t\mathbf{j} + 5\mathbf{k}$.

MISCELLANEOUS PROBLEMS

6.59. Prove that due to the rotation of the earth about its axis the apparent weight of an object of mass m at colatitude λ is $m\sqrt{(g - \omega^2 R \sin^2 \lambda)^2 + (\omega^2 R \sin \lambda \cos \lambda)^2}$ where R is the radius of the earth.

6.60. Prove that the angle β which the apparent vertical at colatitude λ makes with the true vertical is given by $\tan \beta = \dfrac{\omega^2 R \sin \lambda \cos \lambda}{g - \omega^2 R \sin^2 \lambda}$.

6.61. Explain physically why the true vertical and apparent vertical would coincide at the equator and also the north and south poles.

6.62. A stone is twirled in a vertical circle by a string of length 10 ft. Prove that it must have a speed of at least 20 ft/sec at the bottom of its path in order to complete the circle.

6.63. A car C [Fig. 6-8] is to go completely around the vertical circular loop of radius a without leaving the track. Assuming the track is frictionless, determine the height H at which it must start.

6.64. A particle of mass m is constrained to move on a frictionless vertical circle of radius a which rotates about a fixed diameter with constant angular speed ω. Prove that the particle will make small oscillations about its equilibrium position with a frequency given by $2\pi a\omega/\sqrt{a^2\omega^4 - g^2}$.

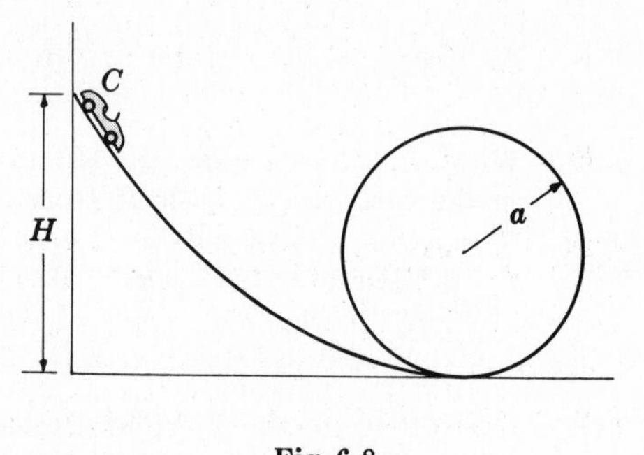

Fig. 6-8

6.65. Discuss what happens in Problem 6.64 if $\omega = \sqrt{g/a}$.

6.66. A hollow cylindrical tube AOB of length $2a$ [Fig. 6-9] rotates with constant angular speed ω about a vertical axis through the center O. A particle is initially at rest in the tube at a distance b from O. Assuming no frictional forces, find (a) the position and (b) the speed of the particle at any time.

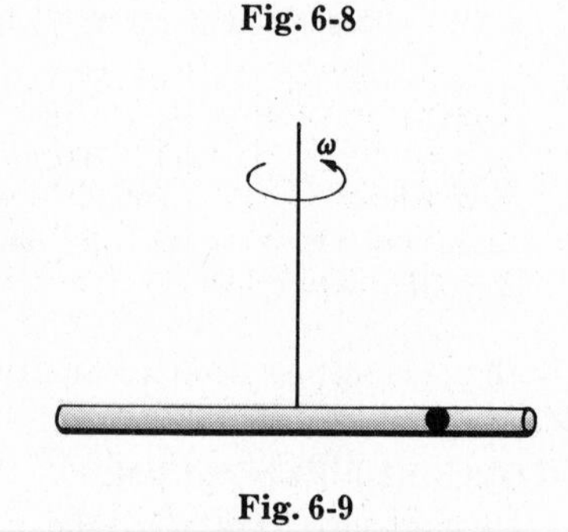

Fig. 6-9

6.67. (a) How long will it take the particle of Problem 6.66 to come out of the tube and (b) what will be its speed as it leaves? *Ans.* (a) $\dfrac{1}{\omega} \ln (a + \sqrt{a^2 - b^2})$

6.68. Find the force on the particle of Problem 6.66 at any position in the tube.

6.69. A mass, attached to a string which is suspended from a fixed point, moves in a horizontal circle having center vertically below the fixed point with a speed of 20 revolutions per minute. Find the distance of the center of the circle below the fixed point. *Ans.* 2.23 meters

6.70. A particle on a frictionless horizontal plane at colatitude λ is given an initial speed v_0 in a northward direction. Prove that it describes a circle of radius $v_0/(2\omega \cos \lambda)$ with period $\pi/(\omega \cos \lambda)$.

6.71. The pendulum bob of a conical pendulum describes a horizontal circle of radius a. If the length of the pendulum is l, prove that the period is given by $4\pi^2\sqrt{l^2 - a^2}/g$.

6.72. A particle constrained to move on a circular wire of radius a and coefficient μ is given an initial velocity $\mathbf{v}_0$. Assuming no other forces act, how long will it take for the particle to come to rest?

6.73. (a) Prove that if the earth were to rotate at an angular speed given by $\sqrt{2g/R}$ where R is its radius and g the acceleration due to gravity, then the weight of a particle of mass m would be the same at all latitudes. (b) What is the numerical value of this angular speed?
Ans. (b) 1.74×10^{-3} rad/sec

6.74. A cylindrical tank containing water rotates about its axis with constant angular speed ω so that no water spills out. Prove that the shape of the water surface is a paraboloid of revolution.

6.75. Work (*a*) Problem 6.16 and (*b*) Problem 6.17, accurate to terms involving ω^2.

6.76. Prove that due to the earth's rotation about its axis, winds in the northern hemisphere traveling from a high pressure area to a low pressure area are rotated in a counterclockwise sense when viewed above the earth's surface. What happens to winds in the southern hemisphere?

6.77. (*a*) Prove that in the northern hemisphere winds from the north, east, south and west are deflected respectively toward the west, north, east and south as indicated in Fig. 6-10. (*b*) Use this to explain the origin of *cyclones.*

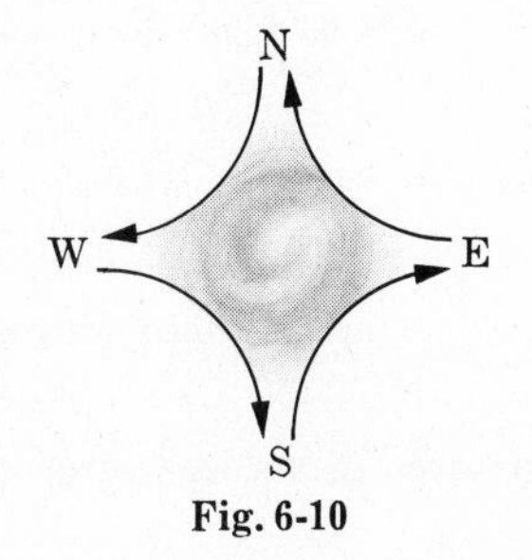

Fig. 6-10

6.78. Find the condition on the angular speed so that a particle will describe a horizontal circle inside of a frictionless vertical cone of angle α.

6.79. Work Problem 6.78 for a hemisphere.

6.80. The period of a simple pendulum is given by P. Prove that its period when it is suspended from the ceiling of a train moving with speed v_0 around a circular track of radius ρ is given by $P\sqrt{\rho g}/\sqrt[4]{v_0^4 + \rho^2 g^2}$.

6.81. Work Problem 6.25 accurate to terms involving ω^2.

6.82. A thin hollow cylindrical tube OA inclined at angle α with the horizontal rotates about the vertical with constant angular speed ω [see Fig. 6-11]. If a particle constrained to move in this tube is initially at rest at a distance a from the intersection O of the tube and the vertical axis of rotation, prove that its distance r from O at any time t is $r = a \cosh(\omega t \cos \alpha)$.

6.83. Work Problem 6.82 if the rod has coefficient of friction μ.

6.84. Prove that the particle of Problem 6.82 is in stable equilibrium between the distances from O given by

$$\frac{g \sin \alpha}{\omega^2}\left(\frac{1 - \mu \tan \alpha}{\tan \alpha + \mu}\right) \quad \text{and} \quad \frac{g \sin \alpha}{\omega^2}\left(\frac{1 + \mu \tan \alpha}{\tan \alpha - \mu}\right)$$

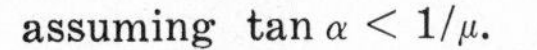

assuming $\tan \alpha < 1/\mu$.

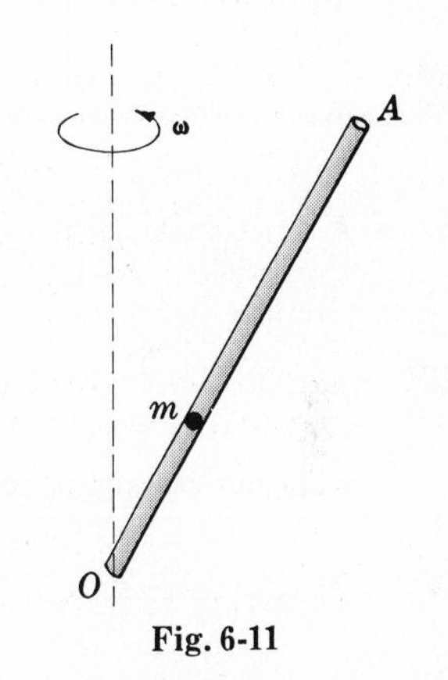

Fig. 6-11

6.85. A train having a maximum speed equal to v_0 is to round a curve with radius of curvature ρ. Prove that if there is to be no lateral thrust on the outer track, then this track should be at a height above the inner track given by $av_0^2/\sqrt{v_0^4 + \rho^2 g^2}$ where a is the distance between tracks.

6.86. A projectile is fired at colatitude λ with velocity v_0 directed toward the west and at angle α with the horizontal. Prove that if terms involving ω^2 are neglected, then the time taken to reach the maximum height is

$$\frac{v_0 \sin \alpha}{g} - \frac{2\omega v_0^2 \sin \lambda \sin \alpha \cos \alpha}{g^2}$$

Compare with the case where $\omega = 0$, i.e. that the earth does not rotate about its axis.

6.87. In Problem 6.86, prove that the maximum height reached is

$$\frac{v_0^2 \sin^2 \alpha}{2g} - \frac{2\omega v_0^3 \sin \lambda \sin^2 \alpha \cos \alpha}{g^2}$$

Compare with the case where $\omega = 0$.

6.88. Prove that the range of the projectile of Problem 6.86 is

$$\frac{v_0^2 \sin 2\alpha}{g} + \frac{\omega v_0^3 \sin\alpha \sin\lambda\,(8\sin^2\alpha - 6)}{3g^2}$$

Thus show that if terms involving ω^2 and higher are neglected, the range will be larger, smaller or the same as the case where $\omega = 0$, according as $\alpha > 60°$, $\alpha < 60°$ or $\alpha = 60°$ respectively.

6.89. If a projectile is fired with initial velocity $v_1\mathbf{i} + v_2\mathbf{j} + v_3\mathbf{k}$ from the origin of a coordinate system fixed relative to the earth's surface at colatitude λ, prove that its position at any later time t will be given by

$$x = v_1 t + \omega v_2 t^2 \cos\lambda$$

$$y = v_2 t - \omega t^2(v_1 \cos\lambda + v_3 \sin\lambda) + \tfrac{1}{3}\omega g t^3 \sin\lambda$$

$$z = v_3 t - \tfrac{1}{2}g t^2 + \omega v_2 t^2 \sin\lambda$$

neglecting terms involving ω^2.

6.90. Work Problem 6.89 so as to include terms involving ω^2 but exclude terms involving ω^3.

6.91. An object of mass m initially at rest is dropped from height h to the earth's surface at colatitude λ. Assuming that air resistance proportional to the instantaneous speed of the object is taken into account as well as the rotation of the earth about its axis, prove that after time t the object is deflected east of the vertical by the amount

$$\frac{2\omega \sin\lambda}{\beta^3}\left[(g - 2h\beta^2)(1 - e^{-\beta t}) + \beta^3 h t e^{-\beta t} - \beta g t + \tfrac{1}{2}g\beta^2 t^2\right]$$

neglecting terms of order ω^2 and higher.

6.92. Work Problem 6.91, obtaining accuracy up to and including terms of order ω^2.

6.93. A frictionless inclined plane of length l and angle α located at colatitude λ is so situated that a particle placed on it would slide under the influence of gravity from north to south. If the particle starts from rest at the top, prove that it will reach the bottom in a time given by

$$\sqrt{\frac{2l}{g \sin\alpha}} + \frac{2\omega l \sin\lambda \cos\alpha}{3g}$$

and that its speed at the bottom is

$$\sqrt{2gl \sin\alpha} - \tfrac{4}{3}\omega l \sin\alpha \cos\alpha \sin\lambda$$

neglecting terms of order ω^2.

6.94. (*a*) Prove that by the time the particle of Problem 6.93 reaches the bottom it will have undergone a deflection of magnitude

$$\frac{2l\omega}{3}\sqrt{\frac{2l}{g \sin\alpha}}\cos(\alpha + \lambda)$$

to the east or west respectively according as $\cos(\alpha + \lambda)$ is greater than or less than zero. (*b*) Discuss the case where $\cos(\alpha + \lambda) = 0$. (*c*) Use the result of (*a*) to arrive at the result of Problem 6.17.

6.95. Work Problems 6.93 and 6.94 if the inclined plane has coefficient of friction μ.

Chapter 7 SYSTEMS of PARTICLES

DISCRETE AND CONTINUOUS SYSTEMS

Up to now we have dealt mainly with the motion of an object which could be considered as a particle or point mass. In many practical cases the objects with which we are concerned can more realistically be considered as *collections* or *systems* of particles. Such systems are called *discrete* or *continuous* according as the particles can be considered as separated from each other or not.

For many practical purposes a discrete system having a very large but finite number of particles can be considered as a continuous system. Conversely a continuous system can be considered as a discrete system consisting of a large but finite number of particles.

DENSITY

For continuous systems of particles occupying a region of space it is often convenient to define a mass per unit volume which is called the *volume density* or briefly *density*. Mathematically, if ΔM is the total mass of a volume $\Delta\tau$ of particles, then the density can be defined as

$$\sigma = \lim_{\Delta\tau \to 0} \frac{\Delta M}{\Delta\tau} \tag{1}$$

The density is a function of position and can vary from point to point. When the density is a constant, the system is said to be of *uniform density* or simply *uniform*.

When the continuous system of particles occupy a surface, we can similarly define a *surface density* or mass per unit area. Similarly when the particles occupy a line [or curve] we can define a mass per unit length or *linear density*.

RIGID AND ELASTIC BODIES

In practice, forces applied to systems of particles will change the distances between individual particles. Such systems are often called *deformable* or *elastic bodies*. In some cases, however, deformations may be so slight that they may for most practical purposes be considered non-existent. It is thus convenient to define a mathematical model in which the distance between any two specified particles of a system remains the same regardless of applied forces. Such a system is called a *rigid body*. The mechanics of rigid bodies is considered in Chapters 9 and 10.

DEGREES OF FREEDOM

The number of coordinates required to specify the position of a system of one or more particles is called the number of *degrees of freedom* of the system.

Example 1.

A particle moving freely in space requires 3 coordinates, e.g. (x, y, z), to specify its position. Thus the number of degrees of freedom is 3.

Example 2.

A system consisting of N particles moving freely in space requires $3N$ coordinates to specify its position. Thus the number of degrees of freedom is $3N$.

A rigid body which can move freely in space has 6 degrees of freedom, i.e. 6 coordinates are required to specify the position. See Problem 7.2.

CENTER OF MASS

Let $\mathbf{r}_1, \mathbf{r}_2, \ldots, \mathbf{r}_N$ be the position vectors of a system of N particles of masses $m_1, m_2, \ldots, m_N$ respectively [see Fig. 7-1].

The *center of mass* or *centroid* of the system of particles is defined as that point C having position vector

$$\bar{\mathbf{r}} = \frac{m_1\mathbf{r}_1 + m_2\mathbf{r}_2 + \cdots + m_N\mathbf{r}_N}{m_1 + m_2 + \cdots + m_N} = \frac{1}{M}\sum_{\nu=1}^{N} m_\nu \mathbf{r}_\nu \tag{2}$$

where $M = \sum_{\nu=1}^{N} m_\nu$ is the *total mass* of the system. We sometimes use $\sum_\nu$ or simply $\sum$ in place of $\sum_{\nu=1}^{N}$.

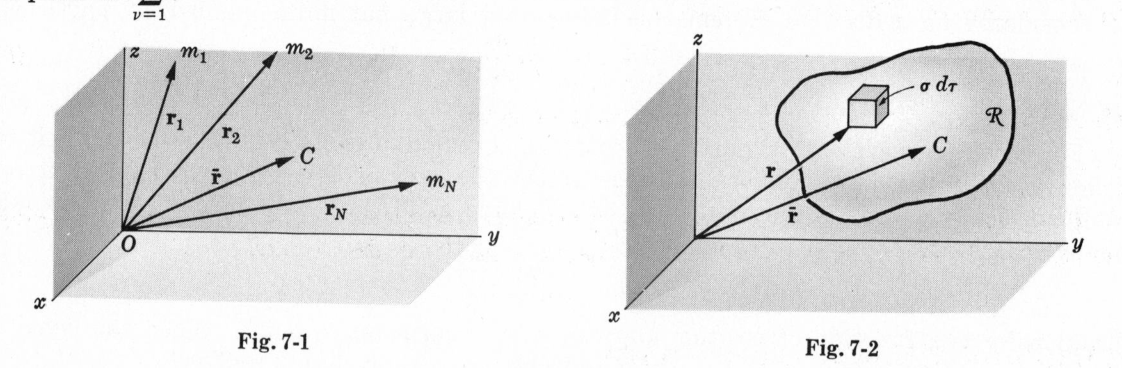

Fig. 7-1

Fig. 7-2

For continuous systems of particles occupying a region $\mathcal{R}$ of space in which the volume density is σ, the center of mass can be written

$$\bar{\mathbf{r}} = \frac{\int_{\mathcal{R}} \sigma \mathbf{r}\, d\tau}{\int_{\mathcal{R}} \sigma\, d\tau} \tag{3}$$

where the integral is taken over the entire region $\mathcal{R}$ [see Fig. 7-2]. If we write

$$\bar{\mathbf{r}} = \bar{x}\mathbf{i} + \bar{y}\mathbf{j} + \bar{z}\mathbf{k}, \qquad \mathbf{r}_\nu = x_\nu\mathbf{i} + y_\nu\mathbf{j} + z_\nu\mathbf{k}$$

then (3) can equivalently be written as

$$\bar{x} = \frac{\sum m_\nu x_\nu}{M}, \qquad \bar{y} = \frac{\sum m_\nu y_\nu}{M}, \qquad \bar{z} = \frac{\sum m_\nu z_\nu}{M} \tag{4}$$

and

$$\bar{x} = \frac{\int_{\mathcal{R}} \sigma x\, d\tau}{M}, \qquad \bar{y} = \frac{\int_{\mathcal{R}} \sigma y\, d\tau}{M}, \qquad \bar{z} = \frac{\int_{\mathcal{R}} \sigma z\, d\tau}{M} \tag{5}$$

where the total mass is given by either

$$M = \sum m_\nu \tag{6}$$

or

$$M = \int_{\mathcal{R}} \sigma\, d\tau \tag{7}$$

The integrals in (3), (5) or (7) can be single, double or triple integrals, depending on which may be preferable.

In practice it is fairly simple to go from discrete to continuous systems by merely replacing summations by integrations. Consequently we will present all theorems for discrete systems.

CENTER OF GRAVITY

If a system of particles is in a uniform gravitational field, the center of mass is sometimes called the *center of gravity*.

MOMENTUM OF A SYSTEM OF PARTICLES

If $\mathbf{v}_\nu = d\mathbf{r}_\nu/dt = \dot{\mathbf{r}}_\nu$ is the velocity of m_ν, the total momentum of the system is defined as

$$\mathbf{p} = \sum_{\nu=1}^{N} m_\nu \mathbf{v}_\nu = \sum_{\nu=1}^{N} m_\nu \dot{\mathbf{r}}_\nu \tag{8}$$

We can show [see Problem 7.3] that

$$\mathbf{p} = M\bar{\mathbf{v}} = M\frac{d\bar{\mathbf{r}}}{dt} = M\dot{\bar{\mathbf{r}}} \tag{9}$$

where $\bar{\mathbf{v}} = d\bar{\mathbf{r}}/dt$ is the velocity of the center of mass.

This is expressed in the following

Theorem 7.1. The total momentum of a system of particles can be found by multiplying the total mass M of the system by the velocity $\bar{\mathbf{v}}$ of the center of mass.

MOTION OF THE CENTER OF MASS

Suppose that the internal forces between any two particles of the system obey Newton's third law. Then if $\mathbf{F}$ is the resultant external force acting on the system, we have [see Problem 7.4]

$$\mathbf{F} = \frac{d\mathbf{p}}{dt} = M\frac{d^2\bar{\mathbf{r}}}{dt^2} = M\frac{d\bar{\mathbf{v}}}{dt} \tag{10}$$

This is expressed in

Theorem 7.2. The center of mass of a system of particles moves as if the total mass and resultant external force were applied at this point.

CONSERVATION OF MOMENTUM

Putting $\mathbf{F} = \mathbf{0}$ in (10), we find that

$$\mathbf{p} = \sum_{\nu=1}^{N} m_\nu \mathbf{v}_\nu = \text{constant} \tag{11}$$

Thus we have

Theorem 7.3. If the resultant external force acting on a system of particles is zero, then the total momentum remains constant, i.e. is conserved. In such case the center of mass is either at rest or in motion with constant velocity.

This theorem is often called the *principle of conservation of momentum.* It is a generalization of Theorem 2-8, page 37.

ANGULAR MOMENTUM OF A SYSTEM OF PARTICLES

The quantity

$$\mathbf{\Omega} \;=\; \sum_{\nu=1}^{N} m_\nu(\mathbf{r}_\nu \times \mathbf{v}_\nu) \tag{12}$$

is called the total *angular momentum* [or *moment of momentum*] of the system of particles about origin O.

THE TOTAL EXTERNAL TORQUE ACTING ON A SYSTEM

If $\mathbf{F}_\nu$ is the external force acting on particle ν, then $\mathbf{r}_\nu \times \mathbf{F}_\nu$ is called the *moment of the force* $\mathbf{F}_\nu$ or *torque* about O. The sum

$$\mathbf{\Lambda} \;=\; \sum_{\nu=1}^{N} \mathbf{r}_\nu \times \mathbf{F}_\nu \tag{13}$$

is called the *total external torque* about the origin.

RELATION BETWEEN ANGULAR MOMENTUM AND TOTAL EXTERNAL TORQUE

If we assume that the internal forces between any two particles are always directed along the line joining the particles [i.e. they are *central forces*], then we can show as in Problem 7.12 that

$$\mathbf{\Lambda} \;=\; \frac{d\mathbf{\Omega}}{dt} \tag{14}$$

Thus we have

Theorem 7.4. The total external torque on a system of particles is equal to the time rate of change of the angular momentum of the system, provided the internal forces between particles are central forces.

CONSERVATION OF ANGULAR MOMENTUM

Putting $\mathbf{\Lambda} = \mathbf{0}$ in *(14)*, we find that

$$\mathbf{\Omega} \;=\; \sum_{\nu=1}^{N} m_\nu(\mathbf{r}_\nu \times \mathbf{v}_\nu) \;=\; \text{constant} \tag{15}$$

Thus we have

Theorem 7.5. If the resultant external torque acting on a system of particles is zero, then the total angular momentum remains constant, i.e. is conserved.

This theorem is often called the *principle of conservation of angular momentum.* It is the generalization of Theorem 2.9, page 37.

KINETIC ENERGY OF A SYSTEM OF PARTICLES

The total *kinetic energy* of a system of particles is defined as

$$T \;=\; \frac{1}{2}\sum_{\nu=1}^{N} m_\nu v_\nu^2 \;=\; \frac{1}{2}\sum_{\nu=1}^{N} m_\nu \dot{\mathbf{r}}_\nu^2 \tag{16}$$

WORK

If $\mathcal{F}_\nu$ is the force (external and internal) acting on particle ν, then the total work done in moving the system of particles from one state [symbolized by 1] to another [symbolized by 2] is

$$W_{12} \;=\; \sum_{\nu=1}^{N} \int_1^2 \mathcal{F}_\nu \cdot d\mathbf{r}_\nu \tag{17}$$

As in the case of a single particle, we can prove the following

Theorem 7.6. The total work done in moving a system of particles from one state where the kinetic energy is T_1 to another where the kinetic energy is T_2, is

$$W_{12} = T_2 - T_1 \tag{18}$$

POTENTIAL ENERGY. CONSERVATION OF ENERGY

When all forces, external and internal, are conservative, we can define a total potential energy V of the system. In such case we can prove the following

Theorem 7.7. If T and V are respectively the total kinetic energy and total potential energy of a system of particles, then

$$T + V = \text{constant} \tag{19}$$

This is the *principle of conservation of energy* for systems of particles.

MOTION RELATIVE TO THE CENTER OF MASS

It is often useful to describe the motion of a system of particles about [or relative to] the center of mass. The following theorems are of fundamental importance. In all cases primes denote quantities relative to the center of mass.

Theorem 7.8. The total linear momentum of a system of particles about the center of mass is zero. In symbols,

$$\sum_{\nu=1}^{N} m_\nu \mathbf{v}'_\nu = \sum_{\nu=1}^{N} m_\nu \dot{\mathbf{r}}'_\nu = 0 \tag{20}$$

Theorem 7.9. The total angular momentum of a system of particles about any point O equals the angular momentum of the total mass assumed to be located at the center of mass plus the angular momentum about the center of mass. In symbols,

$$\mathbf{\Omega} = \bar{\mathbf{r}} \times M\bar{\mathbf{v}} + \sum_{\nu=1}^{N} m_\nu(\mathbf{r}'_\nu \times \mathbf{v}'_\nu) \tag{21}$$

Theorem 7.10. The total kinetic energy of a system of particles about any point O equals the kinetic energy of translation of the center of mass [assuming the total mass located there] plus the kinetic energy of motion about the center of mass. In symbols,

$$T = \frac{1}{2}M\bar{\mathbf{v}}^2 + \frac{1}{2}\sum_{\nu=1}^{N} m_\nu v'^2_\nu \tag{22}$$

Theorem 7.11. The total external torque about the center of mass equals the time rate of change in angular momentum about the center of mass, i.e. equation (*14*) holds not only for inertial coordinate systems but also for coordinate systems moving with the center of mass. In symbols,

$$\mathbf{\Lambda}' = \frac{d\mathbf{\Omega}'}{dt} \tag{23}$$

If motion is described relative to points other than the center of mass, the results in the above theorems become more complicated.

IMPULSE

If $\mathbf{F}$ is the total external force acting on a system of particles, then

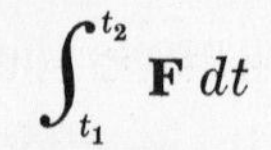

$$\int_{t_1}^{t_2} \mathbf{F}\, dt \tag{24}$$

is called the *total linear impulse* or briefly *total impulse*. As in the case of one particle, we can prove

Theorem 7.12. The total linear impulse is equal to the change in linear momentum.

Similarly if $\mathbf{\Lambda}$ is the total external torque applied to a system of particles about origin O, then

$$\int_{t_1}^{t_2} \mathbf{\Lambda}\, dt \tag{25}$$

is called the total *angular impulse*. We can then prove

Theorem 7.13. The total angular impulse is equal to the change in angular momentum.

CONSTRAINTS. HOLONOMIC AND NON-HOLONOMIC CONSTRAINTS

Often in practice the motion of a particle or system of particles is restricted in some way. For example, in *rigid bodies* [considered in Chapters 9 and 10] the motion must be such that the distance between any two particular particles of the rigid body is always the same. As another example, the motion of particles may be restricted to curves or surfaces.

The limitations on the motion are often called *constraints*. If the constraint condition can be expressed as an equation

$$\phi(\mathbf{r}_1, \mathbf{r}_2, \ldots, \mathbf{r}_N, t) = 0 \tag{26}$$

connecting the position vectors of the particles and the time, then the constraint is called *holonomic*. If it cannot be so expressed it is called *non-holonomic*.

VIRTUAL DISPLACEMENTS

Consider two possible configurations of a system of particles at a particular instant which are consistent with the forces and constraints. To go from one configuration to the other, we need only give the νth particle a displacement $\delta\mathbf{r}_\nu$ from the old to the new position. We call $\delta\mathbf{r}_\nu$ a *virtual displacement* to distinguish it from a *true displacement* [denoted by $d\mathbf{r}_\nu$] which occurs in a time interval where forces and constraints could be changing. The symbol δ has the usual properties of the differential d; for example, $\delta(\sin\theta) = \cos\theta\,\delta\theta$.

STATICS OF A SYSTEM OF PARTICLES. PRINCIPLE OF VIRTUAL WORK

In order for a system of particles to be in equilibrium, the resultant force acting on each particle must be zero, i.e. $\mathbf{F}_\nu = \mathbf{0}$. It thus follows that $\mathbf{F}_\nu \cdot \delta\mathbf{r}_\nu = 0$ where $\mathbf{F}_\nu \cdot \delta\mathbf{r}_\nu$ is called the *virtual work*. By adding these we then have

$$\sum_{\nu=1}^{N} \mathbf{F}_\nu \cdot \delta\mathbf{r}_\nu = 0 \tag{27}$$

If constraints are present, then we can write

$$\mathbf{F}_\nu = \mathbf{F}_\nu^{(a)} + \mathbf{F}_\nu^{(c)} \tag{28}$$

where $\mathbf{F}_\nu^{(a)}$ and $\mathbf{F}_\nu^{(c)}$ are respectively the *actual force* and *constraint force* acting on the νth particle. By assuming that the virtual work of the constraint forces is zero [which is true for rigid bodies and for motion on curves and surfaces without friction], we arrive at

Theorem 7.14. A system of particles is in equilibrium if and only if the total virtual work of the actual forces is zero, i.e. if

$$\sum_{\nu=1}^{N} \mathbf{F}_\nu^{(a)} \cdot \delta \mathbf{r}_\nu = 0 \tag{29}$$

This is often called the *principle of virtual work.*

EQUILIBRIUM IN CONSERVATIVE FIELDS. STABILITY OF EQUILIBRIUM

The results for equilibrium of a particle in a conservative force field [see page 38] can be generalized to systems of particles. The following theorems summarize the basic results.

Theorem 7.15. If V is the total potential of a system of particles depending on coordinates $q_1, q_2, \ldots$, then the system will be in equilibrium if

$$\frac{\partial V}{\partial q_1} = 0, \quad \frac{\partial V}{\partial q_2} = 0, \quad \ldots \tag{31}$$

Since the virtual work done on the system is

$$\delta V = \frac{\partial V}{\partial q_1}\delta q_1 + \frac{\partial V}{\partial q_2}\delta q_2 + \cdots$$

(*31*) is equivalent to the principle of virtual work.

Theorem 7.16. A system of particles will be in *stable equilibrium* if the potential V is a minimum.

In case V depends on only one coordinate, say q_1, sufficient conditions are

$$\frac{\partial V}{\partial q_1} = 0, \quad \frac{\partial^2 V}{\partial q_1^2} > 0$$

Other cases of equilibrium where the potential is not a minimum are called *unstable.*

D'ALEMBERT'S PRINCIPLE

Although Theorem 7.14 as stated applies to the statics of a system of particles, it can be restated so as to give an analogous theorem for dynamics. To do this we note that according to Newton's second law of motion,

$$\mathbf{F}_\nu = \dot{\mathbf{p}}_\nu \quad \text{or} \quad \mathbf{F}_\nu - \dot{\mathbf{p}}_\nu = 0 \tag{30}$$

where $\mathbf{p}_\nu$ is the momentum of the νth particle. The second equation amounts to saying that a moving system of particles can be considered to be in equilibrium under a force $\mathbf{F}_\nu - \dot{\mathbf{p}}_\nu$, i.e. the actual force together with the added force $-\dot{\mathbf{p}}_\nu$ which is often called the *reversed effective force* on particle ν. By using the principle of virtual work we can then arrive at

Theorem 7.17. A system of particles moves in such a way that the total virtual work

$$\sum_{\nu=1}^{N} (\mathbf{F}_\nu^{(a)} - \dot{\mathbf{p}}_\nu) \cdot \delta \mathbf{r}_\nu = 0 \tag{32}$$

With this theorem, which is often called *D'Alembert's principle,* we can consider dynamics as a special case of statics.

Solved Problems

DEGREES OF FREEDOM

7.1. Determine the number of degrees of freedom in each of the following cases: (*a*) a particle moving on a given space curve; (*b*) five particles moving freely in a plane; (*c*) five particles moving freely in space; (*d*) two particles connected by a rigid rod moving freely in a plane.

(*a*) The curve can be described by the parametric equations $x = x(s)$, $y = y(s)$, $z = z(s)$ where s is the parameter. Then the position of a particle on the curve is determined by specifying one coordinate, and hence there is one degree of freedom.

(*b*) Each particle requires two coordinates to specify its position in the plane. Thus $5 \cdot 2 = 10$ coordinates are needed to specify the positions of all 5 particles, i.e. the system has 10 degrees of freedom.

(*c*) Since each particle requires three coordinates to specify its position, the system has $5 \cdot 3 = 15$ degrees of freedom.

(*d*) **Method 1.**

The coordinates of the two particles can be expressed by (x_1, y_1) and (x_2, y_2), i.e. a total of 4 coordinates. However, since the distance between these points is a constant a [the length of the rigid rod], we have $(x_1 - x_2)^2 + (y_1 - y_2)^2 = a^2$ so that one of the coordinates can be expressed in terms of the others. Thus there are $4 - 1 = 3$ degrees of freedom.

Method 2.

The motion is completely specified if we give the two coordinates of the center of mass and the angle made by the rod with some specified direction. Thus there are $2 + 1 = 3$ degrees of freedom.

7.2. Find the number of degrees of freedom for a rigid body which (*a*) can move freely in three dimensional space, (*b*) has one point fixed but can move in space about this point.

(*a*) **Method 1.**

If 3 non-collinear points of a rigid body are fixed in space, then the rigid body is also fixed in space. Let these points have coordinates (x_1, y_1, z_1), (x_2, y_2, z_2), (x_3, y_3, z_3) respectively, a total of 9. Since the body is rigid we must have the relations

$$(x_1 - x_2)^2 + (y_1 - y_2)^2 + (z_1 - z_2)^2 = \text{constant}, \quad (x_2 - x_3)^2 + (y_2 - y_3)^2 + (z_2 - z_3)^2 = \text{constant},$$
$$(x_3 - x_1)^2 + (y_3 - y_1)^2 + (z_3 - z_1)^2 = \text{constant}$$

hence 3 coordinates can be expressed in terms of the remaining 6. Thus 6 independent coordinates are needed to describe the motion, i.e. there are 6 degrees of freedom.

Method 2.

To fix one point of the rigid body requires 3 coordinates. An axis through this point is fixed if we specify 2 ratios of the direction cosines of this axis. A rotation about the axis can then be described by 1 angular coordinate. The total number of coordinates required, i.e. the number of degrees of freedom, is $3 + 2 + 1 = 6$.

(*b*) The motion is completely specified if we know the coordinates of two points, say (x_1, y_1, z_1) and (x_2, y_2, z_2), where the fixed point is taken at the origin of a coordinate system. But since the body is rigid we must have

$$x_1^2 + y_1^2 + z_1^2 = \text{constant}, \quad x_2^2 + y_2^2 + z_2^2 = \text{constant}, \quad (x_1 - x_2)^2 + (y_1 - y_2)^2 + (z_1 - z_2)^2 = \text{constant}$$

from which 3 coordinates can be found in terms of the remaining 3. Thus there are 3 degrees of freedom.

CENTER OF MASS AND MOMENTUM OF A SYSTEM OF PARTICLES

7.3. Prove Theorem 7.1, page 167: The total momentum of a system of particles can be found by multiplying the total mass M of the system by the velocity $\bar{\mathbf{v}}$ of the center of mass.

The center of mass is by definition, $\bar{\mathbf{r}} = \dfrac{\sum m_\nu \mathbf{r}_\nu}{M}$.

Then the total momentum is $\mathbf{p} = \sum m_\nu \mathbf{v}_\nu = \sum m_\nu \dot{\mathbf{r}}_\nu = M\, d\bar{\mathbf{r}}/dt = M\bar{\mathbf{v}}$.

7.4. **Prove Theorem 7.2, page 167: The center of mass of a system of particles moves as if the total mass and resultant external force were applied at this point.**

Let $\mathbf{F}_\nu$ be the resultant external force acting on particle ν while $\mathbf{f}_{\nu\lambda}$ is the internal force on particle ν due to particle λ. We shall assume that $\mathbf{f}_{\nu\nu} = \mathbf{0}$, i.e. particle ν does not exert any force on itself.

By Newton's second law the total force on particle ν is

$$\mathbf{F}_\nu + \sum_\lambda \mathbf{f}_{\nu\lambda} = \frac{d\mathbf{p}_\nu}{dt} = \frac{d^2}{dt^2}(m_\nu \mathbf{r}_\nu) \tag{1}$$

where the second term on the left represents the resultant internal force on particle ν due to all other particles.

Summing over ν in equation (*1*), we find

$$\sum_\nu \mathbf{F}_\nu + \sum_\nu \sum_\lambda \mathbf{f}_{\nu\lambda} = \frac{d^2}{dt^2}\left\{\sum_\nu m_\nu \mathbf{r}_\nu\right\} \tag{2}$$

Now according to Newton's third law of action and reaction, $\mathbf{f}_{\nu\lambda} = -\mathbf{f}_{\lambda\nu}$ so that the double summation on the left of (*2*) is zero. If we then write

$$\mathbf{F} = \sum_\nu \mathbf{F}_\nu \quad \text{and} \quad \bar{\mathbf{r}} = \frac{1}{M}\sum_\nu m_\nu \mathbf{r}_\nu \tag{3}$$

(*2*) becomes

$$\mathbf{F} = M\frac{d^2\bar{\mathbf{r}}}{dt^2} \tag{4}$$

Since $\mathbf{F}$ is the total external force on all particles applied at the center of mass $\bar{\mathbf{r}}$, the required result is proved.

7.5. **A system of particles consists of a 3 gram mass located at $(1, 0, -1)$, a 5 gram mass at $(-2, 1, 3)$ and a 2 gram mass at $(3, -1, 1)$. Find the coordinates of the center of mass.**

The position vectors of the particles are given respectively by

$$\mathbf{r}_1 = \mathbf{i} - \mathbf{k}, \quad \mathbf{r}_2 = -2\mathbf{i} + \mathbf{j} + 3\mathbf{k}, \quad \mathbf{r}_3 = 3\mathbf{i} - \mathbf{j} + \mathbf{k}$$

Then the center of mass is given by

$$\bar{\mathbf{r}} = \frac{3(\mathbf{i}-\mathbf{k}) + 5(-2\mathbf{i}+\mathbf{j}+3\mathbf{k}) + 2(3\mathbf{i}-\mathbf{j}+\mathbf{k})}{3+5+2} = -\frac{1}{10}\mathbf{i} + \frac{3}{10}\mathbf{j} + \frac{7}{5}\mathbf{k}$$

Thus the coordinates of the center of mass are $(-\frac{1}{10}, \frac{3}{10}, \frac{7}{5})$.

7.6. **Prove that if the total momentum of a system is constant, i.e. is conserved, then the center of mass is either at rest or in motion with constant velocity.**

The total momentum of the system is given by

$$\mathbf{p} = \sum m_\nu \mathbf{v}_\nu = \sum m_\nu \dot{\mathbf{r}}_\nu = \frac{d}{dt}\sum m_\nu \mathbf{r}_\nu = M\frac{d}{dt}\left\{\frac{\sum m_\nu \mathbf{r}_\nu}{M}\right\} = M\frac{d\bar{\mathbf{r}}}{dt}$$

Then if $\mathbf{p}$ is constant, so also is $d\bar{\mathbf{r}}/dt$, the velocity of the center of mass.

7.7. **Explain why the ejection of gases at high velocity from the rear of a rocket will move the rocket forward.**

Since the gas particles move backward with high velocity and since the center of mass does not move, the rocket must move forward. For applications involving rocket motion, see Chapter 8.

7.8. Find the centroid of a solid region $\mathcal{R}$ as in Fig. 7-3.

Consider the volume element $\Delta\tau_\nu$ of the solid. The mass of this volume element is

$$\Delta M_\nu = \sigma_\nu \Delta\tau_\nu = \sigma_\nu \Delta x_\nu \Delta y_\nu \Delta z_\nu$$

where σ_ν is the density [mass per unit volume] and $\Delta x_\nu, \Delta y_\nu, \Delta z_\nu$ are the dimensions of the volume element. Then the centroid is given approximately by

$$\frac{\sum \mathbf{r}_\nu \Delta M_\nu}{\sum \Delta M_\nu} = \frac{\sum \mathbf{r}_\nu \sigma_\nu \Delta\tau_\nu}{\sum \sigma_\nu \Delta\tau_\nu} = \frac{\sum \mathbf{r}_\nu \sigma_\nu \Delta x_\nu \Delta y_\nu \Delta z_\nu}{\sum \sigma_\nu \Delta x_\nu \Delta y_\nu \Delta z_\nu}$$

where the summation is taken over all volume elements of the solid.

Fig. 7-3

Taking the limit as the number of volume elements becomes infinite in such a way that $\Delta\tau_\nu \to 0$ or $\Delta x_\nu \to 0, \Delta y_\nu \to 0, \Delta z_\nu \to 0$, we obtain for the centroid of the solid:

$$\bar{\mathbf{r}} = \frac{\int_{\mathcal{R}} \mathbf{r}\, dM}{\int_{\mathcal{R}} dM} = \frac{\int_{\mathcal{R}} \mathbf{r}\,\sigma\, d\tau}{\int_{\mathcal{R}} \sigma\, d\tau} = \frac{\iiint_{\mathcal{R}} \mathbf{r}\,\sigma\, dx\, dy\, dz}{\iiint_{\mathcal{R}} \sigma\, dx\, dy\, dz}$$

where the integration is to be performed over $\mathcal{R}$, as indicated.

Writing $\mathbf{r} = x\mathbf{i} + y\mathbf{j} + z\mathbf{k}$, $\bar{\mathbf{r}} = \bar{x}\mathbf{i} + \bar{y}\mathbf{j} + \bar{z}\mathbf{k}$, this can also be written in component form as

$$\bar{x} = \frac{\iiint_{\mathcal{R}} x\,\sigma\, dx\, dy\, dz}{\iiint_{\mathcal{R}} \sigma\, dx\, dy\, dz}, \quad \bar{y} = \frac{\iiint_{\mathcal{R}} y\,\sigma\, dx\, dy\, dz}{\iiint_{\mathcal{R}} \sigma\, dx\, dy\, dz}, \quad \bar{z} = \frac{\iiint_{\mathcal{R}} z\,\sigma\, dx\, dy\, dz}{\iiint_{\mathcal{R}} \sigma\, dx\, dy\, dz}$$

7.9. Find the centroid of the region bounded by the plane $x + y + z = a$ and the planes $x = 0$, $y = 0$, $z = 0$.

The region, which is a tetrahedron, is indicated in Fig. 7-4. To find the centroid, we use the results of Problem 7.8.

In forming the sum over all volume elements of the region, it is advisable to proceed in an orderly fashion. One possibility is to add first all terms corresponding to volume elements contained in a column such as PQ in the figure. This amounts to keeping x_ν and y_ν fixed and adding over all z_ν. Next keep x_ν fixed but sum over all y_ν. This amounts to adding all columns, such as PQ, contained in a slab RS, and consequently amounts to summing over all cubes contained in such a slab. Finally, vary x_ν. This amounts to addition of all slabs such as RS.

In performing the integration over $\mathcal{R}$, we use these same ideas. Thus keeping x and y constant, integrate from $z = 0$ [base of column PQ] to $z = a - x - y$ [top of column PQ]. Next keep x constant and integrate with respect to y. This amounts to addition of columns having bases in the xy plane [$z = 0$] located anywhere from R [where $y = 0$] to S [where $x + y = a$ or $y = a - x$], and the integration is from $y = 0$ to $y = a - x$. Finally, we add all slabs parallel to the yz plane, which amounts to integration from $x = 0$ to $x = a$. We thus obtain

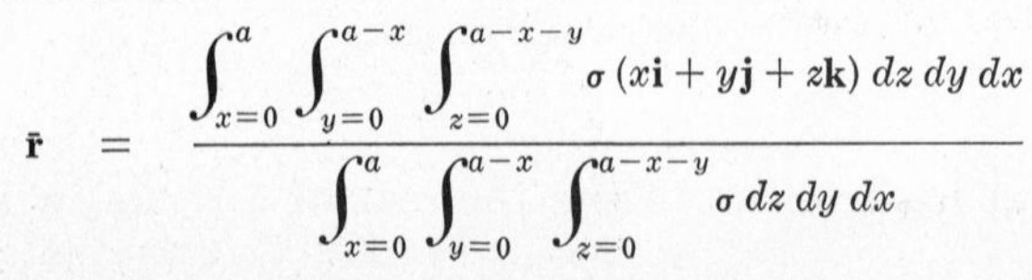

$$\bar{\mathbf{r}} = \frac{\int_{x=0}^{a}\int_{y=0}^{a-x}\int_{z=0}^{a-x-y} \sigma\,(x\mathbf{i} + y\mathbf{j} + z\mathbf{k})\, dz\, dy\, dx}{\int_{x=0}^{a}\int_{y=0}^{a-x}\int_{z=0}^{a-x-y} \sigma\, dz\, dy\, dx}$$

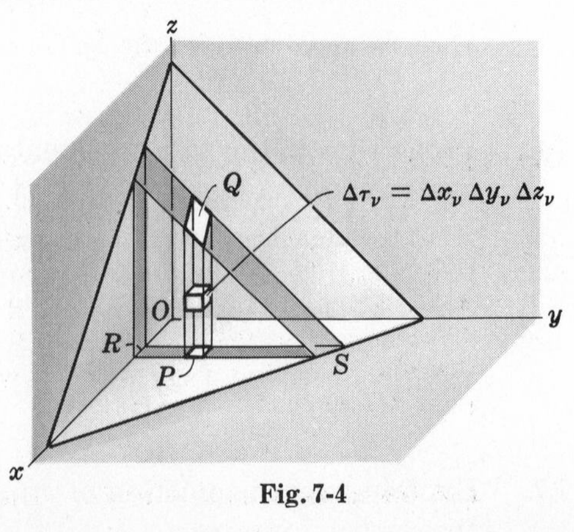

Fig. 7-4

Since σ is constant in this case, it may be cancelled. The denominator without σ is evaluated to be $a^3/6$, and the numerator without σ is $(a^4/24)(\mathbf{i} + \mathbf{j} + \mathbf{k})$. Thus the center of mass is $\bar{\mathbf{r}} = (a/4)(\mathbf{i} + \mathbf{j} + \mathbf{k})$ or $\bar{x} = a/4$, $\bar{y} = a/4$, $\bar{z} = a/4$.

7.10. Find the centroid of a semi-circular region of radius a.

Method 1. Using rectangular coordinates.

Choose the region as in Fig. 7-5. The equation of the circle C is $x^2 + y^2 = a^2$ or $y = \sqrt{a^2 - x^2}$ since $y \geqq 0$.

If σ is the mass per unit area, assumed constant, then the coordinates of the centroid are given by

$$\bar{x} = \frac{\int_{\mathcal{R}} x\,\sigma\,dA}{\int_{\mathcal{R}} \sigma\,dA} = \frac{\iint_{\mathcal{R}} x\,dy\,dx}{\iint_{\mathcal{R}} dy\,dx} = \frac{\int_{x=-a}^{a}\int_{y=0}^{\sqrt{a^2-x^2}} x\,dy\,dx}{\int_{x=-a}^{a}\int_{y=0}^{\sqrt{a^2-x^2}} dy\,dx} = 0$$

$$\bar{y} = \frac{\int_{\mathcal{R}} y\,\sigma\,dA}{\int_{\mathcal{R}} \sigma\,dA} = \frac{\iint_{\mathcal{R}} y\,dy\,dx}{\iint_{\mathcal{R}} dy\,dx} = \frac{\int_{x=-a}^{a}\int_{y=0}^{\sqrt{a^2-x^2}} y\,dy\,dx}{\int_{x=-a}^{a}\int_{y=0}^{\sqrt{a^2-x^2}} dy\,dx} = \frac{2a^3/3}{\pi a^2/2} = \frac{4a}{3\pi}$$

Note that we can write $\bar{x} = 0$ immediately, since by symmetry the centroid is on the y axis. The denominator for $\bar{y}$ can be evaluated without integrating by noting that it represents the semi-circular area which is $\frac{1}{2}\pi a^2$.

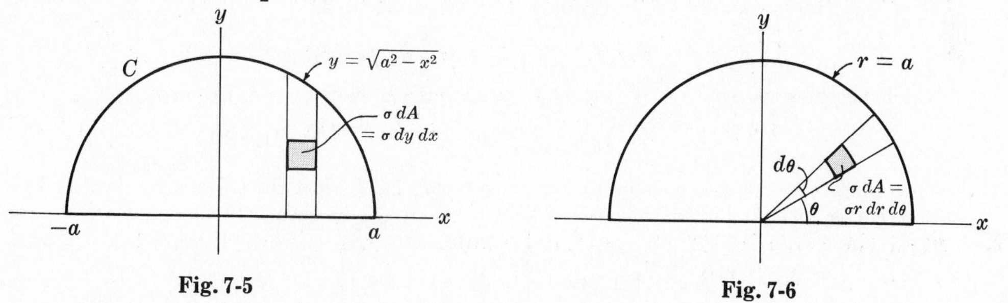

Fig. 7-5

Fig. 7-6

Method 2. Using polar coordinates.

The equation of the circle is $r = a$ [see Fig. 7-6]. As before, we see by symmetry that the centroid must lie on the y axis, so that $\bar{x} = 0$. Since $y = r\sin\theta$ and $dA = r\,dr\,d\theta$ in polar coordinates, we can write

$$\bar{y} = \frac{\int_{\mathcal{R}} y\,\sigma\,dA}{\int_{\mathcal{R}} \sigma\,dA} = \frac{\int_{\theta=0}^{\pi}\int_{r=0}^{a} (r\sin\theta)\,r\,dr\,d\theta}{\int_{\theta=0}^{\pi}\int_{r=0}^{a} r\,dr\,d\theta} = \frac{2a^3/3}{\pi a^2/2} = \frac{4a}{3\pi}$$

7.11. Find the center of mass of a uniform solid hemisphere of radius a.

By symmetry the center of mass lies on the z axis [see Fig. 7-7]. Subdivide the hemisphere into solid circular plates of radius r, such as $ABCDEA$. If the center G of such a ring is at distance z from the center O of the hemisphere, $r^2 + z^2 = a^2$. Then if dz is the thickness of the plate, the volume of each ring is

$$\pi r^2\,dz = \pi(a^2 - z^2)\,dz$$

and the mass is $\pi\sigma(a^2 - z^2)\,dz$. Thus we have

$$\bar{z} = \frac{\int_{z=0}^{a} \pi\sigma z(a^2 - z^2)\,dz}{\int_{z=0}^{a} \pi\sigma(a^2 - z^2)\,dz} = \frac{3}{8}a$$

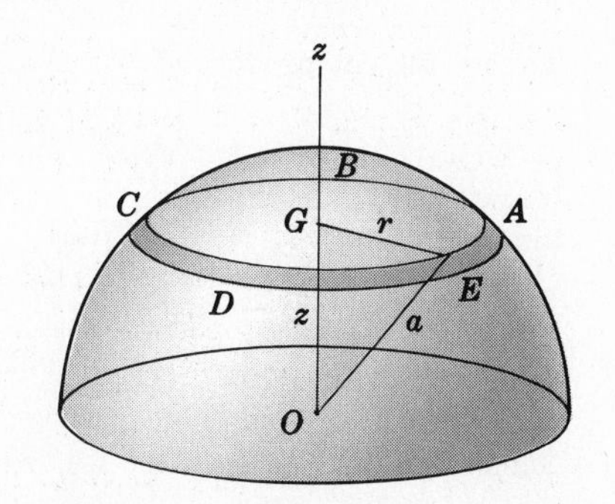

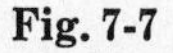
Fig. 7-7

ANGULAR MOMENTUM AND TORQUE

7.12. Prove Theorem 7.4, page 168: The total external torque on a system of particles is equal to the time rate of change of angular momentum of the system, provided that the internal forces between particles are central forces.

As in equation (1) of Problem 7.4, we have

$$\mathbf{F}_\nu + \sum_\lambda \mathbf{f}_{\nu\lambda} = \frac{d\mathbf{p}_\nu}{dt} = \frac{d}{dt}(m_\nu \mathbf{v}_\nu) \tag{1}$$

Multiplying both sides of (1) by $\mathbf{r}_\nu \times$, we have

$$\mathbf{r}_\nu \times \mathbf{F}_\nu + \sum_\lambda \mathbf{r}_\nu \times \mathbf{f}_{\nu\lambda} = \mathbf{r}_\nu \times \frac{d}{dt}(m_\nu \mathbf{v}_\nu) \tag{2}$$

Since
$$\mathbf{r}_\nu \times \frac{d}{dt}(m_\nu \mathbf{v}_\nu) = \frac{d}{dt}\{m_\nu(\mathbf{r}_\nu \times \mathbf{v}_\nu)\} \tag{3}$$

(2) becomes
$$\mathbf{r}_\nu \times \mathbf{F}_\nu + \sum_\lambda \mathbf{r}_\nu \times \mathbf{f}_{\nu\lambda} = \frac{d}{dt}\{m_\nu(\mathbf{r}_\nu \times \mathbf{v}_\nu)\} \tag{4}$$

Summing over ν in (4), we find

$$\sum_\nu \mathbf{r}_\nu \times \mathbf{F}_\nu + \sum_\nu \sum_\lambda \mathbf{r}_\nu \times \mathbf{f}_{\nu\lambda} = \frac{d}{dt}\left\{\sum_\nu m_\nu(\mathbf{r}_\nu \times \mathbf{v}_\nu)\right\} \tag{5}$$

Now the double sum in (5) is composed of terms such as

$$\mathbf{r}_\nu \times \mathbf{f}_{\nu\lambda} + \mathbf{r}_\lambda \times \mathbf{f}_{\lambda\nu} \tag{6}$$

which becomes on writing $\mathbf{f}_{\lambda\nu} = -\mathbf{f}_{\nu\lambda}$ according to Newton's third law,

$$\mathbf{r}_\nu \times \mathbf{f}_{\nu\lambda} - \mathbf{r}_\lambda \times \mathbf{f}_{\nu\lambda} = (\mathbf{r}_\nu - \mathbf{r}_\lambda) \times \mathbf{f}_{\nu\lambda} \tag{7}$$

Then since we suppose that the forces are central, i.e. $\mathbf{f}_{\nu\lambda}$ has the same direction as $\mathbf{r}_\nu - \mathbf{r}_\lambda$, it follows that (7) is zero and also that the double sum in (5) is zero. Thus equation (5) becomes

$$\sum_\nu \mathbf{r}_\nu \times \mathbf{F}_\nu = \frac{d}{dt}\left\{\sum_\nu m_\nu(\mathbf{r}_\nu \times \mathbf{v}_\nu)\right\} \quad \text{or} \quad \mathbf{\Lambda} = \frac{d\mathbf{\Omega}}{dt}$$

where $\mathbf{\Lambda} = \sum_\nu \mathbf{r}_\nu \times \mathbf{F}_\nu$, $\mathbf{\Omega} = \sum_\nu m_\nu(\mathbf{r}_\nu \times \mathbf{v}_\nu)$.

WORK, KINETIC ENERGY AND POTENTIAL ENERGY

7.13. Prove Theorem 7.6, page 169: The total work done in moving a system of particles from one state to another with kinetic energies T_1 and T_2 respectively is $T_2 - T_1$.

The equation of motion of the νth particle in the system is

$$\mathcal{F}_\nu = \mathbf{F}_\nu + \sum_\lambda \mathbf{f}_{\nu\lambda} = \frac{d}{dt}(m_\nu \dot{\mathbf{r}}_\nu) \tag{1}$$

Taking the dot product of both sides with $\dot{\mathbf{r}}_\nu$, we have

$$\mathcal{F}_\nu \cdot \dot{\mathbf{r}}_\nu = \mathbf{F}_\nu \cdot \dot{\mathbf{r}}_\nu + \sum_\lambda \mathbf{f}_{\nu\lambda} \cdot \dot{\mathbf{r}}_\nu = \dot{\mathbf{r}}_\nu \cdot \frac{d}{dt}(m_\nu \dot{\mathbf{r}}_\nu) \tag{2}$$

Since
$$\dot{\mathbf{r}}_\nu \cdot \frac{d}{dt}(m_\nu \dot{\mathbf{r}}_\nu) = \frac{1}{2}\frac{d}{dt}\{m_\nu(\dot{\mathbf{r}}_\nu \cdot \dot{\mathbf{r}}_\nu)\} = \frac{1}{2}\frac{d}{dt}(m_\nu v_\nu^2)$$

(2) can be written

$$\mathcal{F}_\nu \cdot \dot{\mathbf{r}}_\nu = \mathbf{F}_\nu \cdot \dot{\mathbf{r}}_\nu + \sum_\lambda \mathbf{f}_{\nu\lambda} \cdot \dot{\mathbf{r}}_\nu = \frac{1}{2}\frac{d}{dt}(m_\nu v_\nu^2) \tag{3}$$

Summing over ν in equation (3), we find

$$\sum_\nu \mathcal{F}_\nu \cdot \dot{\mathbf{r}}_\nu = \sum_\nu \mathbf{F}_\nu \cdot \dot{\mathbf{r}}_\nu + \sum_\nu \sum_\lambda \mathbf{f}_{\nu\lambda} \cdot \dot{\mathbf{r}}_\nu = \frac{1}{2}\frac{d}{dt}\left(\sum_\nu m_\nu v_\nu^2\right) \tag{4}$$

Integrating both sides of (4) with respect to t from $t = t_1$ to $t = t_2$, we find

$$W_{12} = \sum_\nu \int_{t_1}^{t_2} \boldsymbol{\mathcal{F}}_\nu \cdot \dot{\mathbf{r}}_\nu \, dt = \sum_\nu \int_{t_1}^{t_2} \mathbf{F}_\nu \cdot \dot{\mathbf{r}}_\nu \, dt + \sum_\nu \sum_\lambda \int_{t_1}^{t_2} \mathbf{f}_{\nu\lambda} \cdot \dot{\mathbf{r}}_\nu \, dt$$

$$= \frac{1}{2} \sum_\nu \int_{t_1}^{t_2} \frac{d}{dt}(m_\nu v_\nu^2) \, dt$$

Using the fact that $\dot{\mathbf{r}}_\nu \, dt = d\mathbf{r}_\nu$ and the symbols 1 and 2 for the states at times t_1 and t_2 respectively, this can be written

$$W_{12} = \sum_\nu \int_1^2 \boldsymbol{\mathcal{F}}_\nu \cdot d\mathbf{r}_\nu = \sum_\nu \int_1^2 \mathbf{F}_\nu \cdot d\mathbf{r}_\nu + \sum_\nu \sum_\lambda \int_1^2 \mathbf{f}_{\nu\lambda} \cdot d\mathbf{r}_\nu = T_2 - T_1 \tag{5}$$

where T_1 and T_2 are the total kinetic energies at t_1 and t_2 respectively. Since

$$W_{12} = \sum_\nu \int_1^2 \boldsymbol{\mathcal{F}}_\nu \cdot d\mathbf{r}_\nu \tag{6}$$

is the total work done (by external and internal forces) in moving the system from one state to another, the required result follows.

It should be noted that the double sum in (5) indicating work done by the internal forces, cannot be reduced to zero either by using Newton's third law or the assumption of central forces. This is in contradistinction to the double sums in Problems 7.4 and 7.12 which can be reduced to zero.

7.14. Suppose that the internal forces of a system of particles are conservative and are derived from a potential

$$V_{\lambda\nu}(r_{\lambda\nu}) = V_{\nu\lambda}(r_{\nu\lambda})$$

where $r_{\lambda\nu} = r_{\nu\lambda} = \sqrt{(x_\lambda - x_\nu)^2 + (y_\lambda - y_\nu)^2 + (z_\lambda - z_\nu)^2}$ is the distance between particles λ and ν of the system.

(a) Prove that $\sum_\nu \sum_\lambda \mathbf{f}_{\lambda\nu} \cdot d\mathbf{r}_\nu = -\frac{1}{2} \sum_\nu \sum_\lambda dV_{\lambda\nu}$ where $\mathbf{f}_{\lambda\nu}$ is the internal force on particle ν due to particle λ.

(b) Evaluate the double sum $\sum_\nu \sum_\lambda \int_1^2 \mathbf{f}_{\lambda\nu} \cdot d\mathbf{r}_\nu$ of Problem 7.13.

(a) The force acting on particle ν is

$$\mathbf{f}_{\nu\lambda} = -\frac{\partial V_{\lambda\nu}}{\partial x_\nu}\mathbf{i} - \frac{\partial V_{\lambda\nu}}{\partial y_\nu}\mathbf{j} - \frac{\partial V_{\lambda\nu}}{\partial z_\nu}\mathbf{k} = -\operatorname{grad}_\nu V_{\lambda\nu} = -\nabla_\nu V_{\lambda\nu} \tag{1}$$

The force acting on particle λ is

$$\mathbf{f}_{\lambda\nu} = -\frac{\partial V_{\lambda\nu}}{\partial x_\lambda}\mathbf{i} - \frac{\partial V_{\lambda\nu}}{\partial y_\lambda}\mathbf{j} - \frac{\partial V_{\lambda\nu}}{\partial z_\lambda}\mathbf{k} = -\operatorname{grad}_\lambda V_{\lambda\nu} = -\nabla_\lambda V_{\lambda\nu} = -\mathbf{f}_{\nu\lambda} \tag{2}$$

The work done by these forces in producing the displacements $d\mathbf{r}_\nu$ and $d\mathbf{r}_\lambda$ of particles ν and λ respectively is

$$\mathbf{f}_{\nu\lambda} \cdot d\mathbf{r}_\nu + \mathbf{f}_{\lambda\nu} \cdot d\mathbf{r}_\lambda = -\left\{\frac{\partial V_{\lambda\nu}}{\partial x_\nu} dx_\nu + \frac{\partial V_{\lambda\nu}}{\partial y_\nu} dy_\nu + \frac{\partial V_{\lambda\nu}}{\partial z_\nu} dz_\nu + \frac{\partial V_{\lambda\nu}}{\partial x_\lambda} dx_\lambda + \frac{\partial V_{\lambda\nu}}{\partial y_\lambda} dy_\lambda + \frac{\partial V_{\lambda\nu}}{\partial z_\lambda} dz_\lambda\right\}$$

$$= -dV_{\lambda\nu}$$

Then the total work done by the internal forces is

$$\sum_\nu \sum_\lambda \mathbf{f}_{\lambda\nu} \cdot d\mathbf{r}_\nu = -\frac{1}{2} \sum_\nu \sum_\lambda dV_{\lambda\nu} \tag{3}$$

the factor $\frac{1}{2}$ on the right being introduced because otherwise the terms in the summation would enter twice.

(*b*) By integrating (*3*) of part (*a*), we have

$$\sum_{\nu}\sum_{\lambda}\int_1^2 \mathbf{f}_{\lambda\nu}\cdot d\mathbf{r}_\nu \quad = \quad -\frac{1}{2}\sum_{\nu}\sum_{\lambda}\int_1^2 dV_{\lambda\nu} \quad = \quad V_1^{(\text{int})} - V_2^{(\text{int})} \tag{4}$$

where $V_1^{(\text{int})}$ and $V_2^{(\text{int})}$ denote the total internal potentials

$$\frac{1}{2}\sum_{\nu}\sum_{\lambda} V_{\lambda\nu} \tag{5}$$

at times t_1 and t_2 respectively.

7.15. Prove that if both the external and internal forces for a system of particles are conservative, then the principle of conservation of energy is valid.

If the external forces are conservative, then we have

$$\mathbf{F}_\nu \quad = \quad -\nabla V_\nu \tag{1}$$

from which

$$\sum_{\nu}\int_1^2 \mathbf{F}_\nu\cdot d\mathbf{r}_\nu \quad = \quad -\sum_{\nu}\int_1^2 dV_\nu \quad = \quad V_1^{(\text{ext})} - V_2^{(\text{ext})} \tag{2}$$

where $V_1^{(\text{ext})}$ and $V_2^{(\text{ext})}$ denote the total external potential

$$\sum_{\nu} V_\nu$$

at times t_1 and t_2 respectively.

Using (*2*) and equation (*4*) of Problem 7.14(*b*) in equation (*5*) of Problem 7.13, we find

$$T_2 - T_1 \quad = \quad V_1^{(\text{ext})} - V_2^{(\text{ext})} + V_1^{(\text{int})} - V_2^{(\text{int})} \quad = \quad V_1 - V_2 \tag{3}$$

where

$$V_1 \quad = \quad V_1^{(\text{ext})} + V_1^{(\text{int})} \quad \text{and} \quad V_2 \quad = \quad V_2^{(\text{ext})} + V_2^{(\text{int})} \tag{4}$$

are the respective total potential energies [external and internal] at times t_1 and t_2. We thus find from (*3*),

$$T_1 + V_1 \quad = \quad T_2 + V_2 \quad \text{or} \quad T + V \quad = \quad \text{constant} \tag{5}$$

which is the principle of conservation of energy.

MOTION RELATIVE TO THE CENTER OF MASS

7.16. Let $\mathbf{r}'_\nu$ and $\mathbf{v}'_\nu$ be respectively the position vector and velocity of particle ν relative to the center of mass. Prove that (*a*) $\sum_{\nu} m_\nu \mathbf{r}'_\nu = 0$, (*b*) $\sum_{\nu} m_\nu \mathbf{v}'_\nu = 0$.

(*a*) Let $\mathbf{r}_\nu$ be the position vector of particle ν relative to O and $\bar{\mathbf{r}}$ the position vector of the center of mass C relative to O. Then from the definition of the center of mass,

$$\bar{\mathbf{r}} \quad = \quad \frac{1}{M}\sum_{\nu} m_\nu \mathbf{r}_\nu \tag{1}$$

where $M = \sum_{\nu} m_\nu$. From Fig. 7-8 we have

$$\mathbf{r}_\nu \quad = \quad \mathbf{r}'_\nu + \bar{\mathbf{r}} \tag{2}$$

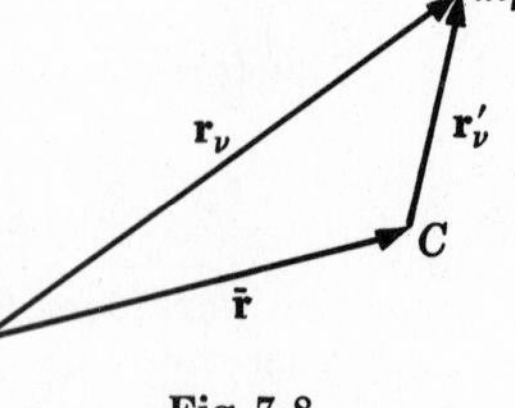

Fig. 7-8

Then substituting (*2*) into (*1*), we find

$$\bar{\mathbf{r}} \quad = \quad \frac{1}{M}\sum_{\nu} m_\nu(\mathbf{r}'_\nu + \bar{\mathbf{r}}) \quad = \quad \frac{1}{M}\sum_{\nu} m_\nu \mathbf{r}'_\nu + \bar{\mathbf{r}}$$

from which

$$\sum_{\nu} m_\nu \mathbf{r}'_\nu \quad = \quad 0 \tag{3}$$

(*b*) Differentiating both sides of (*3*) with respect to t, we have $\sum_{\nu} m_\nu \mathbf{v}'_\nu = 0$.

7.17. Prove Theorem 7.9, page 169: The total angular momentum of a system of particles about any point O equals the angular momentum of the total mass assumed to be located at the center of mass plus the angular momentum about the center of mass.

Let $\mathbf{r}_\nu$ be the position vector of particle ν relative to O, $\bar{\mathbf{r}}$ the position vector of the center of mass C relative to O and $\mathbf{r}'_\nu$ the position vector of particle ν relative to C. Then

$$\mathbf{r}_\nu = \mathbf{r}'_\nu + \bar{\mathbf{r}} \tag{1}$$

Differentiating with respect to t, we find

$$\mathbf{v}_\nu = \dot{\mathbf{r}}_\nu = \dot{\mathbf{r}}'_\nu + \dot{\bar{\mathbf{r}}} = \mathbf{v}'_\nu + \bar{\mathbf{v}} \tag{2}$$

where $\bar{\mathbf{v}}$ is the velocity of the center of mass relative to O, $\mathbf{v}_\nu$ is the velocity of particle ν relative to O, and $\mathbf{v}'_\nu$ is the velocity of particle ν relative to C.

The total angular momentum of the system about O is

$$\begin{aligned}\mathbf{\Omega} &= \sum_\nu m_\nu(\mathbf{r}_\nu \times \mathbf{v}_\nu) = \sum_\nu m_\nu\{(\mathbf{r}'_\nu + \bar{\mathbf{r}}) \times (\mathbf{v}'_\nu + \bar{\mathbf{v}})\} \\ &= \sum_\nu m_\nu(\mathbf{r}'_\nu \times \mathbf{v}'_\nu) + \sum_\nu m_\nu(\mathbf{r}'_\nu \times \bar{\mathbf{v}}) + \sum_\nu m_\nu(\bar{\mathbf{r}} \times \mathbf{v}'_\nu) + \sum_\nu m_\nu(\bar{\mathbf{r}} \times \bar{\mathbf{v}}) \end{aligned} \tag{3}$$

Now by Problem 7.16,

$$\sum_\nu m_\nu(\mathbf{r}'_\nu \times \bar{\mathbf{v}}) = \left\{\sum_\nu m_\nu \mathbf{r}'_\nu\right\} \times \bar{\mathbf{v}} = \mathbf{0}$$

$$\sum_\nu m_\nu(\bar{\mathbf{r}} \times \mathbf{v}'_\nu) = \bar{\mathbf{r}} \times \left\{\sum_\nu m_\nu \mathbf{v}'_\nu\right\} = \mathbf{0}$$

$$\sum_\nu m_\nu(\bar{\mathbf{r}} \times \bar{\mathbf{v}}) = \left\{\sum_\nu m_\nu\right\}(\bar{\mathbf{r}} \times \bar{\mathbf{v}}) = M(\bar{\mathbf{r}} \times \bar{\mathbf{v}})$$

Then (3) becomes, as required,

$$\mathbf{\Omega} = \sum_\nu m_\nu(\mathbf{r}'_\nu \times \mathbf{v}'_\nu) + M(\bar{\mathbf{r}} \times \bar{\mathbf{v}})$$

7.18. Prove Theorem 7.10, page 169: The total kinetic energy of a system of particles about any point O equals the kinetic energy of the center of mass [assuming the total mass located there] plus the kinetic energy of motion about the center of mass.

The kinetic energy relative to O [see Fig. 7-8] is

$$T = \frac{1}{2}\sum_\nu m_\nu v_\nu^2 = \frac{1}{2}\sum_\nu m_\nu(\dot{\mathbf{r}}_\nu \cdot \dot{\mathbf{r}}_\nu) \tag{1}$$

Using equation (2) of Problem 7.16 we find

$$\dot{\mathbf{r}}_\nu = \dot{\bar{\mathbf{r}}} + \dot{\mathbf{r}}'_\nu = \bar{\mathbf{v}} + \mathbf{v}'_\nu$$

Thus (1) can be written

$$\begin{aligned} T &= \frac{1}{2}\sum_\nu m_\nu\{(\bar{\mathbf{v}} + \mathbf{v}'_\nu)\cdot(\bar{\mathbf{v}} + \mathbf{v}'_\nu)\} \\ &= \frac{1}{2}\sum_\nu m_\nu \bar{\mathbf{v}}\cdot\bar{\mathbf{v}} + \sum_\nu m_\nu \bar{\mathbf{v}}\cdot\mathbf{v}'_\nu + \frac{1}{2}\sum_\nu m_\nu \mathbf{v}'_\nu\cdot\mathbf{v}'_\nu \\ &= \frac{1}{2}\left(\sum_\nu m_\nu\right)\bar{v}^2 + \bar{\mathbf{v}}\cdot\left\{\sum_\nu m_\nu \mathbf{v}'_\nu\right\} + \frac{1}{2}\sum_\nu m_\nu v'^2_\nu \\ &= \frac{1}{2}M\bar{v}^2 + \frac{1}{2}\sum_\nu m_\nu v'^2_\nu \end{aligned}$$

since $\sum_\nu m_\nu \mathbf{v}'_\nu = \mathbf{0}$ by Problem 7.16.

IMPULSE

7.19. Prove Theorem 7.12: The total linear impulse is equal to the change in linear momentum.

The total external force by equation (4) of Problem 7.4 is

$$\mathbf{F} = M\frac{d^2\bar{\mathbf{r}}}{dt^2} = M\frac{d\bar{\mathbf{v}}}{dt}$$

Then the total linear impulse is

$$\int_{t_1}^{t_2} \mathbf{F}\,dt = \int_{t_1}^{t_2} M\frac{d\bar{\mathbf{v}}}{dt}\,dt = M\bar{\mathbf{v}}_2 - M\bar{\mathbf{v}}_1 = \mathbf{p}_2 - \mathbf{p}_1$$

where $\mathbf{p}_1 = M\bar{\mathbf{v}}_1$ and $\mathbf{p}_2 = M\bar{\mathbf{v}}_2$ represent the total momenta at times t_1 and t_2 respectively.

CONSTRAINTS. HOLONOMIC AND NON-HOLONOMIC CONSTRAINTS

7.20. In each of the following cases state whether the constraint is holonomic or non-holonomic and give a reason for your answer: (*a*) a bead moving on a circular wire; (*b*) a particle sliding down an inclined plane under the influence of gravity; (*c*) a particle sliding down a sphere from a point near the top under the influence of gravity.

(*a*) The constraint is holonomic since the bead, which can be considered a particle, is constrained to move on the circular wire.

(*b*) The constraint is holonomic since the particle is constrained to move along a surface which is in this case a plane.

(*c*) The constraint is non-holonomic since the particle after reaching a certain location on the sphere will leave the sphere.

Another way of seeing this is to note that if $\mathbf{r}$ is the position vector of the particle relative to the center of the sphere as origin and a is the radius of the sphere, then the particle moves so that $\mathbf{r}^2 \geqq a^2$. This is a non-holonomic constraint since it is not of the form (26), page 170. An example of a holonomic constraint would be $\mathbf{r}^2 = a^2$.

STATICS. PRINCIPLE OF VIRTUAL WORK. STABILITY

7.21. Prove the principle of virtual work, Theorem 7.14, page 171.

For equilibrium, the net resultant force $\mathbf{F}_\nu$ on each particle must be zero, so that

$$\sum_\nu \mathbf{F}_\nu \cdot \delta\mathbf{r}_\nu = 0 \tag{1}$$

But since $\mathbf{F}_\nu = \mathbf{F}_\nu^{(a)} + \mathbf{F}_\nu^{(c)}$ where $\mathbf{F}_\nu^{(a)}$ and $\mathbf{F}_\nu^{(c)}$ are the actual and constraint forces acting on the νth particle, (1) can be written

$$\sum_\nu \mathbf{F}_\nu^{(a)} \cdot \delta\mathbf{r}_\nu + \sum_k \mathbf{F}_\nu^{(c)} \cdot \delta\mathbf{r}_\nu = 0 \tag{2}$$

If we assume that the virtual work of the constraint forces is zero, the second sum on the left of (2) is zero, so that we have

$$\sum_\nu \mathbf{F}_\nu^{(a)} \cdot \delta\mathbf{r}_\nu = 0 \tag{3}$$

which is the principle of virtual work.

7.22. Two particles of masses m_1 and m_2 are located on a frictionless double incline and connected by an inextensible massless string passing over a smooth peg [see Fig. 7-9 below]. Use the principle of virtual work to show that for equilibrium we must have

$$\frac{\sin\alpha_1}{\sin\alpha_2} = \frac{m_2}{m_1}$$

where α_1 and α_2 are the angles of the incline.

Method 1.

Let $\mathbf{r}_1$ and $\mathbf{r}_2$ be the respective position vectors of masses m_1 and m_2 relative to O.

The actual forces (due to gravity) acting on m_1 and m_2 are respectively

$$\mathbf{F}_1^{(a)} = m_1\mathbf{g}, \quad \mathbf{F}_2^{(a)} = m_2\mathbf{g} \tag{1}$$

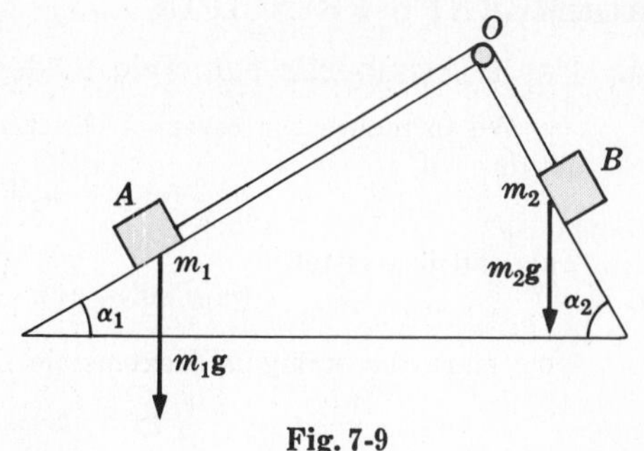

Fig. 7-9

According to the principle of virtual work,

$$\sum \mathbf{F}_\nu^{(a)} \cdot \delta\mathbf{r}_\nu = 0$$

or

$$\mathbf{F}_1^{(a)} \cdot \delta\mathbf{r}_1 + \mathbf{F}_2^{(a)} \cdot \delta\mathbf{r}_2 = 0 \tag{2}$$

where $\delta\mathbf{r}_1$ and $\delta\mathbf{r}_2$ are the virtual displacements of m_1 and m_2 down the incline. Using (1) in (2),

$$m_1\mathbf{g} \cdot \delta\mathbf{r}_1 + m_2\mathbf{g} \cdot \delta\mathbf{r}_2 = 0 \tag{3}$$

or

$$m_1 g\, \delta r_1 \sin\alpha_1 + m_2 g\, \delta r_2 \sin\alpha_2 = 0 \tag{4}$$

Then since the string is inextensible, i.e. $\delta r_1 + \delta r_2 = 0$ or $\delta r_2 = -\delta r_1$, (4) becomes

$$(m_1 g \sin\alpha_1 - m_2 g \sin\alpha_2)\delta r_1 = 0$$

But since δr_1 is arbitrary, we must have $m_1 g \sin\alpha_1 - m_2 g \sin\alpha_2 = 0$, i.e.,

$$\frac{\sin\alpha_1}{\sin\alpha_2} = \frac{m_2}{m_1} \tag{5}$$

Method 2.

When it is not clear which forces are constraint forces doing no work, we can take into account *all* forces and then apply the principle of virtual work. Thus, for example, taking into account the reaction forces $\mathbf{R}_1$ and $\mathbf{R}_2$ due to the inclines on the particles and the tension forces $\mathbf{T}_1$ and $\mathbf{T}_2$, the principle of virtual work becomes

$$(m_1\mathbf{g} + \mathbf{T}_1 + \mathbf{R}_1) \cdot \delta\mathbf{r}_1 + (m_2\mathbf{g} + \mathbf{T}_2 + \mathbf{R}_2) \cdot \delta\mathbf{r}_2 = 0 \tag{6}$$

Now since the inclines are assumed smooth [so that the reaction forces are perpendicular to the inclines] we have

$$\mathbf{R}_1 \cdot \delta\mathbf{r}_1 = 0, \quad \mathbf{R}_2 \cdot \delta\mathbf{r}_2 = 0 \tag{7}$$

Also, since there is no friction at the peg, the tensions $\mathbf{T}_1$ and $\mathbf{T}_2$ have the same magnitude. Thus we have, using the fact that $\delta\mathbf{r}_1$ and $\delta\mathbf{r}_2$ are directed down the corresponding inclines and the fact that $\delta r_2 = -\delta r_1$,

$$\begin{aligned}\mathbf{T}_1 \cdot \delta\mathbf{r}_1 + \mathbf{T}_2 \cdot \delta\mathbf{r}_2 &= -T_1\,\delta r_1 - T_2\,\delta r_2 \\ &= (T_2 - T_1)\delta r_1 = 0 \end{aligned} \tag{8}$$

since $T_1 = T_2$. Then using (7) and (8), (6) becomes

$$m_1\mathbf{g} \cdot \delta\mathbf{r}_1 + m_2\mathbf{g} \cdot \delta\mathbf{r}_2 = 0$$

as obtained in (3).

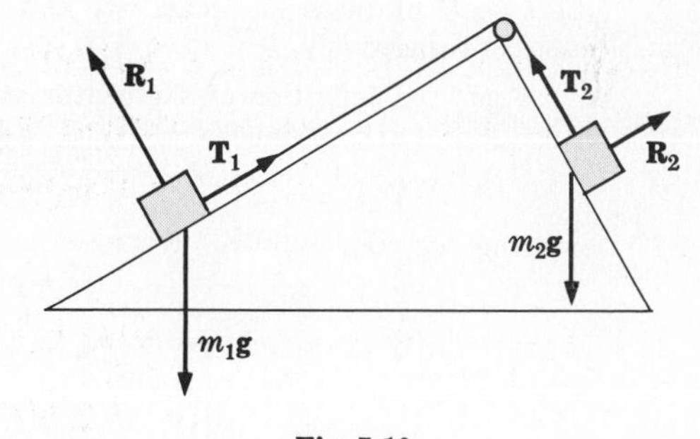

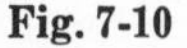

Fig. 7-10

7.23. Use Theorem 7.15, page 171, to solve Problem 7.22.

Let the string have length l and suppose that the lengths of string OA and OB on the inclines [Fig. 7-9] are x and $l - x$ respectively. The total potential energy using a horizontal plane through O as reference level is

$$V = -m_1 g x \sin\alpha_1 - m_2 g(l - x)\sin\alpha_2$$

Then for equilibrium we must have

$$\frac{\partial V}{\partial x} = -m_1 g \sin\alpha_1 + m_2 g \sin\alpha_2 = 0 \quad \text{or} \quad \frac{\sin\alpha_1}{\sin\alpha_2} = \frac{m_2}{m_1}$$

It should be noted that V is not a minimum in this case so that the equilibrium is not stable, as is also evident physically.

D'ALEMBERT'S PRINCIPLE

7.24. Use D'Alembert's principle to describe the motion of the masses in Problem 7.22.

We introduce the reversed effective forces $m_1\ddot{\mathbf{r}}_1$ and $m_2\ddot{\mathbf{r}}_2$ in equation (3) of Problem 7.22 to obtain

$$(m_1\mathbf{g} - m_1\ddot{\mathbf{r}}_1)\cdot\delta\mathbf{r}_1 + (m_2\mathbf{g} - m_2\ddot{\mathbf{r}}_2)\cdot\delta\mathbf{r}_2 = 0 \tag{1}$$

This can be written

$$(m_1 g \sin\alpha_1 - m_1\ddot{r}_1)\delta r_1 + (m_2 g \sin\alpha_2 - m_2\ddot{r}_2)\delta r_2 = 0 \tag{2}$$

Now since the string is inextensible so that $r_1 + r_2 =$ constant, we have

$$\delta r_1 + \delta r_2 = 0, \qquad \ddot{r}_1 + \ddot{r}_2 = 0$$

or $\delta r_2 = -\delta r_1$, $\ddot{r}_2 = -\ddot{r}_1$. Thus (2) becomes, after dividing by $\delta r_1 \neq 0$,

$$m_1 g \sin\alpha_1 - m_1\ddot{r}_1 - m_2 g \sin\alpha_2 - m_2\ddot{r}_1 = 0$$

or

$$\ddot{r}_1 = \frac{m_1 g \sin\alpha_1 - m_2 g \sin\alpha_2}{m_1 + m_2}$$

Thus particle 1 goes down or up the incline with constant acceleration according as $m_1 g \sin\alpha_1 > m_2 g \sin\alpha_2$ or $m_1 g \sin\alpha_1 < m_2 g \sin\alpha_2$ respectively. Particle 2 in these cases goes up or down respectively with the same constant acceleration.

We can also use a method analogous to the second method of Problem 7.22.

MISCELLANEOUS PROBLEMS

7.25. Two particles having masses m_1 and m_2 move so that their relative velocity is $\mathbf{v}$ and the velocity of their center of mass is $\bar{\mathbf{v}}$. If $M = m_1 + m_2$ is the total mass and $\mu = m_1 m_2/(m_1 + m_2)$ is the *reduced mass* of the system, prove that the total kinetic energy is $\frac{1}{2}M\bar{v}^2 + \frac{1}{2}\mu v^2$.

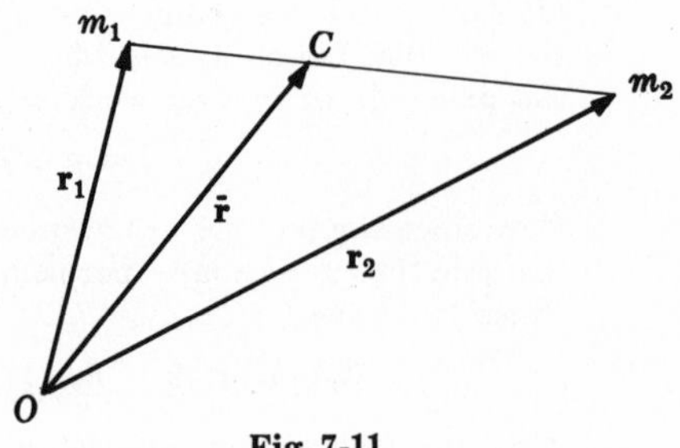

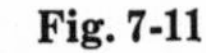

Fig. 7-11

Let $\mathbf{r}_1$, $\mathbf{r}_2$ and $\bar{\mathbf{r}}$ be the position vectors with respect to O of mass m_1, mass m_2 and the center of mass C respectively.

From the definition of the center of mass, we have

$$\bar{\mathbf{r}} = \frac{m_1\mathbf{r}_1 + m_2\mathbf{r}_2}{m_1 + m_2} \qquad \text{and} \qquad \dot{\bar{\mathbf{r}}} = \frac{m_1\dot{\mathbf{r}}_1 + m_2\dot{\mathbf{r}}_2}{m_1 + m_2}$$

or using $\mathbf{v}_1 = \dot{\mathbf{r}}_1$, $\mathbf{v}_2 = \dot{\mathbf{r}}_2$, $\bar{\mathbf{v}} = \dot{\bar{\mathbf{r}}}$,

$$m_1\mathbf{v}_1 + m_2\mathbf{v}_2 = (m_1 + m_2)\bar{\mathbf{v}} \tag{1}$$

If the velocity of m_1 relative to m_2 is $\mathbf{v}$, then

$$\mathbf{v} = \frac{d}{dt}(\mathbf{r}_1 - \mathbf{r}_2) = \dot{\mathbf{r}}_1 - \dot{\mathbf{r}}_2 = \mathbf{v}_1 - \mathbf{v}_2$$

so that

$$\mathbf{v}_1 - \mathbf{v}_2 = \mathbf{v} \tag{2}$$

Solving (1) and (2) simultaneously, we find

$$\mathbf{v}_1 = \bar{\mathbf{v}} + \frac{m_2\mathbf{v}}{m_1 + m_2}, \qquad \mathbf{v}_2 = \bar{\mathbf{v}} - \frac{m_1\mathbf{v}}{m_1 + m_2}$$

Then the total kinetic energy is

$$\begin{aligned} T &= \frac{1}{2}m_1\mathbf{v}_1^2 + \frac{1}{2}m_2\mathbf{v}_2^2 \\ &= \frac{1}{2}m_1\left(\bar{\mathbf{v}} + \frac{m_2\mathbf{v}}{m_1 + m_2}\right)^2 + \frac{1}{2}m_2\left(\bar{\mathbf{v}} - \frac{m_1\mathbf{v}}{m_1 + m_2}\right)^2 \\ &= \frac{1}{2}(m_1 + m_2)\bar{v}^2 + \frac{1}{2}\frac{m_1 m_2}{m_1 + m_2}v^2 = \frac{1}{2}M\bar{v}^2 + \frac{1}{2}\mu v^2 \end{aligned}$$

7.26. Find the centroid of a uniform semicircular wire of radius a.

By symmetry [see Fig. 7-12] the centroid of the wire must be on the y axis, so that $\bar{x} = 0$. If σ is the mass per unit length of the wire, then if ds represents an element of arc, we have $ds = a\,d\theta$ so that

$$\bar{y} = \frac{\int_C y\,\sigma\,ds}{\int_C \sigma\,ds} = \frac{\int_0^\pi (a\sin\theta)(a\,d\theta)}{\int_0^\pi a\,d\theta} = \frac{2a^2}{\pi a} = \frac{2a}{\pi}$$

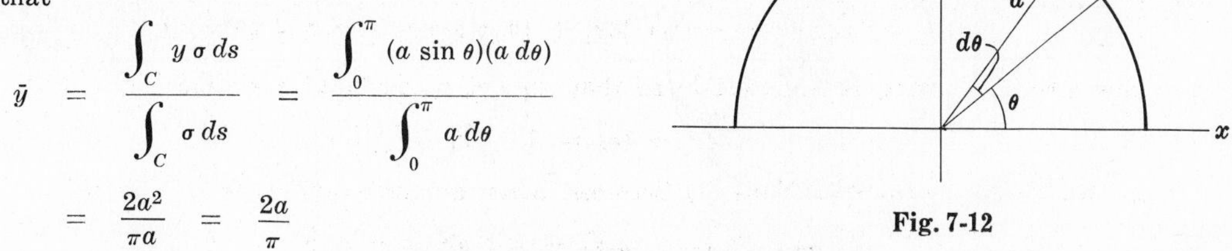

Fig. 7-12

7.27. Suppose that n systems of particles be given having centroids at $\bar{\mathbf{r}}_1, \bar{\mathbf{r}}_2, \ldots, \bar{\mathbf{r}}_n$ and total masses $M_1, M_2, \ldots, M_n$ respectively. Prove that the centroid of all the systems is at

$$\frac{M_1\bar{\mathbf{r}}_1 + M_2\bar{\mathbf{r}}_2 + \cdots + M_n\bar{\mathbf{r}}_n}{M_1 + M_2 + \cdots + M_n}$$

Let system 1 be composed of masses $m_{11}, m_{12}, \ldots$ located at $\mathbf{r}_{11}, \mathbf{r}_{12}, \ldots$ respectively. Similarly let system 2 be composed of masses $m_{21}, m_{22}, \ldots$ located at $\mathbf{r}_{21}, \mathbf{r}_{22}, \ldots$. Then by definition,

$$\bar{\mathbf{r}}_1 = \frac{m_{11}\mathbf{r}_{11} + m_{12}\mathbf{r}_{12} + \cdots}{m_{11} + m_{12} + \cdots} = \frac{m_{11}\mathbf{r}_{11} + m_{12}\mathbf{r}_{12} + \cdots}{M_1}$$

$$\bar{\mathbf{r}}_2 = \frac{m_{21}\mathbf{r}_{21} + m_{22}\mathbf{r}_{22} + \cdots}{m_{21} + m_{22} + \cdots} = \frac{m_{21}\mathbf{r}_{21} + m_{22}\mathbf{r}_{22} + \cdots}{M_2}$$

$$\vdots$$

$$\bar{\mathbf{r}}_n = \frac{m_{n1}\mathbf{r}_{n1} + m_{n2}\mathbf{r}_{n2} + \cdots}{m_{n1} + m_{n2} + \cdots} = \frac{m_{n1}\mathbf{r}_{n1} + m_{n2}\mathbf{r}_{n2} + \cdots}{M_n}$$

But the centroid for all systems is located at

$$\bar{\mathbf{r}} = \frac{(m_{11}\mathbf{r}_{11} + m_{12}\mathbf{r}_{12} + \cdots) + (m_{21}\mathbf{r}_{21} + m_{22}\mathbf{r}_{22} + \cdots) + \cdots + (m_{n1}\mathbf{r}_{n1} + m_{n2}\mathbf{r}_{n2} + \cdots)}{(m_{11} + m_{12} + \cdots) + (m_{21} + m_{22} + \cdots) + \cdots + (m_{n1} + m_{n2} + \cdots)}$$

$$= \frac{M_1\bar{\mathbf{r}}_1 + M_2\bar{\mathbf{r}}_2 + \cdots + M_n\bar{\mathbf{r}}_n}{M_1 + M_2 + \cdots + M_n}$$

7.28. Find the centroid of a solid of constant density consisting of a cylinder of radius a and height H surmounted by a hemisphere of radius a [see Fig. 7-13].

Let $\bar{r}$ be the distance of the centroid of the solid from the base. The centroid of the hemisphere of radius a is at distance $\frac{3}{8}a + H$ from the base of the solid, and its mass is $M_1 = \frac{2}{3}\pi a^3\sigma$ [see Problem 7.11].

The centroid of the cylinder of radius a and height H is at distance $\frac{1}{2}H$ from the base of the solid and its mass is $M_2 = \pi a^2 H\sigma$.

Then by Problem 7.27,

$$\bar{r} = \frac{(\frac{2}{3}\pi a^3\sigma)(\frac{3}{8}a + H) + (\pi a^2 H\sigma)(\frac{1}{2}H)}{\frac{2}{3}\pi a^3\sigma + \pi a^2 H\sigma} = \frac{3a^2 + 8aH + 6H^2}{8a + 12H}$$

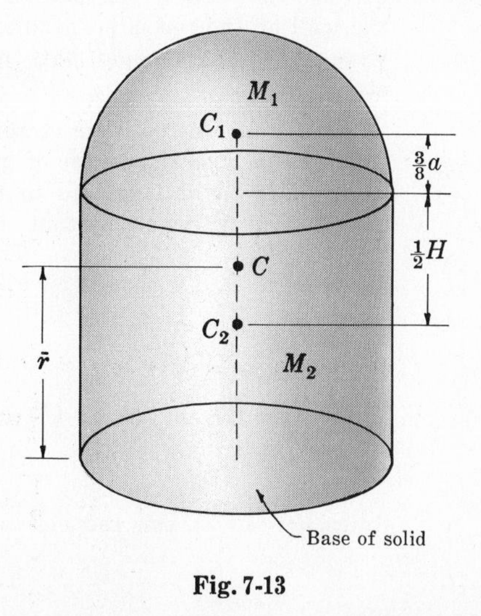

Fig. 7-13

7.29. A circular hole of radius $a/2$ is cut out of a circular region of radius a, as shown in Fig. 7-14. Find the centroid of the shaded region thus obtained.

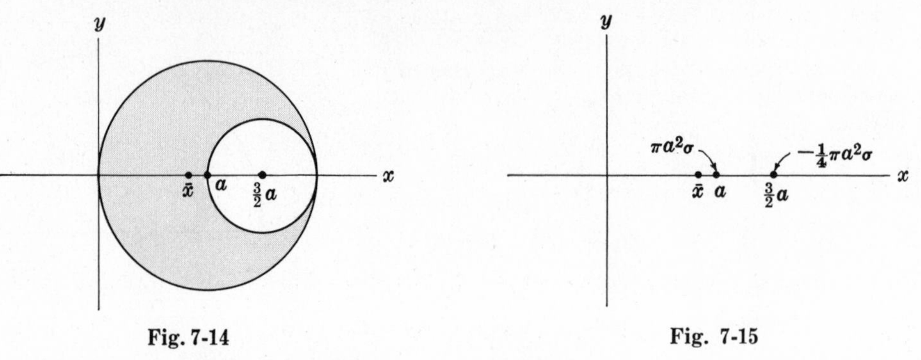

Fig. 7-14 Fig. 7-15

By symmetry the centroid is located on the x axis, so that $\bar{y} = 0$.

We can replace the circular region of radius a by the mass $M_1 = \pi a^2\sigma$ concentrated at its centroid $x_1 = a$ [Fig. 7-15]. Similarly, we can replace the circular hole of radius $a/2$ by the *negative mass* $M_2 = -\frac{1}{4}\pi a^2\sigma$ concentrated at its centroid $x_2 = \frac{3}{2}a$. Then the centroid is located on the x axis at

$$\bar{x} \quad = \quad \frac{M_1x_1 + M_2x_2}{M_1 + M_2} \quad = \quad \frac{(\pi a^2\sigma)(a) + (-\frac{1}{4}\pi a^2\sigma)(\frac{3}{2}a)}{\pi a^2\sigma - \frac{1}{4}\pi a^2\sigma} \quad = \quad \frac{5}{6}a$$

7.30. A uniform rod PQ [see Fig. 7-16] of mass m and length L has its end P resting against a smooth vertical wall AB while its other end Q is attached by means of an inextensible string OQ of length l to the fixed point O on the wall. Assuming that the plane of P, Q and O is vertical and perpendicular to the wall, show that equilibrium occurs if

$$\sin\alpha = \frac{\sqrt{4L^2 - l^2}}{l\sqrt{3}}, \qquad \sin\beta = \frac{\sqrt{4L^2 - l^2}}{L\sqrt{3}}$$

There is only one actual force, i.e. the weight $m\mathbf{g}$ of the rod. Other forces acting are the force of the wall on the rod and the tension in the string. However, these are constraint forces and can do no work. This can be seen since if P were to slide down the wall no work would be done, because the wall is frictionless and thus the force due to the wall on the rod is perpendicular to the wall. Also if Q were to drop, it could only move perpendicular to the string at Q.

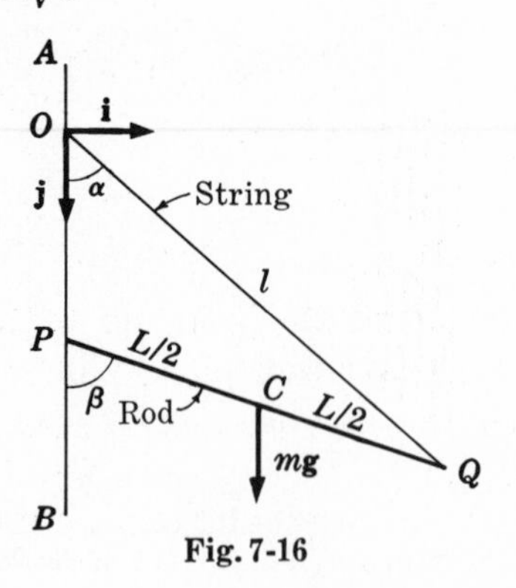

Fig. 7-16

Let $\mathbf{r}$ be the position vector of the center of mass C [in this case also the center of gravity] relative to O. Also let $\mathbf{i}$ and $\mathbf{j}$ be unit vectors in the horizontal and vertical directions respectively so that $\mathbf{r} = x\mathbf{i} + y\mathbf{j}$.

From Fig. 7-16,

$$\mathbf{OQ} = \mathbf{OP} + \mathbf{PQ} \tag{1}$$

$$\mathbf{OQ} = \mathbf{OC} + \mathbf{CQ} \tag{2}$$

Then from (1), on taking the dot product with $\mathbf{i}$,

$$\mathbf{OQ}\cdot\mathbf{i} = \mathbf{OP}\cdot\mathbf{i} + \mathbf{PQ}\cdot\mathbf{i}$$

Since $\mathbf{OP}\cdot\mathbf{i} = 0$, this reduces to

$$\mathbf{OQ}\cdot\mathbf{i} = \mathbf{PQ}\cdot\mathbf{i}$$

or

$$l\sin\alpha = L\sin\beta \tag{3}$$

Similarly on taking the dot product of both sides of (2) with $\mathbf{j}$,

$$\mathbf{OQ}\cdot\mathbf{j} = \mathbf{OC}\cdot\mathbf{j} + \mathbf{CQ}\cdot\mathbf{j}$$

or
$$l\cos\alpha = y + \tfrac{1}{2}L\cos\beta \tag{4}$$

Now a virtual displacement of the center of mass C is given by

$$\delta\mathbf{r} = \delta x\,\mathbf{i} + \delta y\,\mathbf{j} \tag{5}$$

Since $m\mathbf{g}$ is the only actual force, the principle of virtual work becomes

$$m\mathbf{g}\cdot\delta\mathbf{r} = 0 \tag{6}$$

Using (5), this becomes
$$mg\,\delta y = 0 \quad\text{or}\quad \delta y = 0 \tag{7}$$

Now from (3) and (4), we have

$$l\cos\alpha\,\delta\alpha = L\cos\beta\,\delta\beta$$
$$-l\sin\alpha\,\delta\alpha = \delta y - \tfrac{1}{2}L\sin\beta\,\delta\beta$$

since l and L are constants and since δ has the same properties as the differential operator d.

Since $\delta y = 0$ from (7), these equations become

$$l\cos\alpha\,\delta\alpha = L\cos\beta\,\delta\beta \tag{8}$$
$$l\sin\alpha\,\delta\alpha = \tfrac{1}{2}L\sin\beta\,\delta\beta \tag{9}$$

From (8) and (9), we have on division,

$$\frac{\sin\alpha}{\cos\alpha} = \frac{1}{2}\frac{\sin\beta}{\cos\beta} \tag{10}$$

Now from (3),
$$\sin\beta = (l/L)\sin\alpha \tag{11}$$

so that
$$\cos\beta = \sqrt{1-(l^2/L^2)\sin^2\alpha} \tag{12}$$

Thus equation (10) can be written

$$\frac{\sin\alpha}{\sqrt{1-\sin^2\alpha}} = \frac{1}{2}\,\frac{l\sin\alpha}{\sqrt{L^2-l^2\sin^2\alpha}} \tag{13}$$

Dividing by $\sin\alpha$ and squaring both sides, we find

$$\sin\alpha = \frac{\sqrt{4L^2-l^2}}{l\sqrt{3}} \tag{14}$$

and from (11)
$$\sin\beta = \frac{\sqrt{4L^2-l^2}}{L\sqrt{3}} \tag{15}$$

as required.

7.31. A uniform solid consists of a cylinder of radius a and height H on a hemisphere of radius a, as indicated in Fig. 7-17. Prove that the solid is in stable equilibrium on a horizontal plane if and only if $a/H > \sqrt{2}$.

By Problem 7.28 the centroid C is at a distance CB from the center B of the hemisphere given by

$$H - \frac{3a^2+8aH+6H^2}{8a+12H} = \frac{6H^2-3a^2}{8a+12H}$$

Then the distance of the centroid C above the plane is

$$CP = CD + DP = CB\cos\theta + BQ$$
$$= \frac{6H^2-3a^2}{8a+12H}\cos\theta + a$$

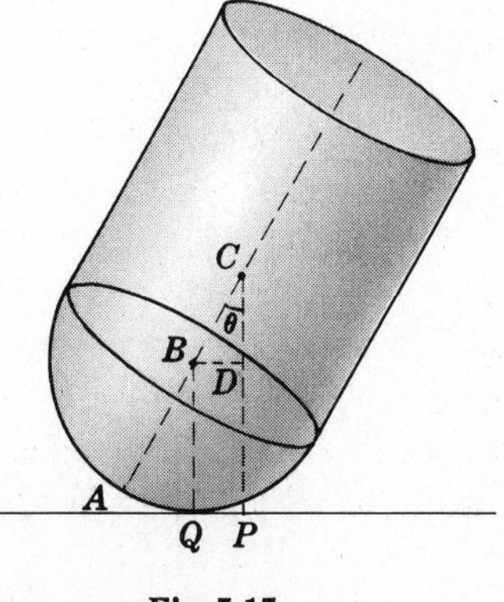

Fig. 7-17

so that the potential energy (or potential) is

$$V = Mg\left(\frac{6H^2 - 3a^2}{8a + 12H}\cos\theta + a\right)$$

Equilibrium occurs where $\frac{\partial V}{\partial \theta} = 0$ or $Mg\left(\frac{3a^2 - 6H^2}{8a + 12H}\right)\sin\theta = 0$, i.e. $\theta = 0$.

Then the equilibrium will be stable if

$$\left.\frac{\partial^2 V}{\partial \theta^2}\right|_{\theta=0} = Mg\left(\frac{3a^2 - 6H^2}{8a + 12H}\right)\cos\theta\bigg|_{\theta=0} = Mg\left(\frac{3a^2 - 6H^2}{8a + 12H}\right) > 0$$

i.e. $3a^2 - 6H^2 > 0$ or $a/H > \sqrt{2}$.

7.32. A uniform chain has its ends suspended from two fixed points at the same horizontal level. Find an equation for the curve in which it hangs.

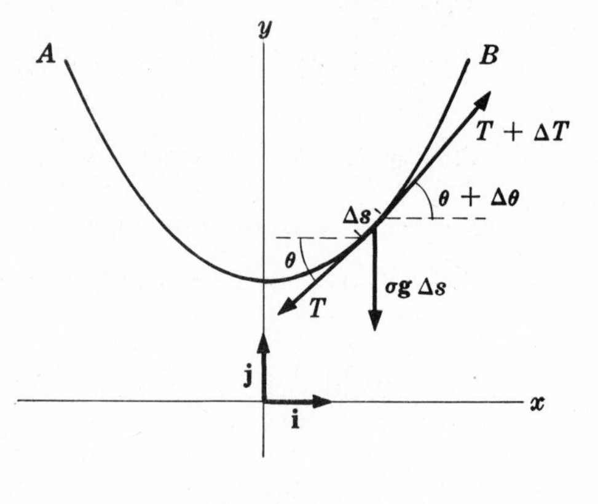

Fig. 7-18

Let A and B [Fig. 7-18] be the fixed points. An element of the chain of length Δs is in equilibrium under the tensions of magnitude T and $T + \Delta T$ due to the rest of the chain and also the weight $\sigma g\,\Delta s$ of the element of chain. Now from Fig. 7-18 if the directions of the vectors corresponding to T and $T + \Delta T$ make angles of θ and $\theta + \Delta\theta$ with the x axis respectively, we have as the condition for equilibrium [neglecting terms of order $(\Delta\theta)^2$ and higher],

$$(T + \Delta T)\cos(\theta + \Delta\theta)\mathbf{i} + (T + \Delta T)\sin(\theta + \Delta\theta)\mathbf{j} - (T\cos\theta\,\mathbf{i} + T\sin\theta\,\mathbf{j}) - \sigma g\mathbf{j}\,\Delta s = \mathbf{0}$$

or

$$(T + \Delta T)\cos(\theta + \Delta\theta) = T\cos\theta \tag{1}$$

$$(T + \Delta T)\sin(\theta + \Delta\theta) - T\sin\theta = \sigma g\,\Delta s \tag{2}$$

Equation (*1*) shows that the horizontal component $T\cos\theta$ must be a constant, which we shall take as T_0 which corresponds to the tension at the lowest point of the chain, where $\theta = 0$. Thus

$$T\cos\theta = T_0 \tag{3}$$

From (*2*) we find on dividing by $\Delta\theta$,

$$\frac{(T + \Delta T)\sin(\theta + \Delta\theta) - T\sin\theta}{\Delta\theta} = \sigma g\frac{\Delta s}{\Delta\theta} \tag{4}$$

Taking the limit of both sides of (*4*) as $\Delta\theta \to 0$, we find

$$\frac{d}{d\theta}(T\sin\theta) = \sigma g\frac{ds}{d\theta} \tag{5}$$

Using (*3*) to eliminate T, (*5*) becomes

$$\frac{d}{d\theta}(T_0\tan\theta) = \sigma g\frac{ds}{d\theta} \tag{6}$$

or

$$\frac{ds}{d\theta} = \frac{T_0}{\sigma g}\sec^2\theta = b\sec^2\theta \tag{7}$$

where $b = T_0/\sigma g$. Now

$$\frac{dx}{ds} = \cos\theta, \quad \frac{dy}{ds} = \sin\theta \tag{8}$$

Thus from (*7*) and (*8*),

$$\frac{dx}{d\theta} = \frac{dx}{ds}\frac{ds}{d\theta} = (\cos\theta)(b\sec^2\theta) = b\sec\theta \tag{9}$$

$$\frac{dy}{d\theta} = \frac{dy}{ds}\frac{ds}{d\theta} = (\sin\theta)(b\sec^2\theta) = b\sec\theta\tan\theta \tag{10}$$

Integrating (*9*) and (*10*) with respect to θ, we find

$$x = b \ln(\sec\theta + \tan\theta) + c_1 \tag{11}$$

$$y = b \sec\theta + c_2 \tag{12}$$

Let us assume that at the lowest point of the chain, i.e. at $\theta = 0$, $x = 0$ and $y = b$. Then from (*11*) and (*12*) we find $c_1 = 0$, $c_2 = 0$. Thus

$$x = b \ln(\sec\theta + \tan\theta) \tag{13}$$

$$y = b \sec\theta \tag{14}$$

From (*13*) we have

$$\sec\theta + \tan\theta = e^{x/b} \tag{15}$$

But

$$\sec^2\theta - \tan^2\theta = (\sec\theta + \tan\theta)(\sec\theta - \tan\theta) = 1 \tag{16}$$

Then dividing (*16*) by (*15*), we find

$$\sec\theta - \tan\theta = e^{-x/b} \tag{17}$$

Adding (*15*) to (*17*) and using (*14*), we find

$$y = \frac{b}{2}(e^{x/b} + e^{-x/b}) = b \cosh\frac{x}{b} \tag{18}$$

This curve is called a *catenary* [from the Latin, meaning chain].

Supplementary Problems

DEGREES OF FREEDOM

7.33. Determine the number of degrees of freedom in each of the following cases: (*a*) a particle moving on a plane curve; (*b*) two particles moving on a space curve and having constant distance between them; (*c*) three particles moving in space so that the distance between any two of them is always constant. *Ans.* (*a*) 1, (*b*) 1, (*c*) 6

7.34. Find the number of degrees of freedom for a rigid body which (*a*) moves parallel to a fixed plane, (*b*) has two points fixed but can otherwise move freely. *Ans.* (*a*) 3, (*b*) 1

7.35. Find the number of degrees of freedom for a system consisting of a thin rigid rod which can move freely in space and a particle which is constrained to move on the rod. *Ans.* 4

CENTER OF MASS AND MOMENTUM OF A SYSTEM OF PARTICLES

7.36. A quadrilateral $ABCD$ has masses 1, 2, 3 and 4 units located at its vertices $A(-1,-2,2)$, $B(3,2,-1)$, $C(1,-2,4)$ and $D(3,1,2)$. Find the coordinates of the center of mass. *Ans.* $(2,0,2)$

7.37. A system consists of two particles of masses m_1 and m_2. Prove that the center of mass of the system divides the line joining m_1 to m_2 into two segments whose lengths are in the ratio m_2 to m_1.

7.38. A bomb dropped from an airplane explodes in midair. Prove that if air resistance is neglected, then the center of mass describes a parabola.

7.39. Three particles of masses 2, 1, 3 respectively have position vectors $\mathbf{r}_1 = 5t\mathbf{i} - 2t^2\mathbf{j} + (3t-2)\mathbf{k}$, $\mathbf{r}_2 = (2t-3)\mathbf{i} + (12-5t^2)\mathbf{j} + (4+6t-3t^3)\mathbf{k}$, $\mathbf{r}_3 = (2t-1)\mathbf{i} + (t^2+2)\mathbf{j} - t^3\mathbf{k}$ where t is the time. Find (*a*) the velocity of the center of mass at time $t = 1$ and (*b*) the total linear momentum of the system at $t = 1$. *Ans.* (*a*) $3\mathbf{i} - 2\mathbf{j} - \mathbf{k}$, (*b*) $18\mathbf{i} - 12\mathbf{j} - 6\mathbf{k}$

7.40. Three equal masses are located at the vertices of a triangle. Prove that the center of mass is located at the intersection of the medians of the triangle.

7.41. A uniform plate has the shape of the region bounded by the parabola $y = x^2$ and the line $y = H$ in the xy plane. Find the center of mass. *Ans.* $\bar{x} = 0,\ \bar{y} = \frac{3}{5}H$

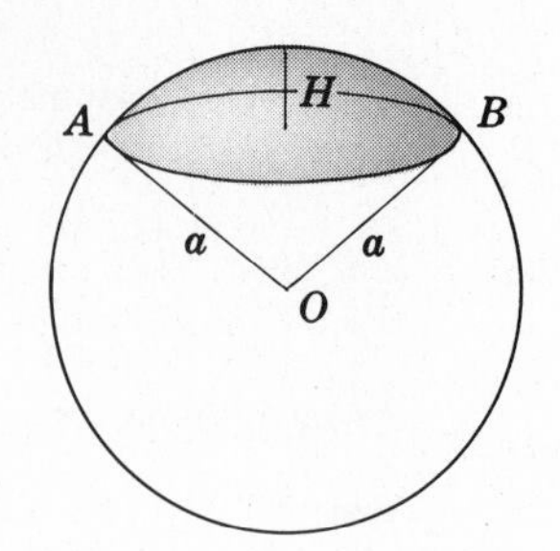

7.42. Find the center of mass of a uniform right circular cone of radius a and height H.
Ans. That point on the axis at distance $\frac{3}{4}H$ from the vertex.

7.43. The shaded region of Fig. 7-19 is a solid spherical cap of height H cut off from a uniform solid sphere of radius a. (*a*) Prove that the centroid of the cap is located at a distance $\frac{3}{4}(2a - H)^2/(3a - H)$ from the base AB. (*b*) Discuss the cases $H = 0$, $H = a$ and $H = 2a$.

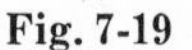
Fig. 7-19

7.44. Find the center of mass of a uniform plate bounded by $y = \sin x$ and the x axis. *Ans.* $\bar{x} = \pi/2,\ \bar{y} = \pi/8$

7.45. Find the center of mass of a rod of length l whose density is proportional to the distance from one end O.
Ans. $\frac{2}{3}l$ from end O

7.46. Find the centroid of a uniform solid bounded by the planes $4x + 2y + z = 8,\ x = 0,\ y = 0,\ z = 0.$
Ans. $\bar{\mathbf{r}} = \frac{1}{16}a(\mathbf{i} + 2\mathbf{j} + 4\mathbf{k})$

7.47. A uniform solid is bounded by the paraboloid of revolution $x^2 + y^2 = cz$ and the plane $z = H$ [see Fig. 7-20]. Find the centroid. *Ans.* $\bar{x} = 0,\ \bar{y} = 0,\ \bar{z} = \frac{2}{3}H$

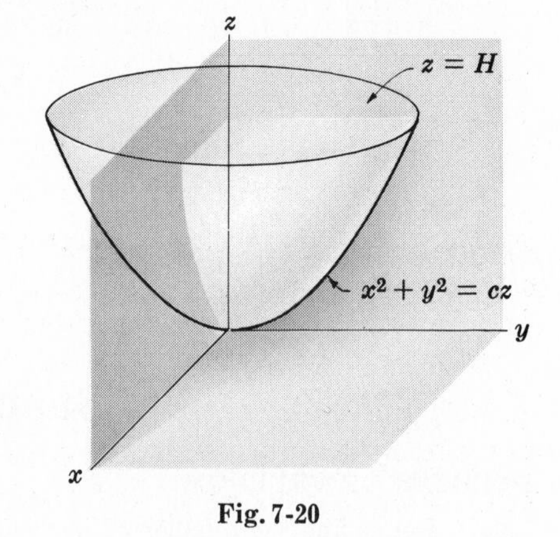

Fig. 7-20

ANGULAR MOMENTUM AND TORQUE

7.48. Three particles of masses 2, 3 and 5 move under the influence of a force field so that their position vectors relative to a fixed coordinate system are given respectively by $\mathbf{r}_1 = 2t\mathbf{i} - 3\mathbf{j} + t^2\mathbf{k}$, $\mathbf{r}_2 = (t+1)\mathbf{i} + 3t\mathbf{j} - 4\mathbf{k}$ and $\mathbf{r}_3 = t^2\mathbf{i} - t\mathbf{j} + (2t-1)\mathbf{k}$ where t is the time. Find (*a*) the total angular momentum of the system and (*b*) the total external torque applied to the system, taken with respect to the origin.
Ans. (*a*) $(31 - 12t)\mathbf{i} + (6t^2 - 10t - 12)\mathbf{j} + (21 + 5t^2)\mathbf{k}$
(*b*) $-12\mathbf{i} + (12t - 10)\mathbf{j} + 10t\mathbf{k}$

7.49. Work Problem 7.48 if the total angular momentum and torque are taken with respect to the center of mass.

7.50. Verify that in (*a*) Problem 7.48 and (*b*) Problem 7.49 the total external torque is equal to the time rate of change in angular momentum.

7.51. In Problem 7.48 find (*a*) the total angular momentum and (*b*) the total external torque taken about a point whose position vector is given by $\mathbf{r} = t\mathbf{i} - 2t\mathbf{j} + 3\mathbf{k}$. Does the total external torque equal the time rate of change in angular momentum in this case? Explain.

7.52. Verify Theorem 7.9, page 169, for the system of particles of Problem 7.48.

7.53. State and prove a theorem analogous to that of Theorem 7.9, page 169, for the total external torque applied to a system.

7.54. Is the angular momentum conserved in Problem 7.38? Explain.

WORK, ENERGY AND IMPULSE

7.55. Find the total work done by the force field of Problem 7.48 in moving the particles from their positions at time $t = 1$ to their positions at time $t = 2$. *Ans.* 42

7.56. Is the work of Problem 7.55 the same as that done on the center of mass assuming all mass to be concentrated there? Explain.

7.57. Find the total kinetic energy of the particles in Problem 7.48 at times (*a*) $t = 1$ and (*b*) $t = 2$. Discuss the connection between your results and the result of Problem 7.55.

Ans. (*a*) 72.5, (*b*) 30.5

7.58. Find the total linear momentum of the system of particles in Problem 7.48 at times (*a*) $t = 1$ and (*b*) $t = 2$. *Ans.* (*a*) $17\mathbf{i} + 4\mathbf{j} + 14\mathbf{k}$, (*b*) $27\mathbf{i} + 4\mathbf{j} + 18\mathbf{k}$

7.59. Find the total impulse applied to the system of Problem 7.48 from $t = 1$ to $t = 2$ and discuss the connection of your result with Problem 7.58. *Ans.* $10\mathbf{i} + 4\mathbf{k}$

7.60. Prove Theorem 7.13, page 170.

7.61. Verify Theorem 7.13, page 170, for the system of particles in Problem 7.48.

CONSTRAINTS, STATICS, VIRTUAL WORK, STABILITY AND D'ALEMBERT'S PRINCIPLE

7.62. In each case state whether the constraint is holonomic or non-holonomic and give a reason for your answer: (*a*) a particle constrained to move under gravity on the inside of a vertical paraboloid of revolution whose vertex is downward; (*b*) a particle sliding on an ellipsoid under the influence of gravity; (*c*) a sphere rolling and possibly sliding down an inclined plane; (*d*) a sphere rolling down an inclined plane parallel to a fixed vertical plane; (*e*) a particle sliding under gravity on the outside of an inverted vertical cone.

Ans. (*a*) holonomic, (*b*) non-holonomic, (*c*) non-holonomic, (*d*) holonomic, (*e*) holonomic

7.63. A lever ABC [Fig. 7-21] has weights W_1 and W_2 at distances a_1 and a_2 from the fixed support B. Using the principle of virtual work, prove that a necessary and sufficient condition for equilibrium is $W_1 a_1 = W_2 a_2$.

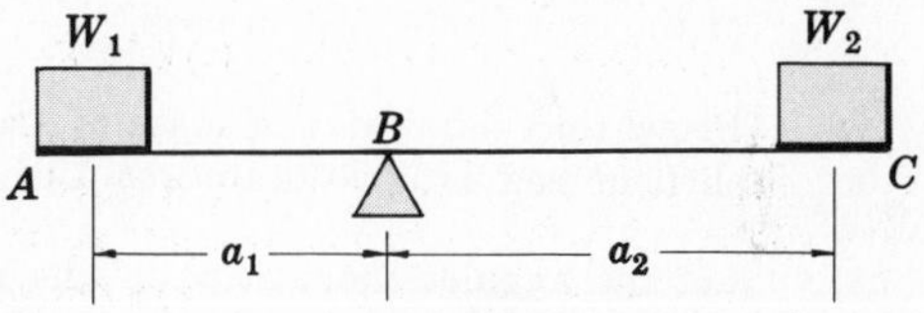

Fig. 7-21

7.64. Work Problem 7.63 if one or more additional weights are placed on the lever.

7.65. An inextensible string of negligible mass hanging over a smooth peg at B [see Fig. 7-22] connects one mass m_1 on a frictionless inclined plane of angle α to another mass m_2. Using D'Alembert's principle, prove that the masses will be in equilibrium if $m_2 = m_1 \sin \alpha$.

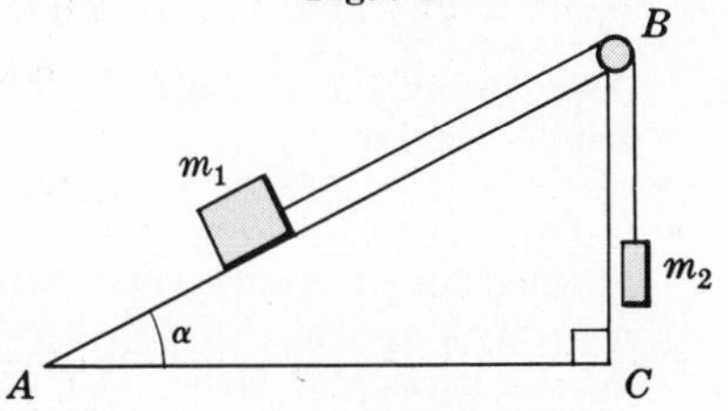

Fig. 7-22

7.66. Work Problem 7.65 if the incline has coefficient of friction μ. *Ans.* $m_2 = m_1(\sin \alpha - \mu \cos \alpha)$

7.67. A ladder AB of mass m has its ends on a smooth wall and floor [see Fig. 7-23]. The foot of the ladder is tied by an inextensible rope of negligible mass to the base C of the wall so that the ladder makes an angle α with the floor. Using the principle of virtual work, find the magnitude of the tension in the rope.

Ans. $\frac{1}{2}mg \cot \alpha$

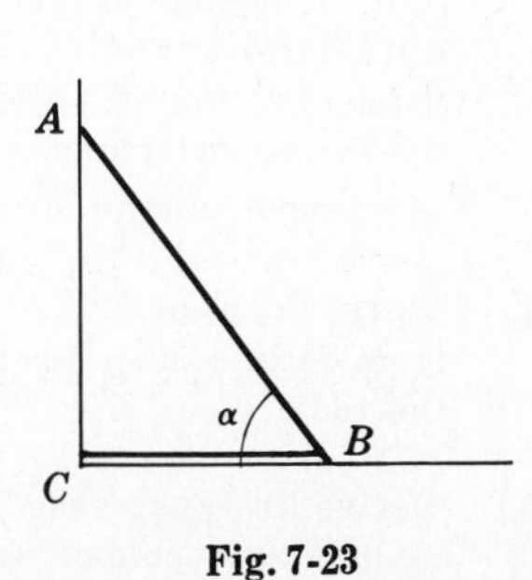

Fig. 7-23

7.68. Work (*a*) Problem 7.63 and (*b*) Problem 7.65 by using the potential energy method. Prove that the equilibrium in each case is unstable.

7.69. A thin uniform rod of length l has its two ends constrained to move on the circumference of a smooth vertical circle of radius a [see Fig. 7-24]. Determine conditions for equilibrium.

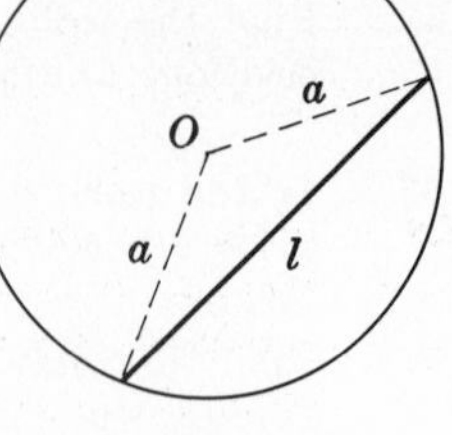

Fig. 7-24

7.70. Is the equilibrium of the rod of Problem 7.69 stable or not? Explain.

7.71. A solid hemisphere of radius a is located on a perfectly rough inclined plane of angle α.

(a) Prove that it is in stable equilibrium if $\alpha < \sin^{-1}(3/8)$.

(b) Are there any other values of α for which equilibrium can occur? Which of these, if any, yield stable equilibrium?

7.72. Use D'Alembert's principle to obtain the equations of motion of masses m_1 and m_2 of Problem 7.65.

7.73. Work Atwood's machine problem [see Problem 7.22, page 180] by using D'Alembert's principle.

7.74. Use D'Alembert's principle to determine the equations of motion of a simple pendulum.

MISCELLANEOUS PROBLEMS

7.75. Prove that the center of mass of a uniform circular arc of radius a and central angle α is located on the axis of symmetry at a distance from the center equal to $(a \sin \alpha)/\alpha$.

7.76. Discuss the cases (a) $\alpha = \pi/2$ and (b) $\alpha = \pi$ in Problem 7.75.

7.77. A circle of radius a is removed from a uniform circular plate of radius $b > a$, as indicated in Fig. 7-25. If the distance between their centers A and B is D, find the center of mass.
Ans. The point at distance $a^2D/(b^2 - a^2)$ below B.

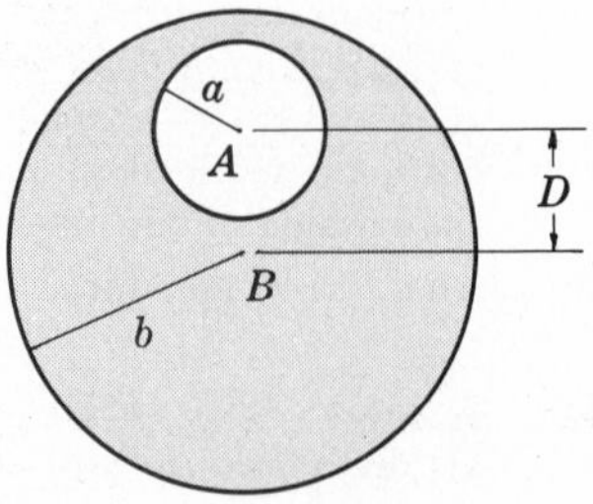

7.78. Work Problem 7.77 if the circles are replaced by spheres.
Ans. The point at distance $a^3D/(b^3 - a^3)$ below B.

7.79. Prove that the center of mass does not depend on the origin of the coordinate system used.

7.80. Prove that the center of mass of a uniform thin hemispherical shell of radius a is located at a distance $\frac{1}{2}a$ from the center.

Fig. 7-25

7.81. Let the angular momentum of the moon about the earth be $\mathbf{A}$. Find the angular momentum of a system consisting of only the earth and the moon about the center of mass. Assume the masses of the earth and moon to be given by M_e and M_m respectively. *Ans.* $M_e\mathbf{A}/(M_e + M_m)$

7.82. Does Theorem 7.13, page 170, apply in case the angular momentum is taken about an arbitrary point? Explain.

7.83. In Fig. 7-26, AD, BD and CD are uniform thin rods of equal length a and equal weight w. They are smoothly hinged at D and have ends A, B and C on a smooth horizontal plane. To prevent the motion of ends A, B and C, we use an inextensible string ABC of negligible mass which is in the form of an equilateral triangle. If a weight W is suspended from D so that the rods make equal angles α with the horizontal plane, prove that the magnitude of the tension in the string is $\frac{1}{9}\sqrt{3}\,(W + 3w) \cot \alpha$.

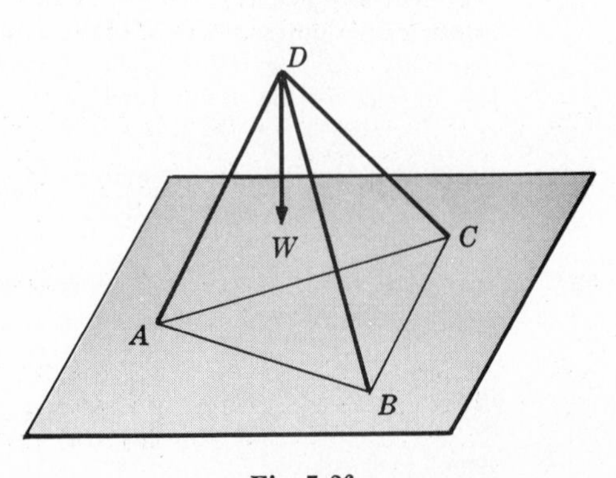

Fig. 7-26

7.84. Work Problem 7.83 if the weight W is removed from D and suspended from the center of one of the rods.

7.85. Derive an expression for (a) the total angular momentum and (b) the total torque of a system about an arbitrary point.

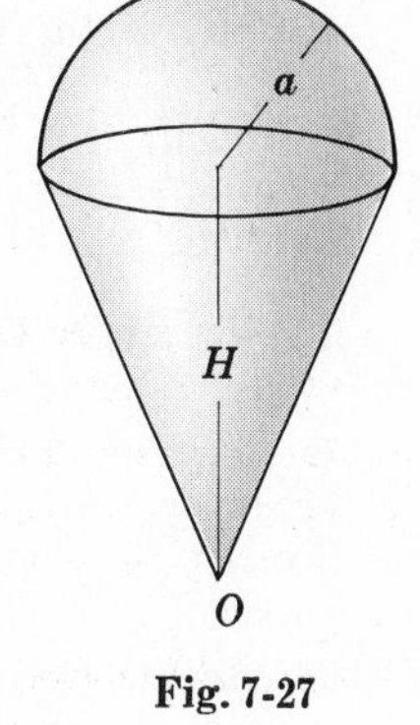

Fig. 7-27

7.86. Prove that the torque about any point P is equal to the time rate of change in angular momentum about P if and only if (a) P is fixed in space, (b) P coincides with the center of mass or (c) P is moving with a velocity which is in the same direction as the center of mass.

7.87. Find the centroid of a solid of constant density consisting of a right circular cone of radius a and height H surmounted by a hemisphere of radius a [see Fig. 7-27].
Ans. At height $\frac{3}{4}(a^2 + H^2)/(2a + H)$ above O.

7.88. Work Problem 7.87 if the density of the cone is twice the density of the hemisphere. *Ans.* At height $\frac{3}{8}(a^2 + 2H^2)/(a + H)$ above O.

7.89. A hemisphere of radius a is cut out of a uniform solid cube of side $b > 2a$ [see Fig. 7-28]. Find the center of mass of the remaining solid.

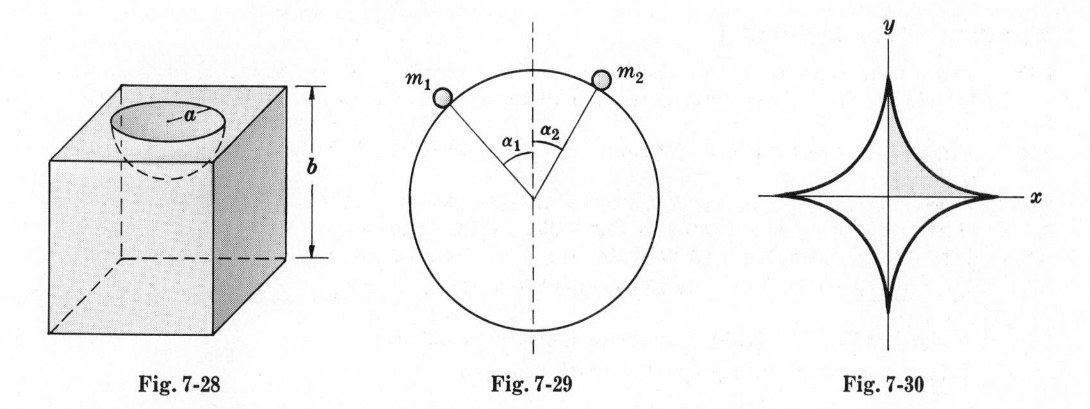

Fig. 7-28 Fig. 7-29 Fig. 7-30

7.90. A uniform chain of 45 kg wt is suspended from two fixed supports 15 meters apart. If the sag in the middle is 20 cm, find the tension at the supports. *Ans.* 450 kg wt

7.91. A chain of length L and constant density σ is suspended from two fixed points at the same horizontal level. If the sag of the chain at the middle is at a distance D below the horizontal line through the fixed points, prove that the tension at the lowest point of the chain is $\sigma(L^2 - 4D^2)/8D$.

7.92. Three particles of masses m_1, m_2, m_3 are located at the vertices of a triangle opposite sides having lengths a_1, a_2, a_3 respectively. Prove that the center of mass lies at the intersection of the angle bisectors of the triangle if and only if $m_1/a_1 = m_2/a_2 = m_3/a_3$.

7.93. Masses m_1 and m_2 are on a frictionless circular cylinder connected by an inextensible string of negligible mass [see Fig. 7-29]. (a) Using the principle of virtual work, prove that the system is in equilibrium if $m_1 \sin \alpha_1 = m_2 \sin \alpha_2$. (b) Is the equilibrium stable? Explain.

7.94. Work Problem 7.93 if friction is taken into account.

7.95. Derive an expression for the total kinetic energy of a system of particles relative to a point which may be moving in space. Under what conditions is the expression mathematically simplified? Discuss the physical significance of the simplification.

7.96. Find the center of mass of a uniform plate shown shaded in Fig. 7-30 which is bounded by the hypocycloid $x^{2/3} + y^{2/3} = a^{2/3}$ and the lines $x = 0$, $y = 0$. [*Hint.* Parametric equations for the hypocycloid are $x = a \cos^3 \theta$, $y = a \sin^3 \theta$.] *Ans.* $\bar{x} = \bar{y} = 256a/315\pi$

7.97. Let m_1, m_2, m_3 be the masses of three particles and $\mathbf{v}_{12}, \mathbf{v}_{23}, \mathbf{v}_{13}$ be their relative velocities.

(a) Prove that the total kinetic energy of the system about the center of mass is

$$\frac{m_1 m_2 v_{12}^2 + m_2 m_3 v_{23}^2 + m_1 m_3 v_{13}^2}{m_1 + m_2 + m_3}$$

(b) Generalize the result in (a).

7.98. A chain of variable density is suspended from two fixed points on the same horizontal level. Prove that if the density of the chain varies as the horizontal distance from a vertical line through its center, then the shape of the chain will be a parabola.

7.99. Discuss the relationship of Problem 7.98 with the shape of a suspension bridge.

7.100. A solid consists of a uniform right circular cone of vertex angle α on a uniform hemisphere of the same density, as indicated in Fig. 7-31. Prove that the solid can be in stable equilibrium on a horizontal plane if and only if $\alpha > 60°$.

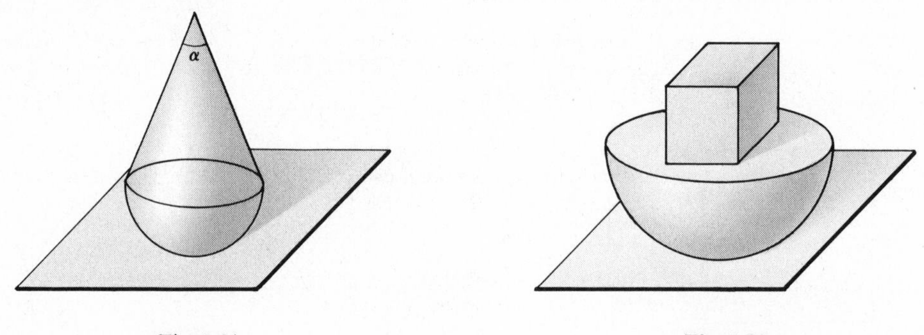

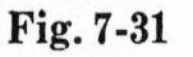
Fig. 7-31

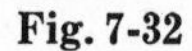
Fig. 7-32

7.101. A uniform solid [see Fig. 7-32] consists of a hemisphere of radius a surmounted by a cube of side b symmetrically placed about the center of the hemisphere. Find the condition on a and b for stable equilibrium. *Ans.* $a/b > \sqrt[4]{2/\pi}$

7.102. Find the centroid of the area bounded by the cycloid

$$x = a(\theta - \sin\theta), \qquad y = a(1 - \cos\theta)$$

and the x axis. *Ans.* $(\pi a, 5a/6)$

7.103. Prove that if the component of the torque about point P in any direction is zero, then the component of angular momentum about P in that direction is conserved if (a) P is a fixed point, (b) P coincides with the center of mass or (c) P is a point moving in the same direction as the center of mass.

7.104. In Problem 7.103, is the angular momentum conserved only if (a), (b) or (c) occurs? Explain.

7.105. Prove that the virtual work due to a force is equal to the sum of the virtual works which correspond to all components of the force.

7.106. Prove that it is impossible for one sphere to be in stable equilibrium on a fixed sphere which is perfectly rough [i.e. with coefficient of friction $\mu = 1$]. Is it possible for equilibrium to occur at all? Explain.

7.107. A uniform solid having the shape of the paraboloid of revolution $cz = x^2 + y^2$, $c > 0$ rests on the xy plane, assumed horizontal. Prove that if the height of the paraboloid is H, then the equilibrium is stable if and only if $H < \frac{3}{4}c$.

7.108. Work Problem 7.107 if the xy plane is inclined at an angle α with the horizontal.

7.109. In Fig. 7-33, AC and BC are frictionless wires in a vertical plane making angles of 60° and 30° respectively with the horizontal. Two beads of masses 3 gm and 6 gm are located on the wires, connected by a thin rod of negligible mass. Prove that the system will be in equilibrium when the rod makes an angle with the horizontal given by $\tan^{-1}(\frac{5}{9}\sqrt{3})$.

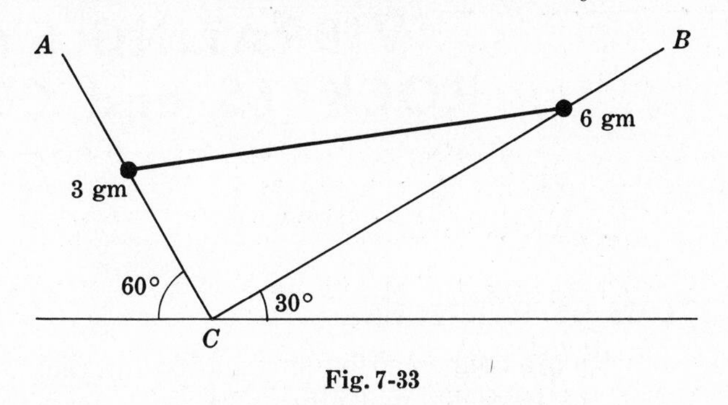

Fig. 7-33

7.110. Prove each of the following theorems due to *Pappus*.

(*a*) If a closed curve C in a plane is revolved about an axis in the plane which does not intersect it, then the volume generated is equal to the area bounded by C multiplied by the distance traveled by the centroid of the area.

(*b*) If an arc of a plane curve (closed or not) is revolved about an axis in the plane which does not intersect it, then the area of the surface generated is equal to the length of the arc multiplied by the distance traveled by the centroid of the arc.

7.111. Use Pappus' theorems to find (*a*) the centroid of a semicircular plate, (*b*) the centroid of a semicircular wire, (*c*) the centroid of a plate in the form of a right triangle, (*d*) the volume of a cylinder.

7.112. Find the (*a*) surface area and (*b*) volume of the doughnut shaped region obtained by revolving a circle of radius a about a line in its plane at a distance $b > a$ from its center.

Ans. (*a*) $4\pi^2 ab$, (*b*) $2\pi^2 a^2 b$

Chapter 8

APPLICATIONS to VIBRATING SYSTEMS, ROCKETS and COLLISIONS

VIBRATING SYSTEMS OF PARTICLES

If two or more particles are connected by springs [or interact with each other in some equivalent manner], then the particles will vibrate or oscillate with respect to each other.

As seen in Chapter 4, a vibrating or oscillating particle such as the simple harmonic oscillator or bob of a simple pendulum, has a single frequency of vibration. In the case of systems of particles, there is generally more than one frequency of vibration. Such frequencies are called *normal frequencies*. The motions of the particles in these cases are often called *multiply-periodic vibrations*.

A *mode of vibration* [i.e. a particular way in which vibration occurs, due to particular initial conditions for example] in which only one of the normal frequencies is present is called a *normal mode of vibration* or simply a *normal mode*. See Problems 8.1-8.3.

PROBLEMS INVOLVING CHANGING MASS. ROCKETS

Thus far we have restricted ourselves to motions of particles having constant mass. An important class of problems involves changing mass. An example is that of a rocket which moves forward by expelling particles of a fuel mixture backward. See Problems 8.4 and 8.5.

COLLISIONS OF PARTICLES

During the course of their motions two or more particles may collide with each other. Problems which consider the motions of such particles are called *collision* or *impact problems*.

In practice we think of colliding objects, such as spheres, as having *elasticity*. The time during which such objects are in contact is composed of a *compression time* during which slight deformation may take place, and *restitution time* during which the shape is restored. We assume that the spheres are smooth so that forces exerted are along the *common normal* to the spheres through the point of contact [and passing through their centers].

A collision can be *direct* or *oblique*. In a *direct collision* the direction of motion of both spheres is along the common normal at the point of contact both before and after collision. A collision which is not direct is called *oblique*.

Fundamental in collision problems is the following principle called *Newton's collision rule* which is based on experimental evidence. We shall take it as a postulate.

Newton's collision rule. Let $\mathbf{v}_{12}$ and $\mathbf{v}'_{12}$ be the relative velocities of the spheres along the common normal before and after impact. Then

$$\mathbf{v}'_{12} = -\epsilon \mathbf{v}_{12}$$

The quantity ϵ, called the *coefficient of restitution*, depends on the materials of which the objects are made and is generally taken as a constant between 0 and 1. If $\epsilon = 0$ the collision is called *perfectly inelastic* or briefly *inelastic*. If $\epsilon = 1$ the collision is called *perfectly elastic* or briefly *elastic*.

In the case of perfectly elastic collisions the total kinetic energy before and after impact is the same.

CONTINUOUS SYSTEMS OF PARTICLES

For some problems the number of particles per unit length, area or volume is so large that for all practical purposes the system can be considered as continuous. Examples are a vibrating violin string, a vibrating drumhead or membrane, or a sphere rolling down an inclined plane.

The basic laws of Chapter 7 hold for such continuous systems of particles. In applying them, however, it is necessary to use integration in place of summation over the number of particles and the concept of density.

THE VIBRATING STRING

Let us consider an elastic string such as a violin or piano string which is tightly stretched between the fixed points $x = 0$ and $x = l$ of the x axis [see Fig. 8-1]. If the string is given some initial displacement [such as, for example, by plucking it] and is then released, it will vibrate or oscillate about the equilibrium position.

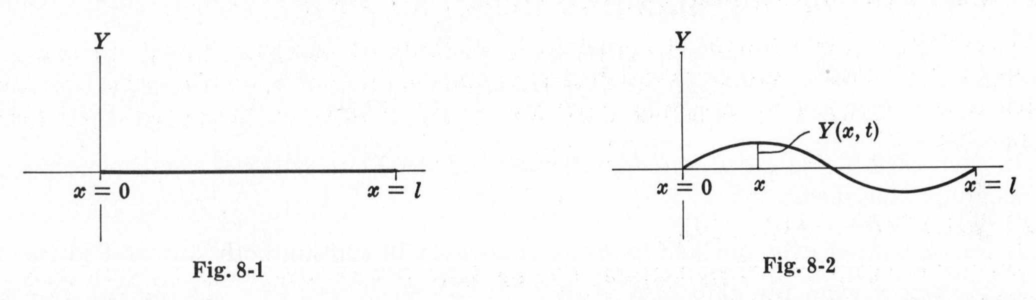

Fig. 8-1

Fig. 8-2

If we let $Y(x, t)$ denote the displacement of any point x of the string from the equilibrium position at time t [see Fig. 8-2], then the equation governing the vibrations is given by the *partial differential equation*

$$\frac{\partial^2 Y}{\partial t^2} = c^2 \frac{\partial^2 Y}{\partial x^2} \tag{1}$$

where if T is the (constant) tension throughout the string and σ is the (constant) density [mass per unit length of string],

$$c^2 = T/\sigma \tag{2}$$

The equation (*1*) holds in case the vibrations are assumed so small that the slope $\partial Y/\partial x$ at any point of the string is much less than one.

BOUNDARY-VALUE PROBLEMS

The problem of solving an equation such as (*1*) subject to various conditions, called boundary conditions, is often called a *boundary-value problem*. An important method for solving such problems makes use of *Fourier series*.

FOURIER SERIES

Under certain conditions [usually satisfied in practice and outlined below] a function $f(x)$, defined in the interval $\gamma < x < \gamma + 2l$ and having period $2l$ outside of this interval, has the series expansion

$$f(x) \quad = \quad \frac{a_0}{2} + \sum_{n=1}^{\infty} \left(a_n \cos \frac{n\pi x}{l} + b_n \sin \frac{n\pi x}{l} \right) \tag{3}$$

where the coefficients in the series, called *Fourier coefficients*, are given by

$$a_n \quad = \quad \frac{1}{l} \int_{\gamma}^{\gamma+2l} f(x) \cos \frac{n\pi x}{l}\, dx \tag{4}$$

$$b_n \quad = \quad \frac{1}{l} \int_{\gamma}^{\gamma+2l} f(x) \sin \frac{n\pi x}{l}\, dx \tag{5}$$

Such a series is called the *Fourier series* of $f(x)$. For many problems $\gamma = 0$ or $-l$.

ODD AND EVEN FUNCTIONS

If $\gamma = -l$, certain simplifications can occur in the coefficients (4) and (5) as indicated below:

1. If $f(-x) = f(x)$,

$$a_n \quad = \quad \frac{2}{l} \int_0^l f(x) \cos \frac{n\pi x}{l}\, dx, \qquad b_n = 0 \tag{6}$$

In such case $f(x)$ is called an *even function* and the Fourier series corresponding to $f(x)$ has only cosine terms.

2. If $f(-x) = -f(x)$,

$$a_n = 0, \qquad b_n \quad = \quad \frac{2}{l} \int_0^l f(x) \sin \frac{n\pi x}{l}\, dx \tag{7}$$

In such case $f(x)$ is called an *odd function* and the Fourier series corresponding to $f(x)$ has only sine terms.

If $f(x)$ is neither even nor odd its Fourier series will contain both sine and cosine terms.

Examples of even functions are x^4, $3x^6 + 4x^2 - 5$, $\cos x$, $e^x + e^{-x}$ and the function shown graphically in Fig. 8-3. Examples of odd functions are x^3, $2x^5 - 5x^3 + 4$, $\sin x$, $e^x - e^{-x}$ and the function shown graphically in Fig. 8-4.

Examples of functions which are neither even nor odd are $x^4 + x^3$, $x + \cos x$ and the function shown graphically in Fig. 8-5.

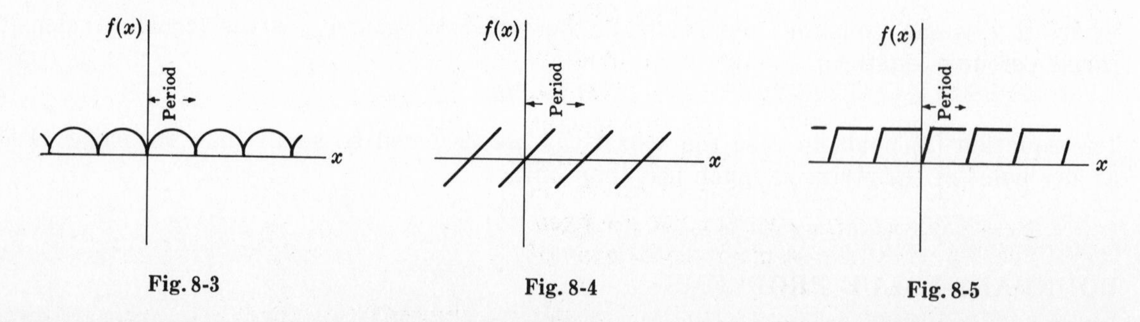

Fig. 8-3 Fig. 8-4 Fig. 8-5

If a function is defined in the "half period" $x = 0$ to $x = l$ and is specified as odd, then the function is known throughout the interval $-l < x < l$ and so the series which contains only sine terms can be found. This series is often called the *half range Fourier sine series*. Similarly, a function defined from $x = 0$ to $x = l$ which is specified as even has a series expansion called the *half range Fourier cosine* series.

CONVERGENCE OF FOURIER SERIES

Let us assume the following conditions on $f(x)$:

1. $f(x)$ is defined in $\gamma < x < \gamma + 2l$.

2. $f(x)$ and its derivative $f'(x)$ are piecewise continuous in $\gamma < x < \gamma + 2l$. [A function is said to be *piecewise continuous* in an interval if the interval can be divided into a finite number of subintervals in each of which the function is continuous and bounded, i.e. there is a constant $B > 0$ such that $-B < f(x) < B$. An example of such a function is indicated in Fig. 8-6.]

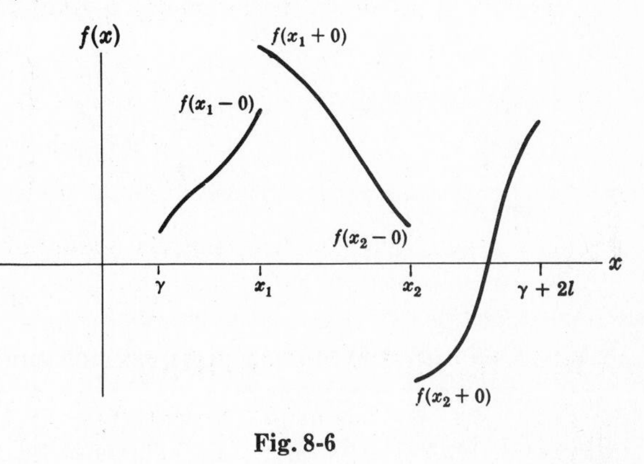

Fig. 8-6

3. At each point of discontinuity, for example, x_1 [or x_2] in Fig. 8-6, $f(x)$ has finite limits from the right and left denoted respectively by $f(x_1 + 0)$ and $f(x_1 - 0)$ [or $f(x_2 + 0)$, $f(x_2 - 0)$].

4. $f(x)$ has period $2l$, i.e. $f(x + 2l) = f(x)$.

These conditions if satisfied are *sufficient* to guarantee the validity of equation (*3*) [i.e. the series on the right side of (*3*) actually converges to $f(x)$] at each point where $f(x)$ is continuous. At each point where $f(x)$ is discontinuous, (*3*) is still valid if $f(x)$ is replaced by $\frac{1}{2}[f(x+0) + f(x-0)]$, i.e. the mean value of the right and left hand limits.

The conditions described above are known as *Dirichlet conditions.*

Solved Problems

VIBRATING SYSTEMS OF PARTICLES

8.1. Two equal masses m are connected by springs having equal spring constant κ, as shown in Fig. 8-7, so that the masses are free to slide on a frictionless table AB. The walls at A and B to which the ends of the springs are attached are fixed. Set up the differential equations of motion of the masses.

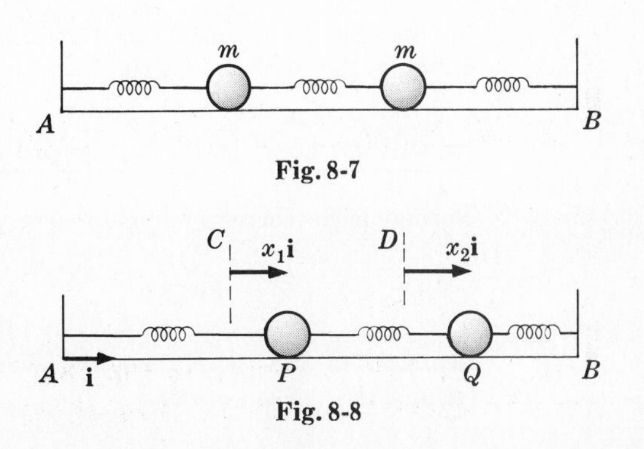

Fig. 8-7

Fig. 8-8

Let $x_1\mathbf{i}$ and $x_2\mathbf{i}$ [Fig. 8-8] denote the displacements of the masses from their equilibrium positions C and D at any time t.

Consider the forces acting on the first mass at P. There will be a force due to the spring on the right given by $\kappa(x_2\mathbf{i} - x_1\mathbf{i}) = \kappa(x_2 - x_1)\mathbf{i}$, and a force due to the spring on the left given by $-\kappa x_1\mathbf{i}$. Thus the net force acting on the first mass at P is

$$\kappa(x_2 - x_1)\mathbf{i} - \kappa x_1\mathbf{i}$$

In the same way the net force acting on the second mass at Q is

$$\kappa(x_1 - x_2)\mathbf{i} - \kappa x_2\mathbf{i}$$

Then by Newton's second law we have

$$m\frac{d^2}{dt^2}(x_1\mathbf{i}) = \kappa(x_2 - x_1)\mathbf{i} - \kappa x_1\mathbf{i}$$

$$m\frac{d^2}{dt^2}(x_2\mathbf{i}) = \kappa(x_1 - x_2)\mathbf{i} - \kappa x_2\mathbf{i}$$

or

$$m\ddot{x}_1 = \kappa(x_2 - 2x_1) \qquad (1)$$

$$m\ddot{x}_2 = \kappa(x_1 - 2x_2) \qquad (2)$$

8.2. Find (*a*) the normal frequencies and (*b*) the normal modes of vibration for the system in Problem 8.1.

(*a*) Let $x_1 = A_1 \cos \omega t$, $x_2 = A_2 \cos \omega t$ in equations (*1*) and (*2*) of Problem 8.1. Then we find, after simplifying,

$$(2\kappa - m\omega^2)A_1 - \kappa A_2 = 0 \qquad (1)$$

$$-\kappa A_1 + (2\kappa - m\omega^2)A_2 = 0 \qquad (2)$$

Now if A_1 and A_2 are not both zero, we must have

$$\begin{vmatrix} 2\kappa - m\omega^2 & -\kappa \\ -\kappa & 2\kappa - m\omega^2 \end{vmatrix} = 0 \qquad (3)$$

or $\quad (2\kappa - m\omega^2)(2\kappa - m\omega^2) - \kappa^2 = 0 \quad$ or $\quad m^2\omega^4 - 4\kappa m\omega^2 + 3\kappa^2 = 0$

Solving for ω^2, we find $\omega^2 = \dfrac{4\kappa m \pm \sqrt{16\kappa^2m^2 - 12\kappa^2m^2}}{2m^2}$ giving

$$\omega^2 = \kappa/m \quad \text{and} \quad \omega^2 = 3\kappa/m \qquad (4)$$

Then the normal (or natural) frequencies of the system are given by

$$f = \frac{1}{2\pi}\sqrt{\frac{\kappa}{m}} \quad \text{and} \quad f = \frac{1}{2\pi}\sqrt{\frac{3\kappa}{m}} \qquad (5)$$

The normal frequencies are also called *characteristic frequencies* and the determinant (*3*) is called the *characteristic determinant* or *secular determinant.*

(*b*) To find the normal mode corresponding to $\omega = \sqrt{\kappa/m}$, let $\omega^2 = \kappa/m$ in equations (*1*) and (*2*). Then we find

$$A_1 = A_2$$

In this case the normal mode of vibration corresponds to the motion of the masses in the *same* direction [i.e. both to the right and both to the left] as indicated in Fig. 8-9.

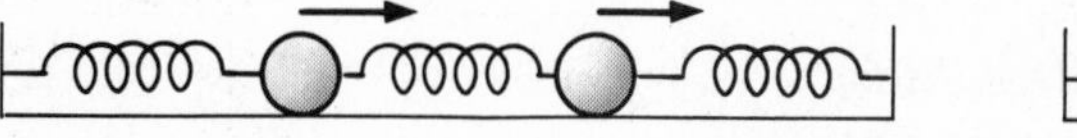

Normal mode corresponding to $\omega = \sqrt{\kappa/m}$

Fig. 8-9

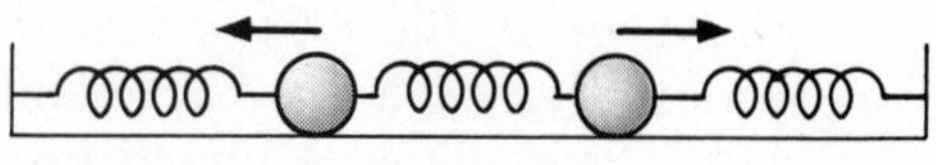

Normal mode corresponding to $\omega = \sqrt{3\kappa/m}$

Fig. 8-10

Similarly to find the normal mode corresponding to $\omega = \sqrt{3\kappa/m}$, let $\omega^2 = 3\kappa/m$ in equations (*1*) and (*2*). Then we find

$$A_1 = -A_2$$

In this case the normal mode of vibration corresponds to the motion of the masses in *opposite* directions [i.e. when one moves to the right the other moves to the left, and vice versa] as indicated in Fig. 8-10 above.

In working this problem we could also just as well have assumed, $x_1 = B_1 \sin \omega t$, $x_2 = B_2 \sin \omega t$ or $x_1 = A_1 \cos \omega t + B_1 \sin \omega t$, $x_2 = A_2 \cos \omega t + B_2 \sin \omega t$ or $x_1 = C_1 e^{i\omega t}$, $x_2 = C_2 e^{i\omega t}$.

8.3. Suppose that in Problem 8.1 the first mass is held at its equilibrium position while the second mass is given a displacement of magnitude $a > 0$ to the right of its equilibrium position. The masses are then released. Find the position of each mass at any later time.

Writing $\omega_1 = \sqrt{\kappa/m}$ and $\omega_2 = \sqrt{3\kappa/m}$, the general motion of both masses is described by

$$x_1 = C_1 \cos \omega_1 t + C_2 \sin \omega_1 t + C_3 \cos \omega_2 t + C_4 \sin \omega_2 t \qquad (1)$$

$$x_2 = D_1 \cos \omega_1 t + D_2 \sin \omega_1 t + D_3 \cos \omega_2 t + D_4 \sin \omega_2 t \qquad (2)$$

where the coefficients are all constants. Substituting these into equation (*1*) or equation (*2*) of Problem 8.1 [both give the same results], we find on equating corresponding coefficients of $\cos \omega_1 t$, $\sin \omega_1 t$, $\cos \omega_2 t$, $\sin \omega_2 t$ respectively,

$$D_1 = C_1, \quad D_2 = C_2, \quad D_3 = -C_3, \quad D_4 = -C_4$$

Thus equations (*1*) and (*2*) can be written

$$x_1 = C_1 \cos \omega_1 t + C_2 \sin \omega_1 t + C_3 \cos \omega_2 t + C_4 \sin \omega_2 t \qquad (3)$$

$$x_2 = C_1 \cos \omega_1 t + C_2 \sin \omega_1 t - C_3 \cos \omega_2 t - C_4 \sin \omega_2 t \qquad (4)$$

We now determine C_1, C_2, C_3, C_4 subject to the initial conditions

$$x_1 = 0, \quad x_2 = a, \quad \dot{x}_1 = 0, \quad \dot{x}_2 = 0 \qquad \text{at } t = 0 \qquad (5)$$

From these conditions we find respectively

$$C_1 + C_3 = 0, \quad C_1 - C_3 = a, \quad C_2\omega_1 + C_4\omega_2 = 0, \quad C_2\omega_1 - C_4\omega_2 = 0$$

From these we find $\qquad C_1 = \frac{1}{2}a, \quad C_2 = 0, \quad C_3 = -\frac{1}{2}a, \; C_4 = 0 \qquad (6)$

Thus equations (*3*) and (*4*) give the required equations

$$x_1 = \tfrac{1}{2}a(\cos \omega_1 t - \cos \omega_2 t) \qquad (7)$$

$$x_2 = \tfrac{1}{2}a(\cos \omega_1 t + \cos \omega_2 t) \qquad (8)$$

where $\omega_1 = \sqrt{\kappa/m}$, $\omega_2 = \sqrt{3\kappa/m}$.

Note that in the motion described by (*7*) and (*8*), both normal frequencies are present. These equations show that the general motion is a *superposition of the normal modes*. This is sometimes called the *superposition principle*.

CHANGING MASS. ROCKETS

8.4. Derive an equation for the motion of a rocket moving in a straight line.

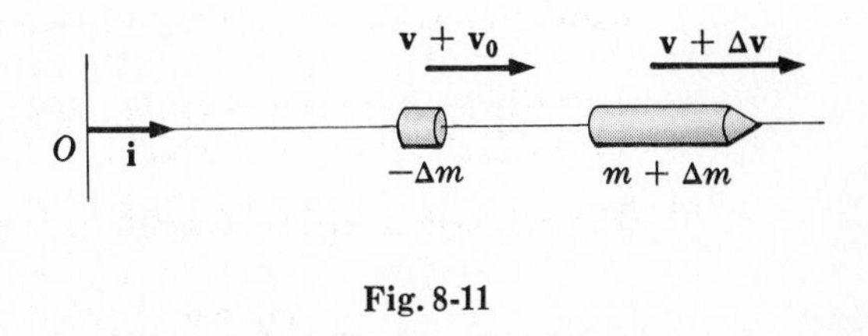

Fig. 8-11

Let m be the total mass of the rocket at time t. At a later time $t + \Delta t$ suppose that the mass is $m + \Delta m$ due to expulsion of a mass $-\Delta m$ of gas through the back of the rocket. Note that $-\Delta m$ is actually a positive quantity since Δm is assumed negative.

Let $\mathbf{v}$ and $\mathbf{v} + \Delta\mathbf{v}$ be the velocities of the rocket at times t and $t + \Delta t$ relative to an inertial system with origin at O. The velocity of the mass of gas ejected from the back of the rocket relative to O is $\mathbf{v} + \mathbf{v}_0$ where $-\mathbf{v}_0$ is the velocity of the gas relative to the rocket.

Since the change in momentum of the system is equal to the impulse, we have

$$\text{Total momentum at } t + \Delta t \; - \; \text{total momentum at } t \quad = \quad \text{impulse}$$

$$\{(m + \Delta m)(\mathbf{v} + \Delta\mathbf{v}) + (-\Delta m)(\mathbf{v} + \mathbf{v}_0)\} - m\mathbf{v} \quad = \quad \mathbf{F}\,\Delta t \tag{1}$$

where $\mathbf{F}$ is the net external force acting on the rocket.

Equation (1) can be written as

$$m\frac{\Delta\mathbf{v}}{\Delta t} - \mathbf{v}_0\frac{\Delta m}{\Delta t} + \frac{\Delta\mathbf{v}}{\Delta t}\Delta m \quad = \quad \mathbf{F}$$

Then taking the limit as $\Delta t \to 0$, we find

$$m\frac{d\mathbf{v}}{dt} - \mathbf{v}_0\frac{dm}{dt} \quad = \quad \mathbf{F} \tag{2}$$

Writing $\mathbf{v} = v\mathbf{i}$, $\mathbf{v}_0 = -v_0\mathbf{i}$, $\mathbf{F} = F\mathbf{i}$, this becomes

$$m\frac{dv}{dt} + v_0\frac{dm}{dt} \quad = \quad F \tag{3}$$

8.5. **Find the velocity of the rocket of Problem 8.4 assuming that gas is ejected at a constant rate and at constant velocity with respect to it and that it moves vertically upward in a constant gravitational field.**

If the gas is ejected at constant rate $\alpha > 0$, then $m = m_0 - \alpha t$ where m_0 is the mass of the rocket at $t = 0$. Since $\mathbf{F} = -mg\mathbf{i}$ (or $F = -mg$) and $dm/dt = -\alpha$, equation (3) of Problem 8.4 can be written

$$(m_0 - \alpha t)\frac{dv}{dt} - \alpha v_0 \quad = \quad -(m_0 - \alpha t)g \qquad \text{or} \qquad \frac{dv}{dt} \quad = \quad -g + \frac{\alpha v_0}{m_0 - \alpha t} \tag{1}$$

Integrating, we find

$$v \quad = \quad -gt - v_0 \ln(m_0 - \alpha t) + c_1 \tag{2}$$

If $v = 0$ at $t = 0$, i.e. if the rocket starts from rest, then

$$0 \quad = \quad 0 - v_0 \ln m_0 + c_1 \qquad \text{or} \qquad c_1 \quad = \quad v_0 \ln m_0$$

Thus (2) becomes

$$v \quad = \quad -gt + v_0 \ln\left(\frac{m_0}{m_0 - \alpha t}\right)$$

which is the speed at any time. The velocity is $\mathbf{v} = v\mathbf{i}$.

Note that we must have $m_0 - \alpha t > 0$, otherwise there will be no gas expelled from the rocket, in which case the rocket will be out of fuel.

COLLISIONS OF PARTICLES

8.6. **Two masses m_1 and m_2 traveling in the same straight line collide. Find the velocities of the particles after collision in terms of the velocities before collision.**

Assume that the straight line is taken to be the x axis and that the velocities of the particles before and after collisions are $\mathbf{v}_1, \mathbf{v}_2$ and $\mathbf{v}_1', \mathbf{v}_2'$ respectively.

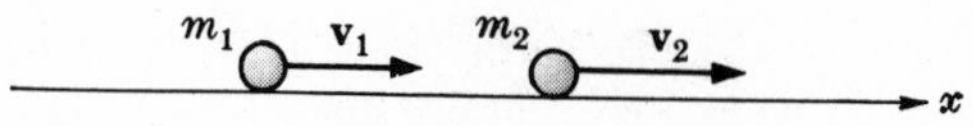

By Newton's collision rule, page 194,

$$\mathbf{v}_1' - \mathbf{v}_2' \quad = \quad \epsilon(\mathbf{v}_2 - \mathbf{v}_1) \tag{1}$$

Fig. 8-12

By the principle of conservation of momentum,

$$\text{Total momentum after collision} \quad = \quad \text{total momentum before collision}$$

$$m_1\mathbf{v}_1' + m_2\mathbf{v}_2' \quad = \quad m_1\mathbf{v}_1 + m_2\mathbf{v}_2 \tag{2}$$

Solving (1) and (2) simultaneously,

$$\mathbf{v}_1' = \frac{(m_1 - \epsilon m_2)\mathbf{v}_1 + m_2(1+\epsilon)\mathbf{v}_2}{m_1 + m_2} \tag{3}$$

$$\mathbf{v}_2' = \frac{m_1(1+\epsilon)\mathbf{v}_1 + (m_2 - \epsilon m_1)\mathbf{v}_2}{m_1 + m_2} \tag{4}$$

8.7. Discuss Problem 8.6 for the case of (a) a perfectly inelastic collision, (b) a perfectly elastic collision.

(a) Here we put $\epsilon = 0$ in (3) and (4) of Problem 8.6 to obtain

$$\mathbf{v}_1' = \frac{m_1\mathbf{v}_1 + m_2\mathbf{v}_2}{m_1 + m_2}, \qquad \mathbf{v}_2' = \frac{m_1\mathbf{v}_1 + m_2\mathbf{v}_2}{m_1 + m_2}$$

Thus after collision the two particles move with the same velocity, i.e. they move as if they were stuck together as a single particle.

(b) Here we put $\epsilon = 1$ in (3) and (4) of Problem 8.6 to obtain

$$\mathbf{v}_1' = \frac{(m_1 - m_2)\mathbf{v}_1 + 2m_2\mathbf{v}_2}{m_1 + m_2}, \qquad \mathbf{v}_2' = \frac{2m_1\mathbf{v}_1 + (m_2 - m_1)\mathbf{v}_2}{m_1 + m_2}$$

These velocities are not the same.

8.8. Show that for a perfectly elastic collision of the particles of Problem 8.6 the total kinetic energy before collision equals the total kinetic energy after collision.

Using the result of Problem 8.7(b), we have

$$\text{Total kinetic energy after collision} = \frac{1}{2}m_1\mathbf{v}_1'^2 + \frac{1}{2}m_2\mathbf{v}_2'^2$$

$$= \frac{1}{2}m_1\left\{\frac{(m_1 - m_2)\mathbf{v}_1 + 2m_2\mathbf{v}_2}{m_1 + m_2}\right\}^2 + \frac{1}{2}m_2\left\{\frac{2m_1\mathbf{v}_1 + (m_2 - m_1)\mathbf{v}_2}{m_1 + m_2}\right\}^2$$

$$= \frac{1}{2}m_1v_1^2 + \frac{1}{2}mv_2^2$$

$$= \text{total kinetic energy before collision}$$

8.9. Two spheres of masses m_1 and m_2 respectively, collide obliquely. Find their velocities after impact in terms of their velocities before impact.

Let $\mathbf{v}_1, \mathbf{v}_2$ and $\mathbf{v}_1', \mathbf{v}_2'$ be the velocities of the spheres before and after impact respectively, as indicated in Fig. 8-13. Choose a coordinate system so that the xy plane is the plane of $\mathbf{v}_1$ and $\mathbf{v}_2$, and so that at the instant of impact the x axis passes through the centers C_1 and C_2 of the spheres.

By the conservation of momentum, we have

$$m_1\mathbf{v}_1 + m_2\mathbf{v}_2 = m_1\mathbf{v}_1' + m_2\mathbf{v}_2' \tag{1}$$

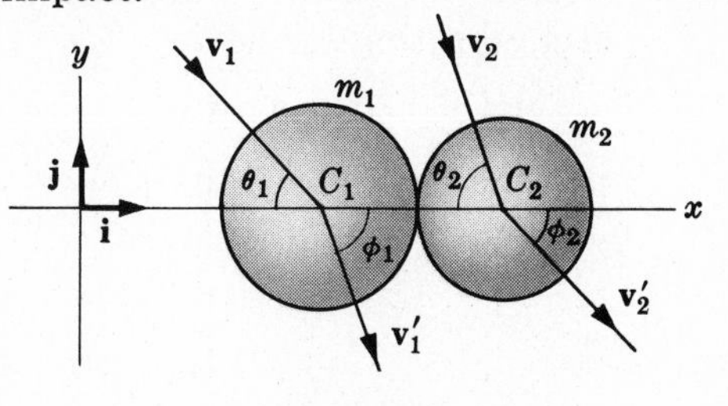

Fig. 8-13

From Fig. 8-13 we see that

$$\mathbf{v}_1 = v_1(\cos\theta_1\,\mathbf{i} - \sin\theta_1\,\mathbf{j}) \tag{2}$$

$$\mathbf{v}_2 = v_2(\cos\theta_2\,\mathbf{i} - \sin\theta_2\,\mathbf{j}) \tag{3}$$

$$\mathbf{v}_1' = v_1'(\cos\phi_1\,\mathbf{i} - \sin\phi_1\,\mathbf{j}) \tag{4}$$

$$\mathbf{v}_2' = v_2'(\cos\phi_2\,\mathbf{i} - \sin\phi_2\,\mathbf{j}) \tag{5}$$

Substituting equations (2)-(5) in (1) and equating coefficients of $\mathbf{i}$ and $\mathbf{j}$, we have

$$m_1 v_1 \cos\theta_1 + m_2 v_2 \cos\theta_2 = m_1 v_1' \cos\phi_1 + m_2 v_2' \cos\phi_2 \quad (6)$$

$$m_1 v_1 \sin\theta_1 + m_2 v_2 \sin\theta_2 = m_1 v_1' \sin\phi_1 + m_2 v_2' \sin\phi_2 \quad (7)$$

By Newton's collision rule, we have

Relative velocity after impact along x axis

$= -\epsilon\{$relative velocity before impact along x axis$\}$

or

$$\mathbf{v}_1' \cdot \mathbf{i} - \mathbf{v}_2' \cdot \mathbf{i} = -\epsilon(\mathbf{v}_1 \cdot \mathbf{i} - \mathbf{v}_2 \cdot \mathbf{i}) \quad (8)$$

which on using equations (2)-(5) becomes

$$v_1' \cos\phi_1 - v_2' \cos\phi_2 = -\epsilon(v_1 \cos\theta_1 - v_2 \cos\theta_2) \quad (9)$$

Furthermore, since the tangential velocities before and after impact are equal,

$$\mathbf{v}_1 \cdot \mathbf{j} = \mathbf{v}_1' \cdot \mathbf{j} \quad (10)$$

$$\mathbf{v}_2 \cdot \mathbf{j} = \mathbf{v}_2' \cdot \mathbf{j} \quad (11)$$

or

$$v_1 \sin\theta_1 = v_1' \sin\phi_1 \quad (12)$$

$$v_2 \sin\theta_2 = v_2' \sin\phi_2 \quad (13)$$

Equation (7) is automatically satisfied by using equations (12) and (13).

From equations (6) and (9) we find

$$v_1' \cos\phi_1 = \frac{(m_1 - m_2\epsilon)v_1 \cos\theta_1 + m_2(1+\epsilon)v_2 \cos\theta_2}{m_1 + m_2}$$

$$v_2' \cos\phi_2 = \frac{m_1(1+\epsilon)v_1 \cos\theta_1 + (m_2 - m_1\epsilon)v_2 \cos\theta_2}{m_1 + m_2}$$

Then using (12) and (13) we find

$$\begin{aligned}\mathbf{v}_1' &= v_1'(\cos\phi_1\,\mathbf{i} - \sin\phi_1\,\mathbf{j}) \\ &= \frac{(m_1 - m_2\epsilon)v_1 \cos\theta_1\,\mathbf{i} + m_2(1+\epsilon)v_2 \cos\theta_2\,\mathbf{i}}{m_1 + m_2} - v_1 \sin\theta_1\,\mathbf{j}\end{aligned}$$

$$\begin{aligned}\mathbf{v}_2' &= v_2'(\cos\phi_2\,\mathbf{i} - \sin\phi_2\,\mathbf{j}) \\ &= \frac{m_1(1+\epsilon)v_1 \cos\theta_1\,\mathbf{i} + (m_2 - m_1\epsilon)v_2 \cos\theta_2\,\mathbf{i}}{m_1 + m_2} - v_2 \sin\theta_2\,\mathbf{j}\end{aligned}$$

CONTINUOUS SYSTEMS OF PARTICLES

8.10. Derive the partial differential equation (1), page 195, for the transverse vibrations of a vibrating string.

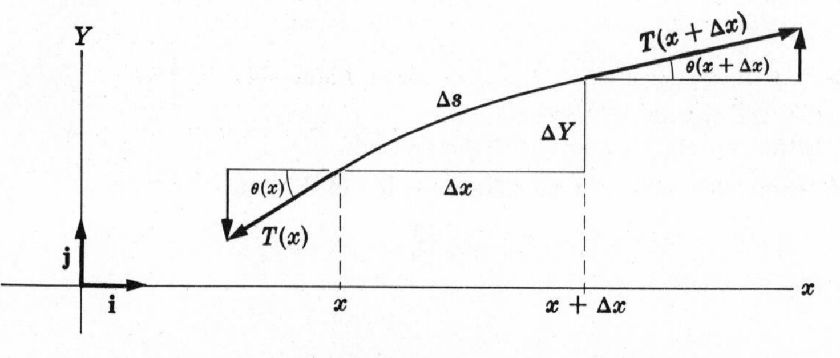

Fig. 8-14

Let us consider the motion of an element of the string of length Δs, greatly magnified in Fig. 8-14.

The forces acting on the element due to the remainder of the string are given by the tensions, as shown in Fig. 8-14, of magnitude $T(x)$ and $T(x + \Delta x)$ at the ends x and $x + \Delta x$ of the element.

The net horizontal force in direction **i** acting on the element is

$$[T(x+\Delta x)\cos\theta(x+\Delta x) - T(x)\cos\theta(x)]\mathbf{i} \tag{1}$$

The net vertical force in direction **j** acting on the element is

$$[T(x+\Delta x)\sin\theta(x+\Delta x) - T(x)\sin\theta(x)]\mathbf{j} \tag{2}$$

If we assume that the horizontal motion in direction **i** is negligible, the net force *(1)* is zero. Using the fact that the acceleration of the element is $\partial^2 Y/\partial t^2$ approximately and that its mass is $\sigma\,\Delta s$ where σ is the mass per unit length, we have from *(2)* and Newton's second law,

$$\sigma\,\Delta s\,\frac{\partial^2 Y}{\partial t^2}\mathbf{j} = [T(x+\Delta x)\sin\theta(x+\Delta x) - T(x)\sin\theta(x)]\mathbf{j} \tag{3}$$

or, dividing by $\Delta x\,\mathbf{j}$,

$$\sigma\frac{\Delta s}{\Delta x}\frac{\partial^2 Y}{\partial t^2} = \frac{T(x+\Delta x)\sin\theta(x+\Delta x) - T(x)\sin\theta(x)}{\Delta x} \tag{4}$$

or

$$\sigma\sqrt{1+\left(\frac{\Delta Y}{\Delta x}\right)^2}\,\frac{\partial^2 Y}{\partial t^2} = \frac{T(x+\Delta x)\sin\theta(x+\Delta x) - T(x)\sin\theta(x)}{\Delta x}$$

Taking the limit as $\Delta x \to 0$, this becomes

$$\sigma\sqrt{1+\left(\frac{\partial Y}{\partial x}\right)^2}\,\frac{\partial^2 Y}{\partial t^2} = \frac{\partial}{\partial x}\{T\sin\theta\} \tag{5}$$

Since

$$\sin\theta = \frac{\tan\theta}{\sqrt{1+\tan^2\theta}} = \frac{\partial Y/\partial x}{\sqrt{1+(\partial Y/\partial x)^2}}$$

equation *(5)* can be written

$$\sigma\sqrt{1+\left(\frac{\partial Y}{\partial x}\right)^2}\,\frac{\partial^2 Y}{\partial t^2} = \frac{\partial}{\partial x}\left\{\frac{T\,\partial Y/\partial x}{\sqrt{1+(\partial Y/\partial x)^2}}\right\} \tag{6}$$

To simplify this equation we make the assumption that vibrations are small so that the slope $\partial Y/\partial x$ is small in absolute value compared with 1. Then we can neglect $(\partial Y/\partial x)^2$ compared with 1 and *(6)* becomes

$$\sigma\frac{\partial^2 Y}{\partial t^2} = \frac{\partial}{\partial x}\left(T\frac{\partial Y}{\partial x}\right) \tag{7}$$

If we further assume that the tension T is constant throughout the string and that σ is also constant, *(7)* becomes

$$\frac{\partial^2 Y}{\partial t^2} = c^2\frac{\partial^2 Y}{\partial x^2} \tag{8}$$

where $c^2 = T/\sigma$. Unless otherwise specified, when we deal with the vibrating string we shall refer to equation *(8)*.

8.11. Derive the equation of Problem 8.10 if the string is horizontal and gravity is taken into account.

In this case we must add to the right hand side of equation *(3)* of Problem 8.10 the force on the element due to gravity

$$-m\mathbf{g} = -\sigma\,\Delta s\,g\mathbf{j}$$

The effect of this is to replace equation *(8)* of Problem 8.10 by

$$\frac{\partial^2 Y}{\partial t^2} = c^2\frac{\partial^2 Y}{\partial x^2} - g$$

FOURIER SERIES

8.12. Graph each of the following functions.

(a) $f(x) = \begin{cases} 3 & 0 < x < 5 \\ -3 & -5 < x < 0 \end{cases}$ Period = 10

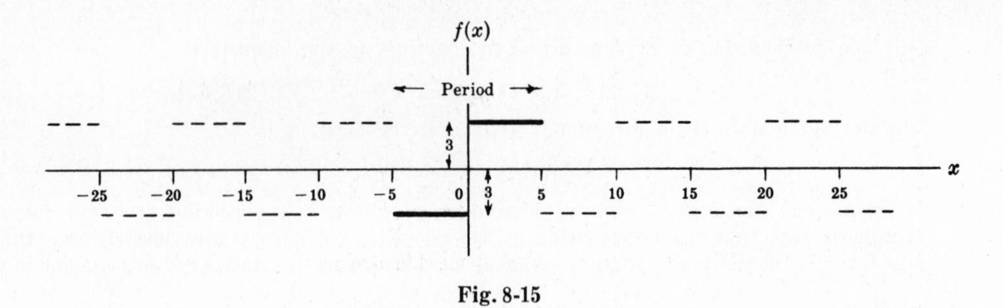

Fig. 8-15

Since the period is 10, that portion of the graph in $-5 < x < 5$ (indicated heavy in Fig. 8-15 above) is extended periodically outside this range (indicated dashed). Note that $f(x)$ is not defined at $x = 0, 5, -5, 10, -10, 15, -15$, etc. These values are the *discontinuities* of $f(x)$.

(b) $f(x) = \begin{cases} \sin x & 0 \leqq x \leqq \pi \\ 0 & \pi < x < 2\pi \end{cases}$ Period $= 2\pi$

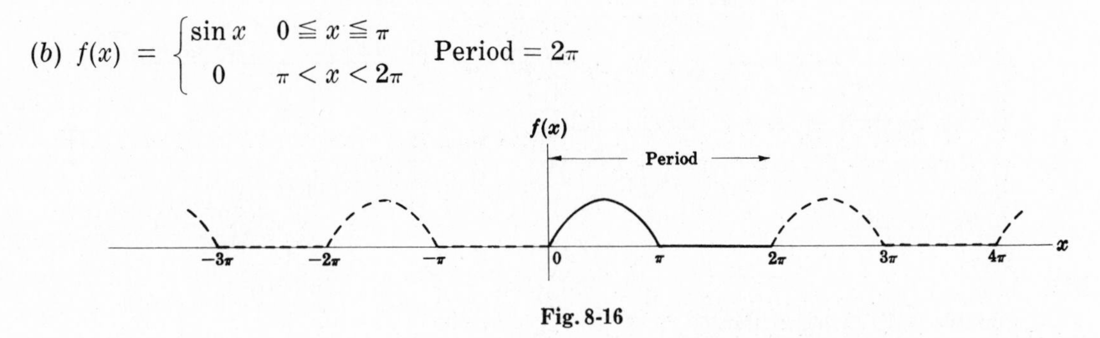

Fig. 8-16

Refer to Fig. 8-16 above. Note that $f(x)$ is defined for all x and is continuous everywhere.

8.13. Prove $\displaystyle\int_{-l}^{l} \sin\frac{k\pi x}{l}\,dx = \int_{-l}^{l} \cos\frac{k\pi x}{l}\,dx = 0$ if $k = 1, 2, 3, \ldots$.

$$\int_{-l}^{l} \sin\frac{k\pi x}{l}\,dx = -\frac{l}{k\pi}\cos\frac{k\pi x}{l}\bigg|_{-l}^{l} = -\frac{l}{k\pi}\cos k\pi + \frac{l}{k\pi}\cos(-k\pi) = 0$$

$$\int_{-l}^{l} \cos\frac{k\pi x}{l}\,dx = \frac{l}{k\pi}\sin\frac{k\pi x}{l}\bigg|_{-l}^{l} = \frac{l}{k\pi}\sin k\pi - \frac{l}{k\pi}\sin(-k\pi) = 0$$

8.14. Prove (a) $\displaystyle\int_{-l}^{l} \cos\frac{m\pi x}{l}\cos\frac{n\pi x}{l}\,dx = \int_{-l}^{l} \sin\frac{m\pi x}{l}\sin\frac{n\pi x}{l}\,dx = \begin{cases} 0 & m \neq n \\ l & m = n \end{cases}$

(b) $\displaystyle\int_{-l}^{l} \sin\frac{m\pi x}{l}\cos\frac{n\pi x}{l}\,dx = 0$

where m and n can assume any of the values $1, 2, 3, \ldots$.

(a) From trigonometry: $\cos A \cos B = \frac{1}{2}\{\cos(A - B) + \cos(A + B)\}$, $\sin A \sin B = \frac{1}{2}\{\cos(A - B) - \cos(A + B)\}$.

Then, if $m \neq n$, by Problem 8.13,

$$\int_{-l}^{l} \cos\frac{m\pi x}{l}\cos\frac{n\pi x}{l}\,dx = \frac{1}{2}\int_{-l}^{l}\left\{\cos\frac{(m-n)\pi x}{l} + \cos\frac{(m+n)\pi x}{l}\right\}dx = 0$$

Similarly if $m \neq n$,

$$\int_{-l}^{l} \sin\frac{m\pi x}{l}\sin\frac{n\pi x}{l}\,dx = \frac{1}{2}\int_{-l}^{l}\left\{\cos\frac{(m-n)\pi x}{l} - \cos\frac{(m+n)\pi x}{l}\right\}dx = 0$$

If $m = n$, we have

$$\int_{-l}^{l} \cos\frac{m\pi x}{l}\cos\frac{n\pi x}{l}\,dx = \frac{1}{2}\int_{-l}^{l}\left(1 + \cos\frac{2n\pi x}{l}\right)dx = l$$

$$\int_{-l}^{l} \sin\frac{m\pi x}{l}\sin\frac{n\pi x}{l}\,dx = \frac{1}{2}\int_{-l}^{l}\left(1 - \cos\frac{2n\pi x}{l}\right)dx = l$$

Note that if $m = n = 0$ these integrals are equal to $2l$ and 0 respectively.

(b) We have $\sin A \cos B = \frac{1}{2}\{\sin(A-B) + \sin(A+B)\}$. Then by Problem 8.13, if $m \neq n$,

$$\int_{-l}^{l} \sin\frac{m\pi x}{l}\cos\frac{n\pi x}{l}\,dx = \frac{1}{2}\int_{-l}^{l}\left\{\sin\frac{(m-n)\pi x}{l} + \sin\frac{(m+n)\pi x}{l}\right\}dx = 0$$

If $m = n$,
$$\int_{-l}^{l} \sin\frac{m\pi x}{l}\cos\frac{n\pi x}{l}\,dx = \frac{1}{2}\int_{-l}^{l} \sin\frac{2n\pi x}{l}\,dx = 0$$

The results of parts (a) and (b) remain valid even when the limits of integration $-l, l$ are replaced by $\gamma, \gamma + 2l$ respectively.

8.15. If

$$f(x) = A + \sum_{n=1}^{\infty}\left(a_n\cos\frac{n\pi x}{l} + b_n\sin\frac{n\pi x}{l}\right)$$

prove that by making suitable assumptions concerning term by term integration of infinite series, that for $n = 1, 2, 3, \ldots$,

(a) $a_n = \dfrac{1}{l}\displaystyle\int_{-l}^{l} f(x)\cos\frac{n\pi x}{l}\,dx$, (b) $b_n = \dfrac{1}{l}\displaystyle\int_{-l}^{l} f(x)\sin\frac{n\pi x}{l}\,dx$, (c) $A = \dfrac{a_0}{2}$.

(a) Multiplying

$$f(x) = A + \sum_{n=1}^{\infty}\left(a_n\cos\frac{n\pi x}{l} + b_n\sin\frac{n\pi x}{l}\right) \tag{1}$$

by $\cos\frac{m\pi x}{l}$ and integrating from $-l$ to l, using Problem 8.14, we have

$$\begin{aligned}\int_{-l}^{l} f(x)\cos\frac{m\pi x}{l}\,dx &= A\int_{-l}^{l}\cos\frac{m\pi x}{l}\,dx \qquad (2)\\ &\quad + \sum_{n=1}^{\infty}\left\{a_n\int_{-l}^{l}\cos\frac{m\pi x}{l}\cos\frac{n\pi x}{l}\,dx + b_n\int_{-l}^{l}\cos\frac{m\pi x}{l}\sin\frac{n\pi x}{l}\,dx\right\}\\ &= a_m l \qquad \text{if } m \neq 0\end{aligned}$$

Thus
$$a_m = \frac{1}{l}\int_{-l}^{l} f(x)\cos\frac{m\pi x}{l}\,dx \qquad \text{if } m = 1, 2, 3, \ldots$$

(b) Multiplying (1) by $\sin\frac{m\pi x}{l}$ and integrating from $-l$ to l, using Problem 8.14, we have

$$\begin{aligned}\int_{-l}^{l} f(x)\sin\frac{m\pi x}{l}\,dx &= A\int_{-l}^{l}\sin\frac{m\pi x}{l}\,dx \qquad (3)\\ &\quad + \sum_{n=1}^{\infty}\left\{a_n\int_{-l}^{l}\sin\frac{m\pi x}{l}\cos\frac{n\pi x}{l}\,dx + b_n\int_{-l}^{l}\sin\frac{m\pi x}{l}\sin\frac{n\pi x}{l}\,dx\right\}\\ &= b_m l\end{aligned}$$

Thus $$b_m = \frac{1}{l}\int_{-l}^{l} f(x)\sin\frac{m\pi x}{l}\,dx \qquad \text{if } m = 1,2,3,\ldots$$

(c) Integration of (1) from $-l$ to l, using Problem 8.13, gives

$$\int_{-l}^{l} f(x)\,dx = 2Al \qquad \text{or} \qquad A = \frac{1}{2l}\int_{-l}^{l} f(x)\,dx$$

Putting $m = 0$ in the result of part (a), we find $a_0 = \frac{1}{l}\int_{-l}^{l} f(x)\,dx$ and so $A = \frac{a_0}{2}$.

The above results also hold when the integration limits $-l, l$ are replaced by $\gamma, \gamma + 2l$.

Note that in all parts above we have assumed interchange of summation and integration. Even when this assumption is not warranted, the coefficients a_m and b_m as obtained above are called *Fourier coefficients* corresponding to $f(x)$, and the corresponding series with these values of a_m and b_m is called the *Fourier series* corresponding to $f(x)$. An important problem in this case is to investigate conditions under which this series actually converges to $f(x)$. Sufficient conditions for this convergence are the *Dirichlet conditions* given on page 197.

8.16. (a) Find the Fourier coefficients corresponding to the function

$$f(x) = \begin{cases} 0 & -5 < x < 0 \\ 3 & 0 < x < 5 \end{cases} \qquad \text{Period} = 10$$

(b) Write the corresponding Fourier series.

(c) How should $f(x)$ be defined at $x = -5$, $x = 0$ and $x = 5$ in order that the Fourier series will converge to $f(x)$ for $-5 \leqq x \leqq 5$?

The graph of $f(x)$ is shown in Fig. 8-17 below.

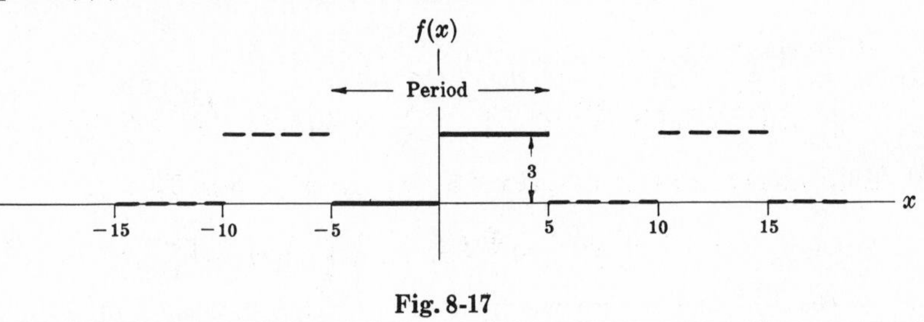

Fig. 8-17

(a) Period $= 2l = 10$ and $l = 5$. Choose the interval γ to $\gamma + 2l$ as -5 to 5, so that $\gamma = -5$. Then

$$a_n = \frac{1}{l}\int_{\gamma}^{\gamma+2l} f(x)\cos\frac{n\pi x}{l}\,dx = \frac{1}{5}\int_{-5}^{5} f(x)\cos\frac{n\pi x}{5}\,dx$$

$$= \frac{1}{5}\left\{\int_{-5}^{0} (0)\cos\frac{n\pi x}{5}\,dx + \int_{0}^{5} (3)\cos\frac{n\pi x}{5}\,dx\right\} = \frac{3}{5}\int_{0}^{5}\cos\frac{n\pi x}{5}\,dx$$

$$= \frac{3}{5}\left(\frac{5}{n\pi}\sin\frac{n\pi x}{5}\right)\Bigg|_{0}^{5} = 0 \qquad \text{if } n \neq 0$$

If $n = 0$, $$a_n = a_0 = \frac{3}{5}\int_{0}^{5}\cos\frac{0\pi x}{5}\,dx = \frac{3}{5}\int_{0}^{5} dx = 3.$$

$$b_n = \frac{1}{l}\int_{\gamma}^{\gamma+2l} f(x)\sin\frac{n\pi x}{l}\,dx = \frac{1}{5}\int_{-5}^{5} f(x)\sin\frac{n\pi x}{5}\,dx$$

$$= \frac{1}{5}\left\{\int_{-5}^{0} (0)\sin\frac{n\pi x}{5}\,dx + \int_{0}^{5} (3)\sin\frac{n\pi x}{5}\,dx\right\} = \frac{3}{5}\int_{0}^{5}\sin\frac{n\pi x}{5}\,dx$$

$$= \frac{3}{5}\left(-\frac{5}{n\pi}\cos\frac{n\pi x}{5}\right)\Bigg|_{0}^{5} = \frac{3(1-\cos n\pi)}{n\pi}$$

(*b*) The corresponding Fourier series is

$$\frac{a_0}{2} + \sum_{n=1}^{\infty}\left(a_n \cos\frac{n\pi x}{l} + b_n \sin\frac{n\pi x}{l}\right) = \frac{3}{2} + \sum_{n=1}^{\infty}\frac{3(1-\cos n\pi)}{n\pi}\sin\frac{n\pi x}{5}$$

$$= \frac{3}{2} + \frac{6}{\pi}\left(\sin\frac{\pi x}{5} + \frac{1}{3}\sin\frac{3\pi x}{5} + \frac{1}{5}\sin\frac{5\pi x}{5} + \cdots\right)$$

(*c*) Since $f(x)$ satisfies the Dirichlet conditions, we can say that the series converges to $f(x)$ at all points of continuity and to $\dfrac{f(x+0)+f(x-0)}{2}$ at points of discontinuity. At $x = -5, 0$ and 5, which are points of discontinuity, the series converges to $(3+0)/2 = 3/2$ as seen from the graph. If we redefine $f(x)$ as follows,

$$f(x) = \begin{cases} 3/2 & x = -5 \\ 0 & -5 < x < 0 \\ 3/2 & x = 0 \\ 3 & 0 < x < 5 \\ 3/2 & x = 5 \end{cases} \qquad \text{Period} = 10$$

then the series will converge to $f(x)$ for $-5 \leqq x \leqq 5$.

8.17. If $f(x)$ is even, show that (*a*) $a_n = \dfrac{2}{l}\displaystyle\int_0^l f(x)\cos\frac{n\pi x}{l}\,dx$, (*b*) $b_n = 0$.

(*a*)
$$a_n = \frac{1}{l}\int_{-l}^{l} f(x)\cos\frac{n\pi x}{l}\,dx = \frac{1}{l}\int_{-l}^{0} f(x)\cos\frac{n\pi x}{l}\,dx + \frac{1}{l}\int_0^l f(x)\cos\frac{n\pi x}{l}\,dx$$

Letting $x = -u$,

$$\frac{1}{l}\int_{-l}^{0} f(x)\cos\frac{n\pi x}{l}\,dx = \frac{1}{l}\int_0^l f(-u)\cos\left(\frac{-n\pi u}{l}\right)du = \frac{1}{l}\int_0^l f(u)\cos\frac{n\pi u}{l}\,du$$

since by definition of an even function $f(-u) = f(u)$. Then

$$a_n = \frac{1}{l}\int_0^l f(u)\cos\frac{n\pi u}{l}\,du + \frac{1}{l}\int_0^l f(x)\cos\frac{n\pi x}{l}\,dx = \frac{2}{l}\int_0^l f(x)\cos\frac{n\pi x}{l}\,dx$$

(*b*)
$$b_n = \frac{1}{l}\int_{-l}^{l} f(x)\sin\frac{n\pi x}{l}\,dx = \frac{1}{l}\int_{-l}^{0} f(x)\sin\frac{n\pi x}{l}\,dx + \frac{1}{l}\int_0^l f(x)\sin\frac{n\pi x}{l}\,dx \qquad (1)$$

If we make the transformation $x = -u$ in the first integral on the right of (*1*), we obtain

$$\frac{1}{l}\int_{-l}^{0} f(x)\sin\frac{n\pi x}{l}\,dx = \frac{1}{l}\int_0^l f(-u)\sin\left(\frac{-n\pi u}{l}\right)du = -\frac{1}{l}\int_0^l f(-u)\sin\frac{n\pi u}{l}\,du$$

$$= -\frac{1}{l}\int_0^l f(u)\sin\frac{n\pi u}{l}\,du = -\frac{1}{l}\int_0^l f(x)\sin\frac{n\pi x}{l}\,dx \qquad (2)$$

where we have used the fact that for an even function $f(-u) = f(u)$ and in the last step that the dummy variable of integration u can be replaced by any other symbol, in particular x. Thus from (*1*), using (*2*), we have

$$b_n = -\frac{1}{l}\int_0^l f(x)\sin\frac{n\pi x}{l}\,dx + \frac{1}{l}\int_0^l f(x)\sin\frac{n\pi x}{l}\,dx = 0$$

8.18. Expand $f(x) = x$, $0 < x < 2$, in a half range (*a*) sine series, (*b*) cosine series.

(*a*) Extend the definition of the given function to that of the odd function of period 4 shown in Fig. 8-18 below. This is sometimes called the *odd extension* of $f(x)$. Then $2l = 4, l = 2$.

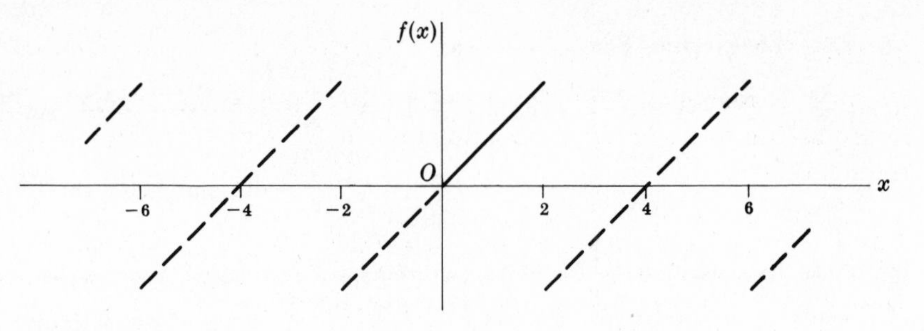

Fig. 8-18

Thus $a_n = 0$ and

$$b_n = \frac{2}{l}\int_0^l f(x)\sin\frac{n\pi x}{l}\,dx = \frac{2}{2}\int_0^2 x\sin\frac{n\pi x}{2}\,dx$$

$$= \left\{(x)\left(\frac{-2}{n\pi}\cos\frac{n\pi x}{2}\right) - (1)\left(\frac{-4}{n^2\pi^2}\sin\frac{n\pi x}{2}\right)\right\}\Bigg|_0^2 = \frac{-4}{n\pi}\cos n\pi$$

Then

$$f(x) = \sum_{n=1}^{\infty}\frac{-4}{n\pi}\cos n\pi\,\sin\frac{n\pi x}{2}$$

$$= \frac{4}{\pi}\left(\sin\frac{\pi x}{2} - \frac{1}{2}\sin\frac{2\pi x}{2} + \frac{1}{3}\sin\frac{3\pi x}{2} - \cdots\right)$$

(*b*) Extend the definition of $f(x)$ to that of the even function of period 4 shown in Fig. 8-19 below. This is the *even extension* of $f(x)$. Then $2l = 4$, $l = 2$.

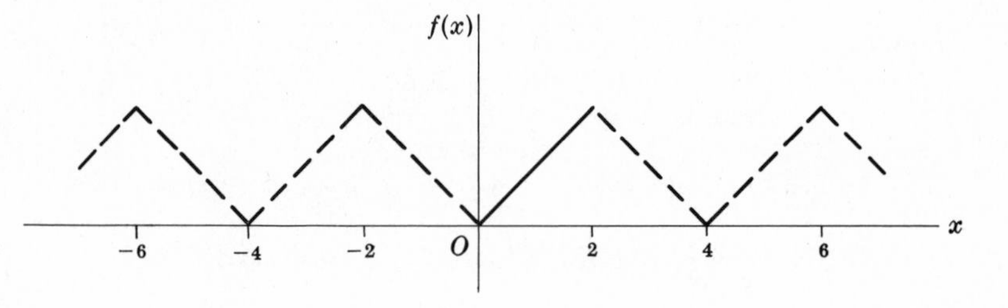

Fig. 8-19

Thus $b_n = 0$,

$$a_n = \frac{2}{l}\int_0^l f(x)\cos\frac{n\pi x}{l}\,dx = \frac{2}{2}\int_0^2 x\cos\frac{n\pi x}{2}\,dx$$

$$= \left\{(x)\left(\frac{2}{n\pi}\sin\frac{n\pi x}{2}\right) - (1)\left(\frac{-4}{n^2\pi^2}\cos\frac{n\pi x}{2}\right)\right\}\Bigg|_0^2$$

$$= \frac{4}{n^2\pi^2}(\cos n\pi - 1) \qquad \text{if } n \neq 0$$

If $n = 0$, $a_0 = \int_0^2 x\,dx = 2$.

Then

$$f(x) = 1 + \sum_{n=1}^{\infty}\frac{4}{n^2\pi^2}(\cos n\pi - 1)\cos\frac{n\pi x}{2}$$

$$= 1 - \frac{8}{\pi^2}\left(\cos\frac{\pi x}{2} + \frac{1}{3^2}\cos\frac{3\pi x}{2} + \frac{1}{5^2}\cos\frac{5\pi x}{2} + \cdots\right)$$

It should be noted that the given function $f(x) = x$, $0 < x < 2$, is represented *equally well* by the two *different* series in (*a*) and (*b*).

SOLUTIONS OF VIBRATING STRING PROBLEMS

8.19. Find the transverse displacement of a vibrating string of length l with fixed endpoints if the string is initially given a displacement $f(x)$ from its equilibrium position and then released.

Let the transverse displacement of any point x of the string at time t be $Y(x,t)$. Since the ends $x=0$ and $x=l$ of the string are fixed, we must have $Y(0,t)=0$ and $Y(l,t)=0$. Since the initial displacement is $f(x)$, we have $Y(x,0)=f(x)$; and since the initial velocity is zero, we have $Y_t(x,0)=0$ where Y_t denotes the partial derivative with respect to t. We must thus solve the boundary-value problem

$$\frac{\partial^2 Y}{\partial t^2} = c^2 \frac{\partial^2 Y}{\partial x^2} \tag{1}$$

$$Y(0,t)=0, \quad Y(l,t)=0, \quad Y(x,0)=f(x), \quad Y_t(x,0)=0 \tag{2}$$

Assume a solution to (1) of the form $Y = XT$ where X depends only on x and T depends only on t. Then substituting into (1), using X'' to denote d^2X/dx^2 and T'' to denote d^2T/dt^2, we have

$$X T'' = c^2 X'' T$$

or

$$\frac{X''}{X} = \frac{T''}{c^2 T} \tag{3}$$

Since one side depends only on x and the other side depends on t while x and t are independent, the only way in which (2) can be valid is if each side is a constant, which we shall take as $-\lambda^2$. Thus

$$\frac{X''}{X} = \frac{T''}{c^2 T} = -\lambda^2$$

or

$$X'' + \lambda^2 X = 0, \quad T'' + \lambda^2 c^2 t = 0$$

These equations have solutions

$$X = A_1 \cos \lambda x + B_1 \sin \lambda x, \qquad T = A_2 \cos \lambda ct + B_2 \sin \lambda ct$$

Thus a solution is given by

$$Y(x,t) = XT = (A_1 \cos \lambda x + B_1 \sin \lambda x)(A_2 \cos \lambda ct + B_2 \sin \lambda ct) \tag{4}$$

From the first condition in (2), we have

$$B_1(A_2 \cos \lambda ct + B_2 \sin \lambda ct) = 0$$

so that $A_1 = 0$ [since if the second factor is zero then the solution is identically zero, which we do not want]. Thus

$$Y(x,t) = B_1 \sin \lambda x\,(A_2 \cos \lambda ct + B_2 \sin \lambda ct)$$

$$= \sin \lambda x\,(b \cos \lambda ct + a \sin \lambda ct) \tag{5}$$

on writing $B_1 A_2 = b$, $B_1 B_2 = a$.

Using the second condition of (2) in (5), we see that $\sin \lambda l = 0$ or $\lambda l = n\pi$ where $n = 1, 2, 3, \ldots$. Thus $\lambda = n\pi/l$ and the solution so far is

$$Y(x,t) = \sin \frac{n\pi x}{l}\left(b \cos \frac{n\pi ct}{l} + a \sin \frac{n\pi ct}{l}\right) \tag{6}$$

By differentiating with respect to t, this becomes

$$Y_t(x,t) = \sin \frac{n\pi x}{l}\left(-\frac{n\pi cb}{l} \sin \frac{n\pi ct}{l} + \frac{n\pi ca}{l} \cos \frac{n\pi ct}{l}\right)$$

so that the fourth condition in (2) gives

$$Y_t(x,0) = \sin \frac{n\pi x}{l}\left(\frac{n\pi ca}{l}\right) = 0$$

from which $a = 0$. Thus (6) becomes

$$Y(x,t) = b \sin \frac{n\pi x}{l} \cos \frac{n\pi ct}{l} \tag{7}$$

To satisfy the third condition of (2) we use the fact that solutions of (7) multiplied by constants as well as sums of solutions are also solutions [the *superposition theorem* or *principle* for linear differential equations]. Thus we arrive at the solution

$$Y(x,t) \quad = \quad \sum_{n=1}^{\infty} b_n \sin\frac{n\pi x}{l} \cos\frac{n\pi ct}{l} \tag{8}$$

Using the third condition of (2) in (8) we must have

$$Y(x,0) \quad = \quad f(x) \quad = \quad \sum_{n=1}^{\infty} b_n \sin\frac{n\pi x}{l} \tag{9}$$

But this is simply the expansion of $f(x)$ in a Fourier sine series and the coefficients are given by

$$b_n \quad = \quad \frac{2}{l}\int_0^l f(x)\sin\frac{n\pi x}{l}\,dx$$

Thus the solution is given by

$$Y(x,t) \quad = \quad \sum_{n=1}^{\infty}\left\{\frac{2}{l}\int_0^l f(x)\sin\frac{n\pi x}{l}\,dx\right\}\sin\frac{n\pi x}{l}\cos\frac{n\pi ct}{l} \tag{10}$$

The method of solution assuming $Y = XT$ is often called the method of *separation of variables.*

8.20. A string with fixed ends is picked up at its center a distance H from the equilibrium position and released. Find the displacement at any position at any time.

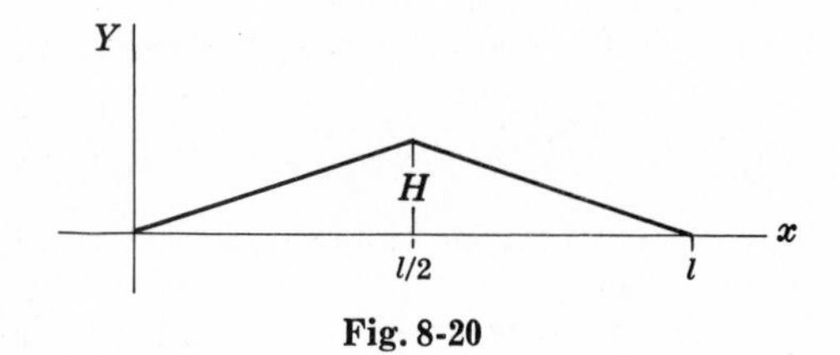

Fig. 8-20

From Fig. 8-20 we see that the initial displacement of the string is given by

$$Y(x,0) \quad = \quad f(x) \quad = \quad \begin{cases} 2Hx/l & 0 < x \leqq l/2 \\ 2H(l-x)/l & l/2 \leqq x \leqq l \end{cases}$$

Now

$$\begin{aligned} b_n \quad &= \quad \frac{2}{l}\int_0^l f(x)\sin\frac{n\pi x}{l}\,dx \\ &= \quad \frac{2}{l}\left\{\int_0^{l/2}\frac{2Hx}{l}\sin\frac{n\pi x}{l}\,dx + \int_{l/2}^{l}\frac{2H}{l}(l-x)\sin\frac{n\pi x}{l}\,dx\right\} \\ &= \quad \frac{8H}{\pi^2}\frac{\sin(n\pi/2)}{n^2} \end{aligned}$$

on using integration by parts to evaluate the integrals. Using this in equation (10) of Problem 8.19, we find

$$\begin{aligned} Y(x,t) \quad &= \quad \frac{8H}{\pi^2}\sum_{n=1}^{\infty}\frac{\sin(n\pi/2)}{n^2}\sin\frac{n\pi x}{l}\cos\frac{n\pi ct}{l} \\ &= \quad \frac{8H}{\pi^2}\left\{\frac{1}{1^2}\sin\frac{\pi x}{l}\cos\frac{\pi ct}{l} - \frac{1}{3^2}\sin\frac{3\pi x}{l}\cos\frac{3\pi ct}{l} + \frac{1}{5^2}\sin\frac{5\pi x}{l}\cos\frac{5\pi ct}{l} - \cdots\right\} \end{aligned}$$

8.21. Find the normal frequencies and normal modes for the vibrating string in Problem 8.20.

The normal mode corresponding to the lowest normal frequency is given by the first term in the solution of Problem 8.20, i.e.,

$$\frac{8H}{\pi^2}\sin\frac{\pi x}{l}\cos\frac{\pi ct}{l}$$

The frequency is given by f_1 where

$$2\pi f_1 \quad = \quad \frac{\pi c}{l} \qquad \text{or} \qquad f_1 \quad = \quad \frac{c}{2l} \quad = \quad \frac{1}{2l}\sqrt{\frac{T}{\sigma}}$$

Since the cosine varies between -1 and $+1$, the mode is such that the string oscillates as in Fig. 8-21 from the heavy curve to the dashed curve and back again.

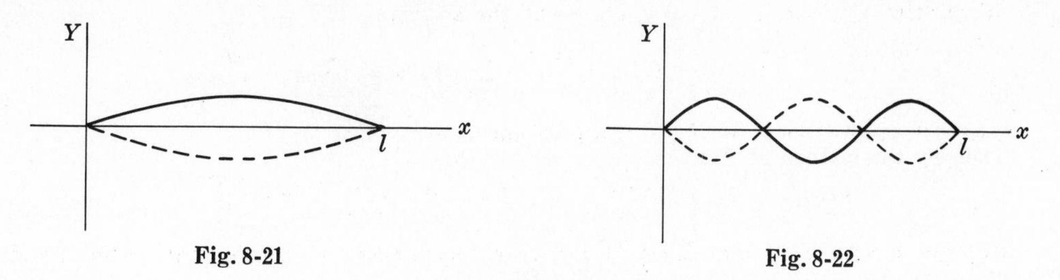

Fig. 8-21 **Fig. 8-22**

The next higher frequency is given by the mode corresponding to the next term in the series which, except for sign, is

$$\frac{8H}{9\pi^2}\sin\frac{3\pi x}{l}\cos\frac{3\pi ct}{l}$$

In this case the frequency is given by

$$2\pi f_3 = \frac{3\pi c}{l} \quad \text{or} \quad f_3 = \frac{3c}{2l} = \frac{3}{2l}\sqrt{\frac{T}{\sigma}}$$

The mode is indicated in Fig. 8-22.

The higher normal frequencies are given by

$$f_5 = \frac{5}{2l}\sqrt{\frac{T}{\sigma}}, \quad f_7 = \frac{7}{2l}\sqrt{\frac{T}{\sigma}}, \quad \ldots$$

The amplitudes of modes corresponding to the even frequencies

$$f_2 = \frac{2}{2l}\sqrt{\frac{T}{\sigma}}, \quad f_4 = \frac{4}{2l}\sqrt{\frac{T}{\sigma}}, \quad \ldots$$

are zero, so that these frequencies are not present. In a general displacement, however, they would be present.

Because of the fact that all higher normal frequencies are integer multiples of the lowest normal frequency, often called the *fundamental frequency*, the vibrating string emits a *musical note*. The higher frequencies are sometimes called *overtones*.

8.22. Find the transverse displacement of a vibrating string of length l with fixed end-points if the string is initially in the equilibrium position and is given a velocity distribution defined by $g(x)$.

In this case we must solve the boundary-value problem

$$\frac{\partial^2 Y}{\partial t^2} = c^2\frac{\partial^2 Y}{\partial x^2} \tag{1}$$

$$Y(0,t) = 0, \quad Y(l,t) = 0, \quad Y(x,0) = 0, \quad Y_t(x,0) = g(x) \tag{2}$$

The method of separation of variables and application of the first two conditions of (2) yields, as in Problem 8.19,

$$Y(x,t) = \sin\frac{n\pi x}{l}\left(b\cos\frac{n\pi ct}{l} + a\sin\frac{n\pi ct}{l}\right)$$

However, in this case if we apply the third condition of (2) we find $b = 0$, so that

$$Y(x,t) = a\sin\frac{n\pi x}{l}\sin\frac{n\pi ct}{l}$$

To satisfy the fourth condition we first note that the superposition principle applies, so that we arrive at the solution

$$Y(x,t) = \sum_{n=1}^{\infty} a_n \sin\frac{n\pi x}{l}\sin\frac{n\pi ct}{l} \tag{3}$$

From this we have by differentiation with respect to t,

$$Y_t(x,t) = \sum_{n=1}^{\infty} \frac{n\pi c a_n}{l} \sin\frac{n\pi x}{l} \cos\frac{n\pi ct}{l}$$

or

$$Y_t(x,0) = g(x) = \sum_{n=1}^{\infty} \frac{n\pi c a_n}{l} \sin\frac{n\pi x}{l}$$

Then by the method of Fourier series we see that

$$\frac{n\pi c a_n}{l} = \frac{2}{l}\int_0^l g(x)\sin\frac{n\pi x}{l}\,dx \quad \text{or} \quad a_n = \frac{2}{n\pi c}\int_0^l g(x)\sin\frac{n\pi x}{l}\,dx \tag{4}$$

Thus the required solution is, on using (4) in (3),

$$Y(x,t) = \sum_{n=1}^{\infty}\left\{\frac{2}{n\pi c}\int_0^l g(x)\sin\frac{n\pi x}{l}\,dx\right\}\sin\frac{n\pi x}{l}\sin\frac{n\pi ct}{l} \tag{5}$$

8.23. **Find the transverse displacement of a vibrating string of length l with fixed end-points if the string initially has a displacement from the equilibrium position given by $f(x)$ and velocity distribution given by $g(x)$.**

The solution to the given problem is the sum of the solutions to the Problems 8.19 and 8.22. Thus the required solution is

$$Y(x,t) = \sum_{n=1}^{\infty}\left\{\frac{2}{l}\int_0^l f(x)\sin\frac{n\pi x}{l}\,dx\right\}\sin\frac{n\pi x}{l}\cos\frac{n\pi ct}{l}$$

$$+ \sum_{n=1}^{\infty}\left\{\frac{2}{n\pi c}\int_0^l g(x)\sin\frac{n\pi x}{l}\,dx\right\}\sin\frac{n\pi x}{l}\sin\frac{n\pi ct}{l}$$

MISCELLANEOUS PROBLEMS

8.24. **A particle is dropped vertically on to a fixed horizontal plane. If it hits the plane with velocity $\mathbf{v}$, show that it will rebound with velocity $-\epsilon\mathbf{v}$.**

The solution to this problem can be obtained from the results of Problem 8.6 by letting m_2 become infinite and $\mathbf{v}_2 = \mathbf{0}$ while $\mathbf{v}_1 = \mathbf{v}$ [where subscripts 1 and 2 refer to the particle and plane respectively]. Then the respective velocities after impact are given by

$$\lim_{m_2\to\infty} \mathbf{v}_1' = \lim_{m_2\to\infty} \frac{\{(m_1/m_2) - \epsilon\}\mathbf{v}}{1 + (m_1/m_2)} = -\epsilon\mathbf{v}$$

$$\lim_{m_2\to\infty} \mathbf{v}_2' = \lim_{m_2\to\infty} \frac{(m_1/m_2)(1+\epsilon)\mathbf{v}}{1 + (m_1/m_2)} = \mathbf{0}$$

Thus the velocity of the particle after impact is $-\epsilon\mathbf{v}$. The velocity of the plane of course remains zero.

8.25. **Suppose that the particle of Problem 8.24 is dropped from rest at a height H above the plane. Prove that the total theoretical distance traveled by the particle before coming to rest is given by $H(1+\epsilon^2)/(1-\epsilon^2)$.**

Let v be the speed of the particle just before it hits the plane. Then by the conservation of energy, $\frac{1}{2}mv^2 + 0 = 0 + mgH$ or $v^2 = 2gH$. Thus by Problem 8.24 the particle rebounds with speed ϵv and reaches a height $(\epsilon v)^2/2g = \epsilon^2 H$. It then travels back to the plane through the distance $\epsilon^2 H$. Thus on the first rebound it travels through the distance $2\epsilon^2 H$.

By similar reasoning we find that on the second, third, ... rebounds it travels through the distance $2\epsilon^4 H, 2\epsilon^6 H, \ldots$. Then the total theoretical distance traveled before coming to rest is

$$H + 2\epsilon^2 H + 2\epsilon^4 H + 2\epsilon^6 H + \cdots = H + 2\epsilon^2 H(1 + \epsilon^2 + \epsilon^4 + \cdots) = H + \frac{2\epsilon^2 H}{1-\epsilon^2} = H\left(\frac{1+\epsilon^2}{1-\epsilon^2}\right)$$

using the result $1 + r + r^2 + r^3 + \cdots = 1/(1-r)$ if $|r| < 1$.

8.26. Two particles having masses m and M are traveling on the x axis (assumed frictionless) with velocities $v_1\mathbf{i}$ and $V_1\mathbf{i}$ respectively. Suppose that they collide and that after the collision (impact) their velocities are $v_2\mathbf{i}$ and $V_2\mathbf{i}$ respectively. Prove that the velocities of the center of mass before and after collision are equal.

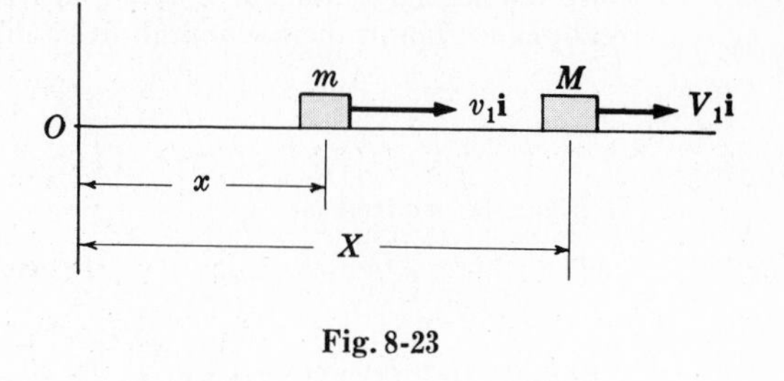

Fig. 8-23

By the conservation of momentum,

$$\text{Total momentum before impact} = \text{total momentum after impact}$$

$$mv_1\mathbf{i} + MV_1\mathbf{i} = mv_2\mathbf{i} + MV_2\mathbf{i}$$

or

$$mv_1 + MV_1 = mv_2 + MV_2$$

Let x and X be the respective coordinates of the particles. Then the center of mass is given by $\bar{r} = (mx + MX)/(m + M)$.

The velocity of the center of mass before impact is

$$\dot{\bar{r}}_1 = (mv_1 + MV_1)/(m + M)$$

The velocity of the center of mass after impact is $\dot{\bar{r}}_2 = (mv_2 + MV_2)/(m + M)$. Thus $\dot{\bar{r}}_1 = \dot{\bar{r}}_2$.

8.27. A particle of mass m slides down a frictionless incline of angle α, mass M and length L which is on a horizontal frictionless plane [see Fig. 8-24]. If the particle starts initially from rest at the top of the incline, prove that the time for the particle to reach the bottom is given by $\sqrt{\dfrac{2L(M + m\sin^2\alpha)}{(M+m)g\sin\alpha}}$.

Choose a fixed vertical xy coordinate system as represented in Fig. 8-24. Let $\mathbf{R}$ be the position vector of the center of mass C of the incline, $\mathbf{A}$ the (constant) vector from C to the top of the incline, and $\mathbf{s}$ the position vector of the particle relative to the top of the incline. Then the position vector of particle m with respect to the fixed coordinate system is $\mathbf{R} + \mathbf{A} + \mathbf{s}$. Since the only force acting on the particle is the weight $m\mathbf{g}$ of the particle, we have by Newton's second law applied to the particle,

$$m\frac{d^2}{dt^2}(\mathbf{R} + \mathbf{A} + \mathbf{s}) = m\mathbf{g} \qquad (1)$$

or

$$\frac{d^2\mathbf{R}}{dt^2} + \frac{d^2\mathbf{s}}{dt^2} = \mathbf{g} \qquad (2)$$

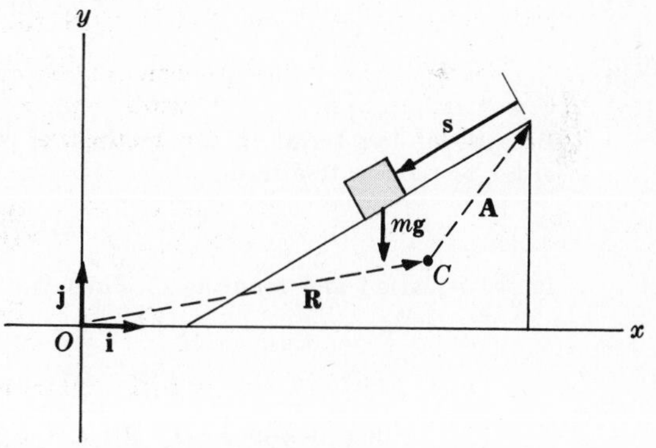

Fig. 8-24

Writing $\mathbf{R} = X\mathbf{i} + Y\mathbf{j}$, $\mathbf{g} = -g\mathbf{j}$ and $\mathbf{s} = s\mathbf{s}_1$, where $\mathbf{s}_1$ is a unit vector down the incline in the direction of $\mathbf{s}$, (2) becomes

$$\frac{d^2X}{dt^2}\mathbf{i} + \frac{d^2s}{dt^2}\mathbf{s}_1 = -g\mathbf{j}$$

Multiplying by $\mathbf{s}_1 \cdot$, this becomes

$$\frac{d^2X}{dt^2}\mathbf{s}_1\cdot\mathbf{i} + \frac{d^2s}{dt^2}\mathbf{s}_1\cdot\mathbf{s}_1 = -g\mathbf{s}_1\cdot\mathbf{j}$$

or

$$-\frac{d^2X}{dt^2}\cos\alpha + \frac{d^2s}{dt^2} = g\sin\alpha \qquad (3)$$

Since the net horizontal force acting on the system consisting of the particle and incline is zero, the total momentum in the horizontal direction before and after the particle starts sliding is zero. Then

$$M\frac{d\mathbf{R}}{dt}\cdot\mathbf{i} + m\frac{d}{dt}(\mathbf{R}+\mathbf{A}+\mathbf{s})\cdot\mathbf{i} = 0$$

This can be written as

$$(M+m)\frac{dX}{dt} - m\frac{ds}{dt}\cos\alpha = 0 \tag{4}$$

Differentiating (4) with respect to t and solving for d^2X/dt^2, we find

$$\frac{d^2X}{dt^2} = \frac{m\cos\alpha}{M+m}\frac{d^2s}{dt^2} \tag{5}$$

Substituting into (3) yields

$$\frac{d^2s}{dt^2} = \frac{(M+m)g\sin\alpha}{M+m-m\cos^2\alpha} = \frac{(M+m)g\sin\alpha}{M+m\sin^2\alpha} \tag{6}$$

Integrating (6) subject to the conditions $s=0$, $ds/dt=0$ at $t=0$, we find

$$s = \frac{1}{2}\left\{\frac{(M+m)g\sin\alpha}{M+m\sin^2\alpha}\right\}t^2$$

which, when $s=L$, yields the required time.

8.28. Solve the vibrating string Problem 8.19 if gravity is taken into account.

The boundary-value problem is

$$\frac{\partial^2Y}{\partial t^2} = c^2\frac{\partial^2Y}{\partial x^2} - g \tag{1}$$

$$Y(0,t)=0,\quad Y(l,t)=0,\quad Y(x,0)=f(x),\quad Y_t(x,0)=0 \tag{2}$$

Because of the term $-g$ the method of separation of variables does not work in this case. In order to remove this term, we let

$$Y(x,t) = Z(x,t) + \psi(x) \tag{3}$$

in the equation and conditions. Thus we find

$$\frac{\partial^2Z}{\partial t^2} = c^2\frac{\partial^2Z}{\partial x^2} + c^2\psi'' - g \tag{4}$$

$$Z(0,t)+\psi(0) = 0,\quad Z(l,t)+\psi(l) = 0,\quad Z(x,0)+\psi(x) = f(x),\quad Z_t(x,0) = 0 \tag{5}$$

The equation (4) and conditions (5) become similar to problems already discussed if we choose ψ such that

$$c^2\psi'' - g = 0,\quad \psi(0) = 0,\quad \psi(l) = 0 \tag{6}$$

In this case (4) and (5) become

$$\frac{\partial^2Z}{\partial t^2} = c^2\frac{\partial^2Z}{\partial x^2} \tag{7}$$

$$Z(0,t) = 0,\quad Z(l,t) = 0,\quad Z(x,0) = f(x)-\psi(x),\quad Z_t(x,0) = 0 \tag{8}$$

Now from (6) we find $\psi''=g/c^2$ or $\psi(x)=gx^2/2c^2+c_1x+c_2$; and since $\psi(0)=0$, $\psi(l)=0$, we obtain $c_2=0$, $c_1=-gl/2c^2$. Thus

$$\psi(x) = \frac{g}{2c^2}(x^2-lx)$$

The solution to equation (7) with conditions (8) is, as in Problem 8.19,

$$Z(x,t) = \sum_{n=1}^{\infty} \left\{\frac{2}{l}\int_0^l [f(x)-\psi(x)] \sin\frac{n\pi x}{l}\,dx\right\} \sin\frac{n\pi x}{l} \cos\frac{n\pi ct}{l}$$

and thus

$$Y(x,t) = \sum_{n=1}^{\infty} \left\{\frac{2}{l}\int_0^l \left[f(x) - \frac{g}{2c^2}(x^2-lx)\right] \sin\frac{n\pi x}{l}\,dx\right\} \sin\frac{n\pi x}{l} \cos\frac{n\pi ct}{l} + \frac{g}{2c^2}(x^2-lx)$$

8.29. Assume that a continuous string, which is fixed at its endpoints and vibrates transversely, is replaced by N particles of mass m at equal distances from each other. Determine the equations of motion of the particles.

We assume that the particles are connected to each other by taut, elastic strings having constant tension T [see Fig. 8-25]. We also assume that the horizontal distances between particles [i.e. in the direction of the unit vector **i**] are equal to a and that the transverse displacement [i.e. in the direction of the unit vector **j**] of particle ν is Y_ν. We assume that there is no displacement of any particle in direction **i** or −**i**.

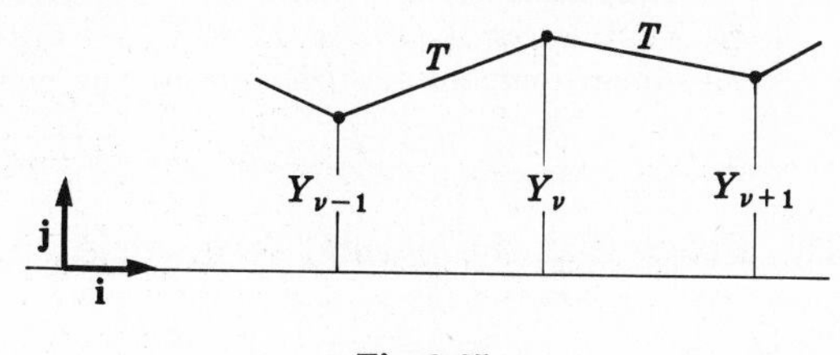

Fig. 8-25

Let us isolate the νth particle. The forces acting on this particle are those due to the $(\nu-1)$st and $(\nu+1)$st particles. We have

$$\text{Transverse force due to } (\nu-1)\text{st particle} = -T\left(\frac{Y_\nu - Y_{\nu-1}}{a}\right)\mathbf{j}$$

$$\text{Transverse force due to } (\nu+1)\text{st particle} = -T\left(\frac{Y_\nu - Y_{\nu+1}}{a}\right)\mathbf{j}$$

Then by Newton's second law the total transverse force acting on particle ν is

$$m\frac{d^2Y_\nu}{dt^2}\mathbf{j} = -T\left(\frac{Y_\nu - Y_{\nu-1}}{a}\right)\mathbf{j} - T\left(\frac{Y_\nu - Y_{\nu+1}}{a}\right)\mathbf{j}$$

or

$$m\frac{d^2Y_\nu}{dt^2} = \frac{T}{a}(Y_{\nu-1} - 2Y_\nu + Y_{\nu+1})$$

i.e.

$$\ddot{Y}_\nu = \frac{T}{ma}(Y_{\nu-1} - 2Y_\nu + Y_{\nu+1}) \tag{1}$$

To take into account the fact that the endpoints are fixed, we assume two particles corresponding to $\nu = 0$ and $\nu = N+1$ for which $Y_0 = 0$, $Y_{N+1} = 0$. Then on putting $\nu = 1$ and $\nu = N$ in equation (1), we find

$$\ddot{Y}_1 = \frac{T}{ma}(-2Y_1 + Y_2), \qquad \ddot{Y}_N = \frac{T}{ma}(Y_{N-1} - 2Y_N) \tag{2}$$

8.30. Obtain the secular determinant condition for the normal frequencies of the system of particles in Problem 8.29.

Let $Y_\nu = A_\nu \cos\omega t$ in equations (1) and (2) of Problem 8.29. Then after simplifying we find

$$-A_{\nu-1} + (2 - ma\omega^2/T)A_\nu - A_{\nu+1} = 0 \qquad \nu = 2, \ldots, N-1 \tag{1}$$

$$(2 - ma\omega^2/T)A_1 - A_2 = 0, \qquad -A_{N-1} + (2 - ma\omega^2/T)A_N = 0 \tag{2}$$

Putting

$$2 - ma\omega^2/T = c \tag{3}$$

these equations can be written

$$cA_1 - A_2 = 0, \quad -A_1 + cA_2 - A_3 = 0, \quad \ldots, \quad -A_{N-1} + cA_N = 0$$

Then if we wish solutions such that $A_\nu \neq 0$, we must require that the Nth order determinant of the coefficients be zero, i.e.,

$$\Delta_N = \begin{vmatrix} c & -1 & 0 & 0 & 0 & \cdots & 0 & 0 & 0 \\ -1 & c & -1 & 0 & 0 & \cdots & 0 & 0 & 0 \\ 0 & -1 & c & -1 & 0 & \cdots & 0 & 0 & 0 \\ \cdots & \cdots & \cdots & \cdots & \cdots & \cdots & \cdots & \cdots & \cdots \\ 0 & 0 & 0 & 0 & 0 & \cdots & -1 & c & -1 \\ 0 & 0 & 0 & 0 & 0 & \cdots & 0 & -1 & c \end{vmatrix} = 0$$

The normal frequencies are obtained by solving this equation for the N values of ω^2.

Although we have used $Y_\nu = A_\nu \cos \omega t$, we could just as well have assumed $Y_\nu = B_\nu \sin \omega t$ or $Y_\nu = A_\nu \cos \omega t + B_\nu \sin \omega t$ or $Y_\nu = C_\nu e^{i\omega t}$. The secular determinant would have come out to be the same [compare the remarks at the end of Problem 8.2(*b*)].

8.31. Prove that the normal frequencies in Problem 8.30 are given by

$$\omega_\alpha^2 = \frac{2T}{ma}\left(1 - \cos\frac{\alpha\pi}{N+1}\right) \qquad \alpha = 1, \ldots, N$$

By expanding the determinant Δ_N of Problem 8.30 in terms of the elements in the first row, we have

$$\Delta_N = c\,\Delta_{N-1} - \Delta_{N-2} \tag{1}$$

Also,

$$\Delta_1 = c, \quad \Delta_2 = c^2 - 1 \tag{2}$$

Putting $N = 2$ in (*1*), we see that equations (*2*) are formally satisfied if we take $\Delta_0 = 1$. Thus conditions consistent with (*1*) and (*2*) are

$$\Delta_0 = 1, \quad \Delta_1 = c \tag{3}$$

To solve the *difference equation* (*1*), assume that $\Delta_N = p^N$ where p is a constant to be determined. Substituting this into (*1*), we find on dividing by p^{N-2},

$$p^2 - cp + 1 = 0 \qquad \text{or} \qquad p = \frac{c \pm \sqrt{c^2-4}}{2}$$

If we call $c = 2\cos\theta$, then

$$p = \cos\theta \pm i\sin\theta = e^{\pm i\theta}$$

Thus solutions of the difference equation are

$$(e^{i\theta})^N = e^{iN\theta} = \cos N\theta + i\sin N\theta \qquad \text{and} \qquad (e^{-i\theta})^N = e^{-Ni\theta} = \cos N\theta - i\sin N\theta$$

Since constants multiplying these solutions and sums of solutions are also solutions [as in the case of linear differential equations], we see that the *general solution* is

$$\Delta_N = G\cos N\theta + H\sin N\theta \tag{4}$$

Now from equations (*3*) we have $\Delta_0 = 1$, $\Delta_1 = 2\cos\theta$ so that $G = 1$, $H = \cot\theta$. Thus

$$\Delta_N = \cos N\theta + \frac{\sin N\theta\cos\theta}{\sin\theta} = \frac{\sin(N+1)\theta}{\sin\theta} \tag{5}$$

This is equal to zero when $\sin(N+1)\theta = 0$ or $\theta = \alpha\pi/(N+1)$, $\alpha = 1, \ldots, N$. Thus using (*3*) of Problem 8.30, we find

$$\omega_\alpha^2 = \frac{2T}{ma}\left(1 - \cos\frac{\alpha\pi}{N+1}\right) \tag{6}$$

8.32. Solve for A_ν in Problem 8.30 and thus find the transverse displacement Y_ν of particle ν.

From equations (1) and (2) of Problem 8.30 we have [on using the normal frequencies (6) of Problem 8.31 and the superscript α to indicate that the A's depend on α],

$$-A_{\nu-1}^{(\alpha)} + 2A_\nu^{(\alpha)} \cos\frac{\alpha\pi}{N+1} - A_{\nu+1}^{(\alpha)} = 0 \qquad (1)$$

together with the end conditions

$$A_0^{(\alpha)} = 0, \qquad A_{N+1}^{(\alpha)} = 0 \qquad (2)$$

The equation (1) subject to conditions (2) can be solved in a manner exactly like that of Problem 8.31, and we find

$$A_\nu^{(\alpha)} = C_\alpha \sin\frac{\alpha\nu\pi}{N+1}$$

where C_α are arbitrary constants. In a similar manner if we had assumed $Y_\nu = B_\nu \sin\omega t$ [see remarks at the end of Problem 8.30] we would have obtained

$$B_\nu^{(\alpha)} = D_\alpha \sin\frac{\alpha\nu\pi}{N+1}$$

Thus solutions are given by

$$C_\alpha \sin\frac{\alpha\nu\pi}{N+1}\cos\omega t \qquad \text{and} \qquad D_\alpha \sin\frac{\alpha\nu\pi}{N+1}\sin\omega t$$

and since sums of solutions are also solutions, we have

$$Y_\nu = \sum_{\alpha=1}^{N} \sin\frac{\alpha\nu\pi}{N+1}(C_\alpha\cos\omega t + D_\alpha \sin\omega t)$$

The constants C_α and D_α are determined from initial conditions.

The analogy with the continuous vibrating string is easily seen.

Supplementary Problems

VIBRATING SYSTEMS OF PARTICLES

8.33. Find the normal frequencies of the vibrations in Problem 8.1, page 197, if the spring constants and masses are all different.

8.34. Two equal masses m on a horizontal frictionless table as shown in Fig. 8-26 are connected by equal springs. The end of one spring is fixed at A and the masses are set into motion. (a) Set up the equations of motion of the system. (b) Find the normal frequencies of vibration. (c) Describe the normal modes of vibration.

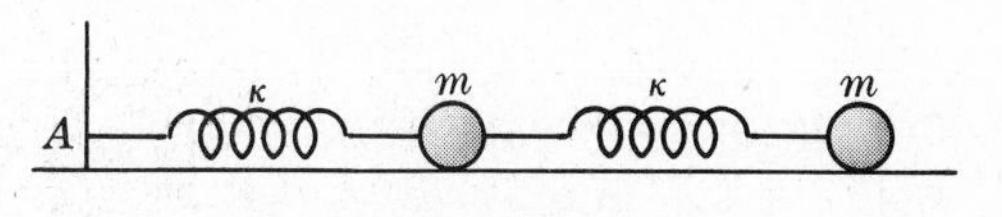

Fig. 8-26

Ans. (b) $f_1 = \dfrac{\sqrt{5}-1}{4\pi}\sqrt{\dfrac{\kappa}{m}}, \quad f_2 = \dfrac{\sqrt{5}+1}{4\pi}\sqrt{\dfrac{\kappa}{m}}$

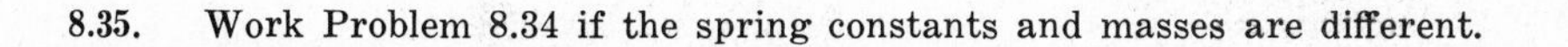

8.35. Work Problem 8.34 if the spring constants and masses are different.

8.36 Two equal masses m are attached to the ends of a spring of constant κ which is on a horizontal frictionless table. If the masses are pulled apart and then released, prove that they will vibrate with respect to each other with period $2\pi\sqrt{m/2\kappa}$.

8.37. Work Problem 8.36 if the masses are different and equal to M_1 and M_2 respectively.
Ans. $2\pi\sqrt{\mu/\kappa}$ where $\mu = M_1M_2/(M_1+M_2)$

8.38. In Fig. 8-27 equal masses m lying on a horizontal frictionless table are connected to each other and to fixed points A and B by means of elastic strings of constant tension T and length l. If the displacements from the equilibrium position AB of the masses are Y_1 and Y_2 respectively, show that the equations of motion are given by

$$\ddot{Y}_1 = \kappa(Y_2 - 2Y_1), \quad \ddot{Y}_2 = \kappa(Y_1 - 2Y_2)$$

where $\kappa = 3T/ml$.

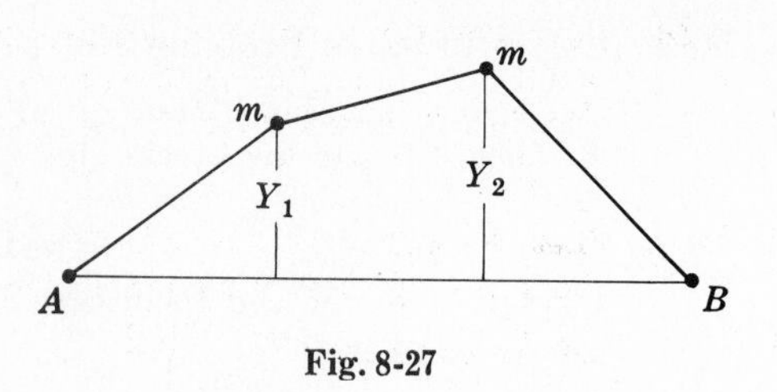

Fig. 8-27

8.39. Prove that the natural frequencies of the vibration in Problem 8.38 are given respectively by

$$\frac{1}{2\pi}\sqrt{\frac{3T}{ml}} \quad \text{and} \quad \frac{1}{2\pi}\sqrt{\frac{9T}{ml}}$$

and describe the modes of vibration.

8.40. Find the normal frequencies and normal modes of vibration for the system of particles of masses m_1 and m_2 connected by springs as indicated in Fig. 8-28.

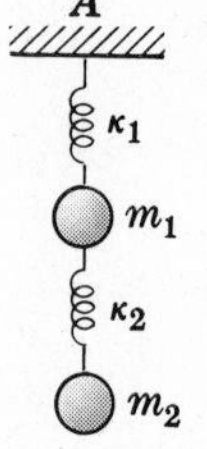

Fig. 8-28

CHANGING MASS. ROCKETS

8.41. (a) Prove that the total distance traveled by the rocket of Problem 8.5 in time t is given by

$$s = v_0\left\{t + \left(\frac{m_0 - \alpha t}{\alpha}\right)\ln\left(\frac{m_0 - \alpha t}{m_0}\right)\right\} - \frac{1}{2}gt^2$$

(b) What is the maximum height which the rocket can reach and how long will it take to achieve this maximum height?

8.42. Suppose that a rocket which starts from rest falls in a constant gravitational field. At the instant it starts to fall it ejects gas at the constant rate α in the direction of the gravitational field and at speed v_0 with respect to the rocket. Find its speed after any time t.

Ans. $gt - v_0 \ln\left(\dfrac{m_0}{m_0 - \alpha t}\right)$

8.43. How far does the rocket of Problem 8.42 travel in time t?

Ans. $\dfrac{1}{2}gt^2 - v_0\left\{t + \left(\dfrac{m_0 - \alpha t}{\alpha}\right)\ln\left(\dfrac{m_0 - \alpha t}{m_0}\right)\right\}$

8.44. Describe how Problem 8.42 can be useful in making a "soft landing" on a planet or satellite?

8.45. Discuss the motion of a two-stage rocket, i.e. one in which one part falls off and the other rocket takes over.

COLLISIONS OF PARTICLES

8.46. A gun fires a bullet of mass m with horizontal velocity $\mathbf{v}$ into a block of wood of mass M which rests on a horizontal frictionless plane. If the bullet becomes embedded in the wood, (a) determine the subsequent velocity of the system and (b) find the loss in kinetic energy.

Ans. (a) $m\mathbf{v}/(M+m)$ (b) $mMv^2/2(M+m)$

8.47. Work Problem 8.46 if the block is moving away from the gun with velocity $\mathbf{V}$.

8.48. A ball which is dropped from a height H onto a floor rebounds to a height $h < H$. Determine the coefficient of restitution. *Ans.* $\sqrt{h/H}$

8.49. A mass m_1 traveling with speed v on a horizontal plane hits another mass m_2 which is at rest. If the coefficient of restitution is ϵ, prove that there is a loss of kinetic energy equal to $m_1m_2(1-\epsilon^2)v^2/2(m_1+m_2)$.

8.50. A billiard ball strikes another billiard ball obliquely at an angle of 45° with their line of centers at the time of impact. If the coefficient of restitution is 1/2, find the angle at which the first ball will "bounce off". *Ans.* $\tan^{-1}(3/5)$

8.51. Let the masses of two colliding particles be m_1, m_2 and their respective velocities before impact be $\mathbf{v}_1, \mathbf{v}_2$. If the coefficient of restitution is ϵ, prove that the loss in kinetic energy as a result of the collision is $\frac{1}{2}\frac{m_1 m_2}{m_1+m_2}(\mathbf{v}_1-\mathbf{v}_2)^2(1-\epsilon^2)$.

8.52. Prove that the momentum which is transferred from the first particle of Problem 8.51 to the second is $\frac{m_1 m_2}{m_1+m_2}(1+\epsilon)(\mathbf{v}_1-\mathbf{v}_2)$.

8.53. A ball is dropped from a height h above a horizontal plane on to an inclined plane of angle α which is resting on the horizontal plane. Prove that if the coefficient of restitution is ϵ, then the ball will next hit the incline at a point which is at a distance $4\epsilon(1+\epsilon)h\sin\alpha$ below the original point of impact.

FOURIER SERIES, ODD AND EVEN FUNCTIONS, FOURIER SINE AND COSINE SERIES

8.54. Graph each of the following functions and find their corresponding Fourier series using properties of even and odd functions wherever applicable.

(a) $f(x) = \begin{cases} 8 & 0 < x < 2 \\ -8 & 2 < x < 4 \end{cases}$ Period 4

(c) $f(x) = 4x,\ 0 < x < 10,$ Period 10

(b) $f(x) = \begin{cases} -x & -4 \leqq x \leqq 0 \\ x & 0 \leqq x \leqq 4 \end{cases}$ Period 8

(d) $f(x) = \begin{cases} 2x & 0 \leqq x < 3 \\ 0 & -3 < x < 0 \end{cases}$ Period 6

Ans. (a) $\frac{16}{\pi}\sum_{n=1}^{\infty}\frac{(1-\cos n\pi)}{n}\sin\frac{n\pi x}{2}$ (c) $20 - \frac{40}{\pi}\sum_{n=1}^{\infty}\frac{1}{n}\sin\frac{n\pi x}{5}$

(b) $2 - \frac{8}{\pi^2}\sum_{n=1}^{\infty}\frac{(1-\cos n\pi)}{n^2}\cos\frac{n\pi x}{4}$ (d) $\frac{3}{2} + \sum_{n=1}^{\infty}\left\{\frac{6(\cos n\pi - 1)}{n^2\pi^2}\cos\frac{n\pi x}{3} - \frac{6\cos n\pi}{n\pi}\sin\frac{n\pi x}{3}\right\}$

8.55. In each part of Problem 8.54, tell where the discontinuities of $f(x)$ are located and to what value the series converges at these discontinuities.

Ans. (a) $x = 0, \pm 2, \pm 4, \ldots;\ 0$ (c) $x = 0, \pm 10, \pm 20, \ldots;\ 20$

(b) no discontinuities (d) $x = \pm 3, \pm 9, \pm 15, \ldots;\ 3$

8.56. Expand $f(x) = \begin{cases} 2 - x & 0 < x < 4 \\ x - 6 & 4 < x < 8 \end{cases}$ in a Fourier series of period 8.

Ans. $\frac{16}{\pi^2}\left\{\cos\frac{\pi x}{4} + \frac{1}{3^2}\cos\frac{3\pi x}{4} + \frac{1}{5^2}\cos\frac{5\pi x}{4} + \cdots\right\}$

8.57. (a) Expand $f(x) = \cos x,\ 0 < x < \pi$, in a Fourier sine series.

(b) How should $f(x)$ be defined at $x = 0$ and $x = \pi$ so that the series will converge to $f(x)$ for $0 \leqq x \leqq \pi$?

Ans. (a) $\frac{8}{\pi}\sum_{n=1}^{\infty}\frac{n\sin 2nx}{4n^2-1}$ (b) $f(0) = f(\pi) = 0$

8.58. (a) Expand in a Fourier series $f(x) = \cos x,\ 0 < x < \pi$ if the period is π; and (b) compare with the result of Problem 8.57, explaining the similarities and differences if any.

Ans. Answer is the same as in Problem 8.57.

8.59. Expand $f(x) = \begin{cases} x & 0 < x < 4 \\ 8 - x & 4 < x < 8 \end{cases}$ in a series of (*a*) sines, (*b*) cosines.

Ans. (*a*) $\dfrac{32}{\pi^2} \sum_{n=1}^{\infty} \dfrac{1}{n^2} \sin \dfrac{n\pi}{2} \sin \dfrac{n\pi x}{8}$ (*b*) $\dfrac{16}{\pi^2} \sum_{n=1}^{\infty} \left(\dfrac{2 \cos n\pi/2 - \cos n\pi - 1}{n^2} \right) \cos \dfrac{n\pi x}{8}$

THE VIBRATING STRING

8.60. (*a*) Solve the boundary-value problem

$$\frac{\partial^2 Y}{\partial t^2} = 4 \frac{\partial^2 Y}{\partial x^2} \qquad 0 < x < \pi,\ t > 0$$

$$Y_x(0, t) = 0, \quad Y(\pi, t) = h, \quad Y(x, 0) = 0, \quad Y_t(x, 0) = 0$$

(*b*) Give a physical interpretation of the problem in (*a*).

Ans. $Y(x, t) = \dfrac{8h}{\pi} \sum_{n=1}^{\infty} \dfrac{(-1)^n}{2n-1} \sin (n - \tfrac{1}{2})x \sin (2n-1)t$

8.61. Solve the boundary-value problem

$$Y_{tt} = Y_{xx} - g \qquad 0 < x < \pi,\ t > 0$$

$$Y(0, t) = 0, \quad Y(\pi, t) = 0, \quad Y(x, 0) = \mu x(\pi - x), \quad Y_t(x, 0) = 0$$

and interpret physically.

Ans. $Y(x, t) = \dfrac{4(2\mu x + g)}{\pi} \sum_{n=1}^{\infty} \dfrac{1}{(2n-1)^3} \sin (2n-1)x \cos (2n-1)t - \tfrac{1}{2} gx(\pi - x)$

8.62. (*a*) Find a solution of the equation $\dfrac{\partial^2 Y}{\partial t^2} = 4 \dfrac{\partial^2 Y}{\partial x^2}$ which satisfies the conditions $Y(0, t) = 0$, $Y(\pi, t) = 0$, $Y(x, 0) = 0.1 \sin x + 0.01 \sin 4x$, $Y_t(x, 0) = 0$ for $0 < x < \pi, t > 0$. (*b*) Interpret physically the boundary conditions in (*a*) and the solution.
Ans. (*a*) $Y(x, t) = 0.1 \sin x \cos 2t + 0.01 \sin 4x \cos 8t$

8.63. (*a*) Solve the boundary-value problem $\dfrac{\partial^2 Y}{\partial t^2} = 9 \dfrac{\partial^2 Y}{\partial x^2}$ subject to the conditions $Y(0, t) = 0$, $Y(2, t) = 0$, $Y(x, 0) = 0.05x(2 - x)$, $Y_t(x, 0) = 0$, where $0 < x < 2, t > 0$. (*b*) Interpret physically.

Ans. (*a*) $Y(x, t) = \dfrac{1.6}{\pi^3} \sum_{n=1}^{\infty} \dfrac{1}{(2n-1)^3} \sin \dfrac{(2n-1)\pi x}{2} \cos \dfrac{3(2n-1)\pi t}{2}$

8.64. Solve Problem 8.63 with the boundary conditions for $Y(x, 0)$ and $Y_t(x, 0)$ interchanged, i.e. $Y(x, 0) = 0$, $Y_t(x, 0) = 0.05x(2 - x)$, and give a physical interpretation.

Ans. $Y(x, t) = \dfrac{3.2}{3\pi^4} \sum_{n=1}^{\infty} \dfrac{1}{(2n-1)^4} \sin \dfrac{(2n-1)\pi x}{2} \sin \dfrac{3(2n-1)\pi t}{2}$

MISCELLANEOUS PROBLEMS

8.65. A spherical raindrop falling in a constant gravitational field grows by absorption of moisture from its surroundings at a rate which is proportional to the instantaneous surface area. Assuming that it starts with radius zero, determine its acceleration. *Ans.* $\tfrac{1}{8} g$

8.66. A cannon of mass M rests on a horizontal plane having coefficient of friction μ. It fires a projectile of mass m with muzzle velocity v_0 in a direction making angle α with the horizontal. Determine how far back the cannon will move due to the recoil.

8.67. A ball is thrown with speed v_0 onto a smooth horizontal plane in a direction making angle α with the plane. If ϵ is the coefficient of restitution, prove that the velocity of the ball after the impact is given by $v_0\sqrt{1 - (1 - \epsilon^2) \sin^2 \alpha}$ in a direction making angle $\tan^{-1} (\epsilon \tan \alpha)$ with the horizontal.

8.68. Prove that the total theoretical time taken for the particle of Problem 8.67 to come to rest is $\sqrt{2H/g}\,(1+\epsilon)/(1-\epsilon)$.

8.69. Prove that while the particle of Problem 8.27, page 213, moves from the top to the bottom of the incline, the incline moves a distance $(mL \cos \alpha)/(M+m)$.

8.70. Prove that the loss of kinetic energy of the spheres of Problem 8.9 is $\frac{1}{2}\mu(v_1 \cos \theta_1 - v_2 \cos \theta_2)^2(1-\epsilon^2)$ where μ is the reduced mass $m_1 m_2/(m_1+m_2)$.

8.71. Prove that the acceleration of a double incline of mass M [Fig. 8-29] which is on a smooth table is given by $\dfrac{(m_1 \sin \alpha_1 \cos \alpha_1 - m_2 \sin \alpha_2 \cos \alpha_2)g}{M + m_1 \sin^2 \alpha_1 + m_2 \sin^2 \alpha_2}$.

8.72. If A is the acceleration of the incline in Problem 8.71, prove that the accelerations of the masses relative to the incline are given numerically by $\dfrac{m_1(A \cos \alpha_1 + g \sin \alpha_1) + m_2(A \cos \alpha_2 - g \sin \alpha_2)}{m_1 + m_2}$.

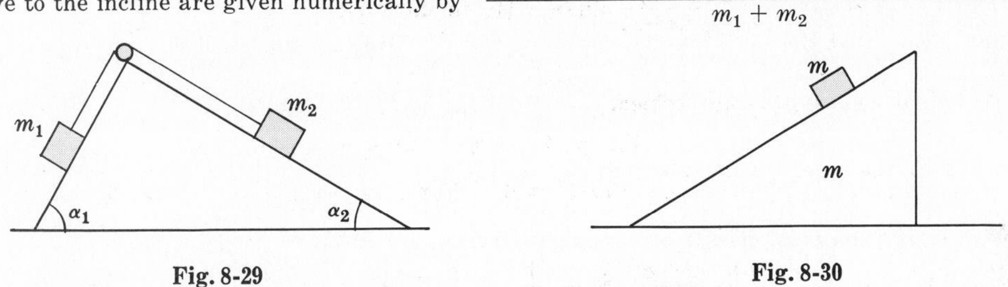

Fig. 8-29

Fig. 8-30

8.73. A mass m slides down an inclined plane of the same mass which is on a horizontal plane with coefficient of friction μ. Prove that the inclined plane moves to the right with acceleration equal to $(1-3\mu)/(3-\mu)$. See Fig. 8-30.

8.74. A gun of mass M is located on an incline of angle α which in turn is on a smooth horizontal plane. The gun fires a bullet of mass m horizontally away from the incline with speed v_0. Find the recoil speed of the gun. *Ans.* $(mv \cos \alpha)/M$ up the incline

8.75. How far up the plane will the gun of Problem 8.74 move before it comes to rest if the incline is (*a*) frictionless, (*b*) has coefficient of friction μ?

8.76. A weight W is dropped from a height H above the plate AB of Fig. 8-31 which is supported by a spring of constant κ. Find the speed with which the weight rebounds.

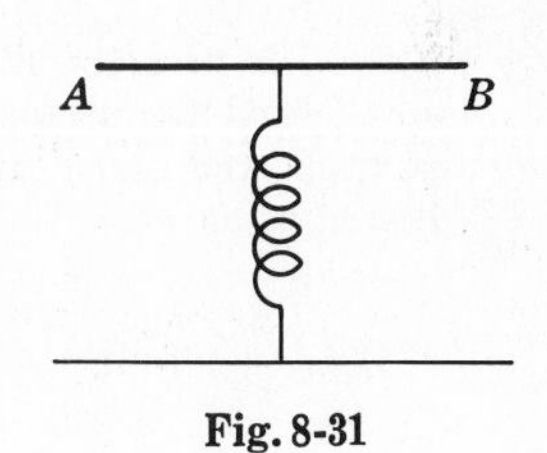

Fig. 8-31

8.77. A ball is thrown with speed v_0 at angle α with a horizontal plane. If it rebounds successively from the horizontal plane, determine its location after n bounces. Assume that the coefficient of restitution is ϵ and that air resistance is negligible.

8.78. Work Problem 8.77 if the horizontal plane is replaced by an inclined plane of angle β and the ball is (*a*) thrown downward, (*b*) thrown upward.

8.79. Obtain the equation (*1*), page 195, for the vibrating string by considering the equations of motion for the N particles of Problem 8.29, page 215, and letting $N \to \infty$.

8.80. Prove that as $N \to \infty$ the normal frequencies as given in Problem 8.31, page 216, approach those for the continuous vibrating string.

8.81. Prove that for $0 \leqq x \leqq \pi$,

(*a*) $$x(\pi - x) = \frac{\pi^2}{6} - \left(\frac{\cos 2x}{1^2} + \frac{\cos 4x}{2^2} + \frac{\cos 6x}{3^2} + \cdots\right)$$

(*b*) $$x(\pi - x) = \frac{8}{\pi}\left(\frac{\sin x}{1^3} + \frac{\sin 3x}{3^3} + \frac{\sin 5x}{5^3} + \cdots\right)$$

8.82. Use Problem 8.81 to show that

$$(a)\ \sum_{n=1}^{\infty} \frac{1}{n^2} = \frac{\pi^2}{6}, \quad (b)\ \sum_{n=1}^{\infty} \frac{(-1)^{n-1}}{n^2} = \frac{\pi^2}{12}, \quad (c)\ \sum_{n=1}^{\infty} \frac{(-1)^{n-1}}{(2n-1)^3} = \frac{\pi^3}{32}.$$

8.83. Prove that $Y = f(x+ct) + g(x-ct)$ is a solution of the equation

$$\frac{\partial^2 Y}{\partial t^2} = c^2 \frac{\partial^2 Y}{\partial x^2}$$

and discuss the connection of this solution with the problem of the vibrating string.

8.84. (a) Prove that the total potential energy of a vibrating string is $V = \frac{T}{2}\int_0^l \left(\frac{\partial Y}{\partial x}\right)^2 dx$.

(b) Thus show that $V = \frac{\pi^2 T}{4l} \sum_{n=1}^{\infty} n^2 \left(a_n \cos\frac{n\pi ct}{l} + b_n \sin\frac{n\pi ct}{l}\right)^2$.

8.85. (a) Prove that the total kinetic energy of the vibrating string is $\text{K.E.} = \frac{1}{2}\sigma \int_0^l \left(\frac{\partial Y}{\partial t}\right)^2 dt$.

(b) Thus show that $\text{K.E.} = \frac{\sigma\pi^2 c^2}{4l} \sum_{n=1}^{\infty} n^2 \left(a_n \cos\frac{n\pi ct}{l} - b_n \sin\frac{n\pi ct}{l}\right)^2$.

(c) Can the kinetic energy be infinite? Explain.

8.86. Prove that the total energy of a vibrating string is $E = \frac{\pi^2 T}{4l} \sum_{n=1}^{\infty} n^2(a_n^2 + b_n^2)$.

8.87. Find the potential energy, kinetic energy and total energy for the string of (a) Problem 8.20, page 210, (b) Problem 8.28, page 214.

8.88. If damping proportional to the instantaneous transverse velocity is taken into account in the problem of the vibrating string, prove that its equation of motion is $\frac{\partial^2 Y}{\partial t^2} + \beta \frac{\partial Y}{\partial t} = c^2 \frac{\partial^2 Y}{\partial x^2}$.

8.89. Prove that the frequencies of vibration for the damped string of Problem 8.88 are given by $\sqrt{n^2\pi^2c^2/l^2 - \beta^2/4}$, $n = 1, 2, 3, \ldots$.

8.90. Solve the problem of the damped vibrating string if the string is fixed at the endpoints $x = 0$ and $x = l$ and the string is (a) given an initial shape $f(x)$ and then released, (b) in the equilibrium position and given an initial velocity distribution $g(x)$, (c) given an initial shape $f(x)$ and velocity distribution $g(x)$.

8.91. Work the problem of the damped vibrating string if gravitation is taken into account.

8.92. Work (a) Problem 8.84(a), (b) Problem 8.85(a), (c) Problem 8.86, (d) Problem 8.88 for the case where the string is replaced by N particles as in Problem 8.29, page 215.

8.93. In Fig. 8-32 the double pendulum system is free to vibrate in a vertical plane. Find the normal frequencies and normal modes assuming small vibrations.

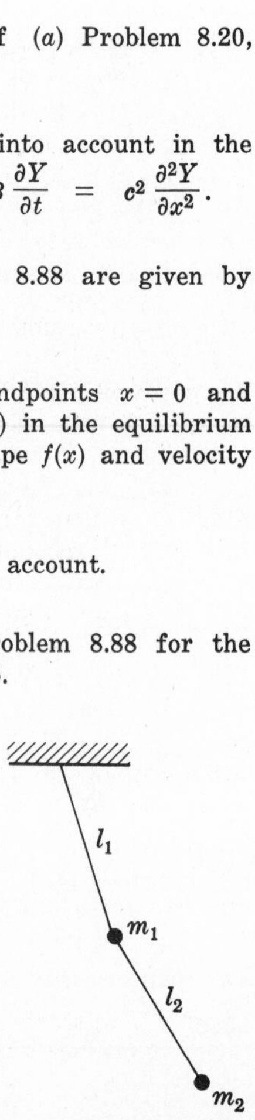

l_1

m_1

l_2

m_2

Fig. 8-32

8.94. Work Problem 8.93 if there is an additional mass m_3 suspended from m_2 by a string of length l_3.

8.95. Generalize the motion of (a) Problem 8.1, (b) Problem 8.34 to N equal particles and springs.

8.96. In Problem 8.95 investigate the limiting case as $N \to \infty$. Discuss the physical significance of the results.

8.97. Solve the boundary-value problem

$$\frac{\partial^2 Y}{\partial t^2} = c^2 \frac{\partial^2 Y}{\partial x^2} + \alpha \sin \omega t$$

$$Y(0, t) = 0, \quad Y(l, t) = 0, \quad Y(x, 0) = f(x), \quad Y_t(x, 0) = 0$$

and give a physical interpretation.

8.98. Work Problem 8.97 if the condition $Y_t(x, 0) = 0$ is replaced by $Y_t(x, 0) = g(x)$.

8.99. Work Problem 8.97 if the partial differential equation is replaced by

$$\frac{\partial^2 Y}{\partial t^2} + \beta \frac{\partial Y}{\partial t} = c^2 \frac{\partial^2 Y}{\partial x^2} + \alpha \sin \omega t$$

and interpret physically.

8.100. Set up the differential equations and initial conditions for the motion of a rocket in an inverse square gravitational field. Do you believe these equations can be solved? Explain.

8.101. Two bodies [such as the sun and earth or earth and moon] of masses m_1 and m_2 move relative to each other under their mutual inverse square attraction according to the universal law of gravitation. If $\mathbf{r}_1$ and $\mathbf{r}_2$ are their position vectors relative to a fixed coordinate system, and $\mathbf{r} = \mathbf{r}_1 - \mathbf{r}_2$, prove that their equations of motion are given by

$$\ddot{\mathbf{r}}_1 = -\frac{Gm_2(\mathbf{r}_1 - \mathbf{r}_2)}{r^3}, \quad \ddot{\mathbf{r}}_2 = -\frac{Gm_1(\mathbf{r}_2 - \mathbf{r}_1)}{r^3}$$

This is called the *problem of two bodies.*

8.102. In Problem 8.101 choose a new origin at the center of mass of the two bodies, i.e. such that $m_1\mathbf{r}_1 + m_2\mathbf{r}_2 = \mathbf{0}$. Thus show that if we let $\mathbf{r}$ be the position vector of m_1 relative to m_2, then

$$\ddot{\mathbf{r}}_1 = -\frac{G(m_1 + m_2)\mathbf{r}_1}{r^3}, \quad \ddot{\mathbf{r}}_2 = -\frac{G(m_1 + m_2)\mathbf{r}_2}{r^3}$$

or, on subtracting,

$$\ddot{\mathbf{r}} = -\frac{G(m_1 + m_2)\mathbf{r}}{r^3}$$

Thus show that the motion of m_1 relative to m_2 is exactly the same as if the body of mass m_2 were fixed and its mass increased to $m_1 + m_2$.

8.103. Using Problem 8.102, obtain the orbit of mass m_1 relative to m_2 and compare with the results of Chapter 5. Are Kepler's first and second laws modified in any way? Explain.

8.104. If P is the period of revolution of m_1 about m_2 and a is the semi-major axis of the elliptical path of m_1 about m_2, prove that

$$\frac{P^2}{a^3} = \frac{4\pi^2}{G(m_1 + m_2)}$$

Compare this result with Kepler's third law: In the case of the earth [or other planet] and sun, does this modified Kepler law have much effect? Explain.

8.105. Set up equations for describing the motion of 3 bodies under a mutual inverse square law of attraction.

8.106. Transform the equations obtained in Problem 8.105 so that the positions of the bodies are described relative to their center of mass. Do you believe these equations can be solved exactly?

8.107. Work Problems 8.105 and 8.106 for N bodies.

Chapter 9 PLANE MOTION of RIGID BODIES

RIGID BODIES

A system of particles in which the distance between any two particles does not change regardless of the forces acting is called a *rigid body*. Since a rigid body is a special case of a system of particles, all theorems developed in Chapter 7 are also valid for rigid bodies.

TRANSLATIONS AND ROTATIONS

A *displacement* of a rigid body is a change from one position to another. If during a displacement all points of the body on some line remain fixed, the displacement is called a *rotation* about the line. If during a displacement all points of the rigid body move in lines parallel to each other the displacement is called a *translation*.

EULER'S THEOREM. INSTANTANEOUS AXIS OF ROTATION

The following theorem, called *Euler's theorem*, is fundamental in the motion of rigid bodies.

Theorem 9.1. A rotation of a rigid body about a fixed point of the body is equivalent to a rotation about a line which passes through the point.

The line referred to is called the *instantaneous axis of rotation*.

Rotations can be considered as finite or infinitesimal. Finite rotations cannot be represented by vectors since the commutative law fails. However, infinitesimal rotations can be represented by vectors.

GENERAL MOTION OF A RIGID BODY. CHASLE'S THEOREM

In the general motion of a rigid body, no point of the body may be fixed. In such case the following theorem, called *Chasle's theorem*, is fundamental.

Theorem 9.2. The general motion of a rigid body can be considered as a translation plus a rotation about a suitable point which is often taken to be the center of mass.

PLANE MOTION OF A RIGID BODY

The motion of a rigid body is simplified considerably when all points move parallel to a given fixed plane. In such case two types of motion, called *plane motion*, are possible.

1. **Rotation about a fixed axis.** In this case the rigid body rotates about a fixed axis perpendicular to the fixed plane. The system has only one degree of freedom [see Chapter 7, page 165] and thus only one coordinate is required for describing the motion.

2. **General plane motion.** In this case the motion can be considered as a translation parallel to the given fixed plane plus a rotation about a suitable axis perpendicular to the plane. This axis is often chosen so as to pass through the center of mass. The number of degrees of freedom for such motion is 3: two coordinates being used to describe the translation and one to describe the rotation.

The axis referred to is the *instantaneous axis* and the point where the instantaneous axis intersects the fixed plane is called the *instantaneous center of rotation* [see page 229].

We shall consider these two types of plane motion in this chapter. The motion of a rigid body in three dimensional space is more complicated and will be considered in Chapter 10.

MOMENT OF INERTIA

A geometric quantity which is of great importance in discussing the motion of rigid bodies is called the *moment of inertia.*

The *moment of inertia of a particle* of mass m about a line or *axis AB* is defined as

$$I = mr^2 \tag{1}$$

where r is the distance from the mass to the line.

The *moment of inertia of a system of particles,* with masses $m_1, m_2, \ldots, m_N$ about the line or axis AB is defined as

$$I = \sum_{\nu=1}^{N} m_\nu r_\nu^2 = m_1 r_1^2 + m_2 r_2^2 + \cdots + m_N r_N^2 \tag{2}$$

where $r_1, r_2, \ldots, r_N$ are their respective distances from AB.

The *moment of inertia of a continuous distribution of mass,* such as the solid rigid body $\mathcal{R}$ of Fig. 9-1, is given by

$$I = \int_{\mathcal{R}} r^2\, dm \tag{3}$$

where r is the distance of the element of mass dm from AB.

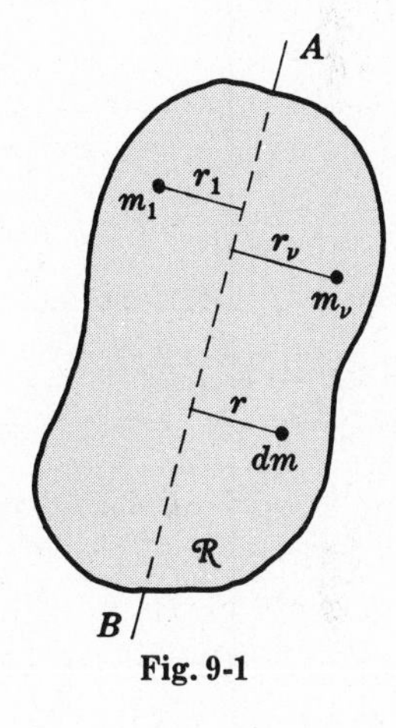

Fig. 9-1

RADIUS OF GYRATION

Let $I = \sum_{\nu=1}^{N} m_\nu r_\nu^2$ be the moment of inertia of a system of particles about AB, and $M = \sum_{\nu=1}^{N} m_\nu$ be the total mass of the system. Then the quantity K such that

$$K^2 = \frac{I}{M} = \frac{\sum_\nu m_\nu r_\nu^2}{\sum_\nu m_\nu} \tag{4}$$

is called the *radius of gyration* of the system about AB.

For continuous mass distributions (4) is replaced by

$$K^2 = \frac{I}{M} = \frac{\int_{\mathcal{R}} r^2\, dm}{\int_{\mathcal{R}} dm} \tag{5}$$

THEOREMS ON MOMENTS OF INERTIA

1. ***Theorem 9.3*: *Parallel Axis Theorem.*** Let I be the moment of inertia of a system about axis AB and let I_C be the moment of inertia of the system about an axis parallel to AB and passing through the center of mass of the system. Then if b is the distance between the axes and M is the total mass of the system, we have

$$I = I_C + Mb^2 \tag{6}$$

2. ***Theorem 9.4*: *Perpendicular Axes Theorem.*** Consider a mass distribution in the xy plane of an xyz coordinate system. Let I_x, I_y and I_z denote the moments of inertia about the x, y and z axes respectively. Then

$$I_z = I_x + I_y \tag{7}$$

SPECIAL MOMENTS OF INERTIA

The following table shows the moments of inertia of various rigid bodies which arise in practice. In all cases it is assumed that the body has uniform [i.e. constant] density.

Rigid Body	Moment of Inertia
1. *Solid Circular Cylinder* of radius a and mass M about axis of cylinder.	$\frac{1}{2}Ma^2$
2. *Hollow Circular Cylinder* of radius a and mass M about axis of cylinder. Wall thickness is negligible.	Ma^2
3. *Solid Sphere* of radius a and mass M about a diameter.	$\frac{2}{5}Ma^2$
4. *Hollow Sphere* of radius a and mass M about a diameter. Sphere thickness is negligible.	Ma^2
5. *Rectangular Plate* of sides a and b and mass M about an axis perpendicular to the plate through the center of mass.	$\frac{1}{12}M(a^2 + b^2)$
6. *Thin Rod* of length a and mass M about an axis perpendicular to the rod through the center of mass.	$\frac{1}{12}Ma^2$

COUPLES

A set of two equal and parallel forces which act in opposite directions but do not have the same line of action [see Fig. 9-2] is called a *couple*. Such a couple has a turning effect, and the *moment* or *torque of the couple* is given by $\mathbf{r} \times \mathbf{F}$.

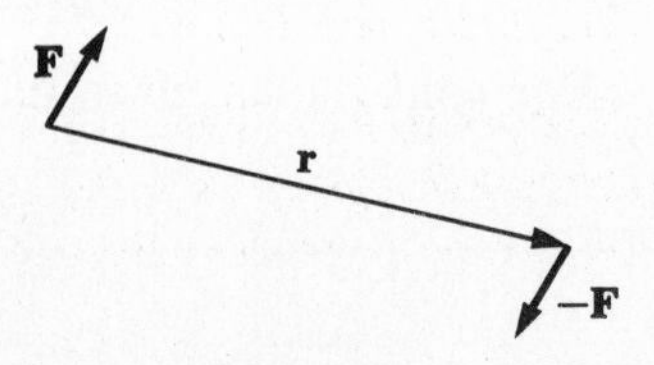

Fig. 9-2

The following theorem is important.

Theorem 9.5. Any system of forces which acts on a rigid body can be equivalently replaced by a single force which acts at some specified point together with a suitable couple.

KINETIC ENERGY AND ANGULAR MOMENTUM ABOUT A FIXED AXIS

Suppose a rigid body is rotating about a fixed axis with angular velocity $\boldsymbol{\omega}$ which has the direction of the axis AB [see Fig. 9-3]. Then the *kinetic energy of rotation* is given by

$$T = \tfrac{1}{2}I\omega^2 \tag{8}$$

where I is the moment of inertia of the rigid body about the axis.

Similarly the *angular momentum* is given by

$$\boldsymbol{\Omega} = I\boldsymbol{\omega} \tag{9}$$

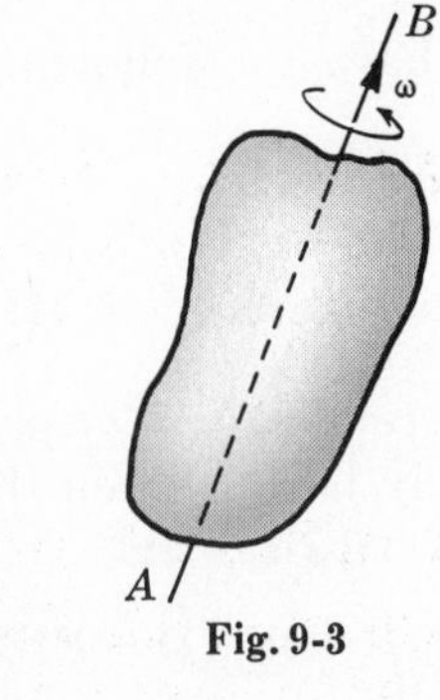

Fig. 9-3

MOTION OF A RIGID BODY ABOUT A FIXED AXIS

Two important methods for treating the motion of a rigid body about a fixed axis are given by the following theorems.

Theorem 9.6**: **Principle of Angular Momentum. If $\boldsymbol{\Lambda}$ is the *torque* or the moment of all external forces about the axis and $\boldsymbol{\Omega} = I\boldsymbol{\omega}$ is the angular momentum, then

$$\boldsymbol{\Lambda} = \frac{d}{dt}(I\boldsymbol{\omega}) = I\dot{\boldsymbol{\omega}} = I\boldsymbol{\alpha} \tag{10}$$

where $\boldsymbol{\alpha}$ is the angular acceleration.

Theorem 9.7**: **Principle of Conservation of Energy. If the forces acting on the rigid body are conservative so that the rigid body has a potential energy V, then

$$T + V = \tfrac{1}{2}I\omega^2 + V = E = \text{constant} \tag{11}$$

WORK AND POWER

Consider a rigid body $\mathcal{R}$ capable of rotating in a plane about an axis O perpendicular to the plane, as indicated in Fig. 9-4. If Λ is the magnitude of the torque applied to the body under the influence of force $\mathbf{F}$ at point $\mathbf{A}$, the *work done in rotating the body* through angle $d\theta$ is

$$dW = \Lambda\, d\theta \tag{12}$$

and the instantaneous *power* developed is

$$\mathcal{P} = \frac{dW}{dt} = \Lambda\omega \tag{13}$$

where ω is the angular speed.

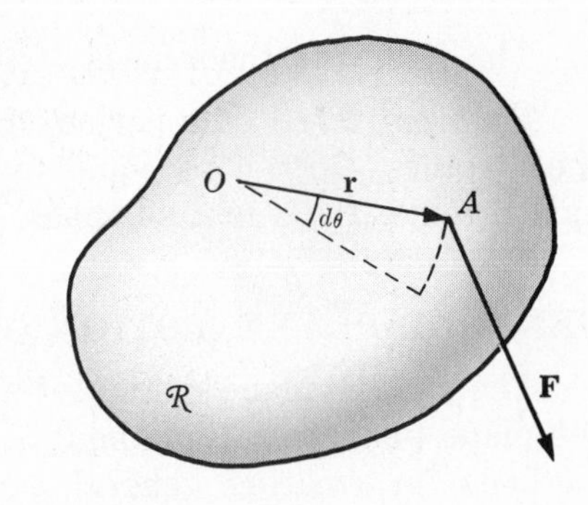

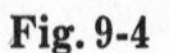
Fig. 9-4

We have the following

Theorem 9.8. The total work done in rotating a rigid body from an angle θ_1 where the angular speed is ω_1 to angle θ_2 where the angular speed is ω_2 is the difference in the kinetic energy of rotation at ω_1 and ω_2. In symbols,

$$\int_{\theta_1}^{\theta_2} \Lambda\, d\theta = \tfrac{1}{2}I\omega_2^2 - \tfrac{1}{2}I\omega_1^2 \tag{14}$$

IMPULSE. CONSERVATION OF ANGULAR MOMENTUM

The time integral of the torque

$$\mathcal{J} = \int_{t_1}^{t_2} \mathbf{\Lambda}\, dt \tag{15}$$

is called the *angular impulse* from time t_1 to t_2.

We have the following theorems.

Theorem 9.9. The angular impulse is equal to the change in angular momentum. In symbols

$$\int_{t_1}^{t_2} \mathbf{\Lambda}\, dt = \mathbf{\Omega}_2 - \mathbf{\Omega}_1 \tag{16}$$

Theorem 9.10*: *Conservation of Angular Momentum. If the net torque applied to a rigid body is zero, then the angular momentum is constant, i.e. is conserved.

THE COMPOUND PENDULUM

Let $\mathcal{R}$ [Fig. 9-5] be a rigid body which is free to oscillate in a vertical plane about a fixed horizontal axis through O under the influence of gravity. We call such a rigid body a *compound pendulum.*

Let C be the center of mass and suppose that the angle between OC and the vertical OA is θ. Then if I_0 is the moment of inertia of $\mathcal{R}$ about the horizontal axis through O, M is the mass of the rigid body and a is the distance OC, we have for the equation of motion,

$$\ddot{\theta} + \frac{Mga}{I_0}\sin\theta = 0 \tag{17}$$

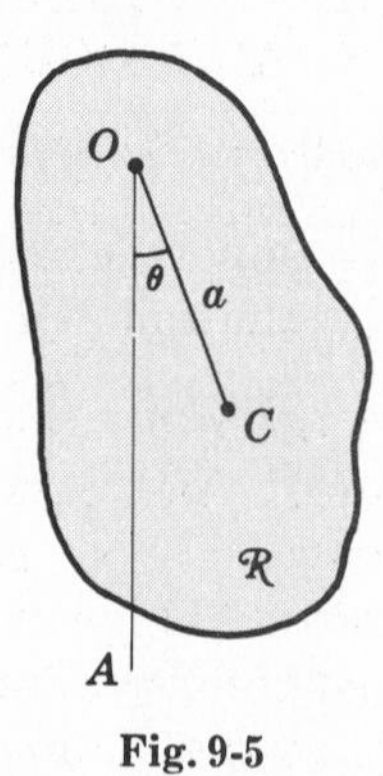

Fig. 9-5

For small oscillations the period of vibration is

$$P = 2\pi\sqrt{I_0/Mga} \tag{18}$$

The length of the equivalent simple pendulum is

$$l = I_0/Ma \tag{19}$$

The following theorem is of interest.

Theorem 9.11. The period of vibration of a compound pendulum is a minimum when the distance $OC = a$ is equal to the radius of gyration of the body about the horizontal axis through the center of mass.

GENERAL PLANE MOTION OF A RIGID BODY

The general plane motion of a rigid body can be considered as a translation parallel to the plane plus a rotation about a suitable axis perpendicular to the plane. Two important methods for treating general plane motion of a rigid body are given by the following theorems.

Theorem 9.12*: *Principle of Linear Momentum. If $\mathbf{r}$ is the position vector of the center of mass of a rigid body relative to an origin O, then

$$\frac{d}{dt}(M\dot{\mathbf{r}}) = M\ddot{\mathbf{r}} = \mathbf{F} \tag{20}$$

where M is the total mass, assumed constant, and $\mathbf{F}$ is the net external force acting on the body.

Theorem 9.13*: *Principle of Angular Momentum. If I_C is the moment of inertia of the rigid body about the center of mass, ω is the angular velocity and Λ_C is the torque or total moment of the external forces about the center of mass, then

$$\Lambda_C = \frac{d}{dt}(I_C \omega) = I_C \dot{\omega} \qquad (21)$$

Theorem 9.14*: *Principle of Conservation of Energy. If the external forces are conservative so that the potential energy of the rigid body is V, then

$$T + V = \tfrac{1}{2}m\dot{r}^2 + \tfrac{1}{2}I_C\omega^2 + V = E = \text{constant} \qquad (22)$$

Note that $\frac{1}{2}m\dot{r}^2 = \frac{1}{2}mv^2$ is the *kinetic energy of translation* and $\frac{1}{2}I_C\omega^2$ is the *kinetic energy of rotation* of the rigid body about the center of mass.

INSTANTANEOUS CENTER. SPACE AND BODY CENTRODES

Suppose a rigid body $\mathcal{R}$ moves parallel to a given fixed plane, say the xy plane of Fig. 9-6. Consider an $x'y'$ plane parallel to the xy plane and rigidly attached to the body.

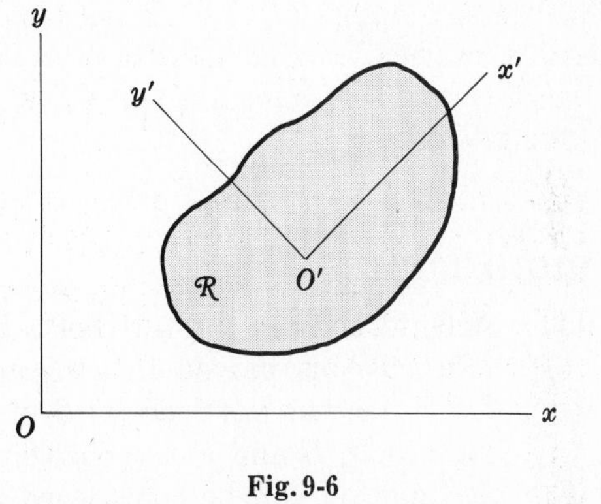

Fig. 9-6

As the body moves there will be at any time t a point of the moving $x'y'$ plane which is instantaneously at rest relative to the fixed xy plane. This point, which may or may not be in the body, is called the *instantaneous center*. The line perpendicular to the plane and passing through the instantaneous center is called the *instantaneous axis*.

As the body moves, the instantaneous center also moves. The locus or path of the instantaneous center relative to the fixed plane is called the *space locus* or *space centrode*. The locus relative to the moving plane is called the *body locus* or *body centrode*. The motion of the rigid body can be described as a rolling of the body centrode on the space centrode.

The instantaneous center can be thought of as that point about which there is rotation without translation. In a pure translation of a rigid body the instantaneous center is at infinity.

STATICS OF A RIGID BODY

The statics or equilibrium of a rigid body is the special case where there is no motion. The following theorem is fundamental.

Theorem 9.15. A necessary and sufficient condition for a rigid body to be in equilibrium is that

$$\mathbf{F} = 0, \quad \Lambda = 0 \qquad (23)$$

where $\mathbf{F}$ is the net external force acting on the body and Λ is the net external torque.

PRINCIPLE OF VIRTUAL WORK AND D'ALEMBERT'S PRINCIPLE

Since a rigid body is but a special case of a system of particles, the principle of virtual work and D'Alembert's principle [see page 171] apply to rigid bodies as well.

PRINCIPLE OF MINIMUM POTENTIAL ENERGY. STABILITY

At a position of equilibrium the net external force is zero, so that if the forces are conservative and V is the potential energy,

$$\mathbf{F} = -\nabla V = \mathbf{0} \tag{24}$$

or in components,

$$\frac{\partial V}{\partial x} = 0, \quad \frac{\partial V}{\partial y} = 0, \quad \frac{\partial V}{\partial z} = 0 \tag{25}$$

In such case V is either a minimum or it is not a minimum. If it is a minimum the equilibrium is said to be *stable* and a slight change of the configuration will restore the body to its original position. If it is not a minimum the body is said to be in *unstable equilibrium* and a slight change of the configuration will move the body away from its original position. We have the following theorem.

Theorem 9.16. A necessary and sufficient condition for a rigid body to be in stable equilibrium is that its potential energy be a minimum.

Solved Problems

RIGID BODIES

9.1. A rigid body in the form of a triangle ABC [Fig. 9-7] is moved in a plane to position DEF, i.e. the vertices A, B and C are carried to D, E and F respectively. Show that the motion can be considered as a translation plus a rotation about a suitable point.

Choose a point G on triangle ABC which corresponds to the point H on triangle DEF. Perform the translation in the direction GH so that triangle ABC is carried to $A'B'C'$. Using H as center of rotation perform the rotation of triangle $A'B'C'$ through the angle θ as indicated so that $A'B'C'$ is carried to DEF. Thus the motion has been accomplished by a translation plus a rotation.

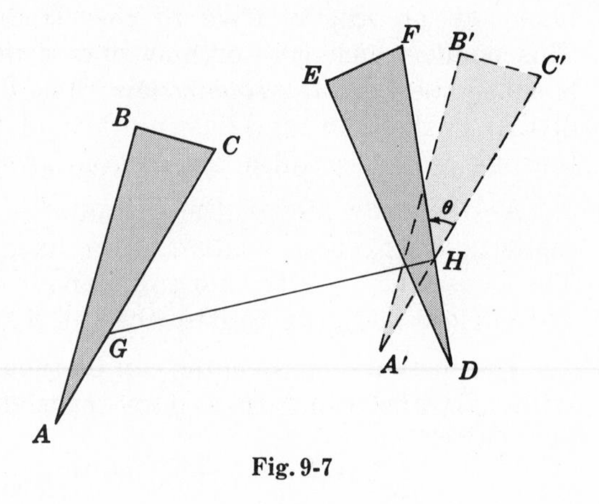

Fig. 9-7

9.2. Give an example to show that finite rotations cannot be represented by vectors.

Let A_x represent a rotation of a body [such as the rectangular parallelepiped of Fig. 9-8(*a*)] about the x axis while A_y represents a rotation about the y axis. We assume that such rotations take place in a positive or counterclockwise sense according to the right hand rule.

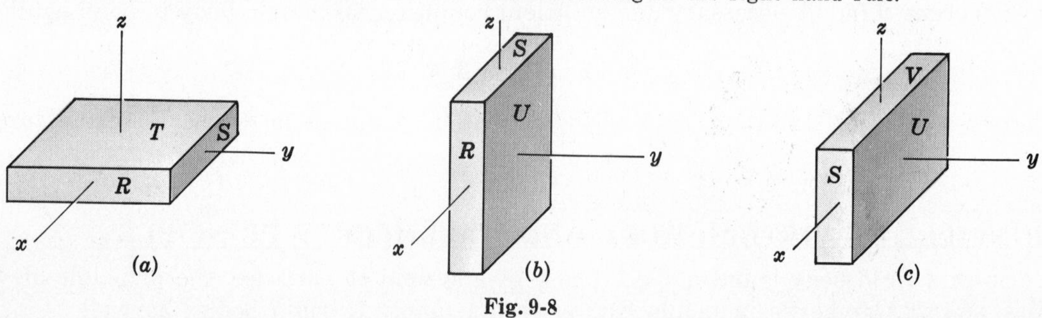

Fig. 9-8

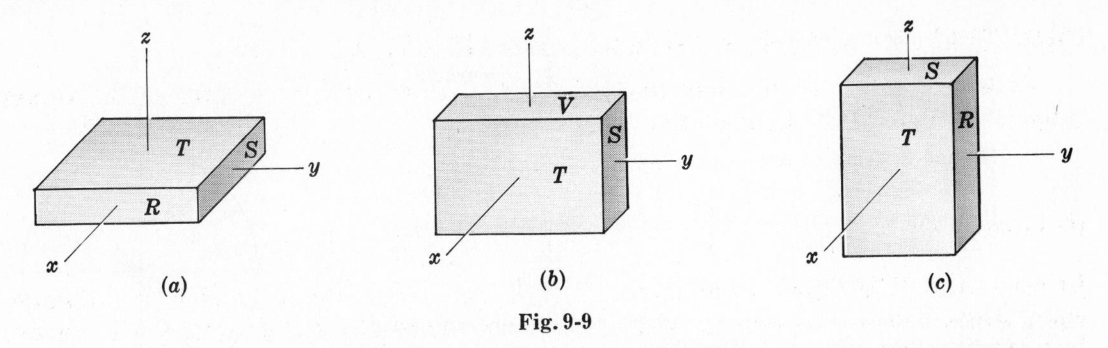

Fig. 9-9

In Fig. 9-8(a) we start with the parallelepiped in the indicated position and perform the rotation A_x about the x axis as indicated in Fig. 9-8(b) and then the rotation about the y axis as indicated in Fig. 9-8(c). Thus Fig. 9-8(c) is the result of the rotation $A_x + A_y$ on Fig. 9-8(a).

In Fig. 9-9(a) we start with the parallelepiped in the same position as in Fig. 9-8(a), but this time we first perform the rotation A_y about the y axis as indicated in Fig. 9-9(b) and then the rotation A_x about the x axis as indicated in Fig. 9-9(c). Thus Fig. 9-9(c) is the result of the rotation $A_y + A_x$ on Fig. 9-9(a).

Since the position of the parallelepiped of Fig. 9-8(c) is not the same as that of Fig. 9-9(c), we conclude that the operation $A_x + A_y$ is not the same as $A_y + A_x$. Thus the commutative law is not satisfied, so that A_x and A_y cannot possibly be represented by vectors.

MOMENTS OF INERTIA

9.3. Two particles of masses m_1 and m_2 respectively are connected by a rigid massless rod of length a and move freely in a plane. Show that the moment of inertia of the system about an axis perpendicular to the plane and passing through the center of mass is μa^2 where the *reduced mass* $\mu = m_1 m_2/(m_1 + m_2)$.

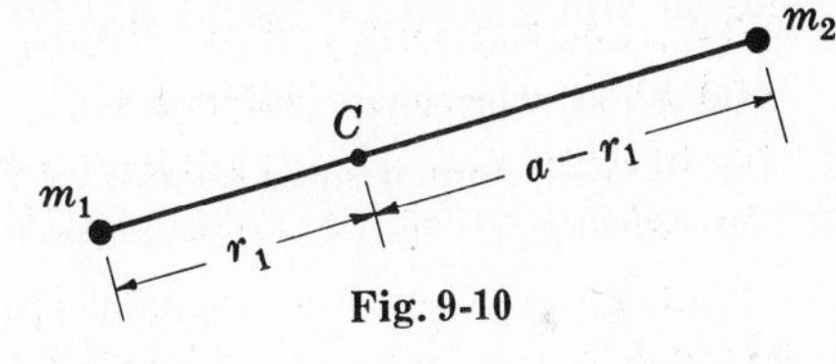

Fig. 9-10

Let r_1 be the distance of mass m_1 from the center of mass C. Then $a - r_1$ is the distance of mass m_2 from C. Since C is the center of mass,

$$m_1 r_1 = m_2(a - r_1) \quad \text{from which} \quad r_1 = \frac{m_2 a}{m_1 + m_2} \quad \text{and} \quad a - r_1 = \frac{m_1 a}{m_1 + m_2}$$

Thus the moment of inertia about an axis through C is

$$m_1 r_1^2 + m_2(a - r_1)^2 = m_1\left(\frac{m_2 a}{m_1 + m_2}\right)^2 + m_2\left(\frac{m_1 a}{m_1 + m_2}\right)^2 = \frac{m_1 m_2}{m_1 + m_2} a^2 = \mu a^2$$

9.4. Find the moment of inertia of a solid circular cylinder of radius a, height h and mass M about the axis of the cylinder.

Method 1, using single integration.

Subdivide the cylinder, a cross section of which appears in Fig. 9-11, into concentric rings one of which is the element shown shaded. The volume of this element is

$$(\text{Area})(\text{thickness}) = (2\pi r\, dr)(h) = 2\pi r h\, dr$$

and the element of mass is $dm = 2\pi\sigma r h\, dr$.

The moment of inertia of dm is

$$r^2\, dm = 2\pi\sigma r^3 h\, dr$$

where σ is the density, and thus the total moment of inertia is

$$I = \int_{r=0}^{a} 2\pi\sigma r^3 h\, dr = \tfrac{1}{2}\pi\sigma h a^4 \qquad (1)$$

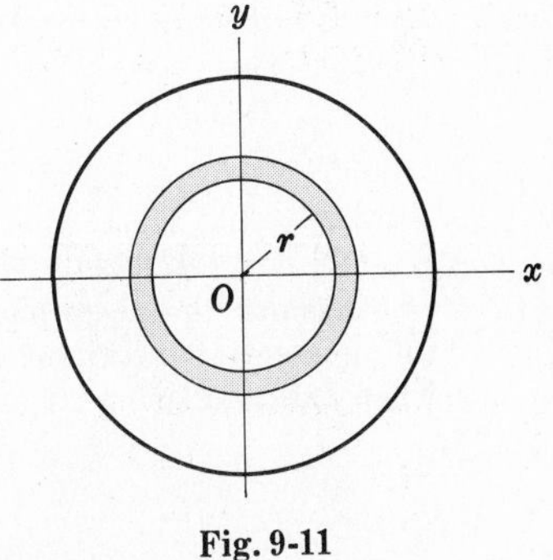

Fig. 9-11

Then since the mass is

$$M = \int_{r=0}^{a} 2\pi\sigma rh\,dr = \sigma\pi a^2 h \qquad (2)$$

we find $I = \frac{1}{2}Ma^2$.

Method 2, using double integration.

Using polar coordinates (r, θ), we see from Fig. 9-12 that the moment of inertia of the element of mass dm distant r from the axis is

$$r^2\,dm = r^2\sigma hr\,dr\,d\theta = \sigma hr^3\,dr\,d\theta$$

since $hr\,dr\,d\theta$ is the volume element and σ is the mass per unit volume (density). Then the total moment of inertia is

$$I = \int_{\theta=0}^{2\pi}\int_{r=0}^{a} \sigma hr^3\,dr\,d\theta = \tfrac{1}{2}\pi\sigma ha^4 \qquad (1)$$

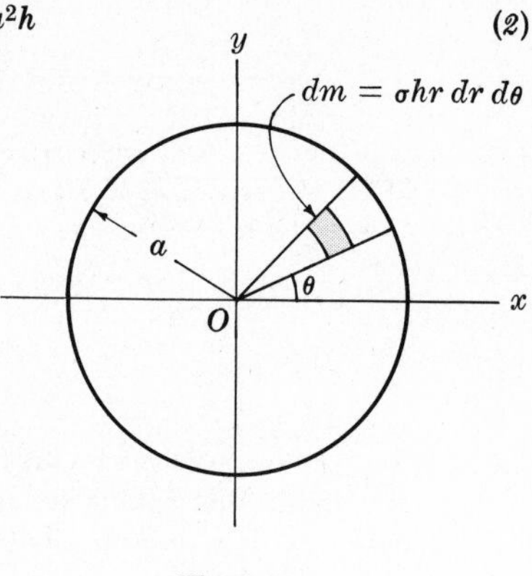

Fig. 9-12

The mass of the cylinder is given by

$$M = \int_{\theta=0}^{2\pi}\int_{r=0}^{a} \sigma hr\,dr\,d\theta = \sigma\pi a^2 h \qquad (2)$$

which can also be found directly by noting that the volume of the cylinder is $\pi a^2 h$. Dividing equation (1) by (2), we find $I/M = \frac{1}{2}a^2$ or $I = \frac{1}{2}Ma^2$.

9.5. Find the radius of gyration, K, of the cylinder of Problem 9.4.

Since $K^2 = I/M = \frac{1}{2}a^2$, $K = a/\sqrt{2} = \frac{1}{2}a\sqrt{2}$.

9.6. Find the (a) moment of inertia and (b) radius of gyration of a rectangular plate with sides a and b about a side.

Method 1, using single integration.

(a) The element of mass shaded in Fig. 9-13 is $\sigma b\,dx$, and its moment of inertia about the y axis is $(\sigma b\,dx)x^2 = \sigma bx^2\,dx$. Thus the total moment of inertia is

$$I = \int_{x=0}^{a} \sigma bx^2\,dx = \tfrac{1}{3}\sigma ba^3$$

Since the total mass of the plate is $M = ab\sigma$, we have $I/M = \frac{1}{3}a^2$ or $I = \frac{1}{3}Ma^2$.

(b) $K^2 = I/M = \frac{1}{3}a^2$ or $K = a/\sqrt{3} = \frac{1}{3}a\sqrt{3}$.

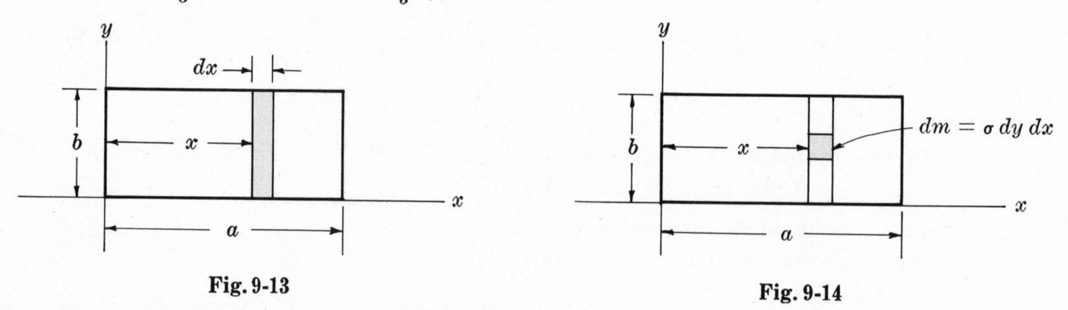

Fig. 9-13 Fig. 9-14

Method 2, using double integration.

Assume the plate has unit thickness. If $dm = \sigma\,dy\,dx$ is an element of mass [see Fig. 9-14], the moment of inertia of dm about the side which is chosen to be on the y axis is $x^2\,dm = \sigma x^2\,dy\,dx$. Then the total moment of inertia is

$$I = \int_{x=0}^{a}\int_{y=0}^{b} \sigma x^2\,dy\,dx = \tfrac{1}{3}\sigma ba^3$$

The total mass of the plate is $M = ab\sigma$. Then, as in Method 1, we find $I = \frac{1}{3}Ma^2$ and $K = \frac{1}{3}a\sqrt{3}$.

9.7. Find the moment of inertia of a right circular cone of height h and radius a about its axis.

Method 1, using single integration.

The moment of inertia of the circular cylindrical disc one quarter of which is represented by PQR in Fig. 9-15 is, by Problem 9.4,

$$\tfrac{1}{2}(\pi r^2 \sigma\, dz)(r^2) \;=\; \tfrac{1}{2}\pi\sigma r^4\, dz$$

since this disc has volume $\pi r^2\, dz$ and radius r.

From Fig. 9-15, $\dfrac{h-z}{h} = \dfrac{r}{a}$ or $r = a\left(\dfrac{h-z}{h}\right)$.

Then the total moment of inertia about the z axis is

$$I \;=\; \tfrac{1}{2}\pi\sigma \int_{z=0}^{h} \left\{a\left(\frac{h-z}{h}\right)\right\}^4 dz \;=\; \tfrac{1}{10}\pi a^4 \sigma h$$

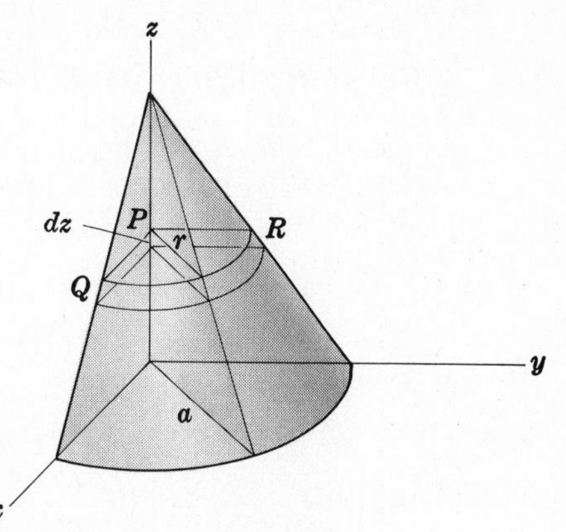

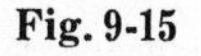
Fig. 9-15

Also,

$$M \;=\; \pi\sigma \int_{z=0}^{h} \left\{a\left(\frac{h-z}{h}\right)\right\}^2 dz \;=\; \tfrac{1}{3}\pi a^2 h\sigma$$

Thus $I = \frac{3}{10}Ma^2$.

Method 2, using triple integration.

Subdivide the cone, one quarter of which is shown in Fig. 9-16, into elements of mass dm as indicated in the figure.

In cylindrical coordinates (r, θ, z) the element of mass dm of the cylinder is $dm = \sigma r\, dr\, d\theta\, dz$ where σ is the density.

The moment of inertia of dm about the z axis is

$$r^2\, dm \;=\; \sigma r^3\, dr\, d\theta\, dz$$

As in Method 1, $\dfrac{h-z}{h} = \dfrac{r}{a}$ or $z = h\left(\dfrac{a-r}{a}\right)$.

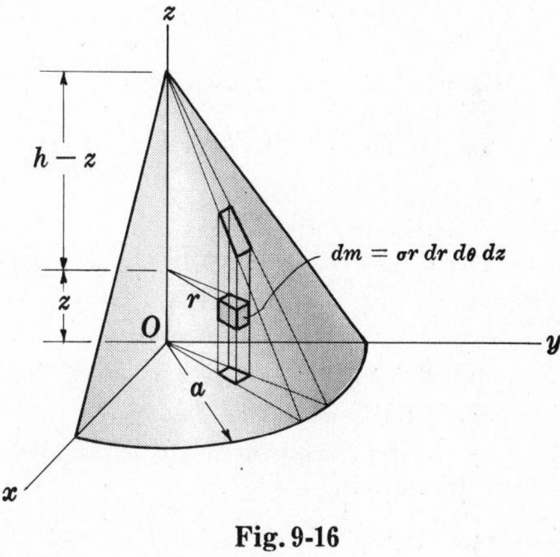

Fig. 9-16

Then the total moment of inertia about the z axis is

$$I \;=\; \int_{\theta=0}^{2\pi}\int_{r=0}^{a}\int_{z=0}^{h(a-r)/a} \sigma r^3\, dr\, d\theta\, dz \;=\; \tfrac{1}{10}\pi a^4 \sigma h$$

The total mass of the cone is

$$M \;=\; \int_{\theta=0}^{2\pi}\int_{r=0}^{a}\int_{z=0}^{h(a-r)/r} \sigma r\, dr\, d\theta\, dz \;=\; \tfrac{1}{3}\pi a^2 h\sigma$$

which can be obtained directly by noting that the volume of the cone is $\frac{1}{3}\pi a^2 h$.

Thus $I = \frac{3}{10}Ma^2$.

9.8. Find the radius of gyration K of the cone of Problem 9.7.

$K^2 = I/M = \frac{3}{10}a^2$ and $K = a\sqrt{\frac{3}{10}} = \frac{1}{10}a\sqrt{30}$.

THEOREMS ON MOMENTS OF INERTIA

9.9. Prove the parallel axis theorem [Theorem 9.3, page 226].

Let OQ be any axis and ACP a parallel axis through the centroid C and distant b from OQ. In Fig. 9-17 below, OQ has been chosen as the z axis so that AP is perpendicular to the xy plane at P.

If $\mathbf{b}_1$ is a unit vector in the direction OP, then the vector OP is given by

$$\mathbf{b} = b\mathbf{b}_1 \tag{1}$$

where b is constant and is the distance between axes.

Let $\mathbf{r}_\nu$ and $\mathbf{r}'_\nu$ be the position vectors of mass m_ν relative to O and C respectively. If $\bar{\mathbf{r}}$ is the position vector of C relative to O then we have

$$\mathbf{r}_\nu = \mathbf{r}'_\nu + \bar{\mathbf{r}} \tag{2}$$

The total moment of inertia of all masses m_ν about axis OQ is

$$I = \sum_{\nu=1}^{N} m_\nu(\mathbf{r}_\nu \cdot \mathbf{b}_1)^2 \tag{3}$$

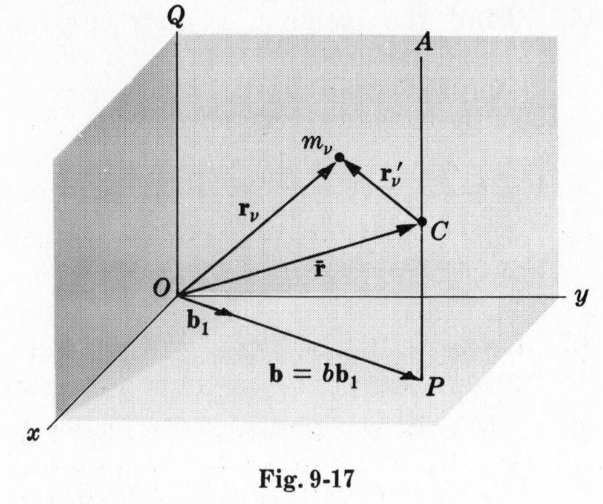

Fig. 9-17

The total moment of inertia of all masses m_ν about axis ACP is

$$I_C = \sum_{\nu=1}^{N} m_\nu(\mathbf{r}'_\nu \cdot \mathbf{b}_1)^2 \tag{4}$$

Then using (2) we find

$$\begin{aligned} I &= \sum_{\nu=1}^{N} m_\nu(\mathbf{r}_\nu \cdot \mathbf{b}_1)^2 = \sum_{\nu=1}^{N} m_\nu(\mathbf{r}'_\nu \cdot \mathbf{b}_1 + \bar{\mathbf{r}} \cdot \mathbf{b}_1)^2 \\ &= \sum_{\nu=1}^{N} m_\nu(\mathbf{r}'_\nu \cdot \mathbf{b}_1)^2 + 2\sum_{\nu=1}^{N} m_\nu(\mathbf{r}'_\nu \cdot \mathbf{b}_1)(\bar{\mathbf{r}} \cdot \mathbf{b}_1) + \sum_{\nu=1}^{N} m_\nu(\bar{\mathbf{r}} \cdot \mathbf{b}_1)^2 \\ &= I_C + 2b\left(\sum_{\nu=1}^{N} m_\nu \mathbf{r}'_\nu\right) \cdot \mathbf{b}_1 + b^2 \sum_{\nu=1}^{N} m_\nu = I_C + Mb^2 \end{aligned}$$

since $\bar{\mathbf{r}} \cdot \mathbf{b}_1 = b$, $\sum_{\nu=1}^{N} m_\nu = M$ and $\sum_{\nu=1}^{N} m_\nu \mathbf{r}'_\nu = \mathbf{0}$ [Problem 7.16, page 178].

The result is easily extended to continuous mass systems by using integration in place of summation.

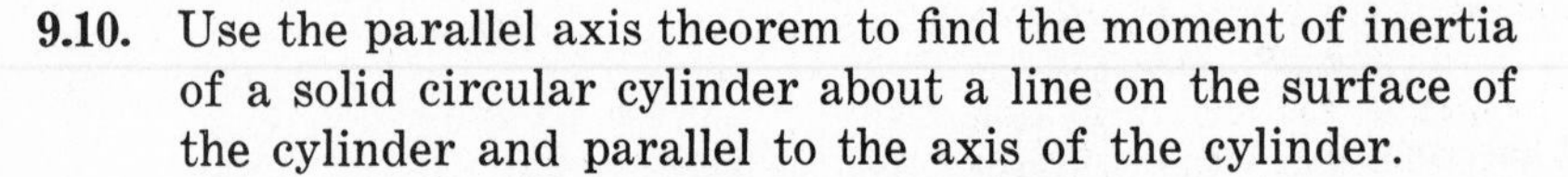

9.10. Use the parallel axis theorem to find the moment of inertia of a solid circular cylinder about a line on the surface of the cylinder and parallel to the axis of the cylinder.

Suppose the cross section of the cylinder is represented as in Fig. 9-18. Then the axis is represented by C, while the line on the surface of the cylinder is represented by A.

If a is the radius of the cylinder, then by Problem 9.4 and the parallel axis theorem we have

$$I_A = I_C + Ma^2 = \tfrac{1}{2}Ma^2 + Ma^2 = \tfrac{3}{2}Ma^2$$

Fig. 9-18

9.11. Prove the perpendicular axes theorem [Theorem 9.4, page 226].

Let the position vector of the particle with mass m_ν in the xy plane be

$$\mathbf{r}_\nu = x_\nu\mathbf{i} + y_\nu\mathbf{j}$$

[see Fig. 9-19]. The moment of inertia of m_ν about the z axis is $m_\nu|\mathbf{r}_\nu|^2$.

Then the total moment of inertia of all particles about the z axis is

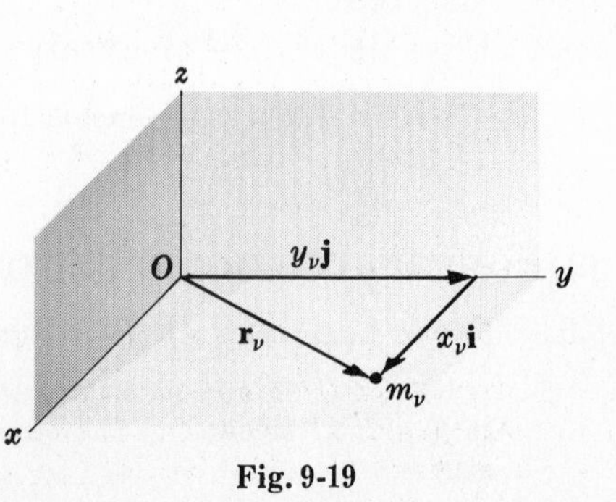

Fig. 9-19

$$I_z = \sum_{\nu=1}^{N} m_\nu |\mathbf{r}_\nu|^2 = \sum_{\nu=1}^{N} m_\nu (x_\nu^2 + y_\nu^2)$$

$$= \sum_{\nu=1}^{N} m_\nu x_\nu^2 + \sum_{\nu=1}^{N} m_\nu y_\nu^2 = I_x + I_y$$

where I_x and I_y are the total moments of inertia about the x axis and y axis respectively.

The result is easily extended to continuous systems.

9.12. Find the moment of inertia of a rectangular plate with sides a and b about an axis perpendicular to the plate and passing through a vertex.

Choose the rectangular plate [see Fig. 9-20] in the xy plane with sides on the x and y axes. Choose the z axis perpendicular to the plate at a vertex.

From Problem 9.6 we have for the moments of inertia about the x and y axes,

$$I_x = \tfrac{1}{3}Mb^2, \qquad I_y = \tfrac{1}{3}Ma^2$$

Then by the perpendicular axes theorem the moment of inertia about the z axis is

$$I_z = I_x + I_y = \tfrac{1}{3}M(b^2 + a^2)$$

$$= \tfrac{1}{3}M(a^2 + b^2)$$

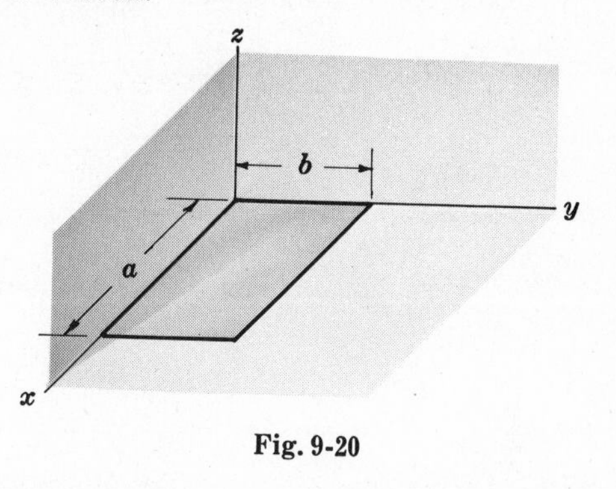

Fig. 9-20

COUPLES

9.13. Prove that a force acting at a point of a rigid body can be equivalently replaced by a single force acting at some specified point together with a suitable couple.

Let the force be $\mathbf{F}_1$ acting at point P_1 as in Fig. 9-21. If Q is any specified point, it is seen that the effect of $\mathbf{F}_1$ alone is the same if we apply two forces $\mathbf{f}_1$ and $-\mathbf{f}_1$ at Q.

In particular if we choose $\mathbf{f}_1 = -\mathbf{F}_1$, i.e. if $\mathbf{f}_1$ has the same magnitude as $\mathbf{F}_1$ but is opposite in direction, we see that the effect of $\mathbf{F}_1$ alone is the same as the effect of the couple formed by $\mathbf{F}_1$ and $\mathbf{f}_1 = -\mathbf{F}_1$ [which has moment $\mathbf{r}_1 \times \mathbf{F}_1$] together with the force $-\mathbf{f}_1 = \mathbf{F}_1$.

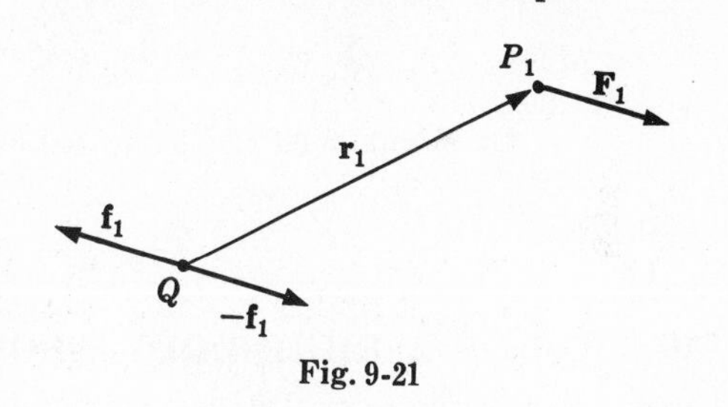

Fig. 9-21

9.14. Prove Theorem 9.5, page 227: Any system of forces which acts on a rigid body can be equivalently replaced by a single force which acts at some specified point together with a suitable couple.

By Problem 9.13 we can replace the force $\mathbf{F}_\nu$ at P_ν by the force $\mathbf{F}_\nu$ at Q plus a couple of moment $\mathbf{r}_\nu \times \mathbf{F}_\nu$. Then the system of forces $\mathbf{F}_1, \mathbf{F}_2, \ldots, \mathbf{F}_N$ at points $P_1, P_2, \ldots, P_N$ can be combined into forces $\mathbf{F}_1, \mathbf{F}_2, \ldots, \mathbf{F}_N$ at Q having resultant

$$\mathbf{F} = \mathbf{F}_1 + \mathbf{F}_2 + \cdots + \mathbf{F}_N$$

together with couples having moments

$$\mathbf{r}_1 \times \mathbf{F}_1, \ \mathbf{r}_2 \times \mathbf{F}_2, \ \ldots, \ \mathbf{r}_N \times \mathbf{F}_N$$

which may be added to yield a single couple. Thus the system of forces can be equivalently replaced by the single force $\mathbf{F}$ acting at Q together with a couple.

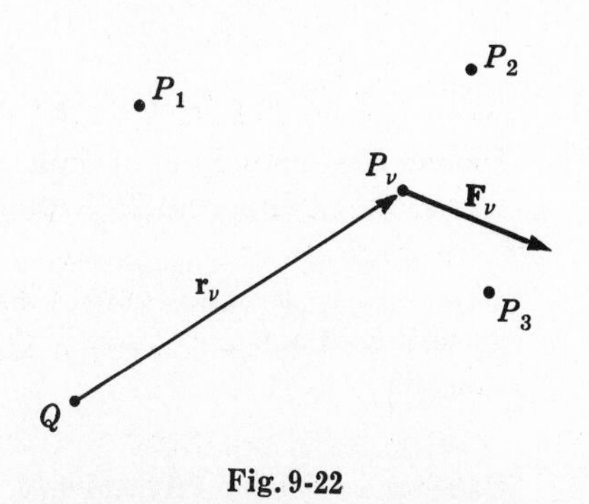

Fig. 9-22

KINETIC ENERGY AND ANGULAR MOMENTUM

9.15. If a rigid body rotates about a fixed axis with angular velocity $\boldsymbol{\omega}$, prove that the kinetic energy of rotation is $T = \frac{1}{2}I\omega^2$ where I is the moment of inertia about the axis.

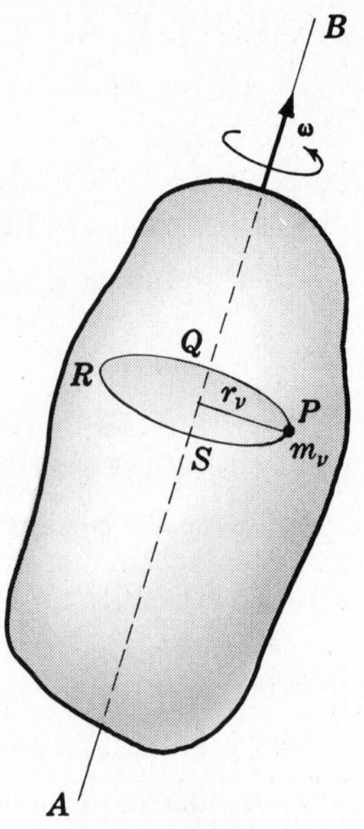

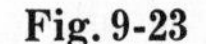
Fig. 9-23

Choose the axis as AB in Fig. 9-23. A particle P of mass m_ν will rotate about the axis with angular speed ω. Then it will describe a circle $PQRSP$ with linear speed $v_\nu = \omega r_\nu$ where r_ν is its distance from axis AB. Thus its kinetic energy of rotation about AB is $\frac{1}{2}m_\nu v_\nu^2 = \frac{1}{2}m_\nu \omega^2 r_\nu^2$, and the total kinetic energy of all particles is

$$T = \sum_{\nu=1}^{N} \tfrac{1}{2}m_\nu \omega^2 r_\nu^2 = \tfrac{1}{2}\left(\sum_{\nu=1}^{N} m_\nu r_\nu^2\right)\omega^2$$
$$= \tfrac{1}{2}I\omega^2$$

where $I = \sum_{\nu=1}^{N} m_\nu r_\nu^2$ is the moment of inertia about AB.

The result could also be proved by using integration in place of summation.

9.16. Prove that the angular momentum of the rigid body of Problem 9.15 is $\boldsymbol{\Omega} = I\boldsymbol{\omega}$.

The angular momentum of particle P about axis AB is $m_\nu r_\nu^2 \boldsymbol{\omega}$. Then the total angular momentum of all particles about axis AB is

$$\boldsymbol{\Omega} = \sum_{\nu=1}^{N} m_\nu r_\nu^2 \boldsymbol{\omega} = \left(\sum_{\nu=1}^{N} m_\nu r_\nu^2\right)\boldsymbol{\omega} = I\boldsymbol{\omega}$$

where $I = \sum_{\nu=1}^{N} m_\nu r_\nu^2$ is the moment of inertia about AB.

The result could also be proved by using integration in place of summation.

MOTION OF A RIGID BODY ABOUT A FIXED AXIS

9.17. Prove the principle of angular momentum for a rigid body rotating about a fixed axis [Theorem 9.6, page 227].

By Problem 7.12, page 176, since a rigid body is a special case of a system of particles, $\boldsymbol{\Lambda} = d\boldsymbol{\Omega}/dt$ where $\boldsymbol{\Lambda}$ is the torque or moment of all external forces about the axis and $\boldsymbol{\Omega}$ is the total angular momentum about the axis.

Since $\boldsymbol{\Omega} = I\boldsymbol{\omega}$ by Problem 9.16, $\boldsymbol{\Lambda} = \dfrac{d}{dt}(I\boldsymbol{\omega}) = I\dfrac{d\boldsymbol{\omega}}{dt} = I\dot{\boldsymbol{\omega}}$.

9.18. Prove the principle of conservation of energy for a rigid body rotating about a fixed axis [Theorem 9.7, page 227] provided the forces acting are conservative.

The principle of conservation of energy applies to any system of particles in which the forces acting are conservative. Hence in particular it applies to the special case of a rigid body rotating about a fixed axis. If T and V are the total kinetic energy and the potential energy, we thus have

$$T + V = \text{constant} = E$$

Using the result of Problem 9.15, this can be written $\frac{1}{2}I\omega^2 + V = E$.

WORK, POWER AND IMPULSE

9.19. Prove equation (*12*), page 227, for the work done in rotating a rigid body about a fixed axis.

Refer to Fig. 9-4, page 227. Let the angular velocity of the body be $\boldsymbol{\omega} = \omega \mathbf{k}$ where $\mathbf{k}$ is a unit vector in the direction of the axis of rotation. The work done by $\mathbf{F}$ is

$$\begin{aligned} dW &= \mathbf{F}\cdot d\mathbf{r} = \mathbf{F}\cdot\frac{d\mathbf{r}}{dt}\,dt = \mathbf{F}\cdot\mathbf{v}\,dt = \mathbf{F}\cdot(\boldsymbol{\omega}\times\mathbf{r})\,dt \\ &= (\mathbf{r}\times\mathbf{F})\cdot\boldsymbol{\omega}\,dt = \mathbf{\Lambda}\cdot\boldsymbol{\omega}\,dt = \Lambda\omega\,dt = \Lambda\,d\theta \end{aligned}$$

where in the last two steps we use $\mathbf{\Lambda} = \Lambda\mathbf{k}$, $\boldsymbol{\omega} = \omega\mathbf{k}$ and $\omega = d\theta/dt$.

9.20. Prove equation (*13*), page 227, for the power developed.

From Problem 9.19 and the fact that $d\theta/dt = \omega$,

$$\mathcal{P} = dW/dt = \Lambda\,d\theta/dt = \Lambda\omega$$

9.21. Prove Theorem 9.8, page 227.

We have $\mathbf{\Lambda} = I\,d\boldsymbol{\omega}/dt$ so that $\Lambda = I\,d\omega/dt$. Then from Problem 9.19 and the fact that $d\theta = \omega\,dt$, we have

$$\text{Work done} = \int_{\theta_1}^{\theta_2} \Lambda\,d\theta = \int_{t_1}^{t_2} I\frac{d\omega}{dt}\omega\,dt = \int_{\omega_1}^{\omega_2} I\omega\,d\omega = \tfrac{1}{2}I\omega_2^2 - \tfrac{1}{2}I\omega_1^2$$

9.22. Prove Theorem 9.9, page 228: The angular impulse is equal to the change in angular momentum.

$$\int_{t_1}^{t_2} \mathbf{\Lambda}\,dt = \int_{t_1}^{t_2} \frac{d\mathbf{\Omega}}{dt}\,dt = \mathbf{\Omega}_2 - \mathbf{\Omega}_1$$

9.23. Prove Theorem 9.10, page 229, on the conservation of angular momentum if the net torque is zero.

From Problem 9.22, if $\mathbf{\Lambda} = \mathbf{0}$ then $\mathbf{\Omega}_2 = \mathbf{\Omega}_1$.

THE COMPOUND PENDULUM

9.24. Obtain the equation of motion (*17*), page 228, for a compound pendulum.

Method 1.

Suppose that the vertical plane of vibration of the pendulum is chosen as the xy plane [Fig. 9-24] where the z axis through origin O is the horizontal axis of suspension.

Let point C have the position vector $\mathbf{a}$ relative to O. Since the body is rigid, $|\mathbf{a}| = a$ is constant and is the distance from O to C.

The only external force acting on the body is its weight $M\mathbf{g} = -Mg\mathbf{j}$ acting vertically downward. Thus we have

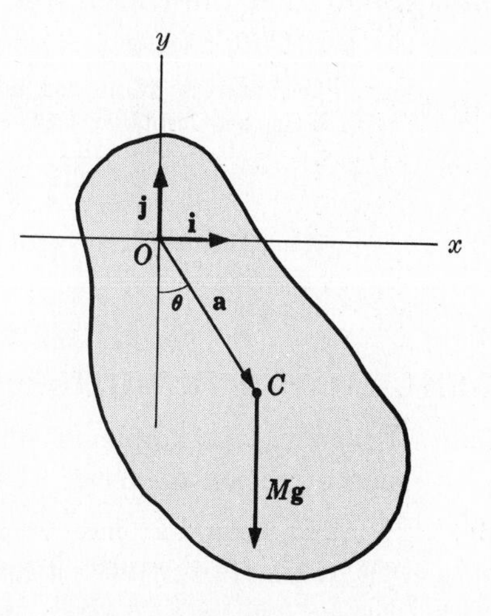

Fig. 9-24

$$\begin{aligned} \mathbf{\Lambda} &= \text{total external torque about } z \text{ axis} \\ &= \mathbf{a}\times M\mathbf{g} = -\mathbf{a}\times Mg\mathbf{j} = aMg\sin\theta\,\mathbf{k} \qquad (1) \end{aligned}$$

where $\mathbf{k}$ is a unit vector in the positive z direction [out of the plane of the paper toward the reader].

Also, the instantaneous angular velocity is

$$\boldsymbol{\omega} = -\omega\mathbf{k} = -\frac{d\theta}{dt}\mathbf{k} = -\dot{\theta}\,\mathbf{k} \qquad (2)$$

so that if I_0 is the moment of inertia about the z axis

$$\boldsymbol{\Omega} = \text{angular momentum about } z \text{ axis} = I_0\boldsymbol{\omega} = -I_0\dot{\theta}\,\mathbf{k}$$

Substituting from (1) and (2) into $\mathbf{\Lambda} = d\boldsymbol{\Omega}/dt$,

$$aMg \sin\theta\,\mathbf{k} = \frac{d}{dt}(-I_0\dot{\theta}\,\mathbf{k}) \qquad \text{or} \qquad \ddot{\theta} + \frac{Mga}{I_0}\sin\theta = 0 \tag{3}$$

Method 2.

The force $M\mathbf{g} = -Mg\mathbf{j}$ is conservative, so that the potential energy V is such that

$$-\nabla V = -\frac{\partial V}{\partial x}\mathbf{i} - \frac{\partial V}{\partial y}\mathbf{j} - \frac{\partial V}{\partial z}\mathbf{k} = -Mg\mathbf{j} \qquad \text{or} \qquad \frac{\partial V}{\partial x} = 0,\ \frac{\partial V}{\partial y} = Mg,\ \frac{\partial V}{\partial z} = 0$$

from which
$$V = Mgy + c = -Mga\cos\theta + c \tag{4}$$

since $y = -a\cos\theta$. This could be seen directly since $y = -a\cos\theta$ is the height of C below the x axis taken as the reference level.

By Problem 9.15, the kinetic energy of rotation is $\frac{1}{2}I_0\omega^2 = \frac{1}{2}I_0\dot{\theta}^2$. Then the principle of conservation of energy gives

$$T + V = \tfrac{1}{2}I_0\dot{\theta}^2 - Mga\cos\theta = \text{constant} = E \tag{5}$$

Differentiating equation (5) with respect to t,

$$I_0\dot{\theta}\ddot{\theta} + Mga\sin\theta\,\dot{\theta} = 0$$

or, since $\dot{\theta}$ is not identically zero, $I_0\ddot{\theta} + Mga\sin\theta = 0$ as required.

9.25. Show that for small vibrations the pendulum of Problem 9.24 has period $P = 2\pi\sqrt{Mga/I_0}$.

For small vibrations we can make the approximation $\sin\theta = \theta$ so that the equation of motion becomes

$$\ddot{\theta} + \frac{Mga}{I_0}\theta = 0 \tag{1}$$

Then, as in Problem 4.23, page 102, we find that the period is $P = 2\pi\sqrt{I_0/Mga}$.

9.26. Show that the length l of a simple pendulum equivalent to the compound pendulum of Problem 9.24 is $l = I_0/Ma$.

The equation of motion corresponding to a simple pendulum of length l suspended vertically from O is [see Problem 4.23, equation (2), page 102]

$$\ddot{\theta} + \frac{g}{l}\sin\theta = 0 \tag{1}$$

Comparing this equation with (1) of Problem 9.25, we see that $l = I_0/Ma$.

GENERAL PLANE MOTION OF A RIGID BODY

9.27. Prove the principle of linear momentum, Theorem 9.12, page 228, for the general plane motion of a rigid body.

This follows at once from the corresponding theorem for systems of particles [Theorem 7-1, page 167], since rigid bodies are special cases.

9.28. Prove the principle of angular momentum, Theorem 9.13, page 229, for general plane motion of a rigid body.

This follows at once from the corresponding theorem for systems of particles [Theorem 7-4, page 168], since rigid bodies are special cases.

9.29. A solid cylinder of radius a and mass M rolls without slipping down an inclined plane of angle α. Show that the acceleration is constant and equal to $\frac{2}{3}g \sin \alpha$.

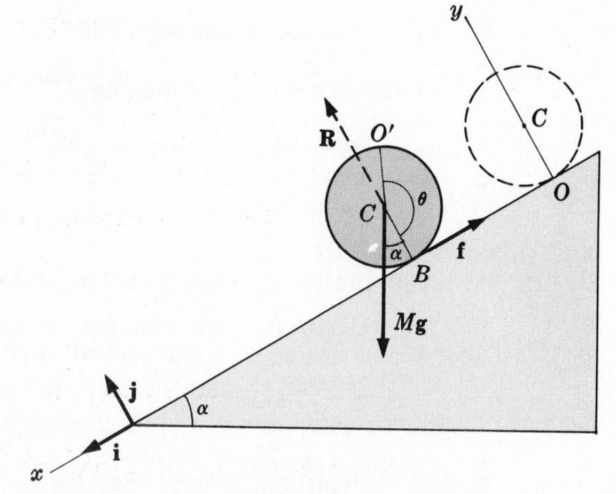

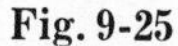
Fig. 9-25

Suppose that initially the cylinder has point O in contact with the plane and that after time t the cylinder has rotated through angle θ [see Fig. 9-25].

The forces acting on the cylinder at time t are: (i) the weight $M\mathbf{g}$ acting vertically downward at the center of mass C; (ii) the reaction $\mathbf{R}$ of the inclined plane acting perpendicular to the plane; (iii) the frictional force $\mathbf{f}$ acting upward along the incline.

Choose the plane in which motion takes place as the xy plane, where the x axis is taken as positive down the incline and the origin is at O.

If $\mathbf{r}$ is the position of the center of mass at time t, then by the principle of linear momentum,

$$M\ddot{\mathbf{r}} = M\mathbf{g} + \mathbf{R} + \mathbf{f} \tag{1}$$

But $\mathbf{g} = g \sin \alpha \, \mathbf{i} - g \cos \alpha \, \mathbf{j}$, $\mathbf{R} = R\mathbf{j}$, $\mathbf{f} = -f\mathbf{i}$. Hence (1) can be written

$$M\ddot{\mathbf{r}} = (Mg \sin \alpha - f)\mathbf{i} + (R - Mg \cos \alpha)\mathbf{j} \tag{2}$$

The total external torque about the horizontal axis through the center of mass is

$$\mathbf{\Lambda} = \mathbf{0} \times M\mathbf{g} + \mathbf{0} \times \mathbf{R} + \mathbf{CB} \times \mathbf{f} = \mathbf{CB} \times \mathbf{f} = (-a\mathbf{j}) \times (-f\mathbf{i}) = -af\,\mathbf{k} \tag{3}$$

The total angular momentum about the horizontal axis through the center of mass is

$$\mathbf{\Omega} = I_C \boldsymbol{\omega} = I_C(-\dot{\theta}\mathbf{k}) = -I_C \dot{\theta}\,\mathbf{k} \tag{4}$$

where I_C is the moment of inertia of the cylinder about this axis.

Substituting (3) and (4) into $\mathbf{\Lambda} = d\mathbf{\Omega}/dt$, we find $-af\mathbf{k} = -I_C\ddot{\theta}\mathbf{k}$ or $I_C\ddot{\theta} = af$.

Using $\mathbf{r} = x\mathbf{i} + y\mathbf{j}$ in (2), we obtain

$$M\ddot{x} = Mg \sin \alpha - f, \qquad M\ddot{y} = R - Mg \cos \alpha \tag{5}$$

Now if there is no slipping, $x = a\theta$ or $\theta = x/a$. Similarly, since the cylinder remains on the incline, $\ddot{y} = 0$; hence from (5), $R = Mg \cos \alpha$.

Using $\theta = x/a$ in $I_C\ddot{\theta} = af$, we have $f = I_C\ddot{x}/a^2$. From Problem 9.4, $I_C = \frac{1}{2}Ma^2$. Then substituting $f = \frac{1}{2}M\ddot{x}$ into the first equation of (5), we obtain $\ddot{x} = \frac{2}{3}g \sin \alpha$ as required.

9.30. Prove that in Problem 9.29 the coefficient of friction must be at least $\frac{1}{3} \tan \alpha$.

The coefficient of friction is $\mu = f/R$.

From Problem 9.29 we have $f = \frac{1}{2}M\ddot{x} = \frac{1}{3}Mg \sin \alpha$ and $R = Mg \cos \alpha$. Thus in order that slipping will not occur, μ must be at least $f/R = \frac{1}{3} \tan \alpha$.

9.31. (*a*) Work Problem 9.29 if the coefficient of friction between the cylinder and inclined plane is μ and (*b*) discuss the motion for different values of μ.

(*a*) In equation (5) of Problem 9.29, substitute $f = \mu R = \mu Mg \cos \alpha$ and obtain

$$\ddot{x} = g(\sin \alpha - \mu \cos \alpha)$$

Note that in this case the center of mass of the cylinder moves in the same manner as a particle sliding down an inclined plane. However, the cylinder may slip as well as roll.

The acceleration due to rolling is $$a\ddot{\theta} = \frac{a^2 f}{I_C} = \frac{a^2 \mu M g \cos\alpha}{\frac{1}{2}Ma^2} = 2\mu g \cos\alpha.$$

The acceleration due to slipping is $$\ddot{x} - a\ddot{\theta} = g(\sin\alpha - 3\mu\cos\alpha).$$

(*b*) If $(\sin\alpha - 3\mu\cos\alpha) > 0$, i.e. $\mu < \frac{1}{3}\tan\alpha$, then slipping will occur. If $(\sin\alpha - 3\mu\cos\alpha) \leqq 0$, i.e. $\mu \geqq \frac{1}{3}\tan\alpha$, then rolling but no slipping will occur. These results are consistent with those of Problem 9.30.

9.32. Prove the principle of conservation of energy [Theorem 9.14, page 229].

This follows from the corresponding theorem for systems of particles, Theorem 7-7, page 169. The total kinetic energy T is the sum of the kinetic energy of translation of the center of mass plus the kinetic energy of rotation about the center of mass, i.e.,

$$T = \tfrac{1}{2}m\dot{\mathbf{r}}^2 + \tfrac{1}{2}I_C\omega^2$$

If V is the potential energy, then the principle of conservation of energy states that if E is a constant,

$$T + V = \tfrac{1}{2}m\dot{\mathbf{r}}^2 + \tfrac{1}{2}I_C\omega^2 + V = E$$

9.33. Work Problem 9.29 by using the principle of conservation of energy.

The potential energy is composed of the potential energy due to the external forces [in this case gravity] and the potential energy due to internal forces [which is a constant and can be omitted]. Taking the reference level as the base of the plane and assuming that the height of the center of mass above this plane initially and at any time t to be H and h respectively, we have

$$\tfrac{1}{2}M\dot{\mathbf{r}}^2 + \tfrac{1}{2}I_C\omega^2 + Mgh = MgH$$

or, using $H - h = x\sin\alpha$ and $\dot{\mathbf{r}}^2 = \dot{x}^2 + \dot{y}^2 = \dot{x}^2$ since $\dot{y} = 0$,

$$\tfrac{1}{2}M\dot{x}^2 + \tfrac{1}{2}I_C\omega^2 = Mgx\sin\alpha$$

Substituting $\omega = \dot{\theta} = \dot{x}/a$ and $I_C = \frac{1}{2}Ma^2$, we find $\dot{x}^2 = \frac{4}{3}gx\sin\alpha$. Differentiating with respect to t, we obtain

$$2\dot{x}\ddot{x} = \tfrac{4}{3}g\dot{x}\sin\alpha \qquad \text{or} \qquad \ddot{x} = \tfrac{2}{3}g\sin\alpha$$

INSTANTANEOUS CENTER. SPACE AND BODY CENTRODES

9.34. Find the position vector of the instantaneous center for a rigid body moving parallel to a given fixed plane.

Choose the XY plane of Fig. 9-26 as the fixed plane and the xy plane as the plane attached to and moving with the rigid body $\mathcal{R}$. Let point P of the xy plane [which may or may not be in the rigid body] have position vectors $\mathbf{R}$ and $\mathbf{r}$ relative to the XY and xy planes respectively. If $\mathbf{v}$ and $\mathbf{v}_A$ are the respective velocities of P and A relative to the XY system,

$$\mathbf{v} = \mathbf{v}_A + \boldsymbol{\omega}\times\mathbf{r} = \mathbf{v}_A + \boldsymbol{\omega}\times(\mathbf{R} - \mathbf{R}_A) \qquad (1)$$

where $\mathbf{R}_A$ is the position vector of A relative to O. If P is to be the instantaneous center, then $\mathbf{v} = \mathbf{0}$ so that

$$\boldsymbol{\omega}\times(\mathbf{R} - \mathbf{R}_A) = -\mathbf{v}_A \qquad (2)$$

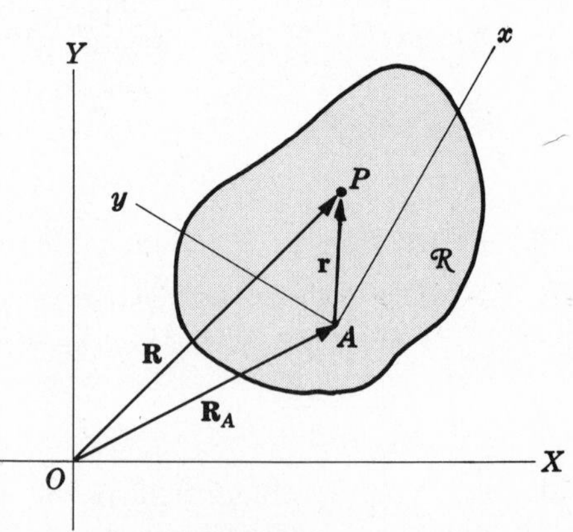

Fig. 9-26

Multiplying both sides of (2) by $\boldsymbol{\omega}\times$ and using (7), page 5,

$$\boldsymbol{\omega}\{\boldsymbol{\omega}\cdot(\mathbf{R} - \mathbf{R}_A)\} - (\mathbf{R} - \mathbf{R}_A)(\boldsymbol{\omega}\cdot\boldsymbol{\omega}) = -\boldsymbol{\omega}\times\mathbf{v}_A$$

Then since $\boldsymbol{\omega}$ is perpendicular to $\mathbf{R} - \mathbf{R}_A$, this becomes

$$(\mathbf{R} - \mathbf{R}_A)\omega^2 = \boldsymbol{\omega}\times\mathbf{v}_A \qquad \text{or} \qquad \mathbf{R} = \mathbf{R}_A + \frac{\boldsymbol{\omega}\times\mathbf{v}_A}{\omega^2} \qquad (3)$$

9.35. A cylinder moves along a horizontal plane. Find the (*a*) space centrode, (*b*) body centrode. Discuss the case where slipping may occur.

(*a*) The general motion is one where both rolling and slipping may occur. Suppose the cylinder is moving to the right with velocity $\mathbf{v}_A$ [the velocity of its center of mass] and is rotating about A with angular velocity $\boldsymbol{\omega}$.

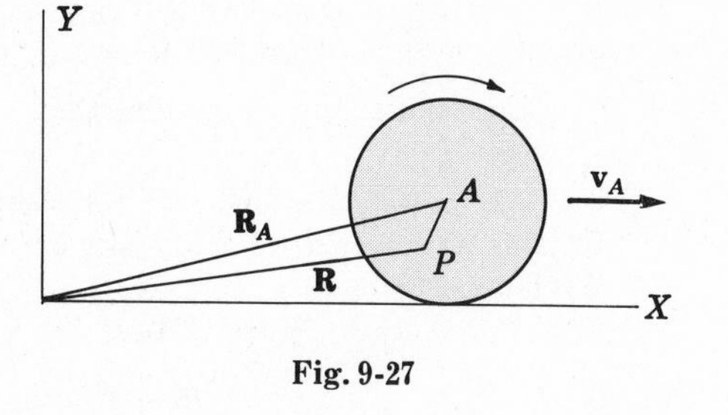

Fig. 9-27

Since $\boldsymbol{\omega} = -\omega\mathbf{k}$ and $\mathbf{v}_A = v_A\mathbf{i}$, we have $\boldsymbol{\omega} \times \mathbf{v}_A = -\omega v_A \mathbf{j}$ so that (*3*) of Problem 9.34 becomes

$$\mathbf{R} = \mathbf{R}_A - \frac{(\omega v_A)\mathbf{j}}{\omega^2} = \mathbf{R}_A - \frac{v_A}{\omega}\mathbf{j}$$

In component form,

$$X\mathbf{i} + Y\mathbf{j} = X_A\mathbf{i} + a\mathbf{j} - (v_A/\omega)\mathbf{j} \qquad \text{or} \qquad X = X_A, \quad Y = a - v_A/\omega$$

Thus the instantaneous center is located vertically above the point of contact of the cylinder with the ground and at height $a - v_A/\omega$ above it.

Then the space centrode is a line parallel to the horizontal and at distance $a - v_A/\omega$ above it. If there is no slipping, then $v_A = a\omega$ and the space centrode is the X axis while the instantaneous center is the point of contact of the cylinder with the X axis.

(*b*) The body centrode is given by $|\mathbf{r}| = v_0/\omega$, or a circle of radius v_0/ω. In case of no slipping, $v_0 = a\omega$ and the body centrode is the circumference of the cylinder.

9.36. Solve Problem 9.29 by using the instantaneous center.

By Problem 9.35, if there is no slipping then the point of contact P of the cylinder with the plane is the instantaneous center. The motion of P is parallel to the motion of the center of mass, so that we can use the result of Problem 7.86(*c*), page 191.

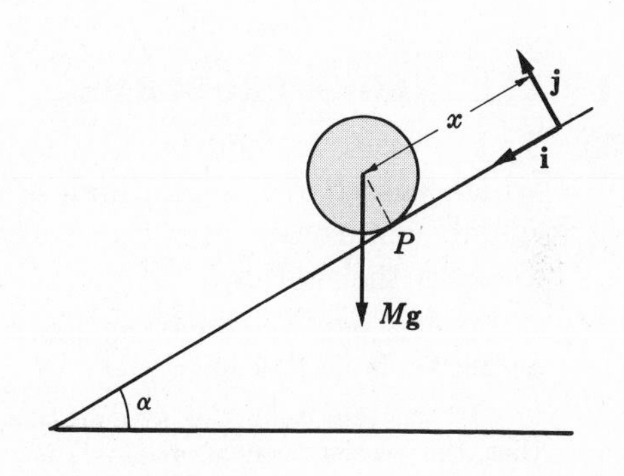

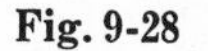
Fig. 9-28

The moment of inertia of the cylinder about P is, by the parallel axis theorem, $\frac{1}{2}Ma^2 + Ma^2 = \frac{3}{2}Ma^2$. The torque about the horizontal axis through P is $Mga \sin\alpha$. Thus

$$\frac{d}{dt}(\tfrac{3}{2}Ma^2\dot{\theta}) = Mga \sin\theta$$

or

$$\ddot{\theta} = \frac{2g}{3a}\sin\theta$$

Since $x = a\theta$, the acceleration is $\ddot{x} = \frac{2}{3}g \sin\theta$.

STATICS OF A RIGID BODY

9.37. A ladder of length l and weight W_l has one end against a vertical wall which is frictionless and the other end on the ground assumed horizontal. The ladder makes an angle α with the ground. Prove that a man of weight W_m will be able to climb the ladder without having it slip if the coefficient of friction μ between the ladder and the ground is at least $\dfrac{W_m + \frac{1}{2}W_l}{W_m + W_l}\cot\alpha$.

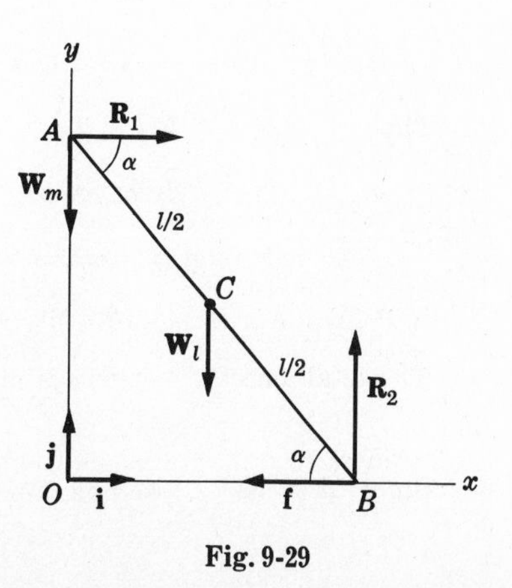

Fig. 9-29

Let the ladder be represented by AB in Fig. 9-29 and choose an xy coordinate system as indicated.

The most dangerous situation in which the ladder would slip occurs when the man is at the top of the ladder. Hence we would require that the ladder be in equilibrium in such case.

The forces acting on the ladder are: (i) the reaction $\mathbf{R}_1 = R_1\mathbf{i}$ of the wall; (ii) the weight $\mathbf{W}_m = -W_m\mathbf{j}$ of the man; (iii) the weight $\mathbf{W}_l = -W_l\mathbf{j}$ of the ladder concentrated at C, the center of gravity; (iv) the reaction $\mathbf{R}_2 = R_2\mathbf{j}$ of the ground; (v) the friction force $\mathbf{f} = -f\mathbf{i}$.

For equilibrium we require that

$$\mathbf{F} = \mathbf{0}, \qquad \mathbf{\Lambda} = \mathbf{0} \tag{1}$$

where $\mathbf{F}$ is the total external force on the ladder and $\mathbf{\Lambda}$ the total external torque taken about a suitable axis which we shall take as the horizontal axis through A perpendicular to the xy plane. We have

$$\mathbf{F} = \mathbf{R}_1 + \mathbf{W}_m + \mathbf{W}_l + \mathbf{R}_2 + \mathbf{f} = (R_1 - f)\mathbf{i} + (-W_m - W_l + R_2)\mathbf{j} = \mathbf{0}$$

if

$$R_1 - f = 0 \qquad \text{and} \qquad -W_m - W_l + R_2 = 0 \tag{2}$$

Also,

$$\begin{aligned}\mathbf{\Lambda} &= (\mathbf{0}) \times \mathbf{R}_1 + (\mathbf{0}) \times \mathbf{W}_m + (\mathbf{AC}) \times \mathbf{W}_l + (\mathbf{AB}) \times \mathbf{R}_2 + (\mathbf{AB}) \times \mathbf{f} \\ &= (\mathbf{0}) \times (R_1\mathbf{i}) + (\mathbf{0}) \times (-W_m\mathbf{j}) + (\tfrac{1}{2}l\cos\alpha\,\mathbf{i} - \tfrac{1}{2}l\sin\alpha\,\mathbf{j}) \times (-W_l\mathbf{j}) \\ &\qquad + (l\cos\alpha\,\mathbf{i} - l\sin\alpha\,\mathbf{j}) \times (R_2\mathbf{j}) + (l\cos\alpha\,\mathbf{i} - l\sin\alpha\,\mathbf{j}) \times (-f\mathbf{i}) \\ &= -\tfrac{1}{2}lW_l\cos\alpha\,\mathbf{k} + lR_2\cos\alpha\,\mathbf{k} - lf\sin\alpha\,\mathbf{k} = \mathbf{0}\end{aligned}$$

if

$$-\tfrac{1}{2}W_l\cos\alpha + R_2\cos\alpha - f\sin\alpha = 0 \tag{3}$$

Solving simultaneously equations (2) and (3), we find

$$f = R_1 = (W_m + \tfrac{1}{2}W_l)\cot\alpha \qquad \text{and} \qquad R_2 = W_m + W_l$$

Then the minimum coefficient of friction necessary to prevent slipping of the ladder is

$$\mu = \frac{f}{R_2} = \frac{W_m + \frac{1}{2}W_l}{W_m + W_l}\cot\alpha$$

MISCELLANEOUS PROBLEMS

9.38. **Two masses m_1 and m_2 are connected by an inextensible string of negligible mass which passes over a frictionless pulley of mass M, radius a and radius of gyration K which can rotate about a horizontal axis through C perpendicular to the pulley. Discuss the motion.**

Choose unit vectors $\mathbf{i}$ and $\mathbf{j}$ in the plane of rotation as shown in Fig. 9-30.

If we represent the acceleration of mass m_1 by $A\mathbf{j}$, then the acceleration of mass m_2 is $-A\mathbf{j}$.

Choose the tensions $\mathbf{T}_1$ and $\mathbf{T}_2$ in the string as shown in the figure. By Newton's second law,

$$m_1A\mathbf{j} = \mathbf{T}_1 + m_1\mathbf{g} = -T_1\mathbf{j} + m_1g\mathbf{j} \tag{1}$$

$$-m_2A\mathbf{j} = \mathbf{T}_2 + m_2\mathbf{g} = -T_2\mathbf{j} + m_2g\mathbf{j} \tag{2}$$

Thus $\quad m_1A = m_1g - T_1, \qquad m_2A = T_2 - m_2g \qquad (3)$

or $\quad T_1 = m_1(g - A), \qquad T_2 = m_2(g + A) \qquad (4)$

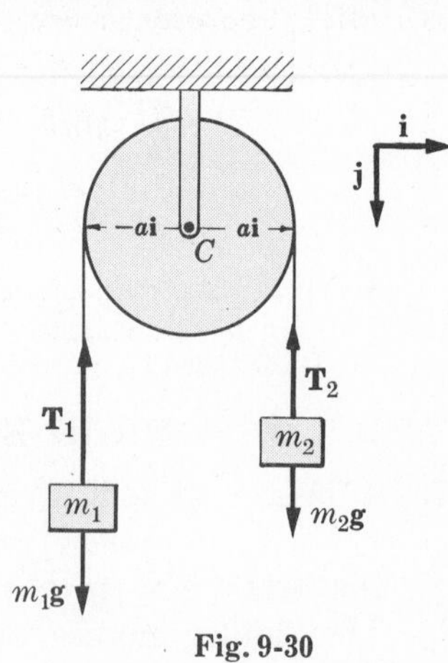

Fig. 9-30

The net external torque about the axis through C is

$$\mathbf{\Lambda} = (-a\mathbf{i}) \times (-T_1\mathbf{j}) + (a\mathbf{i}) \times (-T_2\mathbf{j}) = a(T_1 - T_2)\mathbf{k} \tag{5}$$

The total angular momentum about O is

$$\mathbf{\Omega} = I_C\boldsymbol{\omega} = I_C\omega\mathbf{k} = I_C\dot{\theta}\mathbf{k} \tag{6}$$

Since $\mathbf{\Lambda} = d\mathbf{\Omega}/dt$, we find from (5) and (6),

$$a(T_1 - T_2) = I_C\ddot{\theta} = MK^2\ddot{\theta} \tag{7}$$

If there is no slipping about the pulley, we also have

$$A = a\ddot{\theta} \tag{8}$$

Using (8) in (7),
$$T_1 - T_2 = \frac{MK^2}{a^2}A \tag{9}$$

Using (4) in (9),
$$A = \frac{(m_1 - m_2)g}{m_1 + m_2 + Mk^2/a^2} \tag{10}$$

Thus the masses move with constant acceleration given in magnitude by (10). Note that if $M = 0$, the result (10) reduces to that of Problem 3.22, page 76.

9.39. Find the moment of inertia of a solid sphere about a diameter.

Let O be the center of the sphere and AOB be the diameter about which the moment of inertia is taken [Fig. 9-31]. Divide the sphere into discs such as $QRSTQ$ perpendicular to AOB and having center on AOB at P.

Take the radius of the sphere equal to a, $OP = z$, $SP = r$ and the thickness of the disc equal to dz. Then by Problem 9.4 the moment of inertia of the disc about AOB is

$$\tfrac{1}{2}(\pi r^2 \sigma\, dz)r^2 = \tfrac{1}{2}\pi\sigma r^4\, dz \tag{1}$$

From triangle OSP, $r^2 = a^2 - z^2$. Substituting into (1), the total moment of inertia is

$$I = \int_{z=-a}^{a} \tfrac{1}{2}\pi\sigma(a^2 - z^2)^2\, dz = \tfrac{8}{15}\pi\sigma a^5 \tag{2}$$

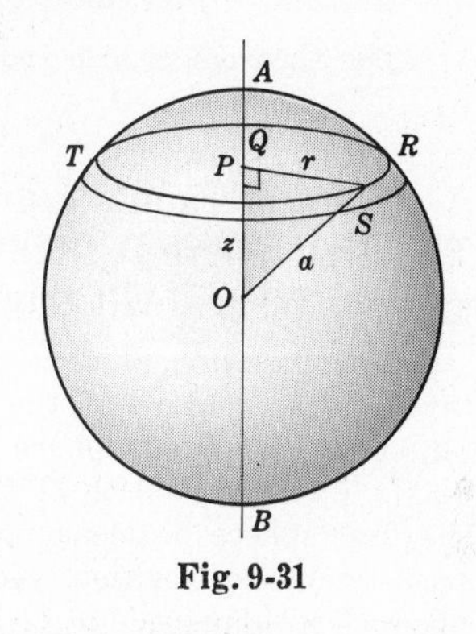

Fig. 9-31

The mass of the sphere is

$$M = \int_{z=-a}^{a} \pi\sigma(a^2 - z^2)\, dz = \tfrac{4}{3}\pi a^3 \sigma \tag{3}$$

which could also be seen by noting that the volume of the sphere is $\frac{4}{3}\pi a^3$.

From (2) and (3) we have $I/M = \frac{2}{5}a^2$ or $I = \frac{2}{5}Ma^2$.

9.40. A cube of edge s and mass M is suspended vertically from one of its edges. (a) Show that the period for small vibrations is $P = 2\pi\sqrt[4]{2}\sqrt{s/3g}$. (b) What is the length of the equivalent simple pendulum?

(a) Since the diagonal of a square of side s has length $\sqrt{s^2 + s^2} = s\sqrt{2}$, the distance OC from axis O to the center of mass is $\frac{1}{2}s\sqrt{2}$.

The moment of inertia I of a cube about an edge is the same as that of a square plate about a side. Thus by Problem 9.6, $I = \frac{1}{3}M(s^2 + s^2) = \frac{2}{3}Ms^2$.

Then the period for small vibrations is, by Problem 9.25,

$$P = 2\pi\sqrt{\tfrac{2}{3}Ms^2/[Mg(\tfrac{1}{2}s\sqrt{2})]} = 2\pi\sqrt[4]{2}\sqrt{s/3g}$$

(b) The length of the equivalent simple pendulum is, by Problem 9.26,

$$l = \tfrac{2}{3}Ms^2/[M(\tfrac{1}{2}s\sqrt{2})] = \tfrac{2}{3}\sqrt{2}\, s$$

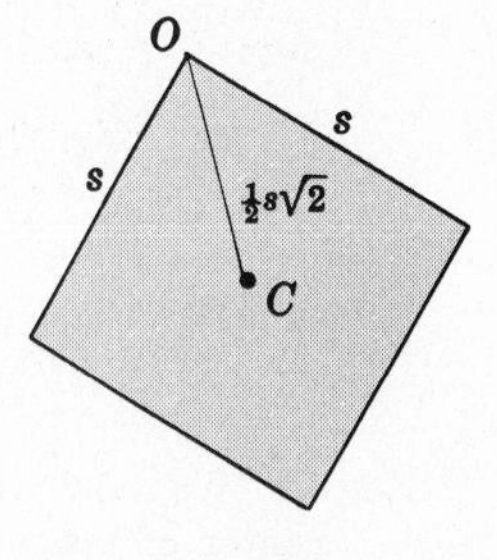

Fig. 9-32

9.41. Prove Theorem 9.11, page 228: The period of small vibrations of a compound pendulum is a minimum when the distance $OC = a$ is equal to the radius of gyration of the body about a horizontal axis through the center of mass.

If I_C is the moment of inertia about the center of mass axis and I_0 is the moment of inertia about the axis of suspension, then by the parallel axis theorem we have

$$I_0 = I_C + Ma^2$$

Then the square of the period for small vibrations is given by

$$P^2 = \frac{4\pi^2 I_0}{Mga} = \frac{4\pi^2}{g}\left(\frac{I_C}{Ma} + a\right) = \frac{4\pi^2}{g}\left(\frac{K_C^2}{a} + a\right)$$

where $K_C^2 = I_C/M$ is the square of the radius of gyration about the center of mass axis.

Setting the derivative of P^2 with respect to a equal to zero, we find

$$\frac{d}{da}(P^2) = \frac{4\pi^2}{g}\left(-\frac{K_C^2}{a^2} + 1\right) = 0$$

from which $a = K_C$. This can be shown to give the minimum value since $d^2(P^2)/da^2 < 0$. Thus the theorem is proved.

The theorem is also true even if the vibrations are not assumed small. See Problem 9.147.

9.42. A sphere of radius a and mass m rests on top of a fixed rough sphere of radius b. The first sphere is slightly displaced so that it rolls without slipping down the second sphere. Where will the first sphere leave the second sphere?

Let the xy plane be chosen so as to pass through the centers of the two spheres, with the center of the fixed sphere as origin O [see Fig. 9-33]. Let the position of the center of mass C of the first sphere be measured by angle θ, and suppose that the position vector of this center of mass C with respect to O is $\mathbf{r}$. Let $\mathbf{r}_1$ and $\boldsymbol{\theta}_1$ be unit vectors as indicated in Fig. 9-33.

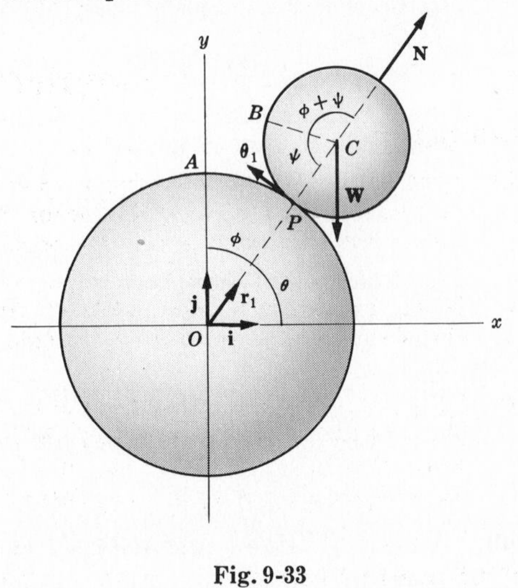

Fig. 9-33

Resolving the weight $\mathbf{W} = -mg\mathbf{j}$ into components in directions $\mathbf{r}_1$ and $\boldsymbol{\theta}_1$, we have [compare Problem 1.43, page 24]

$$\begin{aligned}\mathbf{W} &= (\mathbf{W}\cdot\mathbf{r}_1)\mathbf{r}_1 + (\mathbf{W}\cdot\boldsymbol{\theta}_1)\boldsymbol{\theta}_1 \\ &= (-mg\mathbf{j}\cdot\mathbf{r}_1)\mathbf{r}_1 + (-mg\mathbf{j}\cdot\boldsymbol{\theta}_1)\boldsymbol{\theta}_1 \\ &= -mg\sin\theta\,\mathbf{r}_1 - mg\cos\theta\,\boldsymbol{\theta}_1\end{aligned}$$

The reaction force $\mathbf{N}$ and frictional force $\mathbf{f}$ are $\mathbf{N} = N\mathbf{r}_1$, $\mathbf{f} = f\boldsymbol{\theta}_1$. Using Theorem 9.12, page 228, together with the result of Problem 1.49, page 26, we have

$$\begin{aligned}\mathbf{F} = m\mathbf{a} &= m[(\ddot{r} - r\dot{\theta}^2)\mathbf{r}_1 + (r\ddot{\theta} + 2\dot{r}\dot{\theta})\boldsymbol{\theta}_1] \\ &= \mathbf{W} + \mathbf{N} + \mathbf{f} \\ &= (N - mg\sin\theta)\mathbf{r}_1 + (f - mg\cos\theta)\boldsymbol{\theta}_1\end{aligned}$$

from which $\quad m(\ddot{r} - r\dot{\theta}^2) = N - mg\sin\theta, \qquad m(r\ddot{\theta} + 2\dot{r}\dot{\theta}) = f - mg\cos\theta \qquad (1)$

Since $r = a + b$ [the distance of C from O], these equations become

$$-m(a+b)\dot{\theta}^2 = N - mg\sin\theta, \qquad m(a+b)\ddot{\theta} = f - mg\cos\theta$$

We now apply Theorem 9.13, page 229. The total external torque of all forces about the center of mass C is [since $\mathbf{W}$ and $\mathbf{N}$ pass through C],

$$\boldsymbol{\Lambda} = (-a\mathbf{r}_1)\times\mathbf{f} = (-a\mathbf{r}_1)\times(f\boldsymbol{\theta}_1) = -af\mathbf{k}$$

Also, the angular acceleration of the first sphere about C is

$$\boldsymbol{\alpha} = -\frac{d^2}{dt^2}(\phi + \psi)\mathbf{k} = -(\ddot{\phi} + \ddot{\psi})\mathbf{k}$$

Since there is only rolling and no slipping it follows that arc AP equals arc BP, or $b\phi = a\psi$. Then $\phi = \pi/2 - \theta$ and $\psi = (b/a)(\pi/2 - \theta)$, so that

$$\boldsymbol{\alpha} = -(\ddot{\phi} + \ddot{\psi})\mathbf{k} = -\left(-\ddot{\theta} - \frac{b}{a}\ddot{\theta}\right)\mathbf{k} = \left(\frac{a+b}{a}\right)\ddot{\theta}\mathbf{k}$$

Since the moment of inertia of the first sphere about the horizontal axis of rotation through C is $I = \frac{2}{5}ma^2$, we have by Theorem 9.13,

$$\mathbf{\Lambda} = I\boldsymbol{\alpha}, \qquad -af\mathbf{k} = \tfrac{2}{5}ma^2\left(\frac{a+b}{a}\right)\ddot{\theta}\,\mathbf{k} \qquad \text{or} \qquad f = -\tfrac{2}{5}m(a+b)\,\ddot{\theta}$$

Using this value of f in the second equation of (1), we find

$$\ddot{\theta} = -\frac{5g}{7(a+b)}\cos\theta \tag{2}$$

Multiplying both sides by $\dot{\theta}$ and integrating, we find after using the fact that $\dot{\theta} = 0$ at $t = 0$ or $\theta = \pi/2$,

$$\dot{\theta}^2 = \frac{10g}{7(a+b)}(1 - \sin\theta) \tag{3}$$

Using (3) in the first of equations (1), we find $N = \frac{1}{7}mg(17\sin\theta - 10)$. Then the first sphere leaves the second sphere where $N = 0$, i.e. where $\theta = \sin^{-1} 10/17$.

Supplementary Problems

RIGID BODIES

9.43. Show that the motion of region $\mathcal{R}$ of Fig. 9-34 can be carried into region $\mathcal{R}'$ by means of a translation plus a rotation about a suitable point.

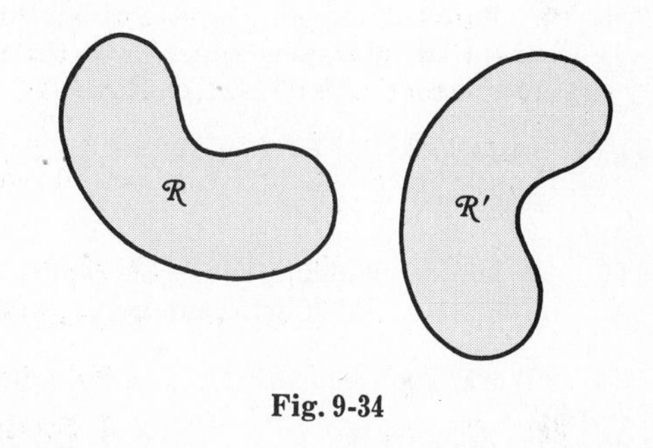

Fig. 9-34

9.44. Work Problem 9.1, page 230, by first applying a translation of the point A of triangle ABC.

9.45. If A_x, A_y, A_z represent rotations of a rigid body about the x, y and z axes respectively, is it true that the associative law applies, i.e. is $A_x + (A_y + A_z) = (A_x + A_y) + A_z$? Justify your answer.

MOMENTS OF INERTIA

9.46. Three particles of masses 3, 5 and 2 are located at the points $(-1, 0, 1)$, $(2, -1, 3)$ and $(-2, 2, 1)$ respectively. Find (a) the moment of inertia and (b) the radius of gyration about the x axis.
Ans. 71

9.47. Find the moment of inertia of the system of particles in Problem 9.46 about (a) the y axis, (b) the z axis. *Ans.* (a) 81, (b) 44

9.48. Find the moment of inertia of a uniform rod of length l about an axis perpendicular to it and passing through (a) the center of mass, (b) an end, (c) a point at distance $l/4$ from an end.
Ans. (a) $\frac{1}{12}Ml^2$, (b) $\frac{1}{3}Ml^2$, (c) $\frac{7}{48}Ml^2$

9.49. Find the (a) moment of inertia and (b) radius of gyration of a square of side a about a diagonal.
Ans. (a) $\frac{1}{12}Ma^2$, (b) $\frac{1}{6}a\sqrt{3}$

9.50. Find the moment of inertia of a cube of edge a about an edge. *Ans.* $\frac{2}{3}Ma^2$

9.51. Find the moment of inertia of a rectangular plate of sides a and b about a diagonal.
Ans. $\frac{1}{6}Ma^2b^2/(a^2 + b^2)$

9.52. Find the moment of inertia of a uniform parallelogram of sides a and b and included angle α about an axis perpendicular to it and passing through its center. *Ans.* $\frac{1}{12}M(a^2 + b^2)\sin^2\alpha$

9.53. Find the moment of inertia of a cube of side a about a diagonal.

9.54. Find the moment of inertia of a cylinder of radius a and height h about an axis parallel to the axis of the cylinder and distant b from its center. *Ans.* $\frac{1}{2}M(a^2 + 2b^2)$

9.55. A solid of constant density is formed from a cylinder of radius a and height h and a hemisphere of radius a as shown in Fig. 9-35. Find its moment of inertia about a vertical axis through their centers. *Ans.* $M(2a^3 + 15a^2h)/(10a + 15h)$

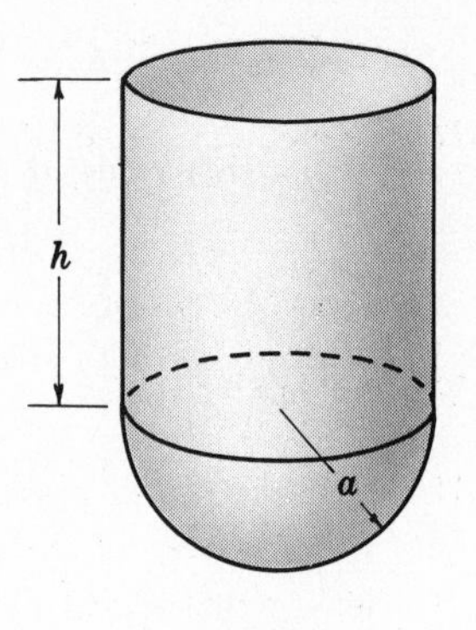

Fig. 9-35

9.56. Work Problem 9.55 if the cylinder is replaced by a cone of radius a and height h.

9.57. Find the moment of inertia of the uniform solid region bounded by the paraboloid $cz = x^2 + y^2$ and the plane $z = h$ about the z axis. *Ans.* $\frac{1}{3}Mch$

9.58. How might you define the moment of inertia of a solid about (a) a point, (b) a plane? Is there any physical significance to these results? Explain.

9.59. Use your definitions in Problem 9.58 to find the moment of inertia of a cube of side a about (a) a vertex and (b) a face. *Ans.* (a) Ma^2, (b) $\frac{1}{3}Ma^2$

KINETIC ENERGY AND ANGULAR MOMENTUM

9.60. A uniform rod of length 2 ft and mass 6 lb rotates with angular speed 10 radians per second about an axis perpendicular to it and passing through its center. Find the kinetic energy of rotation. *Ans.* 100 lb ft^2/sec^2

9.61. Work Problem 9.60 if the axis of rotation is perpendicular to the rod and passes through an end. *Ans.* 400 lb ft^2/sec^2

9.62. A hollow cylindrical disk of radius a and mass M rolls along a horizontal plane with speed v. Find the total kinetic energy. *Ans.* Mv^2

9.63. Work Problem 9.62 for a solid cylindrical disk of radius a. *Ans.* $\frac{3}{4}Mv^2$

9.64. A flywheel having radius of gyration 2 meters and mass 10 kilograms rotates at angular speed of 5 radians/sec about an axis perpendicular to it through its center. Find the kinetic energy of rotation. *Ans.* 1000 joules

9.65. Find the angular momentum of (a) the rod of Problem 9.60 (b) the flywheel of Problem 9.64. *Ans.* (a) 5 lb ft^2/sec, (b) 200 kg m^2/sec

9.66. Prove the result of (a) Problem 9.15, page 236, (b) Problem 9.16, page 236, by using integration in place of summation.

9.67. Derive a "parallel axis theorem" for (a) kinetic energy and (b) angular momentum and explain the physical significance.

MOTION OF A RIGID BODY. THE COMPOUND PENDULUM. WORK, POWER AND IMPULSE

9.68. A constant force of magnitude F_0 is applied tangentially to a flywheel which can rotate about a fixed axis perpendicular to it and passing through its center. If the flywheel has radius a, radius of gyration K and mass M, prove that the angular acceleration is given by F_0a/MK^2.

9.69. How long will it be before the flywheel of Problem 9.68 reaches an angular speed ω_0 if it starts from rest? *Ans.* $MK^2\omega_0/F_0a$

9.70. Assuming that the flywheel of Problem 9.68 starts from rest, find (a) the total work done, (b) the total power developed and (c) the total impulse applied in getting the angular speed up to ω_0. *Ans.* (a) $\frac{1}{2}MK^2\omega_0^2$, (b) $F_0a\omega_0$, (c) $MK^2\omega_0$

9.71. Work (*a*) Problem 9.68, (*b*) Problem 9.69 and (*c*) Problem 9.70 if $F_0 = 10$ newtons, $a = 1$ meter, $K = 0.5$ meter, $M = 20$ kilograms and $\omega_0 = 20$ radians/sec.

Ans. (*a*) 2 rad/sec^2; (*b*) 10 sec; (*c*) 250 joules, 200 joules/sec, 100 newton sec

9.72. Find the period of small vibrations for a simple pendulum assuming that the string supporting the bob is replaced by a uniform rod of length l and mass M while the bob has mass m.

Ans. $2\pi\sqrt{\dfrac{2(M+3m)l}{3(M+2m)g}}$

9.73. Discuss the cases (*a*) $M = 0$ and (*b*) $m = 0$ in Problem 9.72.

9.74. A rectangular plate having edges of lengths a and b respectively hangs vertically from the edge of length a. (*a*) Find the period for small oscillations and (*b*) the length of the equivalent simple pendulum. *Ans.* (*a*) $2\pi\sqrt{2b/3g}$, (*b*) $\frac{2}{3}b$

9.75. A uniform solid sphere of radius a and mass M is suspended vertically downward from a point on its surface. (*a*) Find the period for small oscillations in a plane and (*b*) the length of the equivalent simple pendulum. *Ans.* (*a*) $2\pi\sqrt{7a/5g}$, (*b*) $7a/5$

9.76. A yo-yo consists of a cylinder of mass 80 gm around which a string of length 60 cm is wound. If the end of the string is kept fixed and the yo-yo is allowed to fall vertically starting from rest, find its speed when it reaches the end of the string. *Ans.* 280 cm/sec

9.77. Find the tension in the string of Problem 9.76.
Ans. 19,600 dynes

9.78. A hollow cylindrical disk of mass M moving with constant speed v_0 comes to an incline of angle α. Prove that if there is no slipping it will rise a distance $v_0^2/(g \sin \alpha)$ up the incline.

9.79. If the hollow disk of Problem 9.78 is replaced by a solid disk, how high will it rise up the incline? *Ans.* $3v_0^2/(4g \sin \alpha)$

9.80. In Fig. 9-36 the pulley, assumed frictionless, has radius 0.2 meter and its radius of gyration is 0.1 meter. What is the acceleration of the 5 kg mass? *Ans.* 2.45 m/sec^2

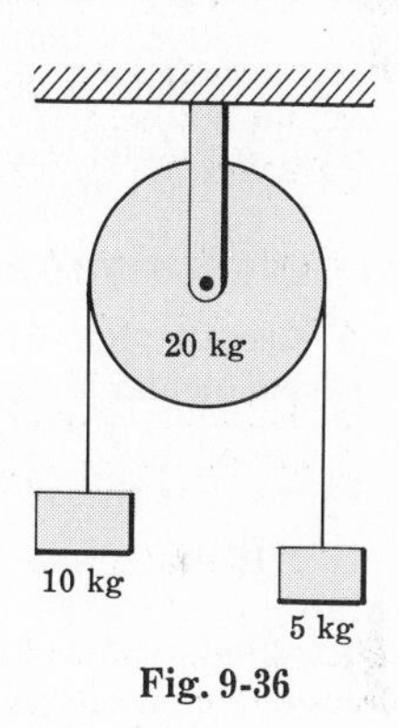

Fig. 9-36

INSTANTANEOUS CENTER. SPACE AND BODY CENTRODES

9.81. A ladder of length l moves so that one end is on a vertical wall and the other on a horizontal floor. Find (*a*) the space centrode and (*b*) the body centrode.

Ans. (*a*) A circle having radius l and center at point O where the floor and wall meet.
(*b*) A circle with the ladder as diameter

9.82. A long rod AB moves so that it remains in contact with the top of a post of height h while its foot B moves on a horizontal line CD [Fig. 9-37]. Assuming the motion to be in one plane, find the locus of instantaneous centers.

9.83. What is the (*a*) body centrode and (*b*) space centrode in Problem 9.82?

9.84. Work Problems 9.82 and 9.83 if the post is replaced by a fixed cylinder of radius a.

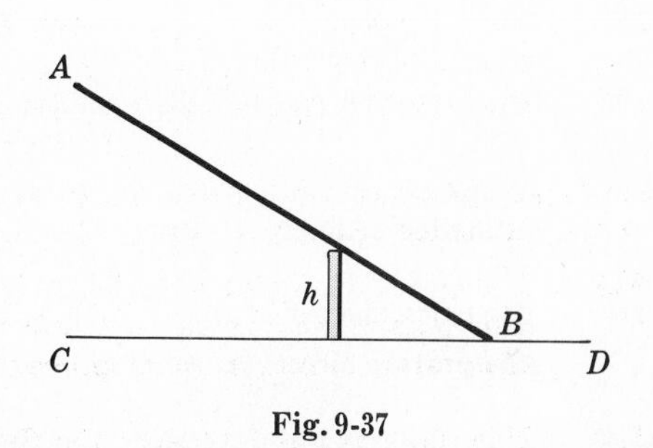

Fig. 9-37

STATICS OF A RIGID BODY

9.85. A uniform ladder of weight W and length l has its top against a smooth wall and its foot on a floor having coefficient of friction μ. (*a*) Find the smallest angle α which the ladder can make with the horizontal and still be in equilibrium. (*b*) Can equilibrium occur if $\mu = 0$? Explain.

9.86. Work Problem 9.85 if the wall has coefficient of friction μ_1.

9.87. In Fig. 9-38, AB is a uniform bar of length l and weight W supported at C. It carries weights W_1 at A and W_2 at D so that $AC = a$ and $CD = b$. Where must a weight W_3 be placed on AC so that the system will be in equilibrium?

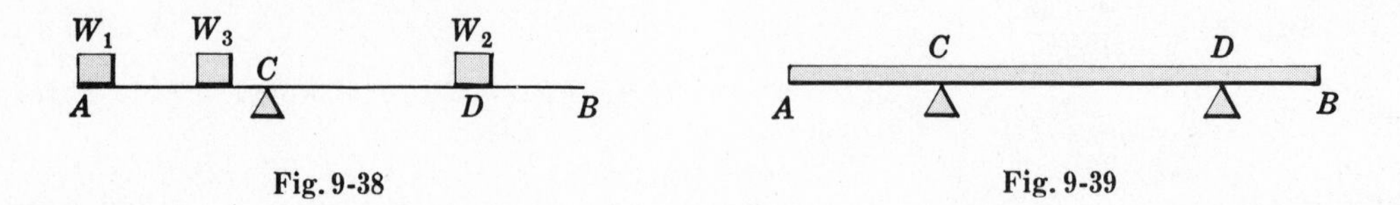

Fig. 9-38 Fig. 9-39

9.88. A uniform triangular thin plate hangs from a fixed point O by strings OA, OB and OC of lengths a, b and c respectively. Prove that the tensions T_1, T_2 and T_3 in the strings are such that $T_1/a = T_2/b = T_3/c$.

9.89. A uniform plank AB of length l and weight W is supported at points C and D distant a from A and b from B respectively [Fig. 9-39]. Determine the reaction forces at C and D respectively.

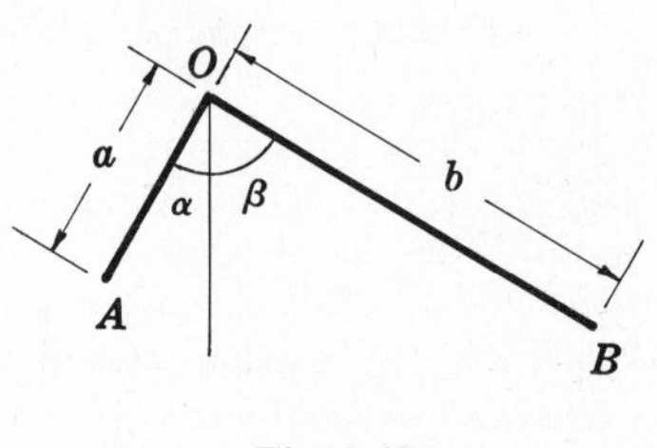

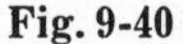

Fig. 9-40

9.90. In Fig. 9-40, OA and OB are uniform rods having the same density and connected at O so that AOB is a right angle. The system is supported at O so that AOB is in a vertical plane. Find the angles α and β for which equilibrium occurs.
Ans. $\alpha = \tan^{-1}(a/b)$, $\beta = \pi/2 - \tan^{-1}(a/b)$

MISCELLANEOUS PROBLEMS

9.91. A circular cylinder has radius a and height h. Prove that the moment of inertia about an axis perpendicular to the axis of the cylinder and passing through the centroid is $\frac{1}{12}M(h^2 + 3a^2)$.

9.92. Prove that the effect of a force on a rigid body is not changed by shifting the force along its line of action.

9.93. A cylinder of radius a and radius of gyration K rolls without slipping down an inclined plane of angle α and length l, starting from rest at the top of the incline. Prove that when it reaches the bottom of the incline is speed will be $\sqrt{(2gla^2 \sin\alpha)/(a^2 + K^2)}$.

9.94. A cylinder resting on top of a fixed cylinder is given a slight displacement so that it rolls without slipping. Determine where it leaves the fixed cylinder.
Ans. $\theta = \sin^{-1} 4/7$ where θ is measured as in Fig. 9-33, page 244.

9.95. Work Problem 9.42 if the sphere is given an initial speed v_0.

9.96. Work Problem 9.94 if the cylinder is given an initial speed v_0.

9.97. A sphere of radius a and radius of gyration K about a diameter rolls without slipping down an incline of angle α. Prove that it descends with constant acceleration given by $(ga^2 \sin\alpha)/(a^2 + K^2)$.

9.98. Work Problem 9.97 if the sphere is (*a*) solid, (*b*) hollow and of negligible thickness.
Ans. (*a*) $\frac{5}{7}g \sin\alpha$, (*b*) $\frac{3}{5}g \sin\alpha$

9.99. A hollow sphere has inner radius a and outer radius b. Prove that if M is its mass, then the moment of inertia about an axis through its center is

$$\frac{2}{5}M\left(\frac{a^4 + a^3b + a^2b^2 + ab^3 + b^4}{a^2 + ab + b^2}\right)$$

Discuss the cases $b = 0$ and $a = b$.

9.100. Wooden plates, all having the same rectangular shape are stacked one above the other as indicated in Fig. 9-41. (*a*) If the length of each plate is $2a$, prove that equilibrium conditions will prevail if the $(n+1)$th plate extends a maximum distance of a/n beyond the nth plate where $n = 1, 2, 3, \ldots$. (*b*) What is the maximum horizontal distance which can be reached if more and more plates are added?

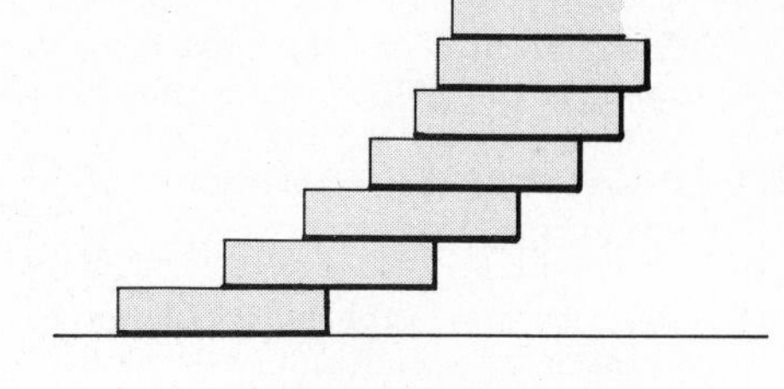

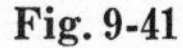

Fig. 9-41

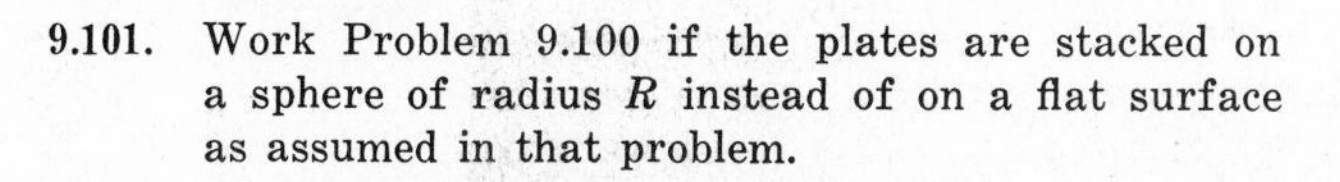

9.101. Work Problem 9.100 if the plates are stacked on a sphere of radius R instead of on a flat surface as assumed in that problem.

9.102. A cylinder of radius a rolls on the inner surface of a smooth cylinder of radius $2a$. Prove that the period of small oscillations is $2\pi\sqrt{3a/2g}$.

9.103. A ladder of length l and negligible weight rests with one end against a wall having coefficient of friction μ_1 and the other end against a floor having coefficient of friction μ_2. It makes an angle α with the floor. (*a*) How far up the ladder can a man climb before the ladder slips? (*b*) What is the condition that the ladder not slip at all regardless of where the man is located?
Ans. (*a*) $\mu_2 l(\mu_1 + \tan\alpha)/(\mu_1\mu_2 + 1)$, (*b*) $\tan\alpha > 1/\mu_2$

9.104. Work Problem 9.103 if the weight of the ladder is not negligible.

9.105. A ladder AB of length l [Fig. 9-42] has one end A on an incline of angle α and the other end B on a vertical wall. The ladder is at rest and makes an angle β with the incline. If the wall is smooth and the incline has coefficient of friction μ, find the smallest value of μ so that a man of weight W_m will be able to climb the ladder without having it slip. Check your answer by obtaining the result of Problem 9.37, page 241, as a special case.

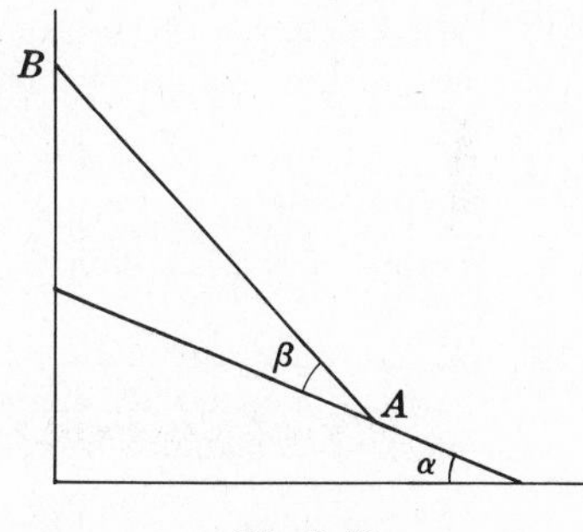

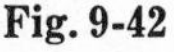

Fig. 9-42

9.106. Work Problem 9.105 if the wall has coefficient of friction μ_1.

9.107. A uniform rod AB with point A fixed rotates about a vertical axis so that it makes a constant angle α with the vertical [Fig. 9-43]. If the length of the rod is l, prove that the angular speed needed to do this is $\omega = \sqrt{(3g \sec\alpha)/2l}$.

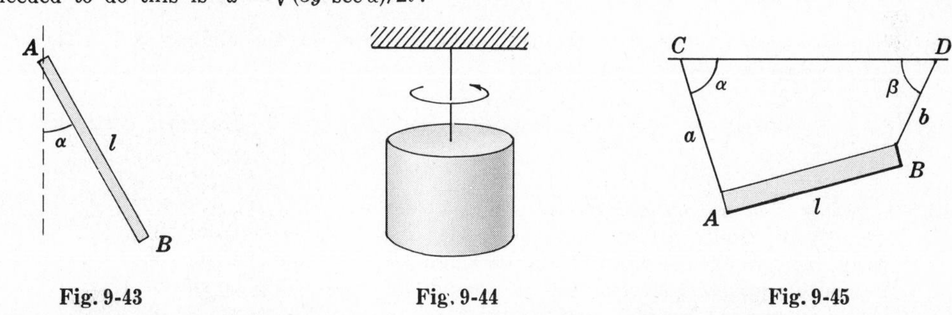

Fig. 9-43 Fig. 9-44 Fig. 9-45

9.108. A circular cylinder of mass m and radius a is suspended from the ceiling by a wire as shown in Fig. 9-44. The cylinder is given an angular twist θ_0 and is then released. If the torque is assumed proportional to the angle through which the cylinder is turned and the constant of proportionality is λ, prove that the cylinder will undergo simple harmonic motion with period $2\pi a\sqrt{m/2\lambda}$.

9.109. Find the period in Problem 9.108 if the cylinder is replaced by a sphere of radius a.
Ans. $2\pi a\sqrt{2m/5\lambda}$

9.110. Work (*a*) Problem 9.108 and (*b*) Problem 9.109 if damping proportional to the instantaneous angular velocity is present. Discuss physically.

9.111. A uniform beam AB of length l and weight W [Fig. 9-45] is supported by ropes AC and BD of lengths a and b respectively making angles α and β with the ceiling CD to which the ropes are fixed. If equilibrium conditions prevail, find the tensions in the ropes.

9.112. In Fig. 9-46 the mass m is attached to a rope which is wound around a fixed pulley of mass M and radius of gyration K which can rotate freely about O. If the mass is released from rest, find (a) the angular speed of the pulley after time t and (b) the tension in the rope.

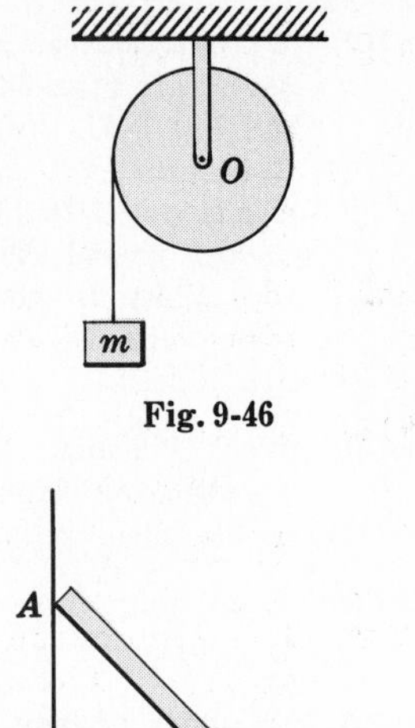

Fig. 9-46

9.113. Prove that the acceleration of the mass m in Problem 9.112 is $ga^2/(a^2+K^2)$.

9.114. Describe how Problem 9.112 can be used to determine the radius of gyration of a pulley.

9.115. A uniform rod AB [Fig. 9-47] of length l and weight W having its ends on a frictionless wall OA and floor OB respectively, slides starting from rest when its foot B is at a distance d from O. Prove that the other end A will leave the wall when the foot B is at a distance from O given by $\frac{1}{3}\sqrt{5l^2+4d^2}$.

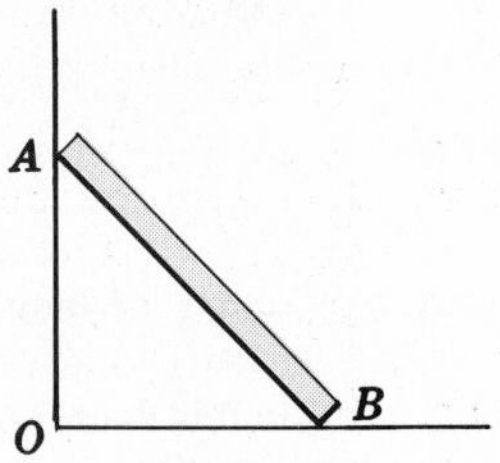

Fig. 9-47

9.116. A cylinder of mass 10 lb rotates about a fixed horizontal axis through its center and perpendicular to it. A rope wound around it carries a mass of 20 lb. Assuming that the mass starts from rest, find its speed after 5 seconds. *Ans.* 128 ft/sec

9.117. What must be the length of a rod suspended from one end so that it will be a seconds pendulum on making small vibrations in a plane? *Ans.* 149 cm

9.118. A solid sphere and a hollow sphere of the same radius both start from rest at the top of an inclined plane of angle α and roll without slipping down the incline. Which one gets to the bottom first? Explain. *Ans.* The solid sphere

9.119. A compound pendulum of mass M and radius of gyration K about a horizontal axis is displaced so that it makes an angle θ_0 with the vertical and is then released. Prove that if the center of mass is at distance a from the axis, then the reaction force on the axis is given by

$$\frac{Mg}{K^2+a^2}\sqrt{([K^2+2a^2]\cos\theta - a^2\cos\theta_0)^2 + (K^2\sin\theta)^2}$$

9.120. A rectangular parallelepiped of sides a, b, and c is suspended vertically from the side of length a. Find the period of small oscillations.

9.121. Find the least coefficient of friction needed to prevent the sliding of a circular hoop down an incline of angle α. *Ans.* $\frac{2}{3}\tan\alpha$

9.122. Find the period of small vibrations of a rod of length l suspended vertically about a point $\frac{1}{3}l$ from one end.

9.123. A pulley system consists of two solid disks of radius r_1 and r_2 respectively rigidly attached to each other and capable of rotating freely about a fixed horizontal axis through the center O. A weight W is suspended from a string wound around the smaller disk as shown in Fig. 9-48. If the radius of gyration of the pulley system is K and its weight is w, find (a) the angular acceleration with which the weight descends and (b) the tension in the string.
Ans. (a) $Wgr_1/(Wr_1^2+wK^2)$, (b) $WwK^2/(Wr_1^2+wK^2)$

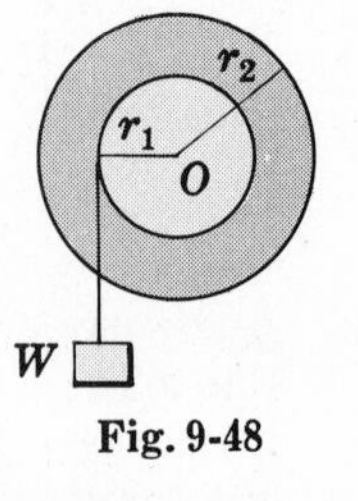

Fig. 9-48

9.124. A solid sphere of radius b rolls on the inside of a smooth hollow sphere of radius a. Prove that the period for small oscillations is given by $2\pi\sqrt{7(a-b)/5g}$.

9.125. A thin circular solid plate of radius a is suspended vertically from a horizontal axis passing through a chord AB [see Fig. 9-49]. If it makes small oscillations about this axis, prove that the frequency of such oscillations is greatest when AB is at distance $a/2$ from the center.

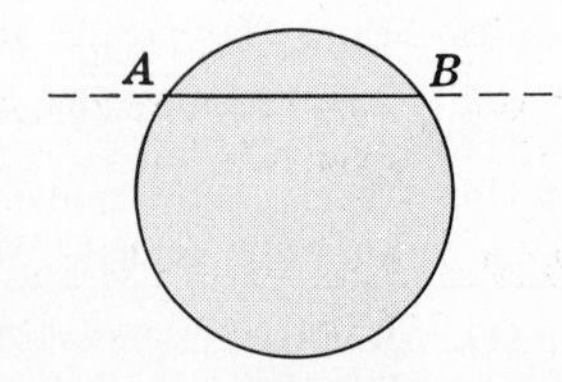

Fig. 9-49

9.126. A uniform rod of length $5l$ is suspended vertically from a string of length $2l$ which has its other end fixed. Prove that the normal frequencies for small oscillations in a plane are $\frac{1}{2\pi}\sqrt{\frac{g}{5l}}$ and $\frac{1}{2\pi}\sqrt{\frac{3g}{l}}$, and describe the normal modes.

9.127. A uniform rod of mass m and length l is suspended from one of its ends. What is the minimum speed with which the other end should be hit so that it will describe a complete vertical circle?

9.128. (*a*) If the bob of a simple pendulum is a uniform solid sphere of radius a rather than a point mass, prove that the period for small oscillations is $2\pi\sqrt{l/g + 2a^2/5gl}$.

(*b*) For what value of l is the period in (*a*) a minimum?

9.129. A sphere of radius a and mass M rolls along a horizontal plane with constant speed v_0. It comes to an incline of angle α. Assuming that it rolls without slipping, how far up the incline will it travel? *Ans.* $10v_0^2/(7g \sin \alpha)$

9.130. Prove that the doughnut shaped solid or *torus* of Fig. 9-50 has a moment of inertia about its axis given by $\frac{1}{4}M(3a^2 + 4b^2)$.

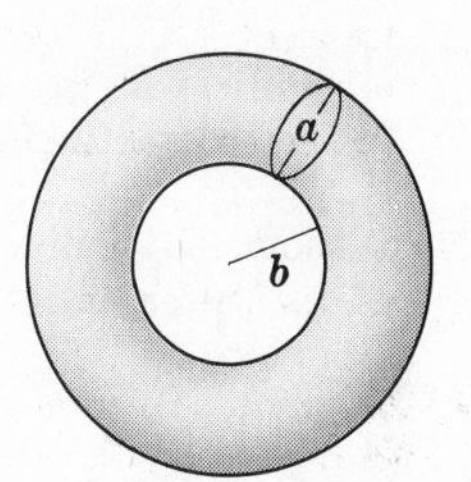

Fig. 9-50

9.131. A cylinder of mass m and radius a rolls without slipping down a 45° inclined plane of mass M which is on a horizontal frictionless table. Prove that while the rolling takes place the incline will move with an acceleration given by $mg/(3M + 2m)$.

9.132. Work Problem 9.131 if the incline is of angle α.
Ans. $(mg \sin 2\alpha)/(3M + 2m - m \cos 2\alpha)$

9.133. Find the (*a*) tension in the rope and (*b*) acceleration of the system shown in Fig. 9-51 if the radius of gyration of the pulley is 0.5 m and its mass is 20 kg.

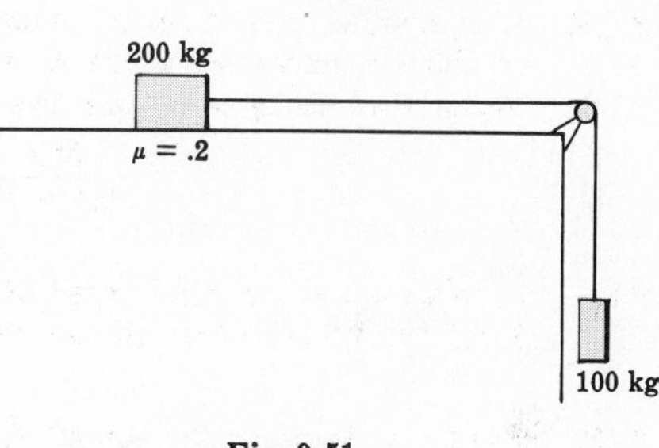

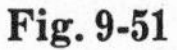
Fig. 9-51

9.134. Compare the result of Problem 9.133 with that obtained assuming the pulley to have negligible mass.

9.135. Prove that if the net external torque about an axis is zero, then it is also zero about any other axis.

9.136. A solid cylindrical disk of radius a has a circular hole of radius b whose center is at distance c from the center of the disk. If the disk rolls down an inclined plane of angle α, find its acceleration. [See Fig. 9-52.]

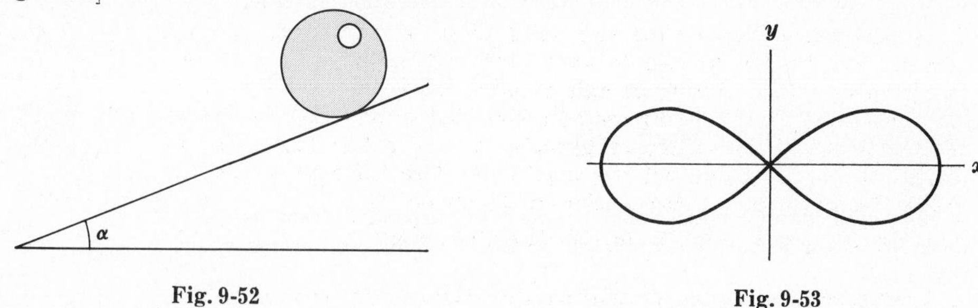

Fig. 9-52

Fig. 9-53

9.137. Find the moment of inertia of the region bounded by the lemniscate $r^2 = a^2 \cos 2\theta$ [see Fig. 9-53] about the x axis. *Ans.* $Ma^2(3\pi - 8)/48$

9.138. Find the largest angle of an inclined plane down which a solid cylinder will roll without slipping if the coefficient of friction is μ.

9.139. Work Problem 9.138 for a solid sphere.

9.140. Discuss the motion of a hollow cylinder of inner radius a and outer radius b as it rolls down an inclined plane of angle α.

9.141. A table top of negligible weight has the form of an equilateral triangle ABC of side s. The legs of the table are perpendicular to the table top at the vertices. A heavy weight W is placed on the table top at a point which is distant a from side BC and b from side AC. Find that part of the weight supported by the legs at A, B and C respectively.

Ans. $\dfrac{2Wa}{s\sqrt{3}},\ \dfrac{2Wb}{s\sqrt{3}},\ W\left(1 - \dfrac{2a+2b}{s\sqrt{3}}\right)$

9.142. Discuss the motion of the disk of Problem 9.136 down the inclined plane if the coefficient of friction is μ.

9.143. A hill has a cross section in the form of a cycloid

$$x = a(\theta + \sin\theta), \quad y = a(1 - \cos\theta)$$

as indicated in Fig. 9-54. A solid sphere of radius b starting from rest at the top of the hill is given a slight displacement so that it rolls without slipping down the hill. Find the speed of its center when it reaches the bottom of the hill.

Ans. $\sqrt{10g(2a-b)/7}$

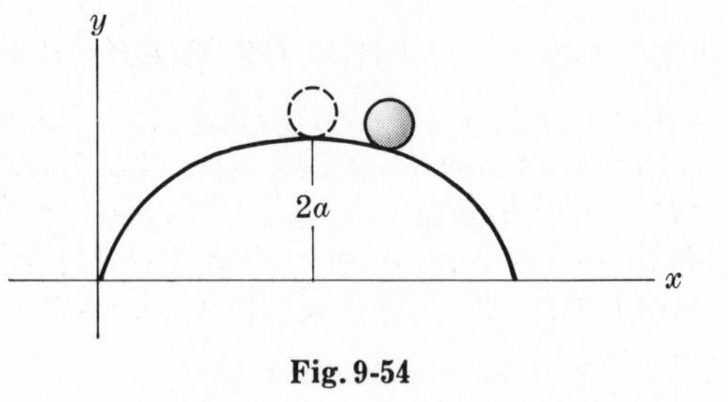

Fig. 9-54

9.144. Work Problem 3.108, page 85, if the masses and moments of inertia of the pulleys are taken into account.

9.145. Work Problem 9.38, page 242, if friction is taken into account.

9.146. A uniform rod of length l is placed upright on a table and then allowed to fall. Assuming that its point of contact with the table does not move, prove that its angular velocity at the instant when it makes an angle θ with the vertical is given in magnitude by $\sqrt{3g(1-\cos\theta)/2l}$.

9.147. Prove Theorem 9.11, page 228, for the case where the vibrations are not necessarily small. Compare Problem 9.41, page 243.

9.148. A rigid body moves parallel to a given fixed plane. Prove that there is one and only one point of the rigid body where the instantaneous acceleration is zero.

9.149. A solid hemisphere of radius a rests with its convex surface on a horizontal table. If it is displaced slightly, prove that it will undergo oscillations with period equal to that of a simple pendulum of equivalent length $4a/3$.

9.150. A solid cylinder of radius a and height h is suspended from axis AB as indicated in Fig. 9-55. Find the period of small oscillations about this axis.

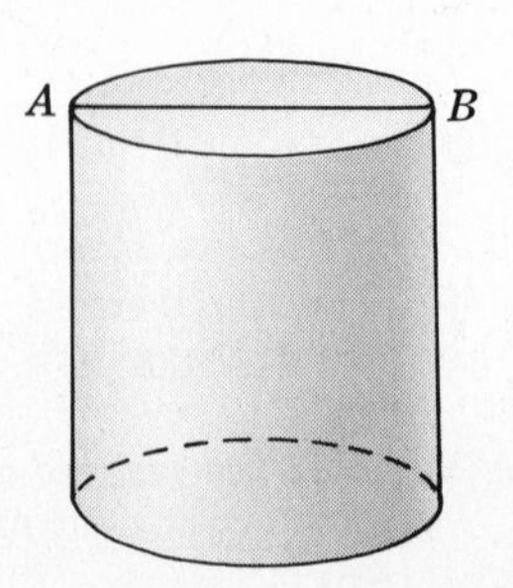

Fig. 9-55

9.151. Prove that a solid sphere will roll without slipping down an inclined plane of angle α if the coefficient of friction is at least $\frac{2}{7}\tan\alpha$.

9.152. Find the least coefficient of friction for an inclined plane of angle α in order that a solid cylinder will roll down it without slipping.

Ans. $\frac{1}{3}\tan\alpha$

Chapter 10

SPACE MOTION of RIGID BODIES

GENERAL MOTION OF RIGID BODIES IN SPACE

In Chapter 9 we specialized the motion of rigid bodies to one of translation of the center of mass plus rotation about an axis through the center of mass and perpendicular to a *fixed* plane. In this chapter we treat the general motion of a rigid body in space. Such general motion is composed of a translation of a fixed point of the body [usually the center of mass] plus rotation about an axis through the fixed point which is not necessarily restricted in direction.

DEGREES OF FREEDOM

The number of degrees of freedom [see page 165] for the general motion of a rigid body in space is 6, i.e. 6 coordinates are needed to specify the motion. We usually choose 3 of these to be the coordinates of a point in the body [usually the center of mass] and the remaining 3 to be angles [for example, the Euler angles, page 257] which describe the rotation of the rigid body about the point.

If a rigid body is constrained in any way, as for example by keeping one point fixed, the number of degrees of freedom is of course reduced accordingly.

PURE ROTATION OF RIGID BODIES

Since the general motion of a rigid body can also be expressed in terms of translation of a fixed point of the rigid body plus rotation of the rigid body about an axis through the point, it is natural for us to consider first the case of pure rotation and later to add the effects of translation. To do this we shall first assume that one point of the rigid body is fixed in space. The effects of translation are relatively easy to handle and can be obtained by using the result *(10)*, page 167.

VELOCITY AND ANGULAR VELOCITY OF A RIGID BODY WITH ONE POINT FIXED

Suppose that point O of the rigid body $\mathcal{R}$ of Fig. 10-1 is fixed. Then at a given instant of time the body will be rotating with *angular velocity* $\boldsymbol{\omega}$ about the instantaneous axis through O. A particle P of the body having position vector $\mathbf{r}_\nu$ with respect to O will have an *instantaneous velocity* $\mathbf{v}_\nu$ given by

$$\mathbf{v}_\nu = \dot{\mathbf{r}}_\nu = \boldsymbol{\omega} \times \mathbf{r}_\nu \qquad (1)$$

See Problem 10.2.

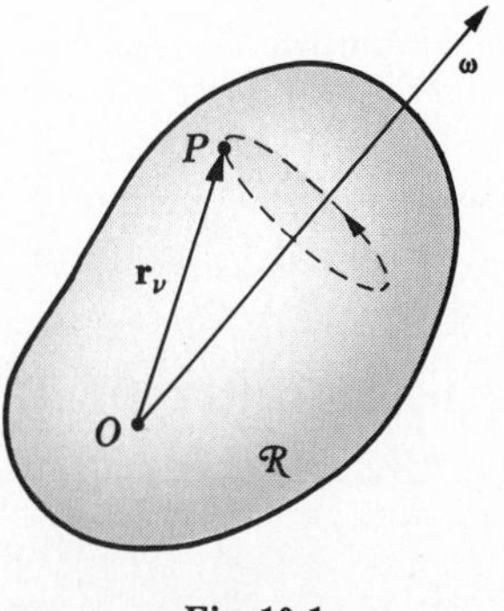

Fig. 10-1

ANGULAR MOMENTUM

The angular momentum of a rigid body with one point fixed about the instantaneous axis through the fixed point is given by

$$\mathbf{\Omega} = \sum m_\nu(\mathbf{r}_\nu \times \mathbf{v}_\nu) = \sum m_\nu\{\mathbf{r}_\nu \times (\boldsymbol{\omega} \times \mathbf{r}_\nu)\} \tag{2}$$

where m_ν is the mass of the νth particle and where the summation is taken over all particles of $\mathcal{R}$.

MOMENTS OF INERTIA. PRODUCTS OF INERTIA

Let us choose a fixed xyz coordinate system having origin O and let us write

$$\begin{gathered}\mathbf{\Omega} = \Omega_x\mathbf{i} + \Omega_y\mathbf{j} + \Omega_z\mathbf{k}, \qquad \boldsymbol{\omega} = \omega_x\mathbf{i} + \omega_y\mathbf{j} + \omega_z\mathbf{k} \\ \mathbf{r}_\nu = x_\nu\mathbf{i} + y_\nu\mathbf{j} + z_\nu\mathbf{k}\end{gathered} \tag{3}$$

Then equation (2) can be written in component form as [see Problem 10.3].

$$\left.\begin{aligned}\Omega_x &= I_{xx}\omega_x + I_{xy}\omega_y + I_{xz}\omega_z \\ \Omega_y &= I_{yx}\omega_x + I_{yy}\omega_y + I_{yz}\omega_z \\ \Omega_z &= I_{zx}\omega_x + I_{zy}\omega_y + I_{zz}\omega_z\end{aligned}\right\} \tag{4}$$

where $$I_{xx} = \sum m_\nu(y_\nu^2 + z_\nu^2), \quad I_{yy} = \sum m_\nu(z_\nu^2 + x_\nu^2), \quad I_{zz} = \sum m_\nu(x_\nu^2 + y_\nu^2) \tag{5}$$

$$\left.\begin{aligned}I_{xy} &= -\sum m_\nu x_\nu y_\nu = I_{yx} \\ I_{yz} &= -\sum m_\nu y_\nu z_\nu = I_{zy} \\ I_{zx} &= -\sum m_\nu z_\nu x_\nu = I_{xz}\end{aligned}\right\} \tag{6}$$

The quantities I_{xx}, I_{yy}, I_{zz} are called the *moments of inertia* about the x, y and z axes respectively. The quantities $I_{xy}, I_{xz}, \ldots$ are called *products of inertia*. For continuous mass distributions these can be computed by using integration.

Note that the products of inertia in (6) have been defined with an associated minus sign. As a consequence minus signs are avoided in (4).

MOMENT OF INERTIA MATRIX OR TENSOR

The nine quantities $I_{xx}, I_{xy}, \ldots, I_{zz}$ can be written in an array often called a *matrix* or *tensor* given by

$$\begin{pmatrix} I_{xx} & I_{xy} & I_{xz} \\ I_{yx} & I_{yy} & I_{yz} \\ I_{zx} & I_{zy} & I_{zz} \end{pmatrix} \tag{7}$$

and each quantity is called an *element* of the matrix or tensor. The diagonal consisting of the elements I_{xx}, I_{yy}, I_{zz} is called the *principal* or *main diagonal*. Since

$$I_{xy} = I_{yx}, \quad I_{xz} = I_{zx}, \quad I_{yz} = I_{zy} \tag{8}$$

it is seen that the elements have symmetry about the main diagonal. For this reason (7) is often referred to as a *symmetric* matrix or tensor.

KINETIC ENERGY OF ROTATION

The kinetic energy of rotation is given by

$$\begin{aligned}T &= \tfrac{1}{2}(I_{xx}\omega_x^2 + I_{yy}\omega_y^2 + I_{zz}\omega_z^2 + 2I_{xy}\omega_x\omega_y + 2I_{xz}\omega_x\omega_z + 2I_{yz}\omega_y\omega_z) \\ &= \tfrac{1}{2}\boldsymbol{\omega}\cdot\mathbf{\Omega}\end{aligned} \tag{9}$$

PRINCIPAL AXES OF INERTIA

A set of 3 mutually perpendicular axes having origin O which are *fixed in the body* and rotating with it and which are such that the products of inertia about them are zero, are called *principal axes of inertia* or briefly *principal axes* of the body.

An important property of a principal axis [which can also be taken as a definition] is that if a rigid body rotates about it the direction of the angular momentum is the same as that of the angular velocity. Thus

$$\mathbf{\Omega} = I\boldsymbol{\omega} \tag{10}$$

where I is a scalar. From this we find [see Problem 10.6] that

$$\left.\begin{aligned} (I_{xx}-I)\omega_x + I_{xy}\omega_y + I_{xz}\omega_z &= 0 \\ I_{yx}\omega_x + (I_{yy}-I)\omega_y + I_{yz}\omega_z &= 0 \\ I_{zx}\omega_x + I_{zy}\omega_y + (I_{zz}-I)\omega_z &= 0 \end{aligned}\right\} \tag{11}$$

In order that (*11*) have solutions other than the trivial one $\omega_x = 0,\ \omega_y = 0,\ \omega_z = 0,$ we require that

$$\begin{vmatrix} I_{xx}-I & I_{xy} & I_{xz} \\ I_{yx} & I_{yy}-I & I_{yz} \\ I_{zx} & I_{zy} & I_{zz}-I \end{vmatrix} = 0 \tag{12}$$

This leads to a cubic equation in I having 3 real roots I_1, I_2, I_3. These are called the *principal moments of inertia*. The directions of the principal axes can be found from (*11*), as shown in Problem 10.6 by finding the ratio $\omega_x : \omega_y : \omega_z$.

An axis of symmetry of a rigid body will always be a principal axis.

ANGULAR MOMENTUM AND KINETIC ENERGY ABOUT THE PRINCIPAL AXES

If we call $\omega_1, \omega_2, \omega_3$ and $\Omega_1, \Omega_2, \Omega_3$ the magnitudes of the angular velocities and angular momenta about the principal axes respectively, then

$$\Omega_1 = I_1\omega_1, \quad \Omega_2 = I_2\omega_2, \quad \Omega_3 = I_3\omega_3 \tag{13}$$

The kinetic energy of rotation about the principal axes is given by

$$T = \tfrac{1}{2}(I_1\omega_1^2 + I_2\omega_2^2 + I_3\omega_3^2) \tag{14}$$

which can be written in vector form as [compare equation (*9*)]

$$T = \tfrac{1}{2}\boldsymbol{\omega}\cdot\mathbf{\Omega} \tag{15}$$

THE ELLIPSOID OF INERTIA

Let $\mathbf{n}$ be a unit vector in the direction of $\boldsymbol{\omega}$. Then

$$\boldsymbol{\omega} = \omega\mathbf{n} = \omega(\cos\alpha\,\mathbf{i} + \cos\beta\,\mathbf{j} + \cos\gamma\,\mathbf{k}) \tag{16}$$

where $\cos\alpha, \cos\beta, \cos\gamma$ are the direction cosines of $\boldsymbol{\omega}$ or $\mathbf{n}$ with respect to the x, y and z axes. Then the kinetic energy of rotation is given by

$$T = \tfrac{1}{2}I\omega^2 \tag{17}$$

where $$I = I_{xx}\cos^2\alpha + I_{yy}\cos^2\beta + I_{zz}\cos^2\gamma + 2I_{xy}\cos\alpha\cos\beta + 2I_{yz}\cos\beta\cos\gamma + 2I_{xz}\cos\alpha\cos\gamma \tag{18}$$

By defining a vector
$$\boldsymbol{\rho} = \mathbf{n}/\sqrt{I} \tag{19}$$
where $\boldsymbol{\rho} = \rho_x\mathbf{i} + \rho_y\mathbf{j} + \rho_z\mathbf{k}$, (18) becomes
$$I_{xx}\rho_x^2 + I_{yy}\rho_y^2 + I_{zz}\rho_z^2 + 2I_{xy}\rho_x\rho_y + 2I_{yz}\rho_y\rho_z + 2I_{zx}\rho_z\rho_x = 1 \tag{20}$$

In the coordinates ρ_x, ρ_y, ρ_z equation (20) represents an *ellipsoid* which is called the *ellipsoid of inertia* or the *momental ellipsoid*.

If the coordinate axes are rotated to coincide with the principal axes of the ellipsoid, the equation becomes
$$I_1\rho_1^2 + I_2\rho_2^2 + I_3\rho_3^2 = 1 \tag{21}$$
where ρ_1, ρ_2, ρ_3 represent the coordinates of the new axes.

EULER'S EQUATIONS OF MOTION

It is convenient to describe the motion of a rigid body relative to a set of coordinate axes coinciding with the principal axes which are fixed in the body and thus rotate as the body rotates. If $\Lambda_1, \Lambda_2, \Lambda_3$ and $\omega_1, \omega_2, \omega_3$ represent the respective components of the external torque and angular velocity along the principal axes, the equations of motion are given by

$$\left.\begin{aligned} I_1\dot{\omega}_1 + (I_3 - I_2)\omega_2\omega_3 &= \Lambda_1 \\ I_2\dot{\omega}_2 + (I_1 - I_3)\omega_3\omega_1 &= \Lambda_2 \\ I_3\dot{\omega}_3 + (I_2 - I_1)\omega_1\omega_2 &= \Lambda_3 \end{aligned}\right\} \tag{22}$$

These are often called *Euler's equations.*

FORCE FREE MOTION. THE INVARIABLE LINE AND PLANE

Suppose that a rigid body is rotating about a fixed point O and that there are no forces acting on the body [except of course the reaction at the fixed point]. Then the total external torque is zero. Thus the angular momentum vector $\boldsymbol{\Omega}$ is constant and so has a fixed direction in space as indicated in Fig. 10-2. The line indicating this direction is called the *invariable line.*

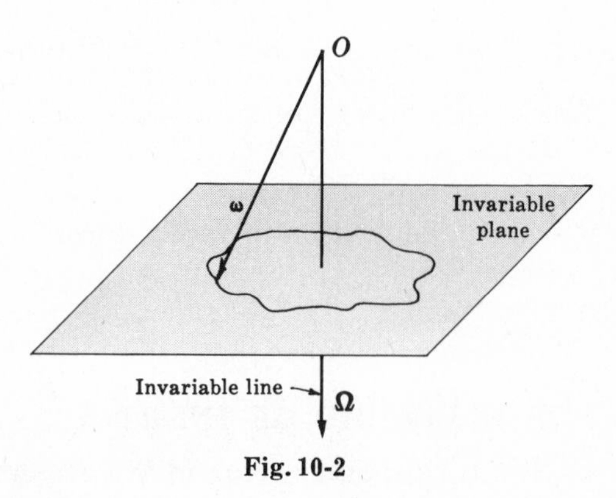

Fig. 10-2

Since the kinetic energy is constant [see Problem 10.34], we have from (15)
$$\boldsymbol{\omega}\cdot\boldsymbol{\Omega} = \text{constant} \tag{23}$$

This means that the projection of $\boldsymbol{\omega}$ on $\boldsymbol{\Omega}$ is constant, so that the terminal point of $\boldsymbol{\omega}$ describes a plane. This plane is called the *invariable plane.*

As the rigid body rotates, an observer fixed relative to the body coordinate axes would see a rotation or *precession* of the angular velocity vector $\boldsymbol{\omega}$ about the angular momentum vector $\boldsymbol{\Omega}$.

POINSOT'S CONSTRUCTION. POLHODE. HERPOLHODE. SPACE AND BODY CONES

As noted by *Poinsot*, the above ideas can be geometrically interpreted as a rolling without slipping of the ellipsoid of inertia corresponding to the rigid body on the invariable plane. The curve described on the invariable plane by the point of contact with the ellipsoid is called the *herpolhode* [see Fig. 10-3]. The corresponding curve on the ellipsoid is called the *polhode*.

To an observer fixed in space it would appear that the vector $\boldsymbol{\omega}$ traces out a cone which is called the *space cone*. To an observer fixed on the rigid body it would appear that $\boldsymbol{\omega}$ also traces out a cone which is called the *body cone*. The motion can then be equivalently described as a rolling without slipping of one cone on the other. See Problem 10.19.

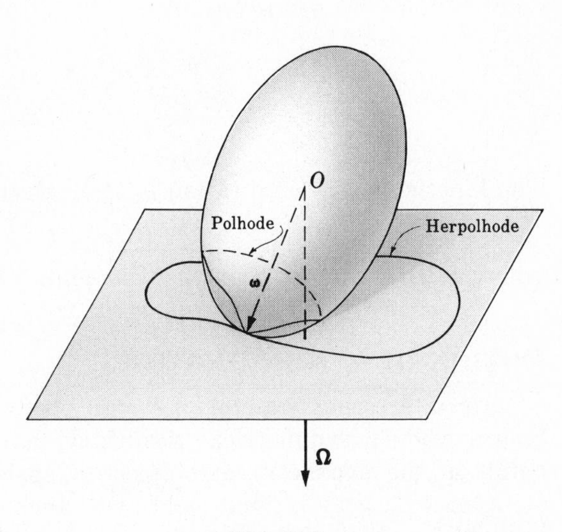

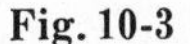
Fig. 10-3

SYMMETRIC RIGID BODIES. ROTATION OF THE EARTH

Simplifications occur in the case of a symmetric rigid body. In such case at least two principal moments of inertia, say I_1 and I_2, are equal and the ellipsoid of inertia is an ellipsoid of revolution. We can then show [see Problem 10.17] that the angular velocity vector $\boldsymbol{\omega}$ precesses about the angular momentum vector $\boldsymbol{\Omega}$ with frequency given by

$$f = \frac{1}{2\pi}\left|\frac{I_3 - I_1}{I_1}\right| A \tag{24}$$

where the constant A is the component of the angular velocity in the direction of the axis of symmetry.

In the case of the earth, which can be assumed to be an ellipsoid of revolution flattened slightly at the poles, this leads to a predicted precession period of about 300 days. In practice, however, the period is found to be about 430 days. The difference is explained as due to the fact that the earth is not perfectly rigid.

THE EULER ANGLES

In order to describe the rotation of a rigid body about a point we use 3 angular coordinates called *Euler angles*. These coordinates denoted by ϕ, θ, ψ are indicated in Fig. 10-4. In this figure the xyz coordinate system can be rotated into the $x'y'z'$ system by successive rotations through the angles ϕ and then θ and then ψ [see Problem 10.20]. The line OA is sometimes called the *line of nodes*.

In practice the x', y', z' axes are chosen as the principal axes or *body axes* of the rigid body while the x, y and z axes or *space axes* are fixed in space.

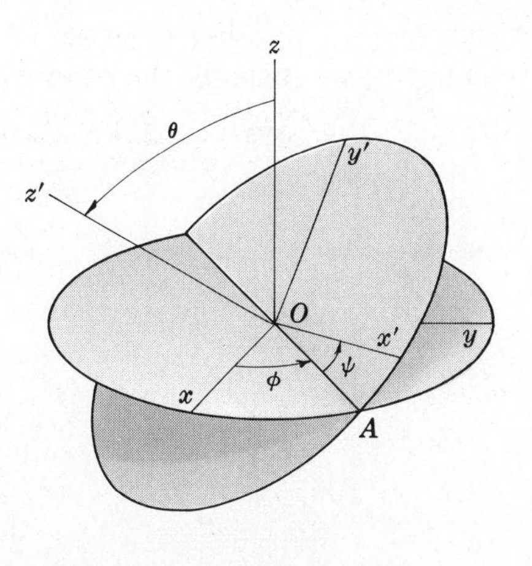

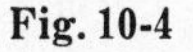
Fig. 10-4

ANGULAR VELOCITY AND KINETIC ENERGY IN TERMS OF EULER ANGLES

In terms of the Euler angles the components $\omega_1, \omega_2, \omega_3$ of the angular velocity along the x', y' and z' axes are given by

$$\left.\begin{array}{lllll} \omega_{x'} & = & \omega_1 & = & \dot{\phi}\sin\theta\sin\psi + \dot{\theta}\sin\psi \\ \omega_{y'} & = & \omega_2 & = & \dot{\phi}\sin\theta\cos\psi - \dot{\theta}\sin\psi \\ \omega_{z'} & = & \omega_3 & = & \dot{\phi}\cos\theta + \dot{\psi} \end{array}\right\} \qquad (25)$$

The kinetic energy of rotation is then given by

$$T = \tfrac{1}{2}(I_1\omega_1^2 + I_2\omega_2^2 + I_3\omega_3^2) \qquad (26)$$

where I_1, I_2, I_3 are the principal moments of inertia.

MOTION OF A SPINNING TOP

An interesting example of rigid body motion occurs when a symmetrical rigid body having one point on the symmetry axis fixed in space is set spinning in a gravitational field. One such example is that of a child's top as shown in Fig. 10-5, where point O is assumed as the fixed point.

For a discussion of the various kinds of motion which can occur, see Problems 10.25-10.32 and 10.36.

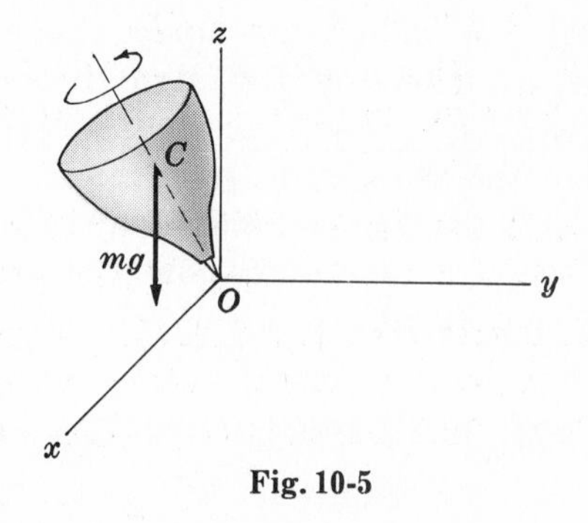

Fig. 10-5

GYROSCOPES

Suppose a circular disk having its axis mounted in *gimbals* [see Fig. 10-6] is given a spin of angular velocity $\boldsymbol{\omega}$. If the *outer gimbal* is turned through an angle, the spin axis of the disk will tend to point in the same direction as previously [see Fig. 10-7]. This assumes of course that friction at the gimbal bearings is negligible.

In general the direction of the spin axis remains fixed even when the outer gimbal, which is attached to some object, moves freely in space. Because of this property the mechanism, which is called a *gyroscope,* finds many applications in cases where maintaining direction [or following some specified course] is important, as for example in navigation and guidance or control of ships, airplanes, submarines, missiles, satellites or other moving vehicles.

A gyroscope is another example of a symmetric spinning rigid body with one point on the symmetry axis [usually the center of mass] taken as fixed.

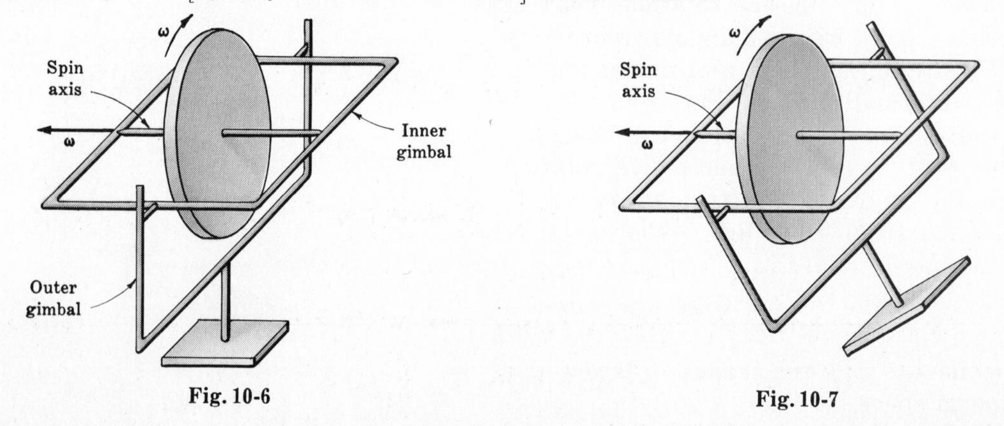

Fig. 10-6

Fig. 10-7

Solved Problems

GENERAL MOTION OF RIGID BODIES IN SPACE

10.1. Find the number of degrees of freedom for a rigid body which (*a*) can move freely in space, (*b*) has one point fixed, (*c*) has two points fixed.

(*a*) 6 [see Problem 7.2(*a*), page 172]

(*b*) 3 [see Problem 7.2(*b*), page 172]

(*c*) If two points are fixed, then the rigid body rotates about the axis joining the two fixed points. Then the number of degrees of freedom is 1, such as for example the angle of rotation of the rigid body about this axis.

10.2. A rigid body undergoes a rotation of angular velocity $\boldsymbol{\omega}$ about a fixed point O. Prove that the velocity $\mathbf{v}$ of any particle of the body having position vector $\mathbf{r}$ relative to O is given by $\mathbf{v} = \boldsymbol{\omega} \times \mathbf{r}$.

This follows at once from Problem 6.1, page 147, on noting that the velocity relative to the moving system is $d\mathbf{r}/dt|_M = d\mathbf{r}/dt|_b = \mathbf{0}$.

ANGULAR MOMENTUM. KINETIC ENERGY. MOMENTS AND PRODUCTS OF INERTIA

10.3. Derive the equations (*4*), page 254, for the components of angular momentum in terms of the moments and products of inertia given by equations (*5*) and (*6*), page 254.

The total angular momentum is given by

$$\boldsymbol{\Omega} \quad = \quad \sum_{\nu=1}^{N} m_\nu(\mathbf{r}_\nu \times \mathbf{v}_\nu) \quad = \quad \sum_{\nu=1}^{N} m_\nu\{\mathbf{r}_\nu \times (\boldsymbol{\omega} \times \mathbf{r}_\nu)\}$$

where we have used Problem 10.2 applied to the νth particle.

Now by equation (*7*), page 5, we have

$$\begin{aligned}
\mathbf{r}_\nu \times (\boldsymbol{\omega} \times \mathbf{r}_\nu) &= \boldsymbol{\omega}(\mathbf{r}_\nu \cdot \mathbf{r}_\nu) - \mathbf{r}_\nu(\boldsymbol{\omega} \cdot \mathbf{r}_\nu) \\
&= (\omega_x\mathbf{i} + \omega_y\mathbf{j} + \omega_z\mathbf{k})(x_\nu^2 + y_\nu^2 + z_\nu^2) \\
&\qquad - (x_\nu\mathbf{i} + y_\nu\mathbf{j} + z_\nu\mathbf{k})(\omega_x x_\nu + \omega_y y_\nu + \omega_z z_\nu) \\
&= \{\omega_x(y_\nu^2 + z_\nu^2) - \omega_y x_\nu y_\nu - \omega_z x_\nu z_\nu\}\mathbf{i} \\
&\qquad + \{\omega_y(x_\nu^2 + z_\nu^2) - \omega_x x_\nu y_\nu - \omega_z y_\nu z_\nu\}\mathbf{j} \\
&\qquad + \{\omega_z(x_\nu^2 + y_\nu^2) - \omega_x x_\nu z_\nu - \omega_y y_\nu z_\nu\}\mathbf{k}
\end{aligned}$$

Then multiplying by m_ν, summing over ν and equating the coefficients of $\mathbf{i}$, $\mathbf{j}$ and $\mathbf{k}$ to Ω_x, Ω_y and Ω_z respectively, we find as required

$$\begin{aligned}
\Omega_x &= \left\{\sum_{\nu=1}^{N} m_\nu(y_\nu^2 + z_\nu^2)\right\}\omega_x + \left\{-\sum_{\nu=1}^{N} m_\nu x_\nu y_\nu\right\}\omega_y + \left\{-\sum_{\nu=1}^{N} m_\nu x_\nu z_\nu\right\}\omega_z \\
&= I_{xx}\omega_x + I_{xy}\omega_y + I_{xz}\omega_z
\end{aligned}$$

$$\begin{aligned}
\Omega_y &= \left\{-\sum_{\nu=1}^{N} m_\nu x_\nu y_\nu\right\}\omega_x + \left\{\sum_{\nu=1}^{N} m_\nu(x_\nu^2 + z_\nu^2)\right\}\omega_y + \left\{-\sum_{\nu=1}^{N} m_\nu y_\nu z_\nu\right\}\omega_z \\
&= I_{xy}\omega_x + I_{yy}\omega_y + I_{yz}\omega_z
\end{aligned}$$

$$\begin{aligned}
\Omega_z &= \left\{-\sum_{\nu=1}^{N} m_\nu x_\nu z_\nu\right\}\omega_x + \left\{-\sum_{\nu=1}^{N} m_\nu y_\nu z_\nu\right\}\omega_y + \left\{\sum_{\nu=1}^{N} m_\nu(x_\nu^2 + y_\nu^2)\right\}\omega_z \\
&= I_{xz}\omega_x + I_{yz}\omega_y + I_{zz}\omega_z
\end{aligned}$$

For continuous mass distributions of density σ, we can obtain the same results by starting with

$$\boldsymbol{\Omega} \quad = \quad \int_{\mathcal{R}} \sigma(\mathbf{r} \times \mathbf{v})\, d\tau \quad = \quad \int_{\mathcal{R}} \sigma\{\mathbf{r} \times (\boldsymbol{\omega} \times \mathbf{r})\}\, d\tau$$

10.4. If a rigid body with one point fixed rotates with angular velocity $\boldsymbol{\omega}$ and has angular momentum $\boldsymbol{\Omega}$, prove that the kinetic energy is given by $T = \frac{1}{2}\boldsymbol{\omega} \cdot \boldsymbol{\Omega}$.

$$\begin{aligned} T &= \tfrac{1}{2}\sum m_\nu v_\nu^2 = \tfrac{1}{2}\sum m_\nu(\dot{\mathbf{r}}_\nu \cdot \dot{\mathbf{r}}_\nu) \\ &= \tfrac{1}{2}\sum m_\nu\{(\boldsymbol{\omega} \times \mathbf{r}_\nu) \cdot (\boldsymbol{\omega} \times \mathbf{r}_\nu)\} = \tfrac{1}{2}\sum m_\nu\{\boldsymbol{\omega} \cdot [\mathbf{r}_\nu \times (\boldsymbol{\omega} \times \mathbf{r}_\nu)]\} \\ &= \tfrac{1}{2}\boldsymbol{\omega} \cdot \sum m_\nu \mathbf{r}_\nu \times (\boldsymbol{\omega} \times \mathbf{r}_\nu) = \tfrac{1}{2}\boldsymbol{\omega} \cdot \boldsymbol{\Omega} \end{aligned}$$

where we have used the abbreviation $\sum$ in place of $\sum_{\nu=1}^{N}$.

10.5. Prove that the kinetic energy in Problem 10.4 can be written

$$T = \tfrac{1}{2}(I_{xx}\omega_x^2 + I_{yy}\omega_y^2 + I_{zz}\omega_z^2 + 2I_{xy}\omega_x\omega_y + 2I_{xz}\omega_x\omega_z + 2I_{yz}\omega_y\omega_z)$$

From Problem 10.4, we have

$$\begin{aligned} T = \tfrac{1}{2}\boldsymbol{\omega} \cdot \boldsymbol{\Omega} &= \tfrac{1}{2}(\omega_x\mathbf{i} + \omega_y\mathbf{j} + \omega_z\mathbf{k}) \cdot (\Omega_x\mathbf{i} + \Omega_y\mathbf{j} + \Omega_z\mathbf{k}) \\ &= \tfrac{1}{2}(\omega_x\Omega_x + \omega_y\Omega_y + \omega_z\Omega_z) \\ &= \tfrac{1}{2}\{\omega_x(I_{xx}\omega_x + I_{xy}\omega_y + I_{xz}\omega_z) \\ &\quad + \omega_y(I_{yx}\omega_x + I_{yy}\omega_y + I_{yz}\omega_z) \\ &\quad + \omega_z(I_{zx}\omega_x + I_{zy}\omega_y + I_{zz}\omega_z)\} \\ &= \tfrac{1}{2}(I_{xx}\omega_x^2 + I_{yy}\omega_y^2 + I_{zz}\omega_z^2 + 2I_{xy}\omega_x\omega_y + 2I_{xz}\omega_x\omega_z + 2I_{yz}\omega_y\omega_z) \end{aligned}$$

using the fact that $I_{xy} = I_{yx}$, $I_{xz} = I_{zx}$, $I_{yz} = I_{zy}$.

PRINCIPAL MOMENTS OF INERTIA AND PRINCIPAL AXES

10.6. Derive equations (*11*), page 255, for the principal moments of inertia and the directions of the principal axes.

Using

$$\boldsymbol{\Omega} = I\boldsymbol{\omega} \qquad (1)$$

together with equations (*3*) and (*4*), page 254, we have

$$\begin{aligned} I_{xx}\omega_x + I_{xy}\omega_y + I_{xz}\omega_z &= I\omega_x \\ I_{yx}\omega_x + I_{yy}\omega_y + I_{yz}\omega_z &= I\omega_y \\ I_{xz}\omega_x + I_{yz}\omega_y + I_{zz}\omega_z &= I\omega_z \end{aligned}$$

or

$$\begin{aligned} (I_{xx} - I)\omega_x + I_{xy}\omega_y + I_{xz}\omega_z &= 0 \\ I_{yx}\omega_x + (I_{yy} - I)\omega_y + I_{yz}\omega_z &= 0 \\ I_{xz}\omega_x + I_{yz}\omega_y + (I_{zz} - I)\omega_z &= 0 \end{aligned} \qquad (2)$$

The principal moments of inertia are found by setting the determinant of the coefficients of $\omega_x, \omega_y, \omega_z$ in (*2*) equal to zero, i.e.,

$$\begin{vmatrix} I_{xx} - I & I_{xy} & I_{xz} \\ I_{yx} & I_{yy} - I & I_{yz} \\ I_{xz} & I_{yz} & I_{zz} - I \end{vmatrix} = 0$$

This is a cubic equation in I leading to three values I_1, I_2, I_3 which are the principal moments of inertia. By putting $I = I_1$ in (*2*) we obtain ratios for $\omega_x : \omega_y : \omega_z$ which yields the direction of $\boldsymbol{\omega}$ or the direction of the principal axis corresponding to I_1. Similarly, by substituting I_2 and I_3 we find the directions of the corresponding principal axes.

10.7. Find the (*a*) moments of inertia and (*b*) products of inertia of a uniform square plate of length a about the x, y and z axes chosen as shown in Fig. 10-8.

(*a*) The moment of inertia of an element $dx\,dy$ of the plate about the x axis if the density is σ is $\sigma y^2\,dx\,dy$. Then the moment of inertia of the entire plate about the x axis is

$$I_{xx} = \int_{x=0}^{a}\int_{y=0}^{a} \sigma y^2\,dx\,dy = \tfrac{1}{3}\sigma a^4 = \tfrac{1}{3}Ma^2 \qquad (1)$$

since the mass of the plate is $M = \sigma a^2$.

Similarly, the moment of inertia of the plate about the y axis is

$$I_{yy} = \int_{x=0}^{a}\int_{y=0}^{a} \sigma x^2\,dx\,dy = \tfrac{1}{3}\sigma a^4 = \tfrac{1}{3}Ma^2 \qquad (2)$$

as is also evident by symmetry.

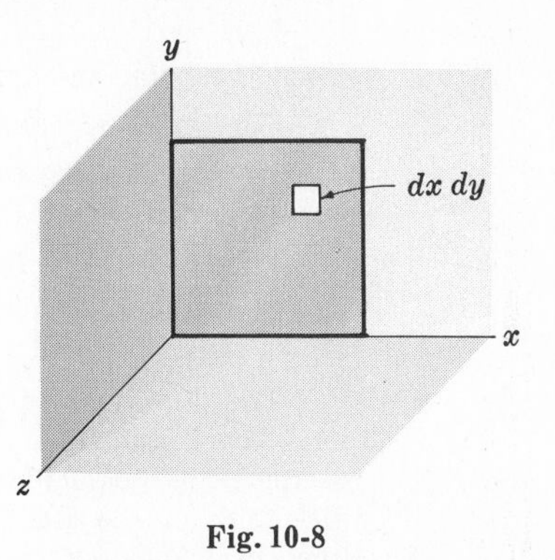

Fig. 10-8

The moment of inertia of $dx\,dy$ about the z axis is $\sigma(x^2+y^2)\,dx\,dy$, and so the moment of inertia of the entire plate about the z axis is

$$I_{zz} = \int_{x=0}^{a}\int_{y=0}^{a} \sigma(x^2+y^2)\,dx\,dy = \tfrac{1}{3}Ma^2 + \tfrac{1}{3}Ma^2 = \tfrac{2}{3}Ma^2 \qquad (3)$$

This also follows from the perpendicular axes theorem [see page 226].

(*b*) The product of inertia of the element $dx\,dy$ of the plate about the x and y axes is $\sigma xy\,dx\,dy$, and so the product of inertia of the entire plate about these axes is

$$I_{xy} = I_{yx} = -\int_{x=0}^{a}\int_{y=0}^{a} \sigma xy\,dx\,dy = -\tfrac{1}{4}\sigma a^4 = -\tfrac{1}{4}Ma^2 \qquad (4)$$

The product of inertia of the element $dx\,dy$ of the plate about the x and z axes is the product of $\sigma\,dx\,dy$ by the distances to the yz and xy planes, which are x and 0 respectively. Thus we must have

$$I_{xz} = I_{zx} = 0, \qquad \text{and similarly} \qquad I_{yz} = I_{zy} = 0 \qquad (5)$$

10.8. Find the (*a*) principal moments of inertia and (*b*) the directions of the principal axes for the plate of Problem 10.7.

(*a*) By Problem 10.6 and the results (*1*)-(*5*) of Problem 10.7, we obtain

$$\begin{vmatrix} \tfrac{1}{3}Ma^2 - I & -\tfrac{1}{4}Ma^2 & 0 \\ -\tfrac{1}{4}Ma^2 & \tfrac{1}{3}Ma^2 - I & 0 \\ 0 & 0 & \tfrac{2}{3}Ma^2 - I \end{vmatrix} = 0 \qquad (1)$$

or
$$[(\tfrac{1}{3}Ma^2 - I)(\tfrac{1}{3}Ma^2 - I) - (-\tfrac{1}{4}Ma^2)(-\tfrac{1}{4}Ma^2)][\tfrac{2}{3}Ma^2 - I] = 0$$

which can be written
$$[I^2 - \tfrac{2}{3}Ma^2 I + \tfrac{7}{144}M^2a^4][\tfrac{2}{3}Ma^2 - I] = 0$$

Setting the first factor equal to zero and using the quadratic formula to solve for I, we find for the three roots of (*1*),

$$I_1 = \tfrac{1}{12}Ma^2, \quad I_2 = \tfrac{7}{12}Ma^2, \quad I_3 = \tfrac{2}{3}Ma^2 \qquad (2)$$

which are the principal moments of inertia.

(*b*) To find the direction of the principal axis corresponding to I_1, we let $I = I_1 = \tfrac{1}{12}Ma^2$ in the equations

$$\left.\begin{aligned} (\tfrac{1}{3}Ma^2 - I)\omega_x - \tfrac{1}{4}Ma^2\omega_y + 0\,\omega_z &= 0 \\ -\tfrac{1}{4}Ma^2\omega_x + (\tfrac{1}{3}Ma^2 - I)\omega_y + 0\,\omega_z &= 0 \\ 0\,\omega_x + 0\,\omega_y + (\tfrac{2}{3}Ma^2 - I)\omega_z &= 0 \end{aligned}\right\} \qquad (3)$$

The first two equations yield $\omega_y = \omega_x$ while the third gives $\omega_z = 0$. Thus the direction of the principal axis is the same as the direction of the angular velocity vector

$$\boldsymbol{\omega} = \omega_x\mathbf{i} + \omega_y\mathbf{j} + \omega_z\mathbf{k} = \omega_x\mathbf{i} + \omega_x\mathbf{j} = \omega_x(\mathbf{i}+\mathbf{j})$$

Then the principal axis corresponding to I_1 is in the direction $\mathbf{i}+\mathbf{j}$.

Similarly, by letting $I = I_2 = \frac{1}{12}Ma^2$ in (3) we find $\omega_y = -\omega_x$, $\omega_z = 0$ so that the direction of the corresponding principal axis is $\boldsymbol{\omega} = \omega_x\mathbf{i} - \omega_x\mathbf{j} = \omega_x(\mathbf{i}-\mathbf{j})$ or $\mathbf{i}-\mathbf{j}$.

If we let $I = I_3 = \frac{2}{3}Ma^2$ in (3) we find $\omega_x = 0$, $\omega_y = 0$ while ω_z is arbitrary. This gives $\boldsymbol{\omega} = \omega_z\mathbf{k}$ which shows that the third principal axis is in the direction $\mathbf{k}$.

The directions of the principal axes are indicated by $\mathbf{i}+\mathbf{j}$, $\mathbf{i}-\mathbf{j}$ and $\mathbf{k}$ in Fig. 10-9. Note that these are mutually perpendicular and that $\mathbf{i}+\mathbf{j}$ and $\mathbf{i}-\mathbf{j}$ have the directions of the diagonals of the square plate which are lines of symmetry.

The principal moments of inertia can also be determined by recognizing the lines of symmetry.

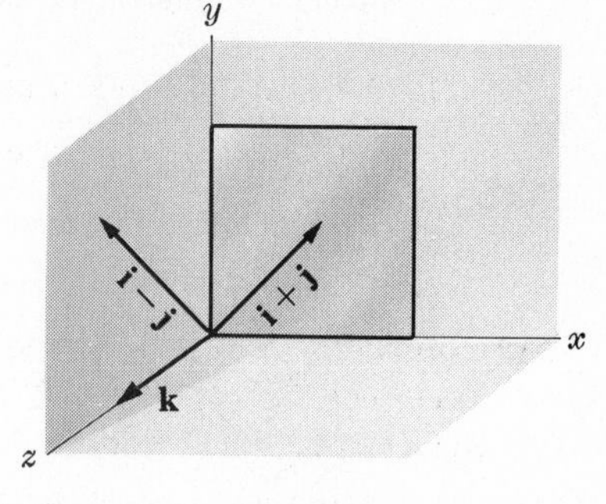

Fig. 10-9

10.9. Find the principal moments of inertia at the center of a uniform rectangular plate of sides a and b.

The principal axes lie along the directions of symmetry and thus must be along the x axis, y axis and z axis [the last of which is perpendicular to the xy plane] as in Fig. 10-10.

By Problems 9.6, 9.9 and 9.11 the principal moments of inertia are found to be $I_1 = \frac{1}{12}Ma^2$, $I_2 = \frac{1}{12}Mb^2$, $I_3 = \frac{1}{12}M(a^2+b^2)$.

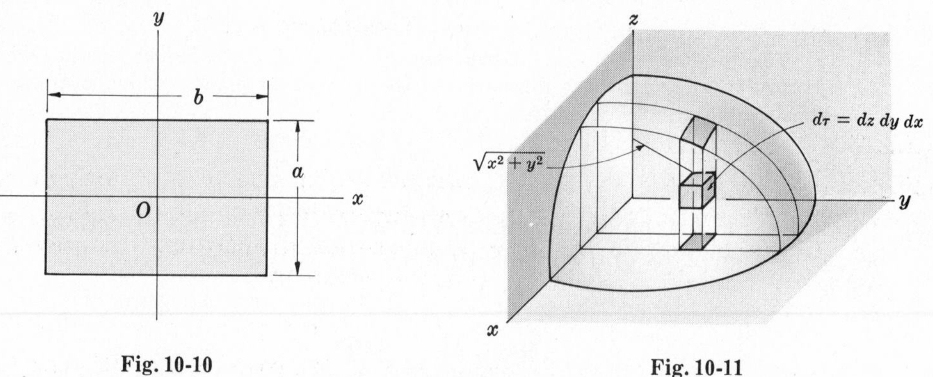

Fig. 10-10

Fig. 10-11

10.10. Find the principal moments of inertia at the center of the ellipsoid

$$\frac{x^2}{a^2} + \frac{y^2}{b^2} + \frac{z^2}{c^2} = 1$$

One eighth of the ellipsoid is indicated in Fig. 10-11. The moment of inertia of the volume element $d\tau$ of mass $\sigma\,d\tau$ about the z axis or "3" axis is $(x^2+y^2)\sigma\,d\tau$, and the total moment of inertia about the z axis is

$$I_3 = 8\int_{x=0}^{a}\int_{y=0}^{b\sqrt{1-x^2/a^2}}\int_{z=0}^{c\sqrt{1-(x^2/a^2+y^2/b^2)}} (x^2+y^2)\sigma\,dz\,dy\,dx$$

Integration with respect to z gives

$$8\sigma c\int_{x=0}^{a}\int_{y=0}^{b\sqrt{1-x^2/a^2}} (x^2+y^2)\sqrt{1-(x^2/a^2+y^2/b^2)}\,dy\,dx$$

To perform this integration let $x = aX$, $y = bY$ where X and Y are new variables. Then the integral can be written

$$8\sigma abc\int_{X=0}^{1}\int_{Y=0}^{\sqrt{1-X^2}} (a^2X^2+b^2Y^2)\sqrt{1-(X^2+Y^2)}\,dY\,dX$$

Introducing polar coordinates R, Θ in this XY plane, this becomes

$$8\sigma abc \int_{R=0}^{1} \int_{\Theta=0}^{\pi/2} (a^2R^2 \cos^2\Theta + b^2R^2 \sin^2\Theta)\sqrt{1-R^2}\, R\, dR\, d\Theta$$

$$= \quad 2\pi\sigma abc(a^2+b^2) \int_{R=0}^{1} R^3\sqrt{1-R^2}\, dR \quad = \quad \tfrac{4}{15}\pi\sigma abc(a^2+b^2)$$

where we use the substitution $1-R^2 = U^2$ in evaluating the last integral.

Since the volume of the ellipsoid is $\frac{4}{3}\pi abc$, the mass is $M = \frac{4}{3}\pi\sigma abc$ and hence $I_3 = \frac{1}{5}M(a^2+b^2)$. By symmetry we find $I_1 = \frac{1}{5}M(b^2+c^2)$, $I_2 = \frac{1}{5}M(a^2+c^2)$.

10.11. Suppose that the ellipsoid of Problem 10.10 is an oblate spheroid such that $a = b$ while c differs slightly from a or b. Prove that to a high degree of approximation, $(I_3 - I_1)/I_1 = 1 - c/a$.

From Problem 10.10, if $a = b$ then $\dfrac{I_3 - I_1}{I_1} = \dfrac{a^2 - c^2}{a^2 + c^2} = \dfrac{(a-c)(a+c)}{a^2+c^2}$. But if c differs only slightly from a then $a + c \approx 2a$ and $a^2 + c^2 \approx 2a^2$. Thus, approximately,

$$(I_3 - I_1)/I \quad = \quad (a-c)(2a)/2a^2 \quad = \quad 1 - c/a$$

10.12. Work Problem 10.11 for the case of the earth assumed to be an oblate spheroid.

Since the polar diameter or distance between north and south poles is very nearly 7900 miles while the equatorial diameter is very nearly 7926 miles, then taking the polar axis as the "3" axis we have $2c = 7900$, $2a = 7926$ or $c = 3950$, $a = 3963$.

Thus by Problem 10.11, $(I_3 - I_1)/I_1 = 1 - 3950/3963 = .00328$.

ELLIPSOID OF INERTIA

10.13. Suppose that the moments and products of inertia of a rigid body $\mathcal{R}$ with respect to an xyz coordinate system intersecting at origin O are $I_{xx}, I_{yy}, I_{zz}, I_{xy}, I_{xz}, I_{yz}$ respectively. Prove that the moment of inertia of $\mathcal{R}$ about an axis making angles α, β, γ with the x, y and z axes respectively is given by

$$I \quad = \quad I_{xx}\cos^2\alpha + I_{yy}\cos^2\beta + I_{zz}\cos^2\gamma$$
$$+\ 2I_{xy}\cos\alpha\cos\beta + 2I_{xz}\cos\alpha\cos\gamma + 2I_{yz}\cos\beta\cos\gamma$$

A unit vector in the direction of the axis is given by

$$\mathbf{n} \quad = \quad \cos\alpha\,\mathbf{i} + \cos\beta\,\mathbf{j} + \cos\gamma\,\mathbf{k}$$

Then if m_ν has position vector $\mathbf{r}_\nu$, its moment of inertia about the axis OA is $m_\nu D_\nu^2$ where $D_\nu = |\mathbf{r}_\nu \times \mathbf{n}|$. But

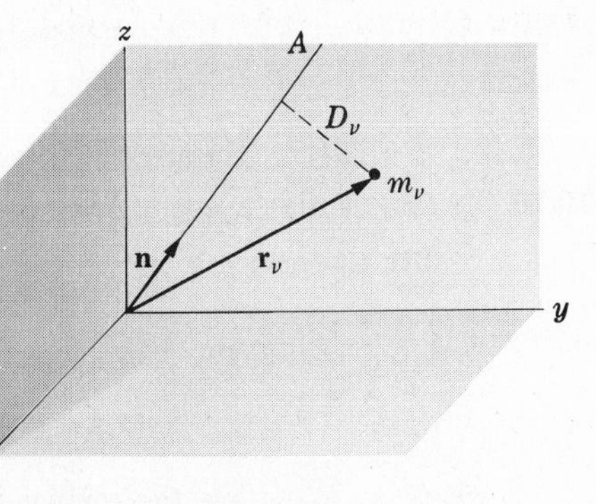

Fig. 10-12

$$\mathbf{r}_\nu \times \mathbf{n} \quad = \quad \begin{vmatrix} \mathbf{i} & \mathbf{j} & \mathbf{k} \\ x_\nu & y_\nu & z_\nu \\ \cos\alpha & \cos\beta & \cos\gamma \end{vmatrix}$$

$$= \quad (y_\nu\cos\gamma - z_\nu\cos\beta)\mathbf{i} + (z_\nu\cos\alpha - x_\nu\cos\gamma)\mathbf{j} + (x_\nu\cos\beta - y_\nu\cos\alpha)\mathbf{k}$$

and

$$|\mathbf{r}_\nu \times \mathbf{n}|^2 \quad = \quad (y_\nu\cos\gamma - z_\nu\cos\beta)^2 + (z_\nu\cos\alpha - x_\nu\cos\gamma)^2 + (x_\nu\cos\beta - y_\nu\cos\alpha)^2$$

$$= \quad (y_\nu^2 + z_\nu^2)\cos^2\alpha + (x_\nu^2 + z_\nu^2)\cos^2\beta + (x_\nu^2 + y_\nu^2)\cos^2\gamma$$
$$-\ 2x_\nu y_\nu\cos\alpha\cos\beta - 2x_\nu z_\nu\cos\alpha\cos\gamma - 2y_\nu z_\nu\cos\beta\cos\gamma$$

Thus the total moment of inertia of all masses m_ν is

$$\begin{aligned} I &= \sum m_\nu D_\nu^2 \\ &= \left\{\sum m_\nu(y_\nu^2+z_\nu^2)\right\}\cos^2\alpha + \left\{\sum m_\nu(x_\nu^2+z_\nu^2)\right\}\cos^2\beta + \left\{\sum m_\nu(x_\nu^2+y_\nu^2)\right\}\cos^2\gamma \\ &\quad + 2\left\{-\sum m_\nu x_\nu y_\nu\right\}\cos\alpha\cos\beta + 2\left\{-\sum m_\nu x_\nu z_\nu\right\}\cos\alpha\cos\gamma \\ &\quad + 2\left\{-\sum m_\nu y_\nu z_\nu\right\}\cos\beta\cos\gamma \\ &= I_{xx}\cos^2\alpha + I_{yy}\cos^2\beta + I_{zz}\cos^2\gamma \\ &\quad + 2I_{xy}\cos\alpha\cos\beta + 2I_{xz}\cos\alpha\cos\gamma + 2I_{yz}\cos\beta\cos\gamma \end{aligned}$$

10.14. **Find an equation for the ellipsoid of inertia corresponding to the square plate of Problem 10.7.**

We have from Problem 10.7,

$$I_{xx} = \tfrac{1}{3}Ma^2,\ I_{yy} = \tfrac{1}{3}Ma^2,\ I_{zz} = \tfrac{2}{3}Ma^2,\ I_{xy} = -\tfrac{1}{4}Ma^2,\ I_{xz} = 0,\ I_{yz} = 0$$

Then the equation of the ellipsoid of inertia is by equation (*20*), page 256,

$$\tfrac{1}{3}Ma^2\rho_x^2 + \tfrac{1}{3}Ma^2\rho_y^2 + \tfrac{2}{3}Ma^2\rho_z^2 - \tfrac{1}{2}Ma^2\rho_x\rho_y = 1$$

or

$$\rho_x^2 + \rho_y^2 + 2\rho_z^2 - \tfrac{3}{2}\rho_x\rho_y = 3/Ma^2$$

EULER'S EQUATIONS OF MOTION

10.15. **Find a relationship between the time rate of change of angular momentum of a rigid body relative to axes fixed in space and in the body respectively.**

If the rigid body axes are chosen as principal axes having directions of the unit vectors $\mathbf{e}_1$, $\mathbf{e}_2$ and $\mathbf{e}_3$ respectively, then the angular momentum becomes

$$\mathbf{\Omega} = I_1\omega_1\mathbf{e}_1 + I_2\omega_2\mathbf{e}_2 + I_3\omega_3\mathbf{e}_3$$

Now by Problem 6.1, page 147, if s and b refer to space (fixed) and body (moving) axes respectively, then

$$\begin{aligned} \left.\frac{d\mathbf{\Omega}}{dt}\right|_s &= \left.\frac{d\mathbf{\Omega}}{dt}\right|_b + \boldsymbol{\omega}\times\mathbf{\Omega} \\ &= I_1\dot{\omega}_1\mathbf{e}_1 + I_2\dot{\omega}_2\mathbf{e}_2 + I_3\dot{\omega}_3\mathbf{e}_3 \\ &\quad + (\omega_1\mathbf{e}_1+\omega_2\mathbf{e}_2+\omega_3\mathbf{e}_3)\times(I_1\omega_1\mathbf{e}_1+I_2\omega_2\mathbf{e}_2+I_3\omega_3\mathbf{e}_3) \\ &= \{I_1\dot{\omega}_1 + (I_3-I_2)\omega_2\omega_3\}\mathbf{e}_1 + \{I_2\dot{\omega}_2 + (I_1-I_3)\omega_1\omega_3\}\mathbf{e}_2 \\ &\quad + \{I_3\dot{\omega}_3 + (I_2-I_1)\omega_2\omega_1\}\mathbf{e}_3 \end{aligned}$$

10.16. **Derive Euler's equations of motion (*22*), page 256.**

By the principle of angular momentum, we have

$$\mathbf{\Lambda} = \left.\frac{d\mathbf{\Omega}}{dt}\right|_s \tag{1}$$

where $\mathbf{\Lambda}$ is the total external torque. Writing

$$\mathbf{\Lambda} = \Lambda_1\mathbf{e}_1 + \Lambda_2\mathbf{e}_2 + \Lambda_3\mathbf{e}_3 \tag{2}$$

where $\Lambda_1, \Lambda_2, \Lambda_3$ are the components of the external torque along the principal axes and making use of (*1*) and Problem 10.15, we find

$$\left.\begin{aligned} I_1\dot{\omega}_1 + (I_3-I_2)\omega_2\omega_3 &= \Lambda_1 \\ I_2\dot{\omega}_2 + (I_1-I_3)\omega_3\omega_1 &= \Lambda_2 \\ I_3\dot{\omega}_3 + (I_2-I_1)\omega_1\omega_2 &= \Lambda_3 \end{aligned}\right\} \tag{3}$$

FORCE FREE MOTION OF A RIGID BODY. ROTATION OF THE EARTH

10.17. A rigid body which is symmetric about an axis has one point fixed on this axis. Discuss the rotational motion of the body, assuming that there are no forces acting other than the reaction force at the fixed point.

Choose the axis of symmetry coincident with one of the principal axes, say the one having direction $\mathbf{e}_3$. Then $I_1 = I_2$ and Euler's equations become

$$I_1\dot{\omega}_1 + (I_3 - I_1)\omega_2\omega_3 = 0 \tag{1}$$

$$I_1\dot{\omega}_2 + (I_1 - I_3)\omega_3\omega_1 = 0 \tag{2}$$

$$I_3\dot{\omega}_3 = 0 \tag{3}$$

From (3), $\omega_3 = \text{constant} = A$ so that (1) and (2) become after dividing by I_1,

$$\dot{\omega}_1 + \left(\frac{I_3 - I_1}{I_1}\right)A\omega_2 = 0 \tag{4}$$

$$\dot{\omega}_2 + \left(\frac{I_1 - I_3}{I_1}\right)A\omega_1 = 0 \tag{5}$$

Differentiating (5) with respect to t and using (4), we find

$$\ddot{\omega}_2 + \left(\frac{I_3 - I_1}{I_1}\right)^2 A^2\omega_2 = 0 \tag{6}$$

or

$$\ddot{\omega}_2 + \kappa^2\omega_2 = 0 \tag{7}$$

where

$$\kappa = \left|\frac{I_3 - I_1}{I_1}\right| A \tag{8}$$

Solving (7), we find $\qquad \omega_2 = B\cos\kappa t + C\sin\kappa t$

If we choose the time scale so that $\omega_2 = 0$ when $t = 0$, then

$$\omega_2 = C\sin\kappa t \tag{9}$$

Then from (5) we have

$$\omega_1 = C\cos\kappa t \tag{10}$$

Thus the angular velocity is

$$\begin{aligned}\boldsymbol{\omega} &= \omega_1\mathbf{e}_1 + \omega_2\mathbf{e}_2 + \omega_3\mathbf{e}_3 \\ &= C\cos\kappa t\,\mathbf{e}_1 + C\sin\kappa t\,\mathbf{e}_2 + A\mathbf{e}_3\end{aligned} \tag{11}$$

From this it follows that the angular velocity is constant in magnitude equal to $\omega = |\boldsymbol{\omega}| = \sqrt{C^2 + A^2}$ and precesses around the "3" axis with frequency

$$f = \frac{\kappa}{2\pi} = \frac{|I_3 - I_1|}{2\pi I_1}A \tag{12}$$

as indicated in Fig. 10-13.

Note that the vector $\boldsymbol{\omega}$ describes a cone about the "3" axis. However, this motion is relative to the body principal axes which are in turn rotating in space with angular velocity $\boldsymbol{\omega}$.

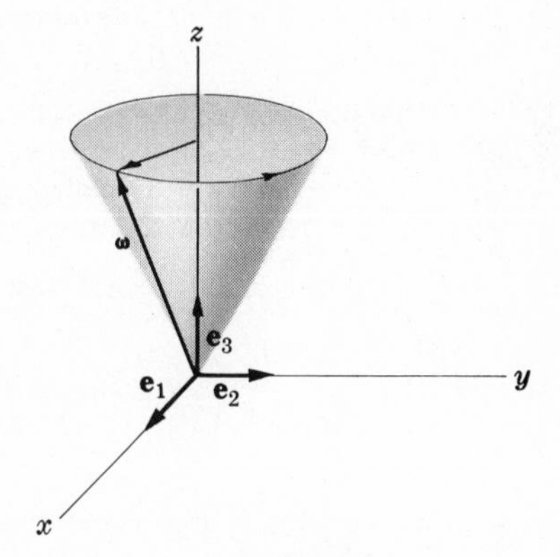

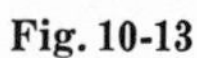

Fig. 10-13

10.18. Calculate the precession frequency of Problem 10.17 in the case of the earth rotating about its axis.

Since the earth rotates about its axis once in a day, we have $\omega_3 = A = 2\pi$ radians/day. Then the precessional frequency is from Problems 10.12 and 10.17,

$$f = \frac{1}{2\pi}\left(\frac{I_3 - I_1}{I_1}\right)A = \frac{1}{2\pi}\left(1 - \frac{c}{a}\right)A = \frac{1}{2\pi}(.00328)(2\pi) = .00328 \text{ radians/day}$$

The period of precession is thus $P = 1/f = 305$ days. The actual observed period is about 430 days and is explained as due to the non-rigidity of the earth.

THE INVARIABLE LINE AND PLANE. POLHODE, HERPOLHODE, SPACE AND BODY CONES

10.19. Describe the rotation of the earth about its axis in terms of the space and body cones.

From Problem 10.17 the angular velocity $\boldsymbol{\omega}$ and angular momentum $\boldsymbol{\Omega}$ are given respectively by

$$\boldsymbol{\omega} = \omega_1\mathbf{e}_1 + \omega_2\mathbf{e}_2 + \omega_3\mathbf{e}_3 = C(\cos\kappa t\,\mathbf{e}_1 + \sin\kappa t\,\mathbf{e}_2) + A\mathbf{e}_3$$

$$\boldsymbol{\Omega} = I_1\omega_1\mathbf{e}_1 + I_1\omega_2\mathbf{e}_2 + I_3\omega_3\mathbf{e}_3 = I_1C\,(\cos\kappa t\,\mathbf{e}_1 + \sin\kappa t\,\mathbf{e}_2) + I_3A\mathbf{e}_3$$

Let α be the angle between $\boldsymbol{\omega}_3 = \omega_3\mathbf{e}_3 = A\mathbf{e}_3$ and $\boldsymbol{\Omega}$. Then

$$\boldsymbol{\omega}_3 \cdot \boldsymbol{\Omega} = |\boldsymbol{\omega}_3|\,|\boldsymbol{\Omega}| \cos\alpha = A\sqrt{I_1^2C^2 + I_3^2A^2}\cos\alpha = I_3A^2$$

and

$$\cos\alpha = \frac{I_3A}{\sqrt{I_1^2C^2 + I_3^2A^2}} \tag{1}$$

Similarly, let β be the angle between $\boldsymbol{\omega}_3$ and $\boldsymbol{\omega}$. Then

$$\boldsymbol{\omega}_3 \cdot \boldsymbol{\omega} = |\boldsymbol{\omega}_3|\,|\boldsymbol{\omega}| \cos\beta = A\sqrt{C^2 + A^2}\cos\beta = A^2$$

and

$$\cos\beta = \frac{A}{\sqrt{C^2 + A^2}} \tag{2}$$

From (1) and (2) we see that

$$\sin\alpha = \frac{I_1C}{\sqrt{I_1^2C^2 + I_3^2A^2}}, \qquad \sin\beta = \frac{C}{\sqrt{C^2 + A^2}} \tag{3}$$

Thus

$$\tan\alpha = \frac{I_1C}{I_3A}, \qquad \tan\beta = \frac{C}{A} \tag{4}$$

or

$$\frac{\tan\alpha}{\tan\beta} = \frac{I_1}{I_3} \tag{5}$$

Now for the earth or any oblate spheroid [flattened at the poles] we have $I_1 < I_3$. It follows that $\alpha < \beta$.

The situation can be described geometrically by Fig. 10-14. The cone with axis in the direction $\boldsymbol{\Omega}$ is fixed in space and is called the *space cone*. The cone with axis $\boldsymbol{\omega}_3 = \omega_3\mathbf{e}_3$ is considered as fixed in the earth and is called the *body cone*. The body cone rolls on the space cone so that the element in common is the angular velocity vector $\boldsymbol{\omega}$. Now

$$\boldsymbol{\omega}_3 \times \boldsymbol{\omega} = \begin{vmatrix} \mathbf{e}_1 & \mathbf{e}_2 & \mathbf{e}_3 \\ 0 & 0 & A \\ I_1C\cos\kappa t & I_1C\sin\kappa t & I_3A \end{vmatrix} = -AI_1C\sin\kappa t\,\mathbf{e}_1 + AI_1C\cos\kappa t\,\mathbf{e}_2$$

Thus

$$\begin{aligned} \boldsymbol{\Omega} \cdot (\boldsymbol{\omega}_3 \times \boldsymbol{\omega}) &= (I_1C\cos\kappa t\,\mathbf{e}_1 + I_1C\sin\kappa t\,\mathbf{e}_2 + I_3A\mathbf{e}_3) \\ &\quad \cdot (-AI_1C\sin\kappa t\,\mathbf{e}_1 + AI_1C\cos\kappa t\,\mathbf{e}_2) \\ &= 0 \end{aligned}$$

It follows from Problem 1.21(*b*), page 16, that $\boldsymbol{\Omega}$, $\boldsymbol{\omega}_3$ and $\boldsymbol{\omega}$ lie in one plane.

An observer fixed in space would see the vector $\boldsymbol{\omega}$ tracing out the space cone [see Fig. 10-14]. An observer fixed in the body would see the vector $\boldsymbol{\omega}$ tracing out the body cone.

In the case of the earth the space cone is inside the body cone due to the fact that $I_1 < I_3$. For the case of a prolate spheroid the reverse situation is true, i.e. $I_1 > I_3$ and the space cone is outside the body cone [see Problem 10.121].

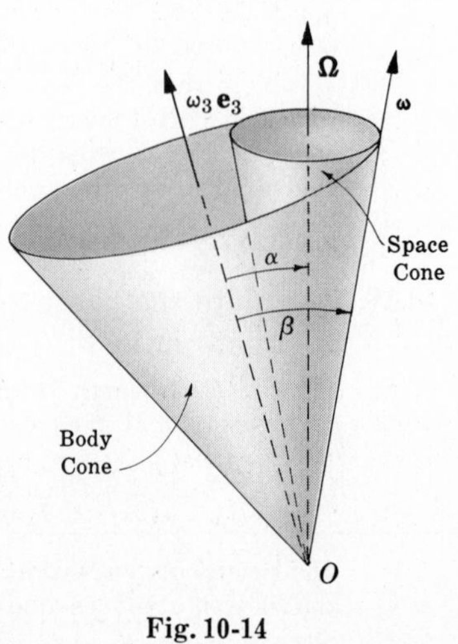

Fig. 10-14

THE EULER ANGLES

10.20. Show by using three separate figures how the xyz coordinate system of Fig. 10-4, page 257, is rotated into the $x'y'z'$ coordinate system by successive rotations through the Euler angles ϕ, θ and ψ.

Refer to Figures 10-15, 10-16 and 10-17. Fig. 10-15 indicates a rotation through angle ϕ of the x and y axes into an X and Y axis respectively while keeping the same z or Z axis.

In Fig. 10-16 a rotation about the X axis through angle θ is indicated so that the Y and Z axes of Fig. 10-15 are carried into the Y' and Z' axes of Fig. 10-16 respectively.

In Fig. 10-17 a rotation about the Z' or z' axis through angle ψ is indicated so that the X' and Y' axes are carried into the x' and y' axes respectively.

In the figures we have indicated unit vectors on the x, y, z axes; X, Y, Z axes; X', Y', Z' axes and x', y', z' axes by $\mathbf{i}, \mathbf{j}, \mathbf{k}$; $\mathbf{I}, \mathbf{J}, \mathbf{K}$; $\mathbf{I}', \mathbf{J}', \mathbf{K}'$ and $\mathbf{i}', \mathbf{j}', \mathbf{k}'$ respectively.

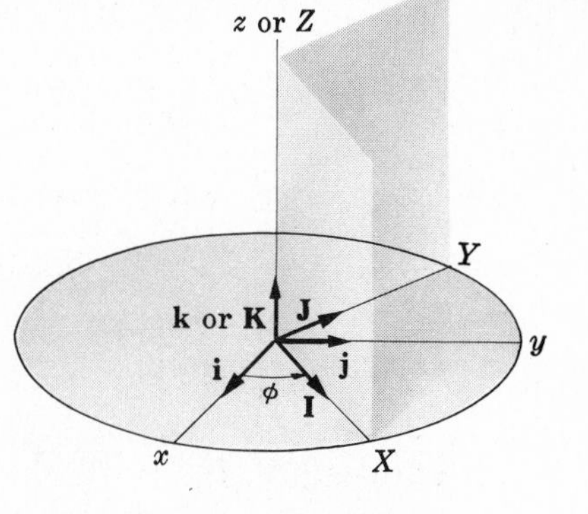

Fig. 10-15

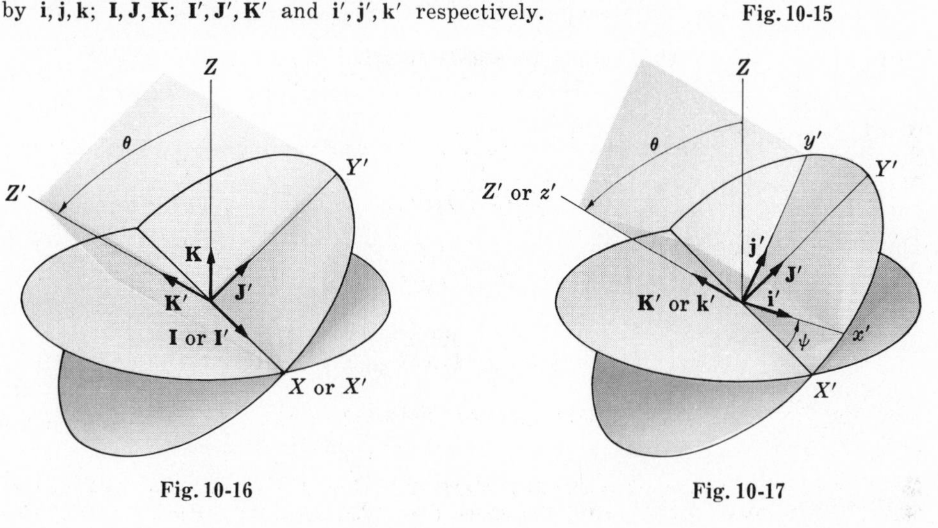

Fig. 10-16

Fig. 10-17

10.21. Find the relationships between the unit vectors (*a*) $\mathbf{i}, \mathbf{j}, \mathbf{k}$ and $\mathbf{I}, \mathbf{J}, \mathbf{K}$ of Fig. 10-15, (*b*) $\mathbf{I}, \mathbf{J}, \mathbf{K}$ and $\mathbf{I}', \mathbf{J}', \mathbf{K}'$ of Fig. 10-16, (*c*) $\mathbf{I}', \mathbf{J}', \mathbf{K}'$ and $\mathbf{i}', \mathbf{j}', \mathbf{k}'$ of Fig. 10-17.

(*a*) From Fig. 10-15,

$$\mathbf{i} = (\mathbf{i}\cdot\mathbf{I})\mathbf{I} + (\mathbf{i}\cdot\mathbf{J})\mathbf{J} + (\mathbf{i}\cdot\mathbf{K})\mathbf{K} = \cos\phi\,\mathbf{I} - \sin\phi\,\mathbf{J}$$
$$\mathbf{j} = (\mathbf{j}\cdot\mathbf{I})\mathbf{I} + (\mathbf{j}\cdot\mathbf{J})\mathbf{J} + (\mathbf{j}\cdot\mathbf{K})\mathbf{K} = \sin\phi\,\mathbf{I} + \cos\phi\,\mathbf{J}$$
$$\mathbf{k} = (\mathbf{k}\cdot\mathbf{I})\mathbf{I} + (\mathbf{k}\cdot\mathbf{J})\mathbf{J} + (\mathbf{k}\cdot\mathbf{K})\mathbf{K} = \mathbf{K}$$

(*b*) From Fig. 10-16,

$$\mathbf{I} = (\mathbf{I}\cdot\mathbf{I}')\mathbf{I}' + (\mathbf{I}\cdot\mathbf{J}')\mathbf{J}' + (\mathbf{I}\cdot\mathbf{K}')\mathbf{K}' = \mathbf{I}'$$
$$\mathbf{J} = (\mathbf{J}\cdot\mathbf{I}')\mathbf{I}' + (\mathbf{J}\cdot\mathbf{J}')\mathbf{J}' + (\mathbf{J}\cdot\mathbf{K}')\mathbf{K}' = \cos\theta\,\mathbf{J}' - \sin\theta\,\mathbf{K}'$$
$$\mathbf{K} = (\mathbf{K}\cdot\mathbf{I}')\mathbf{I}' + (\mathbf{K}\cdot\mathbf{J}')\mathbf{J}' + (\mathbf{K}\cdot\mathbf{K}')\mathbf{K}' = \sin\theta\,\mathbf{J}' + \cos\theta\,\mathbf{K}'$$

(*c*) From Fig. 10-17,

$$\mathbf{I}' = (\mathbf{I}'\cdot\mathbf{i}')\mathbf{i}' + (\mathbf{I}'\cdot\mathbf{j}')\mathbf{j}' + (\mathbf{I}'\cdot\mathbf{k}')\mathbf{k}' = \cos\psi\,\mathbf{i}' - \sin\psi\,\mathbf{j}'$$
$$\mathbf{J}' = (\mathbf{J}'\cdot\mathbf{i}')\mathbf{i}' + (\mathbf{J}'\cdot\mathbf{j}')\mathbf{j}' + (\mathbf{J}'\cdot\mathbf{k}')\mathbf{k}' = \sin\psi\,\mathbf{i}' + \cos\psi\,\mathbf{j}'$$
$$\mathbf{K}' = (\mathbf{K}'\cdot\mathbf{i}')\mathbf{i}' + (\mathbf{K}'\cdot\mathbf{j}')\mathbf{j}' + (\mathbf{K}'\cdot\mathbf{k}')\mathbf{k}' = \mathbf{k}'$$

10.22. Express the unit vectors $\mathbf{i}, \mathbf{j}, \mathbf{k}$ in terms of $\mathbf{i}', \mathbf{j}', \mathbf{k}'$.

From Problem 10.21,

$$\mathbf{i} = \cos\phi\,\mathbf{I} - \sin\phi\,\mathbf{J}, \quad \mathbf{j} = \sin\phi\,\mathbf{I} + \cos\phi\,\mathbf{J}, \quad \mathbf{k} = \mathbf{K}$$

$$\mathbf{I} = \mathbf{I}', \quad \mathbf{J} = \cos\theta\,\mathbf{J}' - \sin\theta\,\mathbf{K}', \quad \mathbf{K} = \sin\theta\,\mathbf{J}' + \cos\theta\,\mathbf{K}'$$

$$\mathbf{I}' = \cos\psi\,\mathbf{i}' - \sin\psi\,\mathbf{j}', \quad \mathbf{J}' = \sin\psi\,\mathbf{i}' + \cos\psi\,\mathbf{j}', \quad \mathbf{K}' = \mathbf{k}'$$

Then

$$\begin{aligned}\mathbf{i} &= \cos\phi\,\mathbf{I} - \sin\phi\,\mathbf{J} = \cos\phi\,\mathbf{I}' - \sin\phi\cos\theta\,\mathbf{J}' + \sin\phi\sin\theta\,\mathbf{K}' \\ &= \cos\phi\cos\psi\,\mathbf{i}' - \cos\phi\sin\psi\,\mathbf{j}' \\ &\qquad - \sin\phi\cos\theta\sin\psi\,\mathbf{i}' - \sin\phi\cos\theta\cos\psi\,\mathbf{j}' + \sin\phi\sin\theta\,\mathbf{k}' \\ &= (\cos\phi\cos\psi - \sin\phi\cos\theta\sin\psi)\mathbf{i}' \\ &\qquad + (-\cos\phi\sin\psi - \sin\phi\cos\theta\cos\psi)\mathbf{j}' + \sin\phi\sin\theta\,\mathbf{k}'\end{aligned}$$

$$\begin{aligned}\mathbf{j} &= \sin\phi\,\mathbf{I} + \cos\phi\,\mathbf{J} = \sin\phi\,\mathbf{I}' + \cos\phi\cos\theta\,\mathbf{J}' - \cos\phi\sin\theta\,\mathbf{K}' \\ &= \sin\phi\cos\psi\,\mathbf{i}' - \sin\phi\sin\psi\,\mathbf{j}' \\ &\qquad + \cos\phi\cos\theta\sin\psi\,\mathbf{i}' + \cos\phi\cos\theta\cos\psi\,\mathbf{j}' - \cos\phi\sin\theta\,\mathbf{k}' \\ &= (\sin\phi\cos\psi + \cos\phi\cos\theta\sin\psi)\mathbf{i}' \\ &\qquad + (-\sin\phi\sin\psi + \cos\phi\cos\theta\cos\psi)\mathbf{j}' - \cos\phi\sin\theta\,\mathbf{k}'\end{aligned}$$

$$\mathbf{k} = \sin\theta\,\mathbf{J}' + \cos\theta\,\mathbf{K}' = \sin\theta\sin\psi\,\mathbf{i}' + \sin\theta\cos\psi\,\mathbf{j}' + \cos\theta\,\mathbf{k}'$$

10.23. Derive equations (*25*), page 258.

$$\begin{aligned}\boldsymbol{\omega} &= \omega_\phi\mathbf{k} + \omega_\theta\mathbf{I}' + \omega_\psi\mathbf{K}' = \dot\phi\mathbf{k} + \dot\theta\mathbf{I}' + \dot\psi\mathbf{K}' \\ &= \dot\phi\sin\theta\sin\psi\,\mathbf{i}' + \dot\phi\sin\theta\cos\psi\,\mathbf{j}' + \dot\phi\cos\theta\,\mathbf{k}' \\ &\qquad + \dot\theta\cos\psi\,\mathbf{i}' - \dot\theta\sin\psi\,\mathbf{j}' + \dot\psi\mathbf{k}' \\ &= (\dot\phi\sin\theta\sin\psi + \dot\theta\cos\psi)\mathbf{i}' \\ &\qquad + (\dot\phi\sin\theta\cos\psi - \dot\theta\sin\psi)\mathbf{j}' + (\dot\phi\cos\theta + \dot\psi)\mathbf{k}'\end{aligned}$$

Then since $\boldsymbol{\omega} = \omega_{x'}\mathbf{i}' + \omega_{y'}\mathbf{j}' + \omega_{z'}\mathbf{k}'$,

$$\omega_{x'} = \omega_1 = \dot\phi\sin\theta\sin\psi + \dot\theta\cos\psi$$

$$\omega_{y'} = \omega_2 = \dot\phi\sin\theta\cos\psi - \dot\theta\sin\psi$$

$$\omega_{z'} = \omega_3 = \dot\phi\cos\theta + \dot\psi$$

10.24. (*a*) Write the kinetic energy of rotation of a rigid body with respect to the principal axes in terms of the Euler angles. (*b*) What does the result in (*a*) become if $I_1 = I_2$?

(*a*) Using Problem 10.23, the required kinetic energy is seen to be

$$\begin{aligned}T &= \tfrac{1}{2}(I_1\omega_1^2 + I_2\omega_2^2 + I_3\omega_3^2) \\ &= \tfrac{1}{2}I_1(\dot\phi\sin\theta\sin\psi + \dot\theta\cos\psi)^2 \\ &\qquad + \tfrac{1}{2}I_2(\dot\phi\sin\theta\cos\psi - \dot\theta\sin\psi)^2 + \tfrac{1}{2}I_3(\dot\phi\cos\theta + \dot\psi)^2\end{aligned}$$

(*b*) If $I_1 = I_2$, the result can be written

$$T = \tfrac{1}{2}I_1(\dot\phi^2\sin^2\theta + \dot\theta^2) + \tfrac{1}{2}I_3(\dot\phi\cos\theta + \dot\psi)^2$$

MOTION OF SPINNING TOPS AND GYROSCOPES

10.25. Set up equations for the motion of a spinning top having fixed point O [see Fig. 10-18].

Let xyz represent an inertial or fixed set of axes having origin O. Let $x'y'z'$ represent principal axes of the top having the same origin. Choose the orientation of the $x'y'$ plane so that Oz, Oz' and Oy' are coplanar. Then the x' axis is in the xy plane. The line ON in the $x'y'$ plane making an angle ψ with the x' axis is assumed to be attached to the top.

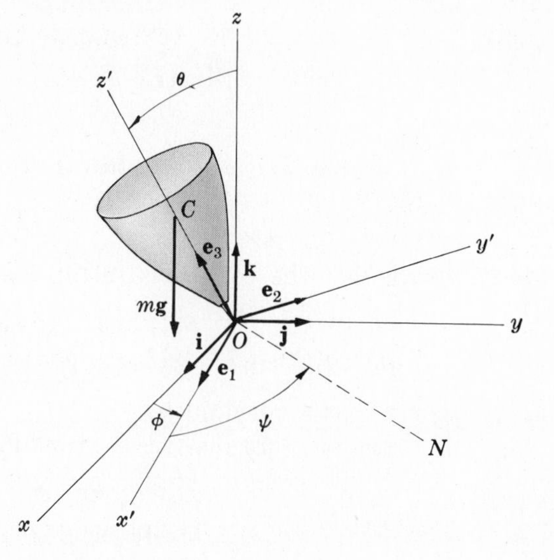

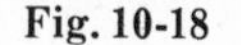
Fig. 10-18

The angular velocity corresponding to the rotation of the $x'y'z'$ axes with respect to the xyz axes is

$$\boldsymbol{\omega} = \omega_1 \mathbf{e}_1 + \omega_2 \mathbf{e}_2 + \omega_3 \mathbf{e}_3 \tag{1}$$

In obtaining the angular momentum we must use the fact that in addition to the component ω_3 due to rotation of the $x'y'z'$ system there is also the component $\mathbf{s} = s\mathbf{e}_3 = \dot{\psi}\mathbf{e}_3$ since the top is spinning about the z' axis. Then the angular momentum is

$$\boldsymbol{\Omega} = I_1\omega_1\mathbf{e}_1 + I_2\omega_2\mathbf{e}_2 + I_3(\omega_3 + s)\mathbf{e}_3 \tag{2}$$

Now if we let subscripts f and b denote the fixed system and body system respectively, we have by Problem 6.1, page 147,

$$\left.\frac{d\boldsymbol{\Omega}}{dt}\right|_f = \left.\frac{d\boldsymbol{\Omega}}{dt}\right|_b + \boldsymbol{\omega} \times \boldsymbol{\Omega} \tag{3}$$

Using (1) and (2) in (3), we find

$$\begin{aligned} \left.\frac{d\boldsymbol{\Omega}}{dt}\right|_f = &\ \{I_1\dot{\omega}_1 + (I_3 - I_2)\omega_2\omega_3 + I_3\omega_2 s\}\mathbf{e}_1 \\ &+ \{I_2\dot{\omega}_2 + (I_1 - I_3)\omega_1\omega_3 - I_3\omega_1 s\}\mathbf{e}_2 \\ &+ \{I_3(\dot{\omega}_3 + \dot{s}) + (I_2 - I_1)\omega_1\omega_2\}\mathbf{e}_3 \end{aligned} \tag{4}$$

The total torque about O is

$$\boldsymbol{\Lambda} = (l\mathbf{e}_3) \times (m\mathbf{g}) = (l\mathbf{e}_3) \times (-mg\mathbf{k}) \tag{5}$$

Since

$$\begin{aligned} \mathbf{k} &= (\mathbf{k}\cdot\mathbf{e}_1)\mathbf{e}_1 + (\mathbf{k}\cdot\mathbf{e}_2)\mathbf{e}_2 + (\mathbf{k}\cdot\mathbf{e}_3)\mathbf{e}_3 = \cos(\pi/2 - \theta)\mathbf{e}_2 + \cos\theta\,\mathbf{e}_3 \\ &= \sin\theta\,\mathbf{e}_2 + \cos\theta\,\mathbf{e}_3 \end{aligned}$$

the torque is

$$\boldsymbol{\Lambda} = -mgl(\mathbf{e}_3 \times \mathbf{k}) = mgl\sin\theta\,\mathbf{e}_1 \tag{6}$$

Then using $\boldsymbol{\Lambda} = \left.\frac{d\boldsymbol{\Omega}}{dt}\right|_f$ with $I_1 = I_2$, we find from (4) and (6),

$$\left.\begin{aligned} I_1\dot{\omega}_1 + (I_3 - I_1)\omega_2\omega_3 + I_3\omega_2 s &= mgl\sin\theta \\ I_1\dot{\omega}_2 + (I_1 - I_3)\omega_1\omega_3 - I_3\omega_1 s &= 0 \\ I_3(\dot{\omega}_3 + \dot{s}) &= 0 \end{aligned}\right\} \tag{7}$$

10.26. Express equations (7) of Problem 10.25 in terms of the Euler angles θ and ϕ of Fig. 10-18.

The components $\omega_1, \omega_2, \omega_3$ can be obtained from Problem 10.23 by formally letting $\psi = 0$. We find

$$\omega_1 = \dot{\theta}, \quad \omega_2 = \dot{\phi}\sin\theta, \quad \omega_3 = \dot{\phi}\cos\theta \tag{1}$$

Then equations (7) of Problem 10.25 become

$$\left.\begin{aligned} I_1\ddot{\theta} + (I_3 - I_1)\dot{\phi}^2 \sin\theta\cos\theta + I_3\dot{\phi}s\sin\theta &= mgl\sin\theta \\ I_1(\ddot{\phi}\sin\theta + \dot{\phi}\dot{\theta}\cos\theta) + (I_1 - I_3)\dot{\theta}\dot{\phi}\cos\theta - I_3\dot{\theta}s &= 0 \\ I_3(\ddot{\phi}\cos\theta - \dot{\phi}\dot{\theta}\sin\theta + \dot{s}) &= 0 \end{aligned}\right\} \tag{2}$$

The quantities $\dot{\phi}$, $\dot{\theta}$ and s are often known as the magnitudes of the *angular velocity of precession*, of *nutation* and of *spin* respectively.

10.27. Prove that the equations (*2*) of Problem 10.26 can be written as

(*a*) $I_1\ddot{\theta} - I_1\dot{\phi}^2\sin\theta\cos\theta + I_3A\dot{\phi}\sin\theta = mgl\sin\theta$

(*b*) $I_1(\ddot{\phi}\sin\theta + 2\dot{\phi}\dot{\theta}\cos\theta) - I_3A\dot{\theta} = 0$

where A is a constant.

From the third equation in (*7*) of Problem 10.25,

$$\omega_3 + s = A \quad \text{or} \quad s = A - \omega_3 \tag{1}$$

Then substitution into the first and second of equations (*7*) yields

$$I_1\dot{\omega}_1 - I_1\omega_2\omega_3 + I_3\omega_2 A = mgl\sin\theta \tag{2}$$

$$I_1\dot{\omega}_2 + I_1\omega_1\omega_3 - I_3\omega_1 A = 0 \tag{3}$$

Using the results (*1*) of Problem 10.26, we find that equations (*2*) and (*3*) reduce to the required equations.

10.28. (*a*) Find the condition for steady precession of a top.
(*b*) Show that two precessional frequencies are possible.

Since θ is constant so that $\ddot{\theta} = 0$, we have from Problem 10.27(*a*),

$$(I_1\dot{\phi}^2\cos\theta - I_3A\dot{\phi} + mgl)\sin\theta = 0$$

or

$$I_1\dot{\phi}^2\cos\theta - I_3A\dot{\phi} + mgl = 0$$

from which

$$\dot{\phi} = \frac{I_3A \pm \sqrt{I_3^2A^2 - 4mglI_1\cos\theta}}{2I_1\cos\theta} \tag{1}$$

Thus there are two frequencies provided that

$$I_3^2A^2 > 4mglI_1\cos\theta \tag{2}$$

If $I_3^2A^2 = 4mglI_1\cos\theta$ only one frequency is possible.

If A is very large, e.g. if the spin of the top is very great, then there are two frequencies, one large and one small, given by

$$I_3A/(I_1\cos\theta), \quad mgl/I_3A \tag{3}$$

10.29. Prove that

(*a*) $\frac{1}{2}I_1(\dot{\theta}^2 + \dot{\phi}^2\sin^2\theta) + \frac{1}{2}I_3A^2 + mgl\cos\theta = \text{constant} = E$

(*b*) $I_1\dot{\phi}\sin^2\theta + I_3A\cos\theta = \text{constant} = K$

and give a physical interpretation of each result.

(*a*) Multiply equations (*7*) of Problem 10.25 by ω_1, ω_2 and $\omega_3 + s$ respectively, and add to obtain

$$I_1(\omega_1\dot{\omega}_1 + \omega_2\dot{\omega}_2) + I_3(\omega_3 + s)(\dot{\omega}_3 + \dot{s}) = mgl\sin\theta\,\dot{\theta}$$

which can be written as

$$\frac{d}{dt}\{\tfrac{1}{2}I_1(\omega_1^2 + \omega_2^2) + \tfrac{1}{2}I_3(\omega_3 + s)^2\} = \frac{d}{dt}(-mgl\cos\theta)$$

Integrating and using $\omega_3 + s = A$ as well as the results $\omega_1 = \dot{\theta}$ and $\omega_2 = \dot{\phi}\sin\theta$, we find

$$\tfrac{1}{2}I_1(\dot{\theta}^2 + \dot{\phi}^2\sin^2\theta) + \tfrac{1}{2}I_3A^2 + mgl\cos\theta = E \tag{1}$$

where E is constant. The result is equivalent to the principle of conservation of energy, since the kinetic energy is

$$T = \tfrac{1}{2}I_1(\dot{\theta}^2 + \dot{\phi}^2 \sin^2\theta) + \tfrac{1}{2}I_3A^2 \quad (2)$$

while the potential energy is $$V = mgl\cos\theta \quad (3)$$

and $T+V=E$ is the total energy.

(b) Multiplying the result of Problem 10.27(b) by $\sin\theta$,

$$I_1\ddot{\phi}\sin^2\theta + 2I_1\dot{\phi}\dot{\theta}\sin\theta\cos\theta - I_3A\dot{\theta}\sin\theta = 0$$

which can be written

$$\frac{d}{dt}(I_1\dot{\phi}\sin^2\theta + I_3A\cos\theta) = 0$$

Integrating, $$I_1\dot{\phi}\sin^2\theta + I_3A\cos\theta = \text{constant} = K \quad (4)$$

To interpret this result physically, we note that the vertical component of the angular momentum is $I_1\dot{\phi}\sin^2\theta + I_3A\cos\theta$ [see Problem 10.123], and this must be constant since the torque due to the weight of the top has zero component in the vertical direction.

10.30. Let $u = \cos\theta$. Prove that:

(a) $$\dot{u}^2 = (\alpha - \beta u)(1-u^2) - (\gamma - \delta u)^2 = f(u)$$

where $\alpha = 2(E - \tfrac{1}{2}I_3A^2)/I_1$, $\beta = 2mgl/I_1$, $\gamma = K/I_1$, $\delta = I_3A/I_1$;

(b) $$t = \int \frac{du}{\sqrt{f(u)}} + \text{constant}$$

(a) From Problem 10.29,

$$\tfrac{1}{2}I_1(\dot{\theta}^2 + \dot{\phi}^2\sin^2\theta) + \tfrac{1}{2}I_3A^2 + mgl\cos\theta = E \quad (1)$$

$$I_1\dot{\phi}\sin^2\theta + I_3A\cos\theta = K \quad (2)$$

From (2), $$\dot{\phi} = \frac{K - I_3A\cos\theta}{I_1\sin^2\theta} \quad (3)$$

Substituting this into (1),

$$\tfrac{1}{2}I_1\dot{\theta}^2 + \frac{(K - I_3A\cos\theta)^2}{2I_1\sin^2\theta} + \tfrac{1}{2}I_3A^2 + mgl\cos\theta = E$$

Letting $u = \cos\theta$ so that $\dot{u} = -\sin\theta\,\dot{\theta}$ and $\sin^2\theta = 1 - u^2$, this becomes

$$\tfrac{1}{2}I_1\frac{\dot{u}^2}{1-u^2} + \frac{(K - I_3Au)^2}{2I_1(1-u^2)} + mglu = E - \tfrac{1}{2}I_3A^2$$

Thus $$\dot{u}^2 + \left(\frac{K - I_3Au}{I_1}\right)^2 + \frac{2mglu(1-u^2)}{I_1} = \frac{2(1-u^2)}{I_1}(E - \tfrac{1}{2}I_3A^2)$$

which can be written as

$$\dot{u}^2 = (\alpha - \beta u)(1-u^2) - (\gamma - \delta u)^2 = f(u) \quad (4)$$

where $$\alpha = (2E - I_3A^2)/I_1, \quad \beta = 2mgl/I_1, \quad \gamma = K/I_1, \quad \delta = I_3A/I_1 \quad (5)$$

Note that with this notation (3) can be written

$$\dot{\phi} = \frac{\gamma - \delta u}{1 - u^2} \quad (6)$$

(*b*) From the result of (*a*) we have, since $\dot{u} > 0$,

$$\dot{u} = \frac{du}{dt} = \sqrt{f(u)} \quad \text{or} \quad dt = \frac{du}{\sqrt{f(u)}}$$

Integrating,

$$t = \int \frac{du}{\sqrt{f(u)}} + c \tag{7}$$

The integral can be evaluated in terms of *elliptic functions* which are periodic.

10.31. (*a*) Prove that $\dot{\theta} = 0$ at those values of u for which

$$f(u) = (\alpha - \beta u)(1 - u^2) - (\gamma - \delta u)^2 = 0$$

(*b*) Prove that the equation in (*a*) has three real roots u_1, u_2, u_3 but that in general not all the angles corresponding to these are real.

(*a*) From Problem 10.30(*a*),

$$\dot{u}^2 = f(u) = (\alpha - \beta u)(1 - u^2) - (\gamma - \delta u)^2 \tag{1}$$

Since $\dot{u} = -\sin\theta\,\dot{\theta}$, it follows that $\dot{\theta} = 0$ where $\dot{u} = 0$ or $f(u) = 0$. Thus $\dot{\theta} = 0$ at the roots of the equation

$$f(u) = (\alpha - \beta u)(1 - u^2) - (\gamma - \delta u)^2 = 0 \tag{2}$$

(*b*) Equation (*1*) can be written as

$$f(u) = \beta u^3 - (\delta^2 + \alpha)u^2 + (2\gamma\delta - \beta)u + \alpha - \gamma^2 \tag{3}$$

Since $\beta > 0$, it follows that

$$f(\infty) = \infty, \quad f(-\infty) = -\infty$$
$$f(1) = -(\gamma - \delta)^2, \quad f(-1) = -(\gamma + \delta)^2$$

Thus there is a change of sign from $-$ to $+$ as u goes from 1 to ∞, and consequently there must be a root, say u_3, between 1 and ∞ as indicated in Fig. 10-19.

Now we know that in order for the motion of the top to take place we must have $f(u) = \dot{u}^2 \geqq 0$. Also, since $0 \leqq \theta \leqq \pi/2$, we must have $0 \leqq u \leqq 1$. It thus follows that there must be two roots u_1 and u_2 between 0 and 1, as indicated in the figure.

It follows that in general there are two corresponding angles θ_1 and θ_2 such that $\cos\theta_1 = u_1$, $\cos\theta_2 = u_2$. In special cases it could happen that $u_1 = u_2$ or $u_2 = u_3 = 1$.

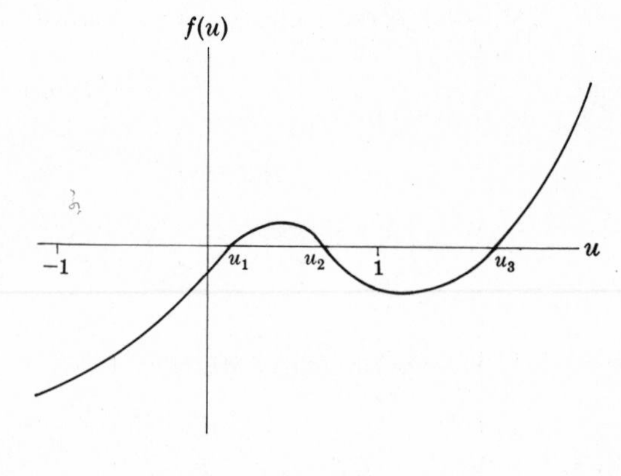

Fig. 10-19

10.32. Give a physical interpretation of the results found in Problem 10.31.

The fact that there are two roots u_1 and u_2 corresponding to θ_1 and θ_2 respectively, shows that the motion of the top is such that its axis always makes an angle θ with the vertical which lies between θ_1 and θ_2. This motion, which is a bobbing up and down of the axis between the limits θ_1 and θ_2, is called *nutation* and takes place at the same time as the *precessional motion* of the axis of the top about the vertical and the *spinning* of the top about its axis. Because the motion can be expressed in terms of elliptic functions [see Problem 10.104], we can show that it is periodic.

In general the tip of the axis of the top will describe one of various types of curves such as indicated in Figs. 10-20, 10-21 and 10-22. The type of curve will depend on the root of the equation [see equation (*6*) of Problem 10.30]

$$\dot{\phi} = \frac{\gamma - \delta u}{1 - u^2} = 0 \tag{1}$$

If this root given by γ/δ is greater than u_2, the curve of Fig. 10-20 occurs. If it is the same as u_2, the curve of Fig. 10-21 obtains. If it is between u_1 and u_2 the curve of Fig. 10-22 occurs. Other cases can arise if the root is the same as u_1 or is less than u_1 [see Problems 10.124].

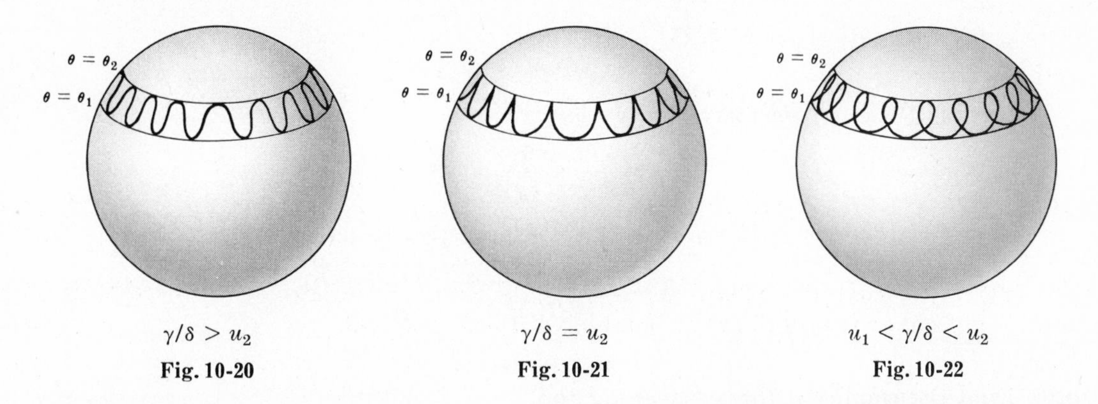

$\gamma/\delta > u_2$ **Fig. 10-20**

$\gamma/\delta = u_2$ **Fig. 10-21**

$u_1 < \gamma/\delta < u_2$ **Fig. 10-22**

Aside from the general motion which is made up of nutation and precession, there are various special cases which can arise. One of these is the case of steady precession with no nutation [see Problem 10.28]. In this case $u_1 = u_2$ so that $\theta_1 = \theta_2$ or $\theta = \text{constant}$. Another case is the "sleeping top" which occurs where $u_2 = u_3 = 1$ and the axis of the spinning top always remains vertical [see Problem 10.36].

MISCELLANEOUS PROBLEMS

10.33. If T is the total kinetic energy of rotation of a rigid body with one point fixed, prove that $dT/dt = \boldsymbol{\omega}\cdot\mathbf{\Lambda}$ where all quantities refer to the body principal axes.

Multiplying both sides of the Euler equations (*3*) of Problem 10.16 by $\omega_1, \omega_2, \omega_3$ respectively and adding, we obtain

$$I_1\omega_1\dot{\omega}_1 + I_2\omega_2\dot{\omega}_2 + I_3\omega_3\dot{\omega}_3 = \omega_1\Lambda_1 + \omega_2\Lambda_2 + \omega_3\Lambda_3 \tag{1}$$

But
$$I_1\omega_1\dot{\omega}_1 + I_2\omega_2\dot{\omega}_2 + I_3\omega_3\dot{\omega}_3 = \frac{1}{2}\frac{d}{dt}(I_1\omega_1^2 + I_2\omega_2^2 + I_3\omega_3^2) = \frac{dT}{dt} \tag{2}$$

and
$$\begin{aligned}\omega_1\Lambda_1 + \omega_2\Lambda_2 + \omega_3\Lambda_3 &= (\omega_1\mathbf{e}_1 + \omega_2\mathbf{e}_2 + \omega_3\mathbf{e}_3)\cdot(\Lambda_1\mathbf{e}_1 + \Lambda_2\mathbf{e}_2 + \Lambda_3\mathbf{e}_3)\\ &= \boldsymbol{\omega}\cdot\mathbf{\Lambda}\end{aligned} \tag{3}$$

Thus (*1*) becomes
$$dT/dt = \boldsymbol{\omega}\cdot\mathbf{\Lambda} \tag{4}$$

10.34. (*a*) Prove that if there are no forces acting on a rigid body with one point fixed, then the total kinetic energy of rotation is constant. (*b*) Thus prove that $\boldsymbol{\omega}\cdot\mathbf{\Omega} = 2T = \text{constant}$.

(*a*) Since there are no forces, $\mathbf{\Lambda} = \mathbf{0}$. Then by Problem 10.33, $dT/dt = 0$ or $T = \text{constant}$.

(*b*) Since $\mathbf{\Omega} = I_1\omega_1\mathbf{e}_1 + I_2\omega_2\mathbf{e}_2 + I_3\omega_3\mathbf{e}_3$ and $\boldsymbol{\omega} = \omega_1\mathbf{e}_1 + \omega_2\mathbf{e}_2 + \omega_3\mathbf{e}_3$,

$$\boldsymbol{\omega}\cdot\mathbf{\Omega} = I_1\omega_1^2 + I_2\omega_2^2 + I_3\omega_3^2 = 2T = \text{constant}$$

10.35. Find the precession frequency of Problem 10.17 in terms of the kinetic energy and angular momentum of the rigid body.

The kinetic energy is

$$T = \tfrac{1}{2}(I_1\omega_1^2 + I_2\omega_2^2 + I_3\omega_3^2) = \tfrac{1}{2}(I_1\omega_1^2 + I_1\omega_2^2 + I_3\omega_3^2) = \tfrac{1}{2}(I_1C^2 + I_3A^2)$$

so that
$$I_1C^2 + I_3A^2 = 2T \tag{1}$$

The angular momentum is

$$\mathbf{\Omega} = I_1\omega_1\mathbf{e}_1 + I_2\omega_2\mathbf{e}_2 + I_3\omega_3\mathbf{e}_3 = I_1\omega_1\mathbf{e}_1 + I_1\omega_2\mathbf{e}_2 + I_3\omega_3\mathbf{e}_3$$
$$= I_1C(\cos\kappa t\ \mathbf{e}_1 + \sin\kappa t\ \mathbf{e}_2) + I_3A\mathbf{e}_3$$

so that $\Omega = |\mathbf{\Omega}| = \sqrt{I_1^2C^2 + I_3^2A^2}$ or

$$I_1^2C^2 + I_3^2A^2 = \Omega^2 \tag{2}$$

Solving (1) and (2) simultaneously, we find

$$C^2 = \frac{2TI_3 - \Omega^2}{I_1(I_3 - I_1)}, \qquad A^2 = \frac{\Omega^2 - 2TI_1}{I_3(I_3 - I_1)} \tag{3}$$

Then from Problem 10.17, equation (12), the precession frequency is

$$f = \frac{1}{2\pi I_1}\sqrt{\frac{|(\Omega^2 - 2TI_1)(I_3 - I_1)|}{I_3}} \tag{4}$$

10.36. Find the condition for a "sleeping top".

For a "sleeping top" we must have $\theta = 0$ and $\dot{\theta} = 0$, since the axis must remain vertical and no nutation can take place. Then from Problem 10.29,

$$I_3A = K, \qquad I_3A^2 = 2(E - mgl)$$

Also, from Problem 10.30 we have $\alpha = 2mgl/I_1$, $\beta = 2mgl/I_1$, $\gamma = I_3A/I_1$, $\delta = I_3A/I_1$. Thus $\alpha = \beta$ and $\gamma = \delta$, and so

$$f(u) = (\alpha - \beta u)(1 - u^2) - (\gamma - \delta u)^2 = \alpha(1 - u)(1 - u^2) - \gamma^2(1 - u)^2 = (1 - u)^2[\alpha(1 + u) - \gamma^2]$$

It follows that $f(u) = 0$ has a double root at $u = 1$, while the third root is given by

$$\frac{\gamma^2}{\alpha} - 1 = \frac{I_3^2A^2}{2mglI_1} - 1$$

Then the top will "sleep" if this root is greater than or at most equal to 1, so that

$$A^2 \geqq 4mglI_1/I_3^2$$

Of course, even though this condition may apply at the beginning, energy will in practice be diminished due to friction at the support so that after some time we will have $A^2 < 4mglI_1/I_3^2$. In such case precession combined with nutation will be introduced. Further loss of energy will ultimately cause the top to fall down.

10.37. Find the torque needed to rotate a rectangular plate of sides a and b [see Fig. 10-23] about a diagonal with constant angular velocity $\boldsymbol{\omega}$.

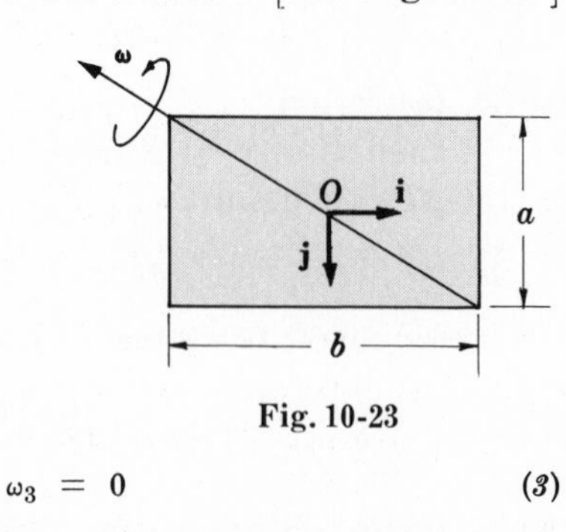

Fig. 10-23

By Problem 10.9 the principal moments of inertia of the plate at the center O are given by

$$I_1 = \tfrac{1}{12}Ma^2, \quad I_2 = \tfrac{1}{12}Mb^2, \quad I_3 = \tfrac{1}{12}M(a^2 + b^2) \tag{1}$$

We have

$$\boldsymbol{\omega} = (\boldsymbol{\omega}\cdot\mathbf{i})\mathbf{i} + (\boldsymbol{\omega}\cdot\mathbf{j})\mathbf{j}$$
$$= -\frac{\omega a}{\sqrt{a^2 + b^2}}\mathbf{i} - \frac{\omega b}{\sqrt{a^2 + b^2}}\mathbf{j} \tag{2}$$

Thus

$$\omega_1 = \frac{-\omega a}{\sqrt{a^2 + b^2}}, \quad \omega_2 = \frac{-\omega b}{\sqrt{a^2 + b^2}}, \quad \omega_3 = 0 \tag{3}$$

Substituting (1) and (3) into Euler's equations

$$I_1\dot{\omega}_1 + (I_3 - I_2)\omega_2\omega_3 = \Lambda_1$$
$$I_2\dot{\omega}_2 + (I_1 - I_3)\omega_3\omega_1 = \Lambda_2$$
$$I_3\dot{\omega}_3 + (I_2 - I_1)\omega_1\omega_2 = \Lambda_3$$

we find $\Lambda_1 = 0, \ \Lambda_2 = 0, \ \Lambda_3 = \dfrac{M(b^2 - a^2)ab\omega^2}{12(a^2 + b^2)}$. Thus the required torque about O is

$$\mathbf{\Lambda} = \frac{M(b^2 - a^2)ab\omega^2}{12(a^2 + b^2)}\mathbf{k} \tag{4}$$

Note that if the rectangular plate is a square, i.e. if $a = b$, then $\mathbf{\Lambda} = \mathbf{0}$.

Supplementary Problems

GENERAL MOTION OF RIGID BODIES IN SPACE

10.38. Find the number of degrees of freedom of (*a*) a sphere free to roll on a plane, (*b*) an ellipsoid free to rotate about a fixed point, (*c*) an airplane moving in space.
Ans. (*a*) 3, (*b*) 3, (*c*) 6

10.39. In Fig. 10-24 a displacement of a tetrahedron in space is indicated. Show directly that the displacement can be accomplished by a translation plus a rotation about a suitable axis, thus illustrating Chasle's theorem [page 224] for space.

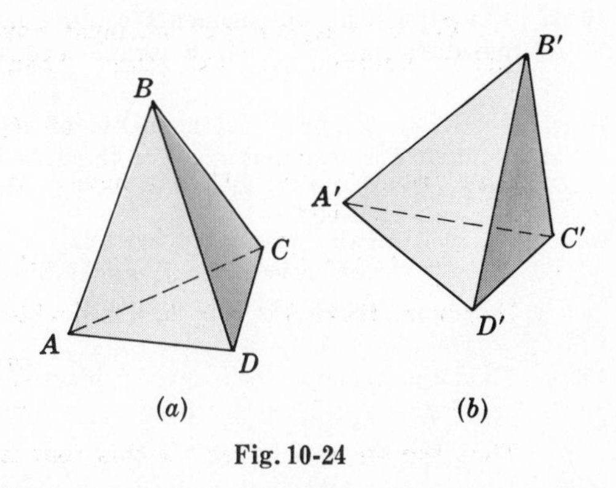

Fig. 10-24

10.40. Give an illustration similar to that of Problem 10.39 involving a rigid body whose surfaces are not plane surfaces.

10.41. Derive the result of Problem 10.2, page 259, without using Problem 6.1, page 147.

ANGULAR MOMENTUM. KINETIC ENERGY. MOMENTS AND PRODUCTS OF INERTIA

10.42. A rigid body consists of 3 particles of masses 2, 1, 4 located at $(1, -1, 1)$, $(2, 0, 2)$, $(-1, 1, 0)$ respectively. Find the angular momentum of the body if it is rotated about the origin with angular velocity $\boldsymbol{\omega} = 3\mathbf{i} - 2\mathbf{j} + 4\mathbf{k}$. *Ans.* $-6\mathbf{j} + 42\mathbf{k}$

10.43. Determine the (*a*) moments of inertia about the x, y and z axes and (*b*) the products of inertia for the rigid body of Problem 10.42.
Ans. (*a*) $I_{xx} = 12$, $I_{yy} = 16$, $I_{zz} = 16$; (*b*) $I_{xy} = 6$, $I_{yz} = 2$, $I_{xz} = -6$

10.44. What is the kinetic energy of rotation for the system of Problem 10.42? *Ans.* 180

10.45. Find the (*a*) moments of inertia and (*b*) products of inertia of a uniform rectangular plate $ABCD$ of sides $AB = a$ and $AD = b$ taken about axes AB, AD and the line perpendicular to the plate at B.
Ans. (*a*) $I_{xx} = \frac{1}{3}Mb^2$, $I_{yy} = \frac{1}{3}Ma^2$, $I_{zz} = \frac{1}{3}M(a^2 + b^2)$
(*b*) $I_{xy} = -\frac{1}{4}Mab$, $I_{yz} = 0$, $I_{xz} = 0$
calling axes through AB and AD the x and y axes respectively.

10.46. Find the (*a*) moments of inertia and (*b*) products of inertia of a cube of side a taken about x, y, z axes coinciding with three intersecting edges of the cube.
Ans. (*a*) $I_{xx} = I_{yy} = I_{zz} = \frac{2}{3}Ma^2$, (*b*) $I_{xy} = I_{yz} = I_{xz} = -\frac{1}{4}Ma^2$

10.47. Find the (*a*) angular momentum and (*b*) kinetic energy of rotation of the cube of Problem 10.46 about the point of intersection O of the three edges if the cube has an angular velocity $\boldsymbol{\omega} = 2\mathbf{i} + 5\mathbf{j} - 3\mathbf{k}$ about O. *Ans.* (*a*) $\frac{1}{12}Ma^2(10\mathbf{i} + 43\mathbf{j} - 45\mathbf{k})$, (*b*) $185Ma^2/12$

10.48. Find the (a) moments of inertia and (b) products of inertia of the uniform solid sphere $x^2 + y^2 + z^2 = a^2$ in the first octant, i.e. in the region $x \geqq 0,\ y \geqq 0,\ z \geqq 0$.
Ans. (a) $I_{xx} = I_{yy} = I_{zz} = \frac{2}{5}Ma^2$, (b) $I_{xy} = I_{yz} = I_{xz} = -2Ma^2/5\pi$

PRINCIPAL MOMENTS OF INERTIA. PRINCIPAL AXES. ELLIPSOID OF INERTIA

10.49. Prove that the principal moments of inertia for a system consisting of two particles of masses m_1 and m_2 connected by a massless rigid rod of length l are $I_1 = I_2 = m_1 m_2 l^2/(m_1 + m_2)$, $I_3 = 0$.

10.50. Find the (a) principal moments of inertia and (b) directions of the principal axes for the system of Problem 10.42.
Ans. (a) $I_1 = 18$, $I_2 = 13 - \sqrt{73}$, $I_3 = 13 + \sqrt{73}$
(b) $\mathbf{j} + \mathbf{k}$, $\frac{1}{6}(I + \sqrt{73})\mathbf{i} - \mathbf{j} + \mathbf{k}$, $\frac{1}{6}(1 - \sqrt{73})\mathbf{i} - \mathbf{j} + \mathbf{k}$,

10.51. Determine the (a) principal moments of inertia and (b) directions of the principal axes for right triangle ABC of Fig. 10-25 about point C.

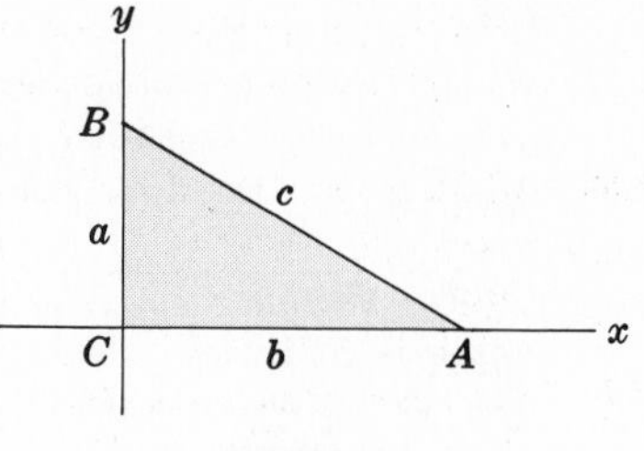

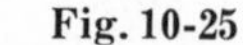
Fig. 10-25

10.52. Find the principal moments of inertia at the center of a parallelogram of sides a and b and acute angle α.

10.53. Find the (a) principal moments of inertia and (b) directions of the principal axes for the cube of Problem 10.46.
Ans. (a) $I_1 = I_2 = \frac{11}{12}Ma^2$, $I_3 = \frac{1}{6}Ma^2$
(b) Axis associated with I_3 is in the direction of the diagonal from the origin. Axes associated with I_1 and I_2 have *any* mutually perpendicular directions in a plane perpendicular to this diagonal.

10.54. Find the principal moments of inertia of a uniform cylinder of radius a and height h.
Ans. $I_1 = I_2 = \frac{1}{12}M(3a^2 + h^2)$, $I_3 = \frac{1}{2}Ma^2$

10.55. Obtain the principal moments of inertia and directions of principal axes for a rectangle of sides a and b by using Problem 10.45 and equations (*11*), page 255. Compare with Problem 10.9, page 262.

10.56. Find the lengths of the axes of the ellipsoid of inertia corresponding to the rectangle of Problem 10.55. *Ans.* $4\sqrt{3/Ma^2}$, $4\sqrt{3/Mb^2}$, $4\sqrt{3/M(a^2 + b^2)}$

10.57. Find the lengths of the axes of the ellipsoid of inertia corresponding to the cube of Problem 10.46.
Ans. $4\sqrt{3/11Ma^2}$, $4\sqrt{3/11Ma^2}$, $2\sqrt{6/Ma^2}$

10.58. Prove that the ellipsoid of inertia for a regular tetrahedron is a sphere and determine its radius.

10.59. If I_1, I_2, I_3 are the principal moments of inertia, prove that

$$I_1 \leqq I_2 + I_3, \quad I_2 \leqq I_1 + I_3, \quad I_3 \leqq I_1 + I_2$$

10.60. Under what conditions do any or all of the equality signs hold in Problem 10.59?

10.61. Prove that if a rigid body is a solid of revolution about a line L, then L is a principal axis corresponding to any part of L.

10.62. Suppose that a rigid body is symmetrical about a plane P. Prove that if L is a line perpendicular to P at point O, then L is a principal axis corresponding to point O.

EULER'S EQUATIONS OF MOTION

10.63. A rigid body having one point O fixed and no external torque about O, has two equal principal axes of inertia. Prove that it must rotate with constant angular velocity.

10.64. Write Euler's equations for the case of plane motion of a rigid body and discuss their physical significance.

10.65. Solve the problem of a compound pendulum by using Euler's equations.

10.66. Describe how Euler's equations can be used to discuss the motion of a solid cylinder rolling down an inclined plane.

10.67. Write Euler's equations in case the axes are not principal axes.

FORCE FREE MOTION. INVARIABLE LINE AND PLANE. POLHODE, HERPOLHODE, SPACE AND BODY CONES

10.68. If two principal moments of inertia corresponding to the fixed point about which a rigid body rotates are equal, prove that (*a*) Poinsot's ellipsoid is an ellipsoid of revolution, (*b*) the polhode is a circle and (*c*) the herpolhode is a circle.

10.69. Discuss the (*a*) invariable line and plane, (*b*) polhode and herpolhode and (*c*) space and body cones for the case of a rigid body which moves parallel to a given plane, i.e. plane motion of a rigid body.

10.70. (*a*) How would you define the instantaneous axis of rotation for space motion of a rigid body? (*b*) What is the relationship between the instantaneous axis of rotation and the space and body cones?

10.71. Prove that relative to its center of mass the axis about which the earth spins in a day will rotate about an axis inclined at 23.5° with respect to it in 25,780 years.

THE EULER ANGLES

10.72. Using the notation of Problem 10.20, page 267, find: (*a*) $\mathbf{I},\mathbf{J},\mathbf{K}$ in terms of $\mathbf{i},\mathbf{j},\mathbf{k}$; (*b*) $\mathbf{I}',\mathbf{J}',\mathbf{K}'$ in terms of $\mathbf{I},\mathbf{J},\mathbf{K}$; (*c*) $\mathbf{i}',\mathbf{j}',\mathbf{k}'$ in terms of $\mathbf{I}',\mathbf{J}',\mathbf{K}'$.

Ans. (*a*) $\mathbf{I} = \cos\phi\,\mathbf{i} + \sin\phi\,\mathbf{j},\quad \mathbf{J} = -\sin\phi\,\mathbf{i} + \cos\phi\,\mathbf{j},\quad \mathbf{K} = \mathbf{k}$

(*b*) $\mathbf{I}' = \mathbf{I},\quad \mathbf{J}' = \cos\theta\,\mathbf{J} + \sin\theta\,\mathbf{K},\quad \mathbf{K}' = -\sin\theta\,\mathbf{J} + \cos\theta\,\mathbf{K}$

(*c*) $\mathbf{i}' = \cos\psi\,\mathbf{I}' + \sin\psi\,\mathbf{J}',\quad \mathbf{J}' = -\sin\psi\,\mathbf{I}' + \cos\psi\,\mathbf{J}',\quad \mathbf{k}' = \mathbf{K}'$

10.73. Prove the results

$$\omega_x = \dot\theta\cos\phi + \dot\psi\sin\theta\sin\phi$$
$$\omega_y = \dot\theta\sin\phi - \dot\psi\sin\theta\cos\phi$$
$$\omega_z = \dot\phi + \psi\cos\theta.$$

10.74. If $I_1 = I_2 = I_3$, prove that the kinetic energy of rotation of a rigid body referred to principal axes is $T = \frac{1}{2}I_1(\dot\phi^2 + \dot\theta^2 + \dot\psi^2 + 2\dot\phi\dot\psi\cos\theta)$.

MOTION OF SPINNING TOPS AND GYROSCOPES

10.75. A top having radius of gyration about its axis equal to 6 cm is spun about its axis. The spinning point is fixed and the center of gravity is on the axis at a distance 3 cm from this fixed point. If it is observed that the top precesses about the vertical at 20 revolutions per minute, find the angular speed of the top about its axis. *Ans.* 3.10 rev/sec or 19.5 rad/sec

10.76. A uniform solid right circular cone of radius a and height h is spun so that its vertex is fixed and its axis is inclined at a constant angle α with the vertical. If the axis precesses about the vertical with period P, determine the angular speed of the cone about its axis.

10.77. Work Problem 10.76 if the cone is surmounted by a uniform solid hemisphere of radius a and the same density.

10.78. Explain physically why the spin axis of the gyroscope of Figures 10-6 and 10-7, page 258, should maintain its direction.

10.79. Explain how a gyroscope can be used to enable a ship, airplane, submarine or missile to follow some specified course of motion.

MISCELLANEOUS PROBLEMS

10.80. A uniform solid cube of side a and mass M has its edges lying on the positive x, y and z axes of a coordinate system with vertex at the origin O. If it rotates about the z axis with constant angular velocity $\boldsymbol{\omega}$, find the angular momentum. *Ans.* $-\frac{1}{12}Ma^2\omega(3\mathbf{i} + 3\mathbf{j} - 8\mathbf{k})$

10.81. Find the moment of inertia of a uniform solid cone of radius a, height h and mass M about (*a*) the base, (*b*) the vertex. *Ans.* (*a*) $\frac{1}{10}Ma^2$, (*b*) $\frac{3}{10}M(2h^2 + a^2)$

10.82. Find the principal moments of inertia at the center of a uniform elliptical plate having semi-major axis a and semi-minor axis b. *Ans.* $I_1 = \frac{1}{4}Mb^2$, $I_2 = \frac{1}{4}Ma^2$, $I_3 = \frac{1}{4}M(a^2 + b^2)$

10.83. A top has the form of a solid circular disk of radius a and mass M with a thin rod of mass m and length l attached to its center [see Fig. 10-26]. Find the angular velocity with which the top should be spun so as to "sleep". Assume that the base point O is fixed.

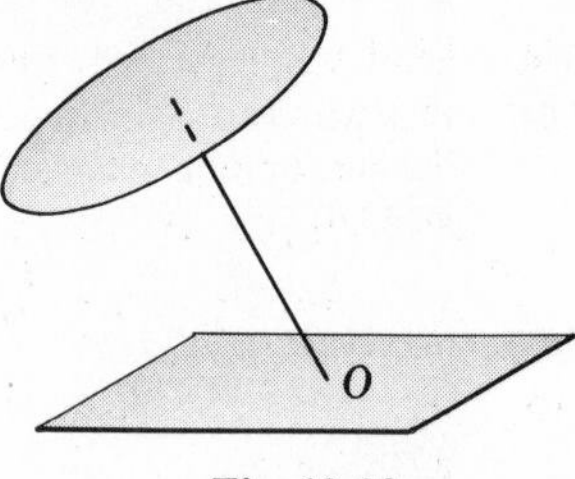

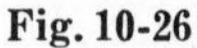
Fig. 10-26

10.84. Work Problem 10.83 for a cone of radius a, height h and mass M.

10.85. Work Problem 10.83 for a cone of radius a, height h and mass M surmounted by a hemisphere of radius a and mass m.

10.86. A coin of radius a is set spinning about a vertical axis with angular velocity $\boldsymbol{\omega}$ [see Fig. 10-27]. Prove that the motion is stable if $\omega^2 > 4g/a$.

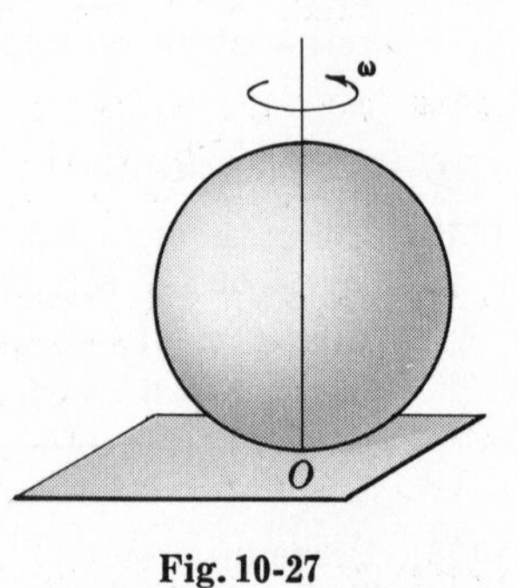

Fig. 10-27

10.87. Suppose that the coin of Problem 10.86 is spun with angular speed s about a diameter which is inclined at an angle α with the vertical and which is fixed at point O. Assuming there is no nutation, find the angular speed with which the coin precesses about the vertical.

10.88. Discuss how gyroscopes can be used to control the motions of a ship on a stormy sea.

10.89. The vertex of a uniform solid cone of radius a, height h and mass M is fixed at point O of a horizontal plane. Prove that if the cone rolls on the plane with angular velocity ω about an axis perpendicular to the plane through O, then the kinetic energy of rotation is $\dfrac{3Mh^2(a^2 + 6h^2)\omega^2}{40(a^2 + h^2)}$.

10.90. Explain how the principal axes of a rigid body can be found if the direction of one of the principal axes is known.

10.91. A uniform solid cone has the radius of its base equal to twice its altitude. Prove that the ellipsoid of inertia corresponding to its vertex is a sphere.

10.92. Explain how a gyroscope can be used as a compass, often called a *gyrocompass*.

10.93. A *dumbbell* consists of two equal masses M attached to a rod ABC of length l and negligible mass [see Fig. 10-28]. The system rotates about a vertical axis DCE with constant angular velocity $\boldsymbol{\omega}$ such that the rod makes a constant angle θ_0 with the vertical. Prove that the angular momentum $\boldsymbol{\Omega}$ of the system describes a cone of angle $\pi/2 - \theta_0$ about $\boldsymbol{\omega}$ and has magnitude $\frac{1}{2}Ml^2\omega \sin\theta_0$.

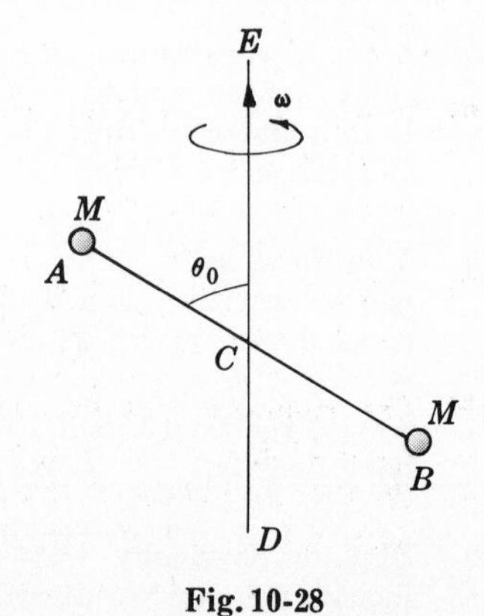

Fig. 10-28

10.94. (a) Prove that the magnitude of the torque needed to keep the system of Problem 10.93 in motion is $\frac{1}{4}Ml^2\omega^2 \sin 2\theta_0$. (b) What is the direction of the torque?

10.95. Work (a) Problem 10.93 and (b) Problem 10.94 if the rod ACB has mass m.

10.96. A thin solid uniform circular plate of radius a has its center attached to the top of a thin fixed vertical rod OA [see Fig. 10-29]. It is spun with constant angular speed ω_0 about an axis which is inclined at angle α with the normal OB to the plate. (a) Prove that the angular velocity vector $\boldsymbol{\omega}$ precesses about the normal OB with period $2\pi/(\omega_0 \cos\alpha)$. (b) Prove that the axis OB describes a space cone with period $2\pi/(\omega_0\sqrt{1+3\cos^2\alpha})$.

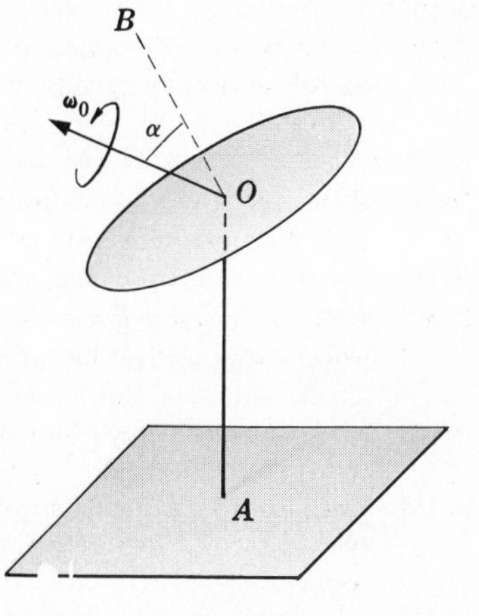

Fig. 10-29

10.97. In Problem 10.96 find the angle through which the plate turns during the time it takes OB to describe the space cone.

10.98. Find the principal moments of inertia of a uniform solid cone of radius a, height h and mass M taken about the (a) vertex, (b) center of mass.

Ans. (a) $I_1 = I_2 = \frac{3}{20}M(a^2+4h^2)$, $I_3 = \frac{3}{10}Ma^2$

(b) $I_1 = I_2 = \frac{3}{80}M(h^2+4a^2)$, $I_3 = \frac{3}{10}Ma^2$

10.99. A compound pendulum of mass M oscillates about a horizontal axis which makes angles α, β, γ with respect to the principal axes of inertia. If the principal moments of inertia are I_1, I_2, I_3 respectively and the distance from the center of mass to the axis of rotation is l, prove that for small oscillations the period is $2\pi\sqrt{Mgl/I}$ where $I = Ml^2 + I_1\cos^2\alpha + I_2\cos^2\beta + I_3\cos^2\gamma$.

10.100. Find the period of small oscillations of a uniform solid cone which rotates about a horizontal axis attached to the vertex of the cone.

10.101. An elliptical plate [see Fig. 10-30] having semi-major and semi-minor axes of lengths a and b respectively is rotated with constant angular speed ω_0 about an axis making a constant angle α with the major axis. Find the torque required to produce this motion.

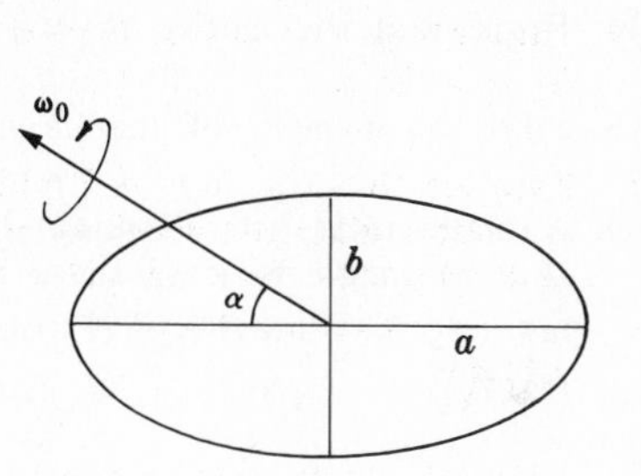

Fig. 10-30

10.102. Work Problem 10.101 if the elliptical plate is replaced by an ellipsoid.

10.103. Given Euler's equations of motion for a rigid body having zero external torque about a fixed point O, i.e.,

$$I_1\dot{\omega}_1 + (I_3 - I_2)\omega_2\omega_3 = 0, \qquad I_2\dot{\omega}_2 + (I_1 - I_3)\omega_3\omega_1 = 0, \qquad I_3\dot{\omega}_3 + (I_2 - I_1)\omega_1\omega_2 = 0$$

prove that

$$I_1\omega_1^2 + I_2\omega_2^2 + I_3\omega_3^2 = \text{constant} = 2T$$

and

$$I_1^2\omega_1^2 + I_2^2\omega_2^2 + I_3^2\omega_3^2 = \text{constant} = H^2$$

10.104. Prove from Problem 10.103 that ω_1, ω_2 and ω_3 satisfy a differential equation of the form $dy/dx = \sqrt{(1-x^2)(1-k^2x^2)}$, and thus show that the angular velocity can be expressed in terms of elliptic functions.

10.105. Find the moment of inertia of a uniform solid cone of radius a, height h and mass M about a line which lies in its surface. *Ans.* $\frac{3}{20}Ma^2(a^2+6h^2)/(a^2+h^2)$

10.106. The moments and products of inertia of a rigid body about the x, y and z axes are $I_{xx} = 3$, $I_{yy} = 10/3$, $I_{zz} = 8/3$, $I_{xy} = 4/3$, $I_{xz} = -4/3$, $I_{yz} = 0$. Find (a) the principal moments of inertia and (b) the directions of the principal axes.

Ans. (a) $I_1 = 3$, $I_2 = 2$, $I_3 = 4$

(b) $\mathbf{e}_1 = \mathbf{i} - 2\mathbf{j} - 2\mathbf{k}$, $\mathbf{e}_2 = -2\mathbf{i} + \mathbf{j} - 2\mathbf{k}$, $\mathbf{e}_3 = -2\mathbf{i} - 2\mathbf{j} + \mathbf{k}$

10.107. A cone having semi-vertical angle α rolls with constant angular speed ω on a horizontal plane with its vertex fixed at a point O. Prove that the axis of the cone rotates about the vertical axis through O with angular speed $\omega \tan \alpha$.

10.108. A horizontal plane rotates about a vertical axis with constant angular velocity $\boldsymbol{\omega}$. A uniform solid sphere of radius a is placed on this plane. Prove that the center describes a circle with angular velocity given in magnitude by $\frac{2}{7}\omega$.

10.109. Work Problem 10.108 if the sphere is not necessarily of constant density.
Ans. $\omega K^2/(K^2 + a^2)$ where K is the radius of gyration about a diameter

10.110. Show how to find the relative maximum and minimum distances from the origin to the ellipsoid $\Phi = Ax^2 + By^2 + Cz^2 + Dxy + Eyz + Fxz = 1$.
[*Hint.* Maximize or minimize the function $\Psi = x^2 + y^2 + z^2$ subject to the condition $\Phi = 1$. To do this use the *method of Lagrange multipliers*, i.e. consider the function $G = \Psi + \lambda\Phi$ where λ is the (constant) Lagrange multiplier and set $\partial G/\partial x$, $\partial G/\partial y$, $\partial G/\partial z$ equal to zero.]

10.111. Explain the relationship of Problem 10.110 to the method of page 255 for obtaining principal moments of inertia and directions of principal axes.

10.112. (*a*) Find the relative maximum and minimum distances from the origin to the ellipsoid $9x^2 + 10y^2 + 8z^2 + 4xy - 4xz = 3$.
(*b*) Discuss the connection of the results of (*a*) with those of Problem 10.106.

10.113. Find the moment of inertia of the system of particles of Problem 10.42 about a line through the point $(2, -1, 3)$ in the direction $3\mathbf{i} - 2\mathbf{j} + 4\mathbf{k}$.

10.114. Prove that the motion of the "sleeping top" of Problem 10.36 is stable if $A^2 \geqq 4mglI_1/I_3^2$.

10.115. Find the moment of inertia of the lemniscate $r^2 = a^2 \cos 2\theta$ about the z axis. *Ans.* $\frac{1}{8}Ma^2$

10.116. A plane rigid body (lamina) has an xy and $x'y'$ coordinate system with common origin O such that the angle between the x and x' axes is α [see Fig. 10-31]. Prove that
(*a*) $I_{x'x'} = I_{xx} \cos^2 \alpha - 2I_{xy} \sin \alpha \cos \alpha + I_{yy} \sin^2 \alpha$
(*b*) $I_{y'y'} = I_{xx} \sin^2 \alpha + 2I_{xy} \sin \alpha \cos \alpha + I_{yy} \sin^2 \alpha$

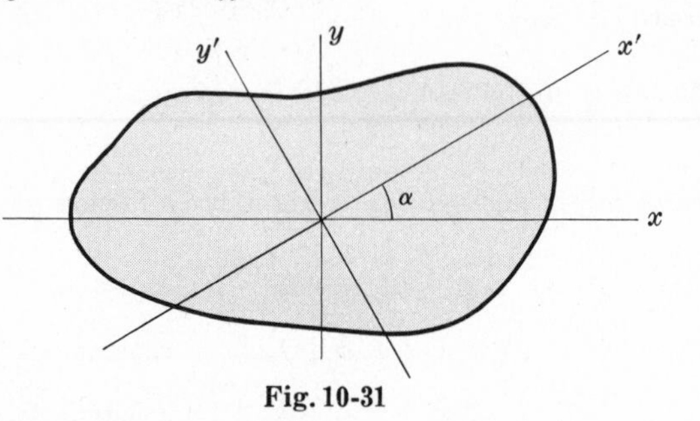

Fig. 10-31

10.117. Use Problem 10.116 to prove that
$$I_{x'x'} + I_{y'y'} = I_{xx} + I_{yy}$$
and give a physical interpretation.

10.118. Referring to Problem 10.116, find an expression for $I_{x'y'}$ in terms of I_{xx}, I_{xy}, I_{yy} and α.

10.119. Use the results of Problems 10.116 and 10.118 to prove that for a plane region having moments and products of inertia defined by I_{xx}, I_{xy}, I_{yy} corresponding to a particular xy coordinate system, the principal axes are obtained by a rotation of these axes through an angle α given by $\tan 2\alpha = I_{xy}/(I_{yy} - I_{xx})$.

10.120. Prove that the lengths of the principal axes in Problem 10.116 are given by
$$\tfrac{1}{2}(I_{xx} + I_{yy}) \pm \sqrt{I_{xy}^2 + \tfrac{1}{4}(I_{xx} - I_{yy})^2}$$

10.121. Discuss Problem 10.19, page 266, if $I_1 > I_3$.

10.122. Find the moment of inertia of a uniform semicircular wire of mass M and radius a about its center. *Ans.* $2M(\pi - 2)a^2/\pi$

10.123. Prove that the expression on the left side of equation *(4)* in Problem 10.29 is the vertical component of the angular momentum.

10.124. Discuss Problem 10.32 if the root of equation *(1)* is *(a)* equal to u_1, *(b)* less than u_1.

10.125. A rigid body consists of 3 particles of masses m_1, m_2 and m_3. The distance between m_1 and m_2; m_2 and m_3; m_3 and m_1 are l_{12}, l_{23} and l_{31} respectively. Prove that the moment of inertia of the system about an axis perpendicular to the plane of the particles through their center of mass is given by

$$\frac{m_1 m_2 l_{12}^2 + m_2 m_3 l_{23}^2 + m_3 m_1 l_{31}^2}{m_1 + m_2 + m_3}$$

10.126. Derive a "parallel axis theorem" for products of inertia and illustrate by means of an example.

10.127. Prove that the principal moments of inertia of a triangle of sides a, b, c and mass M about the center of mass are given by

$$I_1 = I_2 = \frac{M}{72}(a^2 + b^2 + c^2 \pm 2\sqrt{a^4 + b^4 + c^4 - a^2b^2 - b^2c^2 - c^2a^2}), \quad I_3 = \frac{M}{36}(a^2 + b^2 + c^2)$$

10.128. A coin of radius 1.5 cm rolls without slipping on a horizontal table such that the plane of the coin makes an angle of 60° with the table. If the center of the coin moves at a speed of 3 m/sec, prove that the coin moves in a circular path and find its radius. *Ans.* 2.5 m

Chapter 11

LAGRANGE'S EQUATIONS

GENERAL METHODS OF MECHANICS

Up to now we have dealt primarily with the formulation of problems in mechanics by Newton's laws of motion. It is possible to give treatments of mechanics from rather general viewpoints, in particular those due to Lagrange and Hamilton.

Although such treatments reduce to Newton's laws, they are characterized not only by the relative ease with which many problems can be formulated and solved but by their relationship in both theory and application to such advanced fields as quantum mechanics, statistical mechanics, celestial mechanics and electrodynamics.

GENERALIZED COORDINATES

Suppose that a particle or a system of N particles moves subject to possible constraints, as for example a particle moving along a circular wire or a rigid body moving along an inclined plane. Then there will be a minimum number of independent coordinates needed to specify the motion. These coordinates denoted by

$$q_1, q_2, \ldots, q_n \tag{1}$$

are called *generalized coordinates* and can be distances, angles or quantities relating to them. The number n of generalized coordinates is the number of degrees of freedom [see page 165].

Many sets of generalized coordinates may be possible in a given problem, but a strategic choice can simplify the analysis considerably.

NOTATION

In the following the subscript α will range from 1 to n, the number of degrees of freedom, while the subscript ν will range from 1 to N, the number of particles in the system.

TRANSFORMATION EQUATIONS

Let $\mathbf{r}_\nu = x_\nu\mathbf{i} + y_\nu\mathbf{j} + z_\nu\mathbf{k}$ be the position vector of the νth particle with respect to an xyz coordinate system. The relationships of the generalized coordinates (1) to the position coordinates are given by the *transformation equations*

$$\left.\begin{aligned} x_\nu &= x_\nu(q_1, q_2, \ldots, q_n, t) \\ y_\nu &= y_\nu(q_1, q_2, \ldots, q_n, t) \\ z_\nu &= z_\nu(q_1, q_2, \ldots, q_n, t) \end{aligned}\right\} \tag{2}$$

where t denotes the time. In vector form, (2) can be written

$$\mathbf{r}_\nu = \mathbf{r}_\nu(q_1, q_2, \ldots, q_n, t) \tag{3}$$

The functions in (2) or (3) are supposed to be continuous and to have continuous derivatives.

CLASSIFICATION OF MECHANICAL SYSTEMS

Mechanical systems can be classified according as they are *scleronomic* or *rheonomic, holonomic* or *non-holonomic,* and *conservative* or *non-conservative* as defined below.

SCLERONOMIC AND RHEONOMIC SYSTEMS

In many mechanical systems of importance the time t does not enter explicitly in the equations (*2*) or (*3*). Such systems are sometimes called *scleronomic.* In others, as for example those involving moving constraints, the time t does enter explicitly. Such systems are called *rheonomic.*

HOLONOMIC AND NON-HOLONOMIC SYSTEMS

Let $q_1, q_2, \ldots, q_n$ denote the generalized coordinates describing a system and let t denote the time. If all the constraints of the system can be expressed as equations having the form $\phi(q_1, q_2, \ldots, q_n, t) = 0$ or their equivalent, then the system is said to be *holonomic*; otherwise the system is said to be *non-holonomic.* Compare page 170.

CONSERVATIVE AND NON-CONSERVATIVE SYSTEMS

If all forces acting on a system of particles are derivable from a potential function [or potential energy] V, then the system is called *conservative,* otherwise it is *non-conservative.*

KINETIC ENERGY. GENERALIZED VELOCITIES

The total kinetic energy of the system is

$$T = \frac{1}{2}\sum_{\nu=1}^{N} m_\nu \dot{\mathbf{r}}_\nu^2 \tag{4}$$

The kinetic energy can be written as a *quadratic form* in the *generalized velocities* $\dot{q}_\alpha$. If the system is scleronomic [i.e. independent of time t explicitly], then the quadratic form has only terms of the form $a_{\alpha\beta}\dot{q}_\alpha\dot{q}_\beta$. If it is rheonomic, linear terms in $\dot{q}_\alpha$ are also present.

GENERALIZED FORCES

If W is the total work done on a system of particles by forces $\mathbf{F}_\nu$ acting on the νth particle, then

$$dW = \sum_{\alpha=1}^{n} \Phi_\alpha\, dq_\alpha \tag{5}$$

where

$$\Phi_\alpha = \sum_{\nu=1}^{N} \mathbf{F}_\nu \cdot \frac{\partial \mathbf{r}_\nu}{\partial q_\alpha} \tag{6}$$

is called the *generalized force* associated with the generalized coordinate q_α. See Problem 11.6.

LAGRANGE'S EQUATIONS

The generalized force can be related to the kinetic energy by the equations [see Problem 11.10]

$$\frac{d}{dt}\left(\frac{\partial T}{\partial \dot{q}_\alpha}\right) - \frac{\partial T}{\partial q_\alpha} = \Phi_\alpha \tag{7}$$

If the system is conservative so that the forces are derivable from a potential or potential energy V, we can write (7) as

$$\frac{d}{dt}\left(\frac{\partial L}{\partial \dot{q}_\alpha}\right) - \frac{\partial L}{\partial q_\alpha} = 0 \tag{8}$$

where

$$L = T - V \tag{9}$$

is called the *Lagrangian function* of the system, or simply the *Lagrangian.*

The equations (7) or (8) are called *Lagrange's equations* and are valid for holonomic systems which may be scleronomic or rheonomic.

If some of the forces in a system are conservative so as to be derivable from a potential V' while other forces such as friction, etc., are non-conservative, we can write Lagrange's equations as

$$\frac{d}{dt}\left(\frac{\partial L}{\partial \dot{q}_\alpha}\right) - \frac{\partial L}{\partial q_\alpha} = \Phi'_\alpha \tag{10}$$

where $L = T - V'$ and Φ'_α are the generalized forces associated with the non-conservative forces in the system.

GENERALIZED MOMENTA

We define

$$p_\alpha = \frac{\partial T}{\partial \dot{q}_\alpha} \tag{11}$$

to be the *generalized momentum* associated with the generalized coordinate q_α. We often call p_α the momentum *conjugate* to q_α, or the *conjugate momentum.*

If the system is conservative with potential energy depending only on the generalized coordinates, then (11) can be written in terms of the Lagrangian $L = T - V$ as

$$p_\alpha = \frac{\partial L}{\partial \dot{q}_\alpha} \tag{12}$$

LAGRANGE'S EQUATIONS FOR NON-HOLONOMIC SYSTEMS

Suppose that there are m equations of constraint having the form

$$\sum_\alpha A_\alpha\, dq_\alpha + A\, dt = 0, \quad \sum_\alpha B_\alpha\, dq_\alpha + B\, dt = 0, \quad \ldots \tag{13}$$

or equivalently

$$\sum_\alpha A_\alpha \dot{q}_\alpha + A = 0, \quad \sum_\alpha B_\alpha \dot{q}_\alpha + B = 0, \quad \ldots \tag{14}$$

We must of course have $m < n$ where n is the number of coordinates q_α.

The equations (13) or (14) may or may not be integrable so as to obtain a relationship involving the q_α's. If they are not integrable the constraints are *non-holonomic* or *non-integrable*; otherwise they are *holonomic* or *integrable.*

In either case Lagrange's equations can be replaced by

$$\frac{d}{dt}\left(\frac{\partial T}{\partial \dot{q}_\alpha}\right) - \frac{\partial T}{\partial q_\alpha} = \Phi_\alpha + \lambda_1 A_\alpha + \lambda_2 B_\alpha + \cdots \tag{15}$$

where the m parameters $\lambda_1, \lambda_2, \ldots$ are called *Lagrange multipliers* [see Problem 11.18].

If the forces are conservative, (15) can be written in terms of the Lagrangian $L = T - V$ as

$$\frac{d}{dt}\left(\frac{\partial L}{\partial \dot{q}_\alpha}\right) - \frac{\partial L}{\partial q_\alpha} = \lambda_1 A_\alpha + \lambda_2 B_\alpha + \cdots \tag{16}$$

It should be emphasized that the above results are applicable to holonomic (as well as non-holonomic) systems since a constraint condition of the form

$$\phi(q_1, q_2, \ldots, q_n, t) = 0 \tag{17}$$

can by differentiation be written as

$$\sum_\alpha \frac{\partial \phi}{\partial q_\alpha} dq_\alpha + \frac{\partial \phi}{\partial t} dt = 0 \tag{18}$$

which has the form *(13)*.

LAGRANGE'S EQUATIONS WITH IMPULSIVE FORCES

Suppose that the forces $\mathbf{F}_\nu$ acting on a system are such that

$$\lim_{\tau \to 0} \int_0^\tau \mathbf{F}_\nu \, dt = \mathcal{J}_\nu \tag{19}$$

where τ represents a time interval. Then we call $\mathbf{F}_\nu$ *impulsive forces* and $\mathcal{J}_\nu$ are called *impulses.*

If we let the subscripts 1 and 2 denote respectively quantities before and after application of the impulsive forces, Lagrange's equations become [see Problem 11.23]

$$\left(\frac{\partial T}{\partial \dot{q}_\alpha}\right)_2 - \left(\frac{\partial T}{\partial \dot{q}_\alpha}\right)_1 = \mathcal{F}_\alpha \tag{20}$$

where

$$\mathcal{F}_\alpha = \sum_\nu \mathcal{J}_\nu \cdot \frac{\partial \mathbf{r}_\nu}{\partial q_\alpha} \tag{21}$$

If we call $\mathcal{F}_\alpha$ the *generalized impulse,* *(20)* can be written

$$\text{Generalized impulse} = \text{change in generalized momentum} \tag{22}$$

which is a generalization of Theorem 2.6, page 36.

Solved Problems

GENERALIZED COORDINATES AND TRANSFORMATION EQUATIONS

11.1. Give a set of generalized coordinates needed to completely specify the motion of each of the following: *(a)* a particle constrained to move on an ellipse, *(b)* a circular cylinder rolling down an inclined plane, *(c)* the two masses in a double pendulum [Fig. 11-3] constrained to move in a plane.

(a) Let the ellipse be chosen in the xy plane of Fig. 11-1. The particle of mass m moving on the ellipse has coordinates (x, y). However, since we have the transformation equations $x = a \cos \theta$, $y = b \sin \theta$, we can specify the motion completely by use of the generalized coordinate θ.

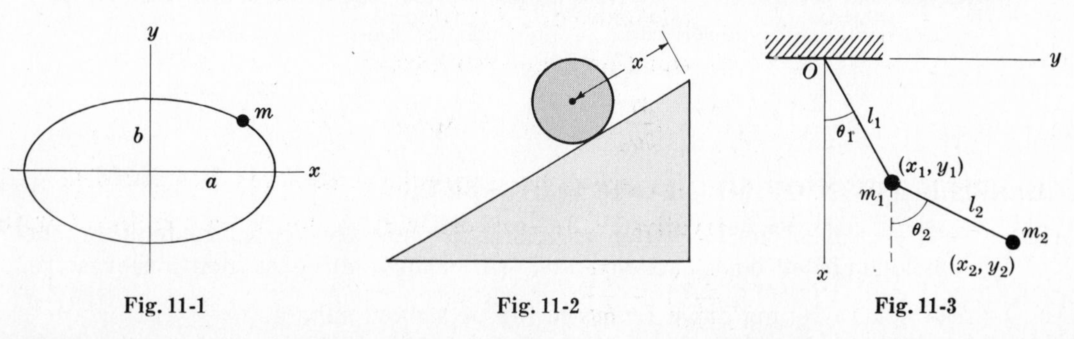

Fig. 11-1 Fig. 11-2 Fig. 11-3

(b) The position of the cylinder [Fig. 11-2 above] on the inclined plane can be completely specified by giving the distance x traveled by the center of mass and the angle θ of rotation turned through by the cylinder about its axis.

If there is no slipping, x is related to θ so that only one generalized coordinate [either x or θ] is needed. If there is slipping, two generalized coordinates x and θ are needed.

(c) Two coordinates θ_1 and θ_2 completely specify the positions of masses m_1 and m_2 [see Fig. 11-3 above] and can be considered as the required generalized coordinates.

11.2. Write the transformation equations for the system in Problem 11.1(c).

Choose an xy coordinate system as shown in Fig. 11-3. Let (x_1, y_1) and (x_2, y_2) be the rectangular coordinates of m_1 and m_2 respectively. Then from Fig. 11-3 we see that

$$x_1 = l_1 \cos\theta_1 \qquad y_1 = l_1 \sin\theta_1$$

$$x_2 = l_1 \cos\theta_1 + l_2 \cos\theta_2 \qquad y_2 = l_1 \sin\theta_1 + l_2 \sin\theta_2$$

which are the required transformation equations.

11.3. Prove that $\dfrac{\partial \dot{\mathbf{r}}_\nu}{\partial \dot{q}_\alpha} = \dfrac{\partial \mathbf{r}_\nu}{\partial q_\alpha}$.

We have $\mathbf{r}_\nu = \mathbf{r}_\nu(q_1, q_2, \ldots, q_n, t)$. Then

$$\dot{\mathbf{r}}_\nu = \frac{\partial \mathbf{r}_\nu}{\partial q_1}\dot{q}_1 + \cdots + \frac{\partial \mathbf{r}_\nu}{\partial q_n}\dot{q}_n + \frac{\partial \mathbf{r}_\nu}{\partial t} \tag{1}$$

Thus

$$\frac{\partial \dot{\mathbf{r}}_\nu}{\partial \dot{q}_\alpha} = \frac{\partial \mathbf{r}_\nu}{\partial q_\alpha} \tag{2}$$

We can look upon this result as a "cancellation of the dots".

11.4. Prove that $\dfrac{d}{dt}\left(\dfrac{\partial \mathbf{r}_\nu}{\partial q_\alpha}\right) = \dfrac{\partial \dot{\mathbf{r}}_\nu}{\partial q_\alpha}$.

We have from (1) of Problem 11.3,

$$\dot{\mathbf{r}}_\nu = \frac{\partial \mathbf{r}_\nu}{\partial q_1}\dot{q}_1 + \cdots + \frac{\partial \mathbf{r}_\nu}{\partial q_n}\dot{q}_n + \frac{\partial \mathbf{r}_\nu}{\partial t} \tag{1}$$

Then

$$\frac{\partial \dot{\mathbf{r}}_\nu}{\partial q_\alpha} = \frac{\partial^2 \mathbf{r}_\nu}{\partial q_\alpha\,\partial q_1}\dot{q}_1 + \cdots + \frac{\partial^2 \mathbf{r}_\nu}{\partial q_\alpha\,\partial q_n}\dot{q}_n + \frac{\partial^2 \mathbf{r}_\nu}{\partial q_\alpha\,\partial t} \tag{2}$$

Now

$$\begin{aligned}\frac{d}{dt}\left(\frac{\partial \mathbf{r}_\nu}{\partial q_\alpha}\right) &= \frac{\partial}{\partial q_1}\left(\frac{\partial \mathbf{r}_\nu}{\partial q_\alpha}\right)\frac{dq_1}{dt} + \cdots + \frac{\partial}{\partial q_n}\left(\frac{\partial \mathbf{r}_\nu}{\partial q_\alpha}\right)\frac{dq_n}{dt} + \frac{\partial}{\partial t}\left(\frac{\partial \mathbf{r}_\nu}{\partial q_\alpha}\right)\\ &= \frac{\partial^2 \mathbf{r}_\nu}{\partial q_1\,\partial q_\alpha}\dot{q}_1 + \cdots + \frac{\partial^2 \mathbf{r}_\nu}{\partial q_n\,\partial q_\alpha}\dot{q}_n + \frac{\partial^2 \mathbf{r}_\nu}{\partial t\,\partial q_\alpha}\end{aligned} \tag{3}$$

Since $\mathbf{r}_\nu$ is assumed to have continuous second order partial derivatives, the order of differentiation does not matter. Thus from (2) and (3) the required result follows.

The result can be interpreted as an interchange of order of the operators, i.e.,

$$\frac{d}{dt}\left(\frac{\partial}{\partial q_\alpha}\right) = \frac{\partial}{\partial q_\alpha}\left(\frac{d}{dt}\right)$$

CLASSIFICATION OF MECHANICAL SYSTEMS

11.5. Classify each of the following according as they are (i) scleronomic or rheonomic, (ii) holonomic or non-holonomic and (iii) conservative or non-conservative.

(a) A sphere rolling down from the top of a fixed sphere.

(*b*) A cylinder rolling without slipping down a rough inclined plane of angle α.

(*c*) A particle sliding down the inner surface, with coefficient of friction μ, of a paraboloid of revolution having its axis vertical and vertex downward.

(*d*) A particle moving on a very long frictionless wire which rotates with constant angular speed about a horizontal axis.

(*a*) **scleronomic** [equations do not involve time t explicitly]
non-holonomic [since rolling sphere leaves the fixed sphere at some point]
conservative [gravitational force acting is derivable from a potential]

(*b*) **scleronomic**
holonomic [equation of constraint is that of a line or plane]
conservative

(*c*) **scleronomic**
holonomic
non-conservative [since force due to friction is not derivable from a potential]

(*d*) **rheonomic** [constraint involves time t explicitly]
holonomic [equation of constraint is that of a line which involves t explicitly]
conservative

WORK, KINETIC ENERGY AND GENERALIZED FORCES

11.6. Derive equations (*5*) and (*6*), page 283, for the work done on a system of particles.

Suppose that a system undergoes increments $dq_1, dq_2, \ldots, dq_n$ of the generalized coordinates. Then the νth particle undergoes a displacement

$$d\mathbf{r}_\nu = \sum_{\alpha=1}^{n} \frac{\partial \mathbf{r}_\nu}{\partial q_\alpha} dq_\alpha$$

Thus the total work done is

$$dW = \sum_{\nu=1}^{N} \mathbf{F}_\nu \cdot d\mathbf{r}_\nu = \sum_{\nu=1}^{N} \left\{ \sum_{\alpha=1}^{n} \mathbf{F}_\nu \cdot \frac{\partial \mathbf{r}_\nu}{\partial q_\alpha} \right\} dq_\alpha = \sum_{\alpha=1}^{n} \Phi_\alpha \, dq_\alpha$$

where

$$\Phi_\alpha = \sum_{\nu=1}^{N} \mathbf{F}_\nu \cdot \frac{\partial \mathbf{r}_\nu}{\partial q_\alpha}$$

We call Φ_α the *generalized force* associated with the generalized coordinate q_α.

11.7. Prove that $\Phi_\alpha = \partial W / \partial q_\alpha$.

We have $dW = \sum \frac{\partial W}{\partial q_\alpha} dq_\alpha$. Also, by Problem 11.6, $dW = \sum \Phi_\alpha \, dq_\alpha$. Then

$$\sum \left(\Phi_\alpha - \frac{\partial W}{\partial q_\alpha} \right) dq_\alpha = 0$$

Thus since the dq_α are independent, all coefficients of dq_α must be zero, so that $\Phi_\alpha = \partial W / \partial q_\alpha$.

11.8. Let $\mathbf{F}_\nu$ be the net external force acting on the νth particle of a system. Prove that

$$\frac{d}{dt}\left\{ \sum_\nu m_\nu \dot{\mathbf{r}}_\nu \cdot \frac{\partial \mathbf{r}_\nu}{\partial q_\alpha} \right\} - \sum_\nu m_\nu \dot{\mathbf{r}}_\nu \cdot \frac{\partial \dot{\mathbf{r}}_\nu}{\partial q_\alpha} = \sum_\nu \mathbf{F}_\nu \cdot \frac{\partial \mathbf{r}_\nu}{\partial q_\alpha}$$

By Newton's second law applied to the νth particle, we have

$$m_\nu \ddot{\mathbf{r}}_\nu = \mathbf{F}_\nu \qquad (1)$$

Then $$m_\nu \ddot{\mathbf{r}}_\nu \cdot \frac{\partial \mathbf{r}_\nu}{\partial q_\alpha} = \mathbf{F}_\nu \cdot \frac{\partial \mathbf{r}_\nu}{\partial q_\alpha} \tag{2}$$

Now by Problem 11.4, $$\frac{d}{dt}\left(\dot{\mathbf{r}}_\nu \cdot \frac{\partial \mathbf{r}_\nu}{\partial q_\alpha}\right) = \ddot{\mathbf{r}}_\nu \cdot \frac{\partial \mathbf{r}_\nu}{\partial q_\alpha} + \dot{\mathbf{r}}_\nu \cdot \frac{d}{dt}\left(\frac{\partial \mathbf{r}_\nu}{\partial q_\alpha}\right)$$
$$= \ddot{\mathbf{r}}_\nu \cdot \frac{\partial \mathbf{r}_\nu}{\partial q_\alpha} + \dot{\mathbf{r}}_\nu \cdot \frac{\partial \dot{\mathbf{r}}_\nu}{\partial q_\alpha} \tag{3}$$

Thus $$\ddot{\mathbf{r}}_\nu \cdot \frac{\partial \mathbf{r}_\nu}{\partial q_\alpha} = \frac{d}{dt}\left(\dot{\mathbf{r}}_\nu \cdot \frac{\partial \mathbf{r}_\nu}{\partial q_\alpha}\right) - \dot{\mathbf{r}}_\nu \cdot \frac{\partial \dot{\mathbf{r}}_\nu}{\partial q_\alpha} \tag{4}$$

Hence from (2) we have, since m_ν is constant,

$$\frac{d}{dt}\left(m_\nu \dot{\mathbf{r}}_\nu \cdot \frac{\partial \mathbf{r}_\nu}{\partial q_\alpha}\right) - m_\nu \dot{\mathbf{r}}_\nu \cdot \frac{\partial \dot{\mathbf{r}}_\nu}{\partial q_\alpha} = \mathbf{F}_\nu \cdot \frac{\partial \mathbf{r}_\nu}{\partial q_\alpha}$$

Summing both sides with respect to ν over all particles, we have

$$\frac{d}{dt}\left\{\sum_\nu m_\nu \dot{\mathbf{r}}_\nu \cdot \frac{\partial \mathbf{r}_\nu}{\partial q_\alpha}\right\} - \sum_\nu m_\nu \dot{\mathbf{r}}_\nu \cdot \frac{\partial \dot{\mathbf{r}}_\nu}{\partial q_\alpha} = \sum_\nu \mathbf{F}_\nu \cdot \frac{\partial \mathbf{r}_\nu}{\partial q_\alpha}$$

11.9. Let T be the kinetic energy of a system of particles. Prove that

(a) $\dfrac{\partial T}{\partial q_\alpha} = \sum_\nu m_\nu \dot{\mathbf{r}}_\nu \cdot \dfrac{\partial \dot{\mathbf{r}}_\nu}{\partial q_\alpha}$, (b) $\dfrac{\partial T}{\partial \dot{q}_\alpha} = \sum_\nu m_\nu \dot{\mathbf{r}}_\nu \cdot \dfrac{\partial \mathbf{r}_\nu}{\partial q_\alpha}$

(a) The kinetic energy is $T = \frac{1}{2}\sum_\nu m_\nu \dot{r}_\nu^2 = \frac{1}{2}\sum_\nu m_\nu \dot{\mathbf{r}}_\nu \cdot \dot{\mathbf{r}}_\nu$. Thus

$$\frac{\partial T}{\partial q_\alpha} = \sum_\nu m_\nu \dot{\mathbf{r}}_\nu \cdot \frac{\partial \dot{\mathbf{r}}_\nu}{\partial q_\alpha}$$

(b) We have by the "cancellation of the dots" [Problem 11.3, page 286],

$$\frac{\partial T}{\partial \dot{q}_\alpha} = \sum m_\nu \dot{\mathbf{r}}_\nu \cdot \frac{\partial \dot{\mathbf{r}}_\nu}{\partial \dot{q}_\alpha} = \sum m_\nu \dot{\mathbf{r}}_\nu \cdot \frac{\partial \mathbf{r}_\nu}{\partial q_\alpha}$$

LAGRANGE'S EQUATIONS

11.10. Prove that $\dfrac{d}{dt}\left(\dfrac{\partial T}{\partial \dot{q}_\alpha}\right) - \dfrac{\partial T}{\partial q_\alpha} = \Phi_\alpha$, $\alpha = 1, \ldots, n$ where $\Phi_\alpha = \sum_\nu \mathbf{F}_\nu \cdot \dfrac{\partial \mathbf{r}_\nu}{\partial q_\alpha}$.

From Problem 11.8,

$$\frac{d}{dt}\left\{\sum_\nu m_\nu \dot{\mathbf{r}}_\nu \cdot \frac{\partial \mathbf{r}_\nu}{\partial q_\alpha}\right\} - \sum_\nu m_\nu \dot{\mathbf{r}}_\nu \cdot \frac{\partial \dot{\mathbf{r}}_\nu}{\partial q_\alpha} = \sum_\nu \mathbf{F}_\nu \cdot \frac{\partial \mathbf{r}_\nu}{\partial q_\alpha} \tag{1}$$

From Problems 11.9(a) and 11.9(b),

$$\sum_\nu m_\nu \dot{\mathbf{r}}_\nu \cdot \frac{\partial \dot{\mathbf{r}}_\nu}{\partial q_\alpha} = \frac{\partial T}{\partial q_\alpha} \tag{2}$$

$$\sum_\nu m_\nu \dot{\mathbf{r}}_\nu \cdot \frac{\partial \mathbf{r}_\nu}{\partial q_\alpha} = \frac{\partial T}{\partial \dot{q}_\alpha} \tag{3}$$

Then substituting (2) and (3) in (1), we find

$$\frac{d}{dt}\left(\frac{\partial T}{\partial \dot{q}_\alpha}\right) - \frac{\partial T}{\partial q_\alpha} = \Phi_\alpha \tag{4}$$

The quantity $$p_\alpha = \frac{\partial T}{\partial \dot{q}_\alpha} \tag{5}$$

is called the *generalized momentum* or *conjugate momentum* associated with the generalized coordinate q_α.

11.11. Suppose that the forces acting on a system of particles are derivable from a potential function V, i.e. suppose that the system is conservative. Prove that if $L = T - V$ is the Lagrangian function, then

$$\frac{d}{dt}\left(\frac{\partial L}{\partial \dot{q}_\alpha}\right) - \frac{\partial L}{\partial q_\alpha} = 0$$

If the forces are derivable from a potential V, then [see Problem 11.7],

$$\Phi_\alpha = \frac{\partial W}{\partial q_\alpha} = -\frac{\partial V}{\partial q_\alpha}$$

Since the potential, or potential energy is a function of only the q's [and possibly the time t],

$$\frac{\partial L}{\partial \dot{q}_\alpha} = \frac{\partial}{\partial \dot{q}_\alpha}(T - V) = \frac{\partial T}{\partial \dot{q}_\alpha}$$

Then from Problem 11.10,

$$\frac{d}{dt}\left(\frac{\partial L}{\partial \dot{q}_\alpha}\right) - \frac{\partial T}{\partial q_\alpha} = -\frac{\partial V}{\partial q_\alpha} \quad \text{or} \quad \frac{d}{dt}\left(\frac{\partial L}{\partial \dot{q}_\alpha}\right) - \frac{\partial L}{\partial q_\alpha} = 0$$

11.12. (a) Set up the Lagrangian for a simple pendulum and (b) obtain an equation describing its motion.

(a) Choose as generalized coordinate the angle θ made by string OB of the pendulum and the vertical OA [see Fig. 11-4]. If l is the length of OB, then the kinetic energy is

$$T = \tfrac{1}{2}mv^2 = \tfrac{1}{2}m(l\dot{\theta})^2 = \tfrac{1}{2}ml^2\dot{\theta}^2 \qquad (1)$$

where m is the mass of the bob.

The potential energy of mass m [taking as reference level a horizontal plane through the lowest point A] is given by

$$V = mg(OA - OC) = mg(l - l\cos\theta)$$
$$= mgl(1 - \cos\theta) \qquad (2)$$

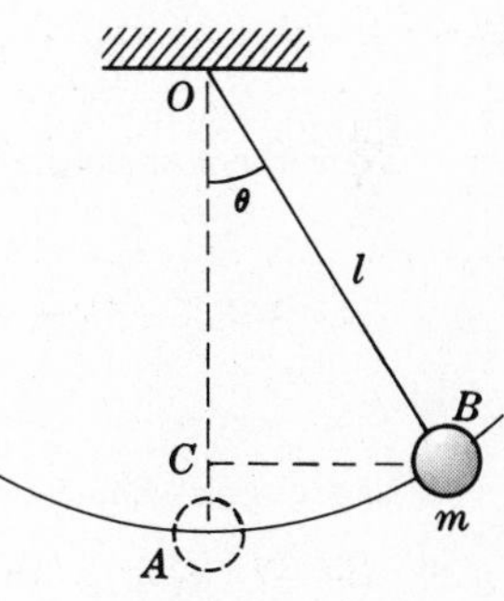
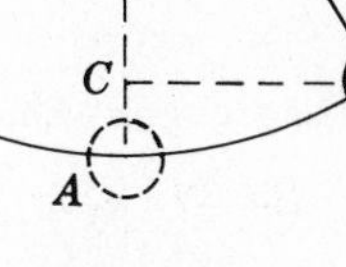

Fig. 11-4

Thus the Lagrangian is

$$L = T - V = \tfrac{1}{2}ml^2\dot{\theta}^2 - mgl(1 - \cos\theta) \qquad (3)$$

(b) Lagrange's equation is

$$\frac{d}{dt}\left(\frac{\partial L}{\partial \dot{\theta}}\right) - \frac{\partial L}{\partial \theta} = 0 \qquad (4)$$

From (3),

$$\frac{\partial L}{\partial \theta} = -mgl\sin\theta, \qquad \frac{\partial L}{\partial \dot{\theta}} = ml^2\dot{\theta} \qquad (5)$$

Substituting these in (4), we find

$$ml^2\ddot{\theta} + mgl\sin\theta = 0 \quad \text{or} \quad \ddot{\theta} + \frac{g}{l}\sin\theta = 0 \qquad (6)$$

which is the required equation of motion [compare Problem 4.23, page 102].

11.13. A mass M_2 hangs at one end of a string which passes over a fixed frictionless non-rotating pulley [see Fig. 11-5 below]. At the other end of this string there is a non-rotating pulley of mass M_1 over which there is a string carrying masses m_1 and m_2. (a) Set up the Lagrangian of the system. (b) Find the acceleration of mass M_2.

Let X_1 and X_2 be the distances of masses M_1 and M_2 respectively below the center of the fixed pulley. Let x_1 and x_2 be the distances of masses m_1 and m_2 respectively below the center of the movable pulley M_1.

Since the strings are fixed in length,

$$X_1 + X_2 = \text{constant} = a, \qquad x_1 + x_2 = \text{constant} = b$$

Then by differentiating with respect to time t,

$$\dot{X}_1 + \dot{X}_2 = 0 \quad \text{or} \quad \dot{X}_2 = -\dot{X}_1$$

and

$$\dot{x}_1 + \dot{x}_2 = 0 \quad \text{or} \quad \dot{x}_2 = -\dot{x}_1$$

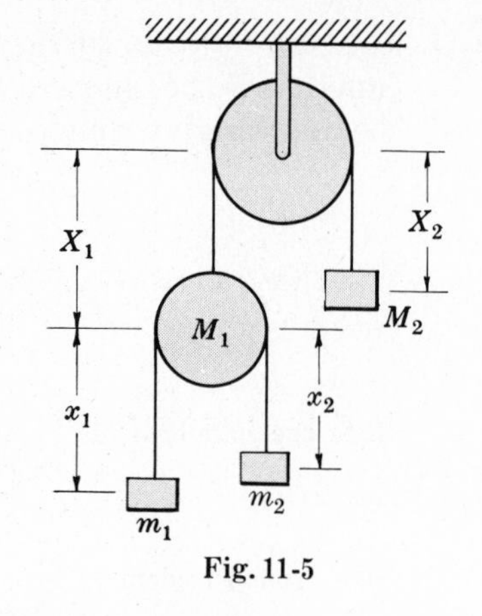

Fig. 11-5

Thus we have

$$\text{Velocity of } M_1 = \dot{X}_1$$

$$\text{Velocity of } M_2 = \dot{X}_2 = -\dot{X}_1$$

$$\text{Velocity of } m_1 = \frac{d}{dt}(X_1 + x_1) = \dot{X}_1 + \dot{x}_1$$

$$\text{Velocity of } m_2 = \frac{d}{dt}(X_1 + x_2) = \dot{X}_1 + \dot{x}_2 = \dot{X}_1 - \dot{x}_1$$

Then the total kinetic energy of the system is

$$T = \tfrac{1}{2}M_1\dot{X}_1^2 + \tfrac{1}{2}M_2\dot{X}_1^2 + \tfrac{1}{2}m_1(\dot{X}_1 + \dot{x}_1)^2 + \tfrac{1}{2}m_2(\dot{X}_1 - \dot{x}_1)^2 \tag{1}$$

The total potential energy of the system measured from a horizontal plane through the center of the fixed pulley as reference is

$$\begin{aligned} V &= -M_1gX_1 - M_2gX_2 - m_1g(X_1 + x_1) - m_2g(X_1 + x_2) \\ &= -M_1gX_1 - M_2g(a - X_1) - m_1g(X_1 + x_1) - m_2g(X_1 + b - x_1) \end{aligned} \tag{2}$$

Then the Lagrangian is

$$\begin{aligned} L &= T - V \\ &= \tfrac{1}{2}M_1\dot{X}_1^2 + \tfrac{1}{2}M_2\dot{X}_1^2 + \tfrac{1}{2}m_1(\dot{X}_1 + \dot{x}_1)^2 + \tfrac{1}{2}m_2(\dot{X}_1 - \dot{x}_1)^2 \\ &\qquad + M_1gX_1 + M_2g(a - X_1) + m_1g(X_1 + x_1) + m_2g(X_1 + b - x_1) \end{aligned} \tag{3}$$

Lagrange's equations corresponding to X_1 and x_1 are

$$\frac{d}{dt}\left(\frac{\partial L}{\partial \dot{X}_1}\right) - \frac{\partial L}{\partial X_1} = 0, \qquad \frac{d}{dt}\left(\frac{\partial L}{\partial \dot{x}_1}\right) - \frac{\partial L}{\partial x_1} = 0 \tag{4}$$

From (3) we have

$$\frac{\partial L}{\partial X_1} = M_1g - M_2g + m_1g + m_2g = (M_1 - M_2 + m_1 + m_2)g$$

$$\frac{\partial L}{\partial \dot{X}_1} = M_1\dot{X}_1 + M_2\dot{X}_1 + m_1(\dot{X}_1 + \dot{x}_1) + m_2(\dot{X}_1 - \dot{x}_1) = (M_1 + M_2 + m_1 + m_2)\dot{X}_1 + (m_1 - m_2)\dot{x}_1$$

$$\frac{\partial L}{\partial x_1} = m_1g - m_2g = (m_1 - m_2)g$$

$$\frac{\partial L}{\partial \dot{x}_1} = m_1(\dot{X}_1 + \dot{x}_1) - m_2(\dot{X}_1 - \dot{x}_1) = (m_1 - m_2)\dot{X}_1 + (m_1 + m_2)\dot{x}_1$$

Thus equations (4) become

$$(M_1 + M_2 + m_1 + m_2)\ddot{X}_1 + (m_1 - m_2)\ddot{x}_1 = (M_1 - M_2 + m_1 + m_2)g$$

$$(m_1 - m_2)\ddot{X}_1 + (m_1 + m_2)\ddot{x}_1 = (m_1 - m_2)g$$

Solving simultaneously, we find

$$\ddot{X}_1 = \frac{(M_1 - M_2)(m_1 + m_2) + 4m_1m_2}{(M_1 + M_2)(m_1 + m_2) + 4m_1m_2}g$$

$$\ddot{x}_1 = \frac{2M_2(m_1 - m_2)}{(M_1 + M_2)(m_1 + m_2) + 4m_1m_2}g$$

Then the downward acceleration of mass M_2 is constant and equal to

$$\ddot{X}_2 = -\ddot{X}_1 = \frac{(M_2 - M_1)(m_1 + m_2) - 4m_1m_2}{(M_1 + M_2)(m_1 + m_2) + 4m_1m_2}g$$

11.14. Use Lagrange's equations to set up the differential equation of the vibrating masses of Problem 8.1, page 197.

Refer to Figs. 8-7 and 8-8 of page 197. The kinetic energy of the system is

$$T = \tfrac{1}{2}m\dot{x}_1^2 + \tfrac{1}{2}m\dot{x}_2^2 \tag{1}$$

Since the stretches of springs AP, PQ and QB of Fig. 8-8 are numerically equal to x_1, $x_2 - x_1$ and x_2 respectively, the potential energy of the system is

$$V = \tfrac{1}{2}\kappa x_1^2 + \tfrac{1}{2}\kappa(x_2 - x_1)^2 + \tfrac{1}{2}\kappa x_2^2 \tag{2}$$

Thus the Lagrangian is

$$L = T - V = \tfrac{1}{2}m\dot{x}_1^2 + \tfrac{1}{2}m\dot{x}_2^2 - \tfrac{1}{2}\kappa x_1^2 - \tfrac{1}{2}\kappa(x_2 - x_1)^2 - \tfrac{1}{2}\kappa x_2^2 \tag{3}$$

Lagrange's equations are

$$\frac{d}{dt}\left(\frac{\partial L}{\partial \dot{x}_1}\right) - \frac{\partial L}{\partial x_1} = 0, \qquad \frac{d}{dt}\left(\frac{\partial L}{\partial \dot{x}_2}\right) - \frac{\partial L}{\partial x_2} = 0 \tag{4}$$

Then since $\dfrac{\partial L}{\partial x_1} = -\kappa x_1 + \kappa(x_2 - x_1) = \kappa(x_2 - 2x_1)$, $\quad \dfrac{\partial L}{\partial \dot{x}_1} = m\dot{x}_1$

$$\frac{\partial L}{\partial x_2} = -\kappa(x_2 - x_1) - \kappa x_2 = \kappa(x_1 - 2x_2), \qquad \frac{\partial L}{\partial \dot{x}_2} = m\dot{x}_2$$

equations (4) become $\qquad m\ddot{x}_1 = \kappa(x_2 - 2x_1), \quad m\ddot{x}_2 = \kappa(x_1 - 2x_2) \qquad (5)$

agreeing with those obtained in Problem 8.1, page 197.

11.15. Use Lagrange's equations to find the differential equation for a compound pendulum which oscillates in a vertical plane about a fixed horizontal axis.

Let the plane of oscillation be represented by the xy plane of Fig. 11-6, where O is its intersection with the axis of rotation and C is the center of mass.

Suppose that the mass of the pendulum is M, its moment of inertia about the axis of rotation is $I_0 = MK^2$ [K = radius of gyration], and distance $OC = h$.

If θ is the instantaneous angle which OC makes with the vertical axis through O, then the kinetic energy is $T = \tfrac{1}{2}I_0\dot{\theta}^2 = \tfrac{1}{2}MK^2\dot{\theta}^2$. The potential energy relative to a horizontal plane through O is $V = -Mgh\cos\theta$. Then the Lagrangian is

$$L = T - V = \tfrac{1}{2}MK^2\dot{\theta}^2 + Mgh\cos\theta$$

Since $\partial L/\partial\theta = -Mgh\sin\theta$ and $\partial L/\partial\dot{\theta} = MK^2\dot{\theta}$, Lagrange's equation is

$$\frac{d}{dt}\left(\frac{\partial L}{\partial \dot{\theta}}\right) - \frac{\partial L}{\partial \theta} = 0$$

i.e.,

$$MK^2\ddot{\theta} + Mgh\sin\theta = 0 \quad \text{or} \quad \ddot{\theta} + \frac{gh}{K^2}\sin\theta = 0$$

Compare Problem 9.24, page 237.

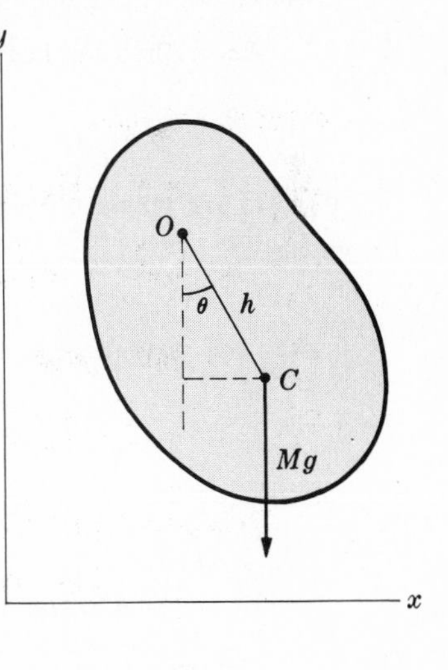

Fig. 11-6

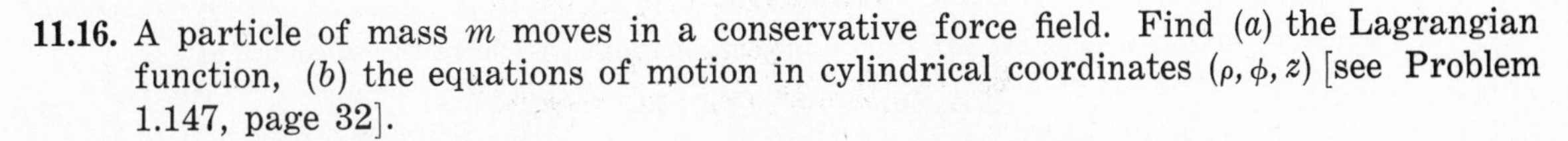

11.16. A particle of mass m moves in a conservative force field. Find (a) the Lagrangian function, (b) the equations of motion in cylindrical coordinates (ρ, ϕ, z) [see Problem 1.147, page 32].

(a) The total kinetic energy $T = \tfrac{1}{2}m[\dot{\rho}^2 + \rho^2\dot{\phi}^2 + \dot{z}^2]$. The potential energy $V = V(\rho, \phi, z)$. Then the Lagrangian function is

$$L = T - V = \tfrac{1}{2}m[\dot{\rho}^2 + \rho^2\dot{\phi}^2 + \dot{z}^2] - V(\rho, \phi, z)$$

(b) Lagrange's equations are

$$\frac{d}{dt}\left(\frac{\partial L}{\partial \dot{\rho}}\right) - \frac{\partial L}{\partial \rho} = 0, \quad \text{i.e.} \quad \frac{d}{dt}(m\dot{\rho}) - \left(m\rho\dot{\phi}^2 - \frac{\partial V}{\partial \rho}\right) = 0 \quad \text{or} \quad m(\ddot{\rho} - \rho\dot{\phi}^2) = -\frac{\partial V}{\partial \rho}$$

$$\frac{d}{dt}\left(\frac{\partial L}{\partial \dot{\phi}}\right) - \frac{\partial L}{\partial \phi} = 0, \quad \text{i.e.} \quad \frac{d}{dt}(m\rho^2\dot{\phi}) + \frac{\partial V}{\partial \phi} = 0 \quad \text{or} \quad m\frac{d}{dt}(\rho^2\dot{\phi}) = -\frac{\partial V}{\partial \phi}$$

$$\frac{d}{dt}\left(\frac{\partial L}{\partial \dot{z}}\right) - \frac{\partial L}{\partial z} = 0, \quad \text{i.e.} \quad \frac{d}{dt}(m\dot{z}) + \frac{\partial V}{\partial z} = 0 \quad \text{or} \quad m\ddot{z} = -\frac{\partial V}{\partial z}$$

11.17. Work Problem 11.16 if the particle moves in the xy plane and if the potential depends only on the distance from the origin.

In this case V depends only on ρ and $z = 0$. Then Lagrange's equations in part (b) of Problem 11.16 become

$$m(\ddot{\rho} - \rho\dot{\phi}^2) = -\frac{\partial V}{\partial \rho}, \qquad \frac{d}{dt}(\rho^2\dot{\phi}) = 0$$

These are the equations for motion in a central force field obtained in Problem 5.3, page 122.

LAGRANGE'S EQUATIONS FOR NON-HOLONOMIC SYSTEMS

11.18. Derive Lagrange's equations (15), page 284, for non-holonomic constraints.

Assume that there are m constraint conditions of the form

$$\sum_\alpha A_\alpha\, dq_\alpha + A\, dt = 0, \quad \sum_\alpha B_\alpha\, dq_\alpha + B\, dt = 0, \quad \ldots \tag{1}$$

where $m < n$, the number of coordinates q_α.

As in Problem 11.10, page 288, we have

$$Y_\alpha \equiv \frac{d}{dt}\left(\frac{\partial T}{\partial \dot{q}_\alpha}\right) - \frac{\partial T}{\partial q_\alpha} = \sum_\nu m_\nu \ddot{\mathbf{r}}_\nu \cdot \frac{\partial \mathbf{r}_\nu}{\partial q_\alpha} \tag{2}$$

If $\delta \mathbf{r}_\nu$ are virtual displacements which satisfy the instantaneous constraints [obtained by considering that time t is a constant], then

$$\delta \mathbf{r}_\nu = \sum_\alpha \frac{\partial \mathbf{r}_\nu}{\partial q_\alpha}\,\delta q_\alpha \tag{3}$$

Now the virtual work done is

$$\delta W = \sum_\nu m_\nu \ddot{\mathbf{r}}_\nu \cdot \delta \mathbf{r}_\nu = \sum_\nu \sum_\alpha m_\nu \ddot{\mathbf{r}}_\nu \cdot \frac{\partial \mathbf{r}_\nu}{\partial q_\alpha}\,\delta q_\alpha = \sum_\alpha Y_\alpha\,\delta q_\alpha \tag{4}$$

Now since the virtual work can be written in terms of the generalized forces Φ_α as

$$\delta W = \sum_\alpha \Phi_\alpha\,\delta q_\alpha \tag{5}$$

we have by subtraction of (4) and (5),

$$\sum_\alpha (Y_\alpha - \Phi_\alpha)\,\delta q_\alpha = 0 \tag{6}$$

Since the δq_α are not all independent, we cannot conclude that $Y_\alpha = \Phi_\alpha$ which would lead to Lagrange's equations as obtained in Problem 11.10.

From (1), since t is constant for instantaneous contraints, we have the m equations

$$\sum_\alpha A_\alpha\,\delta q_\alpha = 0, \quad \sum_\alpha B_\alpha\,\delta q_\alpha = 0, \quad \ldots \tag{7}$$

Multiplying these by the m Lagrange multipliers $\lambda_1, \lambda_2, \ldots$ and adding, we have

$$\sum_\alpha (\lambda_1 A_\alpha + \lambda_2 B_\alpha + \cdots)\,\delta q_\alpha = 0 \tag{8}$$

Subtraction of (6) and (8) yields

$$\sum_{\alpha} (Y_\alpha - \Phi_\alpha - \lambda_1 A_\alpha - \lambda_2 B_\alpha - \cdots)\,\delta q_\alpha = 0 \tag{9}$$

Now because of equations (7) we can solve for m of the quantities δq_α [say $\delta q_1, \ldots, \delta q_m$] in terms of the remaining δq_α [say $\delta q_{m+1}, \ldots, \delta q_n$]. Thus in (9) we can consider $\delta q_1, \ldots, \delta q_m$ as dependent and $\delta q_{m+1}, \ldots, \delta q_n$ as independent.

Let us arbitrarily set the coefficients of the dependent variables equal to zero, i.e.,

$$Y_\alpha - \Phi_\alpha - \lambda_1 A_\alpha - \lambda_2 B_\alpha - \cdots = 0, \qquad \alpha = 1, 2, \ldots, m \tag{10}$$

Then there will be left in the sum (9) only the independent quantities δq_α and since these are arbitrary it follows that their coefficients will be zero. Thus

$$Y_\alpha - \Phi_\alpha - \lambda_1 A_\alpha - \lambda_2 B_\alpha - \cdots = 0, \qquad \alpha = m+1, \ldots, n \tag{11}$$

Equations (2), (10) and (11) thus lead to

$$\frac{d}{dt}\left(\frac{\partial T}{\partial \dot{q}_\alpha}\right) - \frac{\partial T}{\partial q_\alpha} = \Phi_\alpha + \lambda_1 A_\alpha + \lambda_2 B_\alpha + \cdots \qquad \alpha = 1, 2, \ldots, n \tag{12}$$

as required. These equations together with (1) lead to $n+m$ equations in $n+m$ unknowns.

11.19. Derive equations (16), page 284, for conservative non-holonomic systems.

From Problem 11.18,

$$\frac{d}{dt}\left(\frac{\partial T}{\partial \dot{q}_\alpha}\right) - \frac{\partial T}{\partial q_\alpha} = \Phi_\alpha + \lambda_1 A_\alpha + \lambda_2 B_\alpha + \cdots \tag{1}$$

Then if the forces are derivable from a potential, $\Phi_\alpha = -\partial V/\partial q_\alpha$ where V does not depend on $\dot{q}_\alpha$. Thus (1) can be written

$$\frac{d}{dt}\left(\frac{\partial L}{\partial \dot{q}_\alpha}\right) - \frac{\partial L}{\partial q_\alpha} = \lambda_1 A_\alpha + \lambda_2 B_\alpha + \cdots \tag{2}$$

where $L = T - V$.

11.20. A particle of mass m moves under the influence of gravity on the inner surface of the paraboloid of revolution $x^2 + y^2 = az$ which is assumed frictionless [see Fig. 11-7]. Obtain the equations of motion.

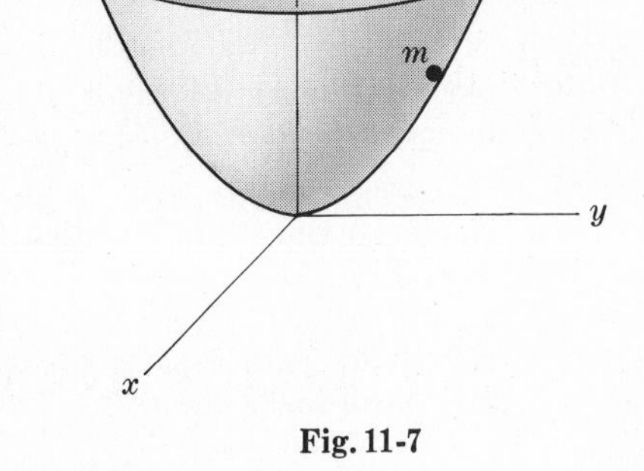

Fig. 11-7

By Problem 11.16, the Lagrangian in cylindrical coordinates is given by

$$L = \tfrac{1}{2}m(\dot{\rho}^2 + \rho^2\dot{\phi}^2 + \dot{z}^2) - mgz \tag{1}$$

Since $x^2 + y^2 = \rho^2$, the constraint condition is $\rho^2 - az = 0$ so that

$$2\rho\,\delta\rho - a\,\delta z = 0 \tag{2}$$

If we call $q_1 = \rho,\ q_2 = \phi,\ q_3 = z$ and compare (2) with the equations (7) of Problem 11.18, we see that

$$A_1 = 2\rho, \quad A_2 = 0, \quad A_3 = -a \tag{3}$$

since only one constraint is given. Lagrange's equations [see Problem 11.19] can thus be written

$$\frac{d}{dt}\left(\frac{\partial L}{\partial \dot{q}_\alpha}\right) - \frac{\partial L}{\partial q_\alpha} = \lambda_1 A_\alpha \qquad \alpha = 1, 2, 3$$

i.e., $$\frac{d}{dt}\left(\frac{\partial L}{\partial \dot{\rho}}\right) - \frac{\partial L}{\partial \rho} = 2\lambda_1\rho, \qquad \frac{d}{dt}\left(\frac{\partial L}{\partial \dot{\phi}}\right) - \frac{\partial L}{\partial \phi} = 0, \qquad \frac{d}{dt}\left(\frac{\partial L}{\partial \dot{z}}\right) - \frac{\partial L}{\partial z} = -\lambda_1 a$$

Using (1), these become

$$m(\ddot{\rho} - \rho\dot{\phi}^2) = 2\lambda_1\rho \tag{4}$$

$$m\frac{d}{dt}(\rho^2\dot{\theta}) = 0 \tag{5}$$

$$m\ddot{z} = -mg - \lambda_1 a \tag{6}$$

We also have the constraint condition

$$2\rho\dot{\rho} - a\dot{z} = 0 \tag{7}$$

The four equations (4), (5), (6) and (7) enable us to find the four unknowns ρ, ϕ, z, λ_1.

11.21. (a) Prove that the particle of Problem 11.20 will describe a horizontal circle in the plane $z = h$ provided that it is given an angular velocity whose magnitude is $\omega = \sqrt{2g/a}$.

(b) Prove that if the particle is displaced slightly from this circular path it will undergo oscillations about the path with frequency given by $(1/\pi)\sqrt{2g/a}$.

(c) Discuss the stability of the particle in the circular path.

(a) The radius of the circle obtained as the intersection of the plane $z = h$ with the paraboloid $\rho^2 = az$ is

$$\rho_0 = \sqrt{ah} \tag{1}$$

Letting $z = h$ in equation (6) of Problem 11.20, we find

$$\lambda_1 = -mg/a \tag{2}$$

Then using (1) and (2) in equation (4) of Problem 11.20 and calling $\dot{\phi} = \omega$, we find $m(-\rho_0\omega^2) = 2(-mg/a)\rho_0$ or $\omega^2 = 2g/a$ from which

$$\omega = \sqrt{2g/a} \tag{3}$$

The period and frequency of the particle in this circular path are given respectively by

$$P_1 = 2\pi\sqrt{\frac{a}{2g}} \quad \text{and} \quad f_1 = \frac{1}{2\pi}\sqrt{\frac{2g}{a}} \tag{4}$$

(b) From equation (5) of Problem 11.20, we find

$$\rho^2\dot{\phi} = \text{constant} = A \tag{5}$$

Assuming that the particle starts with angular speed ω, we find $A = ah\omega$ so that

$$\dot{\phi} = ah\omega/\rho^2 \tag{6}$$

Since the vibration takes place very nearly in the plane $z = h$, we find by letting $z = h$ in equation (6) of Problem 11.20 that

$$\lambda_1 = -mg/a \tag{7}$$

Using (6) and (7) in equation (4) of Problem 11.20, we find

$$\ddot{\rho} - a^2h^2\omega^2/\rho^3 = -2g\rho/a \tag{8}$$

Now if the path departs slightly from the circle, then ρ will depart slightly from ρ_0. Thus we are led to make the transformation

$$\rho = \rho_0 + u \tag{9}$$

in (8), where u is small compared with ρ_0. Then (8) becomes

$$\ddot{u} - \frac{a^2h^2\omega^2}{(\rho_0 + u)^3} = -\frac{2g}{a}(\rho_0 + u) \tag{10}$$

But to a high degree of approximation,

$$\frac{1}{(\rho_0 + u)^3} = \frac{1}{\rho_0^3(1 + u/\rho_0)^3} = \frac{1}{\rho_0^3}\left(1 + \frac{u}{\rho_0}\right)^{-3} = \frac{1}{\rho_0^3}\left(1 - \frac{3u}{\rho_0}\right)$$

by the binomial theorem, where we have neglected terms involving $u^2, u^3, \ldots$. Using the values of ρ_0 and ω given by (1) and (3) respectively, (10) becomes

$$\ddot{u} + (8g/a)u = 0 \tag{5}$$

whose solution is $u = \epsilon_1 \cos \sqrt{8g/a}\, t + \epsilon_2 \sin \sqrt{8g/a}\, t$. Thus

$$\rho = \rho_0 + u = \sqrt{ah} + \epsilon_1 \cos \sqrt{8g/a}\, t + \epsilon_2 \sin \sqrt{8g/a}\, t$$

It follows that if the particle is displaced slightly from the circular path of radius $\rho_0 = \sqrt{ah}$, it will undergo oscillations about the path with frequency

$$f_2 = \frac{1}{2\pi}\sqrt{\frac{8g}{a}} = \frac{1}{\pi}\sqrt{\frac{2g}{a}} \tag{6}$$

or period

$$P_2 = \pi\sqrt{\frac{a}{2g}} \tag{7}$$

It is interesting that the period of oscillation in the circular path given by *(4)* is twice the period of oscillation about the circular path given by *(7)*.

(c) Since the particle tends to return to the circular path when it is displaced slightly from it, the motion is one of *stability*.

11.22. Discuss the physical significance of the Lagrange multipliers $\lambda_1, \lambda_2, \ldots$ in Problem 11.18.

In case there are no constraints the equations of motion are by Problem 11.10,

$$\frac{d}{dt}\left(\frac{\partial T}{\partial \dot{q}_\alpha}\right) - \frac{\partial T}{\partial q_\alpha} = \Phi_\alpha$$

In case there are constraints the equations are by Problem 11.18,

$$\frac{d}{dt}\left(\frac{\partial T}{\partial \dot{q}_\alpha}\right) - \frac{\partial T}{\partial q_\alpha} = \Phi_\alpha + \lambda_1 A_\alpha + \lambda_2 B_\alpha + \cdots$$

It follows that the terms $\lambda_1 A_\alpha + \lambda_2 B_\alpha + \cdots$ correspond to the generalized forces associated with constraints.

Physically, the Lagrange multipliers are associated with the constraint forces acting on the system. Thus when we determine the Lagrange multipliers we are essentially taking into account the effect of the constraint forces without actually finding these forces explicitly.

LAGRANGE'S EQUATIONS WITH IMPULSIVE FORCES

11.23. Derive the equations *(20)*, page 285.

For the case where forces are finite we have by Problem 11.10,

$$\frac{d}{dt}\left(\frac{\partial T}{\partial \dot{q}_\alpha}\right) - \frac{\partial T}{\partial q_\alpha} = \Phi_\alpha \tag{1}$$

where

$$\Phi_\alpha = \sum_\nu \mathbf{F}_\nu \cdot \frac{\partial \mathbf{r}_\nu}{\partial q_\alpha} \tag{2}$$

Integrating both sides of *(1)* with respect to t from $t = 0$ to $t = \tau$,

$$\int_0^\tau \frac{d}{dt}\left(\frac{\partial T}{\partial \dot{q}_\alpha}\right) dt - \int_0^\tau \frac{\partial T}{\partial q_\alpha}\, dt = \int_0^\tau \Phi_\alpha\, dt \tag{3}$$

so that

$$\left(\frac{\partial T}{\partial \dot{q}_\alpha}\right)_{t=\tau} - \left(\frac{\partial T}{\partial \dot{q}_\alpha}\right)_{t=0} - \int_0^\tau \frac{\partial T}{\partial q_\alpha}\, dt = \sum_\nu \left\{\left(\int_0^\tau \mathbf{F}_\nu\, dt\right) \cdot \frac{\partial \mathbf{r}_\nu}{\partial q_\alpha}\right\} \tag{4}$$

Taking the limit as $\tau \to 0$, we have

$$\lim_{\tau \to 0}\left\{\left(\frac{\partial T}{\partial \dot{q}_\alpha}\right)_{t=\tau} - \left(\frac{\partial T}{\partial \dot{q}_\alpha}\right)_{t=0}\right\} - \lim_{\tau\to 0}\int_0^\tau \frac{\partial T}{\partial q_\alpha}\, dt = \sum_\nu \left\{\left(\lim_{\tau\to 0}\int_0^\tau \mathbf{F}_\nu\, dt\right) \cdot \frac{\partial \mathbf{r}_\nu}{\partial q_\alpha}\right\}$$

or
$$\left(\frac{\partial T}{\partial \dot{q}_\alpha}\right)_2 - \left(\frac{\partial T}{\partial \dot{q}_\alpha}\right)_1 = \sum_\nu \mathcal{J}_\nu \cdot \frac{\partial \mathbf{r}_\nu}{\partial q_\alpha} = \mathcal{F}_\alpha$$

using $\lim_{\tau\to 0}\int_0^\tau \frac{\partial T}{\partial q_\alpha}\,dt = 0$ since $\frac{\partial T}{\partial q_\alpha}$ is finite, and $\lim_{\tau\to 0}\int_0^\tau \mathbf{F}_\nu\,dt = \mathcal{J}_\nu$.

11.24. A square $ABCD$ formed by four rods of length $2l$ and mass m hinged at their ends, rests on a horizontal frictionless table. An impulse of magnitude $\mathcal{J}$ is applied to the vertex A in the direction AD. Find the equations of motion.

After the square is struck, its shape will in general be a rhombus [Fig. 11-8].

Suppose that at any time t the angles made by sides AD (or BC) and AB (or CD) with the x axis are θ_1 and θ_2 respectively, while the coordinates of the center M are (x, y). Thus x, y, θ_1, θ_2 are the generalized coordinates.

From Fig. 11-8 we see that the position vectors of the centers E, F, G, H of the rods are given respectively by

$$\mathbf{r}_E = (x - l\cos\theta_1)\mathbf{i} + (y - l\sin\theta_1)\mathbf{j}$$
$$\mathbf{r}_F = (x + l\cos\theta_2)\mathbf{i} + (y - l\sin\theta_2)\mathbf{j}$$
$$\mathbf{r}_G = (x + l\cos\theta_1)\mathbf{i} + (y + l\sin\theta_1)\mathbf{j}$$
$$\mathbf{r}_H = (x - l\cos\theta_2)\mathbf{i} + (y + l\sin\theta_2)\mathbf{j}$$

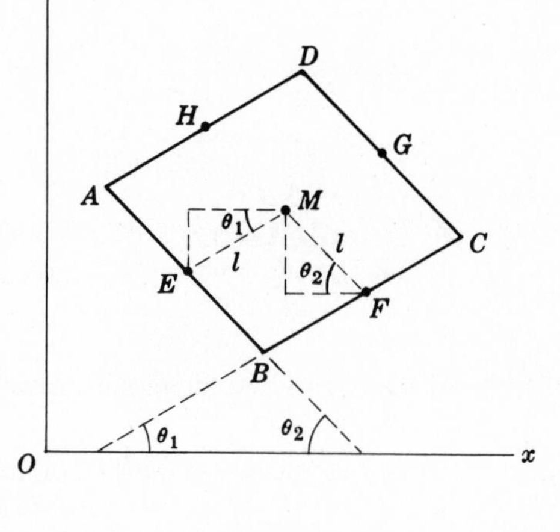

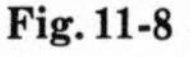
Fig. 11-8

The velocities of E, F, G and H at any time are given by

$$\mathbf{v}_E = \dot{\mathbf{r}}_E = (\dot{x} + l\sin\theta_1\,\dot{\theta}_1)\mathbf{i} + (\dot{y} - l\cos\theta_1\,\dot{\theta}_1)\mathbf{j}$$
$$\mathbf{v}_F = \dot{\mathbf{r}}_F = (\dot{x} - l\sin\theta_2\,\dot{\theta}_2)\mathbf{i} + (\dot{y} - l\cos\theta_2\,\dot{\theta}_2)\mathbf{j}$$
$$\mathbf{v}_G = \dot{\mathbf{r}}_G = (\dot{x} - l\sin\theta_1\,\dot{\theta}_1)\mathbf{i} + (\dot{y} + l\cos\theta_1\,\dot{\theta}_1)\mathbf{j}$$
$$\mathbf{v}_H = \dot{\mathbf{r}}_H = (\dot{x} + l\sin\theta_2\,\dot{\theta}_2)\mathbf{i} + (\dot{y} + l\cos\theta_2\,\dot{\theta}_2)\mathbf{j}$$

The kinetic energy of a rod such as AB is the same as the kinetic energy of a particle of mass m located at its center of mass E plus the kinetic energy of rotation about an axis through E perpendicular to the xy plane. Since the angular velocity has magnitude $\dot{\theta}_2$ and the moment of inertia of a rod of length $2l$ about its center of mass is $I_{AB} = \frac{1}{3}ml^2$, the total energy of rod AB is

$$T_{AB} = \tfrac{1}{2}m\dot{r}_E^2 + \tfrac{1}{2}I_{AB}\dot{\theta}_2^2$$

Similarly, the total kinetic energies of rods BC, CD and AD are

$$T_{BC} = \tfrac{1}{2}m\dot{\mathbf{r}}_F^2 + \tfrac{1}{2}I_{BC}\dot{\theta}_1^2, \quad T_{CD} = \tfrac{1}{2}m\dot{\mathbf{r}}_G^2 + \tfrac{1}{2}I_{CD}\dot{\theta}_2^2, \quad T_{AD} = \tfrac{1}{2}m\dot{\mathbf{r}}_H^2 + \tfrac{1}{2}I_{AD}\dot{\theta}_1^2$$

Thus the total kinetic energy is [using the fact that $I = I_{AB} = I_{BC} = I_{CD} = \frac{1}{3}ml^2$]

$$\begin{aligned} T &= T_{AB} + T_{BC} + T_{CD} + T_{AD} \\ &= \tfrac{1}{2}m(\dot{\mathbf{r}}_E^2 + \dot{\mathbf{r}}_F^2 + \dot{\mathbf{r}}_G^2 + \dot{\mathbf{r}}_H^2) + I(\dot{\theta}_1^2 + \dot{\theta}_2^2) \\ &= \tfrac{1}{2}m(4\dot{x}^2 + 4\dot{y}^2 + 2l^2\dot{\theta}_1^2 + 2l^2\dot{\theta}_2^2) + \tfrac{1}{3}ml^2(\dot{\theta}_1^2 + \dot{\theta}_2^2) \\ &= 2m(\dot{x}^2 + \dot{y}^2) + \tfrac{4}{3}ml^2(\dot{\theta}_1^2 + \dot{\theta}_2^2) \end{aligned}$$

Let us assume that initially the rhombus is a square at rest with its sides parallel to the coordinate axes and its center located at the origin. Then we have

$$x = 0,\ y = 0,\ \ \theta_1 = \pi/2,\ \theta_2 = 0,\ \ \dot{x} = 0,\ \dot{y} = 0,\ \ \dot{\theta}_1 = 0,\ \dot{\theta}_2 = 0$$

If we use the notation $(\ )_1$ and $(\ )_2$ to denote quantities before and after the impulse is applied, we have

$$\left(\frac{\partial T}{\partial \dot{x}}\right)_1 = (4m\dot{x})_1 = 0 \qquad \left(\frac{\partial T}{\partial \dot{y}}\right)_1 = (4m\dot{y})_1 = 0$$
$$\left(\frac{\partial T}{\partial \dot{\theta}_1}\right)_1 = (\tfrac{8}{3}ml^2\dot{\theta}_1)_1 = 0 \qquad \left(\frac{\partial T}{\partial \dot{\theta}_2}\right)_1 = (\tfrac{8}{3}ml^2\dot{\theta}_2)_1 = 0$$

$$\left(\frac{\partial T}{\partial \dot{x}}\right)_2 = (4m\dot{x})_2 = 4m\dot{x} \qquad \left(\frac{\partial T}{\partial \dot{y}}\right)_2 = (4m\dot{y})_2 = 4m\dot{y}$$

$$\left(\frac{\partial T}{\partial \dot{\theta}_1}\right)_2 = (\tfrac{8}{3}ml^2\dot{\theta}_1)_2 = \tfrac{8}{3}ml^2\dot{\theta}_1 \qquad \left(\frac{\partial T}{\partial \dot{\theta}_2}\right)_2 = \tfrac{8}{3}ml^2\dot{\theta}_2 = \tfrac{8}{3}ml^2\dot{\theta}_2$$

Then

$$\left(\frac{\partial T}{\partial \dot{x}}\right)_2 - \left(\frac{\partial T}{\partial \dot{x}}\right)_1 = \mathcal{F}_x \quad \text{or} \quad 4m\dot{x} = \mathcal{F}_x \tag{1}$$

$$\left(\frac{\partial T}{\partial \dot{y}}\right)_2 - \left(\frac{\partial T}{\partial \dot{y}}\right)_1 = \mathcal{F}_y \quad \text{or} \quad 4m\dot{y} = \mathcal{F}_y \tag{2}$$

$$\left(\frac{\partial T}{\partial \dot{\theta}_1}\right)_2 - \left(\frac{\partial T}{\partial \dot{\theta}_1}\right)_1 = \mathcal{F}_{\theta_1} \quad \text{or} \quad \tfrac{8}{3}ml^2\dot{\theta}_1 = \mathcal{F}_{\theta_1} \tag{3}$$

$$\left(\frac{\partial T}{\partial \dot{\theta}_2}\right)_2 - \left(\frac{\partial T}{\partial \dot{\theta}_2}\right)_1 = \mathcal{F}_{\theta_2} \quad \text{or} \quad \tfrac{8}{3}ml^2\dot{\theta}_2 = \mathcal{F}_{\theta_2} \tag{4}$$

where for simplicity we have now removed the subscript $(\ )_2$.

To find $\mathcal{F}_x, \mathcal{F}_y, \mathcal{F}_{\theta_1}, \mathcal{F}_{\theta_2}$ we note that

$$\mathcal{F}_\alpha = \sum_\nu \mathcal{J}_\nu \cdot \frac{\partial \mathbf{r}_\nu}{\partial q_\alpha} \tag{5}$$

where $\mathcal{J}_\nu$ are the impulsive forces. We thus have

$$\mathcal{F}_x = \mathcal{J}_A \cdot \frac{\partial \mathbf{r}_A}{\partial x} + \mathcal{J}_B \cdot \frac{\partial \mathbf{r}_B}{\partial x} + \mathcal{J}_C \cdot \frac{\partial \mathbf{r}_C}{\partial x} + \mathcal{J}_D \cdot \frac{\partial \mathbf{r}_D}{\partial x} \tag{6}$$

$$\mathcal{F}_y = \mathcal{J}_A \cdot \frac{\partial \mathbf{r}_A}{\partial y} + \mathcal{J}_B \cdot \frac{\partial \mathbf{r}_B}{\partial y} + \mathcal{J}_C \cdot \frac{\partial \mathbf{r}_C}{\partial y} + \mathcal{J}_D \cdot \frac{\partial \mathbf{r}_D}{\partial y} \tag{7}$$

$$\mathcal{F}_{\theta_1} = \mathcal{J}_A \cdot \frac{\partial \mathbf{r}_A}{\partial \theta_1} + \mathcal{J}_B \cdot \frac{\partial \mathbf{r}_B}{\partial \theta_1} + \mathcal{J}_C \cdot \frac{\partial \mathbf{r}_C}{\partial \theta_1} + \mathcal{J}_D \cdot \frac{\partial \mathbf{r}_D}{\partial \theta_1} \tag{8}$$

$$\mathcal{F}_{\theta_2} = \mathcal{J}_A \cdot \frac{\partial \mathbf{r}_A}{\partial \theta_2} + \mathcal{J}_B \cdot \frac{\partial \mathbf{r}_B}{\partial \theta_2} + \mathcal{J}_C \cdot \frac{\partial \mathbf{r}_C}{\partial \theta_2} + \mathcal{J}_D \cdot \frac{\partial \mathbf{r}_D}{\partial \theta_2} \tag{9}$$

Now from Fig. 11-8 we find the position vectors of A, B, C, D given by

$$\mathbf{r}_A = (x - l\cos\theta_1 - l\cos\theta_2)\mathbf{i} + (y - l\sin\theta_1 + l\sin\theta_2)\mathbf{j}$$

$$\mathbf{r}_B = (x - l\cos\theta_1 + l\cos\theta_2)\mathbf{i} + (y - l\sin\theta_1 - l\sin\theta_2)\mathbf{j}$$

$$\mathbf{r}_C = (x + l\cos\theta_1 + l\cos\theta_2)\mathbf{i} + (y + l\sin\theta_1 - l\sin\theta_2)\mathbf{j}$$

$$\mathbf{r}_D = (x + l\cos\theta_1 - l\cos\theta_2)\mathbf{i} + (y + l\sin\theta_1 + l\sin\theta_2)\mathbf{j}$$

Since the impulsive force at A is initially in the direction of the positive y axis, we have

$$\mathcal{J}_A = \mathcal{J}\mathbf{j} \tag{10}$$

Thus equations (6)-(9) yield

$$\mathcal{F}_x = 0, \quad \mathcal{F}_y = \mathcal{J}, \quad \mathcal{F}_{\theta_1} = -\mathcal{J}l\cos\theta_1, \quad \mathcal{F}_{\theta_2} = \mathcal{J}l\cos\theta_2 \tag{11}$$

Then equations (1)-(4) become

$$4m\dot{x} = 0, \quad 4m\dot{y} = \mathcal{J}, \quad \tfrac{8}{3}ml^2\dot{\theta}_1 = -\mathcal{J}l\cos\theta_1, \quad \tfrac{8}{3}ml^2\dot{\theta}_2 = \mathcal{J}l\cos\theta_2 \tag{12}$$

11.25. Prove that the kinetic energy developed immediately after application of the impulsive forces in Problem 11.24 is $T = \mathcal{J}^2/2m$.

From equations (12) of Problem 11.24, we have

$$\dot{x} = 0, \quad \dot{y} = \frac{\mathcal{J}}{4m}, \quad \dot{\theta}_1 = -\frac{3\mathcal{J}}{8ml}\cos\theta_1, \quad \dot{\theta}_2 = \frac{3\mathcal{J}}{8ml}\cos\theta_2$$

Substituting these values in the kinetic energy obtained in Problem 11.24, we find

$$T = \frac{\mathcal{J}^2}{8m} + \frac{3\mathcal{J}^2}{8m}(\cos^2\theta_1 + \cos^2\theta_2) \qquad (1)$$

But immediately after application of the impulsive forces, $\theta_1 = \pi/2$ and $\theta_2 = 0$ approximately. Thus *(1)* becomes $T = \mathcal{J}^2/2m$.

MISCELLANEOUS PROBLEMS

11.26. In Fig. 11-9, AB is a straight frictionless wire fixed at point A on a vertical axis OA such that AB rotates about OA with constant angular velocity ω. A bead of mass m is constrained to move on the wire. (*a*) Set up the Lagrangian. (*b*) Write Lagrange's equations. (*c*) Determine the motion at any time.

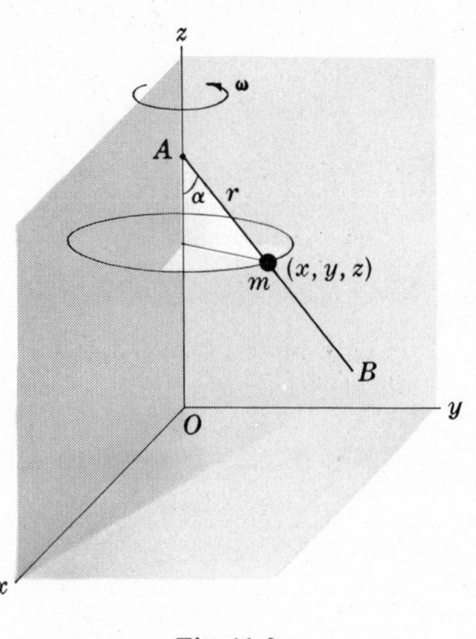

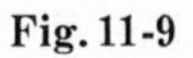
Fig. 11-9

(*a*) Let r be the distance of the bead from point A of the wire at time t. The rectangular coordinates of the bead are then given by

$$x = r\sin\alpha\cos\omega t$$
$$y = r\sin\alpha\sin\omega t$$
$$z = h - r\cos\alpha$$

where it is assumed that at $t = 0$ the wire is in the xz plane and that the distance from O to A is h.

The kinetic energy of the bead is

$$\begin{aligned} T &= \tfrac{1}{2}m(\dot{x}^2 + \dot{y}^2 + \dot{z}^2) \\ &= \tfrac{1}{2}m\{(\dot{r}\sin\alpha\cos\omega t - \omega r\sin\alpha\sin\omega t)^2 \\ &\qquad + (\dot{r}\sin\alpha\sin\omega t + \omega r\sin\alpha\cos\omega t)^2 + (-\dot{r}\cos\alpha)^2\} \\ &= \tfrac{1}{2}m(\dot{r}^2 + \omega^2 r^2\sin^2\alpha) \end{aligned}$$

The potential energy, taking the xy plane as reference level, is $V = mgz = mg(h - r\cos\alpha)$. Then the Lagrangian is

$$L = T - V = \tfrac{1}{2}m(\dot{r}^2 + \omega^2 r^2\sin^2\alpha) - mg(h - r\cos\alpha)$$

(*b*) We have

$$\frac{\partial L}{\partial r} = m\omega^2 r\sin^2\alpha + mg\cos\alpha, \qquad \frac{\partial L}{\partial \dot{r}} = m\dot{r}$$

and Lagrange's equation is $\dfrac{d}{dt}\left(\dfrac{\partial L}{\partial \dot{r}}\right) - \dfrac{\partial L}{\partial r} = 0$ or

$$m\ddot{r} - (m\omega^2 r\sin^2\alpha + mg\cos\alpha) = 0$$

i.e.,

$$\ddot{r} - (\omega^2\sin^2\alpha)r = g\cos\alpha \qquad (1)$$

(*c*) The general solution of equation *(1)* with the right hand side replaced by zero is

$$c_1 e^{(\omega\sin\alpha)t} + c_2 e^{-(\omega\sin\alpha)t}$$

Since the right hand side of *(1)* is a constant, a particular solution is $\dfrac{-g\cos\alpha}{\omega^2\sin^2\alpha}$. Thus the general solution of *(1)* is

$$r = c_1 e^{(\omega\sin\alpha)t} + c_2 e^{-(\omega\sin\alpha)t} - \frac{g\cos\alpha}{\omega^2\sin^2\alpha} \qquad (2)$$

This result can also be written in terms of hyperbolic functions as

$$r = c_3\cosh(\omega\sin\alpha)t + c_4\sinh(\omega\sin\alpha)t - \frac{g\cos\alpha}{\omega^2\sin^2\alpha} \qquad (3)$$

11.27. Suppose that in Problem 11.26 the bead starts from rest at A. How long will it take to reach the end B of the wire assuming that the length of the wire is l?

Since the bead starts from rest at $t=0$, we have $r=0, \dot{r}=0$ at $t=0$. Then from equation (2) of Problem 11.26,

$$c_1 + c_2 = \frac{g\cos\alpha}{\omega^2\sin^2\alpha} \qquad \text{and} \qquad c_1 - c_2 = 0$$

Thus $c_1 = c_2 = \dfrac{g\cos\alpha}{2\omega^2\sin^2\alpha}$ and (2) of Problem 11.26 becomes

$$r = \frac{g\cos\alpha}{2\omega^2\sin^2\alpha}\{e^{(\omega\sin\alpha)t} + e^{-(\omega\sin\alpha)t}\} - \frac{g\cos\alpha}{\omega^2\sin^2\alpha} \tag{1}$$

or

$$r = \frac{g\cos\alpha}{\omega^2\sin^2\alpha}\{\cosh(\omega\sin\alpha)t - 1\} \tag{2}$$

which can also be obtained from equation (3) of Problem 11.26. When $r=l$, (2) yields

$$\cosh(\omega\sin\alpha)t = 1 + \frac{l\omega^2\sin^2\alpha}{g\cos\alpha}$$

so that the required time is

$$t = \frac{1}{\omega\sin\alpha}\cosh^{-1}\left(1 + \frac{l\omega^2\sin^2\alpha}{g\cos^2\alpha}\right)$$

$$= \frac{1}{\omega\sin\alpha}\ln\left\{\left(1 + \frac{l\omega^2\sin^2\alpha}{g\cos^2\alpha}\right) + \sqrt{\left(1 + \frac{l\omega^2\sin^2\alpha}{g\cos^2\alpha}\right)^2 - 1}\right\}$$

11.28. A double pendulum [see Problem 11.1(*c*) and Fig. 11-3, page 285] vibrates in a vertical plane. (*a*) Write the Lagrangian of the system. (*b*) Obtain equations for the motion.

(*a*) The transformation equations given in Problem 11.2, page 286,

$$x_1 = l_1\cos\theta_1 \qquad y_1 = l_1\sin\theta_1$$

$$x_2 = l_1\cos\theta_1 + l_2\cos\theta_2 \qquad y_2 = l_1\sin\theta_1 + l_2\sin\theta_2$$

yield

$$\dot{x}_1 = -l_1\dot{\theta}_1\sin\theta_1 \qquad \dot{y}_1 = l_1\dot{\theta}_1\cos\theta_1$$

$$\dot{x}_2 = -l_1\dot{\theta}_1\sin\theta_1 - l_2\dot{\theta}_2\sin\theta_2 \qquad \dot{y}_2 = l_1\dot{\theta}_1\cos\theta_1 + l_2\dot{\theta}_2\cos\theta_2$$

The kinetic energy of the system is

$$T = \tfrac{1}{2}m_1(\dot{x}_1^2 + \dot{y}_1^2) + \tfrac{1}{2}m_2(\dot{x}_2^2 + \dot{y}_2^2)$$

$$= \tfrac{1}{2}m_1l_1^2\dot{\theta}_1^2 + \tfrac{1}{2}m_2[l_1^2\dot{\theta}_1^2 + l_2^2\dot{\theta}_2^2 + 2l_1l_2\dot{\theta}_1\dot{\theta}_2\cos(\theta_1 - \theta_2)]$$

The potential energy of the system [taking as reference level a plane at distance $l_1 + l_2$ below the point of suspension of Fig. 11-3] is

$$V = m_1g[l_1 + l_2 - l_1\cos\theta_1] + m_2g[l_1 + l_2 - (l_1\cos\theta_1 + l_2\cos\theta_2)]$$

Then the Lagrangian is

$$L = T - V$$

$$= \tfrac{1}{2}m_1l_1^2\dot{\theta}_1^2 + \tfrac{1}{2}m_2[l_1^2\dot{\theta}_1^2 + l_2^2\dot{\theta}_2^2 + 2l_1l_2\dot{\theta}_1\dot{\theta}_2\cos(\theta_1 - \theta_2)] - m_1g[l_1 + l_2 - l_1\cos\theta_1] - m_2g[l_1 + l_2 - (l_1\cos\theta_1 + l_2\cos\theta_2)] \tag{1}$$

(*b*) The Lagrange equations associated with θ_1 and θ_2 are

$$\frac{d}{dt}\left(\frac{\partial L}{\partial\dot{\theta}_1}\right) - \frac{\partial L}{\partial\theta_1} = 0, \qquad \frac{d}{dt}\left(\frac{\partial L}{\partial\dot{\theta}_2}\right) - \frac{\partial L}{\partial\theta_2} = 0 \tag{2}$$

From (*1*) we find

$$\partial L/\partial \theta_1 = -m_2 l_1 l_2 \dot{\theta}_1 \dot{\theta}_2 \sin(\theta_1 - \theta_2) - m_1 g l_1 \sin\theta_1 - m_2 g l_1 \sin\theta_1$$

$$\partial L/\partial \dot{\theta}_1 = m l_1^2 \dot{\theta}_1 + m_2 l_1^2 \dot{\theta}_1 + m_2 l_1 l_2 \dot{\theta}_2 \cos(\theta_1 - \theta_2)$$

$$\partial L/\partial \theta_2 = m_2 l_1 l_2 \dot{\theta}_1 \dot{\theta}_2 \sin(\theta_1 - \theta_2) - m_2 g l_2 \sin\theta_2$$

$$\partial L/\partial \dot{\theta}_2 = m_2 l_2^2 \dot{\theta}_2 + m_2 l_1 l_2 \dot{\theta}_1 \cos(\theta_1 - \theta_2)$$

Thus equations (*2*) become

$$m_1 l_1^2 \ddot{\theta}_1 + m_2 l_1^2 \ddot{\theta}_1 + m_2 l_1 l_2 \ddot{\theta}_2 \cos(\theta_1 - \theta_2) - m_2 l_1 l_2 \dot{\theta}_2 (\dot{\theta}_1 - \dot{\theta}_2) \sin(\theta_1 - \theta_2)$$
$$= -m_2 l_1 l_2 \dot{\theta}_1 \dot{\theta}_2 \sin(\theta_1 - \theta_2) - m_1 g l_1 \sin\theta_1 - m_2 g l_1 \sin\theta_1$$

and

$$m_2 l_2^2 \ddot{\theta}_2 + m_2 l_1 l_2 \ddot{\theta}_1 \cos(\theta_1 - \theta_2) - m_2 l_1 l_2 \dot{\theta}_1 (\dot{\theta}_1 - \dot{\theta}_2) \sin(\theta_1 - \theta_2)$$
$$= m_2 l_1 l_2 \dot{\theta}_1 \dot{\theta}_2 \sin(\theta_1 - \theta_2) - m_2 g l_2 \sin\theta_2$$

which reduce respectively to

$$(m_1 + m_2) l_1^2 \ddot{\theta}_1 + m_2 l_1 l_2 \ddot{\theta}_2 \cos(\theta_1 - \theta_2) + m_2 l_1 l_2 \dot{\theta}_2^2 \sin(\theta_1 - \theta_2) = -(m_1 + m_2) g l_1 \sin\theta_1 \quad (3)$$

and

$$m_2 l_2^2 \ddot{\theta}_2 + m_2 l_1 l_2 \ddot{\theta}_1 \cos(\theta_1 - \theta_2) - m_2 l_1 l_2 \dot{\theta}_1^2 \sin(\theta_1 - \theta_2) = -m_2 g l_2 \sin\theta_2 \quad (4)$$

11.29. Write the equations of Problem 11.28 for the case $m_1 = m_2 = m$ and $l_1 = l_2 = l$.

Letting $m_1 = m_2$, $l_1 = l_2$ in equations (*3*) and (*4*) of Problem 11.28 and simplifying, they can be written

$$2l\ddot{\theta}_1 + l\ddot{\theta}_2 \cos(\theta_1 - \theta_2) + l\dot{\theta}_2^2 \sin(\theta_1 - \theta_2) = -2g \sin\theta_1 \quad (1)$$

$$l\ddot{\theta}_1 \cos(\theta_1 - \theta_2) + l\ddot{\theta}_2 - l\dot{\theta}_1^2 \sin(\theta_1 - \theta_2) = -g \sin\theta_2 \quad (2)$$

11.30. Obtain the equations of Problem 11.29 for the case where the oscillations are assumed to be small.

Using the approximations $\sin\theta = \theta$, $\cos\theta = 1$ and neglecting terms involving $\dot{\theta}^2\theta$, the equations (*1*) and (*2*) of Problem 11.29 become

$$2l\ddot{\theta}_1 + l\ddot{\theta}_2 = -2g\theta_1$$

$$l\ddot{\theta}_1 + l\ddot{\theta}_2 = -g\theta_2$$

11.31. Find the (*a*) normal frequencies and (*b*) normal modes corresponding to the small oscillations of the double pendulum.

(*a*) Let $\theta_1 = A_1 \cos\omega t$, $\theta_2 = A_2 \cos\omega t$ [or $A_1 e^{i\omega t}$, $A_2 e^{i\omega t}$] in the equations of Problem 11.30. Then they can be written

$$\left.\begin{aligned} 2(g - l\omega^2)A_1 - l\omega^2 A_2 &= 0 \\ -l\omega^2 A_1 + (g - l\omega^2)A_2 &= 0 \end{aligned}\right\} \quad (1)$$

In order for A_1 and A_2 to be different from zero, we must have the determinant of the coefficients equal to zero, i.e.,

$$\begin{vmatrix} 2(g - l\omega^2) & -l\omega^2 \\ -l\omega^2 & g - l\omega^2 \end{vmatrix} = 0$$

or $l^2\omega^4 - 4lg\omega^2 + 2g^2 = 0$. Solving, we find

$$\omega^2 = \frac{4lg \pm \sqrt{16l^2g^2 - 8l^2g^2}}{2l^2} = \frac{(2 \pm \sqrt{2})g}{l}$$

or

$$\omega_1^2 = \frac{(2 + \sqrt{2})g}{l}, \quad \omega_2^2 = \frac{(2 - \sqrt{2})g}{l} \quad (2)$$

Thus the normal frequencies are given by

$$f_1 = \frac{\omega_1}{2\pi} = \frac{1}{2\pi}\sqrt{\frac{(2+\sqrt{2})g}{l}} \quad \text{and} \quad f_2 = \frac{\omega_2}{2\pi} = \frac{1}{2\pi}\sqrt{\frac{(2-\sqrt{2})g}{l}} \tag{3}$$

(b) Substituting $\omega^2 = \omega_1^2 = (2+\sqrt{2})g/l$ in equations (1) of Part (a) yields

$$A_2 = -\sqrt{2}\,A_1 \tag{4}$$

This corresponds to the normal mode in which the bobs are moving in *opposite directions.*

Substituting $\omega^2 = \omega_2^2 = (2-\sqrt{2})g/l$ in equations (1) of Part (a) yields

$$A_2 = \sqrt{2}\,A_1 \tag{5}$$

This corresponds to the normal mode in which the bobs are moving in the *same directions.*

11.32. (a) Set up the Lagrangian for the motion of a symmetrical top [see Problem 10.25, page 268] and (b) obtain the equations of motion.

(a) The kinetic energy in terms of the Euler angles [see Problem 10.24, page 268] is

$$T = \tfrac{1}{2}(I_1\omega_1^2 + I_2\omega_2^2 + I_3\omega_3^2) = \tfrac{1}{2}I_1(\dot\phi^2\sin^2\theta + \dot\theta^2) + \tfrac{1}{2}I_3(\dot\phi\cos\theta + \dot\psi)^2 \tag{1}$$

The potential energy is $\qquad V = mgl\cos\theta \qquad (2)$

as seen from Fig. 10-18, page 269, since distance $OC = l$ and the height of the center of mass C above the xy plane is therefore $l\cos\theta$. Thus

$$L = T - V = \tfrac{1}{2}I_1(\dot\phi^2\sin^2\theta + \dot\theta^2) + \tfrac{1}{2}I_3(\dot\phi\cos\theta + \dot\psi)^2 - mgl\cos\theta \tag{3}$$

(b)

$$\begin{aligned}
\partial L/\partial\theta &= I_1\dot\phi^2\sin\theta\cos\theta + I_3(\dot\phi\cos\theta + \dot\psi)(-\dot\phi\sin\theta) + mgl\sin\theta \\
\partial L/\partial\dot\theta &= I_1\dot\theta \\
\partial L/\partial\phi &= 0 \\
\partial L/\partial\dot\phi &= I_1\dot\phi\sin^2\theta + I_3(\dot\phi\cos\theta + \dot\psi)\cos\theta \\
\partial L/\partial\psi &= 0 \\
\partial L/\partial\dot\psi &= I_3(\dot\phi\cos\theta + \dot\psi)
\end{aligned}$$

Then Lagrange's equations are

$$\frac{d}{dt}\left(\frac{\partial L}{\partial\dot\theta}\right) - \frac{\partial L}{\partial\theta} = 0, \quad \frac{d}{dt}\left(\frac{\partial L}{\partial\dot\phi}\right) - \frac{\partial L}{\partial\phi} = 0, \quad \frac{d}{dt}\left(\frac{\partial L}{\partial\dot\psi}\right) - \frac{\partial L}{\partial\psi} = 0$$

or

$$I_1\ddot\theta - I_1\dot\phi^2\sin\theta\cos\theta + I_3(\dot\phi\cos\theta + \dot\psi)\dot\phi\sin\theta - mgl\sin\theta = 0 \tag{4}$$

$$\frac{d}{dt}[I_1\dot\phi\sin^2\theta + I_3(\dot\phi\cos\theta + \dot\psi)\cos\theta] = 0 \tag{5}$$

$$\frac{d}{dt}[I_3(\dot\phi\cos\theta + \dot\psi)] = 0 \tag{6}$$

11.33. Use the results of Problem 11.32 to obtain agreement with the equation of (a) Problem 10.29(b), page 270, and (b) Problem 10.27(a), page 270.

(a) From equations (5) and (6) of Problem 11.32 we obtain on integrating,

$$I_1\dot\phi\sin^2\theta + I_3(\dot\phi\cos\theta + \dot\psi)\cos\theta = \text{constant} = K \tag{1}$$

$$\dot\phi\cos\theta + \dot\psi = A \tag{2}$$

Using (2) in (1), we find $\qquad I_1\dot\phi\sin^2\theta + I_3A\cos\theta = K$

(b) Using (2) in equation (4) of Problem 11.32, we find

$$I_1\ddot\theta - I_1\dot\phi^2\sin\theta\cos\theta + I_3A\dot\phi\sin\theta = mgl\sin\theta$$

11.34. Derive Euler's equations of motion for a rigid body by use of Lagrange's equations.

The kinetic energy in terms of the Euler angles is [see Problem 10.24, page 268]

$$\begin{aligned} T &= \tfrac{1}{2}(I_1\omega_1^2 + I_2\omega_2^2 + I_3\omega_3^2) \\ &= \tfrac{1}{2}I_1(\dot\phi \sin\theta \sin\psi + \dot\theta \cos\psi)^2 + \tfrac{1}{2}I_2(\dot\phi \sin\theta\cos\psi - \dot\theta\sin\psi)^2 + \tfrac{1}{2}I_3(\dot\phi\cos\theta + \dot\psi)^2 \end{aligned}$$

Then

$$\begin{aligned} \partial T/\partial\psi &= I_1(\dot\phi\sin\theta\sin\psi + \dot\theta\cos\psi)(\dot\phi\sin\theta\cos\psi - \dot\theta\sin\psi) \\ &\quad + I_2(\dot\phi\sin\theta\cos\psi - \dot\theta\sin\psi)(-\dot\phi\sin\theta\sin\psi - \dot\theta\cos\psi) \\ &= I_1\omega_1\omega_2 + I_2(\omega_2)(-\omega_1) = (I_1 - I_2)\omega_1\omega_2 \end{aligned}$$

$$\partial T/\partial\dot\psi = I_3(\dot\phi\cos\theta + \dot\psi) = I_3\omega_3$$

Then by Problem 11.10, page 288, Lagrange's equation corresponding to ψ is

$$\frac{d}{dt}\left(\frac{\partial T}{\partial\dot\psi}\right) - \frac{\partial T}{\partial\psi} = \Phi_\psi$$

or

$$I_3\dot\omega_3 + (I_2 - I_1)\omega_1\omega_2 = \Phi_\psi \tag{1}$$

This is Euler's third equation of (*22*), page 256. The quantity Φ_ψ represents the generalized force corresponding to a rotation ψ about an axis and physically represents the component Λ_3 of the torque about this axis [see Problem 11.102].

The remaining equations

$$I_1\dot\omega_1 + (I_3 - I_2)\omega_2\omega_3 = \Lambda_1 \tag{2}$$

$$I_2\dot\omega_2 + (I_1 - I_3)\omega_3\omega_1 = \Lambda_2 \tag{3}$$

can be obtained from symmetry considerations by permutation of the indices. They are not directly obtained by using the Lagrange equations corresponding to θ and ϕ but can indirectly be deduced from them [see Problem 11.79].

11.35. A bead slides without friction on a frictionless wire in the shape of a cycloid [Fig. 11-10] with equations

$$x = a(\theta - \sin\theta), \quad y = a(1 + \cos\theta)$$

where $0 \leqq \theta \leqq 2\pi$. Find (*a*) the Lagrangian function, (*b*) the equation of motion.

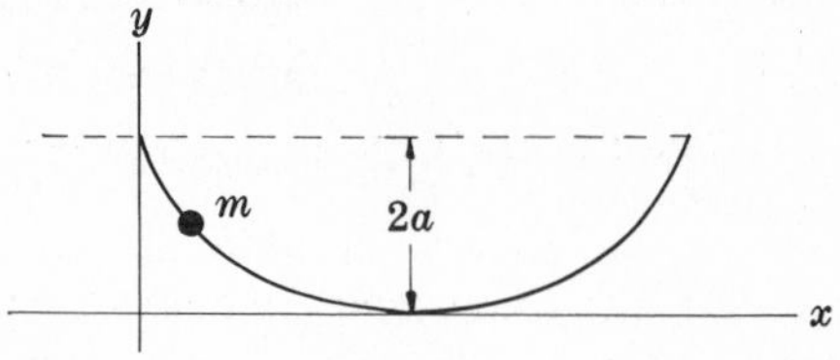

Fig. 11-10

(*a*) Kinetic energy

$$\begin{aligned} T &= \tfrac{1}{2}m(\dot x^2 + \dot y^2) \\ &= \tfrac{1}{2}ma^2\{[(1-\cos\theta)\dot\theta]^2 + [-\sin\theta\,\dot\theta]^2\} \\ &= ma^2(1-\cos\theta)\dot\theta^2 \end{aligned}$$

Potential energy $= V = mgy = mga(1 + \cos\theta)$

Then

Lagrangian $= L = T - V = ma^2(1-\cos\theta)\dot\theta^2 - mga(1+\cos\theta)$

(*b*) $$\frac{d}{dt}\left(\frac{\partial L}{\partial\dot\theta}\right) - \frac{\partial L}{\partial\theta} = 0, \quad \text{i.e.} \quad \frac{d}{dt}[2ma^2(1-\cos\theta)\dot\theta] - [ma^2\sin\theta\,\dot\theta^2 + mga\sin\theta] = 0$$

or

$$\frac{d}{dt}[(1-\cos\theta)\dot\theta] - \tfrac{1}{2}\sin\theta\,\dot\theta^2 - \frac{g}{2a}\sin\theta = 0$$

which can be written

$$(1-\cos\theta)\ddot\theta + \tfrac{1}{2}\sin\theta\,\dot\theta^2 - \frac{g}{2a}\sin\theta = 0$$

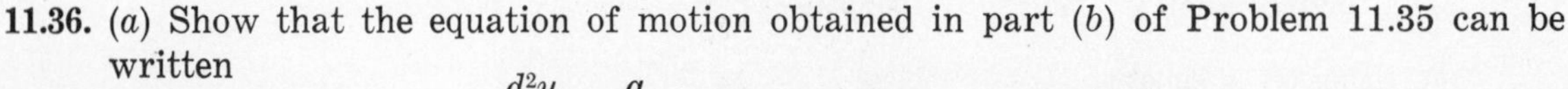

11.36. (*a*) Show that the equation of motion obtained in part (*b*) of Problem 11.35 can be written

$$\frac{d^2u}{dt^2} + \frac{g}{4a}u = 0 \quad \text{where } u = \cos(\theta/2)$$

and thus (*b*) show that the bead oscillates with period $2\pi\sqrt{4a/g}$.

(*a*) If $u = \cos(\theta/2)$, then

$$\frac{du}{dt} = -\tfrac{1}{2}\sin(\theta/2)\dot{\theta}, \qquad \frac{d^2u}{dt^2} = -\tfrac{1}{2}\sin(\theta/2)\ddot{\theta} - \tfrac{1}{4}\cos(\theta/2)\dot{\theta}^2$$

Thus $\dfrac{d^2u}{dt^2} + \dfrac{g}{4a}u = 0$ is the same as

$$-\tfrac{1}{2}\sin(\theta/2)\ddot{\theta} - \tfrac{1}{4}\cos(\theta/2)\dot{\theta}^2 + \frac{g}{4a}\cos(\theta/2) = 0$$

which can be written as

$$\ddot{\theta} + \tfrac{1}{2}\cot(\theta/2)\dot{\theta}^2 - \frac{g}{2a}\cot(\theta/2) = 0 \tag{1}$$

Since

$$\cot(\theta/2) = \frac{\cos(\theta/2)}{\sin(\theta/2)} = \frac{2\sin(\theta/2)\cos(\theta/2)}{2\sin^2(\theta/2)} = \frac{\sin\theta}{1-\cos\theta}$$

it follows that equation (*1*) is the same as that obtained in Problem 11.35(*b*).

(*b*) The solution of the equation is

$$u = \cos(\theta/2) = c_1\cos\sqrt{4a/g}\,t + c_2\sin\sqrt{4a/g}\,t$$

from which we see that $\cos(\theta/2)$ returns to its original value after a time $2\pi\sqrt{4a/g}$ which is the required period. Note that this period is the same as that of a simple pendulum with length $l = 4a$.

An application of this is the *cycloidal pendulum*. See Problem 4.86, page 112.

11.37. Obtain equations for the rolling sphere of Problem 9.42, page 244 by use of Lagrange's equations.

Refer to Fig. 9-33 in which ϕ and ψ represent generalized coordinates. Since the sphere of radius $CP = a$ rolls without slipping on the sphere of radius $OP = b$, we have

$$b\,d\phi/dt = a\,d\psi/dt \quad \text{or} \quad b\dot{\phi} = a\dot{\psi}$$

which shows that if $\phi = 0$ when $\psi = 0$, then

$$b\phi = a\psi \tag{1}$$

Thus ϕ and ψ [and therefore $d\phi$ and $d\psi$ or $\delta\phi$ and $\delta\psi$] are not independent.

The kinetic energy of the rolling sphere is

$$T = \tfrac{1}{2}m(a+b)^2\dot{\phi}^2 + \tfrac{1}{2}I\omega^2$$
$$= \tfrac{1}{2}m(a+b)^2\dot{\phi}^2 + \tfrac{1}{2}(\tfrac{2}{5}ma^2)(\dot{\phi}+\dot{\psi})^2$$

using the fact that $I = \frac{2}{5}ma^2$ is the moment of inertia of the sphere about a horizontal axis through its center of mass.

The potential energy of the rolling sphere [taking the horizontal plane through O as reference level] is

$$V = mg(a+b)\cos\phi$$

Thus the Lagrangian is

$$L = T - V = \tfrac{1}{2}m(a+b)^2\dot{\phi}^2 + \tfrac{1}{5}ma^2(\dot{\phi}+\dot{\psi})^2 - mg(a+b)\cos\phi \tag{2}$$

We use Lagrange's equations (*16*), page 284, for non-holonomic systems. From (*1*) we have

$$b\,\delta\phi - a\,\delta\psi = 0 \tag{3}$$

so that if we call $q_1 = \phi$ and $q_2 = \psi$ and compare with equation (*7*) of Problem 11.18, page 292, we find

$$A_1 = b, \quad A_2 = -a \tag{4}$$

Thus equations (*16*), page 284, become

$$\frac{d}{dt}\left(\frac{\partial L}{\partial\dot{\phi}}\right) - \frac{\partial L}{\partial\phi} = \lambda_1 b \tag{5}$$

$$\frac{d}{dt}\left(\frac{\partial L}{\partial\dot{\psi}}\right) - \frac{\partial L}{\partial\psi} = -\lambda_1 a \tag{6}$$

Substitution of (2) into (5) and (6) yields

$$m(a+b)^2\ddot{\phi} + \tfrac{2}{5}ma^2(\ddot{\phi}+\ddot{\psi}) - mg(a+b)\sin\phi = \lambda_1 b \tag{7}$$

$$\tfrac{2}{5}ma^2(\ddot{\phi}+\ddot{\psi}) = -\lambda_1 a \tag{8}$$

Substituting $\psi = (b/a)\phi$ [from (1)] into (7) and (8), we find

$$m(a+b)^2\ddot{\phi} + \tfrac{2}{5}ma^2(1+b/a)\ddot{\phi} - mg(a+b)\sin\phi = \lambda_1 b \tag{9}$$

$$\tfrac{2}{5}ma^2(1+b/a)\ddot{\phi} = -\lambda_1 a \tag{10}$$

Now from (10) we have $\qquad \lambda_1 = -\tfrac{2}{5}m(a+b)\ddot{\phi}$

and using this in (9) it becomes after simplifying and solving for $\ddot{\phi}$,

$$\ddot{\phi} = \frac{5g}{7(a+b)}\sin\phi$$

This is the same equation as that of (2) in Problem 9.42, page 244, with $\phi = \pi/2 - \theta$. To find the required angle at which the sphere falls off, see Problem 11.104.

11.38. (a) Solve the equations of motion obtained in Problem 11.24, page 296, and (b) interpret physically.

(a) From the first of equations (12) in Problem 11.24 we have

$$x = \text{constant} = 0 \tag{1}$$

since $x = 0$ at $t = 0$. Similarly, from the second of equations (12) we have

$$y = \frac{\mathcal{J}}{4m}t \tag{2}$$

since $y = 0$ at $t = 0$.

From the third of equations (12) we find on separating the variables,

$$\sec\theta_1\,d\theta_1 = -\frac{3\mathcal{J}}{8ml}\,dt$$

or on integrating, $\qquad \ln\cot\left(\frac{\pi}{4} - \frac{\theta_1}{2}\right) = -\frac{3\mathcal{J}t}{8ml} + c_1$

i.e., $\qquad \tan\left(\frac{\pi}{4} - \frac{\theta_1}{2}\right) = c_2 e^{3\mathcal{J}t/8ml}$

Thus since $\theta_1 = \pi/2$ at $t = 0$, we have $c_2 = 0$. This means that for all time we must have $\theta_1 = \pi/2$.

From the fourth of equations (12) in Problem 11.24 we have similarly,

$$\sec\theta_2\,d\theta_2 = \frac{3\mathcal{J}}{8ml}\,dt$$

or on integrating, $\qquad \ln\cot\left(\frac{\pi}{4} - \frac{\theta_2}{2}\right) = \frac{3\mathcal{J}t}{8ml} + c_3$

i.e., $\qquad \tan\left(\frac{\pi}{4} - \frac{\theta_2}{2}\right) = c_4 e^{-3\mathcal{J}t/8ml}$

Now when $t = 0$, $\theta_2 = 0$ so that $c_4 = 1$. Then

$$\tan\left(\frac{\pi}{4} - \frac{\theta_2}{2}\right) = e^{-3\mathcal{J}t/8ml} \quad \text{or} \quad \theta_2 = \frac{\pi}{2} - 2\tan^{-1}(e^{-3\mathcal{J}t/8ml})$$

(b) Equations (1) and (2) show that the center moves along the y axis with constant speed $\mathcal{J}/4m$. The rods AD and BC are always parallel to the y axis while rods AB and CD slowly rotate until finally $[t \to \infty]$ the rhombus collapses, so that all four rods will be on the y axis.

Supplementary Problems

GENERALIZED COORDINATES AND TRANSFORMATION EQUATIONS

11.39. Give a set of generalized coordinates needed to completely specify the motion of each of the following: (*a*) a bead constrained to move on a circular wire; (*b*) a particle constrained to move on a sphere; (*c*) a compound pendulum [see page 228]; (*d*) an Atwood's machine [see Problem 3.22, page 76]; (*e*) a circular disk rolling on a horizontal plane; (*f*) a cone rolling on a horizontal plane.

11.40. Write transformation equations for the motion of a triple pendulum in terms of a suitable set of generalized coordinates.

11.41. A particle moves on the upper surface of a frictionless paraboloid of revolution whose equation is $x^2 + y^2 = cz$. Write transformation equations for the motion of the particle in terms of a suitable set of generalized coordinates.

11.42. Write transformation equations for the motion of a particle constrained to move on a sphere.

CLASSIFICATION OF MECHANICAL SYSTEMS

11.43. Classify each of the following according as they are (i) scleronomic or rheonomic, (ii) holonomic or non-holonomic, and (iii) conservative or non-conservative:

(*a*) a horizontal cylinder of radius a rolling inside a perfectly rough hollow horizontal cylinder of radius $b > a$;

(*b*) a cylinder rolling [and possibly sliding] down an inclined plane of angle α;

(*c*) a sphere rolling down another sphere which is rolling with uniform speed along a horizontal plane;

(*d*) a particle constrained to move along a line under the influence of a force which is inversely proportional to the square of its distance from a fixed point and a damping force proportional to the square of the instantaneous speed.

Ans. (*a*) scleronomic, holonomic, conservative
(*b*) scleronomic, non-holonomic, conservative
(*c*) rheonomic, non-holonomic, conservative
(*d*) scleronomic, holonomic, non-conservative

WORK, KINETIC ENERGY AND GENERALIZED FORCES

11.44. Prove that if the transformation equations are given by $\mathbf{r}_\nu = \mathbf{r}_\nu(q_1, q_2, \ldots, q_n)$, i.e. do not involve the time t explicitly, then the kinetic energy can be written as

$$T = \sum_{\alpha=1}^{n} \sum_{\beta=1}^{n} a_{\alpha\beta}\, \dot{q}_\alpha \dot{q}_\beta$$

where $a_{\alpha\beta}$ are functions of the q_α.

11.45. Discuss Problem 11.44 in case the transformation equations depend explicitly on the time t.

11.46. If $F(\lambda x, \lambda y, \lambda z) = \lambda^n F(x, y, z)$ where λ is a parameter, then F is said to be a *homogeneous function of order n*. Determine which (if any) of the following functions are homogeneous, giving the order in each case:

(*a*) $x^2 + y^2 + z^2 + xy + yz + xz$
(*b*) $3x - 2y + 4z$
(*c*) $xyz + 2xy + 2xz + 2yz$
(*d*) $(x + y + z)/x$
(*e*) $x^3 \tan^{-1}(y/x)$
(*f*) $4 \sin xy$
(*g*) $(x + y + z)/(x^2 + y^2 + z^2)$

Ans. (*a*) homogeneous of order 2, (*b*) homogeneous of order 1, (*c*) non-homogeneous, (*d*) homogeneous of order zero, (*e*) homogeneous or order 3, (*f*) non-homogeneous, (*g*) homogeneous of order -1.

11.47. If $F(x, y, z)$ is homogeneous of order n [see Problem 11.46], prove that

$$x\frac{\partial F}{\partial x} + y\frac{\partial F}{\partial y} + z\frac{\partial F}{\partial z} = nF$$

This is called *Euler's theorem on homogeneous functions.*

[*Hint.* Differentiate both sides of the identity $F(\lambda x, \lambda y, \lambda z) = \lambda^n F(x, y, z)$ with respect to λ and then place $\lambda = 1$.]

11.48. Generalize the result of Problem 11.47.

11.49. Prove that if the transformation equations do not depend explicitly on time t, and T is the kinetic energy, then

$$\dot{q}_1 \frac{\partial T}{\partial \dot{q}_1} + \dot{q}_2 \frac{\partial T}{\partial \dot{q}_2} + \cdots + \dot{q}_n \frac{\partial T}{\partial \dot{q}_n} = 2T$$

Can you prove this directly without the use of Euler's theorem on homogeneous functions [Problem 11.47]?

LAGRANGE'S EQUATIONS

11.50. (a) Set up the Lagrangian for a one dimensional harmonic oscillator and (b) write Lagrange's equations. *Ans.* (a) $L = \frac{1}{2}m\dot{x}^2 - \frac{1}{2}\kappa x^2$, (b) $m\ddot{x} + \kappa x = 0$

11.51. (a) Set up the Lagrangian for a particle of mass m falling freely in a uniform gravitational field and (b) write Lagrange's equations.

11.52. Work Problem 11.51 in case the gravitational force field varies inversely as the square of the distance from a fixed point O assuming that the particle moves in a straight line through O.

11.53. Use Lagrange's equations to describe the motion of a particle of mass m down a frictionless inclined plane of angle α.

11.54. Use Lagrange's equations to describe the motion of a projectile launched with speed v_0 at angle α with the horizontal.

11.55. Use Lagrange's equations to solve the problem of the (a) two-dimensional and (b) three-dimensional harmonic oscillator.

11.56. A particle of mass m is connected to a fixed point P on a horizontal plane by a string of length l. The plane rotates with constant angular speed ω about a vertical axis through a point O of the plane, where $OP = a$. (a) Set up the Lagrangian of the system. (b) Write the equations of motion of the particle.

11.57. The rectangular coordinates (x, y, z) defining the position of a particle of mass m moving in a force field having potential V are given in terms of spherical coordinates (r, θ, ϕ) by the transformation equations

$$x = r \sin\phi \cos\theta, \quad y = r \sin\phi \sin\theta, \quad z = r\cos\phi$$

Use Lagrange's equations to set up the equations of motion.

Ans. $$m[\ddot{r} - r\dot{\phi}^2 - r\dot{\theta}^2 \cos^2\phi] = -\frac{\partial V}{\partial r}$$

$$m\left[\frac{d}{dt}(r^2\dot{\phi}) + r^2\dot{\theta}^2 \sin\phi\cos\phi\right] = -\frac{1}{r}\frac{\partial V}{\partial \theta}$$

$$m\frac{d}{dt}(r^2\dot{\theta}\sin^2\phi) = -\frac{1}{r\sin\theta}\frac{\partial V}{\partial \phi}$$

11.58. Work Problem 11.56 if the particle does not necessarily move in a straight line through O.

11.59. Work Problem 4.23, page 102, by use of Lagrange's equations.

LAGRANGE'S EQUATIONS FOR NON-HOLONOMIC SYSTEMS

11.60. (a) Work Problem 11.20, page 293, if the paraboloid is replaced by the cone $x^2 + y^2 = c^2z^2$. (b) What modification must be made to Problem 11.21, page 294, in this case?

11.61. Use the method of Lagrange's equations for non-holonomic systems to solve the problem of a particle of mass m sliding down a frictionless inclined plane of angle α.

11.62. Work Problem 3.74, page 82 by using the method of Lagrange's equations for non-holonomic systems.

LAGRANGE'S EQUATIONS WITH IMPULSIVE FORCES

11.63. A uniform rod of length l and mass M is at rest on a horizontal frictionless table. An impulse of magnitude $\mathcal{J}$ is applied to one end A of the rod and perpendicular to it. Prove that (a) the velocity given to end A is $4\mathcal{J}/M$, (b) the velocity of the center of mass is $\mathcal{J}/M$ and (c) the rod rotates about the center of mass with angular velocity of magnitude $6\mathcal{J}/Ml$.

11.64. In Fig. 11-11, AB and BC represent two uniform rods having the same length l and mass M smoothly hinged at B and at rest on a horizontal frictionless plane. An impulse is applied at C normal to BC in the direction indicated in Fig. 11-11 so that the initial velocity of point C is $\mathbf{v}_0$. Find (a) the initial velocities of points A and B and (b) the magnitudes of the initial angular velocities of AB and BC about their centers of mass.

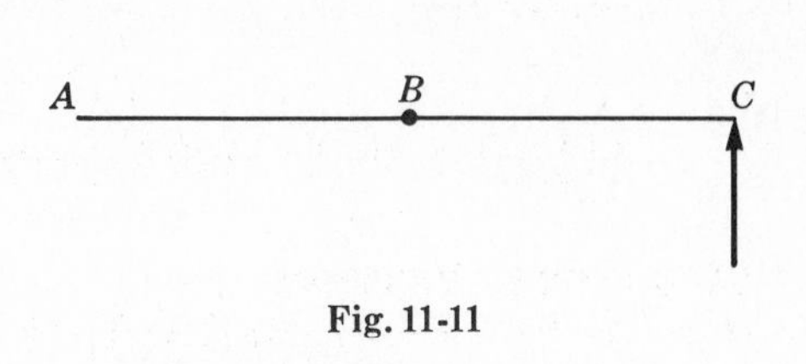

Fig. 11-11

Ans. (a) $\mathbf{v}_0/7,\ -2\mathbf{v}_0/7$; (b) $3v_0/7l,\ -9v_0/7l$

11.65. Prove that the total kinetic energy developed by the system of Problem 11.64 after the impulse is $\frac{1}{7}Mv_0^2$.

11.66. A square of side a and mass M, formed from 4 uniform rods which are smoothly hinged at their edges, rests on a horizontal frictionless plane. An impulse is applied at a vertex in a direction of the diagonal through the vertex so that the vertex is given a velocity of magnitude v_0. Prove that the rods move about their centers of mass with angular speed $3v_0/4a$.

11.67. (a) If $\mathcal{J}$ is the magnitude of the impulse applied to the vertex in Problem 11.66, prove that the kinetic energy developed by the rods is given by $5\mathcal{J}^2/4M$. (b) What is this kinetic energy in terms of v_0? (c) Does the direction of the impulse make any difference? Explain.

11.68. In Problem 11.24, page 296, suppose that the impulse is applied at the center of one of the rods in a direction which is perpendicular to the rod. Prove that the kinetic energy developed is $\mathcal{J}^2/8m$.

MISCELLANEOUS PROBLEMS

11.69. A particle of mass m moves on the inside of a smooth hollow hemisphere of radius a having its vertex on a horizontal plane. With what horizontal speed must it be projected so that it will remain in a horizontal circle at height h above the vertex?

11.70. A particle of mass m is constrained to move inside a thin hollow frictionless tube [see Fig. 11-12] which is rotating with constant angular velocity ω in a horizontal xy plane about a fixed vertical axis through O. Using Lagrange's equations, describe the motion.

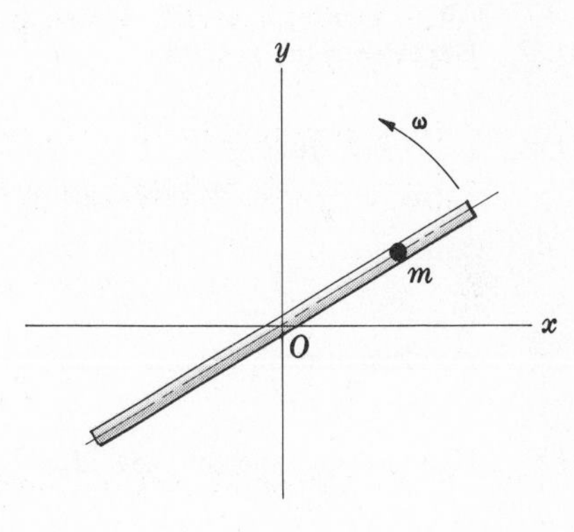

Fig. 11-12

11.71. Work Problem 11.70 if the xy plane is vertical.

11.72. A particle of mass m moves in a central force field having potential $V(r)$ where r is the distance from the force center. Using spherical coordinates, (a) set up the Lagrangian and (b) determine the equations of motion. Can you deduce from these equations that the motion takes place in a plane [compare Problem 5.1, page 121]?

11.73. A particle moves on a frictionless horizontal wire of radius a, acted upon by a resisting force which is proportional to the instantaneous speed. If the particle is given an initial speed v_0, find the position of the particle at any time t.

Ans. $\theta = (mv_0/\kappa)(1 - e^{-\kappa t/ma})$ where θ is the angle which a radius drawn to m makes with a fixed radius such that $\theta = 0$ at $t = 0$, and κ is the constant of proportionality.

11.74. Work Problem 11.73 if the resisting force is proportional to the square of the instantaneous speed.

Ans. $\theta = \dfrac{m}{\kappa a}\ln\left(\dfrac{m + \kappa v_0 t}{m}\right)$

11.75. A spherical pendulum is fixed at point O but is otherwise free to move in any direction. Write equations for its motion.

11.76. Work Problem 9.29, page 239, by use of Lagrange's equations.

11.77. Work Problem 11.20 if the paraboloid of revolution is replaced by the elliptic paraboloid $az = bx^2 + cy^2$ where a, b, c are positive constants.

11.78. Prove that the generalized force corresponding to the angle of rotation about an axis physically represents the component of the torque about this axis.

11.79. (*a*) Obtain Lagrange's equations corresponding to θ and ϕ in Problem 11.34, page 302, and show that these are not the same as equations (*2*) and (*3*) of that problem. (*b*) Show how to obtain equations (*2*) and (*3*) of Problem 11.34 from the Lagrange equations of (*a*).

11.80. Two circular disks, of radius of gyrations K_1, K_2 and masses m_1, m_2 respectively, are suspended vertically on a wire of negligible mass [see Fig. 11-13]. They are set into motion by twisting one or both of the disks in their planes and then releasing. Let θ_1 and θ_2 be the angles made with some specified direction.

(*a*) Prove that the kinetic energy is

$$T = \tfrac{1}{2}(m_1 K_1^2 \dot{\theta}_1^2 + m_2 K_2^2 \dot{\theta}_2^2)$$

(*b*) Prove that the potential energy is

$$V = \tfrac{1}{2}[\tau_1 \theta_1^2 + \tau_2(\theta_2 - \theta_1)^2]$$

where τ_1 and τ_2 are torsion constants, i.e. the torques required to rotate the disks through one radian.

(*c*) Set up Lagrange's equations for the motion.

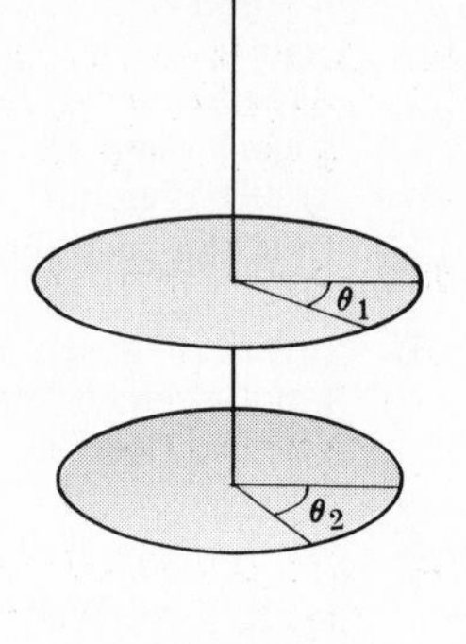

Fig. 11-13

11.81. Solve the vibrating system of Problem 11.80, finding (*a*) the normal frequencies and (*b*) the normal modes of vibration.

11.82. Generalize the results of Problem 11.80 and 11.81 to 3 or more disks.

11.83. (*a*) Prove that if $m_1 \neq m_2$ and $l_1 \neq l_2$ in the double pendulum of Problem 11.28, then the normal frequencies for small oscillations are given by $\omega/2\pi$ where

$$\omega^2 = \frac{(m_1+m_2)(l_1+l_2) \pm \sqrt{(m_1+m_2)[m_1(l_1-l_2)^2 + m_2(l_1+l_2)^2}}{2l_1 l_2 m_1} g$$

(*b*) Discuss the normal modes corresponding to the frequencies in (*a*).

11.84. Examine the special case $l_1 = l_2$, $m_1 \neq m_2$ in Problem 11.83.

11.85. Use Lagrange's equations to describe the motion of a sphere of radius a rolling on the inner surface of a smooth hollow hemisphere of radius $b > a$.

11.86. A particle on the inside surface of a frictionless paraboloid of revolution $az = x^2 + y^2$ at a height H_1 above the vertex is given a horizontal velocity v_0. Find the value of v_0 in order that the particle oscillate between the planes $z = H_1$ and $z = H_2$. *Ans.* $v_0 = \sqrt{2gH_2}$

11.87. Find the period of the oscillation in Problem 11.86.

11.88. A sphere of radius a is given an initial velocity v_0 up a frictionless inclined plane of angle α in a direction which is not along the line of greatest slope. Prove that its center describes a parabola.

11.89. A bead of mass m is constrained to move on a frictionless horizontal circular wire of radius a which is rotating at constant angular speed ω about a fixed vertical axis passing through a point on the wire. Prove that relative to the wire the bead oscillates like a simple pendulum.

11.90. If a particle of mass m and charge e moves with velocity $\mathbf{v}$ in an electric field $\mathbf{E}$ and magnetic field $\mathbf{B}$, the force acting on it is given by

$$\mathbf{F} = e(\mathbf{E} + \mathbf{v} \times \mathbf{B})$$

In terms of a scalar potential Φ and a vector potential $\mathbf{A}$ the fields can be expressed by the relations

$$\mathbf{E} = -\nabla\Phi - \partial\mathbf{A}/\partial t, \qquad \mathbf{B} = \nabla \times \mathbf{A}$$

Prove that the Lagrangian defining the motion of such a particle is

$$L = \tfrac{1}{2}mv^2 + e(\mathbf{A} \cdot \mathbf{v}) - e\Phi$$

11.91. Work Problem 10.86, page 278, by use of Lagrange's equations.

11.92. A uniform rod of length l and mass M has its ends constrained to move on the circumference of a smooth vertical circular wire of radius $a > l/2$ which rotates about a vertical diameter with constant angular speed ω. Obtain equations for the motion of the rod.

11.93. Suppose that the potential V depends on $\dot{q}_\nu$ as well as q_ν. Prove that the quantity

$$T + V - \sum \dot{q}_\nu \frac{\partial V}{\partial \dot{q}_\nu}$$

is a constant.

11.94. Use Lagrange's equations to set up and solve the two body problem as discussed in Chapter 5 [see for example page 121.]

11.95. Find the acceleration of the 5 gm mass in the pulley system of Fig. 11-14. *Ans.* $71g/622$

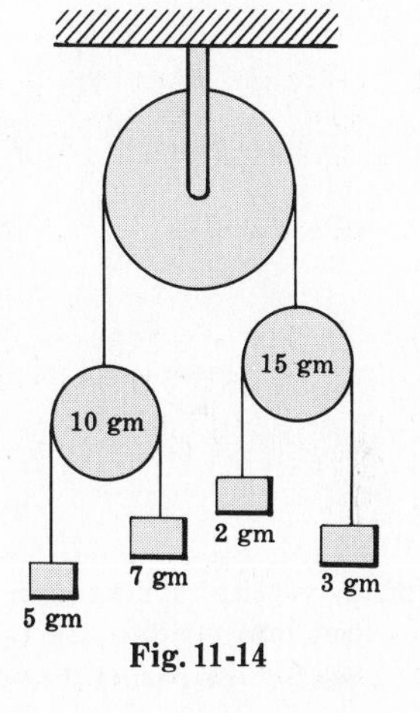

Fig. 11-14

11.96. A circular cylinder of radius a having radius of gyration K with respect to its center, moves down an inclined plane of angle α. If the coefficient of friction is μ, use Lagrange's equations to prove that the cylinder will roll without slipping if $\mu < \dfrac{K^2}{a^2 + K^2}\tan\alpha$. Discuss the cases where μ does not satisfy this inequality.

11.97. Use Lagrange's equations to solve Problem 8.27, page 213.

11.98. Describe the motion of the rods of Problem 11.64 at any time t after the impulse has been applied.

11.99. In Fig. 11-15, AB represents a frictionless horizontal plane having a small opening at O. A string of length l which passes through O has at its ends a particle P of mass m and a particle Q of equal mass which hangs freely. The particle P is given an initial velocity of magnitude v_0 at right angles to string OP when the length $OP = a$. Let r be the instantaneous distance OP while θ is the angle between OP and some fixed line through O.

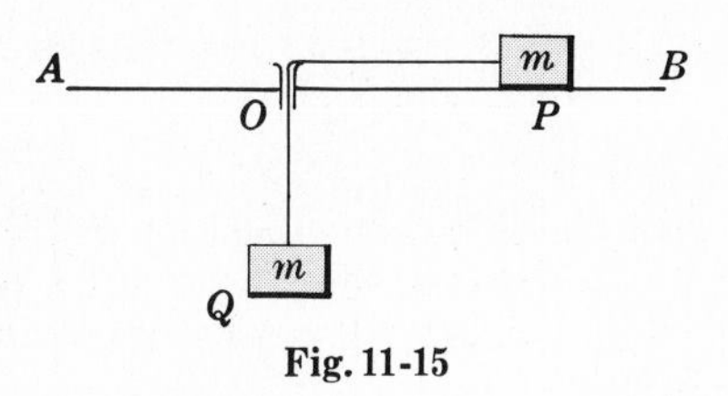

Fig. 11-15

(*a*) Set up the Lagrangian of the system.

(*b*) Write a differential equation for the motion of P in terms of r.

(*c*) Find the speed of P at any position.

Ans. (*a*) $L = \frac{1}{2}m[2\dot{r}^2 + r^2\dot{\theta}^2] + mg(l - r)$

(*b*) $\ddot{r} = a^2v_0^2/r^2 - g$

(*c*) $\dot{r} = \sqrt{2av_0^2 + 2g(a - r) - 2a^2v_0^2/r}$

11.100. Work Problem 11.99 if the masses of particles P and Q are m_1 and m_2 respectively.

11.101. Prove that if $v_0 = \sqrt{ag}$ the particle P of Problem 11.99 remains in stable equilibrium in the circle $r = a$ and that if it is slightly displaced from this equilibrium position it oscillates about this position with simple harmonic motion of period $2\pi\sqrt{2a/3g}$.

11.102. Prove that the quantity Φ_ψ in Problem 11.34, page 302, physically represents the component Λ_3 of the torque.

11.103. Describe the motion of the system of (*a*) Problem 11.63 and (*b*) Problem 11.66 at any time t after the impulse has been applied.

11.104. Show how to find the angle at which the sphere of Problem 11.37, page 303, falls off.

11.105. (*a*) Set up the Lagrangian for the triple pendulum of Fig. 11-16.
(*b*) Find the equations of motion.

11.106. Obtain the normal frequencies and normal modes for the triple pendulum of Problem 11.105 assuming small oscillations.

11.107. Work Problems 11.105 and 11.106 for the case where the masses and lengths are unequal.

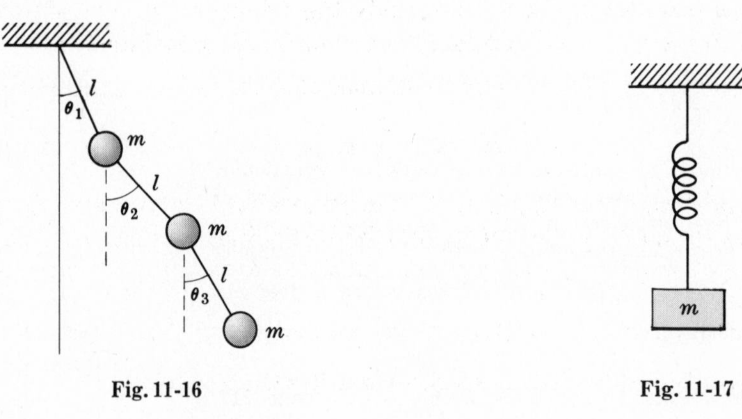

Fig. 11-16

Fig. 11-17

11.108. A vertical spring [Fig. 11-17] has constant κ and mass M. If a mass m is placed on the spring and set into motion, use Lagrange's equations to prove that the system will move with simple harmonic motion of period $2\pi\sqrt{(M + 3m)/3\kappa}$.

Chapter 12 HAMILTONIAN THEORY

HAMILTONIAN METHODS

In Chapter 11 we investigated a formulation of mechanics due to Lagrange. In this chapter we investigate a formulation due to Hamilton known collectively as *Hamiltonian methods* or *Hamiltonian theory*. Although such theory can be used to solve specific problems in mechanics, it develops that it is more useful in supplying fundamental postulates in such fields as quantum mechanics, statistical mechanics and celestial mechanics.

THE HAMILTONIAN

Just as the *Lagrangian function,* or briefly the *Lagrangian,* is fundamental to Chapter 11, so the *Hamiltonian function,* or briefly the *Hamiltonian,* is fundamental to this chapter.

The Hamiltonian, symbolized by H, is defined in terms of the Lagrangian L as

$$H = \sum_{\alpha=1}^{n} p_\alpha \dot{q}_\alpha - L \tag{1}$$

and must be expressed as a function of the generalized coordinates q_α and generalized momenta p_α. To accomplish this the generalized velocities $\dot{q}_\alpha$ must be eliminated from (*1*) by using Lagrange's equations [see Problem 12.3, for example]. In such case the function H can be written

$$H(p_1, \ldots, p_n, q_1, \ldots, q_n, t) \tag{2}$$

or briefly $H(p_\alpha, q_\alpha, t)$, and is also called the *Hamiltonian of the system.*

HAMILTON'S EQUATIONS

In terms of the Hamiltonian, the equations of motion of the system can be written in the symmetrical form

$$\left.\begin{aligned} \dot{p}_\alpha &= -\frac{\partial H}{\partial q_\alpha} \\ \dot{q}_\alpha &= \frac{\partial H}{\partial p_\alpha} \end{aligned}\right\} \tag{3}$$

These are called *Hamilton's canonical equations,* or briefly *Hamilton's equations.* The equations serve to indicate that the p_α and q_α play similar roles in a general formulation of mechanical principles.

THE HAMILTONIAN FOR CONSERVATIVE SYSTEMS

If a system is conservative, the Hamiltonian H can be interpreted as the total energy (kinetic and potential) of the system, i.e.,

$$H = T + V \tag{4}$$

Often this provides an easy way for setting up the Hamiltonian of a system.

IGNORABLE OR CYCLIC COORDINATES

A coordinate q_α which does not appear explicitly in the Lagrangian is called an *ignorable* or *cyclic coordinate*. In such case

$$\dot{p}_\alpha = \frac{\partial L}{\partial q_\alpha} = 0 \tag{5}$$

so that p_α is a constant, often called a *constant of the motion.*

In such case we also have $\partial H/\partial q_\alpha = 0$.

PHASE SPACE

The Hamiltonian formulation provides an obvious symmetry between the p_α and q_α which we call *momentum* and *position coordinates* respectively. It is often useful to imagine a space of $2n$ dimensions in which a *representative point* is indicated by the $2n$ coordinates

$$(p_1, \ldots, p_n, q_1, \ldots, q_n) \tag{6}$$

Such a space is called a $2n$ dimensional *phase space* or a pq phase space.

Whenever we know the state of a mechanical system at time t, i.e. we know all position and momentum coordinates, then this corresponds to a particular point in phase space. Conversely, a point in phase space specifies the state of the mechanical system. While the mechanical system moves in the physical 3 dimensional space, the representative point describes some path in the phase space in accordance with equations (*3*).

LIOUVILLE'S THEOREM

Let us consider a very large collection of conservative mechanical systems having the same Hamiltonian. In such case the Hamiltonian is the total energy and is constant, i.e.,

$$H(p_1, \ldots, p_n, q_1, \ldots, q_n) = \text{constant} = E \tag{7}$$

which can be represented by a surface in phase space.

Let us suppose that the total energies of all these systems lie between E_1 and E_2. Then the paths of all these systems in phase space will lie between the two surfaces $H = E_1$ and $H = E_2$ as indicated schematically in Fig. 12-1.

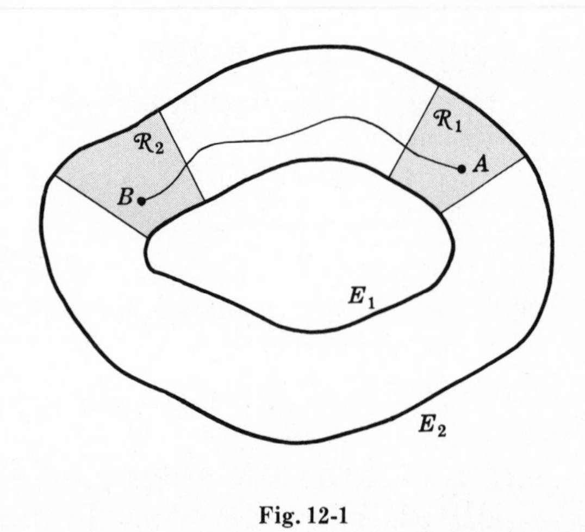

Fig. 12-1

Since the systems have different initial conditions, they will move along different paths in the phase space. Let us imagine that the initial points are contained in region $\mathcal{R}_1$ of Fig. 12-1 and that after time t these points occupy region $\mathcal{R}_2$. For example, the representative point corresponding to one particular system moves from point A to point B. From the choice of $\mathcal{R}_1$ and $\mathcal{R}_2$ it is clear that the number of representative points in them are the same. What is not so obvious is the following theorem called *Liouville's theorem.*

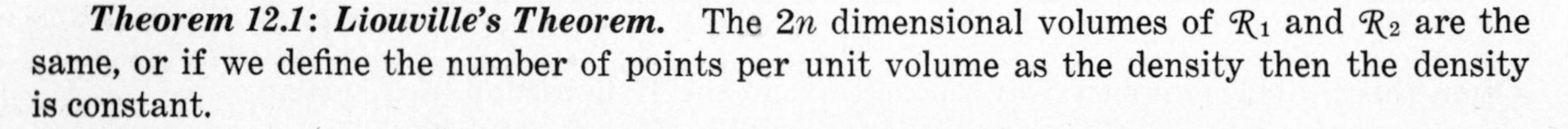

Theorem 12.1*: *Liouville's Theorem. The $2n$ dimensional volumes of $\mathcal{R}_1$ and $\mathcal{R}_2$ are the same, or if we define the number of points per unit volume as the density then the density is constant.

We can think of the points of $\mathcal{R}_1$ as particles of an incompressible fluid which move from $\mathcal{R}_1$ to $\mathcal{R}_2$ in time t.

THE CALCULUS OF VARIATIONS

A problem which often arises in mathematics is that of finding a curve $y = Y(x)$ joining the points where $x = a$ and $x = b$ such that the integral

$$\int_a^b F(x, y, y')\, dx \tag{8}$$

where $y' = dy/dx$, is a maximum or minimum, also called an *extremum* or *extreme value*. The curve itself is often called an *extremal*. It can be shown [see Problem 12.6] that a necessary condition for (*8*) to have an extremum is

$$\frac{d}{dx}\left(\frac{\partial F}{\partial y'}\right) - \frac{\partial F}{\partial y} = 0 \tag{9}$$

which is often called *Euler's equation*. This and similar problems are considered in a branch of mathematics called the *calculus of variations*.

HAMILTON'S PRINCIPLE

The obvious similarity of (*9*) to Lagrange's equations leads one to consider the problem of determining the extremals of

$$\int_{t_1}^{t_2} L(q_1, \ldots, q_n, \dot{q}_1, \ldots, \dot{q}_n, t)\, dt \tag{10}$$

or briefly,

$$\int_{t_1}^{t_2} L\, dt$$

where $L = T - V$ is the Lagrangian of a system.

We can show that a necessary condition for an extremal is

$$\frac{d}{dt}\left(\frac{\partial L}{\partial \dot{q}_\alpha}\right) - \frac{\partial L}{\partial q_\alpha} = 0 \tag{11}$$

which are precisely Lagrange's equations. The result led Hamilton to formulate a general variational principle known as

Hamilton's Principle. A conservative mechanical system moves from time t_1 to time t_2 in such a way that

$$\int_{t_1}^{t_2} L\, dt \tag{12}$$

sometimes called the *action integral*, has an extreme value.

Because the extreme value of (*12*) is often a minimum, the principle is sometimes referred to as *Hamilton's principle of least action*.

The fact that the integral (*12*) is an extremum is often symbolized by stating that

$$\delta \int_{t_1}^{t_2} L\, dt = 0 \tag{13}$$

where δ is the variation symbol.

CANONICAL OR CONTACT TRANSFORMATIONS

The ease in solution of many problems in mechanics often hinges on the particular generalized coordinates used. Consequently it is desirable to examine transformations from one set of position and momentum coordinates to another. For example if we call q_α and p_α the old position and momentum coordinates while Q_α and P_α are the new position and momentum coordinates, the transformation is

$$P_\alpha = P_\alpha(p_1, \ldots, p_n, q_1, \ldots, q_n, t), \qquad Q_\alpha = Q_\alpha(p_1, \ldots, p_n, q_1, \ldots, q_n, t) \tag{14}$$

denoted briefly by

$$P_\alpha = P_\alpha(p_\alpha, q_\alpha, t), \qquad Q_\alpha = Q_\alpha(p_\alpha, q_\alpha, t) \tag{15}$$

We restrict ourselves to transformations called *canonical* or *contact transformations* for which there exists a function $\mathcal{H}$ called the Hamiltonian in the new coordinates such that

$$\dot{P}_\alpha = -\frac{\partial \mathcal{H}}{\partial Q_\alpha}, \qquad \dot{Q}_\alpha = \frac{\partial \mathcal{H}}{\partial P_\alpha} \tag{16}$$

In such case we often refer to Q_α and P_α as *canonical coordinates.*

The Lagrangians in the old and new coordinates are $L(p_\alpha, q_\alpha, t)$ and $\mathcal{L}(P_\alpha, Q_\alpha, t)$ respectively. They are related to the Hamiltonians $H(p_\alpha, q_\alpha, t)$ and $\mathcal{H}(P_\alpha, Q_\alpha, t)$ by the equations

$$H = \sum p_\alpha \dot{q}_\alpha - L, \qquad \mathcal{H} = \sum P_\alpha \dot{Q}_\alpha - \mathcal{L} \tag{17}$$

where the summations extend from $\alpha = 1$ to n.

CONDITION THAT A TRANSFORMATION BE CANONICAL

The following theorem is of interest.

Theorem 12.2. The transformation

$$P_\alpha = P_\alpha(p_\alpha, q_\alpha, t), \qquad Q_\alpha = Q_\alpha(p_\alpha, q_\alpha, t) \tag{18}$$

is canonical if

$$\sum p_\alpha \, dq_\alpha - \sum P_\alpha \, dQ_\alpha \tag{19}$$

is an exact differential.

GENERATING FUNCTIONS

By Hamilton's principle the canonical transformation *(14)* or *(15)* must satisfy the conditions that $\int_{t_1}^{t_2} L\,dt$ and $\int_{t_1}^{t_2} \mathcal{L}\,dt$ are both extrema, i.e. we must simultaneously have

$$\delta \int_{t_1}^{t_2} L\,dt = 0 \quad \text{and} \quad \delta \int_{t_1}^{t_2} \mathcal{L}\,dt = 0 \tag{20}$$

These will be satisfied if there is a function $\mathcal{G}$ such that

$$\frac{d\mathcal{G}}{dt} = L - \mathcal{L} \tag{21}$$

See Problem 12.11. We call $\mathcal{G}$ a *generating function.*

By assuming that $\mathcal{G}$ is a function, which we shall denote by $\mathcal{S}$, of the old position coordinates q_α and the new momentum coordinates P_α as well as the time t, i.e.,

$$\mathcal{G} = \mathcal{S}(q_\alpha P_\alpha, t) \tag{22}$$

we can prove that [see Problem 12.13]

$$p_\alpha = \frac{\partial \mathcal{S}}{\partial q_\alpha}, \quad Q_\alpha = \frac{\partial \mathcal{S}}{\partial P_\alpha}, \quad \mathcal{H} = \frac{\partial \mathcal{S}}{\partial t} + H \tag{23}$$

where

$$\dot{P}_\alpha = -\frac{\partial \mathcal{H}}{\partial Q_\alpha}, \quad \dot{Q}_\alpha = \frac{\partial \mathcal{H}}{\partial P_\alpha} \tag{24}$$

Similar results hold if the generating function is a function of other coordinates [see Problem 12.12].

THE HAMILTON-JACOBI EQUATION

If we can find a canonical transformation leading to $\mathcal{H} \equiv 0$, then we see from (24) that P_α and Q_α will be constants [i.e., P_α and Q_α will be ignorable coordinates]. Thus by means of the transformation we are able to find p_α and q_α and thereby determine the motion of the system. The procedure hinges on finding the right generating function. From the third equation of (23) we see by putting $\mathcal{H} \equiv 0$ that this generating function must satisfy the partial differential equation

$$\frac{\partial \mathcal{S}}{\partial t} + H(p_\alpha, q_\alpha, t) = 0 \tag{25}$$

or

$$\frac{\partial \mathcal{S}}{\partial t} + H\left(\frac{\partial \mathcal{S}}{\partial q_\alpha}, q_\alpha, t\right) = 0 \tag{26}$$

This is called the *Hamilton-Jacobi equation.*

SOLUTION OF THE HAMILTON-JACOBI EQUATION

To accomplish our aims we need to find a suitable solution of the Hamilton-Jacobi equation. Now since this equation contains a total of $n+1$ independent variables, i.e. $q_1, q_2, \ldots, q_n$ and t, one such solution called the *complete solution,* will involve $n+1$ constants. Omitting an arbitrary additive constant and denoting the remaining n constants by $\beta_1, \beta_2, \ldots, \beta_n$ [none of which is additive] this solution can be written

$$\mathcal{S} = \mathcal{S}(q_1, q_2, \ldots, q_n, \beta_1, \beta_2, \ldots, \beta_n, t) \tag{27}$$

When this solution is obtained we can then determine the old momentum coordinates by

$$p_\alpha = \frac{\partial \mathcal{S}}{\partial q_\alpha} \tag{28}$$

Also, if we identify the new momentum coordinates P_α with the constants β_α, then

$$Q_\alpha = \frac{\partial \mathcal{S}}{\partial \beta_\alpha} = \gamma_\alpha \tag{29}$$

where γ_α, $\alpha = 1, \ldots, n$ are constants.

Using these we can then find q_α as functions of β_α, γ_α and t, which gives the motion of the system.

CASE WHERE HAMILTONIAN IS INDEPENDENT OF TIME

In obtaining the complete solution of the Hamilton-Jacobi equation, it is often useful to assume a solution of the form

$$\mathcal{S} = S_1(q_1) + S_2(q_2) + \cdots + S_n(q_n) + F(t) \tag{30}$$

where each function on the right depends on only one variable [see Problems 12.15 and 12.16]. This method, often called the method of *separation of variables,* is especially useful when the Hamiltonian does not depend explicitly on time. We then find that $F(t) = -Et$, and if the time independent part of $\mathcal{S}$ is denoted by

$$S = S_1(q_1) + S_2(q_2) + \cdots + S_n(q_n) \tag{31}$$

the Hamilton-Jacobi equation (*26*) reduces to

$$H\left(\frac{\partial S}{\partial q_\alpha}, q_\alpha\right) = E \tag{32}$$

where E is a constant representing the total energy of the system.

The equation (*32*) can also be obtained directly by assuming a generating function S which is independent of time. In such case equations (*23*) and (*24*) are replaced by

$$p_\alpha = \frac{\partial S}{\partial q_\alpha}, \quad Q_\alpha = \frac{\partial S}{\partial P_\alpha}, \quad \mathcal{H} = H = E \tag{33}$$

where
$$\dot{P}_\alpha = -\frac{\partial \mathcal{H}}{\partial Q_\alpha}, \quad \dot{Q}_\alpha = \frac{\partial \mathcal{H}}{\partial P_\alpha} \tag{34}$$

PHASE INTEGRALS. ACTION AND ANGLE VARIABLES

Hamiltonian methods are useful in the investigation of mechanical systems which are periodic. In such case the projections of the motion of the representative point in phase space on any $p_\alpha q_\alpha$ plane will be closed curves C_α. The line integral

$$J_\alpha = \oint_{C_\alpha} p_\alpha \, dq_\alpha \tag{35}$$

is called a *phase integral* or *action variable.*

We can show [see Problems 12.17 and 12.18] that

$$S = S(q_1, \ldots, q_n, J_1, \ldots, J_n) \tag{36}$$

where
$$p_\alpha = \frac{\partial S}{\partial q_\alpha}, \quad Q_\alpha = \frac{\partial S}{\partial J_\alpha} \tag{37}$$

It is customary to denote the new coordinates Q_α by w_α so that equations (*37*) are replaced by

$$p_\alpha = \frac{\partial S}{\partial q_\alpha}, \quad w_\alpha = \frac{\partial S}{\partial J_\alpha} \tag{38}$$

Thus Hamilton's equations become [see equations (*33*) and (*34*)]

$$\dot{J}_\alpha = -\frac{\partial \mathcal{H}}{\partial w_\alpha}, \quad \dot{w}_\alpha = \frac{\partial \mathcal{H}}{\partial J_\alpha} \tag{39}$$

where $\mathcal{H} = E$ in this case depends only on the constants J_α. Then from the second equation in (*39*),

$$w_\alpha = f_\alpha t + c_\alpha \tag{40}$$

where f_α and c_α are constants. We call w_α *angle variables.* The frequencies f_α are given by

$$f_\alpha = \frac{\partial \mathcal{H}}{\partial J_\alpha} \tag{41}$$

See Problems 12.19 and 12.20.

Solved Problems

THE HAMILTONIAN AND HAMILTON'S EQUATIONS

12.1. If the Hamiltonian $H = \sum p_\alpha \dot{q}_\alpha - L$, where the summation extends from $\alpha = 1$ to n, is expressed as a function of the coordinates q_α and momenta p_α, prove *Hamilton's equations*,

$$\dot{p}_\alpha = -\frac{\partial H}{\partial q_\alpha}, \qquad \dot{q}_\alpha = \frac{\partial H}{\partial p_\alpha}$$

regardless of whether H (*a*) does not or (*b*) does contain the variable time t explicitly.

(*a*) *H does not contain t explicitly.*

Taking the differential of $H = \sum p_\alpha \dot{q}_\alpha - L$, we have

$$dH = \sum p_\alpha \, d\dot{q}_\alpha + \sum \dot{q}_\alpha \, dp_\alpha - \sum \frac{\partial L}{\partial q_\alpha} dq_\alpha - \sum \frac{\partial L}{\partial \dot{q}_\alpha} d\dot{q}_\alpha \tag{1}$$

Then using the fact that $p_\alpha = \partial L / \partial \dot{q}_\alpha$ and $\dot{p}_\alpha = \partial L / \partial q_\alpha$, this reduces to

$$dH = \sum \dot{q}_\alpha \, dp_\alpha - \sum \dot{p}_\alpha \, dq_\alpha \tag{2}$$

But since H is expressed as a function of p_α and q_α, we have

$$dH = \sum \frac{\partial H}{\partial p_\alpha} dp_\alpha + \sum \frac{\partial H}{\partial q_\alpha} dq_\alpha \tag{3}$$

Comparing (*2*) and (*3*) we have, as required,

$$\dot{q}_\alpha = \frac{\partial H}{\partial p_\alpha}, \qquad \dot{p}_\alpha = -\frac{\partial H}{\partial q_\alpha}$$

(*b*) *H does contain t explicitly.*

In this case equations (*1*), (*2*) and (*3*) of part (*a*) are replaced by the equations

$$dH = \sum p_\alpha \, d\dot{q}_\alpha + \sum \dot{q}_\alpha \, dp_\alpha - \sum \frac{\partial L}{\partial q_\alpha} dq_\alpha - \sum \frac{\partial L}{\partial \dot{q}_\alpha} d\dot{q}_\alpha - \frac{\partial L}{\partial t} dt \tag{4}$$

$$dH = \sum \dot{q}_\alpha \, dp_\alpha - \sum \dot{p}_\alpha \, dq_\alpha - \frac{\partial L}{\partial t} dt \tag{5}$$

$$dH = \sum \frac{\partial H}{\partial p_\alpha} dp_\alpha + \sum \frac{\partial H}{\partial q_\alpha} dq_\alpha + \frac{\partial H}{\partial t} dt \tag{6}$$

Then comparing (*5*) and (*6*), we have

$$\dot{q}_\alpha = \frac{\partial H}{\partial p_\alpha}, \qquad \dot{p}_\alpha = -\frac{\partial H}{\partial q_\alpha}, \qquad \frac{\partial H}{\partial t} = -\frac{\partial L}{\partial t}$$

12.2. If the Hamiltonian H is independent of t explicitly, prove that it is (*a*) a constant and is (*b*) equal to the total energy of the system.

(*a*) From equation (*2*) of Problem 12.1 we have

$$\frac{dH}{dt} = \sum \dot{q}_\alpha \dot{p}_\alpha - \sum \dot{p}_\alpha \dot{q}_\alpha = 0$$

Thus H is a constant, say E.

(*b*) By Euler's theorem on homogeneous functions [see Problem 11.47, page 305],

$$\sum \dot{q}_\alpha \frac{\partial T}{\partial \dot{q}_\alpha} = 2T$$

where T is the kinetic energy. Then since $p_\alpha = \partial L/\partial \dot{q}_\alpha = \partial T/\partial \dot{q}_\alpha$ [assuming the potential V does not depend on $\dot{q}_\alpha$], we have $\sum p_\alpha \dot{q}_\alpha = 2T$. Thus as required,

$$H = \sum p_\alpha \dot{q}_\alpha - L = 2T - (T - V) = T + V = E$$

12.3. A particle moves in the xy plane under the influence of a central force depending only on its distance from the origin. (*a*) Set up the Hamiltonian for the system. (*b*) Write Hamilton's equations of motion.

(*a*) Assume that the particle is located by its polar coordinates (r, θ) and that the potential due to the central force is $V(r)$. Since the kinetic energy of the particle is $T = \frac{1}{2}m(\dot{r}^2 + r^2\dot{\theta}^2)$, the Lagrangian is

$$L = T - V = \tfrac{1}{2}m(\dot{r}^2 + r^2\dot{\theta}^2) - V(r) \tag{1}$$

We have
$$p_r = \partial L/\partial \dot{r} = m\dot{r}, \qquad p_\theta = \partial L/\partial \dot{\theta} = mr^2\dot{\theta} \tag{2}$$

so that
$$\dot{r} = p_r/m, \qquad \dot{\theta} = p_\theta/mr^2 \tag{3}$$

Then the Hamiltonian is given by

$$\begin{aligned} H &= \sum_{\alpha=1}^{n} p_\alpha \dot{q}_\alpha - L = p_r\dot{r} + p_\theta\dot{\theta} - \{\tfrac{1}{2}m(\dot{r}^2 + r^2\dot{\theta}^2) - V(r)\} \\ &= p_r\left(\frac{p_r}{m}\right) + p_\theta\left(\frac{p_\theta}{mr^2}\right) - \left\{\frac{1}{2}m\left(\frac{p_r^2}{m^2} + r^2 \cdot \frac{p_\theta^2}{m^2r^4}\right) - V(r)\right\} \\ &= \frac{p_r^2}{2m} + \frac{p_\theta^2}{2mr^2} + V(r) \end{aligned} \tag{4}$$

Note that this is the total energy expressed in terms of coordinates and momenta.

(*b*) Hamilton's equations are $\dot{q}_\alpha = \partial H/\partial p_\alpha, \quad \dot{p}_\alpha = -\partial H/\partial q_\alpha$

Thus
$$\dot{r} = \partial H/\partial p_r = p_r/m, \qquad \dot{\theta} = \partial H/\partial p_\theta = p_\theta/mr^2 \tag{5}$$

$$\dot{p}_r = -\partial H/\partial r = p_\theta^2/mr^3 - V'(r), \qquad \dot{p}_\theta = -\partial H/\partial \theta = 0 \tag{6}$$

Note that the equations (*5*) are equivalent to the corresponding equations (*3*).

PHASE SPACE AND LIOUVILLE'S THEOREM

12.4. Prove Liouville's theorem for the case of one degree of freedom.

We can think of the mechanical system as being described in terms of the motion of representative points through an element of volume in phase space. In the case of a mechanical system with one degree of freedom, we have a two dimensional (p, q) phase space and the volume element reduces to an area element $dp\,dq$ [Fig. 12-2].

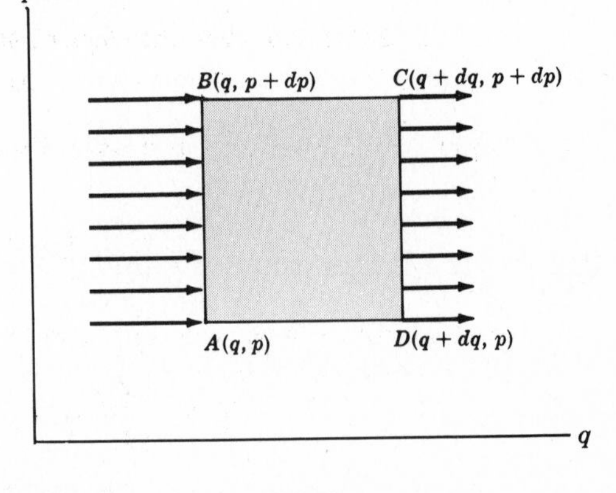

Fig. 12-2

Let $\rho = \rho(p, q, t)$ be the density of representative points, i.e. the number of representative points per unit area as obtained by an appropriate limiting procedure. Since the speed with which representative points enter through AB is $\dot{q}$, the number of representative points which enter through AB per unit time is

$$\rho\dot{q}\,dp \tag{1}$$

The number of representative points which leave through CD is

$$\left\{\rho\dot{q} + \frac{\partial}{\partial q}(\rho\dot{q})\,dq\right\} dp \tag{2}$$

Thus the number which remain in the element is (1) minus (2), or

$$-\frac{\partial}{\partial q}(\rho\dot{q})\,dp\,dq \tag{3}$$

Similarly the number of representative points which enter through AD and leave through BC are respectively

$$\rho\dot{p}\,dq \qquad \text{and} \qquad \left\{\rho\dot{p} + \frac{\partial}{\partial p}(\rho\dot{p})\,dp\right\}dq$$

Thus the number which remain in the element is

$$-\frac{\partial}{\partial p}(\rho\dot{p})\,dp\,dq \tag{4}$$

The increase in representative points is thus [adding (3) and (4)]

$$-\left\{\frac{\partial(\rho\dot{q})}{\partial q} + \frac{\partial(\rho\dot{p})}{\partial p}\right\}dp\,dq$$

Since this is equal to $\frac{\partial\rho}{\partial t}dp\,dq$, we must have

$$\frac{\partial\rho}{\partial t} + \left\{\frac{\partial(\rho\dot{q})}{\partial q} + \frac{\partial(\rho\dot{p})}{\partial p}\right\} = 0$$

or

$$\frac{\partial\rho}{\partial t} + \rho\frac{\partial\dot{q}}{\partial q} + \frac{\partial\rho}{\partial q}\dot{q} + \rho\frac{\partial\dot{p}}{\partial p} + \frac{\partial\rho}{\partial p}\dot{p} = 0 \tag{5}$$

Now by Hamilton's equations $\dot{p} = -\partial H/\partial q$, $\dot{q} = \partial H/\partial p$ so that

$$\frac{\partial\dot{p}}{\partial p} = -\frac{\partial^2 H}{\partial p\,\partial q}, \qquad \frac{\partial\dot{q}}{\partial q} = \frac{\partial^2 H}{\partial q\,\partial p}$$

Thus since we suppose that the Hamiltonian has continuous second order derivatives, it follows that $\partial\dot{p}/\partial p = -\partial\dot{q}/\partial q$. Using this in (5), it becomes

$$\frac{\partial\rho}{\partial t} + \frac{\partial\rho}{\partial q}\dot{q} + \frac{\partial\rho}{\partial p}\dot{p} = 0 \tag{6}$$

But this can be written

$$d\rho/dt = 0 \tag{7}$$

which shows that the density in phase space is constant and thus proves Liouville's theorem.

12.5. Prove Liouville's theorem in the general case.

In the general case the element of volume in phase space is

$$dV = dq_1 \cdots dq_n\,dp_1 \cdots dp_n$$

In exactly the same manner as in Problem 12.4 the increase of representative points in dV is found to be

$$-\left\{\frac{\partial(\rho\dot{q}_1)}{\partial q_1} + \cdots + \frac{\partial(\rho\dot{q}_n)}{\partial q_n} + \frac{\partial(\rho\dot{p}_1)}{\partial p_1} + \cdots + \frac{\partial(\rho\dot{p}_n)}{\partial p_n}\right\}dV$$

and since this is equal to $\frac{\partial\rho}{\partial t}dV$, we must have

$$\frac{\partial\rho}{\partial t} + \frac{\partial(\rho\dot{q}_1)}{\partial q_1} + \cdots + \frac{\partial(\rho\dot{q}_n)}{\partial q_n} + \frac{\partial(\rho\dot{p}_1)}{\partial p_1} + \cdots + \frac{\partial(\rho\dot{p}_n)}{\partial p_n} = 0$$

or

$$\frac{\partial\rho}{\partial t} + \sum_{\alpha=1}^{n}\frac{\partial(\rho\dot{q}_\alpha)}{\partial q_\alpha} + \sum_{\alpha=1}^{n}\frac{\partial(\rho\dot{p}_\alpha)}{\partial p_\alpha} = 0$$

This can be written as

$$\frac{\partial\rho}{\partial t} + \sum_{\alpha=1}^{n}\left(\frac{\partial\rho}{\partial q_\alpha}\dot{q}_\alpha + \frac{\partial\rho}{\partial p_\alpha}\dot{p}_\alpha\right) + \sum_{\alpha=1}^{n}\rho\left(\frac{\partial\dot{q}_\alpha}{\partial q_\alpha} + \frac{\partial\dot{p}_\alpha}{\partial p_\alpha}\right) = 0 \tag{1}$$

Now by Hamilton's equations $\dot{p}_\alpha = -\partial H/\partial q_\alpha$, $\dot{q}_\alpha = \partial H/\partial p_\alpha$ so that

$$\frac{\partial \dot{p}_\alpha}{\partial p_\alpha} = -\frac{\partial^2 H}{\partial p_\alpha \, \partial q_\alpha}, \qquad \frac{\partial \dot{q}_\alpha}{\partial p_\alpha} = \frac{\partial^2 H}{\partial q_\alpha \, \partial p_\alpha}$$

Hence $\partial \dot{p}_\alpha/\partial p_\alpha = -\partial \dot{q}_\alpha/\partial q_\alpha$ and (1) becomes

$$\frac{\partial \rho}{\partial t} + \sum_{\alpha=1}^{n} \left(\frac{\partial \rho}{\partial q_\alpha} \dot{q}_\alpha + \frac{\partial \rho}{\partial p_\alpha} \dot{p}_\alpha \right) = 0 \tag{2}$$

i.e.,

$$d\rho/dt = 0 \tag{3}$$

or ρ = constant.

Note that we have used the fact that if $\rho = \rho(q_1, \ldots, q_n, p_1, \ldots, p_n, t)$ then

$$\frac{d\rho}{dt} = \sum_{\alpha=1}^{n} \left(\frac{\partial \rho}{\partial q_\alpha} \frac{dq_\alpha}{dt} + \frac{\partial \rho}{\partial p_\alpha} \frac{dp_\alpha}{dt} \right) + \frac{\partial \rho}{\partial t} = \sum_{\alpha=1}^{n} \left(\frac{\partial \rho}{\partial q_\alpha} \dot{q}_\alpha + \frac{\partial \rho}{\partial p_\alpha} \dot{p}_\alpha \right) + \frac{\partial \rho}{\partial t}$$

CALCULUS OF VARIATIONS AND HAMILTON'S PRINCIPLE

12.6. Prove that a necessary condition for $I = \int_a^b F(x, y, y')\, dx$ to be an extremum [maximum or minimum] is $\dfrac{d}{dx}\left(\dfrac{\partial F}{\partial y'}\right) - \dfrac{\partial F}{\partial y} = 0$.

Suppose that the curve which makes I an extremum is given by

$$y = Y(x), \qquad a \leqq x \leqq b \tag{1}$$

Then

$$y = Y(x) + \epsilon\, \eta(x) = Y + \epsilon \eta \tag{2}$$

where ϵ is independent of x, is a neighboring curve through $x = a$ and $x = b$ if we choose

$$\eta(a) = \eta(b) = 0 \tag{3}$$

The value of I for this neighboring curve is

$$I(\epsilon) = \int_a^b F(x, Y + \epsilon\eta, Y' + \epsilon\eta')\, dx \tag{4}$$

This is an extremum for $\epsilon = 0$. A necessary condition that this be so is that $\left.\dfrac{dI}{d\epsilon}\right|_{\epsilon=0} = 0$. But by differentiation under the integral sign, assuming this is valid, we find

$$\left.\frac{dI}{d\epsilon}\right|_{\epsilon=0} = \int_a^b \left(\frac{\partial F}{\partial y}\eta + \frac{\partial F}{\partial y'}\eta' \right) dx = 0$$

which can be written on integrating by parts as

$$\int_a^b \frac{\partial F}{\partial x}\eta\, dx + \left.\frac{\partial F}{\partial y'}\eta\right|_a^b - \int_a^b \eta \frac{d}{dx}\left(\frac{\partial F}{\partial y'}\right) dx$$

$$= \int_a^b \eta \left\{ \frac{\partial F}{\partial y} - \frac{d}{dx}\left(\frac{\partial F}{\partial y'}\right) \right\} dx = 0$$

where we have used (3). Since η is arbitrary, we must have

$$\frac{\partial F}{\partial y} - \frac{d}{dx}\left(\frac{\partial F}{\partial y'}\right) = 0 \quad \text{or} \quad \frac{d}{dx}\left(\frac{\partial F}{\partial y'}\right) - \frac{\partial F}{\partial y} = 0$$

which is called *Euler's* or *Lagrange's* equation. The result is easily extended to the integral

$$\int_a^b F(x, y_1, y_1', y_2, y_2', \ldots, y_n, y_n')\, dx$$

and leads to the *Euler's* or *Lagrange's* equations

$$\frac{d}{dx}\left(\frac{\partial F}{\partial y'_\alpha}\right) - \frac{\partial F}{\partial y_\alpha} \;=\; 0 \qquad \alpha = 1, 2, \ldots, n$$

By using a Taylor series expansion we find from (4) that

$$I(\epsilon) - I(0) \;=\; \epsilon \int_a^b \left(\frac{\partial F}{\partial y}\eta + \frac{\partial F}{\partial y'}\eta'\right) dx + \text{higher order terms in } \epsilon^2, \epsilon^3, \text{etc.} \tag{5}$$

The coefficient of ϵ in (5) is often called the *variation* of the integral and is denoted by

$$\delta \int_a^b F(x, y, y')\, dx$$

The fact that $\int_a^b F(x, y, y')\, dx$ is an extremum is thus indicated by

$$\delta \int_a^b F(x, y, y')\, dx \;=\; 0$$

12.7. Discuss the relationship of Hamilton's principle with Problem 12.6.

By identifying the function $F(x, y, y')$ with the Lagrangian $L(t, q, \dot{q})$ where x, y and y' are replaced by $t, q, \dot{q}$ respectively, we see that a necessary condition for the action integral

$$\int_{t_1}^{t_2} L\, dt \tag{1}$$

to be an extremum [maximum or minimum] is given by

$$\frac{d}{dt}\left(\frac{\partial L}{\partial \dot{q}}\right) - \frac{\partial L}{\partial q} \;=\; 0 \tag{2}$$

Since we have already seen that (2) describes the motion of a particle, it follows that such motion can also be achieved by requiring that (1) be an extremum, which is Hamilton's principle.

For systems involving n degrees of freedom we consider the integral (1) where

$$L \;=\; L(t, q_1, \dot{q}_1, q_2, \dot{q}_2, \ldots, q_n, \dot{q}_n)$$

which lead to the Lagrange equations

$$\frac{d}{dt}\left(\frac{\partial L}{\partial \dot{q}_\alpha}\right) - \frac{\partial L}{\partial q_\alpha} \;=\; 0 \qquad \alpha = 1, 2, \ldots, n$$

12.8. A particle slides from rest at one point on a frictionless wire in a vertical plane to another point under the influence of gravity. Find the total time taken.

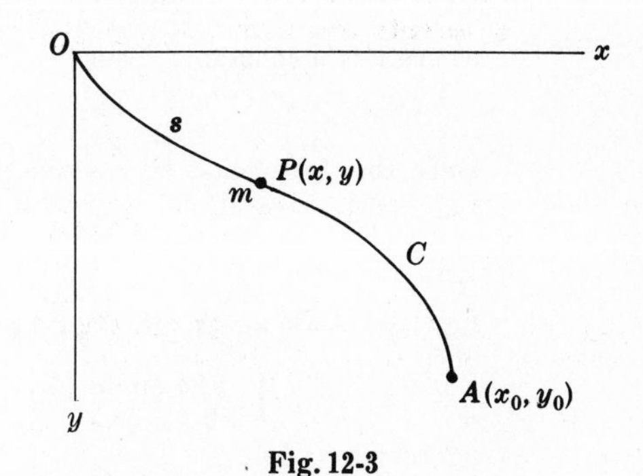

Fig. 12-3

Let the shape of the wire be indicated by curve C in Fig. 12-3 and suppose that the starting and finishing points are taken to be the origin and the point $A(x_0, y_0)$ respectively.

Let $P(x, y)$ denote any position of the particle which we assume has mass m. From the principle of conservation of energy, if we choose the horizontal line through A as reference level, we have

Potential energy at O + kinetic energy at O = potential energy at P + kinetic energy at P

or
$$mgy_0 + 0 \;=\; mg(y_0 - y) + \tfrac{1}{2}m(ds/dt)^2$$

where ds/dt is the instantaneous speed of the particle at time t. Then

$$ds/dt \;=\; \pm\sqrt{2gy} \tag{1}$$

If we measure the arc length s from the origin, then s increases as the particle moves. Thus ds/dt is positive, so that $ds/dt = \sqrt{2gy}$ or $dt = ds/\sqrt{2gy}$.

The total time taken to go from $y = 0$ to $y = y_0$ is

$$\tau = \int_0^{\tau} dt = \int_{y=0}^{y_0} \frac{ds}{\sqrt{2gy}}$$

But $(ds)^2 = (dx)^2 + (dy)^2$ or $ds = \sqrt{1+y'^2}\,dx$. Thus the required time is

$$\tau = \frac{1}{\sqrt{2g}} \int_{y=0}^{y_0} \frac{\sqrt{1+y'^2}}{\sqrt{y}}\,dx \qquad (2)$$

12.9. If the particle of Problem 12.8 is to travel from point O to point A in the least possible time, show that the differential equation of the curve C defining the shape of the wire is $1 + y'^2 + 2yy'' = 0$.

A necessary condition for the time τ given by equation (2) of Problem 12.8 to be a minimum is that

$$\frac{d}{dx}\left(\frac{\partial F}{\partial y'}\right) - \frac{\partial F}{\partial y} = 0 \qquad (1)$$

where
$$F = (1+y'^2)^{1/2}\,y^{-1/2} \qquad (2)$$

Now
$$\partial F/\partial y' = (1+y'^2)^{-1/2}\,y'\,y^{-1/2}, \quad \partial F/\partial y = -\tfrac{1}{2}(1+y'^2)^{1/2}\,y^{-3/2}$$

Substituting these in (1), performing the indicated differentiation with respect to x and simplifying, we obtain the required differential equation.

The problem of finding the shape of the wire is often called the *brachistochrone problem.*

12.10. (a) Solve the differential equation in Problem 12.9 and thus (b) show that the required curve is a *cycloid*.

(a) Since x is missing in the differential equation, let $y' = u$ so that

$$y'' = \frac{du}{dx} = \frac{du}{dy}\frac{dy}{dx} = \frac{du}{dy}y' = u\frac{du}{dy}$$

Then the differential equation becomes

$$1 + u^2 + 2yu\frac{du}{dy} = 0 \quad \text{or} \quad \frac{2u\,du}{1+u^2} + \frac{dy}{y} = 0$$

Integration yields

$$\ln(1+u^2) + \ln y = \ln b \quad \text{or} \quad (1+u^2)y = b$$

where b is a constant. Thus

$$u = y' = \frac{dy}{dx} = \sqrt{\frac{b-y}{y}}$$

since the slope must be positive. Separating the variables and integrating, we find

$$x = \int \sqrt{\frac{y}{b-y}}\,dy + c$$

Letting $y = b\sin^2\theta$, this can be written

$$x = \int \sqrt{\frac{b\sin^2\theta}{b\cos^2\theta}} \cdot 2b\sin\theta\cos\theta\,d\theta + c$$

$$= 2b\int \sin^2\theta\,d\theta + c = b\int (1-\cos 2\theta)\,d\theta + c = \tfrac{1}{2}b(2\theta - \sin 2\theta) + c$$

Thus the parametric equations of the required curve are

$$x = \tfrac{1}{2}b(2\theta - \sin 2\theta) + c, \quad y = b\sin^2\theta = \tfrac{1}{2}b(1-\cos 2\theta)$$

Since the curve must pass through the point $x = 0, y = 0$, we have $c = 0$. Then letting

$$\phi = 2\theta, \quad a = \tfrac{1}{2}b \qquad (1)$$

the required parametric equations are

$$x = a(\phi - \sin\phi), \quad y = a(1-\cos\phi) \qquad (2)$$

(*b*) The equations (*2*) are parametric equations of a *cycloid* [see Fig. 12-4]. The constant a must be determined so that the curve passes through point A. The cycloid is the path taken by a fixed point on a circle as it rolls along a given line [see Problem 12.89].

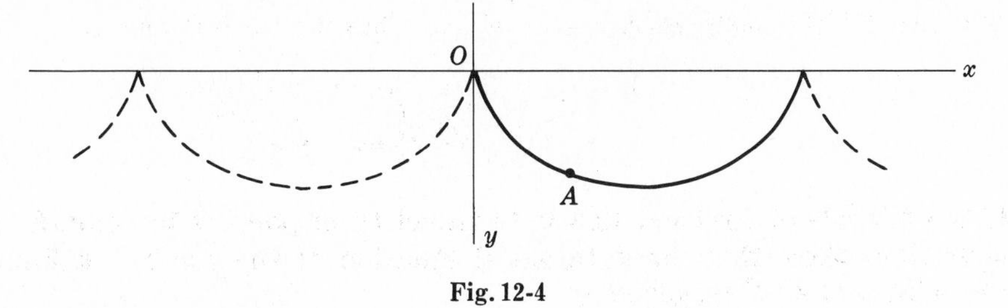

Fig. 12-4

CANONICAL TRANSFORMATIONS AND GENERATING FUNCTIONS

12.11. Prove that a transformation is canonical if there exists a function $\mathcal{G}$ such that $d\mathcal{G}/dt = L - \mathcal{L}$.

The integrals $\int_{t_1}^{t_2} L\,dt$ and $\int_{t_1}^{t_2} \mathcal{L}\,dt$ must simultaneously be extrema so that their variations are zero, i.e.,

$$\delta \int_{t_1}^{t_2} L\,dt = 0 \qquad \text{and} \qquad \delta \int_{t_1}^{t_2} \mathcal{L}\,dt = 0$$

Thus by subtraction,

$$\delta \int_{t_1}^{t_2} (L - \mathcal{L})\,dt = 0$$

This can be accomplished if there exists a function $\mathcal{G}$ such that

$$L - \mathcal{L} = d\mathcal{G}/dt$$

since in such case

$$\delta \int_{t_1}^{t_2} \frac{d\mathcal{G}}{dt}\,dt = \delta\{\mathcal{G}(t_2) - \mathcal{G}(t_1)\} = 0$$

The function $\mathcal{G}$ is called a *generating function.*

12.12. Suppose that the generating function is a function $\mathcal{T}$ of the old and new position coordinates q_α and Q_α respectively as well as the time t, i.e. $\mathcal{T} = \mathcal{T}(q_\alpha, Q_\alpha, t)$. Prove that

$$p_\alpha = \frac{\partial \mathcal{T}}{\partial q_\alpha}, \quad P_\alpha = -\frac{\partial \mathcal{T}}{\partial Q_\alpha}, \quad \mathcal{H} = \frac{\partial \mathcal{T}}{\partial t} + H \qquad \text{where} \qquad \dot{P}_\alpha = -\frac{\partial \mathcal{H}}{\partial Q_\alpha}, \quad \dot{Q}_\alpha = \frac{\partial \mathcal{H}}{\partial P_\alpha}$$

By Problem 12.11,

$$\begin{aligned} \frac{d\mathcal{T}}{dt} &= L - \mathcal{L} = \sum p_\alpha \dot{q}_\alpha - H - \left\{\sum P_\alpha \dot{Q}_\alpha - \mathcal{H}\right\} \\ &= \sum p_\alpha \dot{q}_\alpha - \sum P_\alpha \dot{Q}_\alpha + \mathcal{H} - H \end{aligned}$$

or

$$d\mathcal{T} = \sum p_\alpha\,dq_\alpha - \sum P_\alpha\,dQ_\alpha + (\mathcal{H} - H)\,dt \tag{1}$$

But if $\mathcal{T} = \mathcal{T}(q_\alpha, Q_\alpha, t)$, then

$$d\mathcal{T} = \sum \frac{\partial \mathcal{T}}{\partial q_\alpha}\,dq_\alpha + \sum \frac{\partial \mathcal{T}}{\partial Q_\alpha}\,dQ_\alpha + \frac{\partial \mathcal{T}}{\partial t}\,dt \tag{2}$$

Comparing (*1*) and (*2*), we have as required

$$p_\alpha = \frac{\partial \mathcal{T}}{\partial q_\alpha}, \quad P_\alpha = -\frac{\partial \mathcal{T}}{\partial Q_\alpha}, \quad \mathcal{H} - H = \frac{\partial \mathcal{T}}{\partial t}$$

The equations

$$\dot{P}_\alpha = -\frac{\partial \mathcal{H}}{\partial Q_\alpha}, \quad \dot{Q}_\alpha = \frac{\partial \mathcal{H}}{\partial P_\alpha}$$

follow from the fact that $\mathcal{H}$ is the Hamiltonian in the coordinates P_α, Q_α so that Hamilton's equations hold as in Problem 12.1.

12.13. Let $\mathcal{S}$ be a generating function dependent only on q_α, P_α, t. Prove that

$$p_\alpha = \frac{\partial \mathcal{S}}{\partial q_\alpha}, \quad Q_\alpha = \frac{\partial \mathcal{S}}{\partial P_\alpha}, \quad \mathcal{H} = \frac{\partial \mathcal{S}}{\partial t} + H \qquad \text{where} \qquad \dot{P}_\alpha = -\frac{\partial \mathcal{H}}{\partial Q_\alpha}, \quad \dot{Q}_\alpha = \frac{\partial \mathcal{H}}{\partial P_\alpha}$$

From Problem 12.12, equation (1), we have

$$\begin{aligned} d\mathcal{T} &= \sum p_\alpha \, dq_\alpha - \sum P_\alpha \, dQ_\alpha + (\mathcal{H} - H)\, dt \\ &= \sum p_\alpha \, dq_\alpha - d\left\{\sum P_\alpha Q_\alpha\right\} + \sum Q_\alpha \, dP_\alpha + (\mathcal{H} - H)\, dt \end{aligned}$$

or

$$d\left(\mathcal{T} + \sum P_\alpha Q_\alpha\right) = \sum p_\alpha \, dq_\alpha + \sum Q_\alpha \, dP_\alpha + (\mathcal{H} - H)\, dt \tag{1}$$

i.e.,

$$d\mathcal{S} = \sum p_\alpha \, dq_\alpha + \sum Q_\alpha \, dP_\alpha + (\mathcal{H} - H)\, dt \tag{2}$$

where

$$\mathcal{S} = \mathcal{T} + \sum P_\alpha Q_\alpha \tag{3}$$

But since $\mathcal{S}$ is a function of q_α, P_α, t,

$$d\mathcal{S} = \sum \frac{\partial \mathcal{S}}{\partial q_\alpha} dq_\alpha + \sum \frac{\partial \mathcal{S}}{\partial P_\alpha} dP_\alpha + \frac{\partial \mathcal{S}}{\partial t} dt \tag{4}$$

Comparing (2) and (4),

$$p_\alpha = \frac{\partial \mathcal{S}}{\partial q_\alpha}, \quad Q_\alpha = \frac{\partial \mathcal{S}}{\partial P_\alpha}, \quad \mathcal{H} = \frac{\partial \mathcal{S}}{\partial t} + H$$

The results

$$\dot{P}_\alpha = -\frac{\partial \mathcal{H}}{\partial Q_\alpha}, \quad \dot{Q}_\alpha = \frac{\partial \mathcal{H}}{\partial P_\alpha}$$

follow as in Problem 12.12, since $\mathcal{H}$ is the Hamiltonian.

12.14. Prove that the transformation $P = \frac{1}{2}(p^2 + q^2)$, $Q = \tan^{-1}(q/p)$ is canonical.

Method 1.

Let the Hamiltonians in the coordinates p, q and P, Q be respectively $H(p, q)$ and $\mathcal{H}(P, Q)$ so that $H(p, q) = \mathcal{H}(P, Q)$. Since p, q are canonical coordinates,

$$\dot{p} = -\frac{\partial H}{\partial q}, \quad \dot{q} = \frac{\partial H}{\partial p} \tag{1}$$

But

$$\dot{p} = \frac{\partial p}{\partial P}\dot{P} + \frac{\partial p}{\partial Q}\dot{Q}, \quad \dot{q} = \frac{\partial q}{\partial P}\dot{P} + \frac{\partial q}{\partial Q}\dot{Q} \tag{2}$$

$$\frac{\partial H}{\partial q} = \frac{\partial \mathcal{H}}{\partial P}\frac{\partial P}{\partial q} + \frac{\partial \mathcal{H}}{\partial Q}\frac{\partial Q}{\partial q}, \quad \frac{\partial H}{\partial p} = \frac{\partial \mathcal{H}}{\partial P}\frac{\partial P}{\partial p} + \frac{\partial \mathcal{H}}{\partial Q}\frac{\partial Q}{\partial p} \tag{3}$$

From the given transformation equations we have

$$\frac{\partial P}{\partial p} = p, \quad \frac{\partial P}{\partial q} = q, \quad \frac{\partial Q}{\partial p} = \frac{-q}{p^2 + q^2}, \quad \frac{\partial Q}{\partial q} = \frac{p}{p^2 + q^2}$$

Also, differentiating the transformation equations with respect to P and Q respectively, we find

$$1 = p\frac{\partial p}{\partial P} + q\frac{\partial q}{\partial P}, \quad 0 = \left(p\frac{\partial q}{\partial P} - q\frac{\partial p}{\partial P}\right)\Big/(p^2 + q^2)$$

$$0 = p\frac{\partial p}{\partial Q} + q\frac{\partial q}{\partial Q}, \quad 1 = \left(p\frac{\partial q}{\partial Q} - q\frac{\partial p}{\partial Q}\right)\Big/(p^2 + q^2)$$

Solving simultaneously, we find

$$\frac{\partial p}{\partial P} = \frac{p}{p^2 + q^2}, \quad \frac{\partial q}{\partial P} = \frac{q}{p^2 + q^2}, \quad \frac{\partial p}{\partial Q} = -q, \quad \frac{\partial q}{\partial Q} = p \tag{4}$$

Then equations (1) and (2) become

$$\dot{p} = \frac{p}{p^2+q^2}\dot{P} - q\dot{Q}, \qquad \dot{q} = \frac{q}{p^2+q^2}\dot{P} + p\dot{Q} \tag{5}$$

$$\frac{\partial H}{\partial q} = q\frac{\partial \mathcal{H}}{\partial P} + \frac{p}{p^2+q^2}\frac{\partial \mathcal{H}}{\partial Q}, \qquad \frac{\partial H}{\partial p} = p\frac{\partial \mathcal{H}}{\partial P} - \frac{q}{p^2+q^2}\frac{\partial \mathcal{H}}{\partial Q} \tag{6}$$

Thus from equations (1), (5) and (6) we have

$$\frac{p}{p^2+q^2}\dot{P} - q\dot{Q} = -q\frac{\partial \mathcal{H}}{\partial P} - \frac{p}{p^2+q^2}\frac{\partial \mathcal{H}}{\partial Q}$$

$$\frac{q}{p^2+q^2}\dot{P} + p\dot{Q} = p\frac{\partial \mathcal{H}}{\partial P} - \frac{q}{p^2+q^2}\frac{\partial \mathcal{H}}{\partial Q}$$

Solving these simultaneously we find

$$\dot{P} = -\frac{\partial \mathcal{H}}{\partial Q}, \qquad \dot{Q} = \frac{\partial \mathcal{H}}{\partial P} \tag{7}$$

which show that P and Q are canonical and that the transformation is therefore canonical.

Method 2.

By Theorem 12.2, page 314, the transformation is canonical if

$$\sum p_\alpha\, dq_\alpha - \sum P_\alpha\, dQ_\alpha \tag{8}$$

is an exact differential. In this case (8) becomes

$$p\,dq - P\,dQ = p\,dq - \tfrac{1}{2}(p^2+q^2)\left(\frac{p\,dq - q\,dp}{p^2+q^2}\right)$$

$$= \tfrac{1}{2}(p\,dq + q\,dp) = d(\tfrac{1}{2}pq)$$

an exact differential. Thus the transformation is canonical.

THE HAMILTON-JACOBI EQUATION

12.15. (a) Write the Hamiltonian for the one dimensional harmonic oscillator of mass m. (b) Write the corresponding Hamilton-Jacobi equation. (c) Use the Hamilton-Jacobi method to obtain the motion of the oscillator.

(a) **Method 1.**

Let q be the position coordinate of the harmonic oscillator, so that $\dot{q}$ is its velocity. Since the kinetic energy is $T = \frac{1}{2}m\dot{q}^2$ and the potential energy is $V = \frac{1}{2}\kappa q^2$, the Lagrangian is

$$L = T - V = \tfrac{1}{2}m\dot{q}^2 - \tfrac{1}{2}\kappa q^2 \tag{1}$$

The momentum is

$$p = \partial L/\partial \dot{q} = m\dot{q} \tag{2}$$

so that

$$\dot{q} = p/m \tag{3}$$

Then the Hamiltonian is

$$H = \sum p_\alpha \dot{q}_\alpha - L = p\dot{q} - (\tfrac{1}{2}m\dot{q}^2 - \tfrac{1}{2}\kappa q^2)$$

$$= \tfrac{1}{2}p^2/m + \tfrac{1}{2}\kappa q^2 \tag{4}$$

Method 2.

By Problem 12.2, since the Hamiltonian is the same as the total energy for conservative systems,

$$H = \tfrac{1}{2}m\dot{q}^2 + \tfrac{1}{2}\kappa q^2 = \tfrac{1}{2}m(p/m)^2 + \tfrac{1}{2}\kappa q^2 = \tfrac{1}{2}p^2/m + \tfrac{1}{2}\kappa q^2$$

(b) Using $p = \partial\mathcal{S}/\partial q$ and the Hamiltonian of part (a), the Hamilton-Jacobi equation is [see equation (26), page 315]

$$\frac{\partial\mathcal{S}}{\partial t} + \frac{1}{2m}\left(\frac{\partial\mathcal{S}}{\partial q}\right)^2 + \tfrac{1}{2}\kappa q^2 = 0 \tag{5}$$

(c) Assume a solution to (5) of the form

$$\mathcal{S} = S_1(q) + S_2(t) \tag{6}$$

Then (5) becomes

$$\frac{1}{2m}\left(\frac{dS_1}{dq}\right)^2 + \tfrac{1}{2}\kappa q^2 = -\frac{dS_2}{dt} \tag{7}$$

Setting each side equal to the constant β, we find

$$\frac{1}{2m}\left(\frac{dS_1}{dq}\right)^2 + \tfrac{1}{2}\kappa q^2 = \beta, \qquad \frac{dS_2}{dt} = -\beta$$

whose solutions, omitting constants of integration, are

$$S_1 = \int \sqrt{2m(\beta - \tfrac{1}{2}\kappa q^2)}\, dq, \qquad S_2 = -\beta t \tag{8}$$

so that (6) becomes

$$\mathcal{S} = \int \sqrt{2m(\beta - \tfrac{1}{2}\kappa q^2)}\, dq - \beta t \tag{9}$$

Let us identify β with the new momentum coordinate P. Then we have for the new position coordinate,

$$Q = \frac{\partial\mathcal{S}}{\partial\beta} = \frac{\partial}{\partial\beta}\left\{\int \sqrt{2m(\beta - \tfrac{1}{2}\kappa q^2)}\, dq - \beta t\right\}$$

$$= \frac{\sqrt{2m}}{2}\int \frac{dq}{\sqrt{\beta - \tfrac{1}{2}\kappa q^2}} - t$$

But since the new coordinate Q is a constant γ,

$$\frac{\sqrt{2m}}{2}\int \frac{dq}{\sqrt{\beta - \tfrac{1}{2}\kappa q^2}} - t = \gamma$$

or on integrating,

$$\sqrt{m/\kappa}\,\sin^{-1}\left(q\sqrt{\kappa/2\beta}\right) = t + \gamma$$

Then solving for q,

$$q = \sqrt{2\beta/\kappa}\,\sin\sqrt{\kappa/m}\,(t + \gamma) \tag{10}$$

which is the required solution. The constants β and γ can be found from the initial conditions.

It is of interest to note that the quantity β is physically equal to the total energy E of the system [see Problem 12.92(a)]. The result (9) with $\beta = E$ illustrates equation (31) on page 316.

12.16. Use Hamilton-Jacobi methods to solve Kepler's problem for a particle in an inverse square central force field.

The Hamiltonian is

$$H = \frac{1}{2m}\left(p_r^2 + \frac{p_\theta^2}{r^2}\right) - \frac{K}{r} \tag{1}$$

Then since $p_r = \partial\mathcal{S}/\partial r$, $p_\theta = \partial\mathcal{S}/\partial\theta$, the Hamilton-Jacobi equation is

$$\frac{\partial\mathcal{S}}{\partial t} + \frac{1}{2m}\left\{\left(\frac{\partial\mathcal{S}}{\partial r}\right)^2 + \frac{1}{r^2}\left(\frac{\partial\mathcal{S}}{\partial\theta}\right)^2\right\} - \frac{K}{r} = 0 \tag{2}$$

Let

$$\mathcal{S} = S_1(r) + S_2(\theta) + S_3(t) \tag{3}$$

Then (2) becomes $$\frac{1}{2m}\left\{\left(\frac{dS_1}{dr}\right)^2 + \frac{1}{r^2}\left(\frac{dS_2}{d\theta}\right)^2\right\} - \frac{K}{r} = -\frac{dS_3}{dt}$$

Setting both sides equal to the constant β_3, we find

$$dS_3/dt = -\beta_3 \tag{4}$$

$$\frac{1}{2m}\left\{\left(\frac{dS_1}{dr}\right)^2 + \frac{1}{r^2}\left(\frac{dS_2}{d\theta}\right)^2\right\} - \frac{K}{r} = \beta_3 \tag{5}$$

Integration of (4) yields, apart from a constant of integration,

$$S_3 = -\beta_3 t$$

Multiply both sides of (5) by $2mr^2$ and write it in the form

$$\left(\frac{dS_2}{d\theta}\right)^2 = r^2\left\{2m\beta_3 + \frac{2mK}{r} - \left(\frac{dS_1}{dr}\right)^2\right\}$$

Then since one side depends only on θ while the other side depends only on r, it follows that each side is a constant. Thus

$$dS_2/d\theta = \beta_2 \quad \text{or} \quad S_2 = \beta_2\theta \tag{6}$$

and $$r^2\left\{2m\beta_3 + \frac{2mK}{r} - \left(\frac{dS_1}{dr}\right)^2\right\} = \beta_2^2$$

or $$\frac{dS_1}{dr} = \sqrt{2m\beta_3 + 2mK/r - \beta_2^2/r^2} \tag{7}$$

on taking the positive square root. Then

$$S_1 = \int \sqrt{2m\beta_3 + 2mK/r - \beta_2^2/r^2}\, dr \tag{8}$$

Thus $$\mathcal{S} = \int \sqrt{2m\beta_3 + 2mK/r - \beta_2^2/r^2}\, dr + \beta_2\theta - \beta_3 t \tag{9}$$

Identifying β_2 and β_3 with the new momenta P_r and P_θ respectively, we have

$$Q_r = \frac{\partial\mathcal{S}}{\partial\beta_2} = \frac{\partial}{\partial\beta_2}\int \sqrt{2m\beta_3 + 2mK/r - \beta_2^2/r^2}\, dr + \theta = \gamma_1$$

$$Q_\theta = \frac{\partial\mathcal{S}}{\partial\beta_3} = \frac{\partial}{\partial\beta_3}\int \sqrt{2m\beta_3 + 2mK/r - \beta_2^2/r^2}\, dr - t = \gamma_2$$

since Q_r and Q_θ are constants, say γ_1 and γ_2. On performing the differentiations with respect to β_2 and β_3, we find

$$\int \frac{\beta_2\, dr}{r^2\sqrt{2m\beta_3 + 2mK/r - \beta_2^2/r^2}} = \theta - \gamma_1 \tag{10}$$

$$\int \frac{m\, dr}{\sqrt{2m\beta_3 + 2mK/r - \beta_2^2/r^2}} = t + \gamma_2 \tag{11}$$

The integral in (10) can be evaluated by using the substitution $r = 1/u$, and after integrating we find as the equation of the orbit,

$$r = \frac{\beta_2^2/mK}{1 - \sqrt{1 + 2\beta_3\beta_2^2/mK^2}\cos(\theta + \pi/2 - \gamma_1)} \tag{12}$$

The constant β_3 can be identified with the energy E [see Problem 12.92(b)], thus illustrating equation (31), page 316. If $E = \beta_3 < 0$, the orbit is an ellipse; if $E = \beta_3 = 0$, it is a parabola; and if $E = \beta_3 > 0$, it is a hyperbola. This agrees with the results of Chapter 5.

The equation (11) when integrated yields the position as a function of time.

PHASE INTEGRALS AND ANGLE VARIABLES

12.17. Let $\mathcal{S}$ be a complete solution of the Hamilton-Jacobi equation containing the n constants $\beta_1, \ldots, \beta_n$. Let $J_\alpha = \oint p_\alpha \, dq_\alpha$. Prove that the J_α are functions of the β_α only.

We have
$$\mathcal{S} = S_1(q_1, \beta_1, \ldots, \beta_n) + \cdots + S_n(q_n, \beta_1, \ldots, \beta_n) - \beta_1 t \tag{1}$$
where the constant $\beta_1 = E$, the total energy. Now
$$p_\alpha = \frac{\partial \mathcal{S}}{\partial q_\alpha} = \frac{dS_\alpha}{dq_\alpha} \tag{2}$$
Thus
$$J_\alpha = \oint p_\alpha \, dq_\alpha = \oint \frac{dS_\alpha}{dq_\alpha} \, dq_\alpha \tag{3}$$
But in this integration q_α is integrated out, so that the only quantities remaining are the constants $\beta_1, \ldots, \beta_n$. Thus we have the n equations
$$J_\alpha = J_\alpha(\beta_1, \ldots, \beta_n) \qquad \alpha = 1, \ldots, n \tag{4}$$
Using (4) we can solve for $\beta_1, \ldots, \beta_n$ in terms of $J_1, \ldots, J_n$ and express (1) in terms of the J_α.

12.18. (a) Suppose that the new position and momentum coordinates are taken to be w_α and J_α respectively. Prove that if $\mathcal{H}$ is the new Hamiltonian,
$$\dot{J}_\alpha = -\partial \mathcal{H}/\partial w_\alpha, \qquad \dot{w}_\alpha = \partial \mathcal{H}/\partial J_\alpha$$

(b) Deduce from (a) that
$$J_\alpha = \text{constant} \qquad \text{and} \qquad w_\alpha = f_\alpha t + c_\alpha$$
where f_α and c_α are constants and $f_\alpha = \partial \mathcal{H}/\partial J_\alpha$.

(a) By Hamilton's equations for the canonical coordinates Q_α, P_α,
$$\dot{P}_\alpha = -\partial \mathcal{H}/\partial Q_\alpha, \qquad \dot{Q}_\alpha = \partial \mathcal{H}/\partial P_\alpha \tag{1}$$
Then since the new position and momentum coordinates are taken as $Q_\alpha = w_\alpha$ and $P_\alpha = J_\alpha$, these equations become
$$\dot{J}_\alpha = -\partial \mathcal{H}/\partial w_\alpha, \qquad \dot{w}_\alpha = \partial \mathcal{H}/\partial J_\alpha \tag{2}$$

(b) Since $\mathcal{H} = E$, the new Hamiltonian depends only on the J_α and not on the w_α. Thus from (2) we have
$$\dot{J}_\alpha = 0, \qquad \dot{w}_\alpha = \text{constant} = f_\alpha \tag{3}$$
where $f_\alpha = \partial \mathcal{H}/\partial J_\alpha$. From (3) we find, as required,
$$J_\alpha = \text{constant}, \qquad w_\alpha = f_\alpha t + c_\alpha \tag{4}$$
The quantities J_α are called *action variables* while the corresponding integrals
$$\oint p_\alpha \, dq_\alpha = J_\alpha \tag{5}$$
where the integration is performed over a complete cycle of the coordinate q_α, are called *phase integrals*. The quantities w_α are called *angle variables*.

12.19. (a) Let Δw_α denote the change in w_α corresponding to a complete cycle in the particular coordinate q_r. Prove that
$$\Delta w_\alpha = \begin{cases} 1 & \text{if } \alpha = r \\ 0 & \text{if } \alpha \neq r \end{cases}$$

(*b*) Give a physical interpretation to the result in (*a*).

(*a*)
$$\Delta w_\alpha = \oint \frac{\partial w_\alpha}{\partial q_r} dq_r = \oint \frac{\partial}{\partial q_r}\left(\frac{\partial S}{\partial J_\alpha}\right) dq_r = \oint \frac{\partial}{\partial J_\alpha}\left(\frac{\partial S}{\partial q_r}\right) dq_r$$
$$= \frac{\partial}{\partial J_\alpha}\oint \frac{\partial S}{\partial q_r} dq_r = \frac{\partial J_r}{\partial J_\alpha} = \begin{cases} 1 & \text{if } \alpha = r \\ 0 & \text{if } \alpha \neq r \end{cases}$$

where we have used the fact that $w_\alpha = \partial S/\partial J_\alpha$ [see Problems 12.17 and 12.18] and have assumed that the order of differentiation and integration is immaterial.

(*b*) From (*a*) it follows that w_α changes by one when q_α goes through a complete cycle but that there is no change when any other q goes through a complete cycle. It follows that q_α is a periodic function of w_α of period one. Physically this means that the f_α in equation (*4*) of Problem 12.18 are *frequencies*.

12.20. Determine the frequency of the harmonic oscillator of Problem 12.15.

A complete cycle of the coordinate q [see equation (*10*), Problem 12.15] consists in the motion from $q = -\sqrt{2\beta/\kappa}$ to $q = +\sqrt{2\beta/\kappa}$ and back to $q = -\sqrt{2\beta/\kappa}$. Then the action variable is

$$J = \oint p\,dq = 2\int_{-\sqrt{2\beta/\kappa}}^{\sqrt{2\beta/\kappa}} \sqrt{2m(\beta - \tfrac{1}{2}\kappa q^2)}\,dq = 4\int_0^{\sqrt{2\beta/\kappa}} \sqrt{2m(\beta - \tfrac{1}{2}\kappa q^2)}\,dq$$
$$= 2\pi\beta\sqrt{m/\kappa}$$

Thus
$$\beta = E = \frac{J}{2\pi}\sqrt{\frac{\kappa}{m}} = \mathcal{H} \qquad \text{and} \qquad f = \frac{\partial \mathcal{H}}{\partial J} = \frac{1}{2\pi}\sqrt{\frac{\kappa}{m}}$$

12.21. Determine the frequency of the Kepler problem [see Problem 12.16].

A complete cycle of the coordinate r consists in the motion from $r = r_{\min}$ to $r_{\max}$ and back to $r = r_{\min}$, where $r_{\min}$ and $r_{\max}$ are the minimum and maximum values of r given by the zeros of the quadratic equation [see equation (*10*), Problem 12.16]

$$2m\beta_3 + 2mK/r - \beta_2^2/r^2 = 0 \tag{1}$$

We then have from equations (*6*) and (*7*) of Problem 12.16,

$$J_\theta = \oint p_\theta\,d\theta = \oint \frac{\partial \mathcal{S}}{\partial \theta} d\theta = \oint \frac{dS_2}{d\theta} d\theta = \int_0^{2\pi} \beta_2\,d\theta = 2\pi\beta_2 \tag{2}$$

$$J_r = \oint p_r\,dr = \oint \frac{\partial \mathcal{S}}{\partial r} dr = \oint \frac{dS_1}{dr} dr = 2\int_{r_{\min}}^{r_{\max}} \sqrt{2m\beta_3 + 2mK/r - \beta_2^2/r^2}\,dr$$
$$= 2\pi mK/\sqrt{-2m\beta_3} - 2\pi\beta_2 \tag{3}$$

From (*2*) and (*3*) we have on elimination of β_2,

$$J_\theta + J_r = 2\pi mK/\sqrt{-2m\beta_3} \tag{4}$$

Since $\beta_3 = E$, (*4*) yields

$$E = -\frac{2\pi^2 mK^2}{(J_\theta + J_r)^2} \qquad \text{so that} \qquad \mathcal{H} = -\frac{2\pi^2 mK^2}{(J_\theta + J_r)^2}$$

Then the frequencies are

$$f_\theta = \frac{\partial \mathcal{H}}{\partial J_\theta} = \frac{4\pi^2 mK^2}{(J_\theta + J_r)^3}, \qquad f_r = \frac{\partial \mathcal{H}}{\partial J_r} = \frac{4\pi^2 mK^2}{(J_\theta + J_r)^3}$$

Since these two frequencies are the same, i.e. there is only one frequency, we say that the system is *degenerate*.

MISCELLANEOUS PROBLEMS

12.22. A particle of mass m moves in a force field of potential V. Write (*a*) the Hamiltonian and (*b*) Hamilton's equations in spherical coordinates (r, θ, ϕ).

(*a*) The kinetic energy in spherical coordinates is

$$T = \tfrac{1}{2}m(\dot{r}^2 + r^2\dot{\theta}^2 + r^2 \sin^2\theta\,\dot{\phi}^2) \qquad (1)$$

Then the Lagrangian is

$$L = T - V = \tfrac{1}{2}m(\dot{r}^2 + r^2\dot{\theta}^2 + r^2 \sin^2\theta\,\dot{\phi}^2) - V(r,\theta,\phi) \qquad (2)$$

We have

$$p_r = \partial L/\partial\dot{r} = m\dot{r}, \quad p_\theta = \partial L/\partial\dot{\theta} = mr^2\dot{\theta}, \quad p_\phi = \partial L/\partial\dot{\phi} = mr^2\sin^2\theta\,\dot{\phi} \qquad (3)$$

and

$$\dot{r} = \frac{p_r}{m}, \quad \dot{\theta} = \frac{p_\theta}{mr^2}, \quad \dot{\phi} = \frac{p_\phi}{mr^2\sin^2\theta} \qquad (4)$$

The Hamiltonian is given by

$$\begin{aligned} H &= \sum p_\alpha \dot{q}_\alpha - L \\ &= p_r\dot{r} + p_\theta\dot{\theta} + p_\phi\dot{\phi} - \tfrac{1}{2}m(\dot{r}^2 + r^2\dot{\theta}^2 + r^2\sin^2\theta\,\dot{\phi}^2) + V(r,\theta,\phi) \\ &= \frac{p_r^2}{2m} + \frac{p_\theta^2}{2mr^2} + \frac{p_\phi^2}{2mr^2\sin^2\theta} + V(r,\theta,\phi) \end{aligned} \qquad (5)$$

where we have used the results of equations (*4*).

We can also obtain (*5*) directly by using the fact that for conservative systems the Hamiltonian is the total energy, i.e. $H = T + V$.

(*b*) Hamilton's equations are $\dot{q}_\alpha = \dfrac{\partial H}{\partial p_\alpha}$, $\dot{p}_\alpha = -\dfrac{\partial H}{\partial q_\alpha}$. Then from part (*a*),

$$\dot{r} = \frac{\partial H}{\partial p_r} = \frac{p_r}{m}, \quad \dot{\theta} = \frac{\partial H}{\partial p_\theta} = \frac{p_\theta}{mr^2}, \quad \dot{\phi} = \frac{\partial H}{\partial p_\phi} = \frac{p_\phi}{mr^2\sin^2\theta}$$

$$\dot{p}_r = -\frac{\partial H}{\partial r} = \frac{p_\theta^2}{mr^3} + \frac{p_\phi^2}{mr^3\sin^2\theta} - \frac{\partial V}{\partial r}$$

$$\dot{p}_\theta = -\frac{\partial H}{\partial \theta} = \frac{p_\phi^2\cos\theta}{mr^2\sin^3\theta} - \frac{\partial V}{\partial\theta}$$

$$\dot{p}_\phi = -\frac{\partial H}{\partial\phi} = -\frac{\partial V}{\partial\phi}$$

12.23. A particle of mass m moves in a force field whose potential in spherical coordinates is $V = -(K\cos\theta)/r^2$. Write the Hamilton-Jacobi equation describing its motion.

By Problem 12.22 the Hamiltonian is

$$H = \frac{1}{2m}\left(p_r^2 + \frac{p_\theta^2}{r^2} + \frac{p_\phi^2}{r^2\sin^2\theta}\right) - \frac{K\cos\theta}{r^2} \qquad (1)$$

Writing $p_r = \dfrac{\partial \mathcal{S}}{\partial r}$, $p_\theta = \dfrac{\partial \mathcal{S}}{\partial \theta}$, $p_\phi = \dfrac{\partial \mathcal{S}}{\partial\phi}$, the required Hamilton-Jacobi equation is

$$\frac{\partial\mathcal{S}}{\partial t} + \frac{1}{2m}\left\{\left(\frac{\partial\mathcal{S}}{\partial r}\right)^2 + \frac{1}{r^2}\left(\frac{\partial\mathcal{S}}{\partial\theta}\right)^2 + \frac{1}{r^2\sin^2\theta}\left(\frac{\partial\mathcal{S}}{\partial\phi}\right)^2\right\} - \frac{K\cos\theta}{r^2} = 0 \qquad (2)$$

12.24. (*a*) Find a complete solution of the Hamilton-Jacobi equation of Problem 12.23 and (*b*) indicate how the motion of the particle can be determined.

(*a*) Letting $\mathcal{S} = S_1(r) + S_2(\theta) + S_3(\phi) - Et$ in equation (*2*) of Problem 12.23, it can be written

$$\frac{1}{2m}\left(\frac{dS_1}{dr}\right)^2 + \frac{1}{2mr^2}\left(\frac{dS_2}{d\theta}\right)^2 + \frac{1}{2mr^2\sin^2\theta}\left(\frac{dS_3}{d\phi}\right)^2 - \frac{K\cos\theta}{r^2} = E \qquad (1)$$

Multiplying equation (*1*) by $2mr^2$ and rearranging terms,

$$r^2\left(\frac{dS_1}{dr}\right)^2 - 2mEr^2 = -\left(\frac{dS_2}{d\theta}\right)^2 - \frac{1}{\sin^2\theta}\left(\frac{dS_3}{d\phi}\right)^2 + 2mK\cos\theta$$

Since the left side depends only on r while the right side depends on θ and ϕ, it follows that each side must be a constant which we shall call β_1. Thus

$$r^2\left(\frac{dS_1}{dr}\right)^2 - 2mEr^2 = \beta_1 \qquad (2)$$

and

$$-\left(\frac{dS_2}{d\theta}\right)^2 - \frac{1}{\sin^2\theta}\left(\frac{dS_3}{d\phi}\right)^2 + 2mK\cos\theta = \beta_1 \qquad (3)$$

Multiplying equation (*3*) by $\sin^2\theta$ and rearranging terms,

$$\left(\frac{dS_3}{d\phi}\right)^2 = 2mK\sin^2\theta\cos\theta - \beta_1\sin^2\theta - \sin^2\theta\left(\frac{dS_2}{d\theta}\right)^2 \qquad (4)$$

Since the left side depends only on ϕ while the right side depends only on θ each side must be a constant which we can call β_2. However, since

$$p_\phi = \frac{\partial\mathcal{S}}{\partial\phi} = \frac{dS_3}{d\phi} \qquad (5)$$

we can write $\beta_2 = p_\phi^2$. This is a consequence of the fact that ϕ is a cyclic or ignorable coordinate. Then (*4*) becomes

$$2mK\sin^2\theta\cos\theta - \beta_1\sin^2\theta - \sin^2\theta\left(\frac{dS_2}{d\theta}\right)^2 = p_\phi^2 \qquad (6)$$

By solving equations (*2*), (*6*) and (*5*), we obtain

$$S_1 = \int\sqrt{2mE + \beta_1/r^2}\,dr, \qquad S_2 = \int\sqrt{2mK\cos\theta - p_\phi^2\csc^2\theta - \beta_1}\,d\theta, \qquad S_3 = p_\phi\phi$$

where we have chosen the positive square roots and omitted arbitrary additive constants. The complete solution is

$$\mathcal{S} = \int\sqrt{2mE + \beta_1/r^2}\,dr + \int\sqrt{2mK\cos\theta - p_\phi^2\csc^2\theta - \beta_1}\,d\theta + p_\phi\phi - Et$$

(*b*) The required equations of motion are found by writing

$$\frac{\partial\mathcal{S}}{\partial\beta_1} = \gamma_1, \qquad \frac{\partial\mathcal{S}}{\partial E} = \gamma_2, \qquad \frac{\partial\mathcal{S}}{\partial p_\phi} = \gamma_3$$

and then solving these to obtain the coordinates r, θ, ϕ as functions of time using initial conditions to evaluate the arbitrary constants.

12.25. If the functions F and G depend on the position coordinates q_α, momenta p_α and time t, the *Poisson bracket* of F and G is defined as

$$[F, G] = \sum_\alpha\left(\frac{\partial F}{\partial p_\alpha}\frac{\partial G}{\partial q_\alpha} - \frac{\partial F}{\partial q_\alpha}\frac{\partial G}{\partial p_\alpha}\right)$$

Prove that (*a*) $[F, G] = -[G, F]$, (*b*) $[F_1 + F_2, G] = [F_1, G] + [F_2, G]$, (*c*) $[F, q_r] = \partial F/\partial p_r$, (*d*) $[F, p_r] = -\partial F/\partial q_r$.

(*a*) $$[F, G] = \sum_\alpha\left(\frac{\partial F}{\partial p_\alpha}\frac{\partial G}{\partial q_\alpha} - \frac{\partial F}{\partial q_\alpha}\frac{\partial G}{\partial p_\alpha}\right) = -\sum_\alpha\left(\frac{\partial G}{\partial p_\alpha}\frac{\partial F}{\partial q_\alpha} - \frac{\partial G}{\partial q_\alpha}\frac{\partial F}{\partial p_\alpha}\right) = -[G, F]$$

This shows that the Poisson bracket does not obey the *commutative law of algebra*.

(b)
$$[F_1+F_2, G] = \sum_\alpha \left\{\frac{\partial(F_1+F_2)}{\partial p_\alpha}\frac{\partial G}{\partial q_\alpha} - \frac{\partial(F_1+F_2)}{\partial q_\alpha}\frac{\partial G}{\partial p_\alpha}\right\}$$
$$= \sum_\alpha\left(\frac{\partial F_1}{\partial p_\alpha}\frac{\partial G}{\partial q_\alpha} - \frac{\partial F_1}{\partial q_\alpha}\frac{\partial G}{\partial p_\alpha}\right) + \sum_\alpha\left(\frac{\partial F_2}{\partial p_\alpha}\frac{\partial G}{\partial q_\alpha} - \frac{\partial F_2}{\partial q_\alpha}\frac{\partial G}{\partial p_\alpha}\right)$$
$$= [F_1, G] + [F_2, G]$$

This shows that the Poisson bracket obeys the *distributive law of algebra.*

(c)
$$[F, q_r] = \sum_\alpha\left(\frac{\partial F}{\partial p_\alpha}\frac{\partial q_r}{\partial q_\alpha} - \frac{\partial F}{\partial q_\alpha}\frac{\partial q_r}{\partial p_\alpha}\right) = \frac{\partial F}{\partial p_r}$$

since $\partial q_r/\partial q_\alpha = 1$ for $\alpha = r$ and 0 for $\alpha \neq r$, while $\partial q_r/\partial p_\alpha = 0$ for all α. Since r is arbitrary, the required result follows.

(d)
$$[F, p_r] = \sum_\alpha\left(\frac{\partial F}{\partial p_\alpha}\frac{\partial p_r}{\partial q_\alpha} - \frac{\partial F}{\partial q_\alpha}\frac{\partial p_r}{\partial p_\alpha}\right) = -\frac{\partial F}{\partial q_r}$$

since $\partial p_r/\partial q_\alpha = 0$ for all α, while $\partial p_r/\partial p_\alpha = 1$ for $\alpha = r$ and 0 for $\alpha \neq r$. Since r is arbitrary, the required result follows.

12.26. If H is the Hamiltonian, prove that if f is any function depending on position, momenta and time, then
$$\frac{df}{dt} = \frac{\partial f}{\partial t} + [H, f]$$

$$df = \frac{\partial f}{\partial t}dt + \sum_\alpha\left(\frac{\partial f}{\partial q_\alpha}dq_\alpha + \frac{\partial f}{\partial p_\alpha}dp_\alpha\right) \tag{1}$$

or
$$\frac{df}{dt} = \frac{\partial f}{\partial t} + \sum_\alpha\left(\frac{\partial f}{\partial q_\alpha}\dot{q}_\alpha + \frac{\partial f}{\partial p_\alpha}\dot{p}_\alpha\right) \tag{2}$$

But by Hamilton's equations,
$$\dot{q}_\alpha = \frac{\partial H}{\partial p_\alpha}, \quad \dot{p}_\alpha = -\frac{\partial H}{\partial q_\alpha} \tag{3}$$

Then (*2*) can be written
$$\frac{df}{dt} = \frac{\partial f}{\partial t} + \sum_\alpha\left(\frac{\partial f}{\partial q_\alpha}\frac{\partial H}{\partial p_\alpha} - \frac{\partial f}{\partial p_\alpha}\frac{\partial H}{\partial q_\alpha}\right) = \frac{\partial f}{\partial t} + [H, f]$$

Supplementary Problems

THE HAMILTONIAN AND HAMILTON'S EQUATIONS

12.27. A particle of mass m moves in a force field of potential V. (*a*) Write the Hamiltonian and (*b*) Hamilton's equations in rectangular coordinates (x, y, z).

Ans. (*a*) $H = (p_x^2 + p_y^2 + p_z^2)/2m + V(x, y, z)$

(*b*) $\dot{x} = p_x/m$, $\dot{y} = p_y/m$, $\dot{z} = p_z/m$, $\dot{p}_x = -\partial V/\partial x$, $\dot{p}_y = -\partial V/\partial y$, $\dot{p}_z = -\partial V/\partial z$

12.28. Use Hamilton's equations to obtain the motion of a particle of mass m down a frictionless inclined plane of angle α.

12.29. Work the problem of small oscillations of a simple pendulum by using Hamilton's equations.

12.30. Use Hamilton's equations to obtain the motion of a projectile launched with speed v_0 at angle α with the horizontal.

12.31. Using Hamilton's equations, work the problem of the harmonic oscillator in (*a*) one dimension, (*b*) two dimensions, (*c*) three dimensions.

12.32. Work Problem 3.27, page 78 by using Hamilton's equation.

PHASE SPACE AND LIOUVILLE'S THEOREM

12.33. Explain why the path of a phase point in phase space which represents the motion of a system of particles can never cross itself.

12.34. Carry out the details in the proof of Liouville's theorem for the case of two degrees of freedom.

CALCULUS OF VARIATIONS AND HAMILTON'S PRINCIPLE

12.35. Use the methods of the calculus of variations to find that curve connecting two fixed points in a plane which has the shortest length.

12.36. Prove that if the function F in the integral $\int_a^b F(x, y, y')\,dx$ is independent of x, then the integral is an extremum if $F - y'F_{y'} = c$ where c is a constant.

12.37. Use the result of Problem 12.36 to solve (*a*) Problem 12.9, page 322, (*b*) Problem 12.35.

12.38. It is desired to revolve the curve of Fig. 12-5 having endpoints fixed at $P(x_1, y_1)$ and $Q(x_2, y_2)$ about the x axis so that the area I of the surface of revolution is a minimum.

(*a*) Show that $I = 2\pi \int_{x_1}^{x_2} y\sqrt{1 + y'^2}\,dx$.

(*b*) Obtain the differential equation of the curve.

(*c*) Prove that the required curve is a catenary.

Ans. (*b*) $yy'' = 1 + (y')^2$

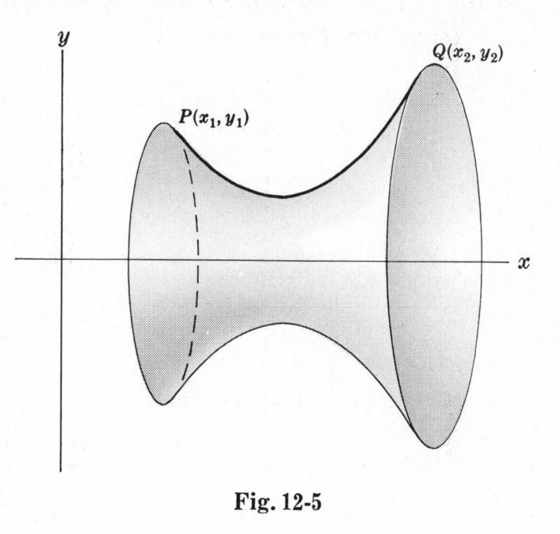

Fig. 12-5

12.39. Two identical circular wires in contact are placed in a soap solution and then separated so as to form a soap film. Explain why the shape of the soap film surface is related to the result of Problem 12.38.

12.40. Use Hamilton's principle to find the motion of a simple pendulum.

12.41. Work the problem of a projectile by using Hamilton's principle.

12.42. Use Hamilton's principle to find the motion of a solid cylinder rolling down an inclined plane of angle α.

CANONICAL TRANSFORMATIONS AND GENERATING FUNCTIONS

12.43. Prove that the transformation $Q = p,\ P = -q$ is canonical.

12.44. Prove that the transformation $Q = q \tan p,\ P = \ln \sin p$ is canonical.

12.45. (*a*) Prove that the Hamiltonian for a harmonic oscillator can be written in the form $H = \frac{1}{2}p^2/m + \frac{1}{2}\kappa q^2$.

(*b*) Prove that the transformation $q = \sqrt{P/\sqrt{\kappa}}\,\sin Q,\ p = \sqrt{mP\sqrt{\kappa}}\,\cos Q$ is canonical.

(*c*) Express the Hamiltonian of part (*a*) in terms of P and Q and show that Q is cyclic.

(*d*) Obtain the solution of the harmonic oscillator by using the above results.

12.46. Prove that the generating function giving rise to the canonical transformation in Problem 12.45(*b*) is $S = \frac{1}{2}\sqrt{\kappa}\, q^2 \cot Q$.

12.47. Prove that the result of two or more successive canonical transformations is also canonical.

12.48. Let $\mathcal{U}$ be a generating function dependent only on Q_α, p_α, t. Prove that

$$P_\alpha = -\frac{\partial \mathcal{U}}{\partial Q_\alpha}, \quad q_\alpha = -\frac{\partial \mathcal{U}}{\partial p_\alpha}, \quad \mathcal{H} = \frac{\partial \mathcal{U}}{\partial t} + H$$

12.49. Let $\mathcal{V}$ be a generating function dependent only on the old and new momenta p_α and P_α respectively and the time t. Prove that

$$q_\alpha = -\frac{\partial \mathcal{V}}{\partial p_\alpha}, \quad Q_\alpha = \frac{\partial \mathcal{V}}{\partial P_\alpha}, \quad \mathcal{H} = \frac{\partial \mathcal{V}}{\partial t} + H$$

12.50. Prove that the generating function $\mathcal{U}$ of Problem 12.48 is related to the generating function $\mathcal{T}$ of Problem 12.12 by $\mathcal{U} = \mathcal{T} - \sum p_\alpha q_\alpha$.

12.51. Prove that the generating function $\mathcal{V}$ of Problem 12.49 is related to the generating function $\mathcal{T}$ Problem 12.12 by $\mathcal{V} = \mathcal{T} + \sum P_\alpha Q_\alpha - \sum p_\alpha q_\alpha$.

THE HAMILTON-JACOBI EQUATION

12.52. Use the Hamilton-Jacobi method to determine the motion of a particle falling vertically in a uniform gravitational field.

12.53. (*a*) Set up the Hamilton-Jacobi equation for the motion of a particle sliding down a frictionless inclined plane of angle α. (*b*) Solve the Hamilton-Jacobi equation in (*a*) and thus determine the motion of the particle.

12.54. Work the problem of a projectile launched with speed v_0 at angle α with the horizontal by using Hamilton-Jacobi methods.

12.55. Use Hamilton-Jacobi methods to describe the motion and find the frequencies of a harmonic oscillator in (*a*) 2 dimensions, (*b*) 3 dimensions.

12.56. Use Hamilton-Jacobi methods to arrive at the generating function of Problem 12.46.

PHASE INTEGRALS AND ANGLE VARIABLES

12.57. Use the method of phase integrals and angle variables to find the frequency of a simple pendulum of length l, assuming that oscillations are small. *Ans.* $\frac{1}{2\pi}\sqrt{\frac{g}{l}}$

12.58. Find the frequencies of (*a*) a 2 dimensional harmonic oscillator, (*b*) a 3 dimensional harmonic oscillator.

12.59. Obtain the frequency of small oscillations of a compound pendulum by using phase integrals.

12.60. Two equal masses m connected by equal springs to fixed walls at A and B are free to slide in a line on a frictionless plane AB [see Fig. 12-6]. Using phase integrals determine the frequencies of the normal modes.

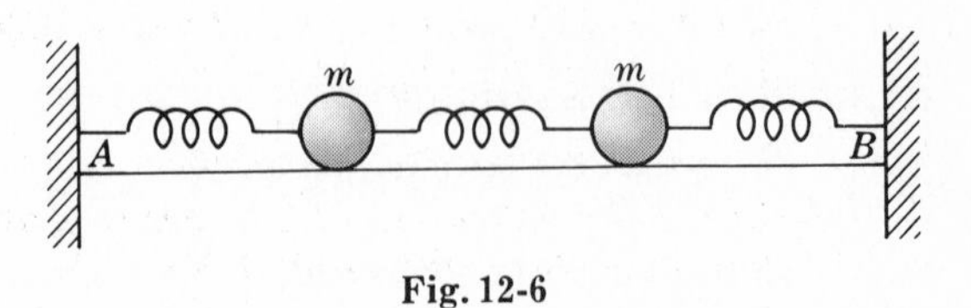

Fig. 12-6

12.61. Discuss Problem 12.57 if oscillations are not assumed small.

MISCELLANEOUS PROBLEMS

12.62. A particle of mass m moves in a force field having potential $V(\rho, \phi, z)$ where ρ, ϕ, z are cylindrical coordinates. Give (*a*) the Hamiltonian and (*b*) Hamilton's equations for the particle.

Ans. (*a*) $H = (p_\rho^2 + p_\phi^2/\rho^2 + p_z^2)/2m + V(\rho, \phi, z)$

(*b*) $\dot{\rho} = p_\rho/m, \ \dot{\phi} = p_\phi/m\rho^2, \ \dot{z} = p_z/m, \ \dot{p}_\rho = p_\phi^2/m\rho^3 - \partial V/\partial\rho, \ \dot{p}_\phi = -\partial V/\partial\phi, \ \dot{p}_z = -\partial V/\partial z$

12.63. A particle of mass m which moves in a plane relative to a fixed set of axes has a Hamiltonian given by the total energy. Find the Hamiltonian relative to a set of axes which rotates at constant angular velocity ω relative to the fixed axes.

12.64. Set up the Hamiltonian for a double pendulum. Use Hamilton-Jacobi methods to determine the normal frequencies for the case of small vibrations.

12.65. Prove that a necessary condition for $I = \int_{t_1}^{t_2} F(t, x, \dot{x}, \ddot{x})\, dt$ to be an extremum is that

$$\frac{\partial F}{\partial x} - \frac{d}{dt}\left(\frac{\partial F}{\partial \dot{x}}\right) + \frac{d^2}{dt^2}\left(\frac{\partial F}{\partial \ddot{x}}\right) = 0$$

Can you generalize this result?

12.66. Work Problem 3.22, page 76, by Hamiltonian methods.

12.67. A particle of mass m moves on the inside of a frictionless vertical cone having equation $x^2 + y^2 = z^2 \tan^2 \alpha$. (*a*) Write the Hamiltonian and (*b*) Hamilton's equations using cylindrical coordinates.

Ans. (*a*) $H = \dfrac{p_\rho^2 \sin^2 \alpha}{2m} + \dfrac{p_\phi^2}{2m\rho^2} + mg\rho \cot \alpha$

(*b*) $\dot{\rho} = \dfrac{p_\rho \sin^2 \alpha}{m}, \quad \dot{p}_\rho = \dfrac{p_\phi^2}{m\rho^3} - mg \cot \alpha$

12.68. Use the results of Problem 12.67 to prove that there will be a stable orbit in any horizontal plane $z = h > 0$, and find the frequency in this orbit.

12.69. Prove that the product of a position coordinate and its canonically conjugate momentum must have the dimension of *action* or energy multiplied by time, i.e. ML^2T^{-1}.

12.70. Perform the integration of equation (*10*) of Problem 12.16 and compare with the solution of the Kepler problem in Chapter 5.

12.71. Verify the integration result (*3*) of Problem 12.21.

12.72. Prove that Euler's equation (*9*), page 313, can be written as

$$y'' \frac{\partial^2 F}{\partial y'^2} + y' \frac{\partial^2 F}{\partial y' \partial y} + \frac{\partial^2 F}{\partial y' \partial x} - \frac{\partial F}{\partial y} = 0$$

12.73. A man can travel by boat with speed v_1 and can walk with speed v_2. Referring to Fig. 12-7, prove that in order to travel from point A on one side of a river bank to a point B on the other side in the least time he must land his boat at point P where angles θ_1 and θ_2 are such that

$$\frac{\sin \theta_1}{\sin \theta_2} = \frac{v_1}{v_2}$$

Discuss the relationship of this result to the refraction of light in the theory of optics.

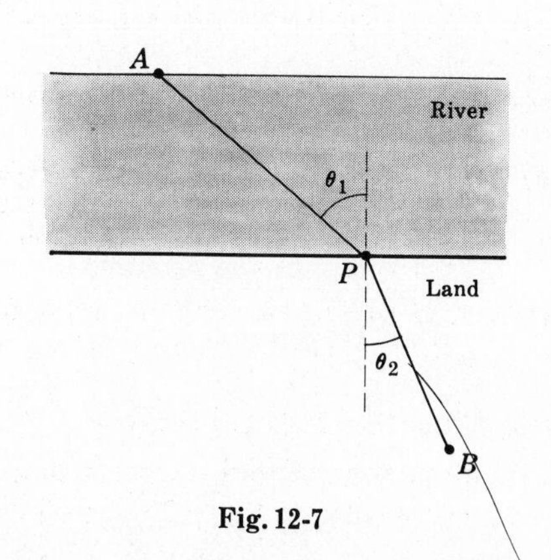

Fig. 12-7

12.74. Prove that if a particle moves under no external forces, i.e. it is a free particle, then the principle of least action becomes one of least time. Discuss the relationship of this result to Problem 12.73.

12.75. Derive the condition for reflection of light in optical theory by using the principle of least time.

12.76. It is desired to find the shape of a curve lying in a plane and having fixed endpoints such that its moment of inertia about an axis perpendicular to the plane and passing through a fixed origin is a minimum.

(a) Using polar coordinates (r, θ), show that the problem is equivalent to minimizing the integral

$$I = \int_{r=r_1}^{r_2} r^2\sqrt{1 + r^2(d\theta/dr)^2}\, dr$$

where the fixed endpoints of the wire are (r_1, θ_1), (r_2, θ_2).

(b) Write Euler's equation, thus obtaining the differential equation of the curve.

(c) Solve the differential equation obtained in (b) and thus find the equation of the curve.

Ans. (c) $r^3 = c_1 \sec(3\theta - c_2)$ where c_1 and c_2 are determined so that the curve passes through the fixed points.

12.77. Use the Hamilton-Jacobi method to set up the equations of motion of a spherical pendulum.

12.78. Use Hamilton-Jacobi methods to solve Problems 11.20, page 293, and 11.21, page 294.

12.79. If $[F, G]$ is the Poisson bracket [see Problems 12.25 and 12.26], prove that

(a) $[F_1 F_2, G] = F_1[F_2, G] + F_2[F_1, G]$

(b) $\dfrac{\partial}{\partial t}[F, G] = \left[\dfrac{\partial F}{\partial t}, G\right] + \left[F, \dfrac{\partial G}{\partial t}\right]$

(c) $\dfrac{d}{dt}[F, G] = \left[\dfrac{dF}{dt}, G\right] + \left[F, \dfrac{dG}{dt}\right]$

12.80. Prove that (a) $[q_\alpha, q_\beta] = 0$, (b) $[p_\alpha, p_\beta] = 0$, (c) $[p_\alpha, q_\beta] = \delta_{\alpha\beta}$

where $\delta_{\alpha\beta} = \begin{cases} 1 & \text{if } \alpha = \beta \\ 0 & \text{if } \alpha \neq \beta \end{cases}$ is called the *Kronecker delta*.

12.81. Evaluate $[H, t]$ where H is the Hamiltonian and t is the time. Are H and t canonically conjugate variables? Explain.

12.82. Prove *Jacobi's identity* for Poisson brackets

$$[F_1, [F_2, F_3]] + [F_2, [F_3, F_1]] + [F_3, [F_1, F_2]] = 0$$

12.83. Illustrate Liouville's theorem by using the one dimensional harmonic oscillator.

12.84. (a) Is the Lagrangian of a dynamical system unique? Explain.

(b) Discuss the uniqueness of the generalized momenta and Hamiltonian of a system.

12.85. (a) Set up the Hamiltonian for a string consisting of N particles [see Problem 8.29, page 215]

(b) Use Hamilton-Jacobi methods to find the normal modes and frequencies.

12.86. Prove that the Poisson bracket is invariant under a canonical transformation.

12.87. Prove that Liouville's theorem is equivalent to the result $\partial\rho/\partial t = [\rho, H]$.

12.88. (a) Let $Q_\alpha = \sum_{\mu=1}^{n} a_{\alpha\mu} q_\mu$, $P_\alpha = \sum_{\mu=1}^{n} b_{\alpha\mu} p_\mu$ where $a_{\alpha\mu}$ and $b_{\alpha\mu}$ are given constants and $\alpha = 1, 2, \ldots, n$. Prove that the transformation is canonical if and only if $b_{\alpha\mu} = \Delta_{\alpha\mu}/\Delta$ where Δ is the determinant.

$$\begin{vmatrix} a_{11} & a_{12} & \cdots & a_{1n} \\ a_{21} & a_{22} & \cdots & a_{2n} \\ \cdots & \cdots & \cdots & \cdots \\ a_{n1} & a_{n2} & \cdots & a_{nn} \end{vmatrix}$$

and $\Delta_{\alpha\mu}$ is the cofactor of the element $a_{\alpha\mu}$ in this determinant.

(b) Prove that the conditions in (a) are equivalent to the condition $\sum P_\alpha Q_\alpha = \sum p_\alpha q_\alpha$.

12.89. Prove that the path taken by a fixed point on a circle as it rolls along a given line is a cycloid.

12.90. (*a*) Express as an integral the total potential energy of a uniform chain whose ends are suspended from two fixed points. (*b*) Using the fact that for equilibrium the total potential energy is a minimum, use the calculus of variations to show that the equation of the curve in which the chain hangs is a *catenary* as in Problem 7.32, page 186. [*Hint.* Find the minimum of the integral subject to the constraint condition that the total length of the chain is a given constant.]

12.91. Use the methods of the calculus of variations to find the closed plane curve which encloses the largest area.

12.92. Prove that the constants (*a*) β in Problem 12.15 and (*b*) β_3 in Problem 12.16 can be identified with the total energy.

12.93. If the theory of relativity is taken into account in the motion of a particle of mass m in a force field of potential V, the Hamiltonian is given by

$$H = \sqrt{p^2c^2 + m^2c^4} + V$$

where c is the speed of light. Obtain the equations of motion for this particle.

12.94. Use Hamiltonian methods to solve the problem of a particle moving in an inverse cube force field.

12.95. Use spherical coordinates to solve Kepler's problem.

12.96. Suppose that m of the n coordinates $q_1, q_2, \ldots, q_n$ are cyclic [say the first m, i.e. $q_1, q_2, \ldots, q_m$]. Let

$$\mathcal{R} = \sum_{\alpha=1}^{m} c_\alpha \dot{q}_\alpha - L \qquad \text{where} \qquad c_\alpha = \partial L/\partial \dot{q}_\alpha$$

Prove that for $\alpha = m+1, \ldots, n$
$$\frac{d}{dt}\left(\frac{\partial \mathcal{R}}{\partial \dot{q}_\alpha}\right) = \frac{\partial \mathcal{R}}{\partial q_\alpha}$$

The function $\mathcal{R}$ is called *Routh's function* or the *Routhian*. By using it a problem involving n degrees of freedom is reduced to one involving $n - m$ degrees of freedom.

12.97. Using the properties
$$\delta L = \frac{\partial L}{\partial y}\delta y + \frac{\partial L}{\partial y'}\delta y', \qquad (\delta y)' = \delta y'$$

of the variational symbol δ [see Problem 12.6] and assuming that the operator δ can be brought under the integral sign, show how Lagrange's equations can be derived from Hamilton's principle.

12.98. Let $P = P(p, q)$, $Q = Q(p, q)$. Suppose that the Hamiltonian expressed in terms of p, q and P, Q are given by $H = H(p, q)$ and $\mathcal{H} = \mathcal{H}(P, Q)$ respectively. Prove that if

$$\dot{q} = \partial H/\partial p, \qquad \dot{p} = -\partial H/\partial q$$

then
$$\dot{Q} = \partial \mathcal{H}/\partial P, \qquad \dot{P} = -\partial \mathcal{H}/\partial Q$$

provided that the *Jacobian determinant* [or briefly *Jacobian*]

$$\frac{\partial(P, Q)}{\partial(p, q)} = \begin{vmatrix} \partial P/\partial p & \partial P/\partial q \\ \partial Q/\partial p & \partial Q/\partial q \end{vmatrix} = 1$$

Discuss the connection of the results with Hamiltonian theory.

12.99. (*a*) Set up the Hamiltonian for a solid cylinder rolling down an inclined plane of angle α.
(*b*) Write Hamilton's equations and deduce the motion of the cylinder from them.
(*c*) Use Hamilton-Jacobi methods to obtain the motion of the cylinder and compare with part (*b*).

12.100. Work Problem 7.22, page 180, by Hamilton-Jacobi methods.

12.101. Write (*a*) the Hamiltonian and (*b*) Hamilton's equations for a particle of charge e and mass m moving in an electromagnetic field [see Problem 11.90, page 309].

Ans. (*a*) $H = \dfrac{1}{2m}(\mathbf{p} - e\mathbf{A})^2 + e\Phi$

(*b*) $\mathbf{v} = \dfrac{1}{m}(\mathbf{p} - e\mathbf{A}), \quad \dot{\mathbf{p}} = -e\nabla\Phi + e\nabla(\mathbf{A}\cdot\mathbf{v})$

12.102. (*a*) Obtain the Hamilton-Jacobi equation for the motion of the particle in Problem 12.101. (*b*) Use the result to write equations for the motion of a charged particle in an electromagnetic field.

12.103. (*a*) Write the Hamiltonian for a symmetrical top and thus obtain the equations of motion. (*b*) Compare the results obtained in (*a*) with those of Chapter 10.

12.104. Prove Theorem 12.2, page 314.

12.105. An atom consists of an electron of charge $-e$ moving in a central force field $\mathbf{F}$ about a nucleus of charge Ze such that

$$\mathbf{F} = -\frac{Ze^2\mathbf{r}}{r^3}$$

where $\mathbf{r}$ is the position vector of the electron relative to the nucleus and Z is the atomic number. In Bohr's quantum theory of the atom the phase integrals are integer multiples of *Planck's constant h*, i.e.,

$$\oint p_r\,dr = n_1 h, \qquad \oint p_\theta\,d\theta = n_2 h$$

Using these equations, prove that there will be only a discrete set of energies given by

$$E_n = -\frac{2\pi^2 m Z^2 e^4}{n^2 h^2}$$

where $n = n_1 + n_2 = 1, 2, 3, 4, \ldots$ is called the *orbital quantum number*.

Appendix A

Units and Dimensions

UNITS

Standardized lengths, times and masses in terms of which other lengths, times and masses are measured are called *units*. For example, a distance can be measured in terms of a standard *foot* or *meter*. A time interval can be measured in terms of *seconds, hours* or *days*. A mass can be measured in terms of *pounds* or *grams*. Many different types of units are possible. However, there are four main types in use at the present time.

1. The CGS or centimeter-gram-second system.
2. The MKS or meter-kilogram-second system.
3. The FPS or foot-pound-second system.
4. The FSS or foot-slug-second system, also called the English gravitational or engineering system.

The first two are sometimes called *metric systems*, while the last two are sometimes called *English systems*. There is an increasing tendency to use metric systems.

The following indicates four consistent sets of units in these systems which can be used with the equation $F = ma$:

CGS System:	F (dynes)	=	m (grams)	×	a (cm/sec^2)
MKS System:	F (newtons)	=	m (kilograms)	×	a (m/sec^2)
FPS System:	F (poundals)	=	m (pounds)	×	a (ft/sec^2)
FSS System:	F (pounds weight)	=	m (slugs)	×	a (ft/sec^2)

In the third through sixth columns of the table on page 340, units of various quantities in these systems are given.

In the table on page 341, conversion factors among units of the various systems are given.

DIMENSIONS

The dimensions of all mechanical quantities may be expressed in terms of the fundamental dimensions of length L, mass M, and time T. In the second column of the table on page 340, the dimensions of various physical quantities are listed.

UNITS AND DIMENSIONS

Physical Quantity	Dimension	CGS System	MKS System	FPS System	FSS System
Length	L	cm	m	ft	ft
Mass	M	gm	kg	lb	slug
Time	T	sec	sec	sec	sec
Velocity	LT^{-1}	cm/sec	m/sec	ft/sec	ft/sec
Acceleration	LT^{-2}	cm/sec^2	m/sec^2	ft/sec^2	ft/sec^2
Force	MLT^{-2}	gm cm/sec^2 = dyne	kg m/sec^2 = newton	lb ft/sec^2 = poundal	slug ft/sec^2 = lbwt
Momentum, Impulse	MLT^{-1}	gm cm/sec = dyne sec	kg m/sec = nt sec	lb ft/sec = pdl sec	slug ft/sec = lbwt sec
Energy, Work	ML^2T^{-2}	gm cm^2/sec^2 = dyne cm = erg	kg m^2/sec^2 = nt m = joule	lb ft^2/sec^2 = ft pdl	slug ft^2/sec^2 = ft lbwt
Power	ML^2T^{-3}	gm cm^2/sec^3 = dyne cm/sec = erg/sec	kg m^2/sec^3 = joule/sec = watt	lb ft^2/sec^3 = ft pdl/sec	slug ft^2/sec^3 = ft lbwt/sec
Volume	L^3	cm^3	m^3	ft^3	ft^3
Density	ML^{-3}	gm/cm^3	kg/m^3	lb/ft^3	slug/ft^3
Angle	—	radian (rad)	rad	rad	rad
Angular velocity	T^{-1}	rad/sec	rad/sec	rad/sec	rad/sec
Angular acceleration	T^{-2}	rad/sec^2	rad/sec^2	rad/sec^2	rad/sec^2
Torque	ML^2T^{-2}	gm cm^2/sec^2 = dyne cm	kg m^2/sec^2 = nt m	lb ft^2/sec^2 = ft pdl	slug ft^2/sec^2 = ft lbwt
Angular momentum	ML^2T^{-1}	gm cm^2/sec	kg m^2/sec	lb ft^2/sec	slug ft^2/sec
Moment of inertia	ML^2	gm cm^2	kg m^2	lb ft^2	slug ft^2
Pressure	$ML^{-1}T^{-2}$	gm/(cm sec^2) = dyne/cm^2	kg/(m sec^2) = nt/m^2	pdl/ft^2	lbwt/ft^2

CONVERSION FACTORS

Length

1 kilometer (km)	=	1000 meters
1 meter (m)	=	100 centimeters
1 centimeter (cm)	=	10^{-2} m
1 millimeter (mm)	=	10^{-3} m
1 micron (μ)	=	10^{-6} m
1 millimicron (mμ)	=	10^{-9} m
1 angstrom (A)	=	10^{-10} m
1 inch (in.)	=	2.540 cm
1 foot (ft)	=	30.48 cm
1 mile (mi)	=	1.609 km
1 mil	=	10^{-3} in.
1 centimeter	=	0.3937 in.
1 meter	=	39.37 in.
1 kilometer	=	0.6214 mile

Area

1 square meter (m^2) = 10.76 ft^2

1 square foot (ft^2) = 929 cm^2

1 square mile (mi^2) = 640 acres

1 acre = 43,560 ft^2

Volume

1 liter (l) = 1000 cm^3 = 1.057 quart (qt) = 61.02 in^3 = 0.03532 ft^3

1 cubic meter (m^3) = 1000 l = 35.32 ft^3

1 cubic foot (ft^3) = 7.481 U.S. gal = 0.02832 m^3 = 28.32 l

1 U.S. gallon (gal) = 231 in^3 = 3.785 l; 1 British gallon = 1.201 U.S. gallon = 277.4 in^3

Mass

1 kilogram (kg) = 2.2046 lb = 0.06852 slug; 1 lb = 453.6 gm = 0.03108 slug

1 slug = 32.174 lb = 14.59 kg

Speed

1 km/hr = 0.2778 m/sec = 0.6214 mi/hr = 0.9113 ft/sec

1 mi/hr = 1.467 ft/sec = 1.609 km/hr = 0.4470 m/sec

Density

1 gm/cm^3 = 10^3 kg/m^3 = 62.43 lb/ft^3 = 1.940 slug/ft^3

1 lb/ft^3 = 0.01602 gm/cm^3; 1 slug/ft^3 = 0.5154 gm/cm^3

Force

1 newton (nt) = 10^5 dynes = 0.1020 kgwt = 0.2248 lbwt

1 pound weight (lbwt) = 4.448 nt = 0.4536 kgwt = 32.17 poundals

1 kilogram weight (kgwt) = 2.205 lbwt = 9.807 nt

1 U.S. short ton = 2000 lbwt; 1 long ton = 2240 lbwt; 1 metric ton = 2205 lbwt

Energy

1 joule = 1 nt m = 10^7 ergs = 0.7376 ft lbwt = 0.2389 cal = 9.481×10^{-4} Btu

1 ft lbwt = 1.356 joules = 0.3239 cal = 1.285×10^{-3} Btu

1 calorie (cal) = 4.186 joules = 3.087 ft lbwt = 3.968×10^{-3} Btu

1 Btu = 778 ft lbwt = 1055 joules = 0.293 watt hr

1 kilowatt hour (kw hr) = 3.60×10^6 joules = 860.0 kcal = 3413 Btu

1 electron volt (ev) = 1.602×10^{-19} joule

Power

1 watt = 1 joule/sec = 10^7 ergs/sec = 0.2389 cal/sec

1 horsepower (hp) = 550 ft lbwt/sec = 33,000 ft lbwt/min = 745.7 watts

1 kilowatt (kw) = 1.341 hp = 737.6 ft lbwt/sec = 0.9483 Btu/sec

Pressure

1 nt/m^2 = 10 dynes/cm^2 = 9.869×10^{-6} atmosphere = 2.089×10^{-2} lbwt/ft^2

1 lbwt/in^2 = 6895 nt/m^2 = 5.171 cm mercury = 27.68 in. water

1 atmosphere (atm) = 1.013×10^5 nt/m^2 = 1.013×10^6 dynes/cm^2 = 14.70 lbwt/in^2
= 76 cm mercury = 406.8 in. water

Angle

1 radian (rad) = 57.296°; 1° = 0.017453 rad

Appendix B

Astronomical Data

THE SUN

Mass	4.4×10^{30} lb	or	2.0×10^{30} kg
Radius	4.32×10^{5} mi	or	6.96×10^{5} km
Mean density	89.2 lb/ft^3	or	1.42 gm/cm^3
Mean surface gravitational acceleration	896 ft/sec^2	or	273 m/sec^2
Escape velocity at surface	385 mi/sec	or	620 km/sec
Period of rotation about axis	25.38 days	or	2.187×10^{6} sec
Universal gravitational constant G	1.068×10^{-9} $ft^3/lb\text{-}sec^2$	or	6.673×10^{-8} $cm^3/gm\text{-}sec^2$

THE MOON

Mean distance from earth	239×10^{3} mi	or	3.84×10^{5} km
Period of rotation about earth	27.3 days	or	2.36×10^{6} sec
Equatorial radius	1080 mi	or	1738 km
Mass	1.63×10^{23} lb	or	7.38×10^{22} kg
Mean density	208 lb/ft^3	or	3.34 gm/cm^3
Mean surface gravitational acceleration	5.30 ft/sec^2	or	1.62 m/sec^2
Escape velocity	1.48 mi/sec	or	2.38 km/sec
Period of rotation about axis	27.3 days	or	2.36×10^{6} sec
Orbital speed	.64 mi/sec	or	1.02 km/sec
Orbital eccentricity	.055		

THE PLANETS

	Mercury	Venus	Earth	Mars	Jupiter	Saturn	Uranus	Neptune	Pluto
Mean distance from sun	36.2×10^6 mi 58.2×10^6 km	67.2×10^6 mi 108×10^6 km	92.9×10^6 mi 150×10^6 km	141×10^6 mi 227×10^6 km	484×10^6 mi 778×10^6 km	887×10^6 mi 1427×10^6 km	1784×10^6 mi 2871×10^6 km	2795×10^6 mi 4498×10^6 km	3670×10^6 mi 5910×10^6 km
Period of rotation about sun	88.0 days 0.241 yr	224.7 days 0.615 yr	365.26 days 1.000 yr	687.0 days 1.88 yr	4333 days 11.86 yr	10,760 days 29.46 yr	30,690 days 84.0 yr	60,180 days 164.8 yr	90,730 days 248.4 yr
Equatorial radius	1500 mi 2420 km	3850 mi 6200 km	3963 mi 6378 km	2110 mi 3400 km	44,370 mi 71,400 km	37,500 mi 60,400 km	14,800 mi 23,800 km	13,900 mi 22,300 km	1850 mi 2960 km
Mass	0.071×10^{25} lb 0.32×10^{24} kg	1.1×10^{25} lb 4.9×10^{24} kg	1.32×10^{25} lb 5.98×10^{24} kg	0.14×10^{25} lb 0.64×10^{24} kg	420×10^{25} lb 1900×10^{24} kg	126×10^{25} lb 570×10^{24} kg	19×10^{25} lb 87×10^{24} kg	22.7×10^{25} lb 103×10^{24} kg	1.2×10^{25} lb 5.4×10^{24} kg
Mean density	330 lb/ft^3 5.3 gm/cm^3	306 lb/ft^3 4.95 gm/cm^3	340 lb/ft^3 5.52 gm/cm^3	247 lb/ft^3 3.95 gm/cm^3	81 lb/ft^3 1.33 gm/cm^3	44 lb/ft^3 0.70 gm/cm^3	100 lb/ft^3 1.56 gm/cm^3	144 lb/ft^3 2.28 gm/cm^3	250 lb/ft^3 4.0 gm/cm^3
Mean surface gravitational acceleration	12 ft/sec^2 3.6 m/sec^2	28 ft/sec^2 8.5 m/sec^2	32.2 ft/sec^2 9.81 m/sec^2	12.5 ft/sec^2 3.8 m/sec^2	85 ft/sec^2 26.0 m/sec^2	37 ft/sec^2 11.2 m/sec^2	31 ft/sec^2 9.4 m/sec^2	49 ft/sec^2 15.0 m/sec^2	26 ft/sec^2 8.0 m/sec^2
Escape velocity	2.6 mi/sec 4.2 km/sec	6.3 mi/sec 10.2 km/sec	7.0 mi/sec 11.2 km/sec	3.1 mi/sec 5.0 km/sec	38 mi/sec 61 km/sec	23 mi/sec 37 km/sec	14 mi/sec 22 km/sec	16 mi/sec 25 km/sec	6 mi/sec 10 km/sec
Period of rotation about axis	88.0 days 7.58×10^6 sec	30 days (estimated)	23.93 hr 86,164 sec	24.61 hr 88,580 sec	9.84 hr 35,430 sec	10.23 hr 36,840 sec	10.8 hr 38,900 sec	15.7 hr 56,400 sec	16.0 hr 57,600 sec
Orbital speed	29.8 mi/sec 47.9 km/sec	21.8 mi/sec 35.1 km/sec	18.5 mi/sec 29.8 km/sec	15.0 mi/sec 24.1 km/sec	8.12 mi/sec 13.1 km/sec	6.00 mi/sec 9.65 km/sec	4.22 mi/sec 6.80 km/sec	3.38 mi/sec 5.44 km/sec	2.95 mi/sec 4.75 km/sec
Orbital eccentricity	0.206	0.0068	0.017	0.093	0.048	0.055	0.047	0.009	0.247

Appendix C

Solutions of Special Differential Equations

DIFFERENTIAL EQUATIONS

An equation which has derivatives or differentials of an unknown function is called a *differential equation.* The *order* of the differential equation is the order of the highest derivative or differential which is present. A *solution* of a differential equation is any relationship between the variables which reduces the differential equation to an identity.

Example 1.

The equation $\frac{dy}{dx} = 2y$ is a differential equation of first order, or order 1. A solution of this equation is $y = ce^{2x}$ where c is any constant, since on substituting this into the given differential equation we have the identity

$$2ce^{2x} = 2ce^{2x}$$

Example 2.

The equation $x^2\,dx + y^3\,dy = 0$ is a differential equation of first order. A solution is $x^3/3 + y^4/4 = c$ where c is any constant, since taking the differential of the solution we have

$$d(x^3/3 + y^4/4) = 0 \qquad \text{or} \qquad x^2\,dx + y^3\,dy = 0$$

Example 3.

The equation $\frac{d^2y}{dx^2} - 3\frac{dy}{dx} + 2y = 4x$ is a differential equation of second order. A solution is $y = c_1e^x + c_2e^{2x} + 2x + 3$ since

$$\frac{d^2y}{dx^2} - 3\frac{dy}{dx} + 2y = (c_1e^x + 4c_2e^{2x}) - 3(c_1e^x + 2c_2e^{2x} + 2) + 2(c_1e^x + c_2e^{2x} + 2x + 3) = 4x$$

In the above examples we have used x as independent variable and y as dependent variable. However, it is clear that any other symbols could just as well have been used. Thus, for instance, the differential equation of Example 3 could be

$$\frac{d^2x}{dt^2} - 3\frac{dx}{dt} + 2x = 4t$$

with independent variable t and dependent variable x and solution $x = c_1e^t + c_2e^{2t} + 2t + 3$.

The above equations are often called *ordinary differential equations* to distinguish them from *partial differential equations* such as $\frac{\partial^2 Y}{\partial t^2} = c^2\frac{\partial^2 Y}{\partial x^2}$ involving two or more independent variables.

ARBITRARY CONSTANTS. GENERAL AND PARTICULAR SOLUTIONS

In the above examples the constants c, c_1, c_2 can take on any values and are called *arbitrary constants.* In practice an nth order differential equation will have a solution involving exactly n independent arbitrary constants. Such a solution is called the *general solution.* All special cases of the general solution obtained by giving the constants special values are then called *particular solutions.* For instance in Example 3 above if we let $c_1 = 5$, $c_2 = -3$ in the general solution $y = c_1e^x + c_2e^{2x} + 2x + 3$, we obtain the particular solution $y = 5e^x - 3e^{2x} + 2x + 3$.

Particular solutions are often found from certain conditions imposed on the problem and sometimes called *boundary* or *initial conditions.* In Example 3 for instance, if we wish to satisfy the conditions $y = 5$ when $x = 0$ and $y' = dy/dx = 1$ when $x = 0$, we obtain $c_1 = 5$, $c_2 = -3$.

A problem in which we are required to solve a differential equation subject to given conditions is often called a *boundary-value problem.*

SOLUTIONS TO SOME SPECIAL FIRST ORDER EQUATIONS

The following list shows some important methods for finding general solutions of first order differential equations.

1. Separation of Variables

If a first order equation can be written as

$$F(x)\,dx + G(y)\,dy = 0 \tag{1}$$

then the variables are said to be *separable* and the general solution obtained by direct integration is

$$\int F(x)\,dx + \int G(y)\,dy = c \tag{2}$$

2. Linear Equations

A first order equation is called *linear* if it has the form

$$\frac{dy}{dx} + P(x)\,y = Q(x) \tag{3}$$

Multiplying both sides by $e^{\int P\,dx}$, this can be written

$$\frac{d}{dx}\{y\,e^{\int P\,dx}\} = Q\,e^{\int P\,dx}$$

Then integrating, the general solution is

$$y\,e^{\int P\,dx} = \int Q\,e^{\int P\,dx}\,dx + c$$

or

$$y = e^{-\int P\,dx}\int Q\,e^{\int P\,dx}\,dx + c\,e^{-\int P\,dx} \tag{4}$$

The factor $e^{\int P\,dx}$ is often called an *integrating factor.*

3. Exact Equation

The equation

$$M\,dx + N\,dy = 0 \tag{5}$$

where M and N are functions of x and y is called an *exact differential equation* if $M\,dx + N\,dy$ can be expressed as an exact differential dU of a function $U(x, y)$. In such case the solution is given by $U(x, y) = c$.

A necessary and sufficient condition that (*5*) be exact is

$$\frac{\partial M}{\partial y} = \frac{\partial N}{\partial x} \tag{6}$$

In some cases an equation is not exact but can be made exact by first multiplying through by a suitably chosen function called an integrating factor as in the case of the linear equation.

4. Homogeneous Equation

If an equation has the form

$$\frac{dy}{dx} = F\left(\frac{y}{x}\right) \tag{7}$$

it is called a *homogeneous equation* and can be solved by the transformation $y = vx$. Using this, (7) becomes

$$v + x\frac{dv}{dx} = F(v) \quad \text{or} \quad \frac{dx}{x} = \frac{dv}{F(v) - v} \tag{8}$$

in which the variables have been separated. Then the general solution is

$$\int \frac{dx}{x} = \int \frac{dv}{F(v) - v} + c \quad \text{where } v = y/x \tag{9}$$

Occasionally other transformations, which may or may not be evident from the form of a given differential equation, serve to obtain the general solution.

SOLUTIONS OF HIGHER ORDER EQUATIONS

The following list shows certain equations of order higher than one which can often be solved.

1. $\dfrac{d^n y}{dx^n} = F(x)$. In this case the equation can be integrated n times to obtain

$$y = \int \cdots \int F(x)\, dx^n + c_1 + c_2 x + c_3 x^2 + \cdots + c_n x^{n-1}$$

2. $\dfrac{d^2 y}{dx^2} = F\left(x, \dfrac{dy}{dx}\right)$. In this case y is missing and if we make the substitution $dy/dx = v$ we find that the equation becomes

$$\frac{dv}{dx} = F(x, v)$$

a first order equation. If this can be solved we replace v by dy/dx, obtaining another first order equation which then needs to be solved.

3. $\dfrac{d^2 y}{dx^2} = F\left(y, \dfrac{dy}{dx}\right)$. Here x is missing and if we make the substitution $dy/dx = v$, noting also that

$$\frac{d^2 y}{dx^2} = \frac{dv}{dx} = \frac{dv}{dy}\frac{dy}{dx} = v\frac{dv}{dy},$$

the given equation can be written as a first order equation

$$v\frac{dv}{dy} = F(y, v)$$

which then needs to be solved.

LINEAR DIFFERENTIAL EQUATIONS OF ORDER HIGHER THAN ONE

We shall consider solutions of linear second order differential equations. The results can easily be extended to linear higher order equations.

A *linear second order equation* has the form

$$\frac{d^2 y}{dx^2} + P(x)\frac{dy}{dx} + Q(x)\, y = R(x) \tag{10}$$

If y_c is the general solution of the equation

$$\frac{d^2y}{dx^2} + P(x)\frac{dy}{dx} + Q(x)\,y = 0 \tag{11}$$

[obtained by replacing the right hand side of (*10*) by zero] and if y_p is any particular solution of (*10*), then the general solution of (*10*) is

$$y = y_c + y_p \tag{12}$$

The equation (*11*) is often called the *complementary equation* and its general solution is called the *complementary solution.*

LINEAR EQUATIONS WITH CONSTANT COEFFICIENTS

The complementary solution is easily obtained when $P(x)$ and $Q(x)$ are constants A and B respectively. In such case equation (*11*) can be written

$$\frac{d^2y}{dx^2} + A\frac{dy}{dx} + By = 0 \tag{13}$$

If we assume as solution $y = e^{\alpha x}$ where α is constant in (*11*), we find that α must satisfy the equation

$$\alpha^2 + A\alpha + B = 0$$

This equation has two roots, and the following cases arise.

1. **Roots are real and distinct, say** $\alpha_1 \neq \alpha_2$.

   In this case solutions are $e^{\alpha_1 x}$ and $e^{\alpha_2 x}$. It also follows that $c_1 e^{\alpha_1 x}$ and $c_2 e^{\alpha_2 x}$ are solutions and that the sum $c_1 e^{\alpha_1 x} + c_2 e^{\alpha_2 x}$ is the general solution.

2. **Roots are real and equal, say** $\alpha_1 = \alpha_2$.

   In this case we find that solutions are $e^{\alpha_1 x}$ and $xe^{\alpha_1 x}$ and the general solution is $c_1 e^{\alpha_1 x} + c_2 x e^{\alpha_1 x}$.

3. **Roots are complex.**

   If A and B are real, these complex roots are conjugate, i.e. $a + bi$ and $a - bi$. In such case solutions are $e^{(a+bi)x} = e^{ax}e^{bix} = e^{ax}(\cos bx + i \sin bx)$ and $e^{(a-bi)x}(\cos bx - i \sin bx)$. The general solution can be written $e^{ax}(c_1 \cos bx + c_2 \sin bx)$.

PARTICULAR SOLUTIONS

To find the general solution of

$$\frac{d^2y}{dx^2} + A\frac{dy}{dx} + By = R(x) \tag{14}$$

we must find a particular solution of this equation and add it to the general solution of (*13*) already obtained above. Two important methods serve to accomplish this.

1. **Method of Undetermined Coefficients.**

   This method can only be used for special functions $R(x)$ such as polynomials and the exponential or trigonometric functions having the form $e^{px}, \cos px, \sin px$ where p is a constant, together with sums and products of such functions. See Problems C.17 and C.18.

2. **Method of Variation of Parameters.**

   In this method we first write the complementary solution in terms of the constants c_1 and c_2. We then replace c_1 and c_2 by functions $f_1(x)$ and $f_2(x)$ so chosen as to satisfy the given equation. The method is illustrated in Problem C.19.

Solved Problems

DIFFERENTIAL EQUATIONS. ARBITRARY CONSTANTS. GENERAL AND PARTICULAR SOLUTIONS

C.1. (*a*) Prove that $y = ce^{-x} + x - 1$ is the general solution of the differential equation

$$\frac{dy}{dx} - x + y = 0$$

(*b*) Find the particular solution such that $y = 3$ when $x = 0$.

(*a*) If $y = ce^{-x} + x - 1$, then $dy/dx = -ce^{-x} + 1$ and so

$$dy/dx - x + y = (-ce^{-x} + 1) - x + (ce^{-x} + x - 1) = 0$$

Thus $y = ce^{-x} + x - 1$ is a solution; and since it has a number of arbitrary constants (namely one) equal to the order of the differential equation, it is the general solution.

(*b*) Since $y = 3$ when $x = 0$, we have from $y = ce^{-x} + x - 1$, $3 = c - 1$ or $c = 4$. Thus $y = 4e^{-x} + x - 1$ is the required particular solution.

C.2. (*a*) Prove that $x = c_1e^t + c_2e^{-3t} + \sin t$ is the general solution of

$$\frac{d^2x}{dt^2} + 2\frac{dx}{dt} - 3x = 2\cos t - 4\sin t$$

(*b*) Find the particular solution such that $x = 2$, $dx/dt = -3$ at $t = 0$.

(*a*) From $x = c_1e^t + c_2e^{-3t} + \sin t$ we have

$$\frac{dx}{dt} = c_1e^t - 3c_2e^{-3t} + \cos t, \qquad \frac{d^2x}{dt^2} = c_1e^t + 9c_2e^{-3t} - \sin t$$

Then
$$\frac{d^2x}{dt^2} + 2\frac{dx}{dt} - 3x = (c_1e^t + 9c_2e^{-3t} - \sin t) + 2(c_1e^t - 3c_2e^{-3t} + \cos t) - 3(c_1e^t + c_2e^{-3t} + \sin t)$$
$$= 2\cos t - 4\sin t$$

Thus $x = c_1e^t + c_2e^{-3t} + \sin t$ is a solution; and since it has *two* arbitrary constants while the differential equation is of order *two*, it is the general solution.

(*b*) From part (*a*), letting $t = 0$ in the expressions for x and dx/dt, we have

$$\begin{cases} 2 = c_1 + c_2 \\ -3 = c_1 - 3c_2 + 1 \end{cases} \qquad \text{or} \qquad \begin{cases} c_1 + c_2 = 2 \\ c_1 - 3c_2 = -4 \end{cases}$$

Solving, we find $c_1 = 1/2$, $c_2 = 3/2$. Then the required particular solution is

$$x = \tfrac{1}{2}e^t + \tfrac{3}{2}e^{-3t} + \sin t$$

SEPARATION OF VARIABLES

C.3. (*a*) Find the general solution of $(x + xy^2)\,dx + (y + x^2y)\,dy = 0$.

(*b*) Find the particular solution such that $y = 2$ when $x = 1$.

(*a*) Write the equation as $x(1 + y^2)\,dx + y(1 + x^2)\,dy = 0$. Dividing by $(1 + x^2)(1 + y^2) \neq 0$ to separate the variables, we find

$$\frac{x\,dx}{1 + x^2} + \frac{y\,dy}{1 + y^2} = 0 \qquad (1)$$

Then we have on integrating,

$$\int \frac{x\,dx}{1 + x^2} + \int \frac{y\,dy}{1 + y^2} = c_1$$

or
$$\tfrac{1}{2}\ln(1 + x^2) + \tfrac{1}{2}\ln(1 + y^2) = c_1$$

This can be written $\frac{1}{2}\ln\{(1+x^2)(1+y^2)\} = c_1$ or

$$(1+x^2)(1+y^2) = c_2 \tag{2}$$

which is the required general solution.

(b) Since $y = 2$ when $x = 1$, we have on substitution in (2), $c_2 = 10$; thus the required particular solution is

$$(1+x^2)(1+y^2) = 10 \qquad \text{or} \qquad x^2+y^2+x^2y^2 = 9$$

C.4. Solve $\dfrac{dR}{dt} = R^2t^2$ if $R = 1$ when $t = 1$.

Separating the variables, we have $\dfrac{dR}{R^2} = t^2\,dt$. Integrating both sides,

$$-\frac{1}{R} = \frac{t^3}{3} + c$$

Substituting $t = 1, R = 1$ we find $c = -4/3$. Thus

$$-\frac{1}{R} = \frac{t^3}{3} - \frac{4}{3} \qquad \text{or} \qquad R = \frac{3}{4-t^3}$$

LINEAR EQUATION

C.5. Solve $\dfrac{dy}{dx} + 2xy = x^3 + x$ if $y = 2$ when $x = 0$.

This is a linear equation of the form (3), page 345, with $P = 2x$, $Q = x^3+x$. An integrating factor is $e^{\int 2x\,dx} = e^{x^2}$. Multiplying the given equation by this factor, we find

$$e^{x^2}\frac{dy}{dx} + 2xye^{x^2} = (x^3+x)e^{x^2}$$

which can be written

$$\frac{d}{dx}(y\,e^{x^2}) = (x^3+x)e^{x^2}$$

Integrating,

$$y\,e^{x^2} = \int (x^3+x)e^{x^2}\,dx + c$$

or, making the substitution $v = x^2$ in the integral,

$$y\,e^{x^2} = \tfrac{1}{2}x^2e^{x^2} + c$$

Thus

$$y = \tfrac{1}{2}x^2 + ce^{-x^2}$$

Since $y = 2$ when $x = 0$, we find $c = 2$. Thus

$$y = \tfrac{1}{2}x^2 + 2e^{-x^2}$$

Check: If $y = \frac{1}{2}x^2 + 2e^{-x^2}$, then $dy/dx = x - 4xe^{-x^2}$. Thus

$$\frac{dy}{dx} + 2xy = x - 4xe^{-x^2} + 2x(\tfrac{1}{2}x^2 + 2e^{-x^2}) = x^3 + x$$

C.6. Solve $\dfrac{dU}{dt} = 3U + 1$ if $U = 0$ when $t = 0$.

Writing the equation in the form

$$\frac{dU}{dt} - 3U = 1 \tag{1}$$

we see that it is linear with integrating factor $e^{\int -3\,dt} = e^{-3t}$. Multiplying (*1*) by e^{-3t}, it can be written as

$$\frac{d}{dt}(Ue^{-3t}) = e^{-3t}$$

Integrating, we have $$Ue^{-3t} = -\tfrac{1}{3}e^{-3t} + c$$

Since $U = 0$ when $t = 0$, we obtain $c = \frac{1}{3}$. Thus

$$Ue^{-3t} = -\tfrac{1}{3}e^{-3t} + \tfrac{1}{3} \quad \text{or} \quad U = \tfrac{1}{3}(e^{3t} - 1) \tag{2}$$

Another method. The equation can also be solved by the method of separation of variables. Thus we have

$$\frac{dU}{3U+1} = dt$$

Integrating, $$\tfrac{1}{3}\ln(3U+1) = t + c$$

Since $U = 0$ when $t = 0$, we find $c = 0$ so that $\frac{1}{3}\ln(3U+1) = t$. Thus $U = \frac{1}{3}(e^{3t} - 1)$.

EXACT EQUATIONS

C.7. Solve $(3x^2 + y\cos x)\,dx + (\sin x - 4y^3)\,dy = 0$.

Comparing with $M\,dx + N\,dy = 0$, we have $M = 3x^2 + y\cos x$, $N = \sin x - 4y^3$. Then $\partial M/\partial y = \cos x = \partial N/\partial x$ and so the equation is exact. Two methods of solution are available.

Method 1.

Since the equation is exact, the left side must be an exact differential of a function U. By grouping terms, we find that the equation can be written

$$3x^2\,dx + (y\cos x\,dx + \sin x\,dy) - 4y^3\,dy = 0$$

i.e., $$d(x^3) + d(y\sin x) + d(-y^4) = 0 \quad \text{or} \quad d(x^3 + y\sin x - y^4) = 0$$

Integration then gives the required solution, $x^3 + y\sin x - y^4 = c$.

Method 2.

The given equation can be written as

$$(3x^2 + y\cos x)\,dx + (\sin x - 4y^3)\,dy = dU = \frac{\partial U}{\partial x}dx + \frac{\partial U}{\partial y}dy$$

Then we must have

$$(1)\quad \frac{\partial U}{\partial x} = 3x^2 + y\cos x \qquad (2)\quad \frac{\partial U}{\partial y} = \sin x - 4y^3$$

Integrating (*1*) with respect to x, keeping y constant, we have

$$U = x^3 + y\sin x + F(y)$$

Then substituting this into (*2*), we find

$$\sin x + F'(y) = \sin x - 4y^3 \quad \text{or} \quad F'(y) = -4y^3$$

where $F'(y) = dF/dy$. Integrating, omitting the constant of integration, we have $F(y) = -y^4$ so that

$$U = x^3 + y\sin x - y^4$$

Then the given differential equation can be written

$$dU = d(x^3 + y\sin x - y^4) = 0$$

and so the solution is $x^3 + y\sin x - y^4 = c$.

HOMOGENEOUS EQUATIONS

C.8. Solve $\dfrac{dy}{dx} = e^{y/x} + \dfrac{y}{x}$.

Let $y = vx$. Then the equation can be written

$$v + x\frac{dv}{dx} = e^v + v \qquad \text{or} \qquad x\frac{dv}{dx} = e^v$$

Separating the variables, $\dfrac{dx}{x} = e^{-v}dv$. Integrating, $\ln x = -e^{-v} + c$. Thus the general solution is $\ln x + e^{-y/x} = c$.

SOLUTIONS OF HIGHER ORDER EQUATIONS

C.9. Solve $\dfrac{d^2U}{dt^2} = 1 + \cos t$ where $U = 2,\ dU/dt = 3$ at $t = 0$.

Integrating once,

$$dU/dt = t + \sin t + c_1$$

Then since $dU/dt = 3$ at $t = 0$, we find $c_1 = 3$. Thus

$$dU/dt = t + \sin t + 3$$

Integrating again, $$U = \tfrac{1}{2}t^2 - \cos t + 3t + c_2$$

Now since $U = 2$ at $t = 0$, we find $c_2 = 3$. The required solution is

$$U = \tfrac{1}{2}t^2 - \cos t + 3t + 3$$

C.10. Solve $xy'' + 2y' = x^2$ where $y' = dy/dx,\ y'' = d^2y/dx^2$.

Since y is missing, let $y' = dy/dx = v$. Then the equation can be written

$$(1)\quad x\frac{dv}{dx} + 2v = x^2 \qquad \text{or} \qquad (2)\quad \frac{dv}{dx} + \frac{2}{x}v = x$$

This is a linear equation in v with integrating factor $e^{\int (2/x)\,dx} = e^{2\ln x} = e^{\ln x^2} = x^2$. Multiplying (2) by x^2, it can be written as

$$\frac{d}{dx}(x^2v) = x^3$$

Then by integration, $x^2v = x^4/4 + c_1$ or

$$v = dy/dx = x^2/4 + c_1/x^2$$

Integrating again, $y = x^3/12 - c_1/x + c_2$.

C.11. Solve $yy'' + (y')^2 = 0$ where $y' = dy/dx,\ y'' = d^2y/dx^2$.

Since x is missing, let $y' = dy/dx = v$. Then

$$y'' = \frac{dv}{dx} = \frac{dv}{dy}\cdot\frac{dy}{dx} = v\frac{dv}{dy}$$

and the given equation can be written

$$yv\frac{dv}{dy} + v^2 = 0 \qquad \text{or} \qquad v\left(y\frac{dv}{dy} + v\right) = 0$$

so that $$(1)\quad v = 0 \qquad \text{or} \qquad (2)\quad y\frac{dv}{dy} + v = 0$$

From (1), $y' = 0$ or $y = c_1$. From (2), $\dfrac{dv}{v} + \dfrac{dy}{y} = 0$, i.e. $\ln v + \ln y = c_2$ or $\ln(vy) = c_2$ so that $vy = c_3$ and

$$v = dy/dx = c_3/y \qquad \text{or} \qquad y\,dy = c_3\,dx$$

Integrating, $$y^2/2 = c_3x + c_4 \qquad \text{or} \qquad y^2 = Ax + B$$

Thus solutions are $y = c_1$ and $y^2 = Ax + B$. Since the first is a special case of the second, the required general solution can be written $y^2 = Ax + B$.

LINEAR EQUATIONS WITH CONSTANT COEFFICIENTS

C.12. Solve $\dfrac{d^2y}{dx^2} - 4\dfrac{dy}{dx} - 5y = 0.$

Letting $y = e^{\alpha x}$ in the equation, we obtain

$$(\alpha^2 - 4\alpha - 5)e^{\alpha x} = 0 \qquad \text{or} \qquad \alpha^2 - 4\alpha - 5 = 0$$

Thus $(\alpha - 5)(\alpha + 1) = 0$ and $\alpha = 5, -1$. Then solutions are e^{5x} and e^{-x} and the general solution is $y = c_1e^{5x} + c_2e^{-x}$.

C.13. Solve $\dfrac{d^2y}{dx^2} + 10\dfrac{dy}{dx} + 25y = 0.$

Letting $y = e^{\alpha x}$, we find $\alpha^2 + 10\alpha + 25 = 0$, i.e. $(\alpha + 5)(\alpha + 5) = 0$, or $\alpha = -5, -5$. Since the root is repeated, solutions are e^{-5x} and xe^{-5x}. Then the general solution is $y = c_1e^{-5x} + c_2xe^{-5x}$.

C.14. Solve $\dfrac{d^2x}{dt^2} + 4\dfrac{dx}{dt} + 4x = 0.$

Letting $x = e^{\alpha t}$, we find $\alpha^2 + 4\alpha + 4 = 0$ or $\alpha = -2, -2$. Then the general solution is $x = c_1e^{-2t} + c_2te^{-2t} = e^{-2t}(c_1 + c_2t)$.

C.15. Solve $\dfrac{d^2y}{dx^2} + 2\dfrac{dy}{dx} + 5y = 0.$

Letting $y = e^{\alpha x}$, we find $\alpha^2 + 2\alpha + 5 = 0$ or $\alpha = -1 \pm 2i$. Then solutions are $e^{(-1+2i)x} = e^{-x}e^{2ix} = e^{-x}(\cos 2x + i \sin 2x)$ and $e^{(-1-2i)x} = e^{-x}e^{-2ix} = e^{-x}(\cos 2x - i \sin 2x)$. The general solution is $y = e^{-x}(c_1 \cos 2x + c_2 \sin 2x)$.

C.16. Solve $d^2y/dx^2 + \omega^2 y = 0.$

Letting $y = e^{\alpha x}$, we find $\alpha^2 + \omega^2 = 0$ or $\alpha = \pm i\omega$. Then solutions are $e^{i\omega x} = \cos \omega x + i \sin \omega x$ and $e^{-i\omega x} = \cos \omega x - i \sin \omega x$. The general solution is thus $y = c_1 \cos \omega x + c_2 \sin \omega x$.

METHOD OF UNDETERMINED COEFFICIENTS

C.17. Solve $\dfrac{d^2y}{dx^2} - 4\dfrac{dy}{dx} - 5y = x^2 + 2e^{3x}.$

By Problem C.12 the complementary solution, i.e. the general solution, of

$$\frac{d^2y}{dx^2} - 4\frac{dy}{dx} - 5y = 0$$

is $$y_c = c_1e^{5x} + c_2e^{-x} \tag{1}$$

Since the right side of the given equation contains a polynomial of the second degree (i.e. x^2) and an exponential ($2e^{3x}$), we are led to the trial particular solution

$$y_p = Ax^2 + Bx + C + De^{3x} \tag{2}$$

where A, B, C, D are constants to be determined.

Substituting (2) for y in the given equation and simplifying, we find

$$(2A - 4B - 5C) + (-8A - 5B)x - 5Ax^2 - 8De^{3x} = x^2 + 2e^{3x}$$

Since this must be an identity, we must have

$$2A - 4B - 5C = 0, \quad -8A - 5B = 0, \quad -5A = 1, \quad -8D = 2$$

Solving, we find $A = -\frac{1}{5}$, $B = \frac{8}{25}$, $C = -\frac{42}{125}$, $D = -\frac{1}{4}$. Then from (2),

$$y_p = -\tfrac{1}{5}x^2 + \tfrac{8}{25}x - \tfrac{42}{125} - \tfrac{1}{4}e^{3x}$$

Thus the required general solution of the given equation is

$$y = y_c + y_p = c_1e^{5x} + c_2e^{-x} - \tfrac{1}{5}x^2 + \tfrac{8}{25}x - \tfrac{42}{125} - \tfrac{1}{4}e^{3x}$$

which can be checked by direct substitution.

C.18. Solve $\dfrac{d^2y}{dx^2} + 10\dfrac{dy}{dx} + 25y = 20\cos 2x$.

The complementary solution [by Problem C.13] is

$$y_c = c_1e^{-5x} + c_2xe^{-5x} \tag{1}$$

Since the right side has the term $\cos 2x$, we are led to the trial solution

$$y_p = A\cos 2x + B\sin 2x \tag{2}$$

Substitution into the given equation yields, after simplifying,

$$(21A + 20B)\cos 2x + (21B - 20A)\sin 2x = 20\cos 2x$$

Equating coefficients of like terms, we have $21A + 20B = 20$, $21B - 20A = 0$. Solving, we find $A = \frac{420}{841}$, $B = \frac{400}{841}$ so that the particular solution is

$$y_p = \tfrac{420}{841}\cos 2x + \tfrac{400}{841}\sin 2x$$

and the general solution of the given equation is

$$y = y_c + y_p = c_1e^{-5x} + c_2xe^{-5x} + \tfrac{420}{841}\cos 2x + \tfrac{400}{841}\sin 2x$$

METHOD OF VARIATION OF PARAMETERS

C.19. Solve $d^2y/dx^2 + y = \tan x$.

The complementary solution is as in Problem C.16 with $\omega = 1$:

$$y_c = c_1\cos x + c_2\sin x \tag{1}$$

We now assume that the solution to the given equation has the form

$$y = f_1\cos x + f_2\sin x \tag{2}$$

where f_1 and f_2 are suitable functions of x. From (2) we have, using primes to denote differentiation with respect to x,

$$dy/dx = -f_1\sin x + f_2\cos x + f_1'\cos x + f_2'\sin x \tag{3}$$

Before finding d^2y/dx^2 let us observe that since there are two functions f_1 and f_2 to be determined and only one relation to be satisfied [namely that the given differential equation must be satisfied] we are free to impose one relation between f_1 and f_2. We choose the relation

$$f_1'\cos x + f_2'\sin x = 0 \tag{4}$$

so as to simplify (3) which then becomes

$$dy/dx = -f_1\sin x + f_2\cos x \tag{5}$$

Another differentiation then leads to

$$d^2y/dx^2 = -f_1\cos x - f_2\sin x - f_1'\sin x + f_2'\cos x \tag{6}$$

From (2) and (6) we see that the given differential equation can be written

$$d^2y/dx^2 + y = -f_1'\sin x + f_2'\cos x = \tan x \tag{7}$$

Thus

$$-f_1'\sin x + f_2'\cos x = \tan x \tag{8}$$

From (4) and (8) we find $f_1' = -\sin^2 x/\cos x$, $f_2' = \sin x$. Thus

$$f_1 = -\int \frac{\sin^2 x}{\cos x}\,dx = -\int \frac{1-\cos^2 x}{\cos x}\,dx = -\int (\sec x - \cos x)\,dx$$

$$= -\ln(\sec x + \tan x) + \sin x + c_1$$

$$f_2 = \int \sin x\,dx = -\cos x + c_2$$

Substituting in (*2*) we find the required general solution,

$$y = c_1 \cos x + c_2 \sin x - \cos x \ln(\sec x + \tan x)$$

Supplementary Problems

DIFFERENTIAL EQUATIONS. ARBITRARY CONSTANTS. GENERAL AND PARTICULAR SOLUTIONS

C.20. Check whether each differential equation has the indicated general solution.

(*a*) $\dfrac{d^2x}{dt^2} - 2\dfrac{dx}{dt} + x = t; \quad x = (c_1 + c_2 t)e^t + t + 2$

(*b*) $t\dfrac{dU}{dt} + U = t^2; \quad t^3 - 3tU = c$

C.21. (*a*) Show that $z = e^{-t}(c_1 \sin t + c_2 \cos t)$ is a general solution of

$$\frac{d^2z}{dt^2} + 2\frac{dz}{dt} + 2z = 0$$

(*b*) Determine the particular solution such that $z = -2$ and $dz/dt = 5$ at $t = 0$.
Ans. (*b*) $z = e^{-t}(3 \sin t - 2 \cos t)$

SOLUTIONS TO SPECIAL FIRST ORDER EQUATIONS

C.22. Solve $dy/dx = -2xy$ if $y = 4$ when $x = 0$. *Ans.* $y = 4e^{-x^2}$

C.23. Solve $\dfrac{dz}{dt} = \dfrac{t\sqrt{1-z^2}}{z\sqrt{1-t^2}}$. *Ans.* $\sqrt{1-t^2} - \sqrt{1-z^2} = c$

C.24. Solve $x\dfrac{dy}{dx} - 2y = x$ if $y(1) = 5$. *Ans.* $y = 6x^2 - x$

C.25. Solve $(x + 2y)\,dx + (2x - 5y)\,dy = 0$ if $y = 1$ when $x = 2$. *Ans.* $x^2 + 4xy - 5y^2 = 7$

C.26. Solve $\dfrac{dy}{dx} = \dfrac{y}{x} + \dfrac{y^2}{x^2}$. *Ans.* $\ln x + (x/y) = c$

C.27. Solve $(ye^x - e^{-y})\,dx + (xe^{-y} + e^x)\,dy = 0$. *Ans.* $ye^x - xe^{-y} = c$

C.28. Solve $(x + xy)\,dx + (xy + y)\,dy = 0$. *Ans.* $(x+1)(y+1) = ce^{x+y}$

C.29. Show that the differential equation $(4y - x^2)\,dx + x\,dy = 0$ has an integrating factor which depends on only one variable and thus solve the equation. *Ans.* $x^4y - \frac{1}{6}x^6 = c$

SOLUTIONS OF HIGHER ORDER EQUATIONS

C.30. Solve $d^2U/dt^2 = t + e^{-t}$ if $U = 3$, $dU/dt = 2$ at $t = 0$. *Ans.* $U = \frac{1}{6}t^3 + e^{-t} + 3t + 2$

C.31. Solve $x\dfrac{d^2y}{dx^2} - 3\dfrac{dy}{dx} = x^2$. *Ans.* $y = -\frac{1}{3}x^3 + c_1x^4 + c_2$

C.32. Solve $U\dfrac{d^2U}{dt^2} + 2\left(\dfrac{dU}{dt}\right)^2 = 0$. *Ans.* $U^3 = c_1t + c_2$

C.33. Solve $\left[1 + \left(\dfrac{dy}{dx}\right)^2\right]^3 = \left(\dfrac{d^2y}{dx^2}\right)^2$. *Ans.* $(x-A)^2 + (y-B)^2 = 1$

LINEAR HIGHER ORDER DIFFERENTIAL EQUATIONS

C.34. Solve $\dfrac{d^2y}{dx^2} - 2\dfrac{dy}{dx} - 8y = 0$. *Ans.* $y = c_1e^{4x} + c_2e^{-2x}$

C.35. Solve $\dfrac{d^2U}{dt^2} + 4\dfrac{dU}{dt} + 4U = 0$ if $U=1, dU/dt=0$ at $t=0$. *Ans.* $U = (1+2t)e^{-2t}$

C.36. Solve $\dfrac{d^2z}{dt^2} + 4\dfrac{dz}{dt} + 5z = 0$. *Ans.* $z = e^{-2t}(c_1 \cos t + c_2 \sin t)$

C.37. Solve $4\dfrac{d^2y}{dx^2} + 25y = 0$ if $y(0) = 10$, $y'(0) = 25$. *Ans.* $y = 10(\cos\frac{5}{2}x + \sin\frac{5}{2}x)$

C.38. Solve $\dfrac{d^2y}{dx^2} - 4\dfrac{dy}{dx} - y = 0$. *Ans.* $y = e^{2x}(c_1e^{\sqrt{5}\,x} + c_2e^{-\sqrt{5}\,x})$

C.39. Solve $4y'' - 20y' + 25y = 0$. *Ans.* $(c_1 + c_2x)e^{5x/2}$

C.40. Solve $\dfrac{d^2y}{dx^2} + 3\dfrac{dy}{dx} + 2y = e^{-3x}$. *Ans.* $y = c_1e^{-x} + c_2e^{-2x} + \frac{1}{2}e^{-3x}$

C.41. Solve $\dfrac{d^2U}{dt^2} + \dfrac{dU}{dt} - 2U = 6t - 10\cos 2t + 5$.
Ans. $U = c_1e^{-2t} + c_2e^t - 3t - 4 + \frac{1}{2}\sin 2t - \frac{3}{2}\cos 2t$

C.42. Solve $y'' + y = \sec x$ by use of the method of variation of parameters.
Ans. $y = c_1 \sin x + c_2 \cos x + x \sin x - \cos x \ln \sec x$

C.43. Solve (*a*) Problem C.40 and (*b*) Problem C.41 by variation of parameters.

C.44. Solve each of the following equations by any method.
(*a*) $y'' - 5y' + 6y = 50 \sin 4x$
(*b*) $y'' + 2y' - 3y = xe^{-x}$
(*c*) $y'' + 4y = \csc 2x$
(*d*) $y'' + 8y' + 25y = 25x + 33 + 18e^{-x}$
Ans. (*a*) $y = c_1e^{2x} + c_2e^{3x} + 2\cos 4x - \sin 4x$
(*b*) $y = c_1e^x + c_2e^{-3x} - \frac{1}{4}xe^{-x}$
(*c*) $y = c_1 \cos 2x + c_2 \sin 2x - \frac{1}{2}x \cos 2x - \frac{1}{4}\sin 2x \ln(\csc 2x)$
(*d*) $y = e^{-4x}(c_1 \cos 3x + c_2 \sin 3x) + x + 1 + e^{-x}$

C.45. Solve $y'' + 4y' + 4y = e^{-2x}$. *Ans.* $y = (c_1 + c_2x)e^{-2x} + \frac{1}{2}x^2e^{-2x}$

C.46. Solve simultaneously: $dx/dt + y = e^t$, $x - dy/dt = t$.
Ans. $x = c_1 \cos t - c_2 \sin t + \frac{1}{2}e^t + t$,
$y = c_1 \sin t + c_2 \cos t + \frac{1}{2}e^t - 1$

C.47. Solve $y'' + y = 4\cos t$ if $y(0) = 2$, $y'(0) = -1$. Is the method of undetermined coefficients applicable? Explain. *Ans.* $y = 2\cos t - \sin t + 2t \sin t$

C.48. Show how to solve linear equations of order higher than two by finding the general solutions of
(*a*) $y''' - 6y'' + 11y' - 6y = 36x$, (*b*) $y^{(iv)} + 2y'' + y = x^2$.
Ans. (*a*) $y = c_1e^x + c_2e^{2x} + c_3e^{3x} - 6x - 11$
(*b*) $y = c_1 \cos x + c_2 \sin x + x(c_3 \cos x + c_4 \sin x) + x^2 - 4$

Appendix D

Index of Special Symbols and Notations

The following list shows special symbols and notations used in this book together with the number of the page on which they first appear. All bold faced letters denote vectors. Cases where a symbol has more than one meaning will be clear from the context.

Symbols

a	length of semi-major axis of ellipse or hyperbola, 38, 118, 119
a_n	Fourier cosine coefficients, 196
$\mathbf{a}$	acceleration, 7
$\mathbf{a}_{P_2/P_1}$	acceleration of particle P_2 relative to particle P_1, 7
A	area, 122
$\mathbf{A}$	vector potential of electromagnetic field, 309
$\dot{\mathbf{A}}$	areal velocity, 122
$\mathcal{A}$	amplitude of steady state oscillation, 90
$\mathcal{A}_{\max}$	maximum amplitude of steady state oscillation, 90
b	length of semi-minor axis of ellipse or hyperbola, 38, 118, 119
b_n	Fourier sine coefficients, 196
$\mathbf{B}$	magnetic intensity, 83
$\mathbf{B}$	unit binormal, 7
c	speed of light, 54
C	curve, 6
D_F, D_M	time derivative operator in fixed and moving systems, 144
$\mathbf{e}_1, \mathbf{e}_2, \mathbf{e}_3$	unit vectors, 72
E	total energy, 36
$\mathbf{E}$	electric intensity, 84
$\mathbf{f}$	force due to friction, 65
$\mathbf{f}_{\nu\lambda}$	internal force on particle ν due to particle λ, 173
$\mathscr{f}$	frequency, 89
$\mathscr{f}_\alpha$	frequencies, 316
$\mathbf{F}$	force, 33
$\mathbf{F}_{12}$	force of particle 1 on particle 2, 33
$\mathbf{F}_{\text{av}}$	average force, 60
$\mathbf{F}_D$	damping force, 87
$\mathbf{F}_\nu$	impulsive forces, 285
$\mathbf{F}_\nu^{(a)}, \mathbf{F}_\nu^{(c)}$	actual and constraint forces acting on particle ν, 170
$\mathscr{F}_\alpha$	generalized impulse, 285
$\mathscr{F}_\nu$	force (external and internal) acting on particle ν of a system of particles, 168

$\mathbf{g}$ acceleration due to gravity, 62
G gravitational constant, 120
$\mathcal{G}$ generating function, 314
h Planck's constant, 338
H Hamiltonian, 311
$\mathcal{H}$ Hamiltonian under a canonical transformation, 314
$\mathbf{i}$ unit vector in direction of positive x axis, 3
I moment of inertia, 225
I_C moment of inertia about axis through center of mass, 226
I_{xx}, I_{yy}, I_{zz} moments of inertia about x, y, z axes, 254
I_{xy}, I_{yz}, I_{xz} products of inertia, 254
I_1, I_2, I_3 principal moments of inertia, 255
$\mathcal{J}_\nu$ impulses, 285
$\mathcal{J}$ angular impulse, 228
$\mathbf{j}$ unit vector in direction of positive y axis, 3
J_α phase integral or action variable, 316
$\mathbf{k}$ unit vector in direction of positive z axis, 3
K radius of gyration, 225
l length, 90
L Lagrangian, 284
$\mathcal{L}$ Lagrangian under a canonical transformation, 314
m mass, 33
m_0 rest mass, 54
M total mass of a system of particles, 166
n number of degrees of freedom, 282
n orbital quantum number, 338
N normal component of reaction force, 65
N number of particles in a system, 166
$\mathbf{N}$ unit normal, 7
p_α generalized or conjugate momenta, 284
$\mathbf{p}$ momentum, 33
P period, 89
P_α new generalized momenta under a canonical transformation, 314
$\mathcal{P}$ power, 34
q electrical charge, 84
q_α generalized coordinates, 282
Q_α new generalized coordinates under a canonical transformation, 314
r spherical coordinate, 32
$\mathbf{r}$ position vector or radius vector, 4
$\bar{\mathbf{r}}$ position vector of center of mass, 166
$\mathbf{r}_1$ unit vector in radial direction, 25
$\mathbf{r}'_\nu$ position vector of particle ν relative to center of mass, 169
R radius of curvature, 8
R range, 75
$R_{\max}$ maximum range, 75
$\mathbf{R}$ resisting force, 64
$\mathbf{R}$ resultant of forces, 47
$\mathcal{R}$ rigid body, 228
$\mathcal{R}$ Routh's function or Routhian, 337
s arc length, 7
$\mathbf{s}$ spin angular velocity, 269
S generating function, 316
$\mathcal{S}$ generating function depending on old position coordinates and new momenta, 314

Greek Symbols

Notations

INDEX